ADVANCES IN MUCOSAL IMMUNOLOGY
Part A

ADVANCES IN EXPERIMENTAL MEDICINE AND BIOLOGY

ADVANCES IN MUCOSAL IMMUNOLOGY

Part A

Edited by

Jiri Mestecky, Michael W. Russell, Susan Jackson, and Suzanne M. Michalek

University of Alabama at Birmingham
Birmingham, Alabama

and

Helena Tlaskalová-Hogenová and Jaroslav Šterzl

Academy of Sciences of the Czech Republic
Prague, Czech Republic

SPRINGER SCIENCE+BUSINESS MEDIA, LLC

Library of Congress Cataloging-in-Publication Data

On file

ISBN 978-1-4613-5796-4   ISBN 978-1-4615-1941-6 (eBook)
DOI 10.1007/978-1-4615-1941-6

Proceedings of the 7th International Congress of Mucosal Immunology,
held August 16–21, 1992, in Prague, Czechoslovakia

PREFACE

The Seventh International Congress of Mucosal Immunology held in Prague, the beautiful old capital of The Czech Republic, 16-20 August 1992, was the first to be sponsored by the Society for Mucosal Immunology, and was the largest since their inception 20 years earlier in Birmingham, Alabama. It was attended by 624 participants who gave 538 presentations, more than 10 times the numbers of the first meeting; these proceedings contain 354 papers that were submitted for publication. The political events in Europe that made it possible to hold this Congress in Prague also allowed for the first time the participation of large numbers of scientists from Eastern Europe, as well as from Asia, and the organizers were truly gratified by this happy circumstance. It is now clear not only that mucosal immunity encompasses the huge area of mucosal surfaces and most physiological organ systems, but also that mucosal immunology extends over the whole global surface and all continents! The sheer size of the Congress and number of manuscripts unfortunately entailed some unexpected problems in editing and assembling the proceedings, partly due to the diversity of linguistic styles not represented at earlier meetings, and we apologize to the authors who have patiently awaited the publication of their contributions.

The past few years have witnessed a growing impact of mucosal immunology on clinical medicine and vaccine development, in part because of the realization that HIV infection, like most other infectious diseases, is predominantly acquired through mucosal surfaces. This is reflected in the number of papers in these volumes and elsewhere that deal with immunity to viral, bacterial, fungal, and parasitic infections, the molecular interactions between these pathogens and the immune system across mucosal epithelia, as well as ingenious strategies to enhance the development of protective immunity at these surfaces, and the regulatory mechanisms involved. These proceedings illustrate this impact.

We wish to record our gratitude to many individuals and groups who contributed to the success of the Congress and preparation of this publication. We thank our administrative and secretarial assistants, especially Ms. Maria Bethune, who has diligently borne the brunt of this effort with remarkable composure. The Congress was supported financially by the following:

ADICO Ltd., Prague, Czechoslovakia
Becton Dickinson, Heidelberg, Germany
Behringwerke Ag. Marburg, Germany
Cheminst, Prague, Czechoslovakia
Chugai Pharmaceutical Co., Ltd., Japan
Daiichi Seiyakau Co., Ltd., Japan
Falk Foundation, Germany
Immuno, Vienna, Austria
Immunotech S.A., Marseille, France
Kabi Pharmacia, Prague, Czechoslovakia
Marion Merrill Dow, U.S.A.
Med Shirotori Co. Ltd., Japan
Medical System Management, Kirkel, Germany

Mucos Pharma, Prague, Czechoslovakia
Nestlé, Vevey, Switzerland
Nippon Boehringer Ingelheim Co., Ltd., Japan
Nissan Prince Osaka Motor Sales Co. Ltd., Japan
Ono Pharmaceutical Co. Ltd., Japan
Procter and Gamble, Cincinnati, U.S.A.
Schöller Pharma, Prague, Czechoslovakia
Secretech, Inc., Birmingham, U.S.A.
Thiemann Therapeutics, Germany
University of Alabama at Birmingham, U.S.A.
University of Texas, Galveston, U.S.A..
Virus Research Institute, Cambridge, U.S.A.
Yamanouchi Pharmaceutical Co. Ltd., Japan

November 1994

The Editors

CONTENTS

PART A

B AND T CELLS OF THE MUCOSAL IMMUNE SYSTEM: TRAFFICKING AND CYTOKINE REGULATION

GNOTOBIOLOGY, ENVIRONMENTAL, NUTRITIONAL, AND INTRINSIC FACTORS IN MUCOSAL IMMUNOLOGY

STRUCTURE, PROTEOLYSIS, AND FUNCTION OF MUCOSAL IMMUNOGLOBULINS: CELLULAR RECEPTORS

PART B

CLINICAL IMMUNOLOGY, IMMUNOPATHOLOGY, IMMUNO-DEFICIENCY, AND ALLERGOLOGY

MICROBIAL, PARASITE, AND HIV MUCOSAL INFECTIONS

IMMUNOLOGY OF THE LIVER

ORAL IMMUNOLOGY AND IMMUNOPATHOLOGY

INDUCTION OF MUCOSAL IMMUNE RESPONSES AND VACCINE
DELIVERY SYSTEMS

INDUCTION AND RECALL OF THE SECONDARY IMMUNE RESPONSE ENTIRELY IN TISSUE CULTURE

Jaroslav Sterzl, Jaroslava Milerova, Lucie Mackova, and Jiri Travnicek

Department of Immunology, Institute of Microbiology, Czech Academy of Sciences, Prague 4, Czech Republic

INTRODUCTION

Biological systems acquire new information which is stored and retrieved through the mechanisms of memory.[1] Immunological memory is defined as the ability of the immune system to react with increased efficiency to antigen encountered in the past. Secondary exposure to the same antigen induces a quicker and more intensive response with higher affinity.[2]

It is difficult to study the character of memory cells in the intact organism because of continous presence of residual antigen, constant recruitment of new waves of cells, and multiple cellular interactions. The difficulties *in vivo* are circumvented by the isolation of memory cells and their adoptive transfer to non-reactive recipients. Although many adoptive transfer studies have been performed (reviewed by Celada)[3], only a limited number of experiments on the transfer of memory cells into tissue cultures have been reported.[4]

The nature of memory cells remains controversial. A widely accepted view holds B-memory cells (B-MC) to be long-lived lymphocyte.[5,6] Contrary to this view is the existence of B-MC with a turnover rate similar to that of most peripheral recirculating lymphocytes[7]; such as B-MC are continually recruited by persisting antigen.

In contrast to the ready recall of secondary immune responses in cultured B-MC, the induction of B-MC under the same conditions in tissue culture with virgin immunocompetent B-lymphocytes (B-ICC) has not been reported; B-MC have been induced only *in vivo* and comparable induction in tissue culture has not been achieved.

The essential condition for generation of immunological memory *in vivo* is the formation of a complex immune network within germinal centers of lymphatic follicules.[8] During primary stimulation, germinal centers are the sites of B-MC generation; the antigen is presented to B lymphocytes on the membranes of follicular dendritic cells in cooperation with T_H cell located in periaterial lymphatic sheets (PALS).[9]

In the present studies we tested the conditions under which secondary responses could be recalled in tissue culture with the goal of initiating the induction of B-MC from virgin B-ICC entirely *in vitro*.

The Marbrook cultivation system with Eagle minimal essential medium (MEM) was supplemented with a mixture of non-essential amino acids and 10% of fetal calf serum (FCS). To maintain the pH during 5-day cultured HEPES solution was added and the cultures were kept in an atmosphere of 96% air and 4% CO_2. After 5 days, the antibody responses were evaluated by enumerating the plaque forming cells (PFC) in agarose by the drop modification[10] of the Jerne method

Secondary Immune Responses Induced in Tissue Culture with B-MC Isolated at Different Intervals after Primary Stimulation *In Vivo*

B-MC were isolated from the spleen of BALB/c mice 10, 80, 115 and 165 days after a primary i v dose of 10^8 SRBC B-MC were restimulated in tissue cultures with increasing doses (10^3-10^9) SRBC (Figs 1 and 2) From Fig 1 showing different days after immunization and Fig 2 illustrating different doses of antigen, the changing dynamics of the secondary response are evident During the active proliferation phase (10 days after stimulation), the IgM secondary response was not inhibited by a large dose of antigen (10^8 SRBC) A small dose (10^5 SRBC) stimulated the secondary response even 165 days after priming, supression was induced by doses of 10^8 and 10^9 SRBC Superposition of results for IgM and IgG responses showed two different populations of memory cells prepared for the secondary response (Fig 3) Taking the response induced by a dose of 10^7 SRBC as 100%, the highest IgM response occurred 150 days after the challenge dose of 10^6, and the SRBC highest IgG response was observed at 80 and 150 days with a challenge dose 10^5 SRBC (Fig 4)

During the entire period followed (10-165 days after priming), the secondary response in tissue culture was preferentially of the IgM type

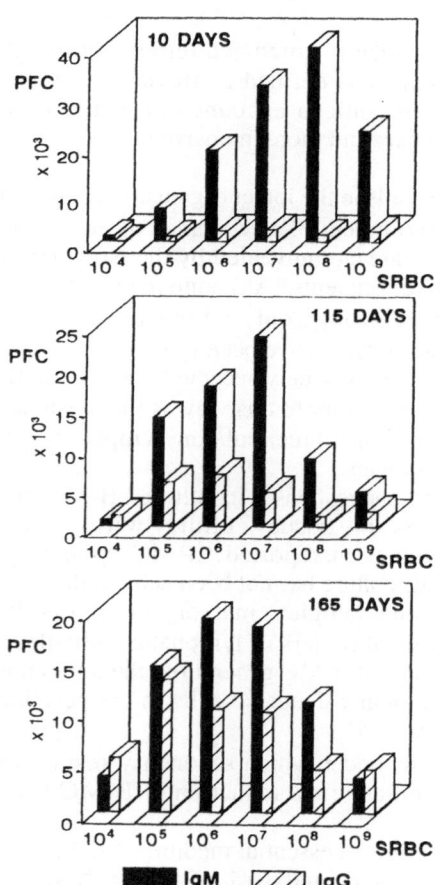

Figure 1. Secondary responses on different days after priming with increasing doses of SRBC

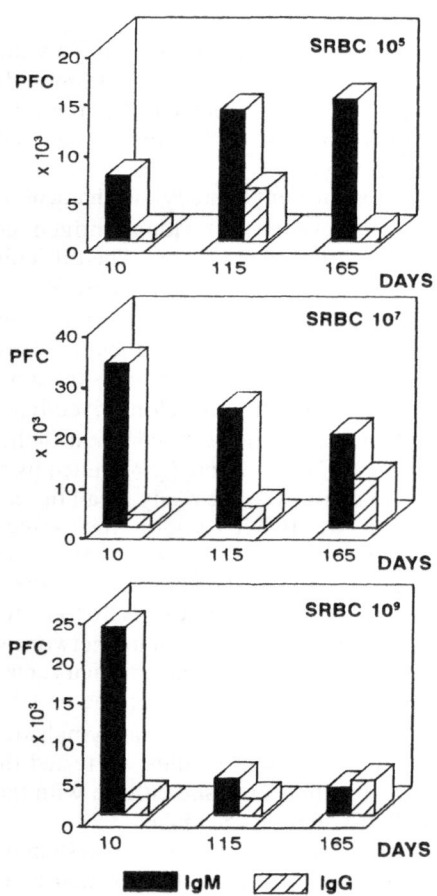

Figure 2. Secondary response with increasing doses of SRBC on different days after priming

2

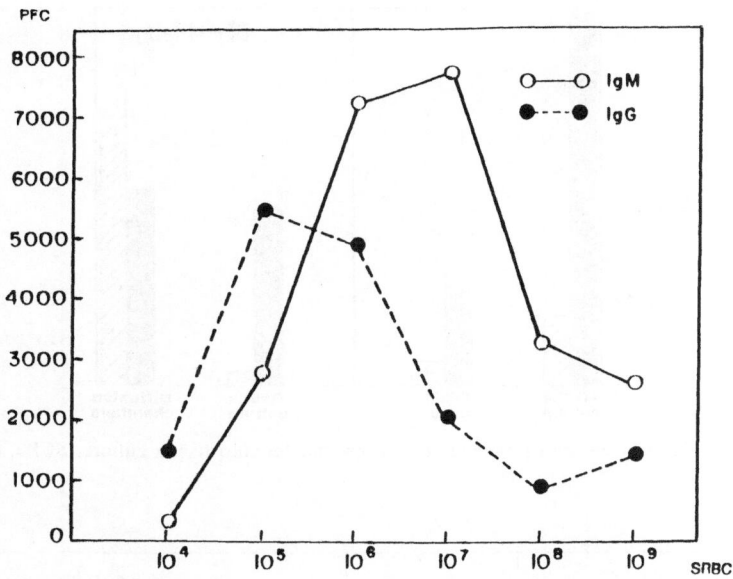

Figure 3. Secondary IgM and IgG responses recalled by different doses of SRBC 80 days after priming.

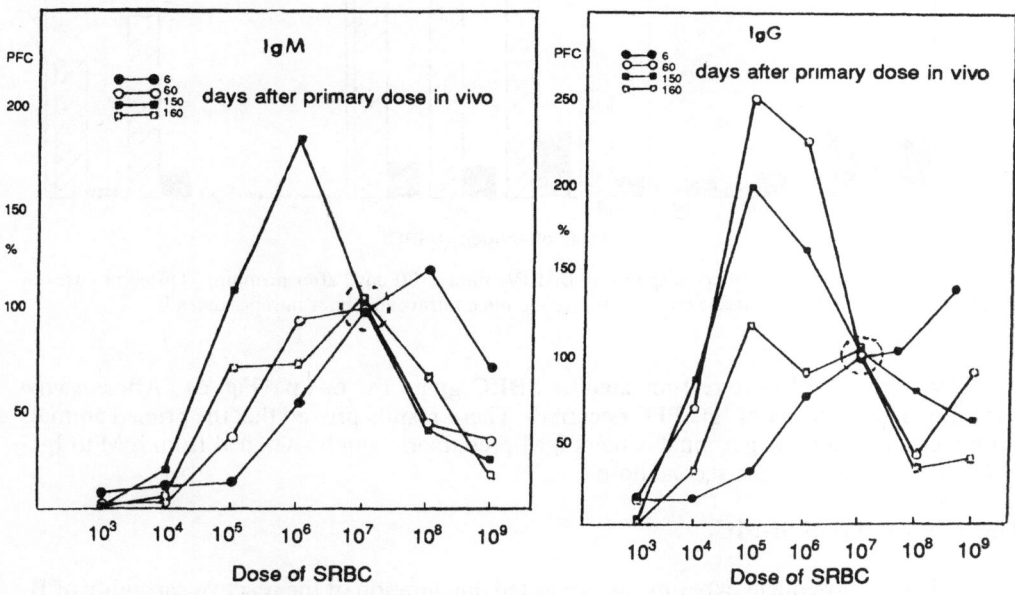

Figure 4. Comparison of IgM and IgG responses: the response to a boosting dose of 10^7 SRBC, was taken as 100%.

IgM and IgG Secondary Responses in Tissue Culture and *In Vivo*

The observation that the IgM secondary response in tissue culture exceeded that number of IgG was a surprising. The cellular basis of this phenomenon was analyzed by transferring isolated B-MC into tissue cultures, into SCID mice, and into diffusion chambers (Fig. 5); under these conditions the IgM response prevailed. The same donors from which

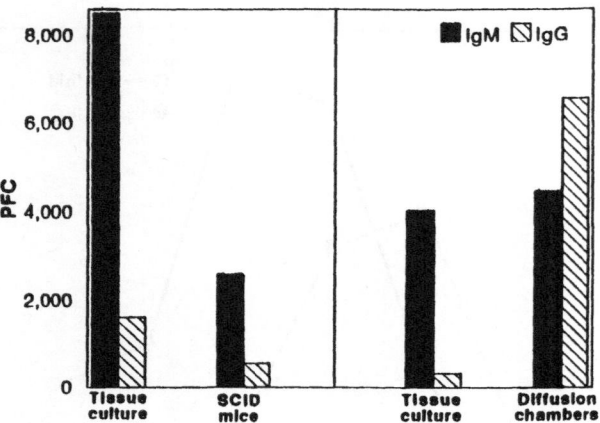

Figure 5. The secondary response of primed cells after transfer into tissue culture, SCID, or diffusion chambers

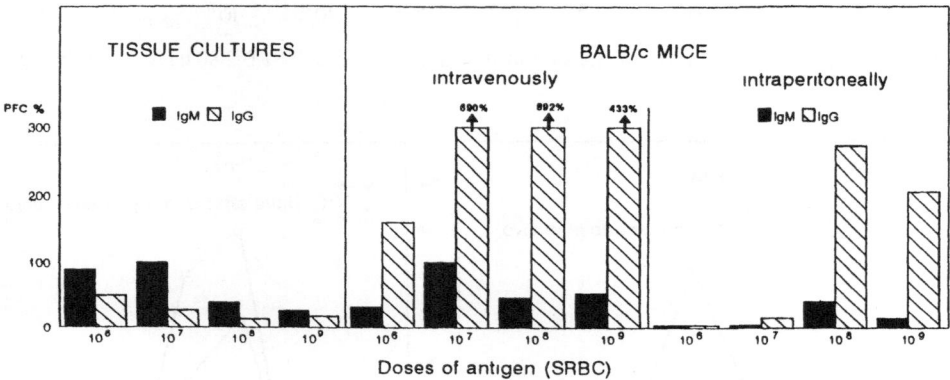

Figure 6. IgM and IgG secondary response in BALB/c mice (130 days after priming) Different doses of SRBC were added to cells in tissue cultures or given to mice intravenously or intraperitoneally

B-MC were isolated were restimulated by SBRC given i v or i p (Fig 6) After *in vivo* boosting, a quick onset of IgG PFC occurred These results proved that the primed animals contained in lymphatic germinal centers IgM precursors which switched from IgM to IgG PFC immediately after the second dose

The Character of B-MC

In our subsequent experiments, we tested the duration of the reactive capability of B-MC and B-ICC in tissue cultures Both types of cells were cultured and the antigen was added on specified days starting 24 h after establishing the cultures PFC responses were determined 5 days after addition of antigen These data demonstrated that both B-MC and B-ICC rapidly lost their ability to be stimulated by antigen, the loss of reactivity had the same rate and slope (Fig 7) This proved the same reactive capacity of virgin and memory cells, they differed only in numbers However, B-MC could be rescued in tissue culture to survive for a long period if they were stimulated by a second dose immediately after isolation (day 0) When the antigen was added after 24 h they lost the ability to respond with IgM and IgG antibody formation very quickly (Fig 8)

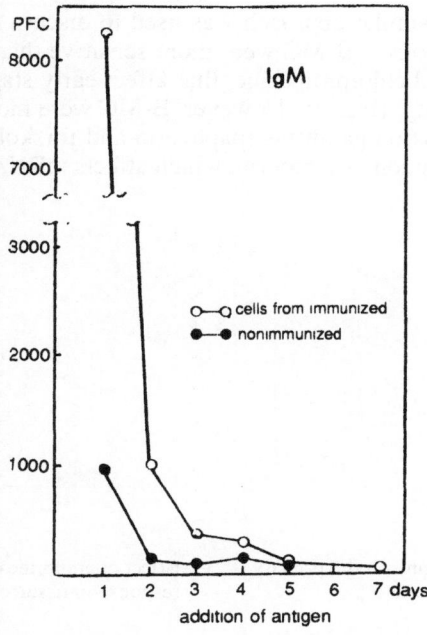

Figure 7. Maintenance of the immunological reactivity of virgin and immunized cells in tissue cultures.

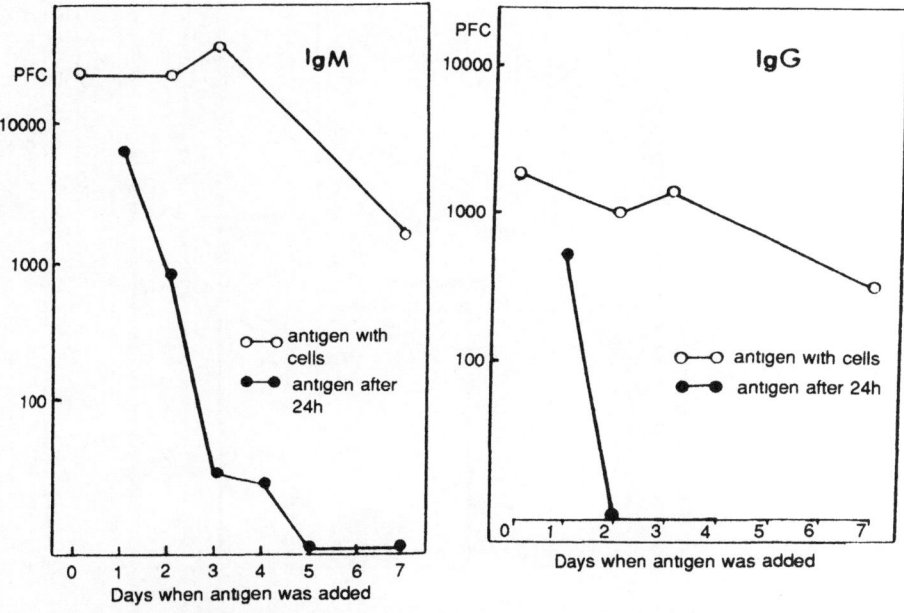

Figure 8. Rescue of immunological reactivity in B-MC by a second dose of antigen immediately after transfer in tissue culture.

Metabolic Changes During B-MC Induction

In the past, we studied the action of drugs that affect ion transport through cell membranes, intracellular transduction pathways, and transcriptional events on lymphocytes activated by antigen [11] A similar approach was used to discern differences between the primary and secondary response B-MC were more sensitive than B-ICC to the action of drugs (such as amiloride and chlorpromazine, that affect early stages of cell activation by acting on membrane signaling) (Fig 9) However, B-MC were more resistant to the action of drugs that affect transduction pathways (papaverin and forskolin, which act on adenyl cyclase), or on gene transcription (azathioprin, which affects mRNA synthesis) (Fig 10)

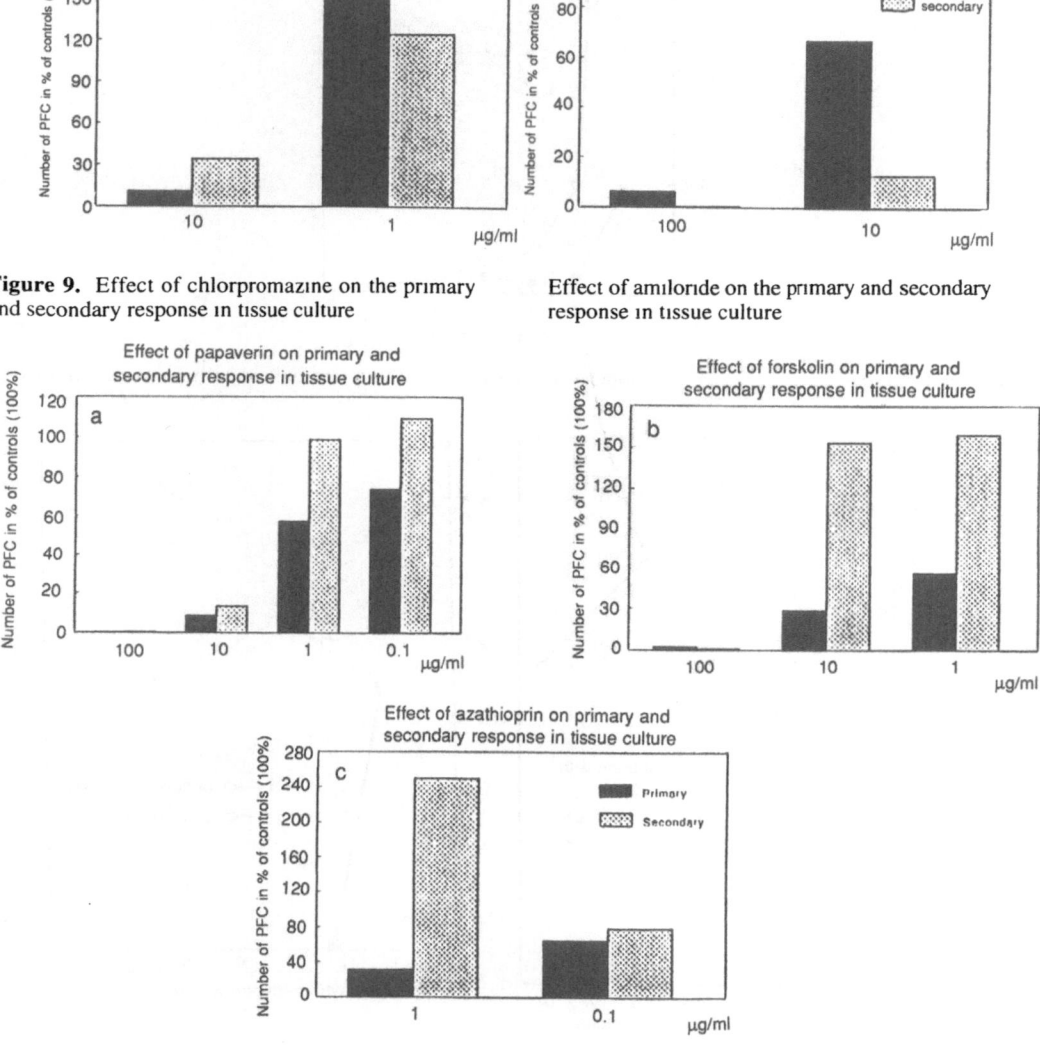

Figure 9. Effect of chlorpromazine on the primary and secondary response in tissue culture

Effect of amiloride on the primary and secondary response in tissue culture

Figure 10. a) Effect of papaverin on the primary and secondary response in tissue culture, b) Effect of forskolin on the primary and secondary response in tissue culture, and c) Effect of azathioprin on the primary and secondary response in tissue culture

Induction of B-MC in BALB/c Mice and in Tissue Culture

Immunological memory is characterized by quantitative and qualitative differences between responses to the first and the second dose of the same antigen B-MC were isolated from spleens of BALB/c mice and from tissue culture 7 days after immunization with a primary dose of 10^5 SRBC. A second dose (10^6 SRBC) was added to the cells immunized *in vivo* or *in vitro*. Virgin B-ICC (controls) were stimulated with the same dose (10^6 SRBC). Both types of cells, mixed with antigen, were injected i.v. into isologous recipients or transferred into tissue culture. A pronounced secondary IgM response was obtained with cells primed *in vivo* or *in vitro* and adoptively transferred into mice when compared with the response elicited in controls. The number of PFC after transfer into tissue culture was significantly increased in comparison with respective controls (Fig. 11).

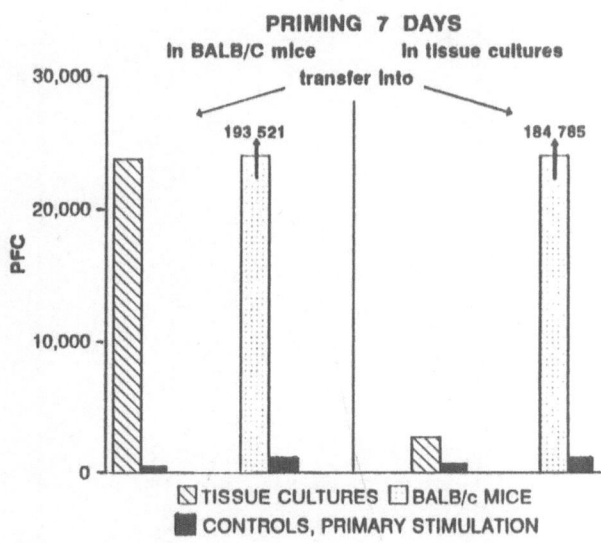

Figure 11. Secondary IgM response of primed cells transferred into tissue culture or adoptively into isologous mice.

CONCLUSION

Early B-MC were successfully induced in tissue cultures and a secondary response was recalled under the same conditions *in vitro*. These results provide a basis for further attempts to induce long-term B-MC in tissue culture.

REFERENCES

1. P. Goelet, V. F. Castellucci, S. Schacher, and E. R. Kandel, *Nature* 322:419 (1986).
2. J. H. Colle, A. M. Le Moal, and P. Truffa-Baci, *Crit. Rev. Immunol.* 10:259 (1990).
3. F. Celada, *Progr. Allergy* 15:223 (1971).
4. D. B. Kotloff and J. J. Cebra., *Molec. Immunol.* 25:147 (1988).
5. B. Schittek and K. Rajewsky, *Nature* 346:749 (1990).
6. S. Strober, *Transplant. Rev.* 24:84 (1975).
7. D. Gray and H. Skarvall, *Nature* 336:70 (1988).
8. I. C. M. MacLennan, Y. J. Liu, S. Oldfield, J. Zhang, P. J. L. Lane, *Curr. Top. Microbiol. Immunol.* 159:37 (1990).
9. H. Buerki, R. R. Craft, M. W. Hess, J. Laissue, H. Cottier, and R. D. Stoner, *Immunol. Lett.* 23:87 (1989).
10. J. Sterzl and L. Mandel, *Folia Microbiol.* 9:173 (1964).
11. J. Sterzl, *in:* "Highlights In Modern Biochemistry", Vol. 2, A. Kotyk *et al.*, eds., p. 1277, VSP, Utrecht (1989).

IN VIVO SWITCHING: IDENTIFICATION OF GERMLINE TRANSCRIPTS FOR HUMAN IgA

C.I.E. Smith,[1] B. Baskin,[1] E. Pattersson,[2] L. Hammarstrom,[1] and K.B. Islam[1]

[1]Center for Biotechnology, Karolinska Institute at Novum, S-141 57 Huddinge and Department of Clinical Immunology
[2]Department of Nephrology, Karolinska Institute at Huddinge Hospital, S-141 86 Huddinge, Sweden

INTRODUCTION

Transforming growth factor β1 (TGF-β1) induces the earliest known steps of isotype switching to IgA in humans. Switching is preceded by transcription of 'I-exons' located 5' of the Ig switch regions. Such mRNA species lack a variable portion and are denoted 'germline' transcripts. We have previously identified I-regions and we have now investigated the *in vivo* expression of human germline transcripts. Decreased levels of germline α transcripts (containing an Iα exon) were found in Ig deficiency diseases and in IgA nephropathy. These findings are compatible with an *in vivo* role of germline transcription in isotype switching and may contribute to the understanding of mechanisms underlaying abnormalities in the immune system.

STRUCTURE OF THE HUMAN IGHC LOCUS

The human IGHC locus contains 2 Cα genes (Cα1 and Cα2) and 7 other functional gene segments encoding the constant region heavy chains of Ig molecules [1,2] (Fig. 1). After activation of the B-lymphocytes, isotype switching may take place enabling cells to synthesize Ig molecules of other isotypes than IgM and IgD. The isotype switch results in a recombination with breakpoints in repetitive sequences being located upstream of all functional C_H genes, with the exception of Cδ[3,4] and the deleted DNA forms circular elements designated 'switch circles'.[5] The existence of Cμ-containing transcripts with a 5' non-translatable exon was demonstrated in 1985.[6] Similar transcripts originating from other C -genes were also identified and a correlation between transcriptional activity and isotype switching was found.[7,8] Such transcripts are normally referred to as germline mRNA; they contain the corresponding CH exons and lack a variable portion. The upstream exon(s) is normally denoted 'I-exon(s)' and the region in which they are located 'I-regions'.[9] The first human I-regions were identified in the Cγ loci[10] in 1989, later followed by Iε[11] and in 1991, Iα1 and Iα2 elements were reported.[12,13]

Advances in Mucosal Immunology, Edited by
J. Mestecky *et al.*, Plenum Press, New York, 1995

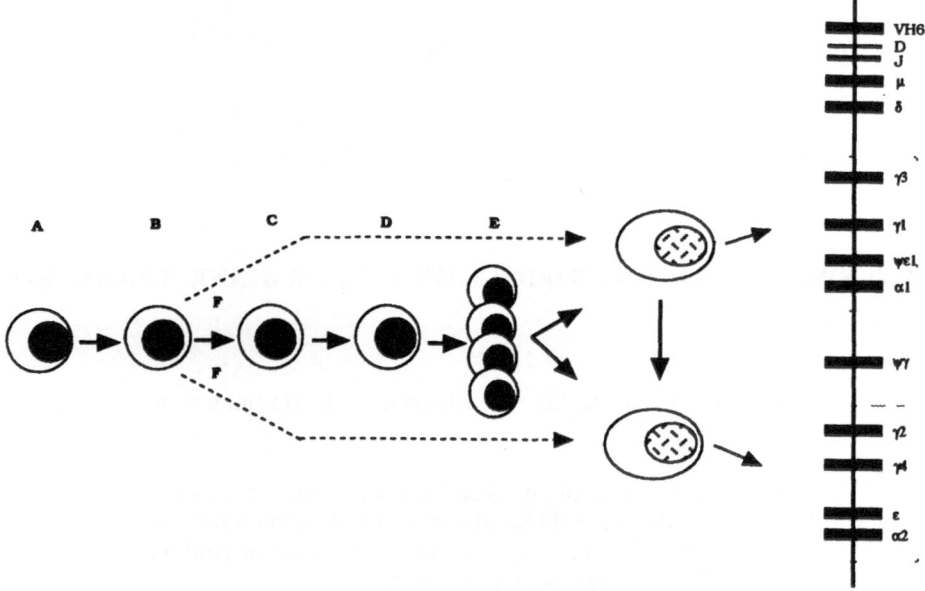

Figure 1. Directed versus stochastic model for isotype switching Solid lines represent the most likely differentiation pathways Dotted lines represent hypothetical pathways (A) Stem cell (B) Early B cell (C) Putative precommitted stage B cells are committed to utilize C_H genes within the first duplication unit (D) Putative stochastic stage committed cells upon stimulation will transcriptionally activate C_H genes within the first duplication unit (E) Putative instructive stage under the influence of exogenous signals delivered by lymphokines or cell to cell interaction, single C_H genes will be utilized in a switch recombination (F) Alternative pathways hypothetical alternative pathways where cells are precommitted to either the first or the second duplication unit

DETECTION OF HUMAN Iα1 and Iα2 TRANSCRIPTS USING RT-PCR

Fig 2 schematically depicts the structure of the human Iα 1 region where the corresponding germline transcripts are initiated Because the Iα exons contain multiple stop codons, it is not feasible to identify the translated product

Furthermore, because we have previously found that the levels of transcripts are rather low[12,14], we have developed a PCR (polymerase chain reaction)- based technique to identify and quantitate the transcripts [14] In short, first strand cDNA was synthesized using *Not* I-d(T) primers followed by amplification of Iα mRNA using primers complementary to the 3' Iα exonic region and to the C_H1 of Cα As Iα mRNA lacks the switch α region, erroneous results caused by contamination of genomic DNA can be excluded on a size basis In order to quantitate Iα mRNA, an irrelevant 56 bp fragment was cloned into the *Pst* I site of PCR amplified Iα mRNA The amount of purified mRNA was assessed and competitive quantitative PCR was carried out using a dilution series of known concentrations of the competitor DNA construct that was co-amplified with cDNA equivalent to 0 33 μg of total RNA For seimquantitave PCR, cDNA was co-amplified using primers for Iα and β-actin

The identity of PCR products was confirmed by size determination of bands on agarose gels as well as by Southern blot hybridization using internal probes The sensitivity was found to be in order of 10 fg of cDNA When human lymphoid organs were analyzed Iα mRNA was detected Higher levels were found in adenoid and tonsil as compared to peripheral blood mononclear cells [14] In order to study mechanisms in immune disease, the amount of Iα message was measured in peripheral blood mononuclear cells from donors with IgA deficiency and in patients IgA nephropathy (Table 1) Using 35 cycles of PCR, Iα

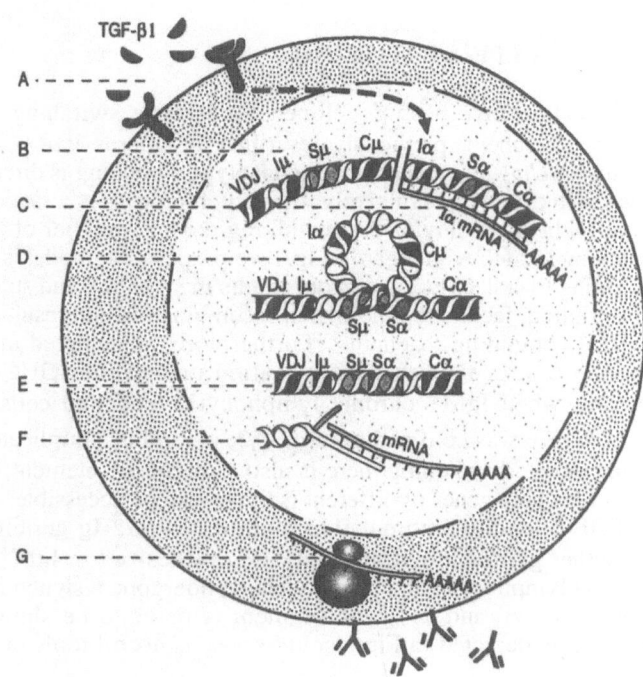

Figure 2. A model showing the role of TGF β1 during IgA class switching (A) Interaction of TGF-β1 with its receptor and the putative signaling pathway leading to the induction of Iα germline mRNA (B) Unrearranged Ig C_H locus (C) Expression of Iα germline mRNA (D) Switch recombination between the Sμ and the Sα region with looping out and subsequent deletion of the intervening sequences (E) Isotype switching to IgA (F) Transcription of α mRNA

Table 1. Expression of germline α mRNA in different disorders

Donors	Expression level[a]
Normal healthy	+
IgA deficiency	( )
IgA nephropathy	

[a]+ denotes Iα mRNA levels detectable after 35 cycles of PCR using cDNA from unstimulated peripheral blood mononuclear cells

mRNA was consistently detectable only in healthy donors Occasionally, low levels were found in IgA deficiency However, the predominant finding was the absence of Iα mRNA[14], indicating that the defect is manifested already in the early stages of isotype switching Analysis of switch recombination break points in our laboratory support this notion (Islam, K B *et al*, in preparation) When cells from patients with IgA nephropathy were analyzed, unexpectedly, decreased levels of Iα mRNA were found (Table 1) [14]

This findings argues in favor of a primary B cell abnormality in IgA nephropathy, as a selective defect in cells being in the process of switching to IgA specific for a certain antigen (sets of antigens) would not be expected to cause a general absence of Iα mRNA

MODELS FOR ISOTYPE SWITCHING

Different models have been developed for isotype switching. Fig. 1 depicts a modified version of the scheme previously described by Sideras et al.[15] The two extremes are *directed* versus *stochastic* models, in which isotype switching is directed by exogenous agents such as interleukins, or, alternatively, the switch process is a random event resulting in the recombination between any of the switch regions. A number of observations favor the directed switching model; with regard to IgA we propose the model is depicted in Fig. 2. Here, TGF-β1 binding to cell surface receptors results in signaling and subsequent activation of Iα-region transcription, facilitating the switch recombination and resulting in the synthesis of full-length (VDJ-containing) α-chains. Furthermore, as depicted in Fig. 1, cells may switch to either the first or the second duplication unit of the IGHC locus. This was particularly evident when IgM[+] chronic lymphocytic leukemia cells were studied, as transcription 3' of Cμ/Cδ was confined to the 5' (γ3, γ1, and α1) duplicated locus.[15] Thus, although we favor directed switching there is also a stochastic element, as a whole region containing constant H chain genes of different isotypes can be accessible.

During TGF-β1-induced stimulation, both α1 and α2 Ig germline transcripts are produced.[12] Whether germline α2 transcripts are synthesized in IgM[+] cells, or whether they are limited to B lymphocytes that already have undergone a switch recombination to a 5' duplicated locus (γ3, γ1, and α1) gene segment remains to be shown. However, we believe that the models depicted in Fig.1 could serve as useful tools in deciphering these alternatives.

ACKNOWLEDGMENTS

This work was supported by the Swedish Medical Research Council, the Ake Wiberg Foundation and the Hesselman Foundation. The skillful technical assistance of Ms. Anita Wirell is gratefully acknowledged.

REFERENCES

1. J. G. Flanagan and T. H. Rabbitts, *Nature* 300:709 (1982).
2. M. H. Hofker, M. A. Walter, and D. W. Cox, *Proc. Natl. Acad. Sci. USA* 86:5567 (1982).
3. H. Sakano, R. Maki, Y. Kurosawa, W. Roeder, and S. Tonegawa, *Nature* 286:676 (1908).
4. U. von Schwedler, H. M. Jack, and M. Wabl, *Nature* 345: 452 (1990).
5. A. Shimizuand and T. Honko, *Cell* 36:801 (1984).
6. G. G. Lennon and R. P. Perry, *Nature* 318:475 (1985).
7. J. Stavnezer-Nordgren and S. Sirlin, *EMBO J.* 5:95 (1986).
8. G. D. Yancopoulos, R. A. DePinho, K. A. Zimmerman, S. G. Lutzker, N. Rosenberg, and F. W. Alt, *EMBO J.* 5:3259 (1986).
9. S. Lutzker and F. W. Alt, *Mol. Cell. Biol.* 8:1849 (1988).
10. P. Sideras, T.-R. Mizuta, H. Kanamori, N. Suzuki, M. Okamoto, K. Kuze, H. Ohno, S. Doi, S. Fukuhara, M. S. Hassan, L. Hammarstrom, C. I. E. Smith, A. Shimizu, and T. Honjo, *Inter. Immunol*. 3:1107 (1991).
11. J.-F. Gauchat, D. A. Lebman, R. L. Coffman, H. Gascan, and J. E. de Vries, *J. Exp. Med.* 172:463 (1990).
12. K. B. Islam, L. Nilsson, P. Sideras, L. Hammarstrom, and C. I. E. Smith, *Inter. Immunol.* 3:1099 (1991).
13. L. Nilsson, K. B. Islam, O. Olafsson, I. Zalcberg, C. Samakovlis, L. Hammarstrom, C. I. E. Smith, and P. Sideras, *Inter. Immunol.* 3: 1107 (1991).

14. K. B. Islam, B. Baskin, B. Christensson, E. Petterson, L. Hammarstrom, and C. I. E. Smith, submitted (1992).
15. P. Sideras, L. Nilsson, K. B. Islam, I. Zalcberg Quintana, L. Freihof, A. Rosen, G. Juliusson, L. Hammarstrom, and C. I. E. Smith, *J. Immunol.* 149:244 (1992).

ANALYSIS OF THE SECOND MESSENGER SYSTEMS INVOLVED IN THE SYNERGISTIC EFFECT OF CHOLERA TOXIN AND INTERLEUKIN-4 ON B CELL ISOTYPE-SWITCHING

Nils Lycke

Department of Medical Microbiology and Immunology
University of Göteborg
S-413 46 Göteborg, Sweden

INTRODUCTION

Cholera toxin (CT) is a powerful immunogen and adjuvant.[1,2] The latter property has attracted much interest from developers of vaccines and especially from researchers with an interest in oral vaccines . We and others have shown that CT used as a mucosal adjuvant together with unrelated antigens greatly enhances the response to these antigens after oral immunization, particularly the serum IgG and mucosal IgA responses.[2-4] These *in vivo* findings suggested the possibility that CT exerted direct immunomodulating effects on B cells. Therefore we undertook a series of *in vitro* studies of the effects of CT on activated murine B cells. We found that CT induces LPS-stimulated membrane IgM-positive B cells to undergo increased switch differentiation to IgG and IgA B cells.[5] Furthermore, we demonstrated that CT acted synergistically with the T cell lymphokine, IL-4, to promote IgG1-switch differentiation which was demonstrable at the cellular- as well as the gene-transcriptional level.[6] The latter effect was evidenced by the strongly enhanced expression of sterile germline γ1-RNA transcripts in the presence of CT and IL-4 as compared to those seen with B cells cultured with IL-4 alone.[6] Our current understanding of Ig-heavy chain gene recombination strongly supports the notion that such germline RNA transcripts, i.e. initiated 5' to switch region of the C-heavy chain gene, preceeds and perhaps directs the final isotype-switching event.[7,8]

In the studies to be described we address the question of which mechanisms are responsible for the isotype-switching effect of CT on B cells and which second messenger systems are involved in this process. In our earlier studies we found that both CT and the non-toxic cholera B-subunit (CT-B) could promote isotype-switching in B cells in the presence of IL-4.[6]. However, these studies were performed with CTB which was purified from whole toxin and might have contained minute contamination with A-subunit/whole toxin[6]. Therefore, for the present study we used a recombinant CTB preparation which was completely free of contaminating A-subunit/whole toxin to compare the effects on IgG1-switching with those stimulated by the intact CT molecule .

MATERIALS AND METHODS

In vitro cultures: Single cell suspensions of normal C57BL/6 mouse B cells from spleen were obtained as described.[6] The cells were cultured at indicated densities in Iscove's complete medium (Gibco, Paisley, Scotland) containing 10% fetal calf serum (FCS) (Gibco), L-glutamine (Gibco), gentamicin, 2-ME (Sigma Chemical Co., St Louis, MO). Cultures were maintained for 6 days at 37° C and 10 % CO_2 in flat-bottomed 96-well plates (Nunclon, Roskilde, Denmark) unless stated otherwise. *Salmonella minnesota*

Advances in Mucosal Immunology, Edited by
J. Mestecky *et al.*, Plenum Press, New York, 1995

lipopolysaccharide (LPS) (Sigma) (10 µg/ml) with CT (List Biological Laboratories Inc. Campbell, CA) or rCTB produced by a derivative of the classical *Vibrio cholerae* 569 Inaba strain lacking the gene for the toxic CTA fragment[9]. In some experiments CT was replaced by dibutyryl-cAMP (dB-cAMP)(Sigma) at indicated concentrations. The IL-4 used for these studies was produced by the X63Ag8-653 myeloma cell line transfected with murine IL-4 cDNA that spontaneously produces IL-4.[10] When indicated, B cell cultures were performed in the presence of the phorbol ester, PMA (Sigma) at 5ng/ml or protein kinase A inhibitors; adenosine-3',5'-cyclic monophosphothioate Rp-isomer (Rp-cAMP)(Biolog, Bremen, Germany) or n-(2-(Methylamino)ethyl)-5-isoquinolinesulfonic dihydrochloride (H-8) (Seikagaku Kogyo, Tokyo, Japan) at 500µM and 15µM respectively.

ELISPOT Technique

Individual cells secreting immunoglobulin of specific isotype were determined by the Elispot technique as described[6].

[^{3}H]TdR Uptake

B-cell enriched spleen cells were cultured for 3 days in medium containing LPS (10 µg/ml) with or without PMA (5 ng/ml)(Sigma) or Rp-cAMP(500µM) (Biolog) or H-8 (15µM)(Seikagaku Kogyo). CT (List) or rCTB at indicated concentrations was added to the cultures in the absence or presence of IL-4 (10.000 U/ml). ^{3}[H]TdR (NEN) (1 mCi/well) was added for the final 6 hours; at harvest, cells were collected on filter paper using a semiautomatic cell harvester (Skatron, Lier, Norway). The filter paper was then placed into a scintillation cocktail (Optofluor, Amersham) and counted in a Beckman counter. Counts per minute (Cpm's) incorporated were expressed as means ± SD of triplicate wells in each experiment.

cAMP-Stimulation Test

A commercial cAMP-test kit was used according to the manufacturer's instructions (Amersham, England).

Northern Blotting of mRNA

The stimulated cells were harvested after 48h and washed repeatedly in cold PBS and stored at -70° C in pellets until RNA extraction was performed as described.[6] After separation in agarose gel, the mRNA was blotted on to Hybond nylon filters. The nylon filters were then hybridized with an Ig-heavy chain constant region-specific DNA probe recognizing γ1 RNA in germline configuration, i.e., the probes[6,7] were specific for the I-region immediately 5' to the switch-region of the γ1 gene (a region that normally is deleted in recombination). The Iγ1 probe was a 2 kb HindIII/PstI fragment from p1/EH10 subcloned into Bluescript (Stratagene, La Jolla, CA).

RESULTS

Effect of CT and rCTB on IgG1 B Cell Differentiation in IL-4 and LPS Stimulated Spleen Cell Cultures

We tested the effect on isotype-switching of various concentrations of a recombinant CTB (rCTB) which was produced in a *Vibrio cholerae* strain lacking the gene for CTA.[9] The rCTB preparation provided enhanced switching to IgG1 in IL-4 (5000 U/ml) stimulated cultures (Fig. 1). The effect of CT and rCTB on isotype-differentiation was further compared by dose-response analysis in cultures containing LPS alone or LPS in combination with IL-4. CT was at least a thousand-fold (on an equimolar basis) more effective compared to rCTB in promoting IgG1-differentiation in LPS and IL-4 stimulated B cell cultures.

Next we investigated whether the synergistic effect on IgG1 differentiation observed with IL-4 plus CT and IL-4 plus rCTB was associated with increases in intracellular cAMP. We found no increase in intracellular cAMP in the presence of IL-4 or rCTB, nor in the combination of these two agents . CT induced significant increases in intracellular cAMP.

Since the whole cholera toxin molecule was more effective as compared to the rCTB in stimulating isotype-switching in B cells in the presence as well as in the absence of IL-4, and because CT but not rCTB stimulated cAMP in these B cells, a critical role for cAMP

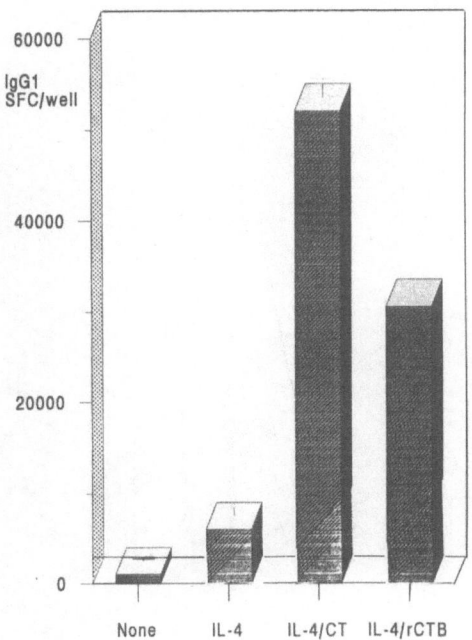

Figure 1. CT and rCTB synergistically enhance IL-4 stimulated IgG1 B cell differentiation. Spleen B cells were cultured in LPS (10µg/ml) and stimulated by a suboptimal concentration of IL-4 (5.000 U/ml) in the presence or absence of CT (0.1 µg/ml) or rCTB (1.0 µg/ml). Single IgG1 secreting cells (SFC) were determined and the numbers are expressed as means of triplicate wells ±SD.

was suggested in IgG1-switching. Therefore, additional experiments were performed and CT was replaced by dBcAMP in the B cell cultures. Strong augmenting effects of dBcAMP were observed on IgG1 SFC formation in the presence of IL-4 and LPS. In subsequent analysis of the effects of rCTB and dBcAMP on isotype-switching in B cells we combined these two agents. The intention was that such an experiment would mimic the effects of the whole CT molecule by providing both B cell membrane GM1-receptor interaction through the rCTB as well as increases in intracellular cAMP by adding dBcAMP. The result from these experiments clearly demonstrated that the combination of the two agents was several-fold more efficient in promoting IgG1 SFC formation in the IL-4 and LPS stimulated B cell cultures as compared to either agent used alone (Fig. 2).

The enhancing effect of CT was tested in the presence of Rp-cAMP or H-8, agents capable of inhibiting cAMP-dependent protein kinases (PKA). These agents significantly abrogated the synergistic effect of CT on IL-4 stimulated IgG1-differentiation . No significant inhibitory or enhancing effect by Rp-cAMP or H-8 on B cell proliferation was observed in the cultures. Thus , CT affects isotype-switching by a mechanism involving the PKA-system.

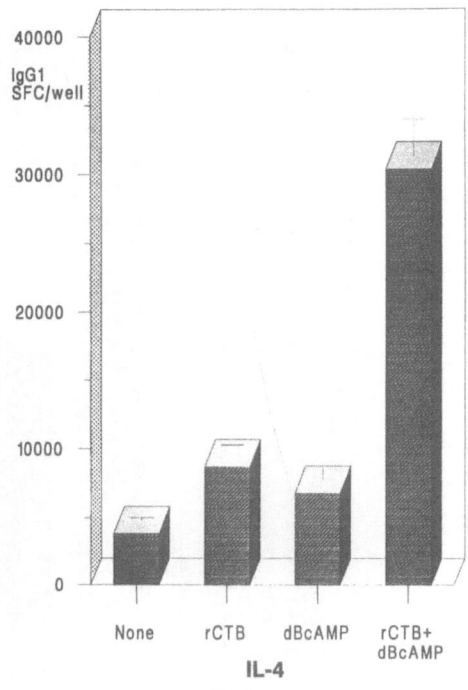

Figure 2. rCTB and dBcAMP synergistically promote IL-4 stimulated IgG1 B cell differentiation. Spleen B cells were cultured in LPS (10µg/ml) and a suboptimal concentration of IL-4 (5.000 U/ml) in the presence or absence of rCTB (1.0 µg/ml) or dBcAMP (100µM) or the combination of these two agents. Single IgG1-secreting cells (SFC) were determined and the numbers were expressed as means of triplicate wells ±SD.

Detection of Germline γ1 RNA Transcripts in B Cells Stimulated by CT, rCTB or dBcAMP

In our previous study we found that CT very strongly promoted an increased expression of such γ1-RNA transcripts in B cells stimulated by IL-4.[6] When rCTB or dBcAMP was allowed to replace CT in similar experiments, the northern blot analysis of isolated mRNA using an I-region γ1-specific probe indicated that similar to CT, dBcAMP enhanced the expression of germline γ1-RNA transcripts as compared to those induced by IL-4 alone (Fig. 3). In contrast, rCTB did not alter this level of expression of germline γ1-RNA transcripts, indicating that rCTB did not affect early events in isotype-switching whereas dBcAMP did (Fig. 3).

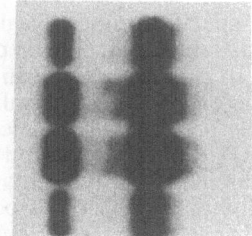

IL-4

IL-4 + CT

IL-4 + dbcAMP

IL-4 + rCTB

Figure 3. Northern blot analysis showing the effect of CT, rCTB or dBcAMP on expression of IL-4 induced germline $\gamma1$ (Iγ1) RNA transcripts. Spleen B cells were cultured for 48h in medium containing LPS (10μg/ml) and an optimal concentration of IL-4 (10.000 U/ml)) in the presence or absence of CT (0.1 μg/ml), rCTB (3.0 μg/ml) or dBcAMP (100 μM).

CT enhances IgG1 Isotype Differentiation in IL-4 Stimulated B Cells in the Presence of Phorbol Esters

Protein kinase C (PKC), was recently shown to be involved in the anti-Ig plus IL-4R activation pathway of resting B cells.[11] Spleen enriched B cells were cultured in the presence or absence of the phorbol ester, PMA. This agent had profound effects on LPS-stimulated B cell differentiation, inhibiting to almost 100% the IgM, IgG1 and IgG3 SFC formation in the cultures without significantly changing the proliferative responses (Table 1).

When IL-4 was added to the LPS and PMA stimulated B cell cultures, it strongly counteracted the inhibitory effect of PMA on B cell differentiation as well as augmented B cell proliferation . In particular IL-4 restored IgM SFC formation while PMA was still very inhibitory to IL-4-induced IgG1-differentiation, which was reduced by 90% as compared to that seen for B cells stimulated with IL-4 and LPS-alone. However, in the presence of CT the IL-4 stimulated IgG1-differentiation was significantly less inhibited by PMA. In contrast, rCTB was much less effective compared to CT in preventing the inhibitory action of PMA on IgG1-differentiation (Table 1). Thus, CT, but not rCTB, exerted a strong enhancing effect on B cell IgG1-differentiation in cultures where IL-4-stimulated IgG1-differentiation was greatly suppressed by PMA.

Table 1. Synergistic effects of CT or rCTB and IL-4 in the presence or absence of the phorbol ester, PMA.

PMA	Agents	Isotype distribution	SFC/well		TdR Uptake
		IgG1	IgG3	IgM	(cpm)
No	None	357±72	1460±288	28350±1964	125806±4802
	IL-4	26649±907	638±189	22680±7091	72489±2421
	IL-4/CT	62937±972	1838±424	24948±2598	68503±2841
	IL-4/rCTB	46134±862	4902±430	18115±2578	27421±3144
Yes	None	84±12	0	1055±436	127280±3239
	IL-4	2772±109	140±12	17577±2553	100906±2460
	IL-4/CT	17577±982	1386±123	23247±982	59490±5689
	IL-4/rCTB	6112±288	420±56	11050±520	28575±1298

Spleen B cells were cultured at 5×10^5 cells/ml for 6 days in LPS (10μg/ml) and in the presence or absence of PMA (5ng/ml) with or without an optimal dose of IL-4 (10.000 U/ml) alone or in combination with CT (0.1 μg/ml) or rCTB (5.0 μg/ml). SFC in of the IgG1,IgG3 and IgM isotypes were determined and expressed as means ±SD of triplicate wells. The data are respresentative of three experiments with identical results. [3][H]TdR was added during the final 6h of the 3 days of incubation. Mean cpm±SD of triplicate cultures.

Northern blot analysis using the Iγ1-specific probe revealed that, despite strong inhibitory effects of PMA on IgG1 SFC formation, IL-4 induced equal (or slightly higher) levels of expression of γ1-RNA transcripts in B cells cultured in the presence as in the absence of PMA. Moreover, CT increased the expression of germline γ1-RNA transcripts in IL-4 stimulated B cells irrespective of the presence or absence of PMA. The strong inhibitory effect by PMA on IL-4-induced IgG1-differentiation and the lack of reduction in γ1-RNA transcipts suggested that PMA did not inhibit early events in isotype-switching in IL-4 stimulated B cells but rather interfered with later, perhaps post-switch, events . CT greatly enhanced γ1 RNA transcripts even in the presence of PMA, suggesting that CT acted early in isotype switch differentiation.

CONCLUSIONS

In this study we provide new information on the mechanisms responsible for the strong enhancing effect of CT on isotype-switching in murine B cells.[5,6] We demonstrated that at least two different mechanisms are involved in the CT-enhanced IL-4-induced IgG1 switch differentiation: 1) increased intracellular cAMP-levels stimulated by the A-subunit potentiates isotype-switching early in differentiation by augmenting the formation of sterile germline γ1-RNA transcripts and 2) the binding of the non-toxic B-subunit to the membrane GM1-ganglioside receptor promotes later stages of differentiation. Moreover, we extend our previous data suggesting that IL-4 and CT affect B cell differentiation by separate pathways by showing that, in contrast to CT, IL-4 does not stimulate cAMP in murine B cells. CT's effect on B cell switch differentiation, on the other hand, can be partially mimicked by dBcAMP and is blocked by inhibitors of cAMP-dependent protein kinases , PKA . The IL-4-pathway, but not the CT-pathway, seemed to be sensitive to changes in protein kinase C activity: because IgG1 SFC formation was almost completely blocked by addition of the phorbol ester, PMA. In contrast, the addition of CT to the PMA-containing cultures resulted in significantly less inhibition of IgG1 production. Northern blot analysis of mRNA using a germline IgG1-specific probe revealed that, despite strong inhibitory effects of PMA on IgG1 SFC formation, IL-4 induced equal (or slightly higher) levels of expression of sterile γ1-RNA transcripts in B cells cultured in the presence as well as in the absence of PMA . This result suggests that PMA blocked later events in IL-4 induced B cell isotype switch differentiation which is further supported by the finding that the whole CT molecule, acting early in differentiation, but not the rCTB, greatly enhances the expression of such transcripts irrespective of the presence or absence of PMA.

The existence of dual mechanisms (cAMP-stimulation and membrane GM1-ganglioside binding) operating together on B cell differentiation help to explain the powerful effect of this molecule on B cell isotype differentiation. Evidence for the need for ADP-ribosylation/cAMP-stimulation for an efficient adjuvant effect *in vivo* is now accumulating.[12] Whether interactions with the cell membrane GM1-ganglioside is required for an adjuvant effect *in vivo* is currently being investigated.

REFERENCES

1. N. Lycke and A.-M. Svennerholm, *In*: New Generation Vaccines. G. C. Woodrow and M. M. Levine eds. Marcel Dekker, New York.p.207 (1990).
2. N. Lycke and J. Holmgren, *Immunology* 59:301 (1986).
3. C.O. Elson and W. Ealding, *J. Immunol.* 132:27 (1984).
4. X. Liang, M. E. Lamm, and J. G. Nedrud, *J. Immunol.* 141:1495 (1988).
5. N. Lycke, and W. Strober, *J. Immunol.* 142:3781 (1989).
6. N. Lycke, E. Severinson, and W. Strober, *J. Immunol.* 145:3316 (1990).
7. J. Stavnezer, G. Radcliffe, Y.-C. Lin, J. Nietupski, L. Berggren, R. Sitia and E. Severinson, *Proc. Natl. Acad. Sci. USA* 85:7704(1988).
8. P. Rothman, Y.-Y. Chen, S. Lutzker, S. C. Li, V. Stewart, R. Coffman, and F.W. Alt, *Mol. Cell. Biol.* 10:1672(1990).
9. J. Sanchez and J. Holmgren, *Proc. Natl. Acad. Sci. USA* 86:481 (1989).
10. H. Karasuyama and F. Melchers, *Eur. J. Immunol.* 18:94 (1988).
11. M. M. Harnett, M. J. Holman, and G. B. Klaus, *J. Immunol.* 147:3831 (1991).
12. N. Lycke, T. Tsuji, and J. Holmgren, *Eur. J. Immunol.* 22:2277 (1992).

TRANSFORMING GROWTH FACTOR BETA (TGFβ) DIRECTS IgA1 AND IgA2 SWITCHING IN HUMAN NAIVE B CELLS

Francine Briere,[1] Thierry Defrance,[2] Beatrice Vanbervliet,[1] Jean-Michel Bridon,[1] Isabelle Durand,[1] Francoise Rousset,[1] and Jacques Banchereau[1]

[1]Schering-Plough, Laboratory for Immunological Research, 27 ch. des Peupliers, B.P. 11, 69571 Dardilly Cedex France; [2]Institut Pasteur, Av. Tony Garnier, 69365 Lyon Cedex 07, France

INTRODUCTION

Until recently, there was no experimental system that permitted the study of soluble factors capable of inducing human B lymphocytes to switch isotypes *in vitro*. Unlike lipopolysaccharide which is currently used for murine B cells, most of human B cell activators are poorly efficient. However, two novel experimental procedures are now available: one is based on the capacity of activated T cells or their membranes to elicit a B cell response which involves interactions in surface molecules.[1] The other one, which is referred to as the CD40 system was developed in our laboratory.[2] In this "CD40 system", polyclonal activation and sustained proliferation of human B cells[3] are obtained by presentation of an anti-CD40 monoclonal antibody on an irradiated mouse fibroblastic L cell line that stably expresses CDw32 (FcγRII). Upon interaction with the CD40 molecule, the Fc portion of the antibody undergoes spatial rearrangement which allows high-affinity interaction with CDw32. One can further enhance the B cell response by coupling the anti-CD40 stimulation with surface Ig (sIg) cross-linking agents such as anti-μ or *S. aureus* Cowan (SAC) particles (Fig. 1).

The CD40 molecule is a 277 amino acid, 48kDa glycosylated phosphoprotein which is expressed on antigen-presenting cells such as B lymphocytes, dendritic cells, and macrophages. The CD40 is a member of the nerve growth factor receptor family, characterized by the presence of 3-4 cysteine-rich motifs in the extracellular domain.[4]

Most studies concerning the differentiation of human B cells have not allowed discrimination between isotype switching and maturation of pre-committed precursors as whole B cell populations have been used. Among the B cells isolated from tonsils, two major subpopulations can be distinguished on the basis of sIgD expression. The sIgD+ B cells represents a homogeneous subpopulation which express sIgM, CD20, CD23, and CD44 in a monotonal fashion.[5]

In contrast, sIgD- B cells do not bind anti-CD23, weakly express sIgM and can be further divided into different subsets of B cells according to CD10, CD20, CD38, and CD44 profiles.[5] Due to the differential expression of these molecules, sIgD+ B cells display phenotypic features of B cells present in the mantle zone of the follicles of secondary lymphoid organs. These cells are considered as being naive i.e. they have not yet encountered the antigen and have not rearranged their heavy chain genes.

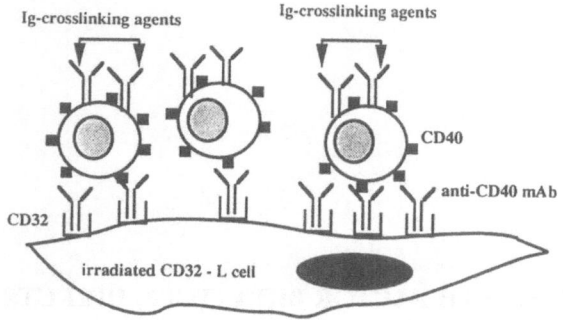

Figure 1. Simultaneous activation of B cells by immobilized anti-CD40 antibodies and Ig-cross-linking reagents: an experimental system.

RESULTS AND DISCUSSION

None of the activators alone (anti-sIg reagents: anti-μ or SAC nor cross-linked anti-CD40) could stimulate Ig synthesis. However, concomitant triggering of sIg and CD40 induced strong IgM secretion from sIgD$^+$ B cells (Table 1). This indicates that dual triggering of antigen receptor and CD40 results in a T cell-independent differentiation of naive B cells.

Table 1. Concomitant triggering of sIgs and CD40 induces IgM secretion from sIgD$^+$ B cells.

Stimulus	IgM	IgG	IgA
		(ug/ml)	
medium	0.08	0.04	0.03
SAC	0.08	0.05	0.03
anti-CD40	0.2	0.05	0.03
SAC + anti-CD40	**29**	0.1	0.07
anti-μ	0.08	0.05	0.04
anti-μ + anti-CD40	**1.8**	0.1	0.04

The pattern of isotypes produced by sIgD$^-$ B cells in this activation system is dramatically different (Table 2). In particular, IgG was reproducibly found to be the major isotype secreted. IgA usually remained a minor component of the Ig response. Thus among the sIgD$^-$ B cell subpopulation, the majority of cells can fully differentiate into IgG-secreting cells upon this activation.

Various cytokines have been tested for their capacity to modify the isotype distribution of sIgD$^+$ B cells (Table 3). IL-10 dramatically enhanced IgM production of sIgD$^+$ B cells, whereas IgG and IgA were produced in a low but reproducible manner (Table 3 and Fig. 2). IL-10 also synergized with SAC and cross-linked anti-CD40 to stimulate IgG, IgM, and IgA production by sIgD$^-$B cells (Fig. 2).

Due to the potentiating effect of IL-10 on IgG and IgA synthesis by sIgD$^-$ B cells, several possibilities may be set forth to explain IgG and IgA production by sIgD$^+$ B cells. First, these culture conditions may result in the expansion and differentiation of a few

Table 2. Concomitant triggering of sIgs and CD40 induces Ig secretion from sIgD⁻ B cells.

Stimulus	IgM	IgG (ug/ml)	IgA
medium	0.1	0.1	0.04
SAC	0.1	0.05	0.05
anti-CD40	0.3	0.3	0.1
SAC + anti-CD40	**4.9**	**7.2**	0.09
anti-μ	0.3	0.3	0.1
anti-μ + anti-CD40	0.2	0.2	0.4

Table 3. Effect of various cytokines on the pattern of Ig isotypes secreted by activated sIgD⁺ B cells.

Stimulus	IgM	IgG (ug/ml)	IgA
medium	9.1	0.1	0.1
IL-1α	16.5	0.2	0.2
IL-2	9.6	0.3	0.2
IL-3	9.2	0.3	0.2
IL-4	5.6	0.05	0.04
IL-5	10	0.2	0.2
IL-6	16.5	0.3	0.2
IL-10	**88.3**	**2.3**	**0.9**
TGF*β*	0.5	0.1	0.1
IFNγ	1.9	0.03	0.1
TNFα	12.9	0.2	0.2

contaminating IgG- and IgA-committed precursors. Second, IL-10 through its potent differentiation effect on B cells, may reveal the switch potential of the CD40 system. Third, IL-10 could be an isotype switch factor. We are presently trying to address this question.

Because TCFβ was previously shown to increase IgA production in lipopolysaccharide-stimulated murine lymphocytes[7], we tested the ability of TGFβ to influence isotype production of human B cells under these culture conditions. As illustrated in Fig. 2, addition of TGFβ strongly potentiated the IL-10-induced IgA synthesis by sIgD⁺cells, whereas TGFβ alone did not have any effect (Table 3). Strikingly, this enhancing effect of TGFβ was restricted to the sIgD⁺ B cell compartment and was isotype-specific. TGFβ suppressed both IgM and IgG production elicited by IL-10 from sIgD⁺B cells. Moreover, IL-10-induced IgG, IgM, and IgA secretion by sIgD⁻B cells were inhibited In a dose-dependent fashion by TGFβ (Fig. 2).

Comparison of the data obtained within the two different B cell subpopulations permits exclusion of the possible expansion of IgA-committed precursors and thus indicates a key role for TGFβ in the induction of IgA isotype commitment.

In this context, Islam *et al.*[8] have recently demonstrated that TGFβ induces germ-line transcripts of both IgA subclasses in *Branhamella catarrhalis*-activated human spleen B cells. Analysis of IgA subclass distribution indicated that both IgA subclasses were produced when sIgD⁺ B cells were activated through CD40 and sIg in the presence of IL-10 and TGFβ (Fig. 3).

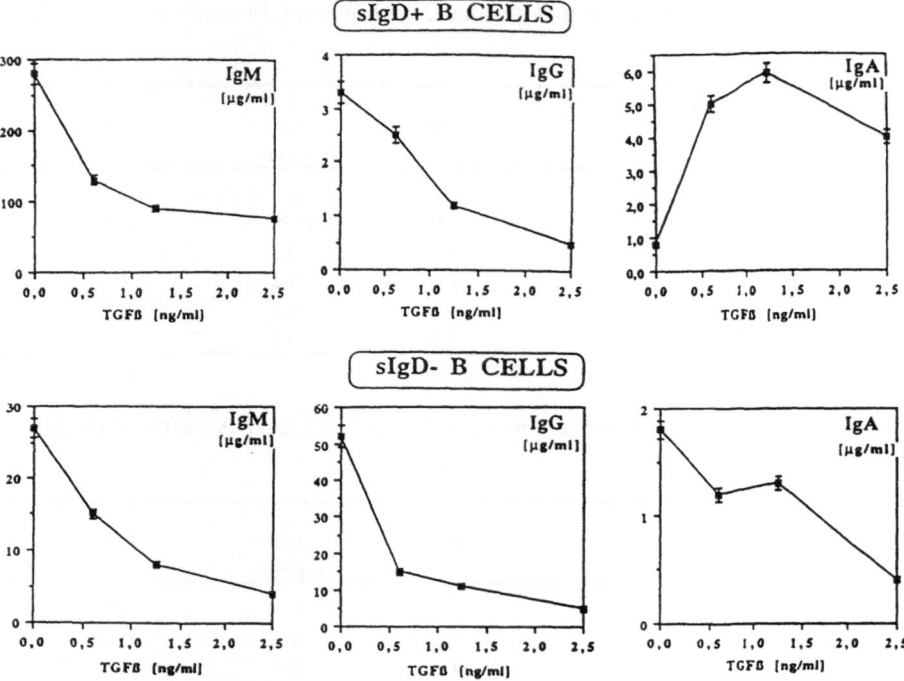

Figure 2. Effect of combinations of IL-10 and TGFβ on the pattern of Ig isotypes secreted by activated sIgD$^+$ and sIgD$^-$ B cells. 5x10^4 purified sIgD$^+$ B cells co-cultured with 5x10^3 irradiated L cells were stimulated with SAC, anti-CD40 in the presence of TGFβ (1.25 ng/ml) and increasing concentrations of IL-10 (3 to 100 ng/ml). IgA1 and IgA2 were measured after a culture period of 10 days. Ig levels represent the mean ± s.d. values of quadruplicate determinations.

Whereas TGFβ increased the secretion of both IgA subclasses, it strongly inhibited the IL-10-induced proliferation of sIgD$^+$ B cells (Fig. 4). This is in keeping with the results that TGFβ does not expand a particular subset of B lymphocytes.

The modification in the pattern of isotype expression induced by IL-10 and TGFβ could be the result of a change in the frequency of switching. To address this question, limiting dilution experiments were performed (Table 4). In the absence of cytokines, virtually no IgA-producing cells could be detected in sIgD$^+$ B cell cultures stimulated by anti-CD40 and SAC or anti-μ. Addition of IL-1- resulted in an increased frequency of IgA-containing cultures. These essentially contained the IgA1 subclass. Addition of TGF β further increased both IgA1 and IgA2 precursor frequencies with an alteration of the IgA subclass ratio in favor of IgA2.

Furthermore, those culture supernatants which had IgA2 always contained IgA1. This suggests that IgA2 production is the result of a sequential isotype switch from IgA1 to IgA2.

CONCLUSION

We have developed an *in vitro* system which allows T cell-independent differentiation of human B cells. This experimental model has permitted us to show that TGFβ can induce the production of both IgA subclasses from naive B cells in the presence of IL-10. TGFβ acts via an increase in the frequency of B cells driven to IgA synthesis by IL-10 with a preferential effect on IgA2.

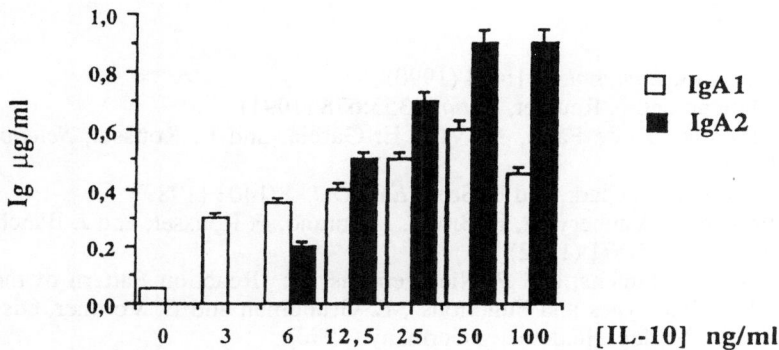

Figure 3. Both IgA subclasses are produced by activated sIgD⁺ B cells in the presence of IL-10 and TGFβ. 5x10⁴ purified sIgD⁺ B cells co-cultured with 5 x10³ irradiated L cells were stimulated with SAC and anti-CD40 in the presence of TGFβ (1.25 ng/ml) and increasing concentrations of IL-10 (3 to 100 ng/ml). IgA1 and IgA2 were measured after a culture period of 10 days. Ig levels represent the mean ± s.d. values of quadruplicate determinations.

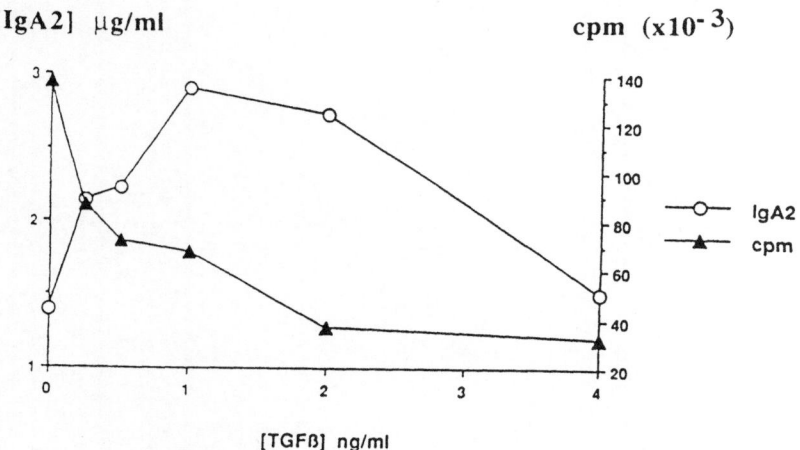

Figure 4. TGFβ inhibits IL-10-induced proliferation of sIgD⁺ B cells. 5x10⁴ purified sIgD⁺ B cells co-cultured with 5 x10³ irradiated L cells were stimulated with SAC and anti CD40 together with 100ng/ml of IL-10 and in the absence or in the presence of TGFβ (0.25, 0.5,1, 2 and 4 ng/ml). IgA2 was measured after a culture period of 10 days and thymidine incorporation was determined on day 6.

Table 4. IgA subclass frequencies upon stimulation of sIgD⁺ B cells with anti-CD40 and SAC.

Cytokine added:	IgA1	IgA2	IgM
IL-10	1:600	1:3000	1:20
IL-10 + TGFβ	1:240	1:550	1:80

Our current hypothesis is that TGFβ direct isotype switching towards IgA and subsequently IL-10 provides the necessary signals for growth and differentiation of the newly IgA-committed B cells.

REFERENCES

1. P. Lipsky, *Res. Immunol.* 141:424 (1990).
2. J. Banchereau and F. Rousset, *Nature* 353:678 (1991).
3. J. Banchereau, P. de Paoli, A. Vlle, E. Garcia, and F. Rousset, *Science* 251:70 (1991).
4. I. Stamenkovic, E. Clark, and B. Seed, *EMBO J.* 8:1403 (1989).
5. T. DeFrance, B. Vanbervliet, F. Briere, I. Durand, F. Rousset, and J. Banchereau, *J. Exp. Med.* 175:671 (1992).
6. F. Kroese, W. Timens, and P. Nieuwenhuis, *in*: "Reaction Pattern of the Lymph Node. Cell Types and Functions", E. Grundman and E. Vollmer, eds., p. 103, Springer-Verlag, Heidelberg, Germany (1990).
7. R. Coffman, D. Lebman, and B. Schrader, *J. Exp. Med.* 170:1039 (1989).
8. K. Islam, L. Nilsson, P. Sideras, L. Hammarstrom, and C. Smith, *Internat. Immunol.* 3:1099 (1991).

LYMPHOKINE mRNA EXPRESSION IN THE HUMAN INTESTINAL MUCOSA AND PBL DETERMINED BY QUANTITATIVE RT/PCR

L. Braun-Elwert,[2] G. E. Mullin,[1] and S. P. James[3]

[1]Mucosal Immunity Section, Laboratory of Clinical Investigation, National Institute of Allergy and Infectious Diseases, Bethesda, MD 20892, USA and [1]Division of Gastroenterology, Department of Medicine, The Johns Hopkins Hospital, Baltimore, MD 21205, USA

[2]present adress: Med. Klinik der FU Berlin, Abt. f. Innere Medizin mit Schwer-punkt Gastroenterologie, Leiter Prof. Dr. E.O. Riecken, Hindenburgdamm 30, 1000 Berlin 45, Germany

[3]present adress: Division of Gastroenterology, University of Maryland at Baltimore, 22S. Greene St., Baltimore, MD 21201, USA

INTRODUCTION

Mucosal T-lymphocytes are unique in that they express predominantly phenotypic markers of activation such as IL-2R[1] and HML-1[2] and the "memory" cell phenotype CD45RO[3-5] but lack the lymph node homing receptor Leu-8.[6] After challenge with specific Ag *in vitro* they do not proliferate but provide help for B-lymphocytes.[7] Furthermore, after in vitro activation they express IL-2, IFNγ, IL-4 and IL-5[8], lymphokines which are thought to be critical for host defense in the normal intestinal mucosa. Since relatively little is known about lymphokine production in the normal human intestine, we wanted to determine the *in situ* expression of these lymphokines. So far, most of the information has been obtained from *in vitro* studies of intestinal lymphocytes that were isolated from surgical specimens using enzymatic methods. In this study, colonoscopic biopsies from normal individuals were used as a source of intestinal lymphocytes. Since only a small amount of total cellular RNA was available for study, we developed a reverse transcription (RT)-PCR method to determine lymphokine mRNA that was both very sensitive and quantitative.

RESULTS

Numerous PCR methods have been developed to measure DNA or mRNA in samples containing as few as a single cell.[9-12] When the efficiency of PCR amplification is the same for different template molecules, quantitation is possible by creating a standard curve that is determined by incorporation of radiolabelled nucleotides into the PCR products after varying numbers of PCR cycles or varying dilutions of the template molecules.[13-18] Quantitative analyses are also possible when the amplification efficiency is the same for different primer pairs.[19]

Recently a quantitative RT-PCR method has been developed that uses synthetic RNA standards that differ in length from the template molecules.[20] Known amounts of synthetic RNA are mixed with total cellular RNA, co reverse transcribed, and coamplified by PCR. The synthetic and template RNA molecules compete in the RT step and subsequently as the respective cDNAs in the PCR amplification step for primers, nucleotides, enzymes etc., thereby controlling for sample to sample variation in the RT and the PCR steps. PCR products from synthetic and template molecules can be distinguished according to their size by gel electrophoresis. Quantitative results are obtained by comparing the respective ethidium stained bands. PCR-products create bands of equal strength only if the RNA in the original samples are equal or nearly equal.

In order to use this method for quantitation of lymphokine mRNA, we developed synthetic RNA-standards containing additional fragments internal to the primer sites and containing a poly (A) tail. The synthetic IL-2 RNA contained the first intron (91 nucleotides), whereas the synthetic RNAs for IFNγ, IL-4 and IL-5 contained inserted fragments that were 133, 240 and 93 nucleotides in length respectively.

Since the PCR products from synthetic RNA are longer than those from cellular RNA, one potential problem was that the latter might have been underestimated because they are amplified with lower efficiency. However, it was found that there was no difference in amplification efficiency for each primer pair tested. Furthermore, there was no difference for cellular RNA from different sources. The amplification efficiency was determined by incorporation of radioactively labelled nucleotides after varying numbers of cycles. Optimal detection of ethidium stained bands was after 30 PCR cycles. Thus, this number of cycles was used throughout all further experiments.

Since each synthetic RNA contained a poly (A) tail, we tested co reverse transcription of total cellular RNA with only one or up to four different synthetic RNAs using oligo d(T) as primer. The PCR step was carried out using the product-specific primers. The results of the PCR products were identical. Therefore, in all subsequent studies, total cellular RNA was co reverse transcribed simultaneously with dilutions of all four synthetic RNAs. This was a particular advantage since there was only a limited amount of cellular RNA available for study.

When the cellular RNA was coamplified with 10-fold dilutions of synthetic RNA, bands of equivalent intensity appeared at the corresponding 10-fold diluted concentrations of synthetic transcripts, demonstrating that the coamplification assay was log-linear over this range. This range was 0.01pg - 1pg for IL-2 mRNA.[21] The range for IFNγ, IL-4, and IL-5 mRNA was 0.1pg - 100pg, 0.1pg - 100pg, 0.01pg - 10pg respectively.[22] The lower number was the minimum amount of synthetic RNA detected using this assay.

After extraction of total cellular RNA the above described method was used to determine the mRNA concentration of IL-2, IFNγ, IL-4, and IL-5 in histologically normal colonoscopic biopsies (4M, 4F) and in PBL from healthy volunteers (2M, 2F). Total cellular RNA from biopsies and PBL was reverse transcribed simultaneously with a mixture of 10-fold dilutions of synthetic RNAs. The PCR amplification was carried out with the cDNA derived from 1mg of total cellular RNA. The values of lymphokine mRNA measured are reported as pg mRNA/mg total cellular RNA. Values between two 10-fold dilutions are reported as intermediate.

Similar IL-2 and IFNγ mRNA concentrations were measured in RNA from biopsies and PBL (Fig. 1a). Values for IFNγ mRNA (1-10pg/mg cellular RNA) were higher than those for IL-2 mRNA (0.01-0.1pg/mg cellular RNA). In contrast IL-4 mRNA was detected in all PBL samples (0.1-1pg/mg cellular RNA) but only in one biopsy (0.01-0.1pg/mg cellular RNA), whereas IL-5 mRNA was detected in two biopsies (0.01pg/mg cellular RNA) but in none of the PBL-samples.

Since most of the lymphokine mRNA detected was expected to be produced by activated T-lymphocytes, additional experiments were performed to estimate the number of T cells present in each sample. This was done indirectly by using a semiquantitative RT/PCR method specific for TCR Cα which is a T-cell specific product. As expected, values for TCR Cα were 100 - 1000 fold higher in the PBL samples than those obtained

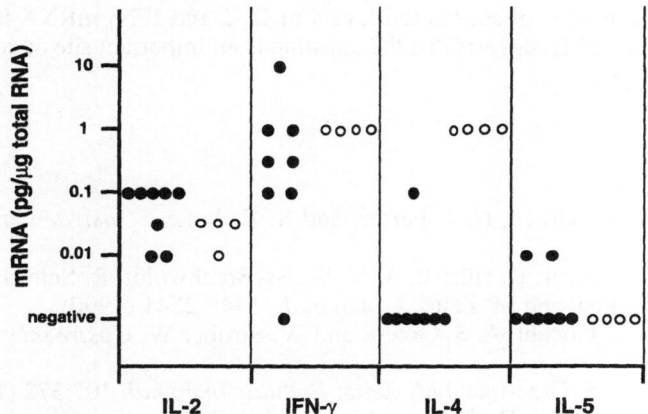

Figure 1a. Lymphokine expression in mucosal biopsies (filled circles) and PBL (open circles).

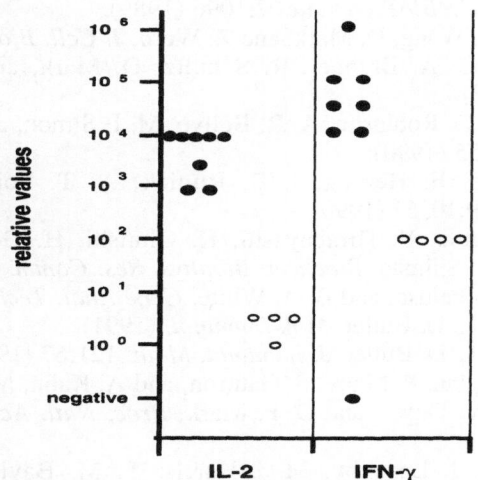

Fgure 1b. Expression of IL-2 and IFN-γ corrected for the concentration of T-cells present in each sample. Shown on the y-axis are relative values. Symbols as in Figure 1a.

from intestinal biopsies demonstrating that mucosal samples compared to PBL contained relatively few T lymphocytes. Considering the concentration of T lymphocytes present in each sample, the expression of IL-2 and IFN-γ is substantially higher in mucosal biopsies compared to PBL (Fig. 1b). Thus, expression of lymphokine mRNA is quantitatively different in PBL and intestinal lymphocytes.

This study demonstrates that quantitative RT/PCR can be used to detect and measure sensitively *in situ* the concentration of lymphokine mRNA in intestinal lymphocytes. Furthermore, the presence of substantial levels of IL-2 and IFNγ mRNA in the intestinal mucosa compared to PBL suggests that the intestine is an important site of constitutive IL-2 and IFNγ production.

REFERENCES

1. M. Zeitz, W. C. Green, N. J. Peffer, and S. P. James, *Gastroenterology* 94:647 (1988).
2. H. L. Schieferdecker, R. Ullrich, A. N. Weiss-Breckwoldt, R. Schwarting, H. Stein, E. O. Riecken, and M. Zeitz, *J. Immunol.* 144: 2541 (1990).
3. S. P. James, C. Fiocchi, A. S. Graeff, and W. Strober W, *Gastroenterology* 91:1483 (1986).
4. S. P. James, A. S. Graeff, and M. Zeitz, *Cellular Immunol.* 107:372 (1987).
5. G. Janossy, M. Bofill, D. Rowe, J. Muir, and P. C. Beverley, *Immunology* 66:517 (1989).
6. M. E. Kanof, W. Strober, C. Fiocchi, M. Zeitz, S. P. James, *J. Immunol.* 141:2029 (1988).
7. M. Zeitz, T. C. Quinn, A. S. Graeff, S. P. James, *Gastroenterology* 94:353.
8. S. P. James, W. C. Kwan, and M. C. Sneller, *J. Immunol.* 144:1251 (1990).
9. E. Razin, K. B. Leslie, and J. W. Schrader, *J. Immunol.* 146:981 (1991).
10. C. A. Brenner, A. W. Tam, P. A. Nelson, E. G. Engleman, N. Suzuki, K. E. Fry, and J. W. Larrick. *BioTechniques* 7:1096 (1989).
11. D. A. Rappolee, A. Wang, D. Mark, and Z. Werb, *J. Cell. Biochem.* 39:1 (1989).
12. D. A. Rappolee, C. A. Brenner, R. Schultz, D. Mark, and Z. Werb, *Science* 241:1823 (1988).
13. J. Singer-Sam, M. O. Robinson, A. R. Bellve, M. I. Simon, and A. D. Riggs, *Nucl. Ac. Res.* 18:1255 (1990).
14. L. D. Murphy, C. E. Herzog, J. B. Rudick, A. T. Fojo, and S. A. Bates, *Biochemistry* 29:10351 (1990).
15. S. Oka, K. Urayama, Y. Hirabayashi, K. Ohnishi, H. Goto, K. Mitamura, S. Kimura, and K. Shimad, *Biochem. Biophys. Res. Comm.* 167:1 (1990).
16. B.C. Delidow, J. J. Peluso, and B. A. White, *Gene Anal. Techn.* 6:120 (1989).
17. K. M. Mohler and L. D. Butler, *Mol. Immunol.* (1991).
18. K. M. Mohler and L. D. Butler, *J. Immunol. Meth.* 121:67 (1989).
19. J. Chelly, J.-C. Kaplan, P. Maire, S. Gautron, and A. Kahn, *Nature* 333:858 (1988).
20. A. M. Wang, M. V. Doyle, and D. F. Mark, *Proc. Natl. Acad. Sci. USA* 86:9717 (1989).
21. G. E. Mullin, A. J. Lazenby, M. L.Harris, T. M. Bayless, and S. P.James, *Gastroenterology* 102:1620 (1992).
22. L. Braun-Elwert , G. E. Mullin, and S. P.James, submitted.

REGULATION OF T CELL REACTIVITIES BY INTESTINAL MUCOSA

Liang Qiao,[1] Guido Schürmann,[2] Stefan C. Meuer,[1]
Reinhard Wallich,[1] Albrecht Schirren,[1] and Frank
Autschbach[1]

[1]Applied Immunology, German Cancer Research Center
[2]Department of Surgery, University of Heidelberg,
Heidelberg, FRG

INTRODUCTION

The large surface area of intestinal mucosa is under constant exposure to pathogens and dietary antigens. The intestinal mucosal immune system together with non-specific barriers like digestive proteases, intestinal mobility, the commensal microflora and mucous coat are believed to provide protection for the host. Mucosal plasma cells produce large amounts of IgA which can traverse the mucosal membrane and prevent the entry of foreign antigens. IgA production needs help by T lymphocytes. Therefore, T cell activation is a prerequisite for IgA production.

T cells can be activated experimentally by binding of monoclonal antibodies (mAb) to CD2, CD3-Ti and CD28 surface molecules, respectively. The second messenger system is not completely elucidated. However, some early events are known.[1] Triggering of TcR-CD3 results in the breakdown of phosphatidylinositol 4,5-bisphosphate by phospholipase C into inositol 1,4,5-trisphosphate and diacylglyceride. Inositol 1,4,5-trisphosphate raises the intracellular free calcium concentration. Diacylglyceride together with calcium translocate the cytosolic protein kinase C (PKC) to the plasma membrane and activate this enzyme.

FUNCTIONAL PROPERTIES OF MUCOSAL T CELLS

Lamina propria T lymphocytes (LPL-T) and intraepithelial T lymphocytes in vitro behave functionally differently from circulating peripheral blood T lymphocytes (PBL-T). Thus, LPL-T isolated from lymphogranuloma venereum-infected nonhuman primates do not proliferate in response to lymphogranuloma venereum stimulation yet provide help for polyclonal immunoglobulin synthesis by B lymphocytes.[2] Human intraepithelial lymphocytes expressing CD2, CD3, TcRαβ and CD8 showed minimal proliferation in response to CD3 antibodies but brisk responses to stimulation of the CD2 receptor.[3] Our recent work demonstrated[4,5,6] that, compared to autologous PBL-T, human LPL-T exhibited a low proliferative response to stimulation by immobilized anti-CD3 mAb in the presence of interleukin 2 (IL-2) whereas these cells largely preserved responses to CD2 and CD28 stimulation *in vitro* (Table 1). The low proliferation to CD3 triggering appear to result from low expression of the IL-2 receptor. This is not a common feature of LPL-T since CD2 and CD28 mAb can induce expression of IL-2 receptor. Signal transduction through TcR/CD3-stimulation is impaired[5] in that triggering by anti-CD3 mAb induces only low amounts of inositol 1,4,5-trisphosphate and no increase of cytoplasmic free calcium.

Moreover, PKC activators PDBu and ionomycin, which respectively mimick the function of diacylglycerides and intracellular free calcium, induced lower proliferation in LPL-T than in PBL-T[6] (Table1).

Table 1. Proliferation of T cell to various stimuli.

	PBL-T	LPL-T	PBL-T+SN[1]
CD2mAb+SRBC	++	++(+++)	++(+++)
CD3mAb+IL-2	++	+(-)	+(-)
CD28mAb+IL-2	++	++	++
PDBu+ionomycin	++	+	+

1: PBL-T preicubated with mucosal supernatant for 2-3 days

REGULATION OF T CELL REACTIVITIES BY INTESTINAL MUCOSA DERIVED SUPERNATANT

It is very interesting to know how T cells are rendered to a state in which they hardly proliferate to antigen receptor stimulation. Functional properties of LPL-T result from local environmental influences as suggested by the studies of co-culture of PBL-T with intestinal mucosa.[5,6] PBL-T were co-cultured with either supernatant of intestinal mucosa or supernatant of smooth muscle tissue or culture medium in a transwell system for 2-3 days. This system consists of two individual compartments which are separated by a semipermeable membrane (pore size 3 μm). Co-culture with supernatant of mucosa tissue induced a similar reactivity in PBL-T as observed in freshly isolated LPL-T (Table 1). In contrast, incubation of PBL-T with supernatant of intestinal muscle tissue or culture medium for the same period did not lead to functional changes.

PROPERTIES OF INTESTINAL MUCOSA DERIVED SUPERNATANT

The soluble factors from intestinal mucosa supernatant were further characterized.[7] The substances which alter T cell reactivities are small molecules since mucosal supernatant completely lost its functional effects when it was dialysed through a semipermeable membrane that allows the passage of molecules of less than 12-14 kilodaltons (Table 2). These molecules are non-protein and non-peptide because mucosal supernatant retained its activities when it was digested with proteinase K overnight at 37°C and boiled at 100°C for 5 min (Table 2). In this way, proteins and peptides are expected to be hydrolysed and proteinase K activity to be subsequently inactivated by heat denaturation.

Intestinal mucosa is endowed with the enzymatic machinery necessary to produce large amounts of reactive oxygen metabolites.[8] Furthermore, human colonic mucosa is relatively deficient in antioxidant enzymes and most of the protective enzyme activity is localised within the epithelium but not in the lamina propria.[9] Oxidants are known to inhibit the growth of mouse T lymphocytes stimulated by Concanavalin A (Con A) or phorbol ester plus calcium ionophore[10] and that of human lymphocytes stimulated by Con A.[11] It is possible that intestinal mucosa contains oxidants which alter T cell reactivities. Our recent data support such an idea since the functional activity of mucosal supernatant was completely lost when mucosal supernatant was co-cultured with PBL-T in the presence of the reducing agents 2-ME (Table 2) or DTT respectively. This finding suggests that the regulatory substances contained in the intestinal mucosa exert oxidative properties. This view is also supported by the finding that pretreatment of T cells with H_2O_2 caused down-regulation of the responses to CD3 and PKC stimulation (Table 2). The response to CD2 stimulation is more resistant to this inhibition. When PBL-T were pretreated with H_2O_2 in the presence of 2-ME, as expected, the effects of H_2O_2 on T cells were reversed by 2-ME.

Table 2. Properties of mucosal supernatant.

Preincubation of PBL-T with	Proliferation to stimulaiton via		
	CD2	CD3	PKC
Medium	++	++	++
SN[1]	++	+(-)	+
SN+Dia[2]	++	++	++
SN+PK[3]	++	+(-)	+
SN+2-ME	++	++	++
H_2O_2	++	+	+

1: mucosal supernatant; 2: dialysed supernatant; 3: supernatant digested with proteinase K.

The unresponsiveness to Con A in hydrogen peroxide treated T cells is supposed to be due to signal transduction impairment in that there is no intracellular free calcium mobilization upon exposure to concanavalin A.[10] Furthermore, oxidants such as N-Chlorosuccinimide, hydrogen peroxide and periodate can inactivate PKC activity.[12-14] This might be the reason why hydrogen peroxide inhibits the proliferation induced by PKC activators. Lamina propria T Lymphocytes have a low responsiveness to stimulation via CD3 and by the PKC activators PBu2 plus ionomycin. Triggering of CD3 in lamina propria T lymphocytes induced no generation of intracellular free calcium. Thus, the behavior of these cells could be the result of oxidant effects. Furthermore, the reducing agent 2-ME can reverse the effects of both hydrogen peroxide and mucosal supernatant. All present data strongly support the notion that the immunoregulatory substances in the intestinal mucosa possess oxidative properties.

IMPORTANCE OF MUCOSAL IMMUNOREGULATORY MECHANISM

Gut mucosal immune tissue can be divided into IgA-inductive sites and mucosal effector sites.[15] The former is represented by Peyer's patches, the appendix, and small solitary lymphoid nodules in which antigens are presented to B and T cells. The antigen specific T and B cells undergo clonal expansion outside of the mucosa e.g. in lymph nodes and in peripheral blood. The expanded cells then recirculate to the mucosal area (mainly lamina propria) where IgA is produced (mucosal effector site). Intestinal mucosa contains substances with oxidative properties which suppress T cell proliferation in response to antigen stimulation. This may prevent harmful effects of proliferating T cells to the local tissue. However, these T cells still may provide help for IgA production. In this way, local immunity is adjusted to be protective but not to be harmful for host self. Abnormalities of this local immunoregulatory mechanism may lead to or enhance pathological conditions. Further identification of substances possessing oxidative properties in the intestinal mucosa may aid the design of novel therapeutic strategies.

REFERENCES

1. S. C. Meuer and K. Resch, *Immunol. Today* 10:523 (1989).
2. M. Zeitz, T. C. Quinn, A. S. Graeff, and S. P. James, *Gastroenterology* 94:353 (1988).
3. E. C. Ebert, *Gastroenterology.* 97:1372 (1989).
4. U. C. Pirzer, G. Schürmann, S. Post, M. Betzler, and S. C. Meuer, *Eur. J. Immunol.* 20:2339 (1990).
5. L. Qiao, G. Schürmann, M. Betzler, and S. C. Meuer, *Gastroenterology* 101:1529 (1991).

6. L. Qiao, G. Schürmann, M. Betzler, and S. C. Meuer, *Eur. J. Immunol.* 21:2385 (1991).
7. L. Qiao, G. Schürmann, F. Autschbach, R. Wallich, and S. C. Meuer, submitted.
8. D. N. Granger, L. A. Hernandez, and M. B. Grisham, *Vien. Dig. Dis.* 18:13 (1986).
9. M. B. Grisham, R. P. MacDermott, and E. A. Deitch, *Inflammation* 14:669 (1990).
10. D. D. Duncan and D. A. Lawrence, *Toxicol. Appl. Pharmacol.* 100:485 (1989).
11. D. C. Zoschke and N. D. Staite, *Clin. Immunol. Immunopathol.* 42:160 (1987).
12. R. Gopalakrishna and W. B. Anderson, *FEBS Lett.* 225:233 (1987).
13. R. Gopalakrishna and W. B. Anderson, *Proc. Natl. Acad. Sci. USA* 86:6758 (1989).
14. R. Gopalakrishna and W. B. Anderson, *Arch. Biochem. Biophys.* 285:382 (1991).
15. J. R. McGhee, J. Mestecky, C. O. Elson, and H. Kiyono, *J. Clin. Immunol.* 9:175 (1989).

VECTOR-ENCODED INTERLEUKIN-5 AND INTERLEUKIN-6 ENHANCE SPECIFIC MUCOSAL IMMUNOGLOBULIN A REACTIVITY *IN VIVO*

Alistair J. Ramsay

Division of Cell Biology, John Curtin School of
Medical Research, Australian National University
Canberra, ACT 2601, Australia

INTRODUCTION

Recent work suggests that T-cell-derived cytokines may influence the development of mucosal immune responses, in particular, the characteristic production of IgA antibody by plasma cells lying beneath epithelial basement membranes and present in secretory glandular tissues. It is now clear that murine interleukin-5 (mIL5) and mIL6 selectively enhance IgA reactivity of activated B-cells *in vitro*,[1-6] alone or in combination with each other or different factors, apparently by enhancing their terminal differentiation.[7-10] Although there is no direct evidence that these factors play a similar role *in vivo*, it has recently been shown that T-cells apparently corresponding to the Th2 type, those secreting a characteristic pattern of cytokines including IL-5 and IL-6,[11] are present at high frequency in IgA effector sites of murine mucosal tissues.[12]

To study the effects of local concentrations of IL5 and IL6 on the develoment of immune responses at mucosae, we have taken the novel approach of encoding the genes for these factors in recombinant vaccinia viruses (rVV) and report here that vector-expressed IL5 and IL6 selectively enhance local antigen-specific IgA reactivity in the lungs of mice immunized with these constructs.

MATERIALS AND METHODS

Plasmids and Viruses

The plasmids pEDFM-5, containing the mIL5 cDNA clone[13] and pCDmIL-6, containing the mIL6 cDNA clone,[14] were provided by Dr H. Campbell (John Curtin School of Medical Research) and Dr K-I. Arai (DNAX), respectively. The plasmid vectors pBCB07, containing the vaccinia P7.5 promoter and pFB-TK were provided by Drs D. Boyle and B. Coupar.[15,16] The thymidine kinase (TK) minus mutant of vaccinia, VV-WR-TK⁻ and the P7.5 promoter were gifts of Dr B. Moss (NIH). The rVV, VV-HA-PR8, which has the hemagglutinin (HA) gene of influenza virus A/PR/8/34 (PR8) inserted in the *HindIII* J region (thus rendering the virus TK minus), and VV-HA-TK, which has

Advances in Mucosal Immunology, Edited by
J. Mestecky *et al.*, Plenum Press, New York, 1995

the TK gene of herpes simplex virus (HSV) inserted as a selectable marker into the *HindIII* F region, were constructed in this laboratory.[17]

Mice and Immunizations

6-9 week old CBA/H mice raised under specific pathogen-free conditions were used in these experiments and were obtained from the Animal Breeding Establishment of the John Curtin School of Medical Research. In the experiments described below, mice were immunized intranasally with 10^7 pfu of rVV.

Lymphoid Cell Preparation

Single cell suspensions of splenocytes were prepared by gentle mincing and erythrocytes and dead cells removed by Ficoll-Paque density gradient separation. Cells at the interphase were washed 3 times and resuspended in RPMI-1640 (Flow Laboratories, McLean, VA) with 10% fetal calf serum and antibiotics (complete medium). Lung cells were similarly prepared, except that the finely minced lung tissue was incubated for 90 min at 37°C with collagenase (Boehringer-Mannheim, Tutzig, F.R.G; cat. no. 103586) at 4mg/lung, Dispase II (Boehringer) at 1.2 units/ml and deoxyribonuclease type II (Calbiochem, La Jolla, CA) at 5 units/ml, in 2ml complete medium per lung. The cell suspension was then passed through 3 layers of sterile cotton gauze to remove particulate matter and washed 3 times, prior to density gradient purification as above. For some experiments, lung cells were depleted of T-cells by treatment with anti-Thy 1.2 antiserum (Serotec, Blackthorn Bicester, England) and of plastic adherent cells prior to separation into mIgA-positive and -negative fractions by panning, as described elsewhere.[7]

Immunoassays

Enumeration of individual antibody-secreting cells (ASC) was accomplished using the ELISPOT assay,[18] with modifications as follows. Initially, 96-well nitrocellulose-bottomed Millititer HA plates (Millipore, Bedford, MA) were filled with 10^3 hemagglutination units of influenza virus PR8 and left overnight at 4°C. After incubation, plates were washed 3 times with phosphate-buffered saline (PBS) to remove unabsorbed material and non-specific binding sites on the nitrocellulose membrane were blocked with PBS/5% BSA for 1 h at 37°C. Well contents were then replaced with cell isolates resuspended in complete medium and added in serial dilutions at 100 μl/well. Plates were left undisturbed for 4 h at 37°C in a humidified atmosphere of 5% CO_2 in air, following which the cells were removed by washing 3 times with PBS then three times with PBS/0.25% Tween 20. Next, biotin-labelled conjugates of goat anti-mouse IgA or IgG (Amersham International, Bucks., England), diluted 1/1000 in PBS/Tween, were added at 100 μl/well and the plates left overnight at 4°C. The wells were then rinsed 3 times with PBS/Tween and incubated with 100 μl streptavidin alkaline phosphatase (Amersham International, 1/2000 dilution in PBS/Tween) for 2 h at 37°C and the plates were developed and spots counted.[18] Levels of IgA in lung fluids, obtained by lavage with 2ml PBS, were determined by standard ELISA using, where appropriate, reagents and procedures described for the ELISPOT assay.

Titration of Virus Growth *In Vivo*

Virus growth in lungs was titrated as described elsewhere.[19]

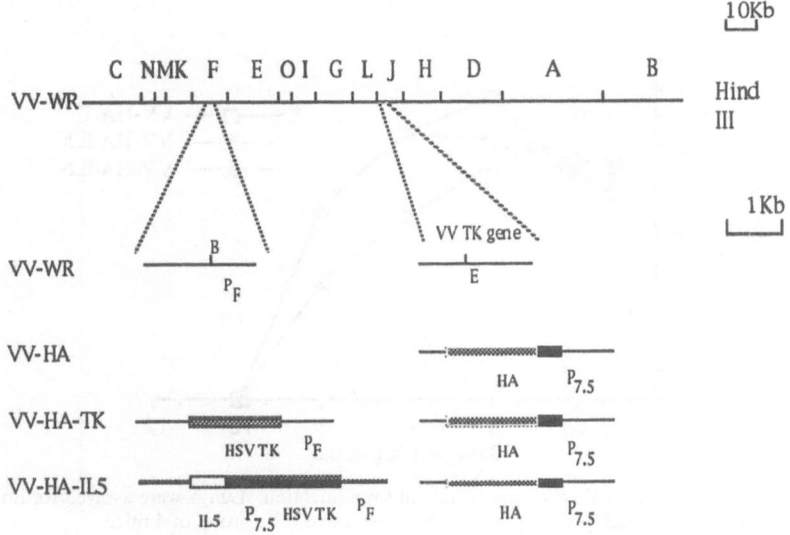

Figure 1. Genomic configuration of VV-HA-IL5. A *HindIII* map of vaccinia virus, strain VV-WR, is shown with insertion points at the *EcoRI* (E) and *BamHI* (B) sites in the J and F fragments, respectively. The positioning of the vaccinia virus thymidine kinase (TK) gene, vaccinia promoters PF and P7.5, the influenza virus HA gene, the herpes simplex virus (HSV) TK gene and the murine IL5 coding sequences are indicated.

RESULTS

Virus Construction and Testing

A *HinPI* fragment, comprising the sequence between nucleotides 24 and 650, was cut from pEDFM-5 and subcloned into the *AccI* site of plasmid vector pBCB07. A *NlaIV/SspI* fragment, comprising the sequence between nucleotides 10 and 958, was cut from pCDmIL-6 and subcloned into the *HincII* site of pBCB07. The resulting recombinant plasmids contained either mIL5 or mIL6 cDNA immediately downstream of a P7.5 VV promoter. These were cut from the respective plasmids in *EcoRI* fragments, ligated into pFB-TK and used in marker rescue as previously described.[15,16] Homologous recombination with VV-HA-PR8 *in vitro* gave rise to VV-HA-IL5 or VV-HA-IL6, which contained the genes for IL5 or IL6 and HSV TK in the *HindIII* F region and the HA gene in the J region. The rVV were plaque purified under methotrexate selection[16] and the presence of inserts in the correct positions confirmed by restriction analysis and Southern blotting (data not shown). Figure 1 shows the genomic configuration of VV-HA-IL5. Expression of HA by rVV was confirmed by immunoprecipitation and production of biologically active cytokine was shown by the ability of 24 h supernatants overlying 143B cell monolayers infected with VV-HA-IL5 or VV-HA-IL6 to stimulate proliferation of T88 cells[20] or 7TD1 cells,[21] respectively (data not shown).

Growth of rVV *In Vivo*

Each of the rVV displayed similar growth kinetics in the lungs of mice when given intranasally (Fig. 2). Growth of the control virus, VV-HA-TK (which encodes the HA gene but no cytokine genes), peaked on the fourth day ($10^{6.9}$ pfu/lung) and that of VV-

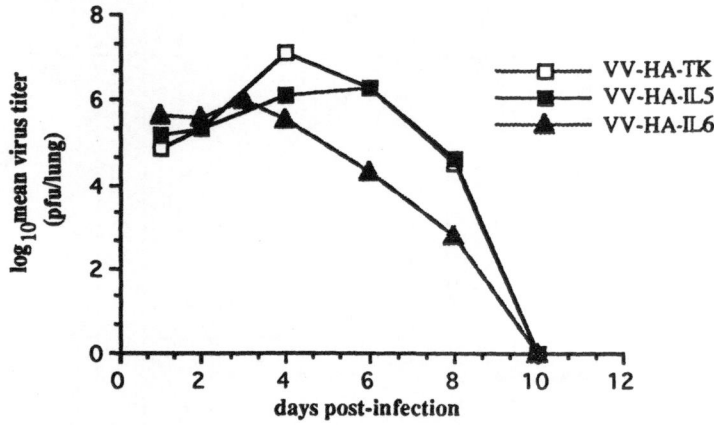

Figure 2. Growth kinetics of rVV following intranasal immunization. Lungs were assayed for infectious virus on the days indicated. Each point represents the mean value for a group of 4 mice.

HA-IL5 on the sixth day ($10^{6.3}$) after infection. Maximum titres of VV-HA-IL6 were found on the third day (10^6), although this virus was recovered at significantly lower titer than the others on day 6 and thereafter. Each of the viruses was last recovered on day 8 and was apparently cleared from the lungs, and from the ovaries and spleen (data not shown), by day 10.

Influence of rVV-Encoded IL5 and IL6 on Antiviral Immunity *In Vivo*

In order to study the influence of IL5 and IL6 on the development of antiviral mucosal antibody responses *in vivo*, we immunised mice intranasally with rVV and monitored the development of immunocytes secreting antibody specific for the co-expressed HA (or "reporter" antigen) in their lungs. As shown in Fig. 3, whereas VV-HA-IL-5 and VV-HA-TK elicited similar specific IgG reactivity, significantly greater numbers of anti-HA IgA ASC were found in the lungs of mice given VV-HA-IL-5 than in those given control virus, despite the remarkably similar growth kinetics exhibited by the two viruses. The elevated response was first detected on day 10 after infection, after virus clearance from the lungs, and peaked on day 14, at 4-fold greater than control levels. Less-pronounced increases in IgA reactivity were found in the lungs of mice given VV-HA-IL-6, however there was no evidence of enhanced systemic reactivity in any of these experiments (data not shown).

We next used ELISA to examine the effects of vector-encoded cytokines on lung responses following secondary immunization of mice primed with control rVV. As shown in Table 1, markedly elevated responses were found in mice boosted at 21 days with either VV-HA-IL-5 or VV-HA-IL-6 when compared to those given a second dose of VV-HA-TK. Anti-HA IgA titers in lung fluid from mice given VV-HA-IL6 were elevated as early as 2 days after boosting, being 4-fold higher than in those given control virus. By 9 days after secondary inoculation, specific lung IgA titers in animals given either VV-HA-IL-5 or VV-HA-IL-6 were 16-fold greater than those measured in immunized controls.

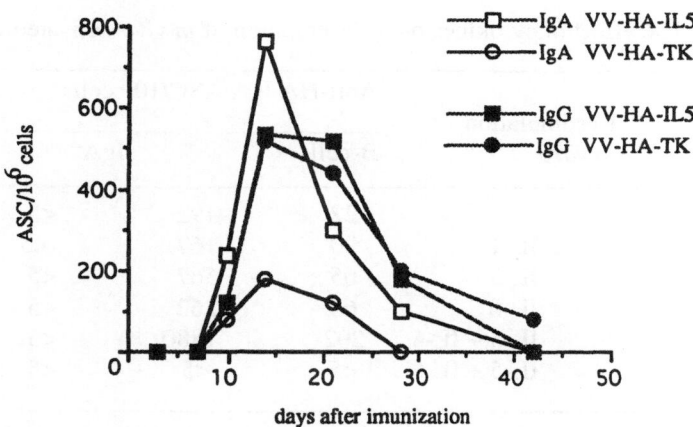

Figure 3. Anti-HA IgA ASC in murine lungs following immunization with rVV. Lung lymphoid cells were assayed by ELISPOT and values represent ASC per million cells.

Table 1. IgA reactivity against HA in lung lavage fluid of mice given rVV

Immunization		.... Reciprocal antibody titre	
Primary	Secondary (d21)	Day 23	Day 30
-	-	<5	<5
VV-HA-TK	-	<5	10
VV-HA-IL5	-	5	20
VV-HA-TK	VV-HA-TK	5	20
VV-HA-TK	VV-HA-IL5	5	320
VV-HA-TK	VV-HA-IL6	20	320

Lung lavage fluid taken as indicated after priming and boosting were assayed for anti-HA IgA by ELISA. Figures represent mean titers for groups of 4 mice.

Response of Antiviral Immunocytes to Cytokines *In Vitro*

The abovementioned results show that IL5 and IL6 enhance specific mucosal IgA reactivity when present at foci of infection. In an attempt to establish which cell populations were responsive to these exogenous factors, we added recombinant cytokines to cultures of B-cells isolated from the lungs of mice infected 7 days earlier with VV-HA-TK. Table 2 shows that, unlike rIL1, both rIL5 and rIL6 increased numbers of anti-HA IgA ASC among unfractionated B-cells after 4 days of culture. A more spectacular increase was seen when rIL5 and rIL6 were added together, suggestive of a synergistic interaction between these factors in promoting IgA reactivity among activated B-cells. The enhanced responses were apparently mediated entirely by the surface IgA-positive cell fraction. Despite their dramatic effects on IgA ASC numbers, neither IL5 nor IL6, unlike the non-specific stimulant, bacterial lipopolysaccharide, stimulated proliferation of these cells (data not shown). The surface IgA-negative cell fraction contained anti-HA IgM ASC, though their numbers were not altered following culture with recombinant cytokines (data not shown).

Table 2. The effect of cytokines on differentiation of *in vivo*-activated B-cells

Immunization	Restimulation *in vitro*	Anti-HA IgA ASC/10^6 cells		
		B-cells	sIgA$^+$	sIgA$^-$
VV-HA-TK	-	22	192	<5
VV-HA-TK	IL-1	20	167	<5
VV-HA-TK	IL-5	65	567	<5
VV-HA-TK	IL-6	68	262	<5
VV-HA-TK	IL-5 + IL-6	202	1880	<5
-	IL-5 + IL-6	<5	<5	<5

Cells were isolated 7d after immunization and B-cell fractions were obtained by panning and cultured for 4d in the presence of cytokines at 10 U/ml . ASC numbers were assayed by ELISPOT.

DISCUSSION

We have made recombinant vaccinia viruses encoding HA of influenza virus in order to generate local anti-HA immune responses in murine lung and have demonstrated specific IgA reactivity in mice given these rVV intranasally. The latter is markedly and selectively enhanced when animals are given constructs which, in addition to HA, express mIL5 or mIL6. It is unlikely that different patterns of growth and persistence of VV-HA-IL5 or VV-HA-IL6 were responsible for the increased responses, as the kinetics of replication of these viruses in lungs were remarkably similar to those of the control virus. The data indicate, therefore, that IL5 and IL6 could play an important role in the development of mucosal immune responses *in vivo*. Such responses are characterized by IgA antibody production by plasma cells which are induced following exposure to antigen in organised mucosal lymphoid structures and which migrate to submucosal tissues and differentiate into IgA-secreting cells.[22] It is likely that the latter process occurs under the influence of activated T helper cells, which are abundant in mucosal tissues,[23] and which presumably secrete factors necessary for differentiation of the IgA B-cells. Certainly, the Th2-type cytokines, IL5 and IL6, significantly enhance IgA synthesis in murine Peyer's patch B cell cultures.[7-9]

The results also indicate that IL5 and IL6 increase terminal differentiation of *in vivo*-activated IgA-committed B-cells, in concordance with existing *in vitro* evidence that these factors act late in B-cell development.[7-10] Others have shown that large, activated B-cells may not require cognate interaction with T helper cells and are able to proliferate and differentiate *in vitro* in response to Th2-derived factors, especially IL5.[24] It is tempting, therefore, to speculate that vector-encoded cytokines in our system, produced at foci of infection, act either directly, or in concert with host-derived factors, to enhance the differentiation of newly-arrived virus-specific IgA plasma cell precursors into antibody-secreting cells. Of possible relevance in this respect is a recent report that mucosal epithelial cells spontaneously produce high levels of IL6.[25] It is also likely that endogenously-produced IL6 would be present locally as the host mounts an inflammatory response to the virus infection. Certainly, our findings provide evidence for synergy between rIL5 and rIL6 in the enhancement of specific IgA reactivity among *in vivo*-activated B-cells. It is also clear that vector-encoded IL6 can act rapidly *in vivo* to promote reactivity of a primed B-cell population, with increased titers of specific IgA found in lung fluids within 2 days of boosting with VV-HA-IL6.

The technology developed and used in these experiments has important implications for the development of vectored vaccines for mucosal or systemic application. Live vectors have great potential for human vaccination, though present problems associated with safety, especially in immunodeficient individuals, and their ability to induce satisfactory immunity against encoded foreign proteins. The findings reported here, and our further data reviewed elsewhere,[26] suggest that the expression of selected cytokines in replicating vectors may allow manipulation of the microenvironment to favour suitable attenuation of the vector as well as the development of appropriate protective immune responses.

ACKNOWLEDGEMENTS

I wish to thank colleagues in the Viral Engineering and Cytokine Research Group at the John Curtin School of Medical Research for their contributions to these studies and numerous others for their kind gifts of plasmids, promoters and reagents. This work received financial support from the National Health and Medical Research Council of Australia, the Commonwealth AIDS Research Grants Committee and the Transdisease Vaccinology Programme of the World Health Organisation.

REFERENCES

1. R. L. Coffman, B. Shrader, J. Carty, T. R. Mosmann, and M. W. Bond, *J. Immunol.* 139:3685 (1987).
2. P. D. Murray, D. T. McKenzie, S. L. Swain, and M. F. Kagnoff, *J. Immunol.* 139:2669 (1987).
3. K. Takatsu, A. Tominaga, N. Harada, S. Mita, M. Matsumo, T. Takahashi, Y. Kikuchi, and N. Yamaguchi, *Immunol. Reviews.* 102:107 (1988).
4. D. Y. Kunimoto, R. P. Nordan, and W. Strober, *J. Immunol.* 143:2230 (1989).
5. R. L. Coffman, D. A. Lebman, and B. Shrader, *J. Exp. Med.* 170:1039 (1989).
6. E. Sonoda, R. Matsumoto, Y. Hitoshi, T. Ishii, M. Sugimoto, S. Araki, A. Tominaga, N. Yamaguchi, and K. Takatsu, *J. Exp. Med.* 170:1415 (1989).
7. G. R. Harriman, D. Y. Kunimoto, J. F. Elliot, V. Paetkau, and W. Strober, *J. Immunol.* 140:3033 (1988).
8. K. W. Beagley, J. H. Eldridge, H. Kiyono, M. P. Everson, W. J. Koopman, T. Honjo, and J. R. McGhee, *J. Immunol.* 141:2035 (1988).
9. K. W. Beagley, J. H. Eldridge, F. Lee, H. Kiyono, M. P. Everson, W. J. Koopman, T. Hirano, T. Kishimoto, and J. R. McGhee, *J. Exp. Med.* 169:2133 (1989).
10. S. Schoenbeck, D. T. McKenzie, and M. F. Kagnoff, *Eur. J. Immunol.* 19:965 (1989).
11. T. R. Mosmann, and R. L. Coffman, *Immunol. Today* 8:223 (1987).
12. T. Taguchi, J. R. McGhee, R. L. Coffman, K. W. Beagley, J. H. Eldridge, K. Takatsu, and H. Kiyono, *J. Immunol.* 145:68 (1990).
13. H. D. Campbell, C. J. Sanderson, Y. Wang, Y. Hort, M. E. Martinson, W. Q. J. Tucker, A. Stellwagen, M. Strath, and I. G. Young, *Eur. J. Biochem.* 174:345 (1990).
14. C-P. Chiu, C. Moulds, R. L. Coffman, D. Rennick, and F. Lee, *Proc. Nat. Acad. Sci. USA* 85:7099 (1988).
15. D. B. Boyle, B. E. H. Coupar, and M. E. Andrew, *Gene* 35:169 (1985).
16. B. E. H. Coupar, M. E. Andrew, and D. B. Boyle, *Gene* 68:1 (1988).

17. M. E. Andrew, B. E. H. Coupar, G. L. Ada, and D. B. Boyle, *Microb. Pathog.* 1:443 (1986).
18. C. Czerkinsky, Z. Moldoveanu, J. Mestecky, L-A. Nilsson, and O. Ouchterlony, *J. Immunol. Meth.* 115:31 (1988).
19. G. Karupiah, B. E. H. Coupar, M. E. Andrew, D. B. Boyle, S. M. Phillips, A. Mullbacher, R. V. Blanden, and I. A. Ramshaw, *J. Immunol.* 144:290 (1990).
20. A. Tominaga, S. Mita, Y. Kikuchi, Y. Hitoshi, K. Takatsu, S-I. Nishikawa, and M. Ogawa, *Growth Factors* 1:135 (1989).
21. J. van Snick, S. Cayphas, A. Vink, C. Uyttenhove, P. Coulie, and R. J. Simpson, *Proc. Nat. Acad. Sci. USA.* 83:9679 (1986).
22. J. Mestecky, and J. R. McGhee, *Adv. Immunol.* 40:153 (1987).
23. M. L. Dunkley, and A. J. Husband, *Reg. Immunol.* 2:213 (1989).
24. L. R. Herron, R. L. Coffman, M. W. Bond, and B. L. Kotzin, *J. Immunol.* 141:842 (1988).
25. K. W. Beagley, J. H. Eldridge, W. K. Aicher, J. Mestecky, S. DiFabio, H. Kiyono, and J. R. McGhee, *Cytokine* 3:107 (1991).
26. I. Ramshaw, J. Ruby, A. Ramsay, G. Ada, and G. Karupiah, *Immunol. Rev.* 127:157 (1992).

PERSISTENT *IN VIVO* ACTIVATION AND TRANSIENT ANERGY TO TCR/CD3 STIMULATION OF NORMAL HUMAN INTESTINAL LYMPHOCYTES

Ruggero DeMaria,[1] Stefano Fais[2] and Roberto Testi[1,2]

[1]Dept. of Experimental Medicine, University of Rome "La Sapienza"
[2]Chair of Gastroenterology, University of Rome "La Sapienza"
Italy

INTRODUCTION

Intestinal lymphocytes represent a unique lymphoid compartment heavily and constantly exposed to antigens. Continuous antigen challenge however, does not normally result in an inflamatory tissue damage, suggesting that cell activation mechanisms are tightly controlled in gut mucosal lymphocytes. Normal intestinal lymphocytes have been found in fact to express *in vivo* some degree of phenotypic activation, yet little if any evidence of in vivo proliferation is detectable. Accordingly, TCR/CD3 mediated *in vitro* proliferative signals are consistently inefficient. This peculiar behavior renders the intestinal immune system an ideal model to unravel in vivo operating mechanisms of suppression and tolerance induction.[1]

In this study we investigated the *in vivo* activation status of both T-IEL and T-LPL from normal human colonic mucosa in terms of expression of surface antigens usually associated with recent or past lymphocyte activation. We also analyzed the ability of both T-IEL and T-LPL to be reactivated in vitro through their TCR/CD3 complex. Our results suggest that most normal gut mucosal lymphocytes show phenotypic evidence of continuous in vivo activation, paralleled by substantial, yet transient, anergy to TCR/CD3 complex stimulation *in vitro*.

T-IEL and T-LPL Are Continuously Activated *In Vivo*

We investigated the in vivo expression of CD69 and CD45R0 on the surface of normal human T-IEL and T-LPL. CD69 is a 28-32 kD disulfide-linked homodimer very quickly and transiently induced on the cell surface upon TCR/CD3 crosslinking in T cells. It therefore represent a useful marker for recent T cell activation.[2] The 180 kD splice isoform of transmembrane tyrosine phosphatase CD45, defined by the epitope CD45R0, is acquired a few days after TCR/CD3 stimulation in T cells and persists on the cell surface for a long time, identifying cells which underwent past activation.[3] It has therefore suggested that most "memory" T cells express high levels of CD45R0. Three color FACS analysis indicated that most T-LPL and virtually all T-IEL express CD69, irrespectively of their CD4, CD8 or γ/δ-TCR phenotype (Fig. 1A), as well as high levels of the CD45R0 epitope (Fig. 1B), suggesting that the vast majority of them are repeatedly and constantly activated *in vivo*.

Most intestinal lymphocytes in humans are likely to be primed by antigen in gut-associated specialized lymphoid areas where they are allowed to proliferate and expand. Afterwards, they are released as "memory" cells into the blood stream through lymphatics, and targeted again to mucosal sites where they may be rechallenged by antigen. *In vivo* CD69 expression by T-LPL and T-IEL provides strong evidence for antigen restimulation at mucosal sites. Such restimulation is likely to be persistent, or very frequent, since CD69

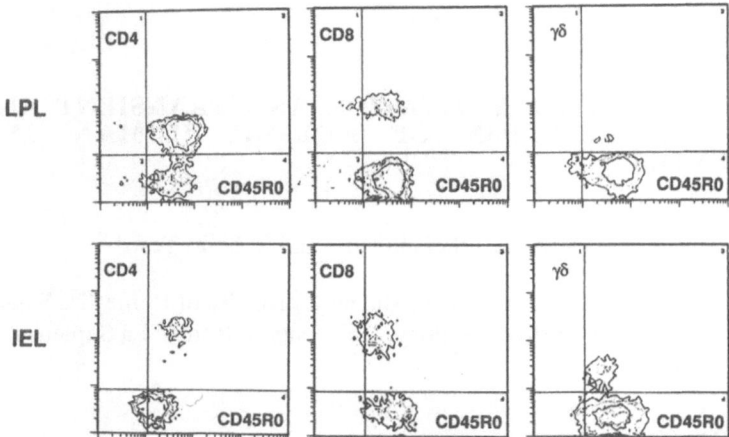

Figure 1A. Simultaneous expression of CD69 and CD4, CD8 or γδ TCR in electronically gated CD3[+] T-LPL (upper panels) and CD3[+] T-IEL (lower panels) by three-color immunofluorescence and FACS analysis, using a FACScan (Becton Dickinson, San Jose, CA). All mAbs were from Becton Dickinson. 98% isotype-matched control mAbs stained cells (not shown) were included in the lower left quadrants.

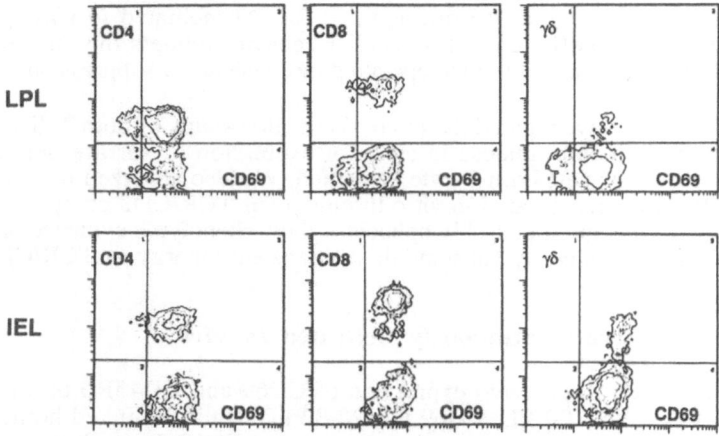

Figure 1B. Simultaneous expression of CD45R0 and CD4, CD8 or γδ TCR in electronically gated CD3[+] T-LPL (upper panels) and CD3[+] T-IEL (lower panels) by three-color immunofluorescence and FACS analysis. >98% isotype-matched control mAbs stained cells (not shown) were included in the lower left quadrants. Transient impairment in TCR/CD3 early signaling in T-IEL and T-LPL.

induction on the cell surface of T cells has been shown to be transient if the stimulus is withdrawn.[2] CD69 on intestinal lymphocytes might play a role in the regulation of cell activation at mucosal sites. It has been shown that CD69 is a signal transducing molecule in activated lymphocytes,[4,5] as well as in other non-lymphoid cells,[6,7] although physiologic ligands have not been identified yet. It is possible, therefore, that CD69 expression may enable gut mucosal T lymphocytes to receive accessory signals which are important for the amplification or modulation of local immune responses. Tyrosine phosphatase activity of CD45 is supposed to play a central role in TCR/CD3-mediated signal transduction.[8] The interaction of CD45R0 with the B cell surface receptor CD22 has been shown to inhibit TCR/CD3-mediated T cell activation.[9] On the other hand, CD22-mediated signals cooperate in B cell activation. T-B interactions via CD45R0/CD22 therefore could be relevant in both controlling T-LPL activation and providing B cell help within the lamina propria enviroment. More difficult to predict, based on present knowledge, is the possible function of CD45R0 expressed by T cells at the epithelial level, where B cells are virtually absent.

Transient Impairment in TCR/CD3 Early Signaling in T-IEL and T-LPL

The possible functional consequences of persistent *in vivo* cell activation on TCR/CD3-mediated signaling were therefore investigated. By analyzing Ca2+ bound Fluor-3 emission in electronically gated CD7+ cells by FACS, we observed that constitutive free cytoplasmic [Ca2+]i in T-IEL and T-LPL was substantially higher (130-150 nM) than that measured in autologous T-PBL (80-100 nM). Moreover, crosslinking of TCR/CD3 by mAbs was essentially ineffective in raising cytoplasmic [Ca2+]i in T-LPL, and the response was greatly impared in T-IEL, compared to autologous T-PBL (Fig. 2A). Similarly, the generation of Ins(1,4,5)P3 levels following TCR/CD3 crosslinking was almost absent in T-LPL and grossly reduced in T-IEL, compared to autologous T-PBL (Fig. 2C). However, when T-LPL were kept as single cell suspension in culture medium for 24 hs at 37 ° C and then stimulated through their TCR/CD3, they almost completely recovered their ability to respond by cytoplasmic [Ca2+]i elevation and by Ins(1,4,5)P3 generation. On the contrary, T-LPL kept at 4° C for 24 hr behaved similarly to freshly isolated T-LPL. Importantly, note that basal cytoplasmic [Ca2+]i in T-LPL kept for 24 hr at 37° C, but not in T-LPL kept at 4° C, returned to levels comparable to T-PBL (Fig. 2B). These results suggested that T-LPL could recover TCR/CD3 responsiveness in a relatively short time, if withdrawn from the site of antigenic challenge and kept metabolically active. Similar experiments with T-IEL could not be performed, due to the massive (>25 %) spontaneous death, likely due to apoptosis, of T-IEL induced in culture at 37° C.[10]

Most T-IEL and T-LPL precursors are therefore likely to enter the mucosal enviroment from the blood stream as conventional CD45R0+ CD69- "memory" T cells, rechallenge antigen with a fully functional TCR/CD3 and quickly express CD69. Prolonged receptor engagement locks intestinal T cells in a CD69+, high cytoplasmic [Ca2+]i status which probably uncouples the TCR/CD3 complex from some downstream signaling effectors. Intestinal lymphocytes may therefore represent a particular type of "memory" T cell, characterized by the simultaneous expression of CD45R0 and CD69, which rather than hyperresponsive to TCR/CD3 triggering, is anergic. Upon receptor disengagement, a certain amount of time is probably needed for recovery of signaling capabilities via the TCR/CD3 by intestinal T cells. *In vivo*, however, due to continuous antigen supply, recently disengaged TCR will be immediately reoccupied, locking again the T cells in the hyporesponsive state. Intestinal T cells are therefore likely to have very little "free time", or a narrow time window for signaling capability, compared to relatively long hyporesponsive and refractory times. It is therefore expected that at any given time point the fraction of cells available for signaling *in vivo* will be extremely small.

Transient anergy through the TCR/CD3 may represent a mechanism to constrain antigen-dependent activation *in vivo* during recirculation of lymphocytes in specialized tissues. Our data suggest that this mechanism may be operating in continuously activated cells of the intestinal mucosa and be regulated at the level of early signaling events.

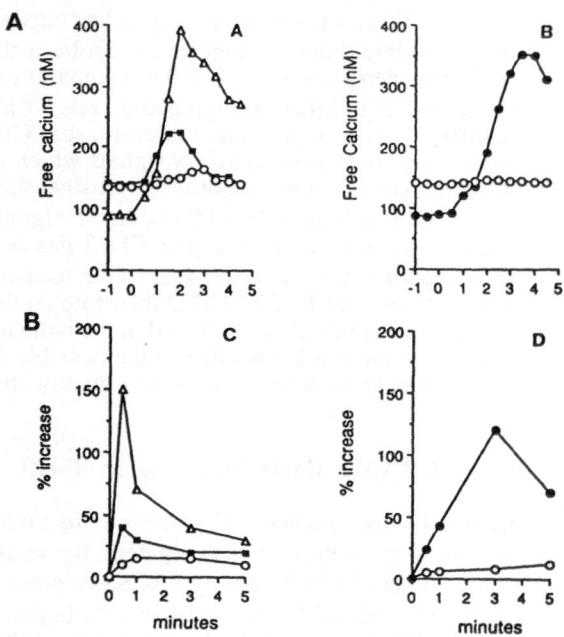

Figure 2A and B. Cytoplasmic [Ca2+]j following TCR/CD3 triggering in CD7+ PBL (A, open triangles), CD7+ LPL (A, open circles) and CD7+ IEL (A, closed squares), CD7+ LPL cultured in medium for 24 hr at 37° C (B, closed circles) or at 4° C (B, open circles). Cells loaded with Fluor-3 and stained with Red613 anti-CD7 were analyzed by FACS before and after stimulation at time 0 with 5 μg/ml of OKT3. 3000 events were recorded every 30 sec and CD7+ cells were analyzed. About 95% of CD7+ LPL and IEL were also CD3+. C and D. Ins (1,4,5) P3 generation following TCR/CD3 triggering in PBL (C, open triangles), LPL (C, open circles) and IEL (C, closed squares), CD7+ LPL cultured in medium for 24 hr at 37° C (D, closed circles) or at 4° C (D, open circles). Equal numbers of CD3+ cells were included in each group. Cells were stimulated at time 0 with 5 mg/ml OKT3 and aliquots were assayed at different times. Ins (1,4,5) P3 extracted from total cell Iysates was measured by RIA (DuPont, Boston).

ACKNOWLEDGMENTS

We are indebt with Drs. M.G. Cifone and M. Silvestri for help in some of the experiments, and with Profs. L. Frati, F. Pallone and A. Santoni for invaluable support and discussion.

REFERENCES

1. P. Brandtzaeg, T. S. Halstesten, K. Kett, P. Krajci, D. Kvale, T. O. Rognum, H. Scott, and L. M. Sollid, *Gastroenterology* 97:1562 (1989).
2. R. Testi, J. H. Phillips, and L. L. Lanier, *J. Immunol.* 142:1854 (1989).
3. A. N. Akbar, L. Terry, A. Timms, P. C. L. Beverley, and G. Janossy. *J. Immunol.* 140:2171 (1988).
4. M. Cebrian, E. Yague, M. Rincon, M. Lopez-Botet, M. O. De Landazuri, and F. Sanchez-Madrid, *J. Exp. Med.* 168:1621. (1988).
5. R. Testi, J. H. Phillips, and L. L. Lanier, *J. Immunol.* 143:1123 (1989).
6. R. Testi, F. Pulcinelli, L. Frati, P. P. Gazzaniga, and A. Santoni, *J. Exp. Med.* 172:701 (1990).
7. R. Gavioli, ,A. Risso, D. Smilovich, I. Baldissarro, M. C. Capra, A. Bargellesi, and E. Cosulich, *Cell. Immunol.* 142:186 (1992).
8. G . A. Koretzky, J. Picus, M. L. Thomas, and A. Weiss, *Nature* 346:66 (1990).
9. I. Stamenkovic, D. Sgroi, A. Aruffo, M. S. Sy, and T. Anderson, *Cell* 66:1133 (1991).
10. J. L. Viney and T. T. MacDonald, *Eur. J. Immunol.* 20:2809 (1990). .

FUNCTIONAL LACTOFERRIN RECEPTORS ON ACTIVATED HUMAN LYMPHOCYTES

Marie-Louise Hammarström, Lucia Mincheva-Nilsson
and Sten Hammarström

Department of Immunology, University of Umeå
S-901 85 Umeå, Sweden

INTRODUCTION

Lactoferrin is an iron binding protein present in external secretions, particularly milk, and in neutrophil granulocytes. It is structurally related to transferrin, which is the major iron binding protein in serum. Transferrin can fully substitute for serum as growth support for activated human lymphocytes and many cell lines[1,2] and cell surface receptors for transferrin are expressed on activated lymphocytes and other proliferating cells.[3,4]

The biological function of lactoferrin is poorly understood. It inhibits bacterial growth by depriving bacteria of iron.[5] It also seems to have a regulatory role in myelopoiesis and in iron absorption in breast feed infants.[6,7] Lactoferrin has been shown to bind to and/or support growth of monocytes, the colon carcinoma cell line HT29 and certain leukemia cell lines.[8,9]

Here we show that activated human peripheral blood lymphocytes express functional lactoferrin receptors. Furthermore, we show that lymphocytes from first trimester human decidua, which contain a large fraction of *in vivo* activated cells, also express lactoferrin receptors.

MATERIALS AND METHODS

Lactoferrin (Lf) and Transferrin (Tr)

Human Lf was purified from colostral skim milk by two alternative procedures. The first involved: precipitation with 80% ammonium sulfate, gel filtration on Sepharose 6B in PBS, pH 7.2, adsorption to rabbit anti-Lf IgG-Sepharose 4B followed by elution with 0.2 M glycine-HCl buffer, pH 2.8.[10] The other procedure involved decaseination at pH 4.7, heparin-Sepharose chromatography[11] followed by gel filtration on Sephadex G 200 in 6 M guanidinium-HCl. Lf was saturated with iron by incubation in 0.07 M $FeCl_3$ in 0.1 M citrate/bicarbonate buffer, pH 8.2 for 18 hours.

Purified human serum Tr (Kabi) was further purified by passage over wheat germ agglutinin Sepharose 6B followed by gel filtration on Sephadex G150.[1]

Anti-Lactoferrin and Anti-Transferrin Antibodies

Rabbit anti-Lf IgG and anti-Tr IgG were immunosorbent purified on Lf-Sepharose 4B and Tr-Sepharose 4B, respectively. F(ab')2-fragments were prepared by pepsin digestion of immunosorbent purified antibodies. FITC conjugated F(ab')2-fragments of immunosorbent purified antibodies were prepared by a dialysis method.

Monoclonal Antibodies (mAbs)

The following mAbs were used: anti-transferrin receptor, anti-CD71 (OKT9) and anti-CD3 (OKT3); purified culture supernatants of hybridoma cell lines from American Type Culture Collection. Anti-CD3 (T3-4B5), anti-CD4 (T3-10), anti-CD8 (DK25), anti-CD20 (B-Ly-1) and anti-CD45RO (UCHL-1) from Dakopatts. Anti-TCRγδ (δTCS1 and TCRδ1) and anti-TCRαβ (BMA 031) from T-cell Sciences. Anti-CD14 (Mf-P9) and anti-CD56 (MY-31) from Becton-Dickinson. Anti-CD20 (B1) from Sera lab and anti-CD45RA-like (D10D11); a kind gift from Dr. B. Axelsson, University of Stockholm, Stockholm, Sweden.

Leukocyte Preparations

Peripheral blood lymphocytes (PBL) were isolated from peripheral blood of healthy adults by gelatin sedimentation, treatment with iron carbonyl and magnet followed by ficoll-isopaque gradient centrifugation.[1] Peripheral blood mononuclear cells (PBMC) were prepared by ficoll-isopaque gradient centrifugation.

Decidua associated mononuclear cells (DMC) were isolated from decidua of healthy women undergoing elective termination of normal pregnancies at the 8 to 12 week of gestation. DMC was prepared by mechanical disruption followed by ficoll-isopaque gradient centrifugation.[12]

Bulk cultures of stimulated PBL and PBMC were established by stimulation with optimal concentrations of leukoagglutinin from *Phaseolus vulgaris* (La, 5 μg/ml), concanavalin A (Con A, 3 μg/ml), OKT3 (50 ng/ml) or a combination of Ionomycin (0.5 μg/ml) and phorbol myristate acetate (PMA; 5 ng/ml) in HEPES buffered RPMI 1640 supplemented with 0.4% human serum albumin (HSA).

Proliferation Assay

PBL or PBMC were incubated in the presence of optimal concentrations of Con A, La or OKT3 in HEPES buffered RPMI 1640 at a concentration of 10^6 cells/ml. Lf, Tr or 0.4% HSA was used as growth support. HSA contains Tr as a major contaminant. The final Tr concentration in HSA supplemented cultures is 20 μg/ml.[1] Proliferation was estimated by [^{3}H]-thymidine incorporation.[1,10] PBL were also used as responder cells in one way MLR with irradiated allogeneic PBMC as stimulator cells (responder/stimulator ratio: 2/1) and incubated in HEPES buffered RPMI 1640 supplemented with 10% human AB+ serum.[10]

Immuno Flow Cytometry

Cells (2×10^6) were pretreated with saturating amounts of Lf or Tr (10 μg) and then incubated with FITC-conjugated F(ab')2 fragments of immunosorbent purified rabbit anti-Lf or rabbit anti-Tr. Determination of other surface markers and two-color staining experiments were performed as described earlier.[12]

RESULTS

Purified Lf supported growth of lymphocytes activated with La (Fig. 1a), Con A or OKT3. Maximal proliferation was reached at Lf concentrations between 5 and 25 µg/ml. The proliferative response was, however, lower than that in standard culture conditions (Fig. 1a). The magnitude of the response in serum-free cultures supplemented with an optimal concentration of Lf was 15 to 80% of that in cultures with Tr. Lf alone did not induce proliferation.

A significant augmentation of the proliferative response was seen if Lf was added to T cells activated by La, Con A or OKT3 (Fig. 1b) in medium containing an optimal concentration of Tr. The proliferative response in one way MLR was also augmented by the addition of Lf to cultures supplemented with serum (Fig. 1c). Augmentation was seen at concentrations between 10 and 100 µg/ml. Lf fully saturated with Fe^{3+} or saturated to 20% gave the same growth supporting or growth augmenting effect.

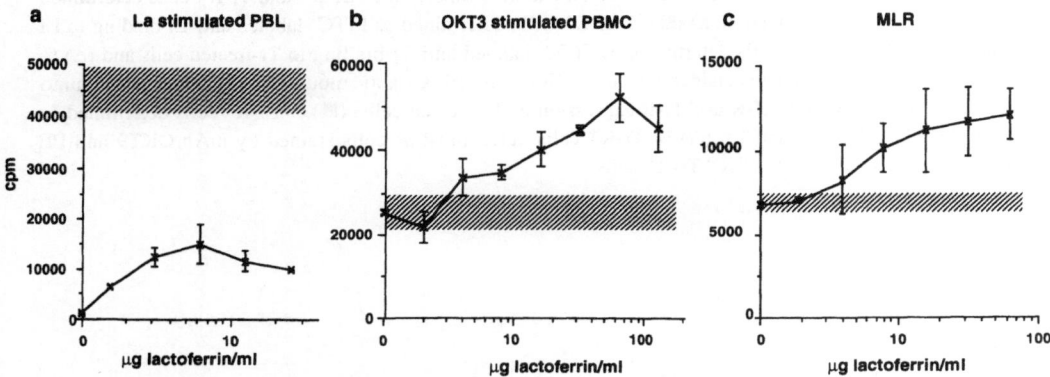

Figure 1. a) PBL stimulated with La in serum-free medium supplemented with increasing amounts of Lf. Shaded area indicates mean cpm ± 1 SD in parallel cultures supplemented with 20 µg Tr/ml and 0.4% HSA. **b)** PBMC stimulated with OKT3 in medium supplemented with 20 µg Tr/ml and 0.4% HSA and increasing amounts of Lf and **c)** PBL incubated with irradiated allogeneic PBMC in medium supplemented with 10% human AB+ serum and increasing amounts of Lf. Shaded areas in **b)** and **c)** indicates cpm ± 1 SD in cultures with no added Lf. Proliferation was measured as [3H]-thymidine uptake after 88 hrs incubation in **(a,b)** and after 112 hrs **(c)** including a final 16 hr pulse.

Freshly isolated PBMC did not bind anti-Lf. After pretreatment with Lf up to 15% of the cells were stained with anti-Lf. The positive cells were found to be large cells expressing the monocyte marker CD14. Thus, in unstimulated PBMC preparations monocytes, but not lymphocytes, expressed Lf-R.

La, Con A, OKT3 and PMA + Ionomycin induced expression of Lf-R on blood lymphocytes. The kinetics of the appearance of Lf-R, and for comparison Tr-R, on the cells are shown in Fig. 2a and b. In standard culture conditions the proportion of Lf-R expressing cells reached a maximum after three days of incubation and then gradually decreased. Tr-R expressing cells appeared at about the same time as Lf-R expressing cells in the cultures and Tr-R+ cells outnumbered the Lf-R+ cells throughout the culture period (Fig. 2a). Cells cultivated in medium supplemented with both 20 µg Lf/ml and 20 µg Tr/ml showed similar kinetics for Lf-R and Tr-R induction but Lf-R expression was prolonged under these culture conditions (Fig. 2b). Up to 70% of the cells in cultures activated for three days expressed Lf-R. When cells from three days activated cultures were divided into small and large cell according to their forward and side scatter characteristics, Lf-R bearing cells were found in both fractions (Fig. 3).

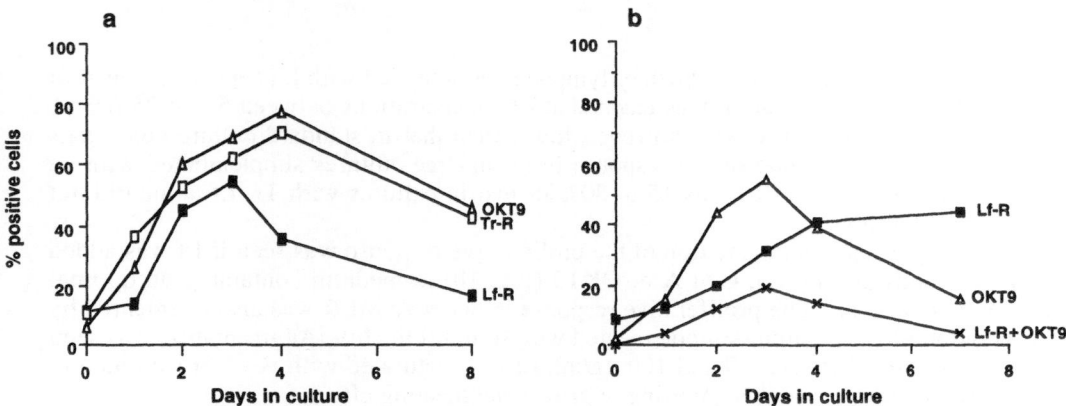

Figure 2. PBMC were stimulated with Con A in medium supplemented with **a)** 20 μg Tr/ml in 0.4% HSA and **b)** 20 μg Lf/ml and 20 μg Tr/ml in 0.4% HSA. and the proportion of Lf-R$^+$ and Tr-R$^+$ cells determined after different times of incubation. **a)** (■) % Lf-R$^+$ cells determined as FITC -labeled anti-Lf binding to Lf treated cells, (□) % Tr-R$^+$ cells determined as FITC -labeled anti-Tr binding to Tr-treated cells and (Δ) % Tr-R$^+$ cells determined as cells stained by mAb OKT9 and FITC- anti- mouse IG **b)** Two color immuno flow cytometric analysis of Lf-R and Tr-R expression on Lf treated cells. (■) % Lf-R$^+$ cells determined as staining by FITC labeled anti-Lf. (Δ) % Tr-R$^+$ cells determined as cells stained by mAb OKT9 and PE labeled anti-mouse IG. (x) % Lf-R$^+$ Tr-R$^+$ cells.

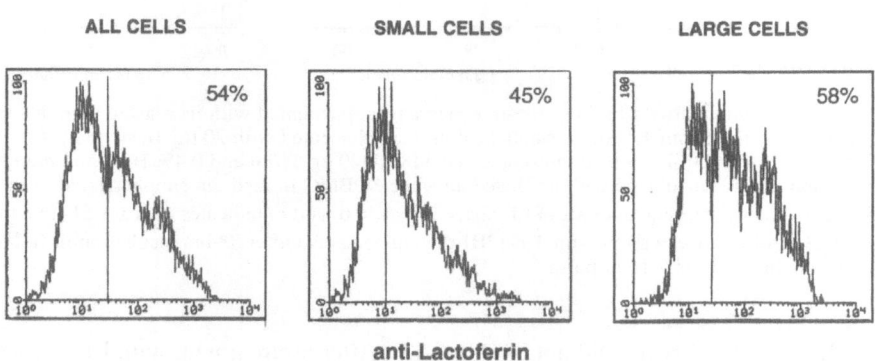

Figure 3. Lf-R determination by immuno flow cytometric analysis of PBMC activated for three days with Con A.

The phenotype of Lf-R expressing cells in three day activated cultures was determined by two color immuno flow cytometry. The majority of the Lf binding cells were T cells (Fig. 4), but Lf-R was also expressed on B cells (CD20$^+$), NK cells (CD56$^+$) and monocytes (CD14$^+$). Lf-R expression was induced on a larger fraction of TCRγδ$^+$ cells than of TCRαβ$^+$ cells (Fig. 4). Large fractions of both CD4$^+$ and CD8$^+$ cells expressed Lf-R (Fig. 4). Tr-R and Lf-R were expressed simultaneously on up to 45% of activated lymphocytes (Fig. 4).

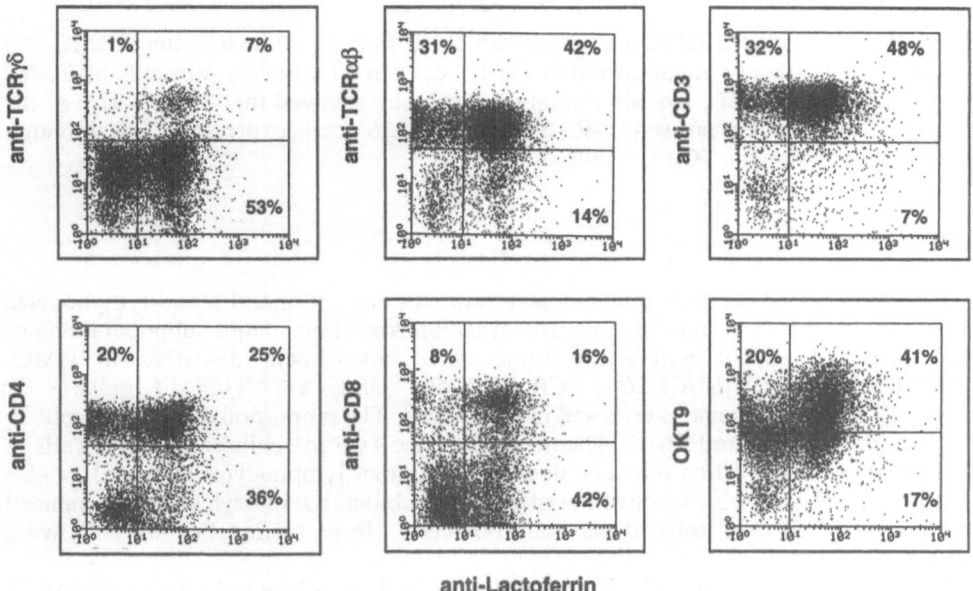

Figure 4. Two color immuno flow cytometric analysis of three day cultures of Con A stimulated PBMC. Percentage of Lf-R$^+$ cells was determined as binding of FITC-labeled anti-Lf to Lf treated cells. Other surface markers were detected by PE- labeled anti-CD4 and anti-CD8 or indirectly using anti-TCRγδ mAbs δTCS1+TCRδ1, anti -TCRαβ, anti-CD3 or OKT9 and PE- labeled anti mouse Ig.

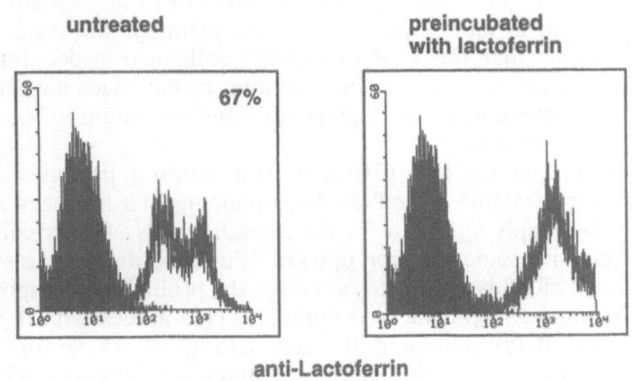

Figure 5 Expression of Lf-R on freshly isolated decidual mononuclear cells

Decidual mononuclear cells (DMC) from the first trimester of normal human pregnancy were isolated and analyzed for Lf-R expression. A large fraction (45-72%) of the cells expressed Lf-R. In contrast to PBL, no pretreatment with Lf was required for detection of Lf-R (Fig. 5), suggesting that the receptors had taken up Lf *in vivo*. However, the mean fluorescence intensity increased ≈ 2.5 fold if the cells were pretreated with Lf before addition of anti-Lf conjugate indicating that the receptors were not fully saturated

(Fig. 5). Freshly isolated DMC also expressed Tr-R. In contrast to activated PBMC, the proportion of Lf-R$^+$ cells outnumbered the Tr-R$^+$ cells in all samples. Virtually all Tr-R$^+$ cells also expressed Lf-R. Double staining experiments showed that the majority of the decidual TCRγδ$^+$ cells expressed Lf-R. Similarly, CD56$^+$ cells expressed Lf-R, although to a lesser extent than the TCRγδ$^+$ cells.

DISCUSSION

Freshly isolated decidual mononuclear cells but not peripheral blood lymphocytes express receptors for the iron binding protein lactoferrin. Four major subpopulations of lymphocytes of approximately the same sizes have been identified in DMC: TCRγδ$^+$/CD56$^+$-, TCRγδ$^+$/CD56$^-$-, TCRγδ$^-$/CD56$^+$- and CD8$^+$/TCRαβ$^+$ cells.[12] In addition, DMC contain some B cells and monocytes.[12] The proportion of TCRγδ$^+$ cells in DMC is high compared to PBMC. The majority of the TCRγδ$^+$ cells and CD56$^+$-cells in decidua expressed activation markers, demonstrating that lymphocytes activated *in vivo* can express Lf-R. Polyclonal activation of peripheral blood lymphocytes *in vitro* induced expression of LF-R on T cells, B cells and NK cells. In particular, the γδT cells were prone to develop Lf-R.

We have characterized the Lf-R biochemically.[13] Two polypeptides of 47 kD and 65 kD molecular weights (reducing conditions) were identified by SDS-PAGE/ autoradiography after passage of lysates from surface labeled activated T cells over Lf-Sepharose. The Tr-R had a molecular weight of 90 kD under the same conditions.

Activated lymphocytes also express Tr-R. The kinetics of receptor expression is similar for both iron binding proteins under standard culture conditions. The levels of Lf-R and Tr-R expression reach their maximum around day 3 of incubation. In these cultures the proportion of Tr-R expressing cells was larger than the proportion of Lf-R expressing cells throughout the incubation period. However, addition of Lf to standard cultures led to increased Lf-R expression at later time points. Interestingly, the proportion of Lf-R expressing cells was larger than the Tr-R expressing cells also in decidual lymphocyte preparations. Lf is produced by cells in the decidual tissue (data not shown). These results suggest that the concentration of Lf present during activation influences induction and/or duration of Lf-R expression.

Interestingly, receptors for two different iron binding proteins are expressed simultaneously on a large fraction of activated lymphocytes. It has been shown that Tr supports proliferation by supplying iron to Tr-R expressing cells. It is possible that T cells actually have two alternative routes for iron uptake. We found that Lf can support growth of polyclonally activated blood lymphocytes although the proliferative response was lower than that obtained in cultures supplemented with Tr. This agrees with the findings that certain human T- and B-lymphoma cell lines can grow in serum-free medium supplemented with Lf only.[9,13] Addition of Lf to cultures supplemented with Tr or serum augmented the proliferative response of both polyclonally activated lymphocytes and lymphocytes responding to alloantigens. This shows that presently used standard culture conditions for *in vitro* activation of human T cells are suboptimal.

Certain T cell subsets may preferentially use Lf as iron support for growth. Indeed, cells expressing Lf-R only were abundant among decidual lymphocytes and were also seen in cultures of activated blood lymphocytes. This implies that one biological function of Lf is to enhance the immune response by providing growth support to selective subsets of lymphocytes such as TCRγδ$^+$ cells.

Lf is present in high concentrations in milk. It is, however, also present in granules of neutrophil granulocytes and in the mucosa. Lf secreted by neutrophil granulocytes and milk-Lf have the same biological activity in a granulocyte - macrophage colony stimulating factor test system.[15] Lf locally secreted by granulocytes at an inflammatory site binds iron. The benefit of this may be twofold: 1) to inhibit bacterial growth by iron deprivation and 2) to provide Lf bound iron that can be utilized by activated T cells.

Taken together our results suggest that Lf is an important growth supporting factor for TCR$\gamma\delta^+$ cells in local immune reactions, especially in the mucosa.

REFERENCES

1. M. -L. Dillner-Centerlind, S. Hammarström, and P. Perlmann, *Eur. J. Immunol.* 9:942 (1979).
2. D.Barnes and G. Sato, *Cell* 22:649 (1980).
3. T. A. Hamilton, H. G. Wada, and H. H. Sussman, *Proc. Natl. Acad. Sci. USA* 76:6406 (1979)
4. I. S. Trowbridge and M. B. Omary, *Proc. Natl. Acad. Sci.. USA* 78:3039 (1981).
5. R. R. Arnold, M. F. Cole, and J. R. McGhee, *Science* 197:263 (1977).
6. H. E. Broxmeyer, M. de Sousa, A. Smithyman, P. Ralph, J. Hamilton, J. I. Kurland, and J. Bognacki, *Blood* 55:324 (1980).
7. G. Spik, B. Brunet, C. Mazurier-Dehaine, G. J. Fontaine, and J. Montreuil, *Acta Paediatr. Scand.* 71:979 (1982).
8. D. Roiron, M. Amouric, J. Marvaldi, and C. Figarella, *Eur. J. Biochem.* 186:367 (1989).
9. Y. Yamada, T. Amagasaki, D. W. Jacobsen, and R. Green, *Blood* 70, 264 (1987).
10. L. Mincheva-Nilsson, M.-L. Hammarström, P. Juto, and S. Hammarström, *Clin. Exp. Immunol.* 79:463 (1990).
11. L. Bläckberg,and O. Hernell. *FEBS Letters* 109:180 (1980).
12. L. Mincheva-Nilsson, S. Hammarström, and M.-L. Hammarström, *J. Immunol.* 149:2203 (1992).
13. L. Mincheva-Nilsson, S. Hammarström, and M.-L. Hammarström, submitted (1992).
14. S. Hashizume, K. Kuroda, and H. Murakami, *Methods Enzymol.* 147:302 (1987).
15. H. E. Broxmeyer, D. C. Bicknell, S. Gillis, E. L. Harris, M. L. Pelus, and G. W. Sledge, Jr, *Blood Cells* 11:429 (1986).

SUBSTANCE P PROMOTES PEYER'S PATCH AND SPLENIC B CELL DIFFERENTIATION

David W. Pascual,[1] Kenneth W. Beagley,[2] Hiroshi Kiyono,[1] and Jerry R. McGhee[3]

[1]Department of Oral Biology; [2]Division of Gastroenterology, Department of Medicine; and [3]Department of Microbiology, University of Alabama at Birmingham. Birmingham, Alabama 35294

INTRODUCTION

Current interpretations of immune processes tends to exclude nonhematopoietic cells and their representative mediators as possible contributors to the immune response. One such excluded element is the nervous system. To date, there is considerable evidence that the sympathetic[1] and the peptidergic[2] nervous systems substantiate neural involvement in immune regulation. This is never more evident than the effects exerted by neuropeptides upon mucosal immune responses. One such neuropeptide is substance P (SP). A member of the tachykinin family, SP is an 11 amino acid peptide[3], and outside the brain, it is found in greatest concentrations in the gut.[3] SP is a product of sensory ganglion cells, and it is transported to peripheral sites where it is stored and released upon noxious stimulation.[3] SP-containing fibers have been localized in the Peyer's patches (PP)[4] suggesting possible influences upon B cell differentiation.

Attempts to delineate possible mechanisms for SP involvement in B cell differentiation have demonstrated SP to be a late-acting differentiation factor.[2] Such differentiation capacity by SP was first shown by Stanisz *et al.*[5] where murine splenic and PP lymphocytes stimulated with concanavalin A resulted in increases in IgA production of 70 and 300%, respectively. Although substantial increases in IgA production were noted, it remained unclear from this study as to which mononuclear cell population or combinations was most greatly affected by SP since each cell subpopulation is known to exhibit SP receptors. To make such evaluations, studies in our own laboratory focused upon examining the effects of SP on defined cell populations. Using subclones of the CD5$^+$ CH12 B lymphoma cell line, these SP receptor$^+$ cells were optimally stimulated with SP in the presence of a co-activation signal.[6] Low levels of lipopolysaccharide (LPS), approximately 50 ng/ml, were required to observe SP-dependent responses. Optimal IgM and IgA production were obtained at subnanomolar concentrations of SP, or at concentrations near the k_D of the receptor. For the IgM-producing cell line, 172% increase in IgM production was observed; for the IgA-producing cell line, 45% increase in IgA production was noted. In a subsequent study[7], SP enhanced IgM and IgG production 500 and 572%, respectively, by LPS-triggered, purified splenic B cell cultures. These SP-dependent increases were specifically inhibited by SP antagonist. LPS co-activation was required since SP by itself exhibited no effect upon Ig production. Further, the observed increases in Ig production were not the result of increased cell proliferation since SP was shown to inhibit their proliferation. Rather, increased Ig production could be attributed to greater numbers of B cells secreting antibodies as demonstrated with an ELISPOT assay.

Thus, these two studies[6,7] support the contention that SP at physiologically relevant concentrations can directly stimulate B cells in their differentiation.

MATERIALS AND METHODS

C3H/HeN mice between the ages of 8 and 12 weeks were obtained from the Frederick Cancer Research Facility (National Cancer Institute, Frederick, MD). Mice were maintained in horizontal laminar flow cabinets, and sterile food and water were provided *ad libitum*.

PP and spleens were aseptically removed. PP were enzymatically dissociated with Dispase® (Boehringer Mannheim Biochemicals, Indianapolis, IN) in Joklik-modified medium (Gibco Laboratories, Grand Island, NY) as previously described.[8] Spleens were mechanically dissociated by pressing through wire sieves as previously described.[7] PP and splenic B cells were positively selected by panning with rabbit $F(ab')_2$ anti-mouse IgM, IgG, and IgA (heavy and light chain specific) antibodies (Zymed Lab., South San Francisco, CA) using established protocols.[7] B cells isolated by this procedure were consistently >98% surface Ig positive.

PP B cells were dispensed in complete media[7] (endotoxin levels less than 0.025 ng/ml) into 96-well microtiter dishes (Costar Corp., Cambridge, MA) at 2×10^5 cells/well. For the interleukin-6 (IL-6) treated PP B cell cultures, 4000 B9 U/ml of recombinant mouse IL-6 was used. Varying concentrations of SP (Peninsula Lab., Belmont, CA) from 10^{-12} to 10^{-8} M were added to quadruplicate cultures treated or untreated with IL-6 in a final culture volume of 100 µl. Cells were cultured for 5 days at 37°C in 5% CO_2 atmosphere, and culture supernatants were subsequently analyzed by isotype-specific sandwich ELISA for IgA, IgG, and IgM production.

Purified splenic B cells were dispensed in complete media into 48-well microtiter dishes (Costar) at 5×10^5 cells/well, and stimulated with 10 µg/ml *Escherichia coli* 055:B5 LPS (tissue culture grade, Sigma Chemical Co., St. Louis, MO). To triplicate cultures, varying concentrations (10^{-13} to 10^{-7} M) of SP or SP agonist, GR73632 (Peninsula Lab.) were added either in the absence or presence of 1000-fold molar excess of SP nonpeptide antagonist, CP 96,345 (kind gift provided by Dr. R. Michael Snider, Pfizer Inc.). Cells in a final volume of 0.5 ml were cultured for 7 days at 37°C in 5% CO_2 atmosphere, and culture supernatants were subsequently analyzed by isotype-specific sandwich ELISA for IgM production.

Isotype-specific ELISAs were used to evaluate IgA, IgG, and IgM levels in culture supernatants. The procedures used were similar to those previously described.[7]

RESULTS AND DISCUSSION

Both murine splenic and PP B lymphocytes have been previously shown to express a single class of high affinity SP receptor with a subnanomolar k_D.[9] In fact, B lymphocytes from the PP express 4- to 5-fold greater number of SP receptors than those from the spleen suggesting that the PP B lymphocytes may be more responsive to SP augmentation. To test such a hypothesis, a suboptimal dose of recombinant murine IL-6 was used to stimulate PP B cell cultures to measure the effects of SP. IL-6 was used since it has been previously shown to be a late-acting differentiation factor for PP B cells whereby maximal levels of IgA can be generated.[8] As shown in Fig. 1, nanomolar concentrations of SP enhanced IgA production. Maximal stimulation was evident at a physiologically relevant dose of 1.0 nM, which represented nearly a 3-fold rise in secreted IgA. In the absence of IL-6 co stimulation, SP was ineffective at stimulating changes in IgA production (Fig. 1).

IgG levels were affected to similar magnitude as IgA levels. IL-6 co-stimulated PP B lymphocytes generated approximately 180% increase in IgG while less than a 30% increase in IgM levels was observed (Fig. 2).

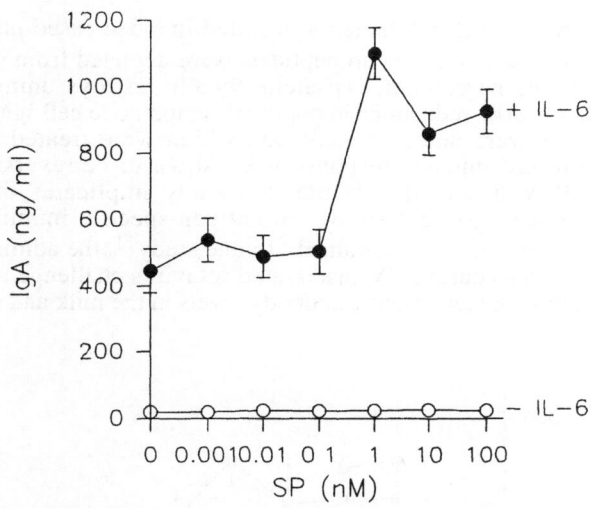

Figure 1. SP synergizes with IL-6 to promote differentiation of Peyer's patch B cells. Purified Peyer's patch B cells (2×10^5) were co-stimulated with 4000 B9 U/ml of recombinant mouse IL-6 (●) and varying doses of SP for 5 days in 100 μl of complete media. Cells stimulated with SP only are also depicted (○). IgA levels were determined by isotype-specific ELISA.

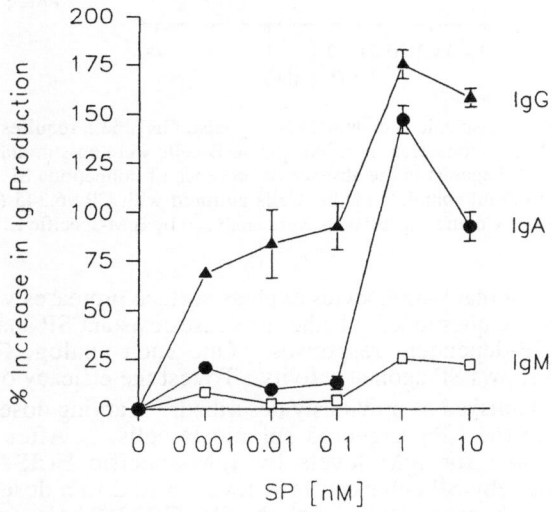

Figure 2. SP synergizes with IL-6 to enhance IgG, but not IgM production by Peyer's patch B cells. Peyer's patch B cell cultures from Fig. 1 were also analyzed for IgG and IgM production using isotype-specific ELISAs. Values depict % change in Ig levels by SP co-stimulated cultures vs cells stimulated with IL-6 only.

These results provide the first evidence that SP can directly affect PP B cells. As indicated by our earlier studies with splenic B cells[7], SP appears to behave as a late-acting differentiation factor.[2] Such evidence implicates a neuronal component, possibly SP, involved in the maturation process of PP B cells. Currently, it is not well understood how SP may be released during antigen challenge. However, it is known that SP is involved in localized immune responses, particularly during inflammatory reactions. This was best evidenced in the study in which SP depletion resulted in a decreased plaque-forming cell response.[10] SP, together with other neuropeptides, were depleted from peripheral sites by the administration of the neurotoxin, capsaicin, which destroys unmyelinated sensory neurons. Approximately 80% reductions in popliteal lymph node cell IgM and IgG plaque-forming cell responses were noted in adult rats which were treated as neonates with capsaicin. This depressed immune response was shown to be reversible upon the co-administration of SP with antigen. While this study implicates the importance of endogenous neuropeptides, particularly SP, in antigen-specific immune responses, the question of SP's adjuvancy was also examined. In one study[11], the administration of SP *in vivo* via miniosmotic pumps during UV-inactivated rotavirus challenge to lactating female mice resulted in enhanced antigen-specific antibody levels in the milk and serum.

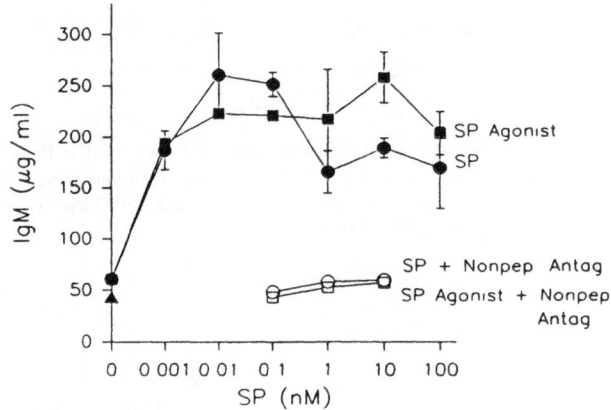

Figure 3. Treatment of murine splenic B cells with SP agonist, GR73632, requires 100-fold more peptide than SP to attain maximal IgM production. Purified splenic B cells were co-stimulated with 10 µg/ml LPS and varying doses of SP or SP agonist in the absence or presence of nonpeptide SP antagonist, CP 96,345, and cultured for 7 days in 0.5 ml complete media. Cells cultured with CP 96,345 (▲) had no effect upon LPS-induced IgM production. Culture supernatants were analyzed by IgM-specific ELISA.

It has been shown that lymphocytes express surface proteases which are capable of inactivating SP. Thus, we questioned whether protease-resistant SP analogs could enhance the potentiation of SP-dependent responses. One such analog, GR73632, is NK-1 receptor-specific, and shows SP agonist activity. To test the efficacy of GR73632, splenic B cell cultures were established as previously described.[7] Varying doses of either SP or SP agonist were added to the LPS triggered splenic B cells. . After 7 days in culture, supernatants were tested for IgM levels by IgM-specific ELISA. Fig. 3 depicts representative data whereby SP enhanced IgM levels 5-fold in a dose-dependent fashion with optimal stimulation between 10 and 100 pM SP. GR73632 also increased IgM levels at similar doses, but maximal effect required a 10 nM dose to achieve a similar 5-fold increase. This contrasts to the finding observed with SP where IgM levels were decreased at these higher doses of peptide yielding only a 3-fold rise in IgM levels. The EC_{50} for this SP agonist appeared similar to SP; consequently, the differences observed reflect greater potentiation of GR73632 at higher peptide doses.

The data presented supports the notion that SP is an extra-immune modifier that participates in the immune process. Inflammation may represent one mechanism by which SP can be introduced into such a process. Since mononuclear cells express receptors for SP, SP (via various routes of induction) can drive the differentiation of B cells by either one or a combination of these possible mechanisms: 1. direct stimulation of B cells in the presence of a co-activation signal, e.g., antigen, IL-6; 2. indirect stimulation of B cells by macrophages exposed to SP to release cytokines, e.g., IL-1, IL-6, or TNF-α; or 3. indirect stimulation of B cells by T cells stimulated with SP to release Th2 type cytokines to drive B cells to IgA differentiation.

ACKNOWLEDGMENT

This work was supported in part by CA54430 (D.W.P.).

REFERENCES

1. D .L. Felten, S. Y. Felten, D. L. Bellinger, S. L. Carlson, K. D. Ackerman, K. S. Madden, J. A. Olschowki, and S. Livnat, *Immunol. Rev.* 100:225 (1987).
2. K. L. Bost and D. W. Pascual, *Amer. J. Physiol.* 262:C537 (1992).
3. B. Pernow, *Pharmacol. Rev.* 35:85 (1983).
4. R. Stead, J. Bienenstock, and A. M. Stainsz, *Immunol. Rev.* 100:333 (1987).
5. A. M. Stanisz, D. Befus, and J. Bienenstock, *J. Immunol.* 136:152 (1986).
6. D. W. Pascual, J. Xu-Amano, H. Kiyono, J. R. McGhee, and K. L. Bost, *J. Immunol.* 146:2130 (1991).
7. D. W. Pascual, J. R. McGhee, H. Kiyono, and K. L. Bost, *Inter. Immunol.* 3:1223 (1991).
8. K. W. Beagley, J. H. Eldridge, J. H. Lee, H. Kiyono, M. P. Everson, W. J. Koopman, T. Hirano, T. Kishimoto, and J. R. McGhee, *J. Exp. Med.* 169:2133 (1989).
9. A. M. Stanisz, R. Scicchitano, P. Dazin, J. Bienenstock, and D. G. Payan, *J. Immunol.* 139:749 (1987).
10. R. D. Helme, A. Eglezos, G. W. Dandie, P. V. Andrews, and R .L. Boyd, *J. Immunol.* 139:3470, (1987).
11. M. K. Ijaz, D. Dent, and L. A. Babiuk, *J. Neuroimmunol.* 26:159 (1990).

THE MIGRATION OF PERITONEAL CELLS TOWARDS THE GUT

Taede Sminia, Marsetyawan Soesatyo, Mohammad
Ghufron, and Theo Thepen

Department of Cell Biology, Section Histology, Vrije
Universiteit, v.d.Boechorststraat 7, Amsterdam,
1081 BT, The Netherlands

INTRODUCTION

Peritoneal cells form a heterogeneous cell population. In the rat about 70% of them are macrophages, 20% consist of granulocytes and mast cells, and the rest comprise dendritic cells and lymphocytes.[1] In microbial infections and after antigen administration, quantitative and qualitative changes occur within these peritoneal cell populations. Peritoneal macrophages efficiently handle antigen locally and transport it via the lymphatics to lymphoid tissues. There are at least two different routes for transport of antigen by peritoneal cells: first, via the lymphatics to the draining lymph nodes[12], and second, across the diaphragm directly to the lung interstitium.[10]

Intraperitoneal immunization can induce a mucosal immune response.[9] The induction of such a response may be related to the migration and homing of (antigen-laden) peritoneal cells to the mucosae. Therefore, we have studied the migration of peritoneal cells and have in particular investigated whether these cells are able to migrate to the gut.

MATERIALS AND METHODS

Animals

Adult female Wistar rats purchased from Harlan Sprague Dawley-CPB (Zeist, The Netherlands) were used throughout this study. They were kept under standard conditions, water and food available *ad libitum*. The animals had a weight of approximately 180-200 grams.

Peritoneal Lavages

After anaesthetization, peritoneal cells were collected by lavage with 10 ml sterile cold (4°C) RPMI 1640 (NPBI, The Netherlands), injected into the peritoneal cavity. After gently shaking the abdomen for a few minutes, the abdominal cavity was opened. The peritoneal cells were washed twice with cold sterile RPMI 1640; erythrocytes were lysed with ammonium chloride. The final concentration used for cell labelling varied between 10^6 and 10^8 per ml.

Cell Labelling

The collected peritoneal cells were labelled with a fluorescent cell linker compound, PKH26 (Zynaxis Cell Science, Inc., Hamburg, Germany). The viable cells, as determined by trypan blue exclusion, were concentrated to 10^6 - 10^8 cells/ml in RPMI 1640 and PKH26; the final concentration of the administrated cells was 2×10^7/ml. After 5 min incubation at 25°C under periodic shaking, the labelling of the cells was stopped by adding an equal volume of sterile fetal calf serum (FCS, 10%) and incubation at 25°C for 1 min. The mixture was then diluted with an equal volume of cold RPMI 1640 and centrifuged at 1200 rpm, 4°C for 7 min. The cell pellet was washed twice with 10 ml RPMI 1640, centrifuged again, and concentrated in saline at a final concentration of 10^6 - 10^8 cells per ml. In some experiments labelled cells were fixed in 1% paraformaldehyde in PBS. Before reinjection, these cells were washed extensively in medium with FCS and finally resuspended in saline at the derived concentration.

Administration of Labelled Cells and Tissue Preparation

10^6 - 10^8 labelled cells were injected into the peritoneal cavity. One, 2, 3 and 5 days after intraperitoneal injection, the animals were sacrificed. Paratracheal and parathymic lymph nodes (PTrLN and PTLN, respectively), a piece of the small and large intestine, Peyer's patches (PP), proximal colonic lymphoid tissue (PCLT), mesenteric lymph nodes (MLN) and spleen were sampled and snap frozen in liquid nitrogen.

Fluorescence Microscopy

Eight μm cryostat sections were air-dried at room temperature for at least 1 h prior to examination. The sections were evaluated with a fluorescence microscope (Axioskop, Zeiss, Germany) equipped with a "TRITC" filter setting. The labelled cells displayed a bright red fluorescent staining.

Enzyme Histochemistry

Acid phosphatase activity was demonstrated using naphthol-AS-BI phosphate as the substrate and hexazotized pararosaniline as the diazonium salt.

RESULTS

Labelling with PKH26 resulted in uniformly red fluorescence of the cell membrane of the peritoneal cells. The viability as determined by trypan blue exclusion was more than 95% in all experiments. The distribution of labelled cells in various tissues at different time intervals is summarized in Table 1.

At day 1 post-labelling numerous labelled cells were found in the subcapsular sinus and the medulla of the PTLN. Furthermore, they were present in the subcapsular sinus of the PTrLN. The labelled cells were brightly red fluorescent and had a round or oval shape. Fluorescent cells were also found in PP. These cells were less brightly labelled and had a granular appearance. Most of them occurred in the dome area and in the follicle; a few were found in the interfollicular area. In spleen, MLN, PCLT and in the lamina propria of the small and large bowel no labelled cells were found.

The number of fluorescent cells were increased in the PTLN and PTrLN two days after peritoneal injection. The majority of these cells was located in the medulla and still a few were found in the subcapsular sinus. Also in PP the number of labelled cells was higher than after day 1. They were located in all 3 compartments: the dome, the B-cell follicle and the interfollicular T-cell area. In the spleen a few labelled cells were found. Hardly any labelled cells were seen in the other tissues studied.

Table 1. Distribution of labelled cells in various tissues.

Tissue	Days post-labelling			
	1	2	3	5
Parathymic lymph node (PTLN)	+	+	+	+
Paratracheal lymph node (PTrLN)	+	+	+	+
Intestinal villi	–	–	–	+/–
Peyer's patch	+	+	+	+/–
Proximal colonic lymphoid tissue	–	+/–	+/–	+/–
Mesenteric lymph node	–	–	–	+/–
Spleen	–	–	–	+/–

+: present; -: absent; +/-: occasionally present

Three Days Post-Labelling

Brightly fluorescent cells were still found in the medulla of the lymph nodes draining the peritoneal cavity (PTLN, PTrLN). Based on their morphology (large oval shaped) and localization (in the medulla) these cells most probably were macrophages. Moreover, acid phosphatase staining showed that these cells display high enzyme activity. In addition, small labelled cells with dendritic cell processes were occasionally detected in the lymph nodes. In the dome and the interfollicular area of PP small labelled cells with cytoplasmic processes were found; large labelled cells occurred in the B-cell follicles. Compared to the fluorescent cells in the PTLN and PTrLN, these cells were less brightly stained. A few labelled cells were found in the spleen and PCLT.

Four Days Post-Labelling

The number of fluorescent cells in the medulla of the PTLN and PTrLN was decreased. In the gut labelled cells were occasionally found in PP, PCLT and in the lamina propria of the small intestine. In the spleen and MLN a few brightly stained fluorescent cells were found.

DISCUSSION

In the present experiment we have shown that fluorescent labelled peritoneal cells injected into the peritoneal cavity migrate to the PTLN and PTrLN, and localize there in the subcapsular sinus and medulla. These findings were expected because these lymph nodes drain the peritoneal cavity.[12] Antigenic material and antigen-laden peritoneal macrophages migrate to these lymph nodes[2], but the migration seems to be independent of the presence of antigen. The present data are in agreement with those of Rosen and Gordon[11], who showed a striking accumulation of labelled mouse macrophages in the outer cortex and medullary cords of the PTLN after injection of these cells into the peritoneal cavity. They suggested that this cell migration is an active process. However, as we found that injected formaldehyde fixed intact macrophages localize at the same sites in the draining lymph nodes, the localization of the injected peritoneal macrophages seems an inactive rather than an active process.

Studies by Thepen[13] on the migration of alveolar macrophages showed that alveolar macrophages injected into the alveoli of rat do not localize in the subcapsular sinus (outer cortex) and medulla but specifically into the T-cell area (paracortical area) of the lung-draining lymph nodes. The cells and factors important to the different migration and localization patterns are not known, but it is likely that these patterns in some way influence the cells of the local microenvironment (peritoneal cavity, alveoli, and draining lymph nodes) The findings demonstrate the important role of the administration route for the outcome of immunization

Fluorescence labelled cells were also detected in the PP Although some of these cells could be granulocytes (about 20% of rat peritoneal cells are granulocytes) the morphology of the labelled cells indicated that they were macrophages and dendritic cells Moreover, acid phosphatase enzyme histochemistry and preliminary double labelling studies supported the presence of injected peritoneal macrophages and dendritic cells in PP

The question arises as to how the peritoneal cells reach PP Lymphocytes enter PP through high endothelial venules, the migration route of macrophages and dendritic cells is unknown Although a direct migration of peritoneal macrophages to PP via lymphatics can not be ruled out, the most likely route for macrophages seems to be the lymph- and blood circulation This notion is sustained by the observation of cells in the spleen (present observations and [11]) These authors also reported that labelled cells were occasionally found in the peripheral blood The specific migration to and localization in PP of peritoneal macrophages and dendritic cells may be governed by receptors on blood cells, factor(s) such as cytokines, and stromal cells (reticulum cells and extra-cellular matrix) that attract and retain macrophages

We realize that the data presented so far are not sufficient to draw definite conclusions on the migration of macrophages and dendritic cells from the peritoneal cavity to the PP But, if such migration and specific homing can be proven, this would be of great importance to demonstrate that intraperitoneal immunization primes for a local mucosal IgA response However, the present findings and current literature point to an important cellular interaction between the peritoneal cavity and the gut The possible interactions are schematically drawn in Fig 1

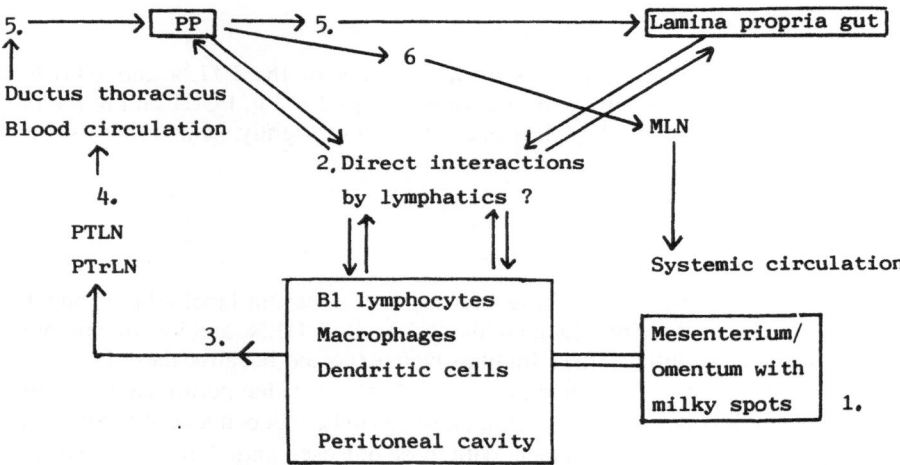

Figure 1. Schematic drawing showing conceivable interactions (cell migration routes) between peritoneal cavity and the gut

   1   Possible origin of peritoneal macrophages and B1 cells, and T cells [6,7,14]

   2   Direct migration of peritoneal cells to the gut or vice versa has not been proven

   3   This route is proven in the present and previous studies [10,11]

   4   Indications for migration from the lymphatics into the blood circulation were among others obtained by Rosen and Gordon[11] and in the present study

   5   Migration of lymphocytes via HEV in PP is proven[3], indication for the influx of immune competent cells into the lamina propria of the gut via HEV-like vessels has been reported[5], the migration route of macrophages and dendritic cells to the gut is as yet unknown

   6   Lymphocytes[4] and macrophages/dendritic cells[8] from the gut primarily migrate to the draining mesenteric lymph nodes

REFERENCES

1. R. H. J. Beelen, D. M. Broekhuis-Fluitsma, C. Korn, and E. C. M. Hoefsmit, *J. Reticuloendothel. Soc.* 23:103 (1988).
2. D. L. Dunn, R. A. Barke, D. C. Ewald, and R. L. Simmons, *Infect. Immun.* 48:287 (1985).
3. J. O. Gowans and E. J. Knight, *Proc. R. Soc. London*, ser.B, 159:257 (1964).
4. S. H. M. Jeurissen, T. Sminia, and G. Kraal, *Cell. Immunol.* 85:264 (1984).
5. S. H. M. Jeurissen, A. M. Duijvestijn, Y. Sonntag, and G. Kraal, *Immunol.* 62:273 (1987).
6. J. W. Koten and W. den Otter, *Lancet* 338:1189 (1991).
7. F. G. M. Kroese, E. C. Butcher, A. M. Stall, P. A. Lalor, S. Adams, and A. Herzenberg, *Int. Immunol.* 1:75 (1989).
8. L. M. Liu and G. G. MacPherson, *Immunology* 73: 281 (1991).
9. N. F. Pierce and F. T. Koster, *J.Immunol.* 124: 307 (1980).
10. M. L. M. Pitt and A. O. Anderson, *J. Adv. Exp. Med. Biol.* 237:627 (1988).
11. H. Rosen and S. Gordon, *Eur. J. Immol.* 20:1251 (1990).
12. N. L. Tilney, *J. Anat.* 109:369 (1971).
13. T. Thepen, Thesis Vrije Universiteit, Amsterdam, the Netherlands (1992).
14. J. F. A. M. Wijffels, R. J. B. M. Hendrickx, J. J. E. Steenbergen, I. L. Eestermans, and R. H. J. Beelen, *Res. Immunol.* 143:401 (1992).

HML-1, A NOVEL INTEGRIN MADE OF THE β7 CHAIN AND OF A DISTINCTIVE α CHAIN, EXERTS AN ACCESSORY FUNCTION IN THE ACTIVATION OF HUMAN IEL VIA THE CD3-TCR PATHWAY

Bernadette Bègue,[1] Sabine Sarnacki,[1] Françoise le Deist,[1] Hélène Buc,[2] Jean Gagnon,[3] Tomaso Méo,[4] Nadine Cerf-Bensussan[1]

[1]INSERM U132, Hôpital Necker-Enfants Malades, Paris
[2]INSERM U75, Faculté Necker, rue Paris
[3]Laboratoire de Biologie Structurale, CEA et CNRS URA 1333, Grenoble
[4]Unité d'immunogénetique et INSERM U276, Institut Pasteur, Paris

INTRODUCTION

Intestinal intraepithelial lymphocytes (IEL) form a large population of T cells located at the interface between the body and the intestinal lumen. They differ from T cells in other lymphoid compartments by their phenotype, mainly CD3+CD8+, the elevated proportion of TcR gd+ cells, their dual thymo-dependent and -independent origin, their cytotoxic properties and their microenvironment in the immediate vicinity of epithelial cells, at some distance from the usual partners of the immune response.[1,2] The hypothesis that IEL possess specific surface receptors enabling interactions with local ligands and controlling their homing, differentiation and/or activation, led us to seek membrane antigens expressed on IEL but not on lymphocytes in other lymphoid organs. One monoclonal antibody, HML-1, raised against human IEL defined a novel membrane antigen preferentially expressed by IEL in the human intestine and in other epithelia (that we will hereafter refer to as MLA for mucosal lymphocyte antigen). Interestingly, in humans, the antibody HML-1 was found to label rare intestinal T cell lymphomas associated with an enteropathy and hyperplasia of IEL, sustaining a previous hypothesis of the intraepithelial origin of such lymphomas[4], and HML-1 labelling was also observed on abnormal cells in mycosis fungoides with marked epidermotropism.[5] However, the expression of this membrane antigen was not entirely restricted to normal or tumoral IEL as it appeared on activated T cells, mainly CD8+.[6,7] Furthermore, HML-1 labelling appeared to be an almost constant phenotypical feature of hairy B cell leukemia.[7-9] In spite of this not entirely specific expression, the antigen defined by HML-1 seemed to be an interesting antigen and a potential candidate for mediating interactions between IEL and enterocytes. This led us to seek more information on its structural and functional properties. Biochemical analysis of the antigen revealed that MLA was a novel member of the integrin family, a large family of molecules which allow either interactions of cells with extracellular matrix proteins or intercellular interactions and which are thereby involved in cell migration, cell differentiation and cell activation.[10] Based on previous studies which demonstrated the role of other lymphocyte-associated integrins, LFA-1 and VLA in the activation of lymphocytes[11,12], we have investigated whether MLA can provide accessory signals enhancing activation of IEL via the CD3-TcR pathway.

MATERIALS AND METHODS

Tissue Specimens

Duodenum or jejunum samples 10 to 15 cm in length were obtained from adult patients undergoing surgery for gastric or pancreatic cancers or chronic pancreatitis, at least 5 cm from pathologic tissue.

Peripheral blood was obtained from healthy donors and from two of the above patients at the time of surgery.

Monoclonal Antibodies

The monoclonal antibodies (mAbs) used for membrane immunofluorescence studies were FITC-conjugated anti-CD25, anti-CD71 (Becton Dickinson, Pont-de-Claix, France), anti-CD11a (IOT16; Immunotech, Luminy, Marseille, France), anti-CD29 (Coulter Immunology, Margency, France), unconjugated or biotinylated HML-1 (3), FITC- or phycoerythrin-conjugated anti-CD3, anti-CD4 and anti-CD8 (Becton-Dickinson). The mAbs used for negative selection of CD8+ PBL have been described elsewhere.[13] The mAbs used for cell stimulation were anti-CD3 (OKT3, Orthodiagnostic, or UCHT1, a kind gift from Dr. P. Beverley), anti-CD29 (4B4; Coulter Immunology), anti-CD11a (IOT16; Immunotech), and anti-MLA. The anti-MLA mAbs were HML-1, and three mAbs recently raised against human IEL and directed to the same antigen.[13,14]

Cells

IEL were isolated using a mechanical procedure.[15] The phenotype of isolated IEL was comparable to that previously described.[6,15,16] Molt 16 cells were a kind gift of Dr. Minowada (Fujisaki Cell Center, Japan). Production of a stable cell line with over 90% HML-1+ cells has been previously described.[14] E-rosette forming (E+) and CD8+ PBL were prepared as described elsewhere.[13]

Procedures for biosynthetic labelling, immunoprecipitation, gel analysis, protein purification and microsequencing have been previously described.[3,14]

Cell Culture

Long term culture of isolated IEL and of the Molt 16 cells has been previously described.[14] For short-term proliferation studies, isolated IEL and enriched peripheral CD8+ T cells were resuspended in culture medium[13] and incubated for 3 to 4 days at $5x10^5$cells/well in flat-bottomed 96-well microtiter plates (Falcon) coated or uncoated with mAbs in PBS overnight at 4°C. Cultures were either pulsed with 1 μCi/well of ^{3}H-thymidine (Amersham, Les Ulis, France) during the last 18 hours or analysed for membrane immunofluorescence.

Membrane Immunofluorescence

Membrane immunofluorescence labeling was performed as described elsewhere[16] and analysed on a FACSCAN (Becton-Dickinson).

Cytoplasmic Free Calcium

E+ PBL and IEL (10×10^6/ml) were loaded with 4mM indo-1 acetomethylester (Indo-1 A.M., Molecular Probes, Eugene, OR, USA) in culture medium for 30 minutes at 37°C, then washed and further incubated for 1 hour at 37°C. The indo-1 fluorescence ratio of individual cells was measured using a FACStar Plus (Becton-Dickinson) before and after adding OKT3 mAb (125 ng/ml) and goat anti-mouse immunoglobulins (Nordic, distributed by Tebu, Perray/Yvelines, France; final dilution 1:100) or the anti-CD2 antibody pair (D66 and T11-1, kindly provided by Dr. A. Bernard; final dilution of ascites: 1/200). UV-excited indo-1 fluorescence was split into high- and low-wavelength emission with a 450 nm long-pass dichroic mirror. Short- and long-wavelength emission were passed through band-pass filters of respectively 405/20 nm and 485/20 nm. The 405nm/485nm fluorescence ratio is displayed as a function of time.

^{32}P Labeling of Phosphoinositides

Lymphocytes were washed, resuspended in 20mM phosphate solution adjusted to pH 7.4 with 25mM Hepes and then incubated in 24-well plates (Falcon) (0.7-0.9 x 10^6 cells/well in 300 μl) with immobilized OKT3 mAb (20 or 50 ng/well coated in PBS overnight at 4°C), a 1:1000 dilution of UCHT1 ascites or buffer alone (control). Twenty μCi of carrier-free ^{32}P-orthophosphate (Amersham) was added to each well. After 15 minutes, the cells were quickly washed with ice-cold buffer and spun down. Phospholipids were extracted with 1ml of cold methanol/chloroform/HCl (200:100:0.75); 300μl of the organic phase was vacuum-concentrated and phospholipids were separated on oxalate-treated silica gel plates (Merck, Chelles, France) with chloroform/acetone/methanol/acetic acid/water (40:15:13:12:8). ^{32}P-labeled phosphoinositides were identified by autoradiography.

RESULTS

MLA is Made of Two Distinct α and β Subunits

Immunoprecipitation of cultured IEL or of an MLA+ cell line (MOLT 16) metabolically labelled with a mixture of ^{35}S methionine and ^{35}S cysteine with the HML-1 antibody showed, after a 6-hour chase, two major bands of approximate MW 120 and 150 kDa in reducing conditions and 105 and 180 kDa in non-reducing conditions (Fig. 1). These two bands, comparable to the two bands precipitated from the cell surface of IEL after iodination, represented two distinct α and β mature polypeptides. First, the α (150 kDa) band but not the b (120 kDa) band was labeled in presence of methionine alone suggesting a different composition in amino acids (data not shown). Second, both chains derived from distinct precursors. Indeed pulse chase experiments combined with deglycosylation studies indicated that the β chain derived from a 88 kDa polypeptide which is decorated with complex N-glycans (30kDa) and a small amount of O-glycans (at least 2 kDa).[14] In contrast, the mature α chain contained approximately 15 kDa complex N-glycans and underwent a postranslational peptidic scission with the loss of a 10 kDa peptide which remained attached to the main core of the protein by a disulfide bond (Ref. 14 and Fig. 1). Finally, HPLC analysis of tryptic digests from both chains revealed quite distinct patterns (see below).

MLA β Chain is the Integrin β7 Chain

Microsequencing of nine HPLC fractions of tryptic digests derived from the MLA β chain allowed identification of about 13 peptides and 150 amino acids representing approximately 20 % of the predictable aminoacid content of the protein. Identified peptides

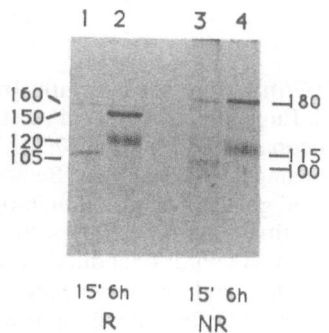

Figure 1. After biosynthetic labelling of MOLT 16 cells with a mixture of ^{35}S methionine and ^{35}S cysteine for 15 minutes (lanes 1 and 3) and a chase for 6 hours (lanes 2 and 4), cell lysates were immunoprecipitated with the HML-1 mAb coupled to protein A-sepharose. Immunoprecipitates were analysed on SDS-polyacrylamide gels in reducing (R: lanes 1 and 2) and non-reducing conditions (NR: lanes 3 and 4).

showed 97% homology with the corresponding sequences of the β7 chain of integrins[14], an integrin subunit recently identified by screening T cell libraries with probes generated by polymerase chain reaction using oligonucleotides derived from the β2 chain.[17,18] The molecular mass and the 8 potential glycosylation sites predicted by the sequence of β7 were compatible with the size and degree of glycosylation of the MLA β chain. Only five residues in the sequence of MLA β differed from the sequence of β7. However, they were detected in fractions which provided two peptides, making it impossible to rule out mistakes in their identification.

MLAα Chain is a Novel Integrin α Chain

Thirteen integrin α chains have been identified in humans. MLAα chain, similarly to a subgroup of α chains, is characterized by the post-translational cleavage of a peptide. Another α chain (VLA- α4) in this subgroup has a MW of 150 kDa[18] comparable to HML-1 and can associate with β7.[19,20] Yet, post-translational cleavage of VLA-α4 results in different cleavage products than that of MLAα.[18] Furthermore, absence of cross-reactivity between VLA-α4 and ML α was confirmed by cross-immunoprecipitation.[14] Finally, sequencing of MLAα tryptic digests confirmed that MLAα is indeed a novel α chain homologous to but different from hitherto cloned integrin α chains (unpublished data).

MLA Can Serve as an Accessory Molecule Favoring Activation of IEL Via the CD3-TcR Pathway

The proliferation index of IEL in response to optimal concentrations of immobilized OKT3 antibody (150ng or 1μg/ well depending on individuals) was variable but not significantly different from that of PBL enriched in CD8+ cells (Fig. 2A). Accordingly, calcium mobilization (Fig. 3) and incorporation of ^{32}P into phosphatidylinositol in response to optimal concentrations of crosslinked OKT3 and/or soluble UCHT-1 (Fig. 4) were not significantly different in human IEL and E+ PBL. At optimal concentrations of anti-CD3 antibody where no accessory signals are required for optimal proliferation of T cells[22], the anti-MLA antibodies exerted minimal or no enhancing effect on the proliferation of IEL.[13]

In contrast, at suboptimal concentrations of OKT3 antibody (10, 50 ng/ well and 150 ng/ well in individuals where this concentration was suboptimal), the anti-MLA antibodies used at a concentration of 1 to 5 μg/ well, exerted a strong synergistic effect on

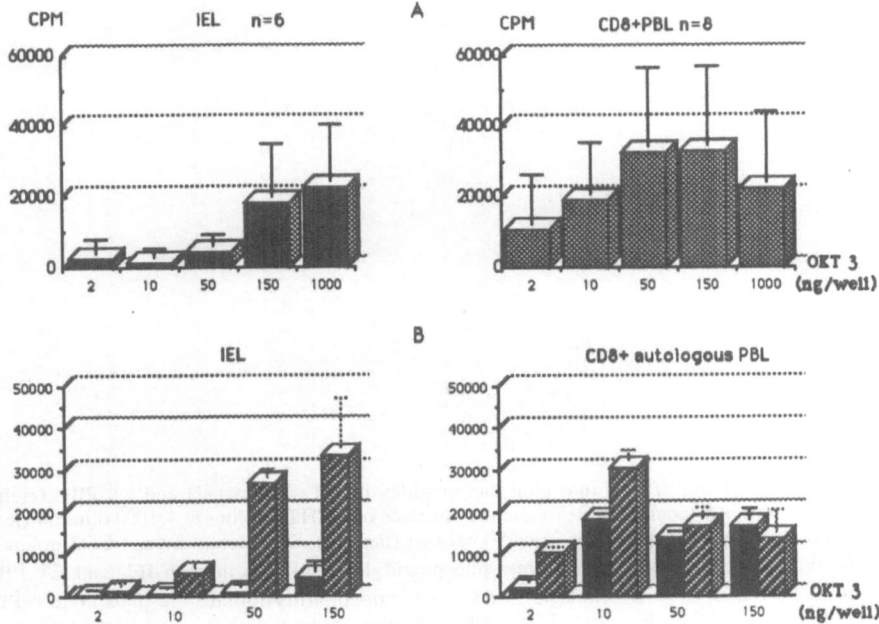

Figure 2. A, the proliferative response of IEL to various concentrations of immobilized OKT3 mAb (left panel) is compared to the proliferative response of PBL enriched in $CD8^+$ cells (right panel) B, the proliferative response to various concentrations of OKT3 either alone (black bars) or coimmobilized with 5µg/well of 6E7 mAb (hatched bars) is compared in IEL (left panel) and autologous PBL enriched in $CD8^+$ cells (right panel) (one representative experiment out of 6 performed)

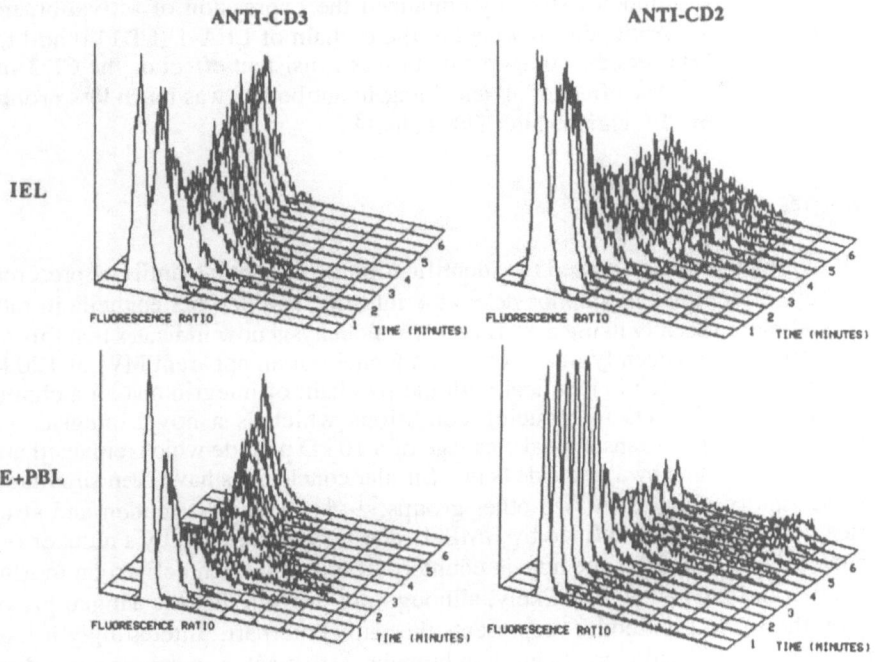

Figure 3 Calcium mobilization in IEL (upper panel) and in E^+ PBL (lower panel) in response to soluble OKT3 mAb crosslinked with rabbit anti-mouse immunoglobulins (left panel anti-CD3) or to soluble D66 and T11 1 (right panel anti-CD2) Results are expressed as the 405nm/485nm fluorescence ratio as a function of time In one representative experiment of four, anti-CD3 and anti-CD2 mAbs mobilized calcium in approximately 65% of IEL and 80 % of E^+ PBL

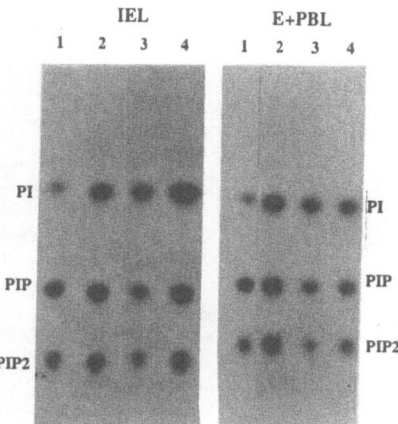

Figure 4. Incorporation of ^{32}P into phosphoinositides in IEL (left panel) and E$^+$ PBL (right panel) incubated without mAb (controls: lanes 1), in the presence of UCHT1 ascites at 1:1000 dilution (lanes 2) or immobilized OKT3, 20ng/well (lanes 3) or 50 ng/well (lanes 4). Stimulation by anti-CD3 mAbs induced markedly increased ^{32}P incorporation into phosphatidylinositol (PI) in both IEL and E$^+$ PBL. The subsequent increase in ^{32}P incorporation into phosphatidylinositol 4-phosphate (PIP) and phosphatidylinositol 4,5-biphosphate (PIP$_2$) is not consistently seen with this technique; however, it was apparent in IEL after stimulation with soluble UCHT1 or 50 ng of immobilized OKT3, and in E$^+$ PBL after stimulation with soluble UCHT1.

the proliferation of IEL (Fig. 2B and Ref. 13). In addition, at OKT3 concentrations of 150 ng/ well, anti-MLA antibodies strongly enhanced the expression of activation markers (CD25 and CD71).[13] Antibodies directed to the α chain of LFA-1 (CD11a) and to the β chain of VLA (CD29) exerted a comparable but less consistent effect on the CD3-induced proliferation of IEL.[13] The effect of all anti-integrin antibodies was much less pronounced on the proliferation of PBL enriched in CD8$^+$ cells.[13]

DISCUSSION

Previous work has allowed the identification of membrane antigens preferentially expressed by intraepithelial lymphocytes in the intestine and in other epithelia in rats[23], in humans[3] and more recently in mice.[20] Biochemical analysis now indicates that this antigen consists of two non-covalently linked chains, a β chain of an apparent MW of 120 kDa in reducing conditions which is identical with the β7 chain of integrin and an a chain of an apparent MW of 150 kDa in reducing conditions which is a novel integrin α chain characterized by the post-translational cleavage of a 10 kD peptide which remained attached to the core of the protein by a disulfide bond. Similar conclusions have been simultaneously and independently reached by two other groups.[21,24] By its distribution and structural properties, the antigen first defined by HML-1 (and now recognized by a number of other antibodies[7,9,13,14,24] appears the human counterpart of the antigen defined on murine IEL by the M290 antibody.[20,25] It is likely, although not proven, that the antigen previously defined by the RGL-1 antibody[23] represents the rat counterpart. Interestingly it has been shown, first in mice[20] and more recently in humans[21], that MLA expression is induced by the TGFβ, a cytokine produced in large amounts by macrophages and enterocytes in the intestinal mucosa as well as by various extraintestinal epithelial cells.[26] Thus, following treatment with TGFβ of lymphocytes expressing the integrin β7-VLA-α4, an integrin involved in lymphocyte homing to Peyer's patches[27,28] expression of the α4 chain was down-regulated whereas expression of the MLA α chain was induced, allowing association of the free β7 chain with the MLA α chain. Endogeneous production of TGFβ by activated T cells[29] and perhaps hairy B cell leukemia may thus explain the expression of MLA at a distance from epithelia.

Integrins are widely distributed surface proteins which mediate cell-cell interactions or interactions of cells with the extracellular matrix. Thereby integrins participate in cell differentiation, cell migration, and cell activation.[10] It is tempting to postulate that MLA, via interactions with a ligand expressed in the epithelial layer, exerts comparable functions. Recently, it has been shown that a large proportion of murine IEL differentiates independently of the thymus and acquires in the gut microenvironment CD3, and the α,β or γ,δ chains of the TcR. In addition, IEL can acquire in the gut the α chain of CD8 and intracytoplasmic granules.[2] The molecular events underlying this differentiation are not known but it is likely that interactions between enterocytes and bone-marrow-derived lymphoid precursors are required. MLA may participate in these interactions in a manner comparable to VLA-4 and LFA-1 which respectively promote interactions of lymphoid precursors with the bone marrow stroma and the thymic epithelium.[30,31] However, investigation of this putative role of MLA will require experimental *in vivo* or *in vitro* models which are not yet readily available in humans. In the past, we have suggested that MLA may favor the migration of IEL into the epithelium. However, the treatment of young rats by antibodies directed to the probable rat counterpart of MLA during the three weeks which follow weaning did not prevent the physiological settlement of epithelium by lymphocytes.[32] Although these data do not definitely rule out the role of MLA in homing of IEL, they strongly argue against this hypothesis. Furthermore, the fact that MLA expression is induced by TGFβ together with the observation of large amounts of TGFβ in the intestinal mucosa suggests that the antigen is induced locally, after migration of lymphocytes. Finally, another possible function of MLA may be its accessory function in the activation of IEL. *In vitro* studies have indeed shown that several lymphocyte-associated integrins can provide, on binding to their respective ligands, accessory signals enhancing the activation of lymphocytes via the TcR-CD3 pathway.[11,12] In the present study, immobilized anti-MLA antibodies, used to mimic the effect of the as yet unidentified MLA ligand, were indeed able to strongly enhance the activation of IEL by anti-CD3 antibodies. At optimal concentrations of OKT3 antibody, activation of IEL judged by the mobilization of calcium, by the incorporation of ^{32}P in phosphatidylinositides, and finally by proliferative index was comparable to that of PBL enriched in CD8+ cells, and anti-MLA antibodies had only a modest or no enhancing effect. In contrast, at suboptimal doses of OKT3 where accessory signals are necessary to obtain maximal proliferation of T cells[22], the anti-MLA antibodies had a marked synergistic effect on the proliferation of IEL and on the expression of activation markers. A comparable but less consistent effect was observed with antibodies directed to two other lymphocyte-associated integrins.[13] Anti-integrin antibodies had less enhancing effect on OKT3-induced proliferation of PBL enriched in CD8+ cells. As discussed elsewhere, this is probably related to the contamination of the latter cells by a small number of accessory cells bypassing any further requirement for accessory signals even at low concentrations of OKT3 antibody.[13] Current *in vitro* studies suggest that activation of T cells requires a first specific signal via the CD3-TcR pathway and a second non-specific signal.[33] It is tempting to speculate that a second signal may be provided to IEL via the interaction of integrins with their respective ligands in the epithelium and that activation of IEL may be modulated by the level of expression of integrin ligands. The absence of ICAM-1, the LFA-1 ligand in normal gut epithelium[34], suggests that LFA-1, an important accessory molecule for peripheral T cells, cannot function as an accessory molecule for IEL in normal conditions. Identification of MLA ligand(s) and of the factors influencing its expression may thus proved to be important for understanding the mechanisms which regulate the response of IEL to intraluminal antigens. It is interesting that preliminary results suggest the presence of an MLA ligand on epithelial cells.[35]

REFERENCES

1. N. Cerf-Bensussan and D.Guy-Grand, *Gastroenterol. Clin. N. Amer.* 20:549 (1991).
2. D. Guy-Grand, N. Cerf-Bensussan, B. Malissen, M. Malassis-Seris, C. Briottet, and P. Vassalli, *J. Exp. Med.* 173:471 (1991).
3. N. Cerf-Bensussan, A. Jarry, N. Brousse, B. Lisowska-Grospierre, D. Guy-Grand, and C.Griscelli, *Eur. J. Immunol.* 17:1279 (1987).
4. J. Spencer, N. Cerf-Bensussan, A. Jarry, N. Brousse, D. Guy-Grand, A. S. Krajewski, and P. Isaacson, *Am. J. Pathol.* 132:1 (1988).
5. M. Sperling, P. Kaudewitz, O. Braun-Falco, H. Stein, *Am. J. Pathol.* 134:955 (1989).
6. H. L. Schieferdecker, R. Ullrich, B. A.Weiss, R. Schwarting, H. Stein, E. O. Riecken, and M. Zeitz, *J. Immunol.* 144:2541 (1990).
7. M. Kruschwitz, G. Fritsche, R. Schwarting, H. Dürkop, K. Micklem, D.Y. Mason, B. Falini, and H. Stein, *J. Clin. Pathol.* 44:636 (1991).
8. P. Möller, B. Mielke, and G. Moldenhauer, *Am. J. Pathol.* 136:509 (1990).
9. L. Flenghi, F. Spinozzi, H. Stein, M. Kruschwitz, S. Pileri, and B. Falini, *Br. J. Haematol.* 76:451 (1990).
10. R.O. Hynes, *Cell* 69:11 (1992).
11. Y. Shimizu, G. A. Van Seventer, K. J. Horgan, and S. Shaw, *Immunol. Rev.* 109:43 (1990).
12. G. A. Van Seventer, W. Newman, Y. Shimizu, T. B. Nutman, Y. Tanaka, K. J. Horgan, T. V. Gopal, E. Ennis, D. O'Sullivan, H. Grey, and S. Shaw, *J. Exp. Med.* 174:901 (1991).
13. S. Sarnacki, B. Bègue, H. Buc, F. Le Deist, and N. Cerf-Bensussan, *Eur. J. Immunol.* 22:2887 (1992).
14. N. Cerf-Bensussan, B. Bègue, J. Gagnon, and T. Méo, *Eur. J. Immunol.* 21:273 and 885 (1992).
15. N. Cerf-Bensussan, D. Guy-Grand, and C. Griscelli, *Gut* 26:81 (1985).
16. A. Jarry, N. Cerf-Bensussan, N. Brousse, F. Selz, and D. Guy-Grand, *Eur. J. Immunol.* 20:1097 (1990).
17. Q. Yuan, W. M. Jiang, G. W. Krissansen, and J. D.Watson, *Int. Immunol.* 2:1097 (1990).
18. D. J. Erle, C. Ruëgg, D. Sheppard, and R. Pytela, *J. Biol. Chem.* 266:11009 (1991).
19. M. E. Hemler, M. J. Elices, C. Parker, and Y. Takada, *Immunol. Rev.* 114:45 (1990).
20. P. J. Kilshaw and S. J. Murrant, *Eur. J. Immunol.* 20:2201 (1990).
21. C. M. Parker, K. L. Cepek, G. J. Russell, S. K. Shaw, D. N. Posnett, R. Schwarting, and M. B. Brenner, *Proc. Natl. Acad. Sci. USA* 89:1924 (1992).
22. F. A. Harding, J. G. McArthur, J. A. Gross, D. H. Raulet, and J. P. Allison, *Nature* 356:607 (1992).
23. N. Cerf-Bensussan, D. Guy-Grand, B. Lisowska-Grospierre, C. Griscelli, and A. K. Bhan, *J. Immunology* 136:76 (1986).
24. K. J. Micklem, Y. Dong, A. Willis, K. A. Pulford, L. Visser, H. Dürkop, S. Poppema, H. Stein, and D. Y. Mason, *Am. J. Pathol.* 139:1297 (1991).
25. Q. Yuan, W.-M. Jiang, E. Leung, D. Hollander, J. D. Watson, G. W. Krissansen, *J. Biol. Chem.* 267:7352 (1992).
26. G. A. Ksander, C. O. Gerhardt, J. R. Dasch, and L. R. Ellinsworth, *J. Histochem. Cytochem.* 38:1831 (1990).
27. B. B. Holzmann, W. McIntyre, and I. L.Weissman, *Cell* 48:549 (1989).
28. B.B. Holzmann and I. L. Weissman, *EMBO J.* 8:735 (1990).
29. C. Lucas, L. N. Bald, B. M. Fendly, M. Mora-Worms, I. S. Figari, E. J. Patzer, and M. A. Palladino, *J. Immunol.* 145:1415 (1990).
30. K. Miyake, I. L. Weissman, J. S. Greenberger, and P. W. Kincade, *J. Exp. Med.* 173:599 (1991).

31. S. Nonoyama, M. Nakayama, T. Shiohara, and J. Yata, *Eur. J. Immunol.* 19:1631 (1989).
32. N. Cerf-Bensussan, A. Jarry, T. Gnéragbé, N. Brousse, B. Lisowska-Grospierre, C. Griscelli, and D. Guy-Grand, *Monogr. Allergy* 24:167 (1988).
33. R. H. Schwartz, *Science* 248:1349 (1990).
34. C. J. Smart, A. Calabrese, D. J. Oakes, P. D. Howdle, and L. K. Tredjdosiewicz, *Scand. J. Immunol.* 34:299 (1991).
35. A. I. Roberts, S. M. O'Connell, and E. C. Ebert, DDW Scientific Sessions San Francisco May 10-13 Abstract 976 (1992).

PHENOTYPE OF HML-1-POSITIVE AND HML-1-NEGATIVE T LYMPHOCYTES IN THE HUMAN INTESTINAL LAMINA PROPRIA

Henrike L. Schieferdecker, Daniela C. Schmidt, Reiner Ullrich, Hans-Ulrich Jahn, Heike Hirseland, and Martin Zeitz

Medical Clinic, Department of Gastroenterology
Klinikum Steglitz, Free University of Berlin
Berlin, FRG

INTRODUCTION

Intestinal lymphocytes encounter antigen in the afferent limb of the gut-associated lymphoid tissue (i.e. the Peyer´s Patches of the small intestine and the lymphoid follicles of the colon and rectum) and then recirculate to the efferent limb, comprising lymphocytes in the epithelium above the basement membrane (intraepithelial lymphocytes) and lymphocytes diffusely spread in the lamina propria (lamina propria lymphocytes, LPL). LPL thus are antigen-activated lymphocytes. In this regard they resemble the so-called memory cells, which differentiate from naive (or virgin) precursor cells upon antigenic stimulation. We therefore postulated that LPL might represent a tissue-specific subpopulation of memory cells.

In recent studies, memory T cells have been identified under *in vitro* conditions, they are characterized by specific functional features and by the bright expression of T cell differentiation antigens as CD45R0, CD2, the leucocyte-function-antigens CD58 (LFA-3) and CD11a/CD18 (LFA-1), and CD29.[1,2] The latter molecules both belong to the integrin family; CD29 is the β-1 chain, while CD11a/CD18 is a β-2 integrin. One hint supporting the hypothesis of LPL being memory cells is, that LPL have bright expression of CD45R0, the surface antigen most commonly used for the definition of memory T cells *in vitro*. In recent studies it has been shown that the phenotype of lamina propria T cells (LP-T) only partially corresponds to the phenotype of the *in vitro* defined memory T cell subpopulation and to the CD45R0 bright subpopulation of peripheral blood T cells (PB-T).[3]

In this study we investigated if these deviations in the phenotype of LP-T can be attributed to a specific subpopulation of LP-T: recently, it has been shown that about 40 % of these cells express HML-1, the human mucosal lymphocyte antigen 1.[4,5] In recent studies, HML-1 has been identified as a novel member of the integrin family,[6,7] consisting of a β-7-chain and a new α-chain. HML-1 is specific for intestinal lymphocytes but its expression can also be induced on peripheral T cells.[5] Similarly to the T cell differentiation antigens[2] and in contrast to activation antigens as for example CD25 and MHC class II molecules, HML-1 remains almost stably expressed after removal of the inducing stimulus. We therefore postulated that HML-1 might be a tissue-specific T cell differentiation antigen and define a subpopulation of memory T cells with peculiar phenotypic features. We therefore investigated the coexpression of HML-1 with further T cell differentiation antigens on LP-T cells and on *in vitro* stimulated PB-T.

Advances in Mucosal Immunology, Edited by
J. Mestecky *et al.*, Plenum Press, New York, 1995

MATERIALS AND METHODS

Cells

Mononuclear cells from the intestinal lamina propria were obtained from macroscopically uninvolved tissue specimens of patients undergoing surgical bowel resection by a method described in Zeitz *et al.*[8] Briefly, the tissue was dissected into small pieces, which were incubated in succesive steps in culture medium containing DTT for the removal of mucus and EDTA for the removal of the epithelium and intraepithelial lymphocytes. LPL were liberated from the remaining tissue by enzymatic tissue degradation using collagenase and purified by density gradient centrifugation over Percoll and Ficoll. Mononuclear cells from the peripheral blood were isolated from heparinized venous blood of healthy volunteers by density gradient centrifugation over Ficoll. For the generation of PHA-blasts, freshly isolated PBL were incubated with PHA for 18 hours, then washed three times to remove the stimulus and incubated for another 5 days.

Staining and Analysis of Cells

Isolated cells were stained with PerCP-conjugated mAb against CD3, biotinylated HML-1 followed by Streptavidin-PE, and FITC-conjugated mAb against various T cell differentiation antigens, and analyzed by flow cytometry [FACScan, Becton Dickinson, LysysII and Consort 30 software]. The cells were first gated for CD3-positive T lymphocytes according to their localization in the foreward/sideward scatter diagram and for the expresssion of CD3 using a histogram gate. Additional gates were then set for HML-1-positive and HML-1-negative T lymphocytes, respectively.

RESULTS AND DISCUSSION

Previous studies have shown that LP-T which are almost exclusively bright for CD45R0, correspond to the CD45R0-bright subpopulation of PB-T in being negative for CD45RA and bright for CD2 and CD58. While CD45R0 bright PB-T were also bright for CD29 and CD11a/CD18, CD45R0 bright LP-T were CD29-dim and contained CD11a/CD18-bright, -dim and -negative cells. Additional differences between CD45R0-bright LP-T and PB-T could be shown with regard to the expression of further surface antigens not associated with T cell differentiation, such as L-selectin, CDw49a, which is the α-chain of VLA-1, and HML-1.[3]

To determine if these deviations can be ascribed to a specific subpopulation of LP-T, we investigated the coexpression of T cell differentiation antigens with HML-1. The results are summarized in Table 1. HML-1-positive and HML-1-negative LP-T were CD45RA-negative and did not differ significantly in the expression profiles of CD2, CD58 and CD29. Whereas HML-1-negative LP-T were also predominantly bright for CD45R0, the surface density of CD45R0 in the HML-1-positive subpopulation was significantly lower. Similarly, CD11a/CD18 was brightly expressed on HML-1-negative LP-T but showed variable surface densities ranging from negative to bright in the HML-1 positive subpopulation.

Thus, the vast majority of HML-1-negative LP-T was homogenous with regard to the expression of T cell differentiation antigens but HML-1-positive cells varied in the expression of CD45R0 and CD11a/CD18. Thus, the phenotype of the HML-1-negative subpopulation of LP-T was shown to correspond to memory T cells *in vitro* except that they had dim expression of CD29. Additional deviations could be detected for the HML-1 positive subpopulation of LP-T: These cells were dim for CD29 and showed variable expression of CD45R0 and CD11a/CD18.

In further studies, we compared the coexpression of various surface antigens with HML-1 on LPL and on *in vitro*-stimulated PBL (PHA-blasts) (Table 1). Both, HML-1-positive LPL and PHA-blasts were CD45RA-negative. Whereas HML-1-positive LPL were

CD29-dim and L-selectin-negative, HML-1-positive PHA-blasts were CD29-bright and contained L-selectin-dim and -bright cells. Therefore, the phenotype of HML-1-positive LPL deviated from the phenotype of the HML-1-positive subpopulation of *in vitro* -stimulated PBL.

Table 1. Expression of T cell differentiation antigens on the HML-1-negative and -positive subpopulations of lamina propria and *in vitro* -stimulated peripheral blood T cells.

Antigen	HML-1⁻ LPL	HML-1⁺ LPL	HML-1⁻ blasts
CD45RA	negative	negative	negative
CD2	bright	bright	n.d.
CD58	bright	bright	n.d.
CD29	dim	dim	**bright**
CD45R0	bright	**dim**	n.d.
CD11a/CD18	bright	**negative to bright**	n.d.
L-selectin	negative	negative	**dim to bright**

Our results support previous notions that memory T cells *in vivo* do not represent a uniform cell population but undergo tissue-specific differentiation. This is reflected by (a) the differential expression of various T cell differentiation antigens on these cells,[3] (b) the expression of the mucosa-specific lymphocyte antigen HML-1, defining a subpopulation of LP-T with specific phenotypic features and (c) the differences in the phenotype of HML-1-positive LPL and PHA-blasts.

The phenotypic characteristics of LP-T presumably correspond to specific functional capabilities of these cells allowing the adaptation of this T cell population to the specific requirements of their tissue localization. HML-1 might be regarded as a tissue-specific T cell differentiation antigen which is expressed on a specialized subpopulation of pre-activated T cells in the human intestine. Further studies will show if the phenotypic differences between HML-1-positive and -negative LP-T are reflected in specific functional capacities of these two subpopulations.

REFERENCES

1. A. N. Akbar, M. Salmon, and G. Janossy, *Immunol. Today* 12:184 (1991).
2. M. E. Sanders, W. Malegapuru, M. W. Makgoba, and S. Shaw, *Immunol. Today* 9:195 (1988).
3. H. L. Schieferdecker, R. Ullrich, H. Hirseland, and M. Zeitz, *J. Immunol.* 149:2816 (1992).
4. N. Cerf-Bensussan, A. Jarry, N. Brousse, B. Lisowska-Grospierre, D. Guy-Grand, and C. Griscelli, *Eur.J. Immunol.* 17:1279 (1987).
5. H. L. Schieferdecker, R. Ullrich, A. N. Weiss-Breckwoldt, R. Schwarting, H. Stein, E.-O. Riecken, and M. Zeitz, *J. Immunol.* 144:2541 (1990).
6. N. Cerf-Bensussan, B. Bègue, J. Gagnon, and T. Meo, *Eur. J. Immunol.* 22:273 and 885 (1992).
7. C. M Parker, K. L. Cepek, G. J. Russell, S. K. Shaw, D. N. Posnett, R. Schwarting, and M. B. Brenner, *Proc. Natl. Acad. Sci. USA.* 89:1924 (1992).
8. M. Zeitz, T. C. Quinn, A. S. Graeff, and S. P. James, *Gastroenterology.* 94:353 (1988).

ADHERENCE OF PORCINE PEYER'S PATCH, PERIPHERAL BLOOD AND LYMPH NODE LYMPHOCYTES TO PEYER'S PATCH LAMINA PROPRIA

Aras Kadioglu and Peter Sheldon

Department of Microbiology, Medical Sciences Building
University of Leicester, PO Box 138, LE1 9HN, England

INTRODUCTION

A vital element of the immune system is the ability of mature lymphocytes to circulate between blood and lymphoid organs in a continuous fashion. This recirculation not only allows detection and immune response to antigen, but also the communication of lymphocytes with cells at other sites within the body. There appear to be at least two separate lymphocyte recirculation patterns , one specific for mucosal tissue and the other for peripheral lymph node.[1,2,3] More recently a third migratory pathway specific for inflamed synovium has been suggested.[4,5]

Tissue specificity is mediated by receptors found on both the recirculating lymphocyte and tissue (endothelium) for which it is selective. These interactions are achieved by means of adhesion molecules, termed "homing receptors" on the lymphocyte and "vascular addressins" on endothelium.[6]

Our aim in this study was to investigate and compare adherence to Mucosa-Associated Lymphoid Tissue of lymphocytes obtained from peripheral blood, lymph node and Peyer's patch (PP) and relate our findings to current opinion on lymphocyte circulation to the gut.

MATERIALS AND METHODS

Preparation of Lymphocyte Suspensions

Source/Animals. Porcine blood, peripheral lymph node, and sections of small intestine were obtained fresh from the local abattoir. The blood was transported in Dextran (Sigma) sedimentation mixture with heparin (4 ml for every 20 ml of blood) in order to stop coagulation and to sediment the red blood cells. The lymph node and gut segments were transported in Hank's balanced salt solution (HBSS) (Gibco).

Porcine PP and Lymph Node Lymphocytes. PPs were identified as raised, dome like areas on the outer epithelium of the small intestine. Whole lengths of porcine small intestine were collected from the abattoir, the segments of Peyer's patch excised from the rest of the tissue and cut into small pieces (1-2 cm^2 in size). A total of 20-25 PP sections were cut. In order to obtain the lymphocytes the freshly cut sections were washed in phosphate-buffered saline twice and added to 100 ml of HBSS which also contained 150 mg of protease (Dispase Neutral protease from *Bacillus polymyxa* grade 2, Boehringer), and then stirred at 37ºC for 45 min. The supernatant, containing the dissociated lymphocytes was aspirated, washed twice and resuspended in supplemented Iscove's culture medium (sICM from Flow) at a concentration of 5×10^6. The same procedure was used for lymph node.

Porcine Peripheral Blood Lymphocytes. These were obtained by density gradient centrifugation using Lymphocyte separation medium (Flow).

In Vitro Adherence Assay

PP tissue was embedded in Tissue Tek OCT compound (Miles), snap-frozen in liquid nitrogen and stored at -80°C. Before sectioning the tissues were kept at -20°C overnight. 10 μm sections were cut from frozen tissue, using a cryostat and the adherence assay[7] was carried out. A known concentration (5×10^6) and volume (50 μL) of lymphocytes was layered onto the tissue section, encircled by a wax pen, agitated and incubated at 70 rpm for 30 min in the 7°C cold room.

After incubation the nonadherent cells were decanted from the tissue by gentle tapping and with the aid of a Q-tip applicator. The tissue sections were then fixed in 2% glutaraldehyde (Sigma) for 10 min and repeatedly washed in PBS. Once dried they were then stained with Haematoxylin and Eosin. Lymphocytes appeared blue while the tissue was pink/red. This enabled easy visual quantification of adherent lymphocytes. After staining, the sections were fixed with DPX mountant (BDH) for permanent storage.

Quantification

Quantification was carried out by bright field microscopy, x 250 magnification, by one observer enumerating lymphocytes overlaying 1 mm^2 of lamina propria (using a graticule) and subtracting the number adherent over similiar areas elsewhere e.g., follicular area of PP.

RESULTS AND DISCUSSION

Our previous studies had shown that adherence in this system was metabolically active (complete inhibition by sodium azide) calcium-dependant (70% inhibition by EDTA) and involved endothelium (80% inhibition by mannose 6-phosphate).

The results shown below indicate that the adherence of PPL was greater than that of PBL or LNL. This would agree with previous work indicating that there are distinct populations of lymphocytes with different organ specificities.

Table 1. Cell adherence to lamina propria/mm^2.

Source of lymphocytes	n	Mean	SEM
Peripheral blood	12	166	31
Lymph node	5	142	33
Peyer's patch	12	498	54

Adherence was restricted to lamina propria and very little adherence was actually observed over the follicular areas of the Peyer's patch tissue. A greater proportion of PPL adhered onto lamina propria than did PBL or LNL. This would suggest that PPL has a greater proportion of lamina propria "homing" lymphocytes than do PBL or LNL.

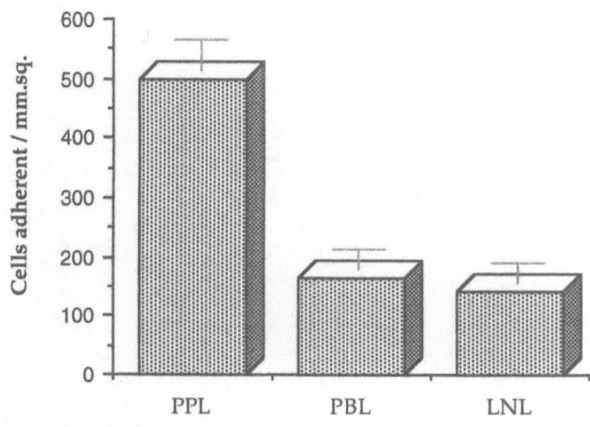

Figure 1. Adherence of Procine Lymphocyte Populations to Lamina Propria of Peyer's Patch.

It remains to be established whether or not the adherence as observed is morphologically between lymphocyte and endothelium. Though the inhibition studies mentioned would be in keeping with this. The existance of endothelial structures within lamina propria has been demonstrated.[8]

CONCLUSION

Porcine PP, lymph node and peripheral blood lymphocytes were layered onto frozen sections of porcine intestine that included PP. Adherence of cells to lamina proria was assayed, there being very little adherence elsewhere. A greater proportion of PP lymphocytes adhered than those derived from peripheral blood or lymph node, suggesting that PP contains a greater proportion of lamina propria seeking lymphocytes than other tissues.

REFERENCES

1. E. C. Butcher, R. G. Scollay, and I. L. Weissman I.L., *Eur. J. Immunol.* 10:556 (1980).
2. W. M. Gallatin, I. L. Weissman , and E. C. Butcher E.C., *Nature* 119:1603 (1983).
3. P. R. Streeter, E. L. Berg, B. R. N. Rouse, R. F. Bargatze, and E. C. Butcher, *Nature* 331:41 (1988).
4. S. Jalkanen, *Springer. Semin. Immunopathol.* 11:187 (1989).
5. S. Jalkanen, R. F. Bargatze, J. Toyos, and E. C. Butcher E.C., *J. Cell. Biol.* 105:983 (1987).
6. T. A. Yednock and S. D. Rosen, *Adv. Immunol.* 44:313 (1989).
7. H. B. Stamper and J. J. Woodruff, *J. Exp. Med.* 144:828 (1976).
8. S. H. M. Jeurissen, A. M. Duijvestijn, Y. Sontag, and G. Kraal, *Immunology* 62:273 (1987).

ASSESSMENT OF ENGRAFTMENT AND FUNCTION OF HUMAN TONSILLAR AND BLOOD MONONUCLEAR CELLS IN IMMUNODEFICIENT MICE

B. Albini,[1] D. Nadal,[2] C. Chen,[3] E. Schlapfer,[2] B.K. Mookerjee,[4] T. Stulnig,[5] , P.L. Ogra,[6] G. Wick,[5] and S.A. Cohen[4]

[1]Departments of Microbiology and Medicine, State University of New York at Buffalo, Buffalo, NY, USA; [2]Kinderspital, Zurich, Eleonorenstiftung Universitats-Kinderklinik, Zurich, Switzerland; [3]Hoffman-LaRoche Inc., Nutley, New Jersey, USA; [4]Veterans Administration Hospital of Buffalo and Department of Medicine, State University of New York at Buffalo, Buffalo, NY, USA; [5]Institute for General and Experimental Pathology, University of Innsbruck, Innsbrusck, Austria; [6]Department of Pediatrics, The University of Texas Medical Branch at Galveston, Galveston, Texas, USA

In 1988, two groups of investigators reported the successful adoptive transfer of nonneoplastic human cells into mice with severe combined immunodeficiency (SCID).[1-3] These mice originated from the CB-17 lcr strain by a spontaneous autosomal recessive mutation[4] affecting the VDJ recombinase system.[5] SCID mice lack functional B and T cells, but have a normal arsenal of macrophages and NK cells. The survival of human immunoreactive cells in SCID mice allows for the study of the human immune system in an animal model. Thus the human immune system is available for experimental protocols feasible up to now only for the study of animal immune systems. SCID mice engrafted with human immunologically reactive cells were called "human-mouse chimeras," "SCID-hu mice" and, hu-SCID mice. These mice are of interest for the study of maturation and differentiation of human hematopoietic cells,[6] tumor immunology,[7,8] human immune responses to infectious agents, neoplasms and autoantigens,[9,10] and drug effects on the human immune system. The hu-SCID mouse also seems an exciting model for the direct study of components of human mucosa-associated lymphoid tissue (MALT), their interactions, functions, and traffic *in vivo*.[11,12]

Before hu-SCID mice can be used for these purposes, however, it seems important to characterize this experimental model as to the success of engraftment, distribution of human cells in murine tissues, functional capacity of human cells, murine cell responses to the xenografted human constituents as well as immune responses of the human graft to murine antigens, and to develop a protocol for the evaluation of the human immune system transferred to SCID mice. In the following, a summary of our experience with human peripheral blood and tonsillar mononuclear cells engrafted in immunodeficient mice is given and criteria are proposed for the routine evaluation of hu-SCID mice.

Advances in Mucosal Immunology, Edited by
J. Mestecky *et al.*, Plenum Press, New York, 1995

The success and extent of engraftment of human cells in SCID mice is determined by characteristics of the donor cell population and the status of the recipient. There is a minimum number requirement for cells in the inoculum, below which engraftment does not occur in otherwise untreated SCID mice.[13] This number seems to be roughly 10^7. Engraftment of human tonsillar and blood mononuclear cells can be achieved consistently only when the donors are positive for antibodies to Epstein Barr virus (EBV).[11,12] This EBV-dependent or EBV-mediated enhancement of engraftment may reflect activating effects of the virus on human mononuclear cells; more specifically, one could speculate that the homology between the encoding sequence for IL-10 and the EBV genome could be involved in human T helper cell type 2 (T_{H2})-stimulating activity;[14] alternatively, the ability of EBV genes to inhibit apopoptosis may enhance the chances for survival and engraftment of human cells in SCID mice.[15]

Some SCID mice produce functional B and T cells. This phenomenon has been termed "leakiness."[16] The frequency of leakiness increases with the age of the mice. Usually, 10-20% of SCIDs are leaky, when tested at 6-14 wks of age. Leaky SCID mice seem to have a diminished potential to allow engraftment of human cells. In addition, the age of SCID mice seems to influence engraftment; in our hands, optimal results are obtained with animals at 6-8 wks of age. Infections in SCID mice preceding cell transfer make engraftment of human cells virtually impossible. This may reflect activation of murine NK cells and macrophages.

Engraftment also depends on the route of inoculation. Intraperitoneal administration leads to engraftment, whereas inocula given i.v. usually do not survive.[2]

It is crucial to trace engrafted human cells in murine tissues and to obtain quantitative data on their engraftment in various organs. It seemed obvious to use CD 3 reagents, which are specific for the human T cell surface molecule[2], but Mosier et al.[17] suggested in 1989 that murine cells may express antigens that cross-react with human CD3 in hu-SCID mice. To explore this possibility, we have used a supravital nuclear dye, H33342,[18] to label the human mononuclear cells in vitro, prior to cell transfer into SCID mice.[11,19] Using this method, it is possible to follow human cells over a period of up to 4-5 wk in various organs of the SCID mouse. The largest number of cells is retained in the peritoneal cavity, but aggregates of cells can be found 2 wk after inoculation in various tissues of the mouse. Murine cells observed after injection of H33342-stained, lethally irradiated human cells into SCID mice, suggesting absence of any significant uptake of dye by murine cells after death of human H33342-labelled cells. In addition to this simple method of cell tracing, it is possible to use in situ hybridization with human chromosome markers[e.g.20] or immuno-histology or immunocytology with an antiserum to human HLA determinants.[21] Finally, an elegant multi-staining procedure for FACS analysis using antisera to several surface markers was developed by Bankert et al. (personal communication), which allows for definition of human cells in SCID mice. Using the supravital dye and human CD 3 and Ig-specific serological reagents in double staining experiments, cells carrying human CD 3 or Ig were stained consistently with H33342.[19] Interestingly, migration of cells into spleen, liver, bone marrow, and other organs was enhanced upon immunological stimulation of the hu-SCID mice (unpublished data).

Tonsillar cells showed migration into lungs, whereas peripheral blood cells did not.[11] Neither of these populations migrated into the *lamina propria* of the intestines.[11] On the other hand, it has been shown that pokeweed mitogen (PWM)-stimulated blood mononuclear cells do populate SCID lamina propria.[22] In our own preliminary experiments, untreated human Peyer's patch cells migrated to the intestines (unpublished results).

Inocula containing low numbers of human cells (below 10^7) do not engraft. Indeed, soon after i.p. administration, human blood and tonsillar mononuclear cells are taken up by murine peritoneal macrophages. At the same time, there is activation and increase in numbers of murine NK cells in the peritoneum and, subsequently, in the spleen (Nadal, D., Schlapfer, E., Ogra, P.L., Cohen, S.A. and Albini, B., manuscript in preparation). This demonstrates that SCID mice indeed do mount a xenograft rejection reaction involving NK cells and macrophages.

Engraftment of human blood mononuclear cells can be enhanced by administration of human growth hormone (hGH).[21] This hormone also enhances migration into other visceral organs of SCID mice.[21] The mechanisms by which hGH enhances the engraftment are not well understood, and require further study. Furthermore, treatment of SCID mice with monoclonal antibodies to asialoGM-determinants eliminate or reduce the number of murine NK cells. It seems that this treatment enhances engraftment of human cells[21], especially when the latter are given in low numbers.

Originally, engraftment of human blood mononuclear cells was assessed primarily by monitoring the presence and increase over time of circulating human Ig in SCID mice.[2] This is best accomplished by ELISA for human total Ig or IgG, which, in most instances, is the predominant isotype present in serum. In addition, it is possible to demonstrate specific antibodies in serum, e.g., to tetanus toxoid[2] or to streptococcal antigens.[23] The immune responses to challenge antigens are almost exclusively of human IgG isotype but follow more protracted kinetics than those seen in the "classical" secondary response.[23]

Another method to monitor B cell function in hu-SCID mice is the use of ELISPOT assays for human Ig, their isotypes, or for specific antibody production.[11,12,24] This approach gives good estimates of human B cell frequencies in various organs of hu-SCID and has been used to evaluate antibody production of various isotypes to respiratory syncycial virus (RSV).[11,12]

The participation of human T cells in immune responses of the hu-SCID mouse has been demonstrated in the case of *Streptococcus pyogenes*-induced hepatic granulomas.[23] These granulomas are T cell dependent[23] and thus do not develop in SCID mice. Upon transfer of human blood mononuclear cells into SCID mice, however, hepatic granulomas develop in the wake of immunization with *S. pyogenes* antigens.[23] Human T cell function in SCID mice can be estimated using reverse plaque assays[25] or ELISA for human cytokines. Lymphocyte proliferation assays with appropriate mitogens may also be used.

In our experience, human Ig produced in the hu-SCID mice are polyclonal. Only after neoplastic transformation of human cells carrying EBV does oligo- or monoclonality appear.[20]

Even though very little is known about graft versus host reactions (GvHR) across xenogeneic barriers, it was somewhat surprising to observe virtually no GvHR in hu-SCID mice.[2] Classical signs of GvHR were observed by us in only 3 out over one hundred hu-SCID mice sacrificed prior to wk 15 after human cells transfer. On the other hand, Bankert et al.[21] have found typical lesions of GvHR in the livers of their hu-SCID mice, especially late after transfer of human cells. Bankert et al.[26] have also described the production of human antibodies reactive with murine erythrocytes in hu-SCID mice; this seems to be a rather frequent phenomenon, and some of the mice developed severe hemolytic crises. These fascinating aspects of hu-SCID mice remain to be studied in more detail.

PROPOSED PROTOCOL FOR THE EVALUATION OF HUMAN DONORS AND RECIPIENT SCID MICE

In the table, we propose a flowchart of assessments necessary to characterize hu-SCID mice and to assess engraftment and functionality of engrafted human cells. The proposed parameters reflect our own experience with experiments using human peripheral blood and tonsillar mononuclar cells and from reports published in the literature.

Table 1. Assessment of hu-SCID mice

Test object	Activity monitored	Procedure
SCID Mouse	Leakiness	Murine serum Ig (before and at conclusion of experiment, ELISA)
	Infections	Bacteriologic and virologic screening of colony (at bi-monthly intervals)
Human Cell Donor	EBV antibody status	Standard immunofluorescence test for EBV antibodies
	Allergy to murine antigenic determinants	Donors' disease history RAST and other allergen tests
	Immunization status	Vaccination record, serum antibodies to bacterial components
	Acute or chronic diseases, age, sex, present medication proliferation and engraftment of B cells	Donor's disease history
		Human serum Ig (at weekly intervals, ELISA, ELISPOT) antibodies of human isotypes (when appropriate, ELISA, ELISPOT)
Hu-SCID Mouse	Proliferation and engraftment of T cells (participation in cell-mediated immune reactions)	
	Proliferation and engraftment of human cells	Composition of peritoneal washings and tissues (as required by experimental protocol, FACS analysis, cytokine reverse plaque assay, ELISA for cytokines; lymphocyte proliferation)

Human blood mononuclear cells engraft also in athymic nude mice, but the engrafting cells are predominantly T cells, and very little human Ig is produced.[23] However, human Ig continued to be produced,[23] suggesting that human B cells also engraft, but that their numbers and activity remain down-regulated. Several other immunodeficient inbred strains of mice have become available and will be tested for transfer of human mononuclear cell grafts, including MALT components; prominent among them are the RAG-deficient mice.

REFERENCES

1. J. M. McCune, R. Namikawa, H. Kaneshima, L. D. Schultz, M. Lieberman, and I. L. Weissman, *Science* 241:1632 (1988).
2. D. E. Mosier, R. J. Gulizia, S. M. Baird, and D. B. Wilson, *Nature* 335:256 (1988).
3. P. C. Taylor, *Int. J. Exp. Pathol.* 73:251 (1992).
4. G. C. Bosma, R. P. Custer, and M. J. Bosma, *Nature* 301:527 (1983).

5. W. Schuler, I. J. Weiler, A. Schuler, R. A. Phillips, N. Rosenberg, T. Mak, J. F. Kearney, R. P. Perry, and M. J. Bosma, *Cell* 46:963 (1986).

6. R. Namikawa, K. N. Weilbaecher, H. Kaneshima, E. J. Yee, and J. M. McCune, *J. Exp. Med.* 171:1055 (1990).

7. R. B. Bankert, T. Umemoto, Y. Sugiyama, F. A. Chen, E. Repasky, and S. Yokota. *Curr. Top. Microbiol. Immunol.* 152:201 (1989).

8. D. T. Purtilo, K. Falk, S. J. Pirrucello, H. Nakamine, K. Kleveland, J. R. David, M. Okano, Y. Yaguchi, W. G. Sanger, and K. W. Beisel, *Int. J. Cancer* 47:510 (1991).

9. J. McClune, H. Kaneshima, J. Krowka, R. Namikawa, H.Outzen, B. Peault, L. Rabin, C. Shih, and E. Yee, *Annu. Rev. Immunol.* 9:399 (1991).

10. H. Tighe, G. Silverman, F. Kozin, R. Tucker, R. Gulizia, C. Peebles, M. Lotz, G. Rhodes, K. Machold, D. E. Mosier, and D. A. Carson, *Eur. J. Immunol.* 20:1843 (1990).

11. D. Nadal, B. Albini, C. Chen, E. Schlapfer, J. M. Bernstein, and P. L. Ogra, *Int. Arch. Allergy Appl. Immunol.* 95:341 (1991).

12. D. Nadal, B. Albini, E. Schlapfer, L. Brodsky, and P. L. Ogra, *Clin. Exp. Immunol.* 85:358 (1991).

13. D. E. Mosier, *J. Clin. Immunol.* 10:185 (1990).

14. K. W. Moore, P. Vieira, D. F. Fiorentino, M. L. Trounstine, T. A. Kahn, and T. R. Mossmann, *Science* 248:1230 (1990).

15. C. D. Gregory, C. Dive, S. Henderson, C. A. Smith, G. T. Williams, J. Gordon, and A. B. Rickinson, *Nature* 349:612 (1991).

16. G. C. Bosma, M. Fried, R. P. Custer, A. Carroll, D. M. Gibson, and M. J. Bosma, *J. Exp. Med.* 167:1016 (1988).

17. D. E. Mosier, R. J. Gulizia, S. M. Bard, and D. B. Wilson, *Nature* 338:211 (1989).

18. M. Brenan and C. R. Parish, *J. Immunol. Methods* 74:31 (1984).

19. C. Chen, D. Nadal, S. A. Cohen, E. Schlapfer, B. K. Mookerjee, A. Vladutiu, M. W. Stinson, P. L. Ogra, and B. Albini, *Int. Arch. Allergy Appl. Immunol.* 97:295 (1992).

20. D. Nadal, B. Albini, E. Schlapfer, J. M. Bernstein, and P. L. Ogra, *J. Gen. Virol.* 73:113 (1992).

21. W. J. Murphy, S. K. Durum, and D. L. Longo, *Proc. Natl. Acad. Sci. USA* 89:4481 (1992).

22. C. Lue, H. Kiyono, K. Fujihashi, J. R. McGhee, and J. Mestecky, *Regional Immunol.* 4:86 (1992).

23. C. Chen, B. Albini, D. Nadal, B. K. Mookerjee, P. L. Ogra, M. W. Stinson, and S. A. Cohen, *Amer. J. Pathol.*(Submitted).

24. C. Czerkinsky, Z. Moldoveanu, J. Mestecky, and L. A. Ouchterlony, *J. Immunol. Methods* 115:31 (1988).

25. C. E. Lewis, *J. Immunol. Methods* 127:51 (1990).

26. S. S. Williams, T. Umemoto, H. Kida, E. A. Repasky, and R. B. Bankert, *J. Immunol.* (in press, 1992).

27. Y. Shinkae, G. Rathbun, K.-P. Lam, E. M. Oltz, V. Steward, M. Mendelsohn, J. Charron, M. Datta, F. Young, A. M. Stall, and F. W. Alt, *Cell* 68:855 (1992).

28. P. Mombaerts, J. Iacomini, R. S. Johnson, K. Herrup, S. Tonegawa, and V. E. Papaionnou, *Cell* 68:869 (1992).

EXPRESSION OF VLA-4 AND L-SELECTIN IN HUMAN GUT-ASSOCIATED LYMPHOID TISSUE (GALT)

Inger Nina Farstad,[1] Trond S. Halstensen,[1] Dag Kvale1,[2]
Olav Fausa,[2] and Per Brandtzaeg[1]

[1]Laboratory for Immunohistochemistry and Immuno-
pathology (LIIPAT), Institute of Pathology, and [2]Medical
Department A, University of Oslo, The National Hospital,
Rikshospitalet, Oslo, Norway

INTRODUCTION

The preferential homing of lymphocytes to certain tissue sites has been studied in several species[19,12]. Peyer's patches (PP) and the appendix have been considered primarily as B-cell structures. However, a recent study of lymphocyte homing in rats revealed no special prefence for thoracic duct-derived B or T cells to bind to PP high endothelial venules (HEVs)[22]. In mice, two heterodimeric integrins are assumed to mediate lymphocyte homing to PP: LPAM-1 consisting of α4 coupled to a β chain designated βp, and LPAM-2 (α4β1), which is homologous to the human very late antigen-4 (VLA-4)[7,8]. However, in humans α4 may also be coupled to β7, homologous to mouse βp[14,9,16]. The endothelial counter-receptor (ligand) for α4β1, vascular adhesion molecule-1 (VCAM-1), is expressed in two isoforms with 6 or 7 immunoglobulin-like domains (6D and 7D): VCAM-7D is the dominant form found on endothelium of inflamed tissues while VCAM-6D (defined by monoclonal antibody (mAb) E1/6) apparently is expressed earlier during inflammation but scarcely on human GALT HEVs[21,13,15]. The ligand for α4β7 is as yet unknown, although stimulated lymphocytes expressing this heterodimer can bind to VCAM-1[3]. However, the efficiency of binding to VCAM-1 was much better for lymphocytes expressing α4β1 than α4β7[3]. A mAb to β7 inhibited lymphocyte binding to PP but not peripheral lymph node (PLN) HEVs in mice[9], suggesting that β7 might be responsible for PP-specific lymphocyte migration in the normal state.

L-selectin (Leu-8; LECAM-1), homologous to murine MEL-14, is highly conserved among various species and is recognized as the major PLN homing receptor[2,18]. Its endothelial ligand is a mucin-like glycoprotein (Sgp50) designated GlyCAM-1[11]. In mice, the ligand for L-selectin or Mel-14 is recognized by mAb MECA-79[20]. This mAb blocks the binding of human L-selectin+ lymphocytes to PLN HEVs and recognizes Sgp50 and other glycoproteins apparently expressed by HEVs in human PLN and tonsils[1,20]. A role for MEL-14 in lymphocyte binding to PP HEVs has also been supported *in vivo*[5].

The aim of this study was to compare the expression of the α4 and β1 chains of VLA-4 as well as L-selectin on peripheral blood (PBL) and GALT lymphocytes to evaluate the possible roles of these adhesion molecules in lymphocyte homing to human PP.

MATERIALS AND METHODS

Primary Antibodies

Monoclonal murine and polyclonal rabbit antibody reagents to the following human antigens were used: CD3 (Leu-4, murine IgG1), L-selectin (Leu-8, murine IgG2a), and CD45RA (Leu-18, murine IgG1), all from Becton Dickinson, Mountain View, CA; CD3 (BMA030, murine IgG2a; Behringwerke, Marburg, Germany); CD19 (HD37, murine IgG1), CD20 (L26, murine IgG2a), and rabbit antisera to CD3 and von Willebrand factor, all from Dakopatts, Glostrup, Denmark; CD45R0 (UCHL-1, murine IgG2a; gift from Dr. P.C.L. Beverly, London, U. K.); α4 (B-5G10, murine IgG1; gift from Dr. Hemler, Dana-Farber Cancer Inst., Harvard Med. School, Boston, MA); β1 (4B4, murine IgG1; Coulter Company, Hialeah, FL); and VCAM-1 (BBA 5, murine IgG1; British Bio-Technology, Abingdon, U.K.).

Lymphocyte Samples

Lymphocytes were obtained from uninflamed appendices (n=5) and PP biopsy specimens (n=2). Isolation procedures were principally as described by Fujihashi *et al.*[4] Dispase (1.5 mg/ml) dissolved in buffer (pH 7.4) containing 0.14M NaCl, 0.004M KCl, 0.001M Na$_2$PO$_4$ and 0.011M glucose was used to dissociate cells from the tissues.

PBL were obtained from healthy donors (n=4) with Ficoll-Hypaque (Pharmacia, Sweden) and subjected to incubations with Dispase for 30 and 60 min to evaluate the proteolytic effect on leucocyte surface molecules. Only the expression of L-selectin was dramatically affected by Dispase and at least 10 h were needed for reconstitution (medium with 10% fetal calf serum, 37°C). Isolation procedures were therefore followed by 10 h incubation before immunostaining.

Flow Cytometry

Cells were single- and double-stained in round-bottomed microtitre plates by conventional methods with primary murine mAbs (see above). Secondary reagents were phycoerythrin-conjugated goat anti-mouse IgG1 and FITC-conjugated goat anti-mouse IgG2a (Southern Biotechnology, Birmingham, AL). Samples were then analysed in a fluorescence-activated cell scanner (FACScan, Becton-Dickinson) and lymphocytes were selected (gated) in the forward/side scatter dot plot. Negative controls consisted of samples incubated with irrelevant primary antibody or only secondary antibody.

Immunohistochemistry

Tissue specimens from PP (n=5) and appendix (n=5) were embedded in OCT compound (Tissue-Tek; Miles Laboratories, Elkhart, IN), oriented on a thin slice of carrot, snap-frozen in isopentane cooled in liquid nitrogen, and stored at -70°C. Cryosections cut at 4 μm were air-dried, fixed in acetone (10 min) at room temperature and stored at -20°C until staining. Mixes of primary antibodies were incubated for 1 h, second step antibodies (mixes of biotinylated or FITC-conjugated goat anti-mouse IgG subclass and unlabelled rabbit antibody to a human antigen) for 1.5 h, and the final incubations (30 min) were accomplished with Streptavidin-Texas Red (BRL, Gaithersburg, MD) and AMCA-conjugated goat anti-rabbit IgG (Vector Laboratories, Burlingame, CA). Negative controls consisted of sections incubated with irrelevant primary antibody.

RESULTS

Flow Cytometry

PBL: Median 70% (range, 65-70%) of lymphocytes expressed L-selectin and CD45RA, 90% (range, 80-95%) expressed α4, and 90% (range, 90-95%) β1. Expression of β1 divided T cells into a β1^{high} and a β1^{low} subpopulation corresponding to the proportions of CD45R0+ vs. CD45RA$^+$ (=R0$^-$) lymphocytes (Fig. 1a).

GALT: Median 25% (range, 15-70%) of isolated cells expressed L-selectin. By paired staining, most L-selectin$^+$ cells were also shown to be CD45RA$^+$, and L-selectin was expressed by both CD3$^+$ and CD19$^+$ cells (Figs. 1b and 2). Most T and B cells (90%) expressed α4 and β1, but the fluorescence intensity for β1 was, in the two experiments performed, comparable only to the β1^{low} fraction of PBL (Fig. 1). The proportions of T and B cells were approximately equal, median 46% (range, 42-55%) CD3+ and 45% (range, 33-55%) CD19$^+$ (Fig. 2).

Immunohistochemistry

PP and appendix cells were evaluated in various tissue compartments: the dome area underneath the specialized follicle-associated epithelium; lymphoid follicle centres; lymphoid follicle mantle zones; and interfollicular zones (Table 1).

α4 was expressed by lymphocytes in the domes, mantle zones, interfollicular areas, and lamina propria adjacent to PP. Two-colour staining for α4 and CD20 or CD3 showed that most mature B and T cells expressed this marker.

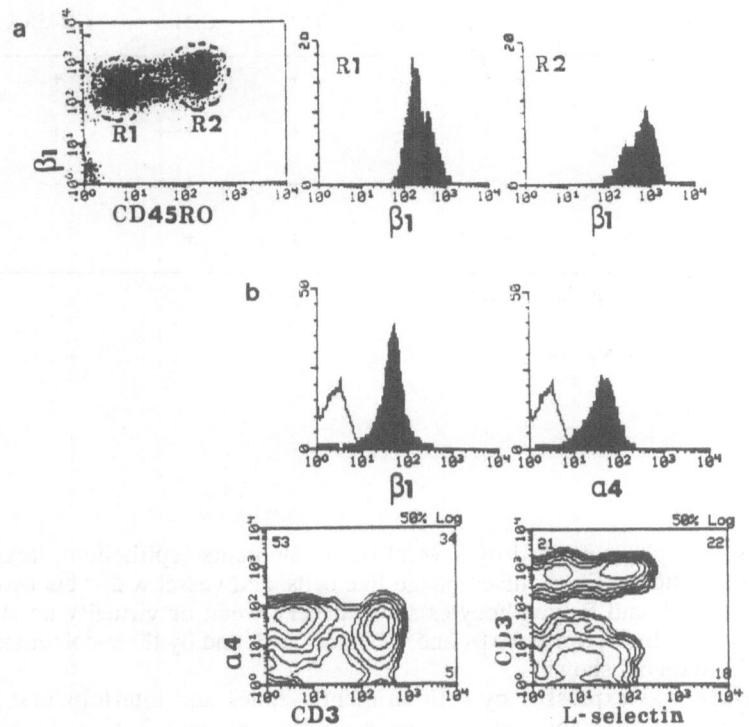

Figure 1. Flow-cytometric comparison of β1 expression by (**a**) PBL and (**b**, top) GALT lymphocytes, and (**b**, bottom) expression of α4 and L-selectin by CD3+ cells in GALT.

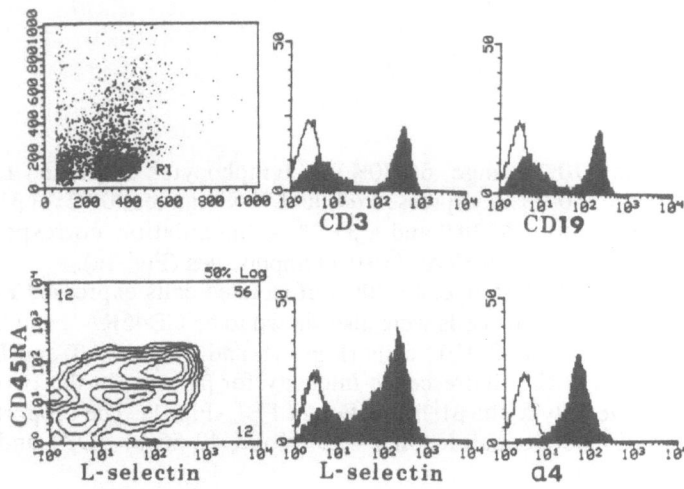

Figure 2. GALT lymphocytes from appendix studied by flow cytometry. Note high co-expression of L-selectin and CD45RA, and equal proportions of CD3+ and CD19+ cells (45%).

Table 1. Expression of L-selectin, α4, β1, CD3, CD19 and CD45RA on GALT lymphocytes

	L-selectin	α4	β1	CD3	CD19	CD45RA
Dome	+*	+	(+)	+	+	+
Follicle	-	-	-	+	+++	+
Mantle zone	++	++	(+)	+	+++	++
Interfollicular area	++	++	+	+++	+	++

* Key for immunofluorescence grading *in situ*:

-        = negative
(+)   = weakly or doubtfully positive
+   = some positive cells
++   = most cells positive
+++   = virtually all cells positive

β1 was strongly expressed by several tissue elements (epithelium, large cells in germinal centres, reticular fibres, macrophage-like cells, and vessel walls) but by only very few T cells. Most T and B lymphocytes showed very weak or virtually no staining as confirmed by paired fluorescence for β1 and CD20 or CD3, and by three-colour staining for α4, β1 and CD3 (data not shown).

L-selectin was expressed by cells in mantle zones and interfollicular areas but rarely in domes and adjacent lamina propria. Thus, α4 and L-selectin expression overlapped topographically in GALT, but the latter marker was found on a much lower proportion of lymphocytes. Two-colour staining for CD45RA and L-selectin revealed almost identical marker distribution except in follicles where some cells with only CD45RA were observed.

VCAM-1 expression was seen in some HEVs at the outer periphery of the endothelial lining. The nature of these positive elements is unknown, but the luminal and cytoplasmic aspects of the endothelial cells were negative.

Reproducibility. Two appendices were evaluated both by flow cytometry and immunohistochemistry; the proportions of cells expressing the respective markers were comparable by the two methods (data not shown).

DISCUSSION

The major findings in this study were: (i) L-selectin was expressed in GALT by both T and B cells in a distribution similar to CD45RA, except in follicle centres; (ii) PBL and GALT lymphocytes expressed the α4 chain with similar fluorescence intensity, while β1 expression was relatively low on PP and appendix lymphocytes; and (iii) VCAM-1 (ligand for α4β1) was not expressed on the luminal side of HEVs in GALT.

Our results were in agreement with a previous study in humans[10] showing that very few gut lamina propria lymphocytes but a sizeable fraction of GALT lymphocytes express L-selectin. Lasky *et al.*[11] proposed that functional interactions of L-selectin with its endothelial ligand might take place in murine PP under certain circumstances, as low levels of Sgp50 mRNA were found in this compartment. This could explain why mAb to MEL-14 was shown to inhibit PP lymphocyte binding by 50-60% in mice[5]. One of our subjects showed high expression of both L-selectin and CD45RA in his appendix but his colon was not examined (20-year old male, organ donor). This observation could support a functional L-selectin-ligand interaction as referred to above or, alternatively, it suggests that shedding of L-selectin occurs rather late after lymphocyte extravasation[2]. The mAb MECA-79, which inhibits L-selectin-mediated lymphocyte binding to PLN HEVs, also reacts slightly with human PP HEVs[1], supporting a role for L-selectin in GALT homing. However, L-selectin might be generally involved in endothelial adherence rather than being a specific homing molecule.

The findings regarding VLA-4 need some further considerations. GALT epithelium, smooth muscle, macrophage-like cells, vessels walls, and fibrous elements surrounding the mantle zones, all stained intensely for β1; scattered T cells were weakly positive by immunohistochemistry while most other lymphocytes were negative (or by FACS analysis equal to the β^{low} fraction of PBL). However, VLA-4 expressed by lymphocytes might be bound to its alternative ligand, fibronectin, thereby explaining the apparent absence of β1 by masking of this epitope *in situ*.

The α4 chain of VLA-4 may be associated with another β subunit on a proportion of circulating lymphocytes because there is a two-fold excess of α4 compared to β1 on freshly isolated B cells[6]. Unstimulated peripheral blood T cells do not express β7, but β7 mRNA was observed 14 days after activation by CD3 antibody[16]. Lack of VCAM-1 expression on HEVs in this study does not support α4β1-mediated lymphocyte binding in GALT. Our results are, rather, compatible with the notion that α4β7 mediates lymphocyte binding in PP, perhaps primarily involving B cells via an as yet unknown endothelial ligand. L-selectin may also be involved as an adhesion molecule in GALT.

REFERENCES

1   E. L. Berg, M. K. Robinson, R. A. Warnock, and E. C. Butcher, *J. Cell. Biol.* 114:343 (1991).

2.   D. Camerini, S. P. James, I. Stamenkovic, and B. Seed, *Nature* 342:48 (1989).

3.   B. M. C. Chan, M. J. Elices, E. Murphy, and M. Hemler, *J. Biol. Chem.* 267:8366 (1992).

4.   K. Fujihashi, J. R. McGhee, C. Lue, K. W. Beagley, T. Taga, T. Hirano, T. Kishimoto, J. Mestecky, and H. Kiyono, *J. Clin. Invest.* 88:248 (1991).

5. A. Hamann, D. Jablonski-Westrich, P. Jonas, and H.-G. Thiele, *Eur. J. Immunol.* 21:2925 (1991).
6. M. E. Hemler, *Ann. Rev. Immunol.* 8:365 (1990).
7. B. Holzmann and I. L. Weissman, *Immunol. Rev.* 108 (1988).
8. B. Holzmann and I. L. Weissman, *Embo J.* 8: 1735 1989).
9. M. C. -T. Hu, D. T. Crowe, I. L. Weissman, and B. Holzmann, *Proc. Natl. Acad. Sci. USA* 89:8254 (1992).
10. M. E. Kanoff, W. Strober, C. Fiocchi, M. Zeitz, and S. P. James, *J. Immunol.* 140:3029 (1988).
11. L. A. Lasky, M. S. Singer, D. Dowbenko, Y. Imai, W. J. Henzel, C. Grimley, C. Fennie, N. Gillet, S. R. Watson, and S. D. Rosen, *Cell* 69:927 (1992).
12. C. R. Mackay, *Immunol. Today* 12:189 (1991).
13. L. Osborn, C. Vassallo, and C.D. Benjamin, *J. Exp. Med.* 176:99 (1992).
14. C. M. Parker, K. L. Cepek, G. J. Russel, S. K. Shaw, D. N. Posnett, R. Schwartin and M. Brenner, *Proc. Natl. Acad. Sci. USA* 89:1924 (1991).
15. G. E. Rice, J. M. Munro, C. Corless, and M. P. Bevilacqua, *Am. J. Pathol.* 138:385 (1991).
16. C. Ruegg, A. A. Postigo, E. E. Sikorski, E. C. Butcher, R. Pytela and D. J. Erle, *J. Cell. Biol.* 117:179 (1992).
17. Y. Shimizu, W. Newman, Y. Tanaka, and S. Shaw, *Immunol. Today* 13:106 (1992).
18. O. Spertini, G. S. Kansas, K. A. Reimann, C. R. Mackay, and T. F. Tedder, *J. Immunol.* 147:942 (1991).
19. L. M. Stoolman, *Cell* 56:907 (1989).
20. P. R. Streeter, B. T. N. Rouse, and E. Butcher. *J. Cell. Biol,* 107:1853 (1988).
21. R. H. Vonderheide and T. H. Springer, *J. Exp. Med.* 175:1433 (1992).
22. J. Westermann, V. Blaschke, G. Zimmermann, U. Hirschfeld, and R. Pabst, *Eur. J. Imunol.* 22:2219 (1992).

COMPARTMENTALIZATION WITHIN THE COMMON MUCOSAL IMMUNE SYSTEM

Zina Moldoveanu,[1,2] Michael W. Russell,[1] Hong
Yin Wu,[1] Wen-Qiang Huang,[2] Richard W.
Compans,[3] and Jiri Mestecky[1,2]

[1]Department of Microbiology, University of
Alabama at Birmingham, Birmingham, AL, U.S.A.
[2]Secretech, Inc., Birmingham, AL, U.S.A.
[3]Department of Microbiology and Immunology,
Emory University, Atlanta, GA, U.S.A.

INTRODUCTION

The mucosal tissues and external secretory glands of humans and animals reveal a remarkable preponderance of IgA-producing plasma cells[1] which, as demonstrated in animal experiments, are derived from precursors in the organized gut- and bronchus-associated lymphoid tissues (GALT; BALT).[2] Despite the general pre-occupation with GALT, several investigators have considered that cells from other sources may contribute to the pool of IgA precursors.[3-6] Surgical removal of all visible GALT or thoracic duct drainage in experimental animals results in only a 50% decrease in the intestinal lamina propria plasma cells, implying that alternative sources contribute significantly to the pool of IgA precursor cells. It is plausible that GALT, BALT, and other sources (peritoneum, tonsils, and rectal lymphoid follicles) may preferentially supply IgA-committed, antigen-sensitized cells to restricted (frequently adjacent) mucosal regions.[7-10] This possibility has considerable practical impact: inhalation or intranasal immunization might stimulate immune responses preferentially in the upper respiratory and digestive tracts rather than in the genital or lower intestinal secretions, whereas ingestion of antigens or rectal immunization would stimulate responses preferentially in corresponding areas and less, for example, in nasal secretions or tears. Although the induction of antibodies in saliva, tears, and milk by mucosal immunization indicates that different inductive sites overlap in their ability to seed IgA-secreting cells to various effector sites, the extent to which the common mucosal immune system (CMIS) is compartmentalized is not clear. As many pathogens display distinct tropism for certain mucosal sites, exploitation of compartmentalization within the CMIS to direct a immune response to a particular site where the infection can be effectively countered would not only be efficient, but also might avoid undesirable effects elsewhere, and this information may be valuable in the design of vaccines and appropriate immunization routes.

MATERIAL AND METHODS

Animals: BALB/c mice, approx. 6 weeks old (5 mice/group) were immunized by intraperitoneal (IP), intragastric (IG), intranasal (IN), intrarectal (IR), or intravaginal (IV) routes with 10 µg cholera toxin (CT), or with *Streptococcus mutans* AgI/II-CTB conjugate [15µg+ 5µg CT (IN), 30µg (IP), and 30µg+ 5µg CT (IG)], or with 10^5 pfu of vaccinia virus expressing gp160 of SIV (VV-SIV 239 env).

Biological samples were collected before and at selected times after immunization, as follows: blood was obtained with heparinized glass capillary tubes from the tail vein; stimulated saliva was collected with capillary tubes after IP injection of carbamyl-choline chloride (1-2µg/mouse); and intestinal lavages were performed according to the method described by Elson *et al.*[11] To determine the isotype and specificity of intestinal antibodies at selected time points during the experiment, we used as an alternative less stressful for animals, the method described by deVos and Dick.[12] Cervico-vaginal secretions were obtained by the instillation and aspiration of 50µl of saline solution.

The tissues analyzed included spleen, salivary glands (SG); Peyer's patches (PP), intestinal lamina propria (LP), mesenteric lymph nodes (MLN), and superficial cervical lymph nodes (CLN). Single cell suspensions were obtained from spleen and lymph nodes by mechanical dissociation, and from SG, PP and LP by enzymatic digestion (Dispase), followed by Percoll gradient centrifugation.[13]

The induced antibodies were measured at the single cell level by ELISPOT assay[14], and in biological fluids by ELISA.

RESULTS AND DISCUSSION

Immunization of mice by alternative mucosal routes such as IN, IV, or IR compared with IG and systemic routes was evaluated by measuring antibodies in serum and secretions and by enumerating cells secreting specific antibodies.

At the single cell level differences were found among routes (IN vs IG vs IP) and antigens (*S. mutans* AgI/II vs CT), regarding the level, isotype, and site of maximum response, as illustrated in Fig.1. IgA antibodies predominated at secretory sites, especially after IG or IN immunization. IgA responses were higher in LP when the antigen was given IG, and in SG after IN immunization, especially when the immunogen displayed cell-binding properties (CT). IgG was the dominant isotype, not only in spleen but also in secretory sites, after systemic immunization (IP).

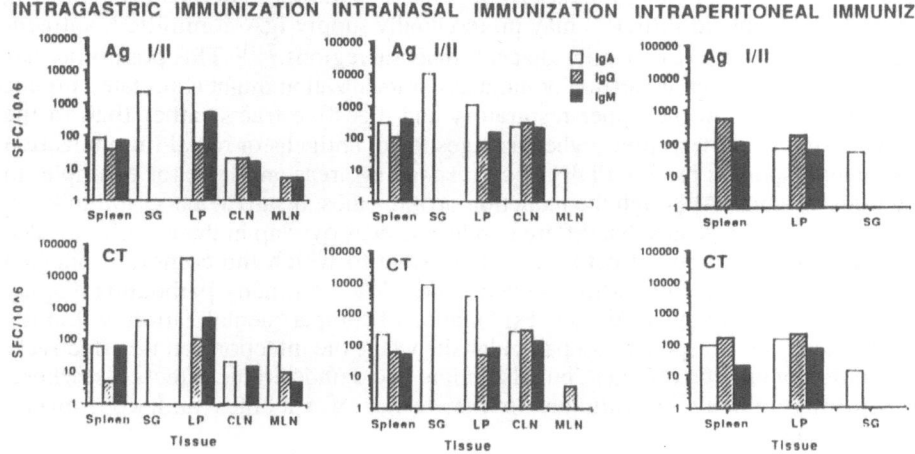

Figure 1. Antibody-secreting cell responses to AgI/II and CT in various tissues 7 days after immunization with AgI/II-CTB conjugate, 3 times at 10-day intervals. SFC-spot-forming cells in the ELISPOT assay.

All isotypes were secreted by cells isolated from CLN after IG or IN immunization, and the numbers were comparable with those obtained from spleens. Antibodies measured in secretions after IG or IN immunization with AgI/II-CTB were consistent with the numbers of antibody-secreting cells in corresponding tissues.

Mice primed and boosted IG, IN, IV, or IR with CT exhibited in serum as well as in secretions, similar levels of IgG and IgM anti-CT antibodies independent of the immunization route. However, the levels of IgA anti-CT differed according to the route of administration: IG administration was the most efficient for inducing antibodies in serum and intestinal lavage, IN for salivary antibodies, and IV and IR induced equal amounts of CT-specific IgA in vaginal washes that were 3 times higher than those measured after IG or IN immunization (data not shown).

IG priming of mice with CT resulted in the production of antigen-specific antibodies in both serum and secretions. Boosting of IG-primed animals with CT via three different mucosal routes, IG, IN or IR, elevated antibodies in the serum, whereas the levels in the secretions differed markedly depending on the route (Fig.2). Thus, salivary antibody levels were highest in the animals boosted IN, while vaginal antibodies were highest in those boosted IR.

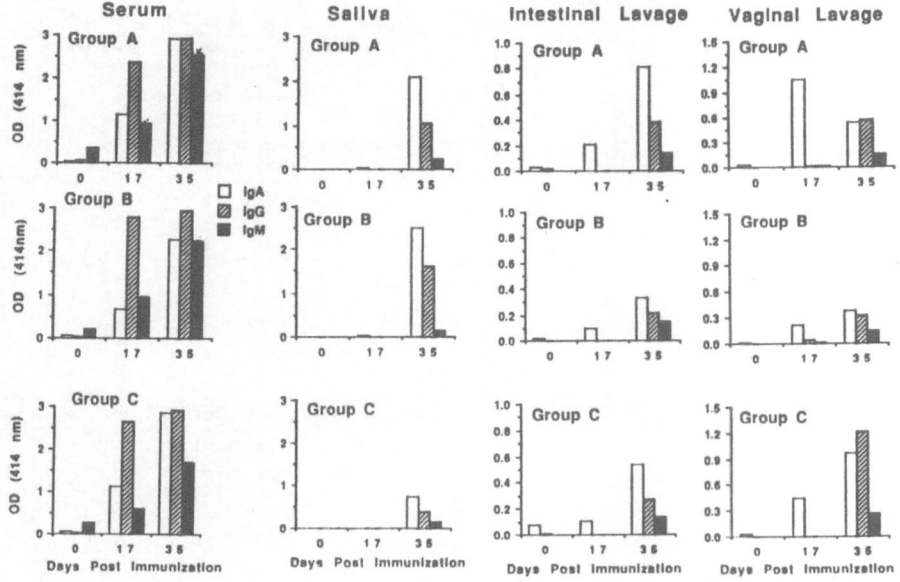

Figure 2. CT-specific antibodies in sera and secretions of mice immunized IG with 10µg/mouse CT, and boosted 17 days later by different routes A - IG, B - IN; C - IR

IR, IV or IG immunization of mice with vaccinia virus expressing gp160 of SIV induced antibodies against the virus, and at a lower level against gp160 in serum and secretions. Although IgM and IgG were the antibody isotypes found in sera, IgA dominated in all analyzed secretions, showing a slight increase at the immunization site (Fig. 3).

The appearance of antibodies in serum and various secretions after immunization at a specific site supports the concept of the common mucosal immune system. However, the results suggested a degree of compartmentalization which may be advantageously exploited in the design of mucosal vaccination strategies.

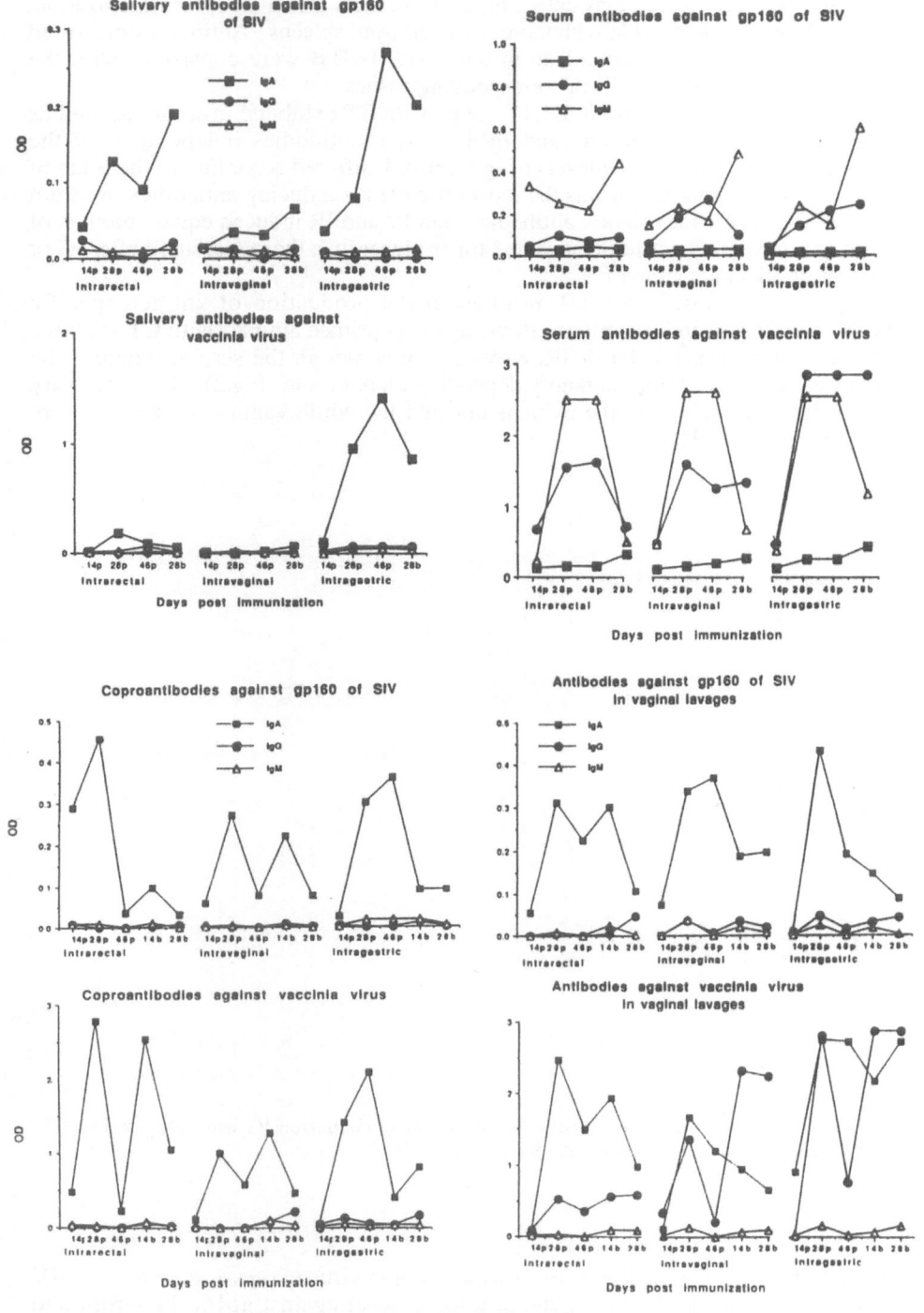

Figure 3. Antibodies specific to SIV and to vaccinia virus measured in sera and secretions of mice primed IR, IV, or IG, and 48 days later boosted by the same routes with VV SIV 239 env.

ACKNOWLEDGMENTS

Supported by US-PHS grants AI-18745, DE-06746, AI-28147, DE-08182.

REFERENCES

1. P. Brandtzaeg, *Curr. Topics Microbiol. Immunol.* 146: 13 (1989).
2. A. J. Husband, ed., "Migration and Homing of Lymphoid Cells", Vol. II, CRC Press, Boca Raton, FL (1988).
3. R. V. Heatley, J. M. Stark, P. Horsewood, E. Bandouvas, F. Cole, and J. Bienenstock, *Immunology* 44: 543 (1981).
4. D. Y. E. Perey, D. Frommel, R. Hong, and R. A. Good, *Lab. Invest.* 22: 212 (1970).
5. G. Mayerhofer and R. Fisher, *Eur. J. Immunol.* 9: 85 (1979).
6. G. A. Enders, S. Balhaus, and W. Brendel, *Immunology* 63: 411 (1988).
7. P. Brandtzaeg, *in*: "Immunology of the Lung and Upper Respiratory Tract", J. Bienenstock, ed., p 28-95, McGraw-Hill Book Co., New York (1984).
8. A. J. Husband and J. L. Gowans, *J. Exp. Med.* 148: 1146 (1978).
9. M. R. McDermott and J. Bienenstock, *J. Immunol.* 122: 1892 (1979).
10. N. F. Pierce and W. C. Cray, *J. Immunol.* 128: 1311 (1982).
11. C. O. Elson, W. Ealding, and J. Lefkowitz, *J. Immunol. Methods.* 67: 101 (1984).
12. T. de Vos, and T. A. Dick, *J. Immunol. Methods* 141: 285 (1991).
13. H. Kiyono, J. R. McGhee, M. J. Wannemuehler, M. V. Frangakis, D. M. Spalding, S. M. Michalek, and W. J. Koopman, *Proc. Natl. Acad. Sci. USA* 79: 596 (1982).
14. C. Czerkinsky, *in*: "Methods of Enzymatic Analysis", H.U. Bergmeyer, J. Bergmeyer, and M. Grassl, eds., p 23, VCH, Weinheim (1986).

INTRAPERITONEAL ADMINISTRATION OF TETANUS TOXOID ELICITS A SPECIFIC RESPONSE OF ANTIBODY-SECRETING CELLS IN THE PERITONEAL CAVITY

Cummins Lue,[1,2] A. Warmold L. Van den Wall Bake,[1]
Shirley J. Prince,[1] Bruce A. Julian,[2] Mei-ling Tseng,[1]
Charles O. Elson III,[2] Hollie H. Hale,[1] and Jiri Mestecky[1,2]

Departments of [1]Microbiology and [2]Medicine
University of Alabama at Birmingham
Birmingham, Alabama 35294, USA

INTRODUCTION

Anatomical studies suggested that the mammalian peritoneum plays an important role in immunological processes. Attention has primarily focussed on the greater omentum where "milky spots" cells play a role in the immunological defense of the peritoneal cavity.[1,2,3,4,5] The peritoneal route of immunization has been used in experimental animals as an effective site for induction of both systemic and mucosal immune responses.[6,7,8] Kroese et al.[9] showed that B cells from the murine peritoneal cavity repopulated the intestinal lamina propria of recipient mice with IgA-secreting cells. It was estimated that up to 50% of murine intestinal IgA-secreting cells were derived from surface IgA-negative precursor in the peritoneal cavity. Recently, Solvason et al.[10] have demonstrated that human fetal omentum may serve as an additional site of B cell generation. These findings prompted us to study the potential of human peritoneal B cells to differentiate into antibody-secreting cells (AbSC). Patients on continuous ambulatory peritoneal dialysis (CAPD) represent a group of human subjects with a permanent access to the peritoneal cavity through an indwelling catheter.[11] We immunized patients on CAPD intraperitoneally (i.p.) with tetanus toxid (TT), a well established protein antigen, to examine whether it can elicit specific antibody production by peritoneal B cells.

MATERIALS AND METHODS

Immunization

Approval by the Institutional Review Board at the University of Alabama was obtained before the study. Nine patients, seven females and two males, with end-stage renal disease and on CAPD were recruited with informed consent. Five subjects were administered one 0.5 ml dose of TT (Tetanus Toxoid Ultrafined, Wyeth-Ayerst, Marietta, PA) each through the indwelling peritoneal catheter after complete drainage of the overnight dialysis fluid; 250 ml of isotonic saline was instilled through the indwelling peritoneal catheter to ensure the deposition of the vaccine into the peritoneal cavity. The regular CAPD schedule was resumed after 8 h. Four subjects served as controls and received the same

Advances in Mucosal Immunology, Edited by
J. Mestecky *et al.,* Plenum Press, New York, 1995

dose of the TT vaccine each into the deltoid muscle (i.m.). Blood, saliva, and peritoneal effluents (the overnight dialysate) were obtained before, 7, 14, and 21 days after immunization.

Cells

Peripheral blood lymphocytes were isolated by centrifugation on a Ficoll-Hypaque (Sigma Chemical Co., St. Louis, MO) density gradient. Peritoneal lymphocytes were obtained by centrifugation of the fresh effluent at 2500 rpm for 15 min.[12,13] After discarding the supernatant, the cell pellet was washed twice with Dulbecco's PBS. Cells were then resuspended in complete medium. Viability was checked by trypan blue exclusion. A differential cell count was carried out by Wright's stain.

Enzyme-Linked Immunospot (ELISPOT) Assay

Peripheral blood and peritoneal mononuclear cells were assayed for total IgSC or AbSC in an ELISPOT assay, as described previously[14,15,16]. For the detection of TT-specific AbSC, individual wells of the Millititer HA plate (Millipore, Bedford, MA) were coated with TT (kindly provided by Wyeth Laboratories, Marietta, PA) overnight at 4°C. Biotinylated goat anti-human IgA, IgG or IgM antibodies (Tago, Burlingame, CA) were used as secondary antibody. For the enumeration of Ig-secreting cells, the wells were coated with F(ab')$_2$ fragements of goat anti-human IgA (Pel-Freez, Rogers, AR), goat anti-human IgG (Jackson, West Grove, PA) or goat anti-human IgM antibodies (Pel-Freez). The secondary antibodies were the same as in the antigen-specific assay.

Fluorescence-Activated Cell Sorting (FACS)

Mononuclear cell samples were stained for lymphocyte markers with phycoerythrine (PE)-conjugated anti-CD3, PE-anti-CD4, FITC-conjugated anti-CD8, PE-anti-CD19 (all Becton Dickinson) monoclonal antibody. The cells were then subjected to two-color analysis by using a FACStar (Becton Dickinson) equipped with an argon laser.

Statistics

The results were expressed as arithmetic means with the SEM. The ANOVA with repeated measurements was used to determine significant differences among the two immunization routes (i.p. versus i.m.). A significance level of 0.05 was chosen in all statistical tests.

RESULTS

Phenotype of Peritoneal Cells

The majority of peritoneal cells were lymphocytes, monocyte/macrophages, polymorphonuclear leukocyte, and mesothelial cells. The phenotype of peritoneal lymphocytes was determined by FACS analysis. CD3$^+$ T cells represented the largest fraction, while CD19$^+$ B cells accounted for less than 15% of all lymphocytes.

Antigen-Specific AbSC

Before immunization, low frequencies of AbSC were detectable in the peripheral blood and peritoneal effluents of most patients. Seven days after i.p. or i.m. immunization with TT, significant numbers of antigen-specific AbSC were found in the peripheral blood and peritoneal effluents. This AbSC response was significant for IgA and IgG both in peripheral blood and peritoneal cavity. The isotype distribution was IgG>IgA. The frequency of AbSC declined when measured 14 and 21 days postimmunization. Intraperitoneal immunization elicited significantly higher numbers of IgA- and IgG- AbSC in peritoneal effluents than the i.m. route. There was no significant difference between the two routes with regard to the AbSC frequencies in peripheral blood, although i.m. immunization tended to induce higher numbers of AbSC in the circulation. Peritoneal effluents were also analyzed for total IgSC. IgG dominated followed by IgA and IgM (Fig. 1).

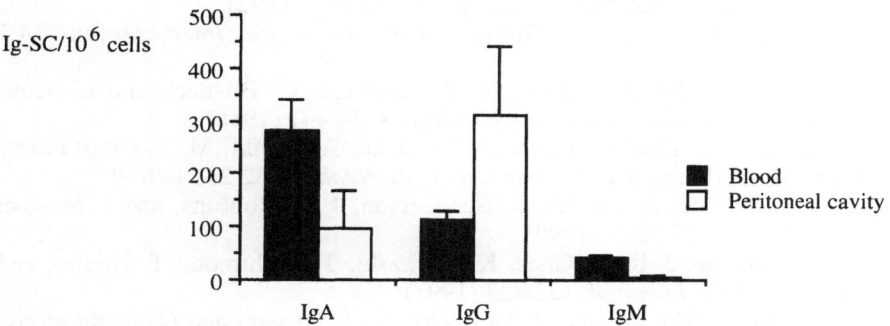

Figure 1.   Immunoglobulin-secreting cells among PBMC and peritoneal lymphocytes (arithmetic means ± SEM).

DISCUSSION

Our study agrees with the findings by several other investigators[12,13,17,18] that peritoneal effluents from patients on CAPD contain in the majority lymphocytes and macrophages. We demonstrated that peritoneal B cells actively secrete IgA, IgG, or IgM. The i.p. administration of the purified protein antigen TT induced the appearance of antigen-specific AbSC in peritoneal cavity and the circulation. In comparison with the control group, the i.p. route of immunization elicited higher numbers of peritoneal AbSC suggesting that i.p. immunization led to the local response of resident B cells, as manifested by the appearance of specific AbSC. The i.m. immunization induced specific AbSC in peritoneal effluents although they were found at lower frequencies than the i.p. route. The AbSC were probably derived from the circulation through migration to milky spots via omental stomata.[2] The peritoneal cavity may play a role as a reservoir for precursors of the B cell capable of differentiation into AbSC upon stimulation by i.p. microbial antigens such as LPS or capsular polysaccharides.

ACKNOWLEDGMENTS

This study was supported by US-PHS grants AI 18745 and DE 08182.

REFERENCES

1.   Y. Hamazaki, *Folia Anat. Japon.* 3:243 (1925).
2.   V. A. Mironov, S. A. Gusev, and A. F. Baradi, *Cell Tissue Res.* 201:327 (1979).
3.   M. Shimotsuma, M. Kawata, A. Hagiwara, and T. Takahashi, *Acta Anatom.* 136:211 (1989).

4   M Shimotsuma, T Takahashi, M Kawata, and K Dux, *Cell Tissue Res* 264 599 (1991)

5   I Trebichavsky, M Holub, L Jaroskova, L Mandel, and F Kovaru, *Cell Tissue Res* 215 437 (1981)

6   K J Beh, A J Husband, and A K Lascelles, *Immunology* 37 385 (1979)

7   N F Pierce and J L Gowans, *J Exp Med* 142 1550 (1975)

8   M A Thapar, E L Parr, and M B Parr, *Immunology* 70 121 (1990)

9   F G M Kroese, E C Butcher, A M Stall, P A Lalor, S Adams, and L A Herzenberg, *Int Immunol* 1 75 (1989)

10   N Solvason and J F Kearney, *J Exp Med* 175 397 (1992)

11   H Tenckhoff and H Schechter, *Transac Amer Soc Artific Intern Organs* 14 181 (1968)

12   B Brando, R Galato, M Seveso, E Sommaruga, G Busnach and L Minetti, *Transact Amen Soc Artific Intern Organs* 34 441 (1988)

13   M T Valle, M L Degl'Innocenti, P Giordano, A Kunkl, M T Constantini, F Perfumo, F Manca, and R Gusmano, *Clin Nephrol* 32 235 (1989)

14   C Lue, S J Prince, A Fattom, R Schneerson, J B Robbins, and J Mestecky, *Infect Immun* 58 2547 (1990)

15   C Lue, H Kiyono, J R McGhee, K Fujihashi, T Kishimoto, T Hirano, and J Mestecky, *Cell Immunol* 132 423 (1991)

16   C Czerkinsky, Z Moldoveanu, J Mestecky, L -Å Nilsson and O Ouchterlony, *J Immunol Methods* 115 31 (1988)

17   T Cichocki, Z Hanicki, W Sulowicz, and O Smolenski, *Nephron* 35 175 (1983)

18   S J Davies, J Suassuna, C S Ogg, and J S Cameron, *Kidney Int* 36 661 (1989)

THE COLON AND RECTUM AS INDUCTOR SITES FOR LOCAL AND DISTANT MUCOSAL IMMUNITY

Bjørn Haneberg,[1,2] Donna Kendall,[1] Helen M. Amerongen,[1] Felice M. Apter,[1] and Marian R. Neutra[1]

[1]Department of Pediatrics, Harvard Medical School and GI Cell Biology Research Laboratory, Children's Hospital, Boston, Massachusetts 02115, USA; [2]Vaccine Department, National Institute of Public Health, Oslo, Norway

INTRODUCTION

The principles governing the induction of local mucosal immune responses are not completely understood. This is due in part to the lack of practical methods for the quantitation of specific IgA antibodies in secretions on local mucosal surfaces. Most current techniques, which rely upon the use of washes of the excised gut[1], the products of intestinal gavage *in vivo*[2], or peripheral blood lymphocytes[3], give only a summary of these responses. To evaluate the effect of vaccines intended for mucosal application, it is important that secretory immune responses be measured at the specific sites which are relevant to protection.

We have developed a method for retrieval of secretions directly from various mucosal surfaces with specially designed wicks which combine a large absorbent capacity with very low protein binding. We summarize here results of a study on the induction of local secretory IgA (S-IgA) antibodies to cholera toxin after immunization either orally, gastrically, by the rectal route, or vaginally.

MATERIALS AND METHODS

Groups of young adult female mice were immunized and boosted over a period of 25 days with a total of 4 doses of cholera toxin (CT, type Inaba 569B). CT was delivered by the oral or vaginal route via a micropipette, or by the gastric or rectal route via a blunt-end steel feeding tube. Samples of various secretions, intestinal contents and blood were obtained as follows. Salivary secretions were collected in wicks after stimulation of salivation by pilocarpine. Blood and feces were collected, and total intestinal contents were obtained by luminal flushing of isolated small intestine.

Secretions adherent to local mucosal surfaces were harvested directly using custom-made cylindrical wicks composed of a mixture of synthetic fibers and cellulose. Specially made glass tubes were used as applicators for insertion of the wicks into the lower colon/rectum or vagina, or into segments of the excised small intestine.

After removal of the wicks from the various mucosal sites, secretions were extracted from the wicks in microcentrifuge tubes by adding a standard volume of PBS with 5% non-

Advances in Mucosal Immunology, Edited by
J. Mestecky *et al.*, Plenum Press, New York, 1995

fat dry milk and protease inhibitors, vortexing, and centrifugation. Similarly, extracts were made of freeze-dried feces by adding a standard volume of PBS containing 5% dry milk and inhibitors, vortexing, and centrifugation.

IgA and IgG antibodies specific for CT in sera, intestinal washes, extracts of feces, and material eluted from absorbent wicks were determined by ELISA. Ninety-six-well plates were coated with CT and samples were applied. Horseradish peroxidase conjugates of goat antibodies directed against mouse IgA and against mouse IgG (Sigma) were used as secondary antibodies. Known amounts of monoclonal IgA and IgG antibodies against CT were used as standard references for determinations of anti-CT activity.[4] Antibody concentrations were calculated based on the weights of the undiluted secretions captured, corrected for dilutions of the samples, and expressed in weight/volume units.

RESULTS AND DISCUSSION

Serum IgG antibodies to CT were present in all mice after this antigen had been delivered to every mucosal site, i.e. after oral, gastric, rectal or vaginal delivery. Antibodies of the IgA class were present in serum after oral, gastric and intestinal immunization, but could not be demonstrated in serum after vaginal delivery. In washes of the small intestine and in extracts of feces, highest levels of IgA antibodies specific to CT could be demonstrated after gastric and colo-rectal immunization; IgA antibody levels were low after oral immunization, and were practically non-existent after vaginal antigen delivery. A positive correlation of anti-CT IgA antibodies in feces with those in intestinal washes was found, confirming that feces collection is of value for non-invasive assessment of intestinal S-IgA responses.[5]

The concentrations of specific anti-CT IgA antibodies which were picked up from local mucosal surfaces by the absorbent wicks were highly dependent on the site of collection. In general, we found low levels in saliva, somewhat higher levels in vaginal secretions, high levels in duodenal and ileal mucus, and very high levels in colo-rectal mucus. It should be noted that collections from the mucosal surface of the small intestine were made after the lumenal contents had been cleared by washes of saline. To some extent, therefore, these IgA concentration differences reflected the relative dilution of the native secretions (saliva) as well as dilution by saline during collection (vagina and small intestine).

The pattern of specific IgA concentrations on mucosal surfaces reflected the site of immunization. For example, a response in saliva was seen only after oral delivery of the antigen. The strongest responses in duodenum and ileum were seen after gastric delivery. It was striking that a very strong response was seen in the lower colon and rectum exclusively after rectal delivery (Table 1). This confirms that the magnitude of regional S-IgA responses in the gut are strongly influenced by the site of uptake of antigen.[6-8]

Table 1. IgA antibodies to cholera toxin in stimulated saliva and in secretions from mucosal surfaces of the small intestine, lower colon and rectum, and vagina, after a series of immunizations by the oral, gastric, colo-rectal, or vaginal route.

	IgA Response			
Antigen delivery	Saliva	Samll intestine	Colon/rectum	Vagina
Oral	+	+	–	+
Gastric	–	++	–	+
Colo-rectal	–	+	++	+
Vaginal	–	–	–	–

Although we found vaginal IgA antibodies after oral, gastric, and colo-rectal delivery of CT as an antigen, antigen delivery to the vagina and cervix did not lead to any S-IgA response in vagina, nor at other mucosal sites (Table 1). This indicates that the oral, small intestinal and colo-rectal mucosae are good inductive sites for a distant mucosal immune response in cervix/vagina, but the lower female reproductive tract is a poor inductive site. This finding is consistent with previous observations of specialized antigen-transporting epithelia and organized mucosal lymphoid tissues in the tonsils[9], small intestine[10], and colon/rectum[11] and with the apparent absence of these structures in cervix or vagina.

Specific anti-CT IgA antibody concentrations along segments of the small intestine also seemed to reflect the site of prior immunization. Thus, high levels were found in duodenal segments after both gastric and colo-rectal antigen delivery, while a second concentration peak in the lower ileum was found only after gastric delivery. It is likely that the a major portion of the first concentration peak represented antibodies secreted via the bile. This suggests that a major portion of the small intestinal antibody response following colo-rectal immunizations is mediated by hepatic secretion.

CONCLUSION

In summary, the wick method has allowed us to obtain quantitative information about the concentrations of specific S-IgA present in secretions associated with local mucosal surfaces. This information may be especially useful in future design of mucosal vaccines for immune protection of the lower gastrointestinal tract and the vagina.

ACKNOWLEDGMENTS

This work was supported by Research Grants HD 17557 and AI 29378 from the National Institutes of Health.

REFERENCES

1. A. A. Merchant, W. S. Groene, E. H. Cheng, and R. D. Shaw, *J. Clin. Microbiol.* 29:1693 (1991).
2. C. O. Elson, W. Ealding, and J. Lefkowitz, *J. Immunol. Methods* 67:101 (1984).
3. B. D. Forrest, *Infect. Immun.* 60:2023 (1992).
4. F. M. Apter, W. I. Lencer, J. J. Mekalanos, and M. R. Neutra, *J. Cell Biol.* 115:399a (1991).
5. B. Haneberg and D. Aarskog, *Clin. Exp. Immunol.* 22:210 (1975).
6. P. L. Ogra and D. T. Karzon, *J. Immunol.* 102:1423 (1969).
7. N. F. Pierce and W. C. Cray, Jr., *J. Immunol.* 128:1311 (1982).
8. P. Brandtzaeg, *J. Infect. Dis.* 165 (suppl 1):s167 (1992).
9. P. Brandtzaeg, *in*: "Immunology of the Ear", J. M. Bernstein, and P. L. Ogra, eds., Raven Press, New York (1987).
10. M. R. Neutra and J. P. Kraehenbuhl, *Trends Cell Biol.* 2:134 (1992).
11. R. L. Owen, A. J. Piazza, and T. H. Ermak, *Am. J. Anat.* 190:10 (1991).

A BASEMENT MEMBRANE MOLECULE PREFERENTIALLY ASSOCIATED WITH MUCOSAL POST-CAPILLARY VENULES

Tracy Hussell, Peter G. Isaacson and Jo Spencer

Department of Histopathology
UCL Medical School
Rockefeller Building
University St
London, WC1E 6JJ

INTRODUCTION

B cell lymphomas of mucosa associated lymphoid tissue (MALT-type lymphomas) are a distinct group of tumours which arise in extranodal sites including the stomach, salivary gland, thyroid and lung.[1] The onset of lymphomas in the salivary gland and thyroid is known to be preceeded by autoimmune disease, though it is not clear whether the tumour cells themselves in such lymphomas recognise autoantigens.[2, 3] We have recently shown that low grade MALT-type lymphomas of the gastrointestinal tract recognise autoantigens.[4] In this study we have investigated an autoantigen recognised by one such case of low grade B cell gastric lymphoma which we believe may itself have a significant role in mucosal immunity.

MATERIALS AND METHODS

Tissues

A low grade B cell gastric lymphoma of mucosa associated lymphoid tissue (MALT-type) was received fresh in the laboratory. Cells were teased from the tumour and frozen down in aliquots for future experiments. The following specimens were received fresh in the laboratory and were snap frozen and stored at -70oC for immunohistochemical studies: 10 normal tonsils, 3 peripheral lymph nodes, 4 specimens of terminal ileum (from right hemicolectomies for carcinoma of the colon) containing Peyer's patches, 10 specimens of appendix, 3 hemicolectomy specimens from patients with Crohn's disease.

Antibodies and Immunohistochemistry

Murine monoclonal antibodies to type IV collagen, Factor VIII (both from Dako UK Ltd, High Wycombe), and vascular endothelium but not lymphatic endothelium (PAL-E, Bio-nuclear services Ltd, Bude) were used to compare with the reactivity of the tumour immunoglobulin in serial sections. Binding of these antibodies was detected on acetone fixed frozen sections using indirect immunoperoxidase with rabbit anti-mouse secondary antiserum conjugated to peroxidase (Dako UK Ltd).

Advances in Mucosal Immunology, Edited by
J. Mestecky *et al.*, Plenum Press, New York, 1995

Preparation of Anti-Idiotypic Antibody

Anti-idiotypic antibodies were produced and characterised as described previously.[5] Briefly, mice were immunised with tumour cells and clones screened using immunohistochemistry for reactivity with tumour but not tonsil tissue. ELISA assays with pooled human serum were used to confirm that these were anti-idiotypic antibodies. The anti-idiotypic antibody produced against the lymphoma in this study cross reacted with a small population of normal T cells.

Preparation of Tumour Immunoglobulin

Tumour cells were fused with the murine myeloma cell line NSO, as described previously, and clones secreting tumour immunoglobulin (Ig) were detected using the anti-idiotypic antibody in an ELISA assay. Positive clones were repeatedly cloned out and the supernatant from these cells used as a source of tumour Ig.[4, 5]

Detection of Tumour Ig Reactivity

The reactivity of the tumour immunoglobulin was detected in tissues from individuals other than the patient from which the lymphoma was taken. Frozen sections were cut, air dried, fixed in acetone and incubated in the tumour Ig for 60 minutes. Sections were then washed and incubated in the murine anti-idiotypic antibody for 60 minutes. The complexes were then detected with rabbit anti-mouse peroxidase conjugate, followed by the diaminobenzidine reagent.

Reactivity of Tumour Ig in Enzyme Digested Tissue Sections

Frozen sections were air dried but not fixed. They were then digested with dilutions of up to 100Êg/ml of either neuraminidase, hyaluronidase, elastase, trypsin, pronase or collagenase (all from Sigma Ltd., Poole) for 30 minutes at room temperature. Sections were then washed and stained to detect the reactivity of the tumour immunoglobulin as described above.

RESULTS

Reactivity of Tumour Ig in Organised Gut-Associated Lymphoid Tissue

The vessels recognised by the tumour Ig were a subset of the total vessels present in organised gut-associated lymphoid tissue. The anti- endothelial cell antibody PAL-E recognised many vessels including capillaries in the follicle centre, the sub-epithelial region of the follicle and the lamina propria. The antigen recognised by tumour Ig however was restricted in its distribution to the HEV's and some larger vessels in the serosa (Fig. 1).

Differential Expression of Basement Membrane Molecule in Different Anatomical Sites

The tumour immunoglobulin recognised the basement membrane area of high endothelial venules in all lymphoid tissues studied. The reactivity in gut-associated lymphoid tissue in the appendix and Peyer's patches exceeded that observed in peripheral lymph node and tonsil (Fig. 2A & B). The most intense expression was observed in association with the vessels in lamina propria of resected bowel from patients with Crohn's disease (Fig. 2C). As shown in Fig. 1, vessels in normal lamina propria do not express the antigen recognized by tumour Ig.

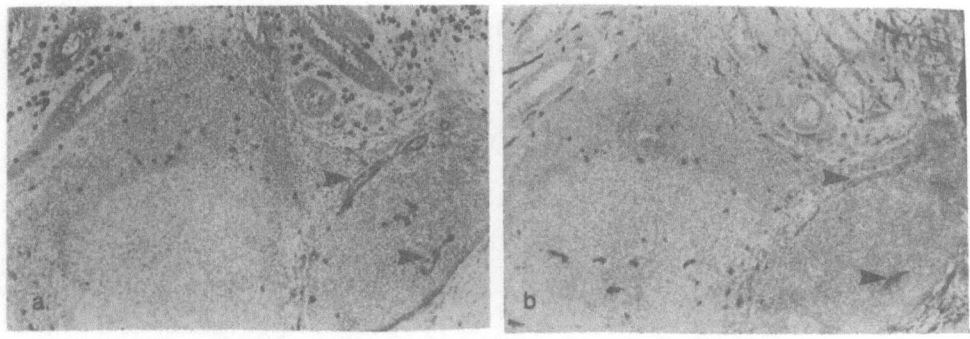

Figure 1. Comparison of the distribution of the basement membrane molecule recognised by the tumour Ig (a) and vessels recognised by anti-endothelial cell antibody PAL-E (b) in frozen sections of appendix. Tumour Ig recognises basement membrane associated with HEV and some larger vessels in the serosa (arrows), but not vessels in the lamina propria or the follicle centre. (Immunoperoxidase x 32)

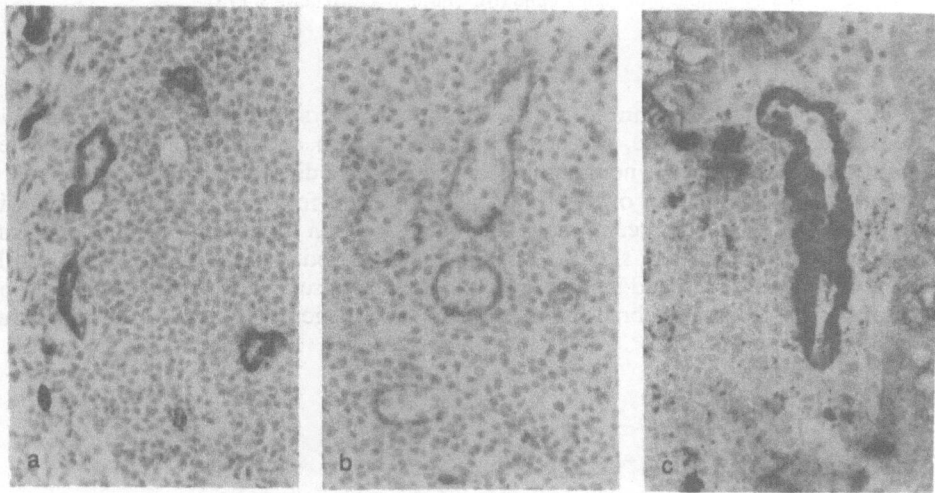

Figure 2. Frozen sections of appendix (a), tonsil (b) and lamina propria of bowel resected for Crohn's disease (c) stained using tumour immunoglobulin There is a quantitative difference in the staining observed in the HEV's at different sites; the staining of HEV in gut-associated lymphoid tissue exceeds the staining in peripheral lymphoid tissues including tonsil Expression of the basment membrane molecule recognised by tumour Ig is highest in the lamina propria in Crohn's disease.

The antigen recognised by tumour Ig was observed in the walls of some medium sized and larger vessels and also in the lymphatics (identified by reactivity with factor VIII but not PAL-E, and their content of mostly small lymphocytes) in Crohn's disease. Lymphocytes could be seen to closely associate with the antigen recognised by tumour Ig in lymphatic structures (Fig. 3).

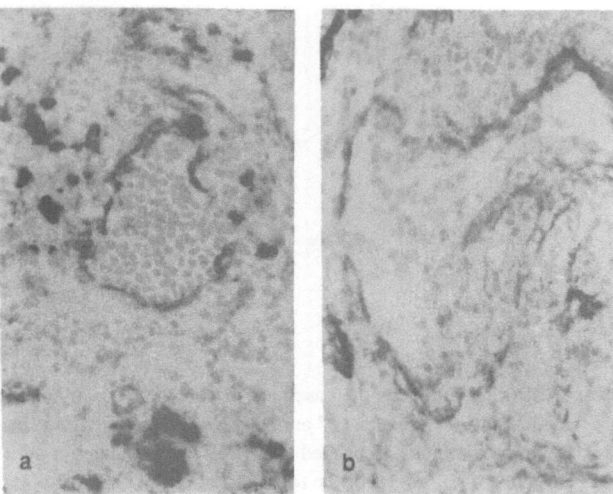

Figure 3. Frozen sections of bowel resected for Crohn's disease showing lymphatics expressing the antigen recognised by the tumour immunoglobulin (a) and associated areas showing lymphocytes closely associated with fibres recognised by tumour immunoglobulin (b). (Immunoperoxidase x 125)
Digestion of Tissue Sections

The reactivity of tumour Ig in tissue sections following digestion with hyaluronidase, neuraminidase, elastase, pronase, trypsin and collagenase was tested. Digestion with hyaluronidase, neuraminidase and elastase did not reduce the reactivity of the tumour Ig. No reduction of reactivity was observed following mild digestion with pronase or trypsin. Higher concentrations or longer digestion times with pronase or trypsin removed the section from the slide and could not be tested. The antigen recognised by tumour Ig was cleanly removed by digestion with collagenase. The possibility that tumour Ig reacted with basement membrane collagen (type IV collagen) was tested in ELISA assay. No reactivity was observed.

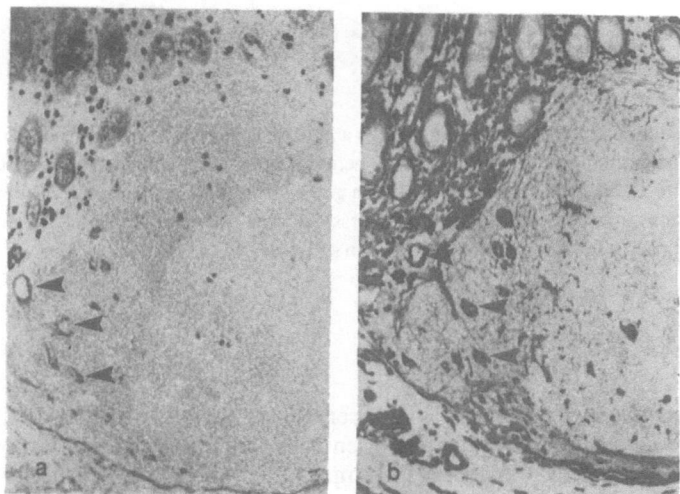

Figure 4. Frozen serial sections of appendix stained with tumour immunoglobulin (a) and anti-type IV collagen (b). Tumour immunoglobulin recognises a subset of the vessels expressing type IV collagen (arrows). Immunoperoxidase x 32.

114

Comparison of the Immunohistologic Reactivity with Anti-Type IV Collagen and Tumour Ig

The distribution of type IV collagen and the antigen recognised by tumour Ig were compared. The distribution of type IV collagen was more extensive than the distribution of tumour Ig, but the antigen recognised by the tumour Ig was expressed by a subset of the vessels expressing type IV collagen (Fig. 4).

DISCUSSION

We have described a low grade B cell lymphoma of MALT type with specificity for a basement membrane molecule which appears to be either collagen, or collagen associated. This molecule is preferentially associated with HEV's in gut associated lymphoid tissue in comparison to HEV's in peripheral lymphoid tissues. In the gut, basement membrane positivity is exclusively associated with HEV's in the T cell zones and also some larger vessels in the serosa. This distribution correlates with the distribution of the smooth muscle pericytes[6] or myoid cells,[7] but we are not aware of any reports of higher numbers of these cells associated with HEV in human gut-associated lymphoid tissue compared to peripheral lymphoid tissues.

Another antibody recognising a basement membrane molecule with a distinctive distribution within the lymphoid microenvironment has been described.[8] This molecule is associated with capillaries in the follicles centre and the distribution appears to be reciprocal to that described here. There is therefore a precedent to suggest that a groups of basement membrane molecules exist which are specifically associated with vessels in certain lymphoid microenvironements. Although the basement membrane molecule recognised by the tumour Ig is not normally expressed in association with endothelium in intestinal lamina propria, there is high expression in the lamina propria in Crohn's disease. Fine basement membrane positivity in association in lymphatics in Crohn's disease, and positive processes between lymphocytes in association with lymphatics was also observed. Lymphocytes which traffic through high endothelial venules, or which enter lymphatics must almost certainly cross the basement membrane structure associated with the endothelium. The role of the molecule described here is not known but it is possible that it contacts, or even interacts with mucosal lymphocytes as they move in and out of gut associated lymphoid tissue through the HEV's and lymphatics.

REFERENCES

1.   P. G. Isaacson and J. Spencer, *Histopathology*, 11:445 (1987)
2.   E. Hyjeck, W. J. Smith, and P. G. Isaacson, *Human Pathol.* 19:766 (1988)
3.   E. Hyjeck and P. G. Isaacson, *Human Pathol.* 19:1315 (1988)
4.   T. Hussell, P. G. Isaacson, J. E. Crabtree, A. Dogan, and J. Spencer, *Am. J. Pathol.* (in press)
5.   J. Spencer, T. Diss, and P. G. Isaacson, *J. Pathol.* 160:231 (1990)
6.   A. O. Anderson, N. D. Anderson, *Am. J. Pathol.* 80: 387 (1975)
7.   M.-F. Tocannier- Pelte, O. Skalli, Y. Kapanci, and G. Gabbiani, *Am. J. Pathol.* 129:109 (1987)
8.   E. H. Jaspars, J. C. van der Linden, R. J. Scheper, L. M. Meijer, *J. Pathol.*, Published Proceedings of the 165th Meeting of the Pathological Society of Great Britain and Ireland. Abstract 137 (1992)

ANALYSIS OF INTRAEPITHELIAL LYMPHOCYTES AND PEYER'S PATCH LYMPHOID TISSUE IN TCR-ALPHA KNOCKOUT MICE

Jo Viney, Karen Philpott, and Mike Owen

Laboratory for Lymphocyte Molecular Biology
Imperial Cancer Research Fund
London WC2A 3PX

INTRODUCTION

Despite numerous studies, attributing function to the distinct T cell receptor (TCR) $\alpha\beta^+$ and TCR $\gamma\delta^+$ IEL has proved difficult. TCR $\alpha\beta^+$ cells are induced by the gut flora, have potent cytotoxic effector activity, and are capable of proliferation and elaborating lymphokines after stimulation *in vitro*.[1-4] In contrast the biological function of TCR $\gamma\delta^+$ cells is still largely unknown, although these cells do have weak cytotoxic activity.[1-4] The stimulus which induces activity and the MHC restriction, if any, of TCR $\gamma\delta^+$ IEL is unclear. Until now it has proved difficult to deplete IEL of all TCR $\alpha\beta^+$ and yet retain sufficient TCR $\gamma\delta^+$ IEL for functional assays at meaningful cell numbers. Several approaches have been taken with the aim of disrupting the T cell repertoire with varying degrees of success. The depletion of $\alpha\beta$ T cells by injecting mice with antibodies to the $\alpha\beta$ TCR, and the expression of prearranged TCR genes in transgenic mice are two examples.[5,6] Instead, we have used the approach of genetically deleting the capacity of mice to make TCR $\alpha\beta^+$ T cells using the technique of homologous recombination.[7] This system yields the cleanest model mouse host for studying the biological function of TCR $\gamma\delta^+$ cells and in particular the role of TCR $\gamma\delta^+$ cells in mucosal immunity.

MATERIALS AND METHODS

Construct and ES Cells

The coding potential of the first $C\alpha$ exon was disrupted by insertion of the neomycin resistance gene, as previously described.[7] ES cells were electroporated with the linearised construct. 10 days after electroporation G418 resistant clones were trypsinised. Half of the cells were taken for PCR analysis and the remainder cultured further.

Production of Chimaeric Mice

ES clones (129) containing the disrupted TCR α allele were injected into blastocysts (BALB/c). After the birth of chimaeric mice, offspring with germline transmission of the mutated TCR α gene were sibling mated to homozygosity. Status of the mice was determined by Southern blot analysis (Fig 1).

$$+ \quad + \quad - \quad + \quad + \quad - \quad + \quad - \quad + \quad + \quad + \quad - \quad + \quad + \quad + \quad + \quad +$$
$$+ \quad - \quad - \quad - \quad + \quad - \quad + \quad - \quad - \quad - \quad - \quad - \quad + \quad - \quad - \quad - \quad -$$

Figure 1. DNA extracted from mouse tails was digested with ECOR1 and analysed by Southern blot hybridisation - status of the mice was determined by the presence of a 4kb fragment for wild type DNA (+/+), a 2.8kb fragment for the disrupted TCRα gene (-/-), or both fragments in heterozygous mice (+/-).

Immunohistochemistry

6μm frozen sections of Peyer's patches and small intestine were stained using the immunoperoxidase technique. Antibodies to CD3, TCRαβ, TCR γδ, CD4 or CD8 were used. Positive cells were identified by the presence of a brown reaction product.

Proliferation Assay

Cells were cultured at a density of 2×10^5 cells/well with Concanavalin A (ConA, 5μg/ml), Phytohaemagglutinin (PHA, 5μg/ml) anti-CD3 mAb (145.2CII, 100μl/ml), anti-β TCR mAb (H57.597, 100μl/ml) and IL2 (lymphocult, 100μl/ml). After 3 days, cultures were pulsed with 1μCi tritiated thymidine for 6 h, harvested and analysed on an automated liquid scintillation counter.

RESULT AND DISCUSSION

TCRα -/- mice are outwardly indistinguishable from TCRα +/- or wild type (TCRα +/+) littermates. The thymus and spleen of TCRα -/- mice are of a normal size and have a similar number of cells to TCRα +/- and TCRα +/+ thymus and spleen. However TCRα -/- mice have barely discernible Peyer's patches, although by immunohistology, the Peyer's patches (Fig. 2) of TCRα -/- mice have grossly normal anatomy. There is also no evidence of disregulation of the small intestine gross architecture. There are, however, fewer IEL in TCRα -/- mice compared to heterozygous TCRα +/- littermates (Fig. 3).

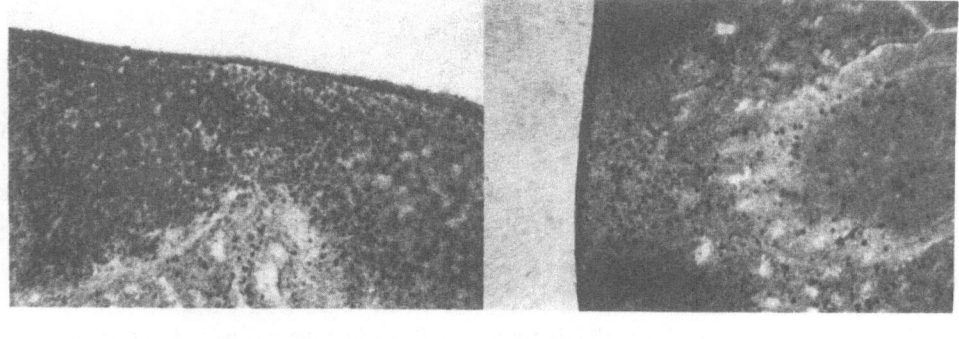

+/-   -/-

Figure 2. Frozen sections of PP were stained for CD3 expression using the immunoperoxidase techniques. CD3+ cells can be seen in reduced numbers in the interfollicular T cell areas of the PP from TCRα -/- mice. There is no evidence of a compensatory increase in TCR γδ+ or sIg+ cells.

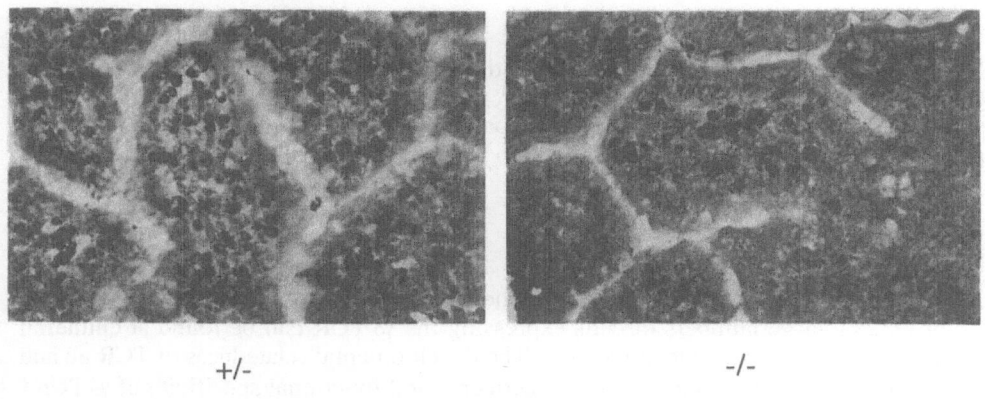

+/-   -/-

Figure 3. Frozen sections of small intestine were stained for CD3 expression using the immunoperoxidase technique. CD3$^+$ lymphocytes can be detected in the intraepithelial compartment of TCRα -/- mice (19.0 ± 2.4 CD3$^+$ IEL/100 epithelial cells in TCRα +/- mice; 9.2 ± CD3$^+$IEL/100 epithelial cells in TCRα −/− mice; n = 4).

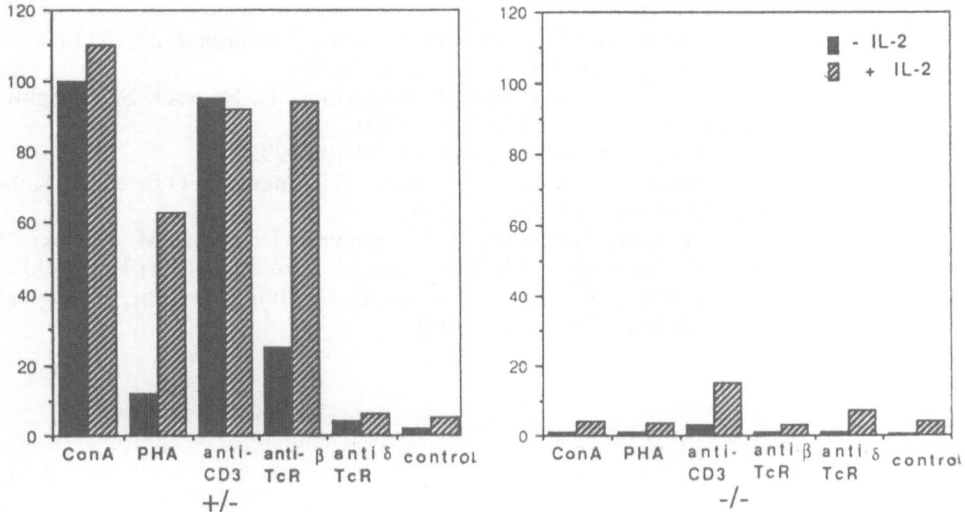

Figure 4. Spleen cells from TCRα +/- and TCRα -/- mice were stimulated with mitogens or with antibodies to the CD3 TCR complex and the proliferative response determined. Results shown are mean cpm of tritiated thymidine incorporation. Cells from TCRα +/- mice proliferated well to all stimuli. In contrast, cells from TCRα-/- showed minimal ability to proliferate except after stimulation with anti-CD3 mAb in the presence of IL2.

In the absence of functional Cα, essentially no mature CD3$^+$ cells can develop in TCRα -/- mice other than CD3$^+$ TCRγδ$^+$ cells. Interestingly, there is no expansion of the γδ T cell population to compensate for elimination of the αβ population of T cells in any of the lymphoid compartments studied. Furthermore, the absence of αβ T cells does not prevent B cell development.

Analysing the proliferative capability of cells after stimulation indicates that TCRα -/- lymphocytes from the spleen are non-responsive to the conventional T cell mitogens ConA and PHA (Fig. 4). Small proliferative responses were obtained after stimulation with anti-CD3 mAb and minimal proliferative responses after stimulation with anti-δ TCR mAb when cells were cultured in the presence of IL2 (Fig. 4). Similar proliferative responses were obtained for PPL from TCRα -/- mice (data not shown). In contrast, IEL failed to respond to any of the stimuli tested. However, it should now be possible to analyse the *in vivo* γδ T cell immune response to challenge, particularly to gut pathogens, in the absence of any TCRαβ cells.

CONCLUSIONS

Although the majority of mature peripheral T cells in rodents express the αβ T cell receptor (TCR), large numbers of cells expressing the γδ TCR can be found at epithelial surfaces, particularly in the gut epithelium. The developmental relatedness of TCR αβ and TCR γδ cells is unresolved, and the precise antigenic and functional specificity of γδ TCR+ cells has so far remained elusive. To investigate the effect of αβ T cells on the development of γδ T cells in immune surveillance and defence, we have generated mice congenitally deficient in TCR αβ expressing cells using the technique of homologous recombination.

REFERENCES

1. J. L. Viney, P. J. Kilshaw, and T. T. MacDonald, *Eur. J. Immunol.* 20:1623 (1990).
2. L. Lefrancois and T. Goodman, *Science* 243:1716 (1989).
3. A. Bandeira, T. Mota-Santos, S. Itohara, S. Degerman, C. Heusser, S. Tonegawa, and A. Coutinho, *J. Exp. Med.* 172:239 (1990).
4. J. L. Viney and T. T. MacDonald, *Immunology* 77:19 (1992).
5. A. Carbone, R. Harbeck, A. Dallas, D. Nemazee, T. Finkel, R. O'Brien, R. Kubo, and W. Born. *Immunol. Rev.* 120:35 (1990).
6. M. Bonneville, S. Itohara, E.Krecko, P. Mombaerts, I. Ishida, M. Katsuki, A. Berns, A. Farr, C. Janeway, and S. Tonegawa, *J. Exp. Med.* 171:1015 (1990).
7. K. L. Philpott, J. L. Viney, G. Kay, S. Rastan, E. Gardiner, S. Chae, H. Hayday, and M.J. Owen, *Science* 256:1448 (1992).

INTESTINAL T CELLS IN CD8α KNOCKOUT MICE AND T CELL RECEPTOR TRANSGENIC MICE

Wai-Ping Fung-Leung,[1,2] Kenji Kishihara,[1] Dawn Gray,[1]
Hung-Sia Teh,[3] Catherine Y Lau,[2] and Tak W Mak[1]

[1]Ontario Cancer Institute, Department of Medical Biophysics
and Immunology, University of Toronto, 500 Sherbourne
St, Toronto, Ontario M4X 1K9, Canada, [2]The R W
Johnson Pharmaceutical Research Institute (Canada), 19
Green Belt drive, Don Mills, Ontario M3C 1L9, Canada,
and [3]Department of Microbiology, University of British
Columbia, Vancouver, B C V6T 1W5, Canada

INTRODUCTION

T cells distributed at the epithelium of the intestines are called intraepithelial lymphocytes (IEL) IEL are heterogenous populations comprising of αβ and γδ T cells The majority of IEL express CD8 mainly as homodimers of the α subunit (CD8αα), whereas T cells in other peripheral lymphoid organs express CD8 as heterodimers of the α and β subunits (CD8αβ) (reviewed in Ref 3) Autoreactive T cells have been reported to be present in IEL population[4] and therefore IEL are suggested to mature through an extra-thymic pathway In the CD8α knockout mice, the CD8 co-receptor was shown to be required for thymic ontogeny of cytotoxic T cells but not helper T cells [1] In the H-Y transgenic mice, CD8+ T cells with the male H-Y antigen specific T cell receptor are thymically deleted in the male but selected in the female mice [2] In this report, the criteria for the ontogeny and function of IEL were studied in the CD8α knockout mice and in the H-Y T cell receptor transgenic mice   In these transgenic and knockout models, the antigen specificity of IEL can be defined by the H-Y specific T cell receptor and the defects observed for IEL in CD8α knockout mice would be a reflection of the indispensable role of CD8α in the ontogeny and function of IEL

RESULTS AND DISCUSSION

IEL Subpopulations in Normal Mice and CD8 Knockout Mice

IEL were collected from C57BL/6 mice and compared with CD8α knockout mice of C57BL/6 background The phenotype of IEL, according to the expression of CD8α, CD8β, CD4 and T cell receptors αβ or γδ, were studied by staining with monoclonal antibodies followed by flow cytometric analysis As shown in Figure 1, CD8α+ T cells were the major subset which was 62% of the total population Percentage of the CD8β+ T cell subset was only 21% of total IEL, indicating that two-third of the CD8α+ T cells express CD8 mainly as homodimers of the α chain (CD8αα). Both αβ and γδ T cells were present in the CD8αα⁺ population. A population of CD8αα⁺ T cells, which co- express CD4 and have a mature level of αβ T cell receptors, was also found in IEL from normal mice. Similar results were reported by others.[5]

CD8α knockout mice with the disrupted CD8α gene were generated by homologous recombination in embryonic stem cells.[1] There is no cell surface expression of CD8α or subunits in these knockout mice. Cytotoxic T cells, which usually are CD8⁺ T cells, cannot mature in the thymus of these mice.[1] However, the population size of IEL in these mice was found to be indifferent from control normal mice. The majority of IEL in CD8α knockout mice did not express CD8 or CD4 (Fig. 1). A prominent population of CD4CD8⁻ cells expressing αβ (15%) or γδ (52%) T cell receptors was found in the knockout mice (unpublished data), whereas the CD4⁻ CD8⁻ αβ and γδ T cells in IEL from normal mice were minimal, 2% and 7% respectively. The results suggest that the lack of CD8α expression results in a phenotypic change of IEL from CD8⁺ to CD4⁻ CD8⁻, but has no obvious effect on the accumulation and the distribution of IEL in the intestines.

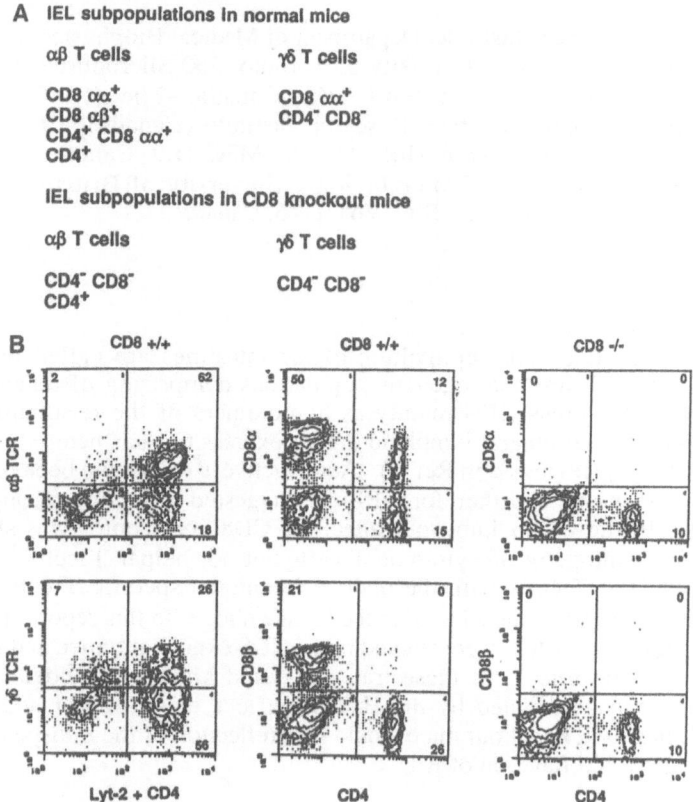

Figure 1. A summary (A) and the phenotypic analyses (B) of IEL subpopulations in normal mice and CD8α knockout mice. IEL were purified from small intestines[6], triple stained with T cell specific antibodies and analysed with the FACScan program

IEL In H-Y T Cell Receptor Transgenic Mice in a Normal and a CD8α-Null Background

Maturation of T cells in the thymus involves the processes of positive and negative selection. Thymic derived T cells are therefore restricted to self-major histocompatibility complex molecules and are tolerant to self-antigens. For the H-Y transgenic model, CD8⁺ T cells with the male H-Y antigen specific T cell receptor are deleted in the thymus of the male, but selected in the female H-2^b mice [2] IEL have been suggested to be generated extra-thymically [4] To study the response of IEL in recognition of antigens, IEL from the H-Y transgenic mice in a normal and in a CD8α-null background were analysed Surprisingly, a pronounced population of CD8$\alpha\alpha^+$ transgenic IEL (60% of total IEL) was found in the male but not in the female transgenic mice (Fig 2) CD8$\alpha\beta$+ transgenic IEL, similar to the CD8$\alpha\beta^+$ transgenic T cells in lymph nodes and the spleen, were found mainly in female mice as a small population (8% of total IEL), and were absent in male transgenic mice Similar results were also obtained by others [7] The presence of CD8$\alpha\alpha^+$ T cells carrying 'forbidden' T cell receptors in the intestines, strongly indicates that this T cell lineage, characterized by the expression of CD8α but not β chains, is not subject to thymic selection The data here supports the notion that CD8$\alpha\alpha^+$ IEL are derived from precursor cells through an extra-thymic pathway The accumulation of CD8$\alpha\alpha^+$ H-Y specific T cells at the intestinal epithelium of transgenic male mice also suggests that the autoreactive IEL respond to antigens by clonal expansion However, no pathological symptoms were observed in the intestines of these transgenic male mice (unpublished data) This model demonstrates that in an *in vivo* system, IEL do not mediate a cytotoxic effect towards target cells upon antigen recognition The function of IEL is therefore likely to be of a regulatory role, probably on the activation of other lymphoid cells at the lamina propria of the intestines

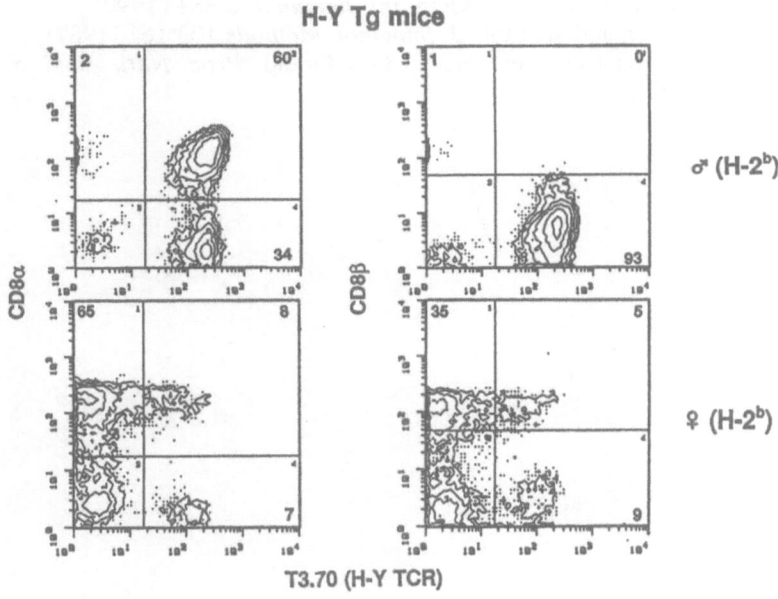

Figure 2 IEL in H Y transgenic mice IEL were collected from small intestines of male and female transgenic mice of H 2^b background IEL were triple stained with antibodies specific for CD8α, CD8β and the H Y transgenic T cell receptor α chain (T3 70) The FACScan program was used for analysis

SUMMARY

Intraepithelial lymphocytes (IEL) refer to the T cells located at the epithelium of the intestines Unlike the T cells in other peripheral lymphoid organs, the majority of IEL express the CD8 cell surface protein To study the role of CD8 in the ontogeny and the function of IEL, phenotypic analysis of IEL from CD8α knockout mice and normal mice was performed The CD8α gene in CD8α knockout mice was disrupted by homologous recombination [1] These mice are defective in thymic maturation of cytotoxic T cells [1] In normal mice, $\alpha\beta$ T cells that were CD8$\alpha\alpha^+$ or CD4$^+$ CD8$\alpha\alpha^+$, and $\gamma\delta$ T cells that were CD8$\alpha\alpha^+$, were the distinct populations found only in IEL In CD8α knockout mice, the population size of IEL remained normal, but the majority of IEL were CD4⁻ CD8⁻ T cells expressing αβ or γδ T cell receptors. IEL from the H-Y transgenic mice[2], which express the male H-Y antigen specific T cell receptor in a normal and in a CD8α-null background, were also studied. In contrast to thymic derived T cells, CD8αα⁺ IEL with the autoreactive ransgenic T cell receptor were not deleted, but clonally expanded in the male transgenic mice. Interestingly, no pathological symptoms were observed in the intestines of these mice. In the absence of CD8α expression, the H-Y specific autoreactive IEL did not accumulate in the intestines. The results suggest that CD8αα⁺ IEL are derived extra-thymically and their responses towards antigens require the CD8 accessory molecule.

REFERENCES

1. W.-P. Fung-Leung, M. W. Schilham, A. Rahemtulla, T. M. KÅndig, M. Vollenweider, J. Potter, W. van Ewijk, and T. W. Mak, *Cell* 65:443 (1991).
2. H. S. Teh, P. Kisielow, B. Scott, H. Kishi, Y. Uematsu, H. BlÅthmann, and H. von Boehmer, *Nature* 335:229 (1988).
3. P. B. Ernst, A. D. Befus, and J. Bienenstock, *Immunol. Today* 6:50 (1985).
4. R. Benedita, P. Vassalli, and D. Guy-Grand, *J. Exp. Med.* 173:483 (1991).
5. R. L. Mosley, D. Styre, and J. R. Klein, *Int. Immunol.* 2:361 (1990).
6. P. J. van der Heijden and W. Stok, *J. Immunol. Methods* 103:161 (1987).
7. R. Benedita, H. von Boehmer, and D. Guy-Grand, *Proc. Natl. Acad. Sci.* USA 89:5336 (1992).

ANALYSIS OF INTESTINAL INTRAEPITHELIAL LYMPHOCYTE (IEL) T CELLS IN MICE EXPRESSING ANTI-CD8 IMMUNOGLOBULIN TRANSGENES

Wilhelm K. Aicher,[1,2] Hermann Eibel,[2] Kohtaro Fujihashi,[1] Tilman Boehm,[2] Kenneth W. Beagley,[1] Jerry R. McGhee[1] and Hiroshi Kiyono[1]

[1]The University of Alabama at Birmingham, Birmingham, AL USA; and [2]The University of Freiburg and MPI for Immunobiology Freiburg, Germany

INTRODUCTION

In mouse, intraepithelial lymphocytes (IEL) of the small intestine contain a large number of CD4-, CD8+ T cells. Smaller numbers of IEL are CD4+, CD8-, CD4-, CD8- (double negative, DN) or CD4+, CD8+ (double positive, DP). An unusual feature of IEL is their predominant expression of γ/δ TCR (up to 75%) and rather small numbers of α/β TCR+ populations.[1,2] Several lines of evidence suggested that IEL develop in a thymus independent pathway (for review see[3]). The γ/δ TCR+ IEL express an α/α homodimer CD8 molecule, which does not react with Ly-3, an antibody that stains virtually all other peripheral CD8+ T cells, including the CD4+, CD8+ thymic T cells.[4] Experiments with thymectomized and irradiated mice which were reconstituted with bone marrow or fetal liver resulted in regeneration of IEL T cells and suggested their thymus independent development.[5,6]

To analyze the development of the CD8+ subset in IEL, we used a transgenic mouse model which is expressing a monoclonal anti-CD8, anti-α-chain IgG_{2a} antibody (manuscript in preparation). The expression of the transgene was expected to modulate the development of thymic T cells. Here we report on the effects of the anti-CD8 expression on CD8+ cells in IEL.

In a second approach, we compared T cell maturation in thymus and in IEL, analyzing the transcription rates of mRNA coding for recombinase activator genes (RAG)-1 and RAG-2 [7]. Expression of both, RAG-1 and RAG-2 genes is required for activation of the V(D)J rearrangement of the TCR genes.[8,9] RAG message and protein is therefore prominently found in immature T cells such as the CD3-, DN and $CD3^{low}\alpha$ β^{low} DP T precursors of the thymic cortex area[10] (D. Schatz, personal communication). Therefore, if immature T cells enter IEL and if a thymus independent T cell maturation occurred in this tissue, we expected to find mRNAs coding for both RAG-1 and RAG-2.

RESULTS

Analysis of Thymocytes in Anti-CD8 Transgenic Mice

Thymocytes isolated from transgenic mice producing an CD8.2 α-chain specific antibody and control littermates were analyzed for development of α/β TCR$^+$ and CD4$^+$ or CD8$^+$ T cell subsets (Fig. 1). The transgenic mice were characteristic of severely reduced numbers of total T cells in thymus (Table 1). Subset analysis of thymocytes revealed that the number of α/β$^+$ T cells in the transgenic thymus was drastically reduced (Fig. 1, Table 1). Analyzing CD4 and CD8 expressing T cells, a relative increase of DN T cells was observed. In parallel, the total numbers of CD4$^+$ and CD8$^+$ T cells was reduced about ten fold (Fig.1), whereas numbers of immature DN T cell precursors remained normal (Fig. 1, Table 1). Therefore, the modulation of T cell numbers and development in the anti-CD8 transgenes was dependent of the expression of CD8 markers which occurs in the DP T cell population passing through the deep thymic cortex.

Analysis of Lymphocytes Isolated from Peripheral Lymph Nodes (LN)

Analysis of T cells isolated from LN of the transgenic mice showed virtually no effect on the CD4$^+$ T cell subset (Fig. 1). T cells expressing CD8 molecules were not detected in LN of the transgenic mice (Fig. 1). The severe reduction of mature CD8$^+$ thymocytes resulted in a complete loss of CD8$^+$ T cells in the systemic secondary lymphoid compartments such as lymph nodes. A repopulation CD8$^+$ cells as observed in case of CD4$^+$ T cells in LN was obviously hindered by the anti-CD8 antibody, but a relative increase of DN T cells was found (not shown).

Effect of Transgenic Anti-CD8 Antibody Expression on IELs

In normal mice the majority of IEL belong to the α/β CD8$^+$ or γ/δ CD8$^+$ subset. Therefore - in contrast to LN - in IEL α/β$^+$ T cells were reduced substantially (Fig. 1). Interestingly, as in LN, the CD4$^+$ T cell remained normal. Again, in IEL of the transgenic mice, the CD8$^+$ T cells were missing (Fig. 1). A detailed analysis of IEL T cell subsets showed that in transgenic mice the absolute numbers of α/β$^+$ DN and α/β$^+$, CD4$^+$, CD8$^-$ T cells were close to the numbers found in controls (Table 1). In the γ/δ$^+$ subset the CD8$^-$ subset was relatively increased, the elimination of the CD8$^+$ was obvious (not shown).

Table 1. Relative elimination rate of T cell subsets in anti-CD8 transgenes compared to controls.

Phenotype	DN	DP,α/β^{high}	CD4$^+$, CD8$^-$	CD4$^-$, CD8$^+$
Thymocytes	0.7	550	12	300
IEL	0.9	75	13	575

Numbers T cells in each subset have been computed from numbers of total T cells/tissue and T cell numbers in each subset as analyzed by flow cytometry. The elimination rate is calculated as the difference of cells obtained in controls compared to the transgenic mice.

Search for Recombinase Activator Gene Message in IEL

TCR gene rearrangement requires activation of both RAG-1 and RAG-2. Total RNA was isolated from thymocytes and gastrointestinal (GI) lymphoid compartments of BALB/c mice 6 weeks of age. PCR amplification (35 cycles) of thymic cDNA with RAG-1

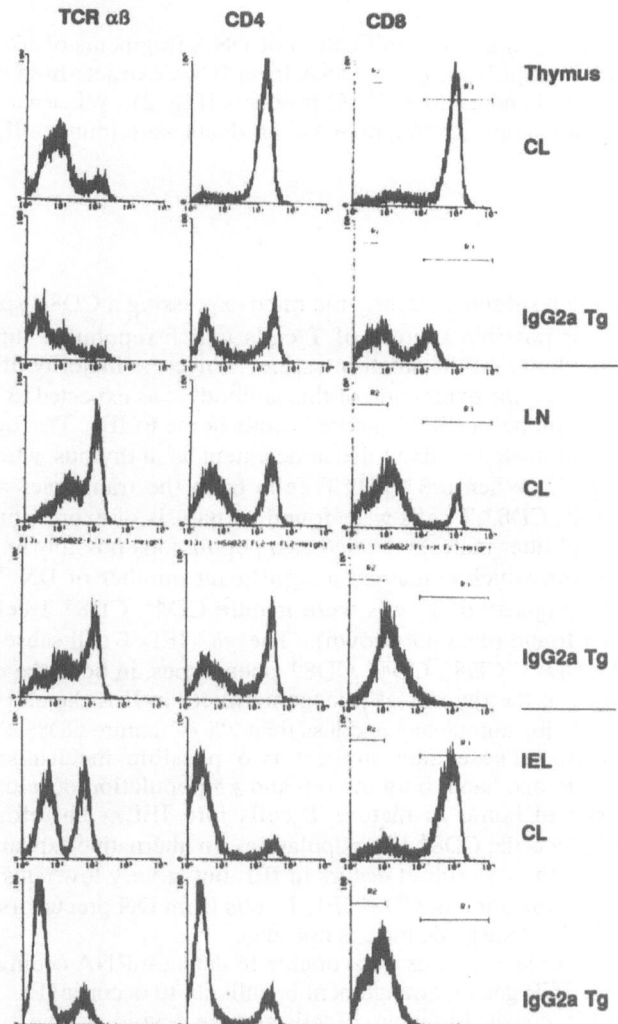

Figure 1. Analysis of T cell subsets in thymus, lymph nodes (LN) and intestinal intraepithelial lymphocytes (IEL) of control litter mates (CL) and anti-CD8 transgenic (IgG_{2a} Tg) mice. T cells were isolated and stained for expression of TCR, CD4 and CD8 markers followed by flow cytometry.

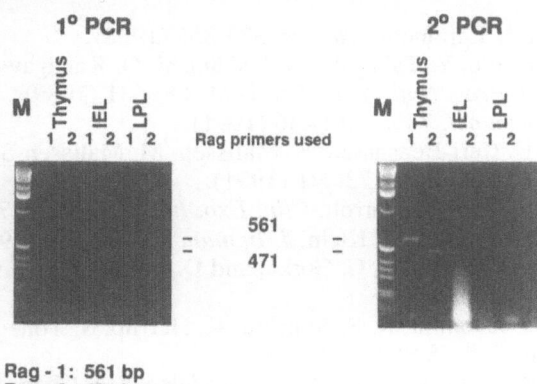

Rag - 1: 561 bp
Rag - 2 : 471 bp

Figure 2. Search for RAG-1 and RAG-2 coding message in thymus and IEL of normal mice. T cell were isolated and total RNA was prepared. cDNA synthesis and RAG-PCR were performed ans described.[7] RAG-1 and RAG-2 expression was detected in thymus but not IEL.

127

and RAG-2 primer pairs resulted in amplification of DNA fragments of 561 bp and 472 bp of length (Fig. 2). PCR amplification of cDNA from RNA extracts from total IEL T cell populations did not result in detectable. RAG products (Fig. 2). When a second PCR was performed from a primary amplification, no RAG products were found in IEL RNA/cDNA preparations (Fig. 2).

DISCUSSION

In analyzing T cell subsets of transgenic mice expressing a CD8.2 specific antibody IgG_{2a}, we searched for possible sources of T cells which repopulate and build up the intraepithelial IEL population of the small intestine. Since the majority of T cells in IEL belongs to the CD8+ subset, the expression of this antibody was expected to modulate the T cell populations in this compartment, if mature T cells home to IEL. We found that in IEL the CD8+ subset was eliminated to about the same extent as in thymus, whereas the CD4+ subset remained normal. When α/β+ IEL T cells from the transgenes were analyzed, virtually no DP or CD4-, CD8+ T cells were found. Analysis of expression of γ/δ TCR in the normal mice (control litter mates) revealed two populations possibly resembling a less mature γ/δ^{low} population which contained a significant number of DN T cells. In the γ/δ^{high} population the majority of T cells were mature CD4-, CD8+ T cells; γ/δ+ CD4+, CD8- T cells were not found (data not shown). The γ/δ+ IEL T cell subsets were further analyzed for DN, PD, CD4+, CD8-, CD4-, CD8+ phenotypes, in both, the α/β+ and γ/δ+ T cell subset. In contrast to the thymus of transgenic mice, in IEL the anti-CD8 antibody eliminated the DP population completely and less than 2% of mature CD8+ T cells remained detectable (not shown). These data suggest two possible mechanisms for T cell development in IEL. On one hand, both the α/β and γ/δ population develop from thymus dependent precursors and home as mature T cells into IEL. Therefore, the thymic elimination of T cells affects the CD8+ IEL population. An alternative explanation proposes, that T cell maturation of the γ/δ+ subset occurs in IEL but at very low rates and with slow kinetics. Therefore, a repopulation of CD8+ IEL T cells from DN precursors as obtained in thymocytes of the anti-CD8 transgenic mice is not seen.

Since we and other laboratories were unable to detect mRNA coding for RAG-2 in IEL T cell populations, TCR gene rearrangement is unlikely to occur in IEL. In conclusion, our data suggest that IEL consist of mature T cells and are devoid of immature precursor T cells.

REFERENCES

1. T. Goodman and L. Lefrancois, *Nature* 333:855 (1988).
2. K. Ito, M. Bonneville, Y. Takagaki, N. Nakanishi, O. Kanagawa, E. G. Krecko, and S. Tonegawa, *Proc. Natl. Acad. Sci. USA* 68: 631 (1989).
3. L. Lefrancois, *Immunol. Today* 12:436 (1991).
4. D. Guy-Grand, N. Cerf-Bensussan, B. Malissen, M. Malissen-Seris, C. Briottet, and P. Vassalli, *J. Exp. Med.* 173:471 (1991).
5. A. Fergusson and D. V. M. Parrott, *Clin. Exp. Immunol.* 12:477 (1972).
6. R. L. Mosley, D. Styre, and J. Klein, *J. Immunol.* 145:1369 (1990).
7. M. A. Oettinger, D. G. Schatz, C. Gorka, and D. Baltimore, *Science* 248:1517 (1990).
8. R. Mombaerts, J. Iacomini, R. S. Johnson, K. Herrup, S. Tonegawa, and V. E. R. Papaioannou, *Cell* 68:869 (1992).
9. Y. Shinkai, G. Rathbun, V. Steward, M. Mendelsohn, J. Charron, M. Datta, F.Young, A.M. Stall, and F. W. Alt, *Cell* 68:855 (1992).
10. D. Guy-Grand, C. Vanden Broecke, C. Briottet, M. Malassis-Seris, F. Selz, and P.Vassalli, *Eur. J. Immunol.* 22:505 (1992).

EFFECT OF NEONATAL THYMECTOMY ON MURINE SMALL INTESTINAL INTRAEPITHELIAL LYMPHOCYTES EXPRESSING T CELL RECEPTOR αβ AND "CLONALLY FORBIDDEN Vβs"

T. Lin, H. Takimoto, G. Matsuzaki, and K. Nomoto

Department of Immunology
Medical Institute of Bioregulation
Kyushu University, Fukuoka Japan

ABSTRACT

Neonatal thymectomy NTX performed on Day 3 of C3H mice causes over a 50% reduction in small intestinal intraepithelial lymphocytes (IEL) expressing the αβ T cell receptor (TCR) when analyzed at 10 weeks of age. Furthermore, this reduction is most notable in αβ TCR IEL expressing the CD8α homodimer and no Thy 1 marker. This is in direct contradiction to the present theory that αβ TCR IEL expressing these markers are thymic independent. Evaluation of Vβ expression shows a relative increase in Vβs which are clonally forbidden (Vβs 3, 5, 11), and which is consistent with the extrathymic origin of these particular IEL.

INTRODUCTION

Despite the near absence of αβ TCR IEL in nude mice,[1,2] many investigators feel that a portion of αβ TCR IEL are thymus independent[3-5] and are distinguished from their thymus dependent counterpart as they phenotypically bear the CD8α homodimer and/or do not have the Thy 1 marker.[3] Furthermore, these thymus-independent αβ TCR IEL are loaded with Vβs which are "clonally forbidden".[5] In the peripheral lymphoid organs, these clonally forbidden Vβs are found in minimal amount and are felt to be deleted by the thymus.[5]

Thymectomy on day 3 has been shown to increase a few of these forbidden clones.[6-8] The exact mechanism is unknown. Furthermore, NTX is associated with a systemic autoimmune like disease which may involve the gastrointestinal tract.[9,10] The mechanism for this is also unknown, and whether there is an association between the forbidden Vβs and autoimmune disease is debatable.[7,10]

In the present study, we have performed NTX on C3H mice to; 1) study the effect on αβ TCR IEL and its CD8α homodimer and Thy 1⁻ subsets, and 2) study the effect on the clonally deleted Vβs.

MATERIAL AND METHODS

C3H mice (IE^k, Mls 1^{b}2^a) were obtained from Kuroda (Japan) and bred and maintained under SPF conditions at the Kyushu University Animal Center. On day 3 after birth, newborn mice underwent thymectomy by suction. Sham operated mice were used as controls. The mice were sacrificed at weeks, and IEL were obtained by mechanical shaking of tissue in 199 media containing DTT, followed by purification with Percoll density centrifugation.[9,10] Upon sacrifice, the mice were closely examined for any gross evidence of thymus. Those with possible remnants were discarded.

The phenotype of the cell preparations were examined by flow cytometer (Becton Dickinson and FACScan) using propidium iodide to gate for dead cells. Antibodies used were FITC-conjugated Vβ3 (KJ25a), Vβ5 (MR 9-4), Vβ6 (44-22-1), Vβ11 (KT-11), pan TCR αβ (H57-597); PE-conjugated CD8α (Lyt 2) and Thy 1.2 ; biotin-conjugated pan TCR αβ (H57-597) and CD8β (Lyt 3). Statistical significance was determined by Student t test.

RESULTS

NTX causes over a 50% reduction in αβ TCR IEL (Table 1). A similar decrease was seen in the peripheral lymph nodes (data not shown). However, when αβ TCR IEL subsets were examined, the Thy 1⁻ and CD8αα subset was most dramatically reduced (Table 1). This is shown by the %Thy⁺ and % Thy 1⁻ ratio, and %CD8αα and %CD8αβ ratio among αβ TCR positive IEL.

Table 1. Effect of NTX on αβ positive Thy 1.2 negative CD8α homodimer. Values represent mean ± SD of at least five separate experiments.

	αβ %	αβ⁺ (Thy1⁺/Thy1⁻)	αβ⁺ (CD8αα/CD8αβ)
C3H SHAM	54.0 ± 5.3[a]	1.5 ± 1.2[b]	3.2 ± 0.9[c]
C3H NTX	1.8 ± 5.3[b]	3.7 ± 2.5[b]	0.3 ± 0.3[c]

Effects on the forbidden clones Vβ3 (Mls 2^a), Vβ5(IE) and Vβ11 (IE) (as a percentage of αβ TCR positive IEL) show a moderate relative increase, but not significant change, in the non-forbidden Vβ6. (Table 2). The total percentage of IEL Vβ3, Vβ5 and Vβ11 are actually decreased with NTX, but not to the extent seen with αβ TCR IEL and Vβ6. Hence, these forbidden clones are relatively increased.

Table 2. Respective Vβs as a percentage of αβ TCR IEL. Values represent mean ± SD of at least 3 separate experiments.

	Vβ3	Vβ5	Vβ6	Vβ11
C3H SHAM	0.4 ± 0.1	2.6 ± 0.3	14.6 ± 9.2	4.2 ± 1.6
C3H NTX	1.8 ± 0.7	5.2 ± 0.7	13.2 ± 9.6	7.8 ± 3.4

DISCUSSION

Though the decrease in αβ TCR IEL with NTX was expected, we were surprised to find that αβ TCR IEL bearing the so-called extrathymic markers (Thy 1⁻ and/or CD8αα) also decreased. Moreover, this decrease was most prominent in these subsets. This data presents considerable doubt to whether these subsets are truly thymus independent. Thus, evidence for the thymic independence of a portion of αβ TCR IEL comes mostly from adult thymectomized, lethally irradiated and bone marrow reconstructed murine models.[4,5,11] There remains a possibility that IEL are more radioresistant than its peripheral counterpart. Another possibility, one in which we are currently investigating and have reason to believe, is that these αβ TCR IEL that are Thy 1⁻ and/or CD8αα are not thymus independent in the true sense. They do develop extrathymically, but the thymus exerts an influence extrathymically, perhaps through some factor(s) produced by the thymus that allows precursors of these αβ TCR IEL to mature. On the other hand, the relative increase in the clonally forbidden Vβs is consistent with the fact that αβ TCR IEL may be thymic independent. However, how NTX increases some "forbidden clones" in the peripheral lymphoid organs is unknown. A similar influence on IEL forbidden clones is also possible. We are presently investigating this possibility.

REFERENCES

1. J. L. Viney, T. T. MacDonald, and P. J. Kilshaw, *Immunology* 66:583 (1989).
2. B. D. Gues, M. V. Enden, C. Coolen, L. Nagelkerken, P.V. Heijden, and J. Rozing, *Eur. J. Immunol.* 20:291 (1990).
3. D. Guy-Grand, N. Cerf-Bensussan, B. Malissen, M. Malassis-Seris, C. Briottet, and P. Vassalli, *J. Exp. Med.* 173:471 (1991).
4. A. Bandeira, S. Itohara, M. Bonneville, O. Burlen-Defranoux, T. Mota-Santos, A. Coutinho, and S. Tonegawa, *Proc. Natl. Acad. Science* 88:43 (1991).
5. B. Rocha, P. Vassalli, and D. Guy- Grand, *J. Exp. Med.* 173:483 (1991).
6. H. Smith, I. M. Chen, R. Kubo, and K. S. K. Tung, *Science* 245:749 (1989).
7. L. A. Jones, L .T. Chin, G. R. Merria, L. M. Nelson, and A. M. Kruisbeck, *J. Exp. Med.* 172:1277 (1990).
8. J. L. Andreu-Sanchez, I. M. de Alboran, M. A. R. Marcos, A. Sanchez-Movilla, C. Martinez-A, and G. Kroener, *J. Exp. Med.* 173:1323 (1991).
9. K. S. K. Tung, S. Smith, C. Teuscher, C. Cook, and R. E. Anderson, *Am. J. Pathology* 126:303 (1987).
10. A. Kojima, Y. Tanaka-Kojima, T. Sakura, and Y. Nishizuka. *Lab. Invest.* 34:550 (1976).
11. R. L. Mosley, D. Styre, and J. R. Klein, *J. Immunol.* 145: 1369 (1990).

DESIALYLATION OF RAT INTRAEPITHELIAL LYMPHOCYTES: EFFECTS ON PROLIFERATION AND DIPEPTIDYL PEPTIDASE IV ACTIVITY

Prosper N. Boyaka, Michel Coste, and Daniel Tomé

Unit of Human Nutrition and Intestinal Physiology
I.N.R.A. 4, Av. de l'Observatoire, 75006 Paris, France

INTRODUCTION

Some intestinal bacterial species, such as *Clostridium*, are known to exhibit extracellular neuraminidase (Nase) activity.[1] This enzyme removes sialic acids from glycoconjugates and could act by participating in the degradation of the protective mucin layer and by enhancing the effects of bacterial toxins. Although the physiological role of bacterial neuraminidase on the intestinal lymphocytes remains undetermined, it is well known that immunological reactions are induced after neuraminidase treatment of lymphocytes. These effects may include the stimulation of lectin-induced blastogenesis,[2] the increase of dead lymphocyte phagocytosis,[3] and the ablation of suppressor activity of antigen-primed T cells.[4]

Interactions between intestinal microbe products and local lymphocyte populations are of great importance. It is well known that small quantities of intact proteins may cross the intestinal barrier[5] and that this phenomenon may be increased in pathological conditions. For example, in *Clostridium difficile* infections, bacterial toxin A is responsible for a massive loss of mucosal integrity due to the disruption of the tight junctional complexes.[6] Therefore, the particular localization of intestinal intraepithelial lymphocytes (IELs) suggests that their functions could be influenced by exogenous components, particularly when abnormalities of intestinal transport occur in various diseases.

In the present report, we hypothesized that uncoupling sialic acids from the membrane might modulate their *in vitro* proliferation. In addition, dipeptidyl peptidase IV (DPIV or CD26) is associated with T-cell proliferation.[7,8,9] We characterized DPIV activity on purified IELs and investigated a possible relationship between DPIV activity and proliferation of IELs. Here we report that, in contrast to spleen cells, Nase treatment of rat IELs was associated with a significant decrease of their proliferation in response to concanavalin A stimuli. Such an effect does not appear to be associated with the variation of membrane DPIV activity during the culture.

MATERIALS AND METHODS

Purification of IELs

Small longitudinally opened pieces (1 to 2 cm) of Wistar-Furth rat upper small intestine were incubated on a tridimensional stirrer 3 times, 20 min at 37° C in RPMI 1640, 2mM L-glutamine, 100 U/ml penicillin, 100 µg/ml streptomycin 5 x 10^{-5} M 2-mercaptoethanol and 2 % heat-inactivated fetal calf serum (FCS). Microscopic examination showed a complete removal of the epithelium without affecting the structure of villi after the third incubation period. Supernatants were pooled and filtered over sterile gauze to remove debris. IELs were then purified on a discontinuous Percoll gradient and they were collected at 1.050-1.070 and 1.070-1.120 $g.ml^{-1}$ density interfaces.

Lymphocyte Culture System

Cells were stimulated with Con A ($1 \mu g.ml^{-1}$) and cultured for 3 days (10% FCS) in the presence of 20% syngeneic adherent spleen cells. The same culture conditions were used for spleen cells. The cultures were pulsed for the last 24 hours with 2.77×10^4 becquerels of $[6-^3H]$thymidine per well. Cells were harvested on glass filter papers and the incorporation of ^3H into DNA was measured with the automatic filter counting system INB-384 (Inotech, Switzerland). The INB-384 detected ^3H radioactivity on dried filters and quantitation is as accurate as with conventional liquid scintillation counting. The comparison with liquid scintillation showed a linear correlation (ki2 > 0.99) and the following relation : cpm scintillator = $3.50^{(a)}$ (cpm INB-384 detector) ($^{(a)}$ Standard Error = 0.03). Therefore, in our experiments the means of Con A-induced proliferation of 5×10^4 cells per round-bottomed well were 5000 cpm for spleen cells and 800 cpm for IELs. The probabilities (p) were calculated by t-test of Student-Fisher.

Neuraminidase Treatment

IEL and spleen cells (5×10^5 to 10^6 in 500 μl of culture medium without FCS) were treated for 30 min at 37° C with 1 $UI.ml^{-1}$ (final concentration) of neuraminidase from *Clostridium perfringens* (Sigma). Cells were then washed 3 times. The viability, determined by trypan-blue exclusion, was superior to 96 % before and after Nase treatment.

Enzyme Activities

The DPIV activity of purified IEL was compared to the enzymatic activities preferentially expressed by enterocytes ; the sucrase-isomaltase (SI) and the alkaline phosphatase (AP). Enzyme substrates (AP; 5mM p-nitrophenyl phosphate and DPIV; 1mM Gly-Pro-p-nitroanilide, final concentrations) were added to cells suspended in quadriplicate in microtiter plates, in a final volume of 200 μl. The absorbance (A) was recorded (AP;15 min and DPIV;120 min) at 405 nm. The units (U) were expressed as $A_{405\ nm}$/min for AP and as nanomoles of p-nitroaniline formed/10 min for DPIV. The activity of sucrase isomaltase was performed according to Dahlqvist[10] and units were expressed as $\mu g.ml^{-1}$ of glucose formed/60 min. The accessory cells and the cell-free control mixtures were tested in parallel for all enzymes assays. Residual activities were expressed according to the following relation : (activity after purification/activity before purification) x 100.

RESULTS AND DISCUSSION

In order to assess the purity of IEL preparations, we measured DPIV, SI and AP activities before and after the Percoll gradient as it is well known that these enzymes are expressed by epithelial cells. The results (Figure 1) showed that the AP and SI activities decreased drastically and were respectively 8.4 ± 1.5 % and 2.5 ± 3.6 % of the unpurified IEL preparation. Surprisingly, the residual DPIV activity was significantly higher (47.2 ± 16.4 %) than that observed for AP (p<0.01, n=4) and SI (p<0.05, n=3) activities. DPIV is a serine exopeptidase, and it has been shown to be identical to CD26 in the immune system.[7] This surface marker was shown to be present on the majority of CD4+ cells and on a portion of CD8+ cells[8] it is able to transduce an activating signal in lymphocytes and seems to be associated with the ability of T cells to produce IL-2.[9] CD26 expression by IELs had never been reported and our observations suggest that rat IELs express this cell surface antigen. Moreover, the measurement of SI and AP activities after IEL purification appears to be a useful method to evaluate epithelial cell contamination. In our experiments, the microscopic evaluation of enterocyte contamination ranged between 0 and 10%.

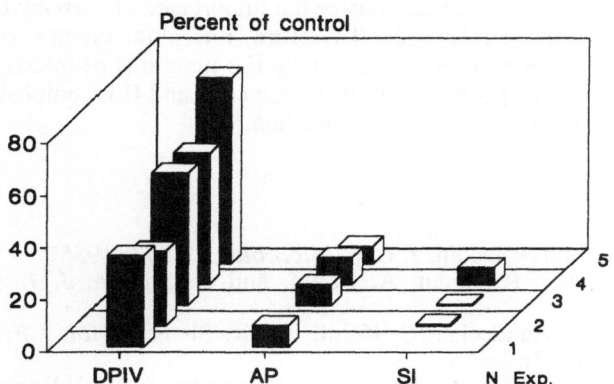

Figure 1. Residual enzyme activities after IEL purification on a Percoll gradient. The control values of enzyme activities before purification were ; DPIV : 0.32 ± 0.07 U/10⁵ cells, AP : 0.12 ± 0.04 U/10⁵ cells and SI : 49 ± 19 U/10⁵ cells (SE for 3 to 5 experiments with intact cells).

The aim of this study was to investigate the effect of Nase treatment on IEL proliferation because it has been demonstrated that removing sialic acids from lymphocytes may modify immune responses. The effect of Nase depends on the concentration used and higher concentrations may induce cell death.[11] In our study, we used conditions that failed to decrease the viability of cells as measured by dye exclusion. Thus, we confirmed that Con A-induced proliferation of spleen cells was significantly increased (256 ± 76%, n=6, p<0.01) after Nase treatment. In contrast, IEL proliferation was inhibited. Inhibition was 47 ± 10% (p<0.01, n=6) of the untreated IELs. When compared to T cells from other compartments, IELs exhibit poor proliferative capacities and the mechanisms involved are unknown. Removing sialic acids might enhance the production of suppressive factors which inhibit proliferation of IEL subsets. Another possibility for the observed inhibitory effect might be the induction of apoptosis because such a phenomenon does not immediately affect cell viability.[12]

Having established the presence of high DPIV activities in purified IEL preparations, it was of interest to explore the relationship between the Nase-induced inhibition of proliferation and the DPIV activity of IELs. As shown in Table 1, DPIV activity, measured on entire cells, remained unchanged during the culture time. Although DPIV is inducible upon activation of T cells, the regulatory pathway involved for different T cell subpopulations is still unknown. Therefore, using specific enzyme inhibitors might provide more insight into the role of this enzyme in IEL proliferation.

Table 1. Effect of Nase treatment on dipeptidyl peptidase IV activity on IELs in culture.[1]

Treatment		Time culture (hours)		
Nase	Con A	0	24	72
-	-	0.324 ± 0.054 [2]	0.357 ± 0.089	0.326 ± 0.146
+	-	0.376 ± 0.156	0.357 ± 0.068	0.409 ± 0.222
-	+	0.310 ± 0.155	0.353 ± 0.095	0.287 ± 0.089
+	+	0.433 ± 0.063	0.344 ± 0.063	0.404 ± 0.163

[1] Values from 3 experiments. [2] DPIV activities are expressed as $U/10^5$ intact whole cells.

In conclusion, our results emphasize the importance of carbohydrate modulation in the regulation of immune responses. Therefore, microbial factors, such as exogenous neuraminidase, might be expected to modulate the functions of intestinal IEL in disease states. The contrasting responses between spleen cells and IELs emphasize the functional peculiarity of this intestinal lymphocyte population.

REFERENCES

1. M. R. Popoff and A. Dodin, *J. Clin. Microbiol.* 22:873 (1985).
2. A. Novogrodsky, R. Lotan, A. Ravid, and N. Sharon, *J. Immunol.* 115:1243 (1975).
3. S. Jibril, B. V. Gandecker, S. Kelm, and R. Shauer, *Biol. Chem. Hoppe-Seyler* 368:819 (1987).
4. M. Nakamura, T. Yoshida, K. Isobe, T. Iwamoto, S. M. J. Rahman, Y. Zhang, T. Hasegawa, M. Ichihara, and I. Nakashima, *Immunol. Lett.* 29:235 (1991).
5. L. S Smith, D. A. Wall, C. H. Gochoco, and G. Wilson, *Adv. Drug Deliv. Rev.* 8:253 (1992).
6. M. Heyman, G. Corthier, F. Lucas, J. C. Meslin, and J. F. Desjeux, *Gut* 30:108 (1989).
7. E. Schön, and S. Ansorge, *Biol. Chem. Hoppe-Seyler* 371:699 (1990).
8. Z. Lojda, *Czechoslov. Med.* 11:181 (1988).
9. E. Schön, H. Demuth, E. Eichmann, H. Horst, I. Körner, J. Kopp, T. Mattern, K. Neubert, F. Noll, A. Ulmer, A. Barth, and S. Ansorge, *Scand. J. Immunol.* 29:127 (1989).
10. Dahlqvist, *Anal. Biochem.* 22:99 (1968).
11. C. Ogier, A. M. Sjögren, and P. Reizenstein, *Biomedicine* 31:250 (1979).
12. W. Swat, L. Ignatowicz, and P. Kisielow, *J. Immunol. Methods*, 137:79 (1991).

MINIMAL SIGNAL REQUIREMENTS FOR SUPERANTIGEN-SPECIFIC T CELL-DRIVEN B CELL PROLIFERATION

Gayle D. Wetzel

Miles Inc.
P.O. Box 1986
Berkeley, CA 94701 USA

INTRODUCTION

A foreign minor lymphocyte stimulating (Mls) antigen recognized in the context of MHC class II products on B cells and macrophages will induce T-cell responses.[1,2] We now appreciate that Mls products function as superantigens[3] and are encoded by mouse tumor viruses.[4] T cell dependent B cell activation involves several signaling steps, which despite some clarification, remain to be elucidated.[5] Some adhesins appear to be involved as well as class II molecules, surface Ig and lymphokines.[5] We employ anti-IgM and Mls[a]- specific cloned lymphokine non-secreting T-cell lines[6-9] to examine the requirements for this type of activation to B-cell proliferation.

RESULTS AND DISCUSSION

Using cultures containing between 1 and 5 B cells per well, limiting dilution analysis allowed the determintation of proliferating B-cell frequencies, expressed as percentages of input cells. A well established paradigm of B-cell proliferation induction uses anti-IgM and IL-4. We show, in Fig. 1, that under stringent conditions, very few resting B cells (less than 1%) actually proliferate with these stimuli alone. We used the H3 cell line kindly provided by Dr. Polly Matzinger,which is H-Y specific and cross-reacts on Mls[a] but secretes no interleukin which we could detect, in an attempt to model a T cell membrane-derived activation signal. The three component system of anti-IgM, IL-4, and lethally irradiated membrane antigen-specific T cells provided sufficient activation signaling to B cells. In fact, Fig. 1 shows that this signaling combination induced as much B-cell proliferation as the combination of powerful B cell mitogens LPS+DXS. Interestingly, T cells and sIg cross-linking were insufficient for optimal B-cell proliferation.

Fig. 2 shows similar data from another experiment, also using a second Mls[a]-specific interleukin non-secreting cloned T-cell line kindly provided by Dr. Harald Prahl, in which the effects of IL 1-5 were shown to be inadequate. To date, only IL-4 has been found to complement anti-IgM and T cells to activate profound B-cell proliferation under these stringent conditions (data not shown). We should caution that we have not yet tested IL 10-12, however.

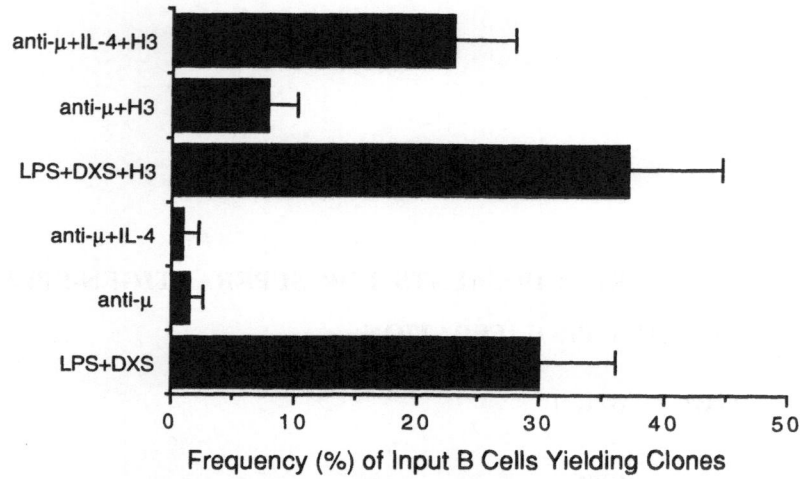

Figure 1. Limiting dilution analysis of B-cell growth with T Cells, anti-IgM and IL-4.

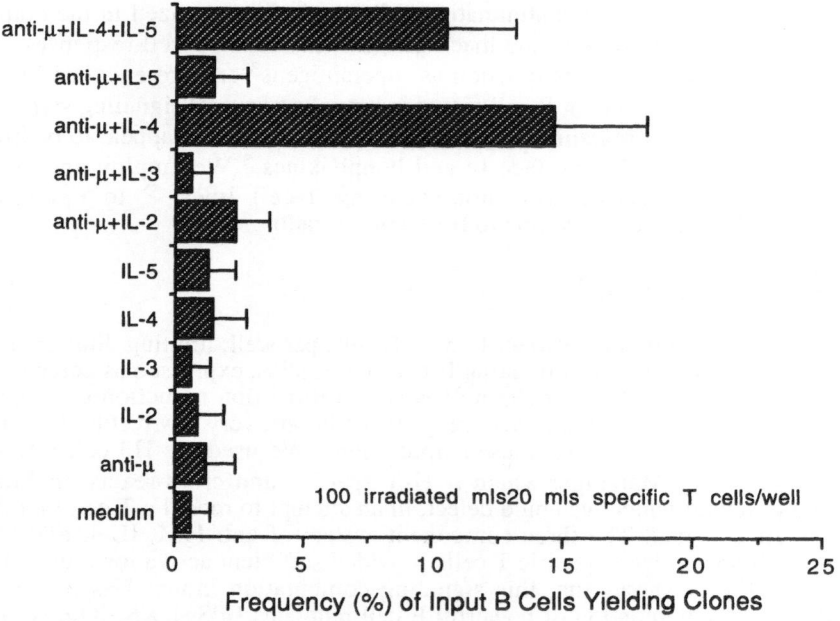

Figure 2. Comparison of the ability of different interleukins to support T cell and anti-IgM Induced B-cell proliferation.

Fig. 3 shows the analyses of clone sizes of B cells stimulated to proliferate by three different activation conditions. With anti-IgM+IL-4 and no T cells, very litte activation was observed, as mentioned above, and clone sizes tended to be small with a mode of 5-8 cells. Addition of T cell-derived membrane activated signals in the form of irradiated

Mls^{a}specific cells shifted the clone size distribution so that many more of the larger clones were observed. This was true although the mode remained equivalent to that seen in the absence of T cells. In fact, the three signal system resulted in more clones of larger size than observed with the combination of LPS+DXS, previously presumed to provide optimal B-cell activation.

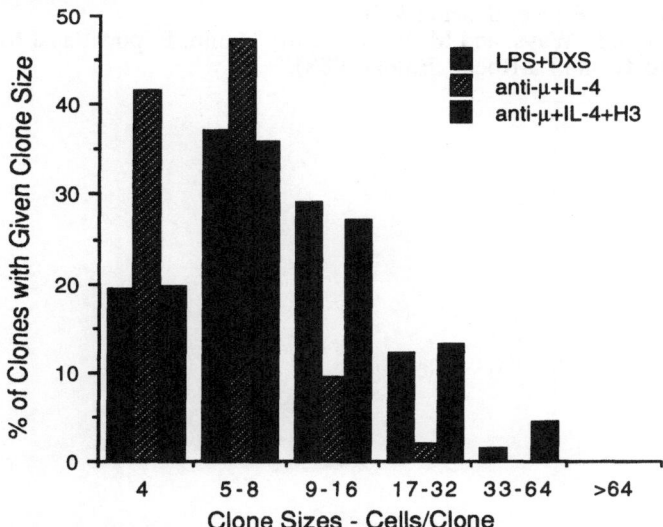

Figure 3. Comparison of clone sizes of B-cells stimulated in the presence and absence of T cells.

These data suggest a minimal model of three signals for optimal B cell proliferation. Obviously, the T-cell clones may deliver more than a single signal, but as a minimal hypothesis, we can identify sIg cross-linking (by anti-IgM and presumably by antigen), amplification by IL-4 (known to enhance some gene expression e.g., MHC class II expression) and an ill-defined third signal possibly delivered by the T-cell surface. We have not been able to replace T cells with anti-class II, anti-LFA1 nor anti-Lyb2 (CD72) antibodies. Antibodies to other B-cell antigens may, however, replace T-cell function.

CONCLUSIONS

The mechanism of T-cell help to B cells is a longstanding focus of immunological investigation. Evidence is accumulating which suggests that secreted T cell-derived lymphokines are insufficient to co-stimulate antigen-induced B-lymphocyte proliferation. We present a minimal system in which the proliferation of limiting numbers of B cells, one in the extreme case, can be examined. Anti-IgM is used to provide surface IgM cross-linking signals. The B cells used in this system spontaneously present endogenous murine mammary tumor virus 7 superantigen to specific T cells. The cloned T-cell lines used were chosen since they secrete no identified interleukins and may, therefore, provide only membrane delivered signals. Although absolutely required by the B cells, these membrane-derived signals, together with anti-IgM remain insufficient to stimulate B cell proliferation. Only when the combination of anti-IgM, irradiated T cells and IL-4 is perceived by the B cell does proliferation ensue. Although the nature of the membrane-derived T-cell signals are undefined and remain the subject of inquiry, this system defines a 3 signal minimal model for T cell-driven B-cell proliferation.

REFERENCES

1. H. Festenstein, *Transplant. Rev.* 15:62 (1973).
2. A. B. Peck, C. A. Janeway, and H. Wigzell, *Nature* 266:840 (1977).
3. J. W. Kappler, U. Staerz, J. White, and P. C. Marrack, *Nature*, 332:35 (1988).
4. Y. Choi, P. Marrack, and J. W. Kappler, *J. Exp. Med.* 175:847-52 (1992).
5. R. J. Noelle and E. C. Snow. *FASEB J.* 5:2770 (1991).

6.   G. D. Wetzel and H. Prahl, *in*: "Annu. Report Basel Inst. Immunol." p.80, Hoffmann La-Roche, Basel (1989).
7.   G. D. Wetzel, P. Matzinger, and H. Prahl, *in:* "Annu. Report Basel Inst. Immunol." p. 62, Hoffmann La-Roche, Basel (1988)
8.   P. Matzinger and J. P. Ways, *in:* "Annu. Report Basel Inst. Immunol." p.53, Hoffmann La-Roche, Basel (1987).
9.   P. Matzinger, J. P. Ways, and M. V. Wiles, *in:* "Annu. Report Basel Inst. Immunol." p. 56, Hoffmann La-Roche, Basel (1988).

EXPRESSION OF CELL ADHESION MOLECULES IN THE FETAL GUT

Ahmet Dogan, Thomas T. MacDonald, and Jo Spencer

Department of Histopathology
UCL Medical School; Rockefeller Building
University Street, London, WC1E 6JJ
United Kingdom

INTRODUCTION

The cell adhesion molecules ICAM-1 and VCAM-1 when expressed on endothelium are thought to be important in modulating lymphocyte movement in areas of inflammation through their ligands LFA-1 and VLA-4 on lymphocytes. They are also expressed by accessory cells and provide costimulatory signals for T cell activation.[1] ELAM-1 is a cell adhesion molecule expressed by the endothelium at sites of acute inflammation and is significant in neutrophil migration to the site of injury.[2,3] Cytokines like IL1, IFN-γ and TNFα upregulate or induce expression of these molecules.[1,2] In intestinal inflammation such as Crohn's disease or coeliac disease expression is increased on endothelium and accessory cells within the lymphoid tissues.[4,5] In this study we examined the expression of ICAM-1, VCAM and ELAM-1 in human fetal intestine from 11 to 22 weeks old fetuses. We also studied used an organ culture model of intestinal inflammation[6] to study the time course of adhesion molecule expression in an invitro fetal gut organ culture model.

METHODS

Human small intestine specimens from 11-22 week old fetuses were either frozen in liquid nitrogen (day 0) or processed for organ culture. At the beginning of cultures, to stimulate the T cells, pokeweed mitogen (PWM) (7 μg/ml) was added to half of the explants, while the other half were kept as controls. Samples of stimulated and control fetal gut were then frozen at 4, 8, 12, 18, 24, 36, 60, 96 hours after beginning of the culture. Frozen section (8 μm) were cut from frozen specimens, fixed in acetone and stained with monoclonal antibodies to CD3, ELAM-1, ICAM-1 and VCAM-1 using an immunoalkaline phosphatase (APAAP) and an immunoperoxidase method. In time course experiments, intensity of staining was scored from 1 to 5 using a semiquantitative system.

RESULTS

Ontogeny of Adhesion Molecule Expression

Primitive Peyer's patches, including those present at 11 weeks expressed ICAM-1 intensely (Figure 1). In other parts of lamina propria ICAM-1 was constitutively expressed on venular endothelium only. VCAM-1 like ICAM-1 was strongly expressed in the primitive Peyer's patches (Figure 1). In other parts of the lamina propria VCAM-1 diffusely and weakly expressed in the upper lamina propria stroma. No constitutive expression of ELAM-1 was observed in fetal intestine of any age.

Advances in Mucosal Immunology, Edited by
J. Mestecky *et al.*, Plenum Press, New York, 1995

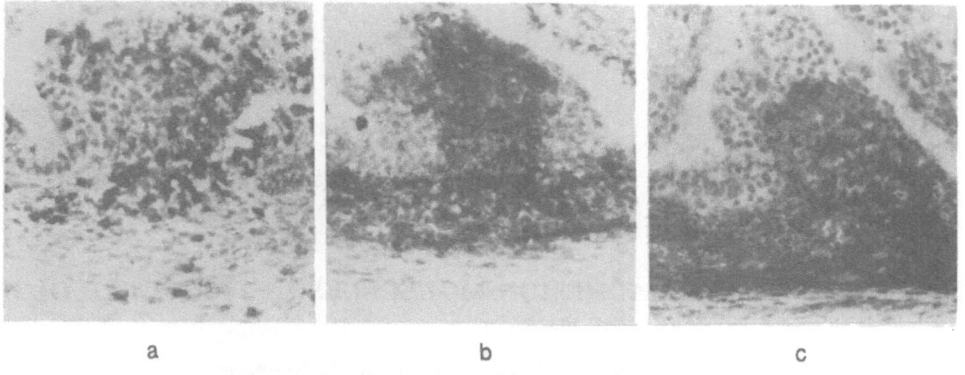

Figure 1. Serial frozen sections of a primitive Peyer's patch from a 16 weeks old fetus, stained with antibodies against CD3 (a), ICAM-1 (b) and VCAM-1 (c). Immunoperoxidase x 100.

Induction of Adhesion Molecule Expression

In the fetal gut organ culture, expression of all three cell adhesion molecules that was studied, were induced or upregulated in the lamina propria. Polyclonal T cell activation strongly enhanced this effect. ELAM - 1 expression on the endothelium reached its peak by 6 - 8 hours and disappeared by 24 hours. ICAM-1 and VCAM-1 were upregulated in the stroma and the endothelium reaching a plateau by 60 hours (Fig. 2).

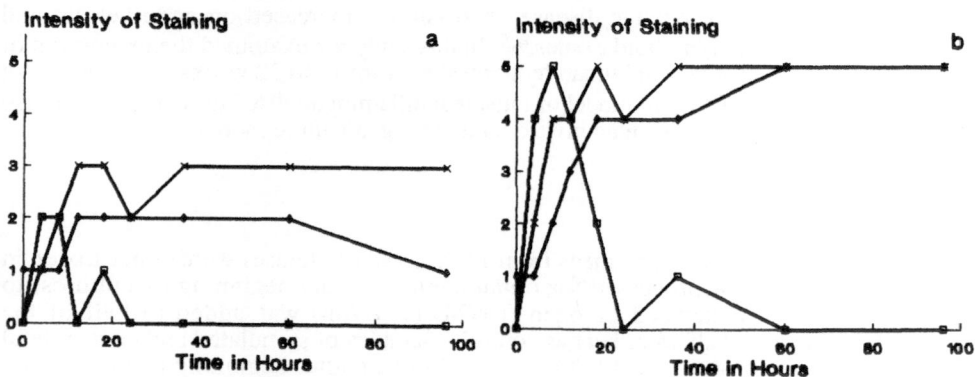

Figure 2. Time course of induction of ICAM-1 (◊), VCAM-1 (X) and ELAM-1 (□) in control (a) and PWM stimulated (b) fetal gut organ cultures.

DISCUSSION

In the fetal intestinal lamina propria, ICAM-1 was expressed on the endothelium and VCAM-1 reactivity was weakly observed in the stroma. In comparison, primitive Peyer's patches strongly expressed these molecules. No constitutive ELAM-1 positivity was present. Considering the role of ICAM-1 and VCAM-1 in lymphocyte migration and stimulation,[1] high levels of expression in the Peyer's patches may indicate an important part for these molecules in arrangement and development of lymphoid tissue within the mucosal microenvironment. In the fetal gut organ culture system we observed strong induction of all adhesion molecules, local T cells were stimulated with mitogens. This was probably due to release of inflammatory mediators. In addition, there was induction in the control cultures without T cell stimulation. This has been reported in other organ culture models[7] and could be caused by relative hypoxia or physical trauma. Time course of induction of ICAM-1, VCAM-1 and ELAM-1 was similar to other in vitro and in vivo studies where cytokines were used as the stimulant.[7,8] Our findings show that when chronic polyclonal T cell activation is used as the initiating event instead of cytokines the dynamics of cell adhesion molecule expression does not change. In conclusion this study suggests an important role for ICAM-1 and VCAM- 1 in the ontogeny of mucosal immune system in the human intestine and implies that T cell activation in the small intestinal microenvironment induces expression of cell adhesion molecules.

REFERENCES

1. T. A. Springer, *Nature* 346:425 (1990).
2. J. S. Pober, M. P. Bevilacqua, D. L. Mendrick, L. A. Lapierre, W. Fiers, and M. A. Gimbrone, *J. Immunol.* 136:1680 (1986).
3. F. W. Luscinscas, M. I. Cybulsky, J. M. Kiely, C. S. Peckins, V. M. Davis, and M. A. Gimbrone, *J. Immunol.* 142:2257 (1989).
4. R. P. Sturgess, J. C. Macartney, M. W. Makgoba, C. H. Hung, D. O. Haskard, and P. J. Ciclitira, *Clin. Exp. Immunol.* 82:489 (1990).
5. G. Malizia, A. Calabrese, M. Cottone, M. Raimondo, L. K. Trejdosiewicz, C. J. Smart, L. Oliva, and L. Pagliaro, *Gastroenterology* 100:150 (1991).
6. T. T. MacDonald, and J. Spencer, *J. Exp. Med.* 167:1341 (1988).
7. D. V. Messadi, J.S. Pober, W. Fiers, M. A. Gimbrone, and G. F. Murphy, *J. Immunology* 139:1557 (1987).
8. J. M. Munro, J. S. Pober, and R. S. Cotran, *Am. J. Pathol.* 135:121 (1989).

A ROLE FOR ANTIGEN IN THE EVOLUTION OF GASTROINTESTINAL MALT-TYPE B CELL LYMPHOMA

Tracy Hussell, P. G. Isaacson and Jo Spencer

Department of Histopathology, UCMSM
University Street, London
WC1E 6JJ
United Kingdom

INTRODUCTION

Low grade B cell lymphomas of mucosa associated lymphoid tissue (MALT) are a discrete group of malignancies occurring most commonly at mucosal sites which are normally devoid of lymphoid tissue.[1] These sites have been shown to acquire reactive lymphoid tissue resembling normal MALT seen in the Peyer's patch of the terminal ileum[1,2] before or at the onset of lymphoma development. Several characteristic features such as the persistence of follicular dendritic cells, the presence of reactive follicles and a range in cellular morphology within these `MALT`-type lymphomas suggest that they may arise in the course of an immune response to antigen.[1,3]

We have investigated the specificity of the tumour immunoglobulin (TIg) in 3 low grade and 1 high grade B cell lymphoma of MALT`-type. We have also studied the effect of cross-linking TIg with anti-idiotypic antibodies (‡-Id) in combination with mitogens *in vitro*.

METHODS

Tissues

The following fresh tissues were received fresh in the laboratory: Three low grade primary B cell gastric lymphomas of MALT type (cases 1-3), 1 high grade case (case 4) and tonsil. Suspensions of the cells and blocks of all tissues were frozen down for subsequent use.

Protocol for Detecting the Reactivity of TIg by Immunohistochemistry

A TIg source was first obtained by fusing tumour cells to the mouse myeloma cell line NSO to give a TIg secreting heterohybrid cell line. We then made a murine anti-idiotypic antibody that specifically recognized the TIg using conventional mouse fusion protocols.[4] Sections from unrelated individuals were then incubated with the TIg source and binding detected using the appropriate anti-id followed by rabbit anti-mouse peroxidase conjugate and the diaminobenzidine reagent. Sections were extensively washed between each addition.

Advances in Mucosal Immunology, Edited by
J. Mestecky *et al.*, Plenum Press, New York, 1995

The Proliferation of MALT-Type Lymphoma B Cells *In vitro*

The proliferative activity of tumour cells in low and high grade cases, in medium alone, ‡-Id alone and with optimal doses of B cell activators Staphylococcus aureus Cowan 1 (SAC), poke weed mitogen (PWM) and phorbol 12-myristate 13-acetate (TPA) was examined. Proliferative activity was quantitated by measuring incorporation of 125IUDR (cpm/1000) Ò standard errors.[5]

RESULTS

Immunohistochemical Reactivity of Tumour Immunoglobulin

By using ‡-Id antibodies immunohistochemically to detect binding of TIg in unrelated tissue sections we have shown that the TIg from 3 low grade `MALT` type lymphomas (cases 1-3) recognises normal tissue components. The antigens in two cases (cases 1-2) had a very restricted tissue distribution and the clarity of the reactivity implies that the TIg in these cases had high affinity for a single antigen (Figs. 1A-B). Broader reactivity was observed in the remaining low grade case (case 3) (Fig. 1C). We were unable to investigate the specificity of the TIg in the high grade case due to the lack of a source of TIg (case 4).

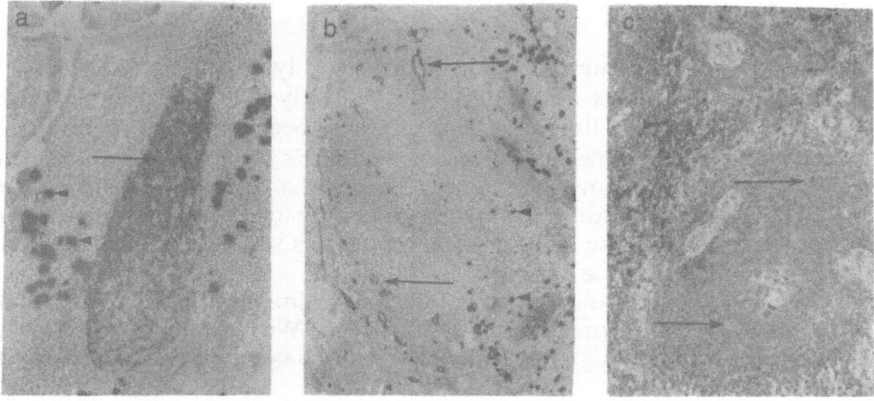

Figure 1. a) section of an appendix showing how the TIg from case 2 recognises an antigen present on follicular dendritic cells (arrow). b) Section of an appendix showing that the TIg from case 1 recognises an antigen present in the smooth muscle surrounding mucosal venules (arrow). c) Section of spleen showing that the TIg from case 3 reacts with a wide variety of cell types (arrow). Non-specific endogenous peroxidase is shown with arrowheads.

The Effect of Cross-Linking Surface Ig With ‡-Id and Mitogens

Of the four cases studied, one low grade case (Fig. 2C) showed enhanced proliferation in response to ‡-Id alone. This could not be increased by the addition of mitogens. In the remaining two low grade cases mitogen responsiveness was observed (Figs. 2A &B) which was affected by ‡-Id either by an enhancement or a reduction in the proliferative response. The high grade case failed to respond to the stimuli studied (Fig. 2D).

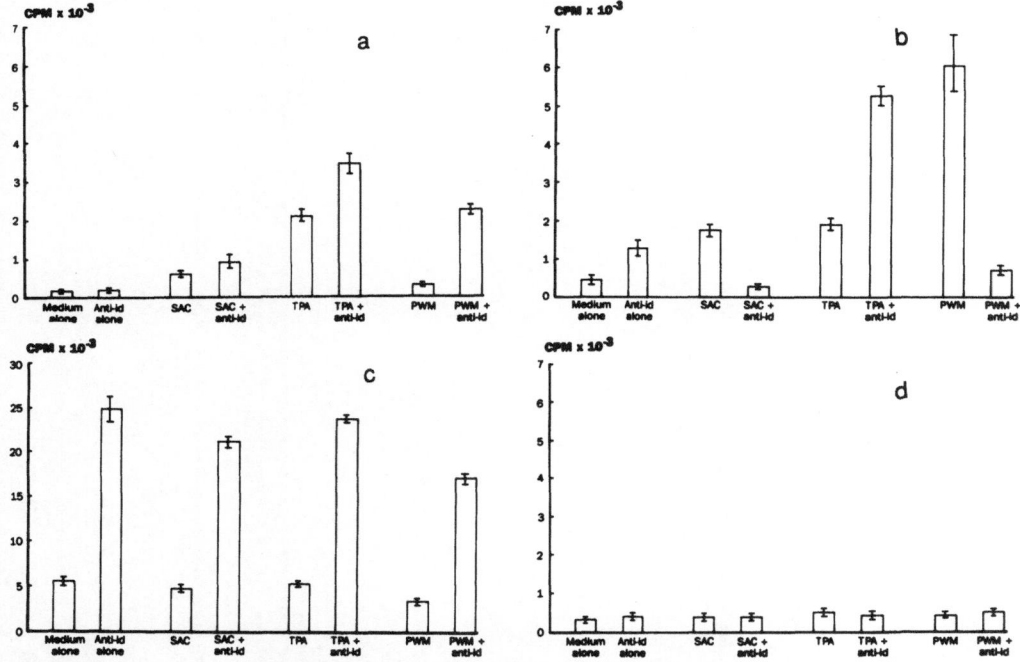

Figure 2. a) Case 1: ‡-Id alone had no effect on the tumour cells. There was a variable response to PWM and stimulation in the presence SAC and TPA. Co-stimulation with ‡-Id enhanced the response of all mitogens. b) Case 2: Note the low response to ‡-Id and proliferation in the presence of SAC, TPA & PWM. In this case ‡-Id synergised with TPA but reduced the response of SAC & PWM. c) Case 3: There was spontaneous proliferation of tumour cells from this case which was substantially increased in the presence of ‡-Id. No additional response to mitogen was observed. d) Case 4: There was no response to ‡-Id alone or to polyclonal B cell activators in the presence or absence of ‡-Id. Further high grade cases will need to be investigated before conclusions can be drawn as to the significance of this result.

DISCUSSION

We have observed that the TIg from 3 cases of low grade ʼMALTʼ-type B cell gastric lymphoma which are characteristically CD5- is specific for normal tissue components. The association between lymphoma of MALT type and preceding autoimmunity has been described before in the salivary gland and thyroid in association with Sjîgrens syndrome and Hashimotoʼs thyroiditis, respectively.[6,7]

We have shown that cross-linking TIg with ‡-Id antibodies in combination with mitogens in vitro caused the proliferation of 3 low grade, but not an individual high grade lymphoma. The tumour cells from case 3 spontaneously proliferated and appeared to have overcome the dependency on signals required for growth and differentiation shown by other lymphomas.

This study suggests that autoimmunity may play a role in the pathogenesis of gastric lymphoma. Furthermore, binding of autoantigens may affect lymphoma cell proliferation and the evolution of the tumours.

REFERENCES

1. P. G. Isaacson and J. Spencer, *Histopathology,* 11:445 (1987).
2. J. Spencer, T. Finn and P. G. Isaacson, Gut, 27:405 (1986).
3. J. K. C. Chan, C. S. Ng, and P. G. Isaacson, *Am. J. Pathol.,* 136:1153 (1990).
4. T. Hussell and P. G. Isaacson, J. Spencer, *Am. J. Pathol.* (in press).
5. T. Hussel and P. G. Isaacson, J. Spencer, *J. Pathol.* (in press).
6. E. Hyjeck and W. J. Smith, P. G. Isaacson, *Human Pathol,* 19:766 (1988).
7. E. Hyjeck and P. G. Isaacson, *Human Pathol.,* 19:1315 (1988).

Figure 3. ...

DISCUSSION

...

REFERENCES

...

IN VIVO PROLIFERATION OF T AND B LYMPHOCYTES IN THE EPITHELIUM AND LAMINA PROPRIA OF THE SMALL INTESTINE

Hermann J. Rothkötter, Timm Kirchhoff and Reinhard Pabst

Centre of Anatomy, Medical School of Hannover
PO Box 610180, D-3000 Hannover 61, Germany

INTRODUCTION

Intraepithelial lymphocytes (IEL) and lymphoid cells in the lamina propria (LPL) of the gastrointestinal (GI) tract are effector or regulator cells for an effective mucosal immune response. Little is known about the proliferation of lymphocytes in these compartments.[1] The thymidine analog bromodesoxyuridine (BrdU) is incorporated during the DNA synthesis of the cell cycle. BrdU can easily be detected using an immunocytological double stain for incorporated BrdU and lymphocyte subsets. This method was used to determine the proliferation of lymphocyte subsets within IEL or LPL suspensions from the jejunum and ileum. During the purification of the isolated IEL and LPL, high numbers of cells are often lost. Therefore in the present study the suspensions were analyzed without further purification steps.

METHODS

The experiments were carried out in six Goettingen minipigs (10 months old). The animals received a single i.v. injection of 20 mg BrdU/kg body weight (BrdU, Sigma, Deisenhofen, Germany), 24 h later parts of the jejunum and ileum were sampled. Cells from the epithelium and lamina propria were isolated according to the method described by Zeitz et al.[2] The stripped mucosa was incubated twice for 1 h in Ca^{++} and Mg^{++} free Hanks balanced salt solution containing 1.26 mmol EDTA and 0.94 mmol Dithiotreitol to isolate IEL. The remaining tissue was incubated two consective times in RPMI 1640 containing 0.15 mg/ml collagenase (Boehringer, Mannheim, Germany) and 0.1 mg/ml DNAse, 4 h and 12 h, respectively, to harvest the LPL. After washing the cell suspensions, lymphocyte counts were performed in a hemocytometer using a phase contrast microscope at a 500 magnification. The lymphocyte yield per gram of tissue was calculated. In differential cell counts, to determine the cell composition, were performed on Giemsa stained cytospot preparations. Without further purification the lymphocyte subpopulations in the cell suspensions were determined by flow cytometry (FACScan, Becton Dickinson, Heidelberg, Germany) using monoclonal antibodies directed against pig lymphocyte subsets. The lymphoid cells were gated based on their forward and side scatter properties.

On cytospot preparations of IEL or LPL an immunocytological double staining method was used to detect incorporated BrdU in lymphocyte subsets. A modification of the method described by Westermann et al.,[3] was used. The subsets were determined using the alkaline phosphatase anti-alkaline phosphatase (APAAP) technique with fast blue dye. Following denaturation of DNA with formamide/sodium hydroxide, the incorporated BrdU was detected using a monoclonal antibody coupled to peroxidase and directed against BrdU (Anti-BrdU, Becton Dickinson, Heidelberg, Germany). On the cytospots the subset positive cells were counted and each checked for BrdU staining.

Advances in Mucosal Immunology, Edited by
J. Mestecky *et al.*, Plenum Press, New York, 1995

RESULTS

Composition of Cell Suspensions and Lymphocyte Yield

The differential counts of Giemsa stained cytospots showed high numbers of granular lymphocytes (17.0 ± 13.6 %) in the IEL suspensions. The plasma cells were mainly detected in the second LPL fraction i.e., after 12 h of incubation (34.3 ± 12.3 %). The LPL suspension of the ileum contained many eosinophils (24.2 ± 12.0 %). In all LP-suspensions a marked number of mast cells was observed (1.1 ± 0.5 x 10^6/g tissue).

The lymphocyte yield/g was higher in the jejunum (33.4 ± 15.4 x 10^6/g tissue) than in the ileum (23.6 ± 6.4 x 10^6). In the forward/side scatter dot plot the lymphocytes could be gated as a well defined population. Most of the cells in these cell suspensions were T cells, each containing a major proportion of $CD8^+$ cells. Ig^+ cells were mainly detected in the second LPL fraction (jejunum: 18.6 ± 15.2 %, ileum: 10.4 ± 6.9 %).

Proliferation

As the very small number of Ig^+ cells among the IEL is probably due to a contamination by LP cells, only the proliferation of total T cells and the $CD8^+$ T cell subset in this fraction were analyzed. In the jejunal IEL suspensions, 3.6 ± 1.4 % $BrdU^+$ T cells were observed. The BrdU-index was two times higher in ileal intraepithelial T cells (7.2 ± 3.9 %). T cell proliferation in the LP of jejunum and ileum was lower than in the IEL T cell fraction (T cells, jejunum: 1.1 ± 0.5 %, ileum: 2.1 ± 1.0 %). Only small numbers of Ig^+ cells were $BrdU^+$ (IgA^+: 1.2 ± 0.6 %, IgM^+: 0.8 ± 0.5 %).

DISCUSSION

It was possible to analyze the isolated IEL and LPL cell suspensions without purification by gradient centrifugation. Using the forward/side scatter properties of the flow cytometer the lymphocyte population could be gated. This is advantageous for studies of the cell phenotype in mucosal suspensions. In this manner cells can be analyzed without cell loss occurring during purification. Using the phase contrast microscopy it is possible to enumerate the lymphocyte numbers in these cell suspensions. This enables one to calculate the numbers of cells per isolated g of mucosa. This is important for analysis of suspensions from normal or inflamed gut tissue. Giemsa stained cytospots are an easy method to analyze the proportion of eosinophils and mast cells within mucosal cell suspensions.

The immunocytological double staining for subsets and incorporated BrdU showed interesting results. So far, no explanation for the higher number of $BrdU^+$ cells in the ileum is available. The $BrdU^+$ cells in the gut epithelium was comparable to those observed using ^{3}H-thymidine.[1] In the skin of pigs a comparable number of $BrdU^+$ cells was observed.[4] So far, little is known about cell proliferation within the lamina propria. But evidence exists that lymphoid cells proliferate in this compartment occurs.[5] So far it is unknown whether the $BrdU^+$ cells proliferate locally or migrate to the epithelium or LP after entering the cell cycle in other lymphoid or non-lymphoid tissues.

REFERENCES

1. C. Röpke and N. B. Everett, *Anat. Rec.,* 185:101 (1976).
2. M. Zeitz, W. C. Greene, N. J. Peffer, and S. P. James, *Gastroenterology* 94:647 (1988).
3. J. Westermann, S. Ronneberg, F. J. Fritz, and R. Pabst, *Eur. J. Immunol.,* 19:1087 (1989).
4. F. J. Fritz, R. Pabst, and R. M. Binns, *Immunology* 71: 508 (1990).
5. H. J. Rothkötter, H. Ulbrich, and R. Pabst, *Pediatr. Res.,* 29:237 (1991).

CYTOTOXIC ACTIVITY OF T CELLS EXPRESSING DIFFERENT T-CELL RECEPTOR VARIABLE GENE PRODUCTS IN THE INTESTINAL MUCOSA

Maria E. Baca-Estrada and Kenneth Croitoru

Intestinal Disease Research Program, Department of Medicine
McMaster University, Hamilton, Ontario, L8N 3Z5 Canada

INTRODUCTION

T cells recognize foreign antigen via the heterodimeric T-cell receptor (TCR).[1,2] The peripheral TCR repertoire is shaped via positive and negative selection processes that influence T-cell maturation in the thymus.[3,4] There is increasing evidence that there are modifications of this repertoire by extrathymic events. In the mouse there is evidence that there exists an extrathymic pathway of T lymphocyte development and that this lineage acquires its TCR repertoire via selection processes that occur outside of the thymus.[5-7] Recent studies examining TCR expression by PCR analysis of mRNA in human mucosal T cells, have found skewing of TCR variable gene usage and oligoclonal expansion when compared to peripheral blood T cells.[8,9] These findings suggest that local factors in the intestinal mucosa influence the not only selective homing, but also expansion or selection of T cells bearing certain TCR variable regions genes. The aim of this study was to examine and compare the functional activity of the different TCR V-gene bearing subsets of mucosal and peripheral T cells from the same individual. Our results showed that the cytolytic activity measured in a redirected cytotoxicity assay, using a panel of 7 monoclonal antibodies specific for different TCR V gene products in mucosal T cells differs from that detected in the peripheral blood T cells from the same individual.

SUBJECTS AND METHODS

Patients

Mucosal and peripheral T cells were isolated from surgical specimens and peripheral blood samples from seven patients undergoing intestinal resections for various inflammatory and noninflammatory illnesses.

Cell Isolation

Peripheral blood lymphocytes were isolated by centrifugation over Ficoll-Paque (Pharmacia, Uppsala). Intraepithelial lymphocytes (IEL) were isolated by dissociation of the intestinal epithelium with Ca^{+2} and Mg^{+2} free Hanks Balanced Salt Solution (Gibco, Grand Island, NY) containing 0.75 mM EDTA as previously described.[10] Lamina propria lymphocytes (LPL) were isolated by further enzymatic digestion of the mucosa with collagenase (20 U/ml) and DNAase (0.01%) (Sigma, St. Louis, MO.). Both lymphocyte populations were further purified by centrifugation in 40% Percoll (Pharmacia) followed by centrifugation over Ficoll-Paque. Cell viability was greater than 95%.

Redirected Cytotoxicity Assay

Effector cells were cultured in a 24 well plate at a concentration of 1 x10^6 cells/ml in

Advances in Mucosal Immunology, Edited by
J. Mestecky et al., Plenum Press, New York, 1995

RPMI media (Gibco, Grand Island, NY) containing 10% fetal calf serum and ConA (10 μg/ml). After 2 days in culture human recombinant IL-2 (5 U/ml, Cetus Corp, CA) was added to the cultures. Cells were cultured for 6 to 8 days then tested for cytolytic activity. ^{51}Cr-labelled P815 mouse mastocytoma cells were pre-incubated with 200 ng of MAb specific to different TCR variable regions (T Cell Diagnostics Inc. Cambridge, MA,), anti-CD3 (OKT3, Ortho Diagnostics) or anti-CD4 (T4, Coulter, Hialeah, FL.) for 30 min at room temperature. Target cells were then resuspended to the appropriate volume and incubated in triplicate with cultured effector cells at a 50:1 effector:target ratio. After 4 h, ^{51}Cr released into the supernatant was measured on a LKB gamma counter. The specific cytotoxicity was calculated as the (measured cpm -spontaneous release)/ (total cpm released - spontaneous release) x 100. The spontaneous release was measured in the presence of an irrelevant antibody and was consistently less than 20% of total release and equivalent to that seen in the absence of antibody. Freshly isolated PBL or mucosal lymphocytes had no detectable cytolytic activity in this assay.

RESULTS AND DISCUSSION

The redirected cytotoxic activity of IEL was analyzed in 2 patients. The pattern of cytotoxicity detected with the panel of anti-TCR MAb was different than the pattern of cytotoxicity seen in the PBL T cells from these patients.

The cytolytic activity of the lamina propria T cells expressing the different TCR V gene also differed from the pattern seen with peripheral T cells from the same patient in 3 out of 5 patients tested. In 2 patients, however, the pattern of LPL cytolytic activity was similar to that seen in cultured peripheral blood T cells. No consistent increase or decrease in function of T cell expressing different TCR variable gene was observed in either the IEL or LPL T-cell populations.

These results suggest that the mucosal microenvironment influences the cytotoxic activity of different TCR V-gene expressing T-cell subsets in a manner that is different from that seen in PBL. These differences may represent the accumulation or expansion of a restricted set of cytotoxic T cells, or alternatively the functional inactivation of a subset of T cells. This may reflect a process of TCR repertoire selection that occurs within the intestine as has been shown for the extrathymic lineage of murine IEL.[5,6] These findings illustrate a possible mechanism by which the cellular reactivity generated in the mucosal compartment can differ from those in the peripheral immune system.

ACKNOWLEDGMENTS

Funded by the Canadian Foundation for Ileitis and Colitis. The technical assistance of Darlene Steele-Norwood is gratefully acknowledged. Maria Baca-Estrada is a recipient of a Fellowship from the Medical Research Council of Canada.

REFERENCES

1.   T. Mak, N. Caccia, N. Kimura, R. Spolski, A. Iwamoto, P. Ohashi, M. D. Reis, and B. Toyonaga, *Cold Spring Harbor Symp. Quant. Biol.* LI:797 (1986).
2.   J. P. Allison, *Annu. Rev. Immunol.* 5:503 (1987).
3.   M. M. Davis, *Annu. Rev. Biochem.* 59:475 (1990).
4.   M. M. Davis, L. J. Berg, A. Y. Lin, B. Fazekas de St.Groth, B. Devaux, C. G. Sagerström, P. J. Bjorkman, and J. F. Elliott, *Cold Spring Harbor Symp. Quant. Biol.* 54:119 (1989).
5.   B. Rocha, *Eur. J. Immunol.* 20:919 (1990).
6.   B. Rocha and H. von Boehmer, *Nature* 251:1225 (1991).
7.   C. M. Parker, V. Groh, H. Band, S. A. Porcelli, C. Morita, M. Fabbi, D. Glass, J. L. Strominger, and M. B. Brenner, *J. Exp. Med.* 171:1597 (1990).
8.   C. Van Kerckhove, G. J. Russell, K. Deusch, K. Reich, A. K. Bhan, H. DerSimonian, and M. B. Brenner, *J. Exp. Med.* 175:57 (1992).
9.   S. P. Balk, E. C. Ebert, R. L. Blumenthal, F. V. McDermott, K. W. Wucherpfennig, S. B. Landau, and R. S. Blumberg, *Science* 253:1411 (1991).
10. E. C. Ebert, *Gastroenterology* 97:1372 (1989).

B-CELL PROMOTING ACTIVITY OF HUMAN COLOSTRUM

Igor I. Slukvin,[1] Valentina V. Pilipenko,[1] Victor P.
Chernyshov,[1] and Alexey A. Philchenkov[2]

[1]Laboratory of Immunology, Ukrainian Institute of
Pediatrics, Obstetrics and Gynecology, 252050
Kiev, Ukraine; and [2]Institute for Oncology and
Radiology Problems, 252022 Kiev, Ukraine

INTRODUCTION

It is well known that human colostrum and milk contain different factors which provide passive immunization for newborn infants.[1-3] However, recently it has been shown that human milk may serve as a mechanism by which active immunity can be transferred to the recipient newborn. Free secretory component (SC), lactoferrin, and S-IgA were described as suppressor factors in human colostrum.[4-6] At the same time, clinical data suggest the presence in human colostrum and milk of factors, that enhance Ig production. In this paper we have assessed the effect of human colostrum on B cell function of mouse splenocytes. A high-performance liquid chromatography (HPLC) gel-filtration was employed to identify the fraction that associated with B cell-promoting activity of human colostrum.

MATERIALS AND METHODS

Human colostrum samples were provided by healthy mothers at 2-3 days of lactation. Colostral whey was prepared from pooled colostrum by centrifugation at 20, 000 g to remove fat, and then acidified to pH 4.6 with 1N HCl to precipitate casein. Then colostral whey was separated by HPLC gel filtration on Protein Pak 300 SW column (Waters) with 0.2M NaH_2PO_4 buffer, pH 7.4. Principal proteins of all fractions were analyzed by PAG-electrophoresis.

Bovine colostrum was obtained during 24-48 h postparturition. Purification procedure consisted of centrifugation to remove cream, acidification of the skim fraction to remove casein, and salting out of the globulin fraction from the neutralized whey.

Spleens were removed from CBA mice by aseptic technique. Mouse splenocytes were cultured at $1 \times 10^5/0.2$ ml of RPMI 1640 medium supplemented with 5% fetal calf serum, 2mM 1-glutamine, 10^{-5}M 2- mercaptoetanol and 10 µg/ml gentamycin in 96-well cluster plates (Linbro) in an atmosphere of 5% CO_2 at 37°C. Proliferation was assessed after 72 h of culture by incorporation of ^{3}H-thymidine during 4 h exposure. Total Ig content in culture supernatants was determined using conventional ELISA technique and polyclonal rabbit anti-mouse Ig antibodies for preparation of solid-phase immunoadsorbent and peroxidase-labeled rabbit anti-mouse Ig antibodies. Color intensity was measured in Multiskam MCC/340 (Labsystems).

Numbers of IgA-, IgM-, IgG1-, IgG2a- and IgG2b-secreting cells were examined by ELISPOT assay[7] using rabbit antibodies to mouse Ig isotypes (Calbiochem) and nitrocellulose-backed 96-well plates.

LPS from *E. coli* 055:B5 (Difco) were used at suboptimal concentration 50 ng/ml. Human colostrum and bovine colostrum or their different fractions were added to cell culture simultaneously with LPS.

RESULTS

Human colostrum stimulated proliferation of mouse splenocytes induced by suboptimal doses of LPS. Stimulation of lymphocyte by human colostrum in the absence of LPS was also observed but to a lesser degree. There was a dose-dependent proliferation of lymphocytes related to the milk concentration present in culture Fig. 1.

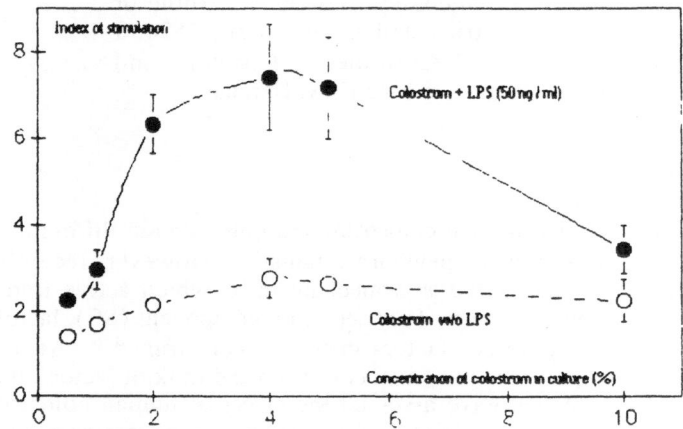

Figure 1. Dose-dependent stimulation by LPS (50ng/ml) and spontaneous mouse spleen cell proliferation by different concentrations of human colostrum. Index of stimulation = c.p.m. stimulated cells/c.p.m. control cells.

The rise in total Ig content in splenocytes culture supernatants was observed after addition of human colostrum. The effect of human colostrum on Ig synthesis was dose-dependent and had a maximum at 1-2% concentrations in medium.

In contrast to human colostrum, bovine colostrum exerted suppressive effect on mouse splenocyte proliferation and Ig synthesis induced by suboptimal doses of LPS. Bovine casein did not significantly affect the Ig synthesis by mouse splenocytes. Salting out of the globulin fraction from bovine colostral whey with 35% ppt of ammonium sulfate showed that suppession of Ig synthesis was associated with bovine colostral globulins.

Using ELISPOT method we found that rise in Ig level in culture after treatment with human colostrum was due to increase number of IgG2a- secreting cells Fig. 2.

After HPLC gel-filtration using Protein Pak 300 SW column, B cell-promoting activity was found in descended part of the second peak Fig. 3. As determined by gel-filtration and PAG-electrophoresis the molecular weight of this fraction was 70-75 kDa.

DISCUSSION

The present study clearly demonstrated that human colostrum contained a factor with B cell promoting activity. This factor enhanced mouse splenocyte proliferation and

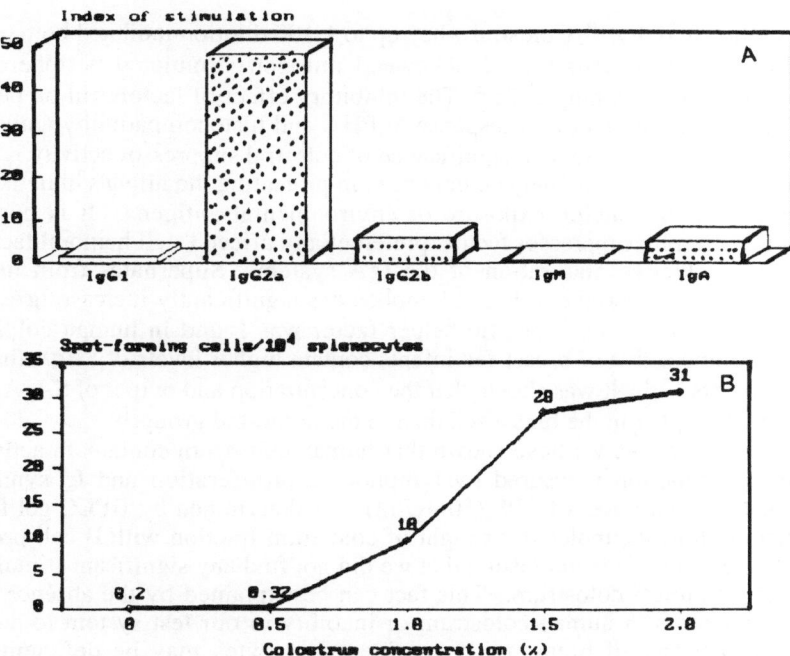

Figure 2. Effect of human colostrum on level of Ig-secreting cells. Mouse splenocytes were cultured with LPS (50ng/ml) and human colostrum whey and after 4 days number of Ig-secreting cells were examined by ELISPOT technique. A. Ig-secreting cells in cultures with 2% of human colostrum whey. B. Dose-dependent stimulation of IgG2a spot-forming cells by human colostrum. The results shown are the mean of 3 separate experiments (3 human colostrum specimens were tested in each experiment)

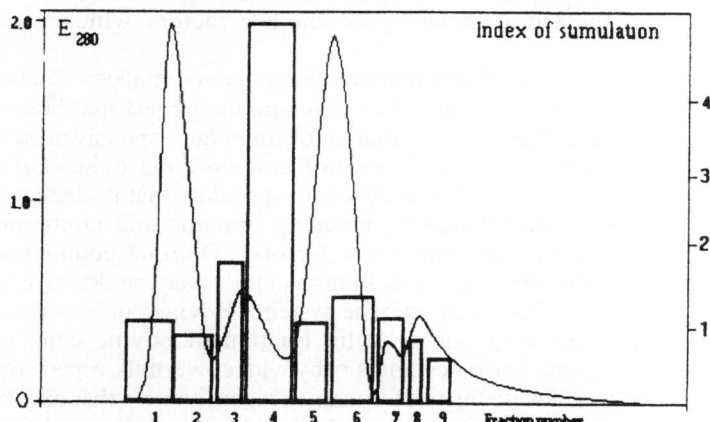

Figure 3. The effect of human colostrum whey fractions (obtained by HPLC gel-filtration) on splenocytes proliferation induced by suboptimal doses of LPS. The line depicts the elution profile of human colostrum from Protein Pak 300SW column. Open bars indicate the effect exerted by fractions on the splenocyte proliferation induced by suboptimal doses of LPS (50ng/ml).

Ig synthesis induced by suboptimal doses of LPS. As determined by the ELISPOT method the rise in Ig level was due to the increased numbers of the IgG2a-secreting cells.

The potential to suppress lymphocyte stimulation was described previously for human colostrum. Preincubation of peripheral blood mononuclear cells with colostrum caused a striking fall of their proliferative response to PHA[3,8], natural killer activity[9], and

155

monocytes motion activity.[10] Colostral whey up to 1:100 dilution inhibited both uptake of ^{3}H-thymidine and differentiation of pokeweed mitogen stimulated peripheral blood lymphocytes into Ig-containing cells.[4] The inhibitory effect of lactoferrin on peripheral blood lymphocytes proliferation in response to PHA and histocompatibility antigens was also demonstrated.[5] The biological significance of colostral suppressor activity is unclear. It seems that suppressor factors may be important in protecting the infant's immune system for overstimulating by sudden exposure of environmental antigens. It is possible to suggest that human milk suppressor factors preferentially affect T cell humoral factors that may promote a selective maturation of the IgA system. Supernates from incubated colostral cells added to peripheral blood lymphocytes significantly increase IgA, but not IgG or IgM synthesis.[11] IgA-specific helper factor was found in human colostrum.[12] Nasal secretions and saliva of breast-fed infants contain higher levels of S-IgA than those of bottle-fed infants.[13-15] It was shown that the concentration and output of S-IgA in urine were significantly higher in the breast-fed than in the bottle-fed group.[16]

In our experiments we have shown that human colostrum contain an activity that enhances B cell function measured by lymphocyte proliferation and Ig synthesis in response to suboptimal doses of LPS (50 ng/ml). As determined by HPLC gel-filtration and PAG-electrophoresis, molecular weight of colostrum fraction with B cell-promoting activity was 70-75 kDa. It is interesting that we did not find any significant stimulation of IgA-synthesis by human colostrum. This fact can be explained by the absence of IgA-specific helper factors in human colostrum or inability in our test system to determine IgA-enhancing activity of breast milk (murine splenocytes may be deficient in IgA precursors). The major effect of human colostrum treatment on splenocyte culture was a marked increase in IgG2a production. The most likely explanation could be that human colostrum contains factors which are involved in the differential control of Ig isotype secretion. IL-1 - IL-6 are able to support the growth of LPS-preactivated B cells.[17] But any of these factors are capable to induce IgG2a synthesis. IFN-γ has been shown to stimulate *in vitro* IgG2a secretion and inhibit IgG1 and IgE secretion by LPS-activated B lymphocytes.[18] Thus, it is possible to suggest that B cell-promoting activity of human colostrum can be connected with IFN-γ or another factors which promote IFN-γ production.

Newborn infants have an altered immune reactivity and mucosal immune system is immature in neonatal period.[3] It is therefore tempting to suggest that human colostrum plays a major role in the initiation and ontogeny of immune responsiveness in neonates. Breast milk can supply baby with B cell growth factors in order to support the immune responses of his immature B cells. It is tempting to speculate that protection afforded by human milk is due to a complex factors including immune and nonimmune defense factors, B cell growth factors and suppressor factors. Optimal combination of these defense and immunoregulatory components in human milk may be decisive and essential factors in the development of neonatal immune system. Bovine colostrum lacked B cell-stimulating activity. More over, the globulin fraction of bovine colostrum showed inhibitory activity. Therefore, bottle-feeding a baby with cow's milk especially on the first day after birth can alter the development of immune responsiveness that may be one of the major causes of increased levels of mortality and morbidity from intestinal and respiratory infections in bottle-fed as compared to breast-fed neonates.

REFERENCES

1. L. A. Hanson, S. Ahelstedt, B. Andersson, B. Carlsson, S. Fallstrom, L. Mellander, O. Porras, T. Soderstrom, and C. Svanborg, *Pediatrics* 75, suppl. 1:172 (1985).
2. P. L. Ogra, M. Fishaut, and C. Theodore, *in* : "Human Milk, its Biological and Social Value", S. Freier and A. I. Edelman, eds., p. 115, Shaare Zedek Medical Center, Jersualem (1980).
3. V. Chernishov and I. Slukvin, *Arch. Immunol. Therap. Exp.* 38:145 (1990).

4. S. S. Crago, R. Kulhavy, S. J. Prince, and J. Mestecky, *Clin. Exp. Immunol.* 45:386 (1981).
5. E. Richie, J. Hillard, R. Gilmore, and D. Gillespie, *J. Reprod. Immunol.* 12:137 (1987).
6. K. Komiyama, S. S. Crago, K. Itoh, I. Moro, and J. Mestecky, *Cell. Immunol.* 101:143 (1986).
7. C. Czerkinsky, L. Nilsson, H. Nygren, O. Ouchterlony, and A. Tarkowski, *J. Immunol. Methods* 65:109 (1983).
8. S. S. Ogra and P. L. Ogra, *J. Pediatr.* 92:550 (1978).
9. I. Moro, T. Abo, S. S. Crago, K. Komiyama, and J. Mestecky, *Cell. Immunol.* 93:467 (1985).
10. J. Clemente, N. Clerizi, M. Espinosa, and F. Leyva-Cobian, *Immunol. Lett.* 12:271 (1986).
11. W. Pittard and K. Bill, *Cell. Immunol.* 42:437 (1977).
12. H. Shinmoto, H. Kawakami, S. Dosako, and Y. Sogo, *Clin. Exp. Immunol.* 66:223 (1986).
13. S. Roberts and D. Freed, *Lancet ii*:1131 (1977).
14. S. Gross and R. Buckley, *Lancet ii*:543 (1980).
15. S. Stephens, *Arch. Dis. Child.* 61:263 (1986).
16. A. Prentice, *Arch. Dis. Child.* 62:792 (1987).
17. W. Lerchardt, H. Karashuyama, A. Rolink, and F. Melchers, *Immunol. Rev.* 99:241 (1987).
18. C. Snapper, C. Peschel, and W. Paul, *J. Immunol.* 140:2121 (1988).

EXPRESSION OF ADHESION MOLECULES ON THE SURFACE OF MALIGNANT CELLS AND CELL LINES

Juri Marinov,[1] Kristian Koubek,[1] and Jan Starý[2]

[1]Institute of Hematology and Blood Transfusion
[2]II Paediatric Department, Faculty Hospital
Prague, Czech Republic

INTRODUCTION

Cell-cell adhesion molecules (CAMs) have been implicated not only in intercellular recognition (adhesion of various types of leukocytes to other cells and to a component of extracellular matrix), but also in morphogenetic events, regeneration, tumor invasion and metastasis. Some CAMs may primarily establish stable bonds between cells, whereas others may be expected to be involved in signalling between cells. Based on nucleic acid sequence data, the majority of known CAMs can be grouped into the following families: the integrin superfamily (VLA antigens, leukocyte integrins, cytoadhesins); the immunoglobulin superfamily; the selectin family and the cadherin family.[1-5]

Glycoprotein LFA-1, belonging to the Integrin superfamily is a non-covalent dimer composed of the Q2 (CD18, 95 kDa) and P (CD11A, 180 kDa) subunits. The Q2 integrin subunit can be alternatively associated with two other P chains: CD11b and CD11c. LFA-1 is a major leukocyte adhesion molecule that binds its broadly expressed ligand ICAM-1 (CD54, 90 kDa) or the alternative ligand ICAM-2. LFA-3 (CD58, 40-65 kDa), broadly expressed on the surface of hematopoietic cells is a ligand for CD2 (LFA-3 Rc, 50 kDa) and together they form the CD2-LFA 3 adhesion system. Adhesion of leukocytes to specific endothelial cells and the subsequent triggering of the process of directing leukocytes to sites of inflammation is mediated by the so-called homing receptors. One of them indispensible for all types of homing is CD44 (80-95 kDa). CD31 (140 kDa) is an adhesion molecule of LAK cells along with CD56 and CD57. Finally CD59 (18 kDa) has been recently shown to play an important role in the leukocyte adhesion process. In 71 patients with blood malignancies (non T-ALL, T-ALL, AMixL, ALL, CML, CLL, HCL, PLL, MDS, B splenic lymphoma, AUL) we have studied the expression of several adhesion molecules: CD2, CD5, CD11a/CD18, CD11c/CD18, CD54, CD58, CD44, CD31, CD59.

MATERIALS AND METHODS

In 71 patients with leukemias and lymphomas, following clinical and cytological examination and morphological typing by standard techniques on Wright-Giemsa stained smears, neoplastic cells (>30%) were obtained from the lymph nodes, peripheral blood, bone marrow and spleen, using sodium metrizoate at a density of 1.077 g/ml. The expression of the adhesion molecules on the surface of leukemic cells was determined by immunofluorescence method using Zeiss microscope and a Flow Cytometer (FACScan,

Becton Dickinson) The monoclonal antibodies (MoAbs) used for immunophenotyping are listed in Table 1

RESULTS AND DISCUSSION

The cases of leukemias and lymphomas we have studied were heterogeneous with respect to the expression of the screened adhesion molecules Important variations were observed between the various subclasses of leukemias and within the same subclass, between individual malignancies Correlation to a certain extent existed with the cell lineage and the maturation stage The results are summarized in Tables 2 and 3 In cases with non T-ALL, LFA-1 (CD11a/CD18) expression varied considerably between individual tumors of a given group A total of 7 of 17 ALLs were LFA-1 negative Another 6 cases displayed low expression and 4 cases were positive No clear correlation was made between LFA-1 expression and stage of cell maturation CD11c was expressed only in one case of non T-

Table 1 Monoclonal antibodies used in this study

MoAbs	CD	Origin	MoAbs	CD	Origin
MT 910	CD2	Rieber	G26	CD58	Huning
Leu 1	CD5	B Dickinson	YTH53 1	CD59	Waldmann
L29	CD11a	Lanier	LAK1	CD31	Zacchi
BL4H4	CD11c	Fiebig	F10442	CD44	Dalhau
MHM23	CD18	McMichael	7F7	CD54	Schulz

ALL ICAM (CD54) was negative in 10 of 16 and expressed at relatively low density in other cases ICAM-1 (CD54) showed a tendency to be expressed in cases with more mature phenotype than in cases of immature phenotype CD58 and CD59 were expressed only in 2/4 and 2/11 cases, the homing receptor CD44 in only 6/15 cases, mostly in immature (I type) and more mature (IV and V types) In AML, LFA-1 expression also varied considerably, missing in 5/30 cases, low in 4 cases and strong expression without a clear correlation between immunophenotypic subgroups in the remainder CD11c showed low or no expression (5/30) ICAM-1(CD54) molecule was present on the surface of malignant cells from 14/30 patients and missing or lowly expressed in 16 cases CD58, CD59 also demonstrated low expression in only 9/30 and 1/24 cases The homing receptor was missing in 17/30 cases, while CD31 molecule was largely expressed in 20/24 In patients with CLL, all molecules showed low or no expression LFA1 in 2/7, CD11c in 0/7, CD54 in 3/7, CD58 in 4/7, CD44 in 3/7 and CD59 in 0/4 cases In patients with CML, LFA1 expression was high, while the rest of the molecules showed variable but low expression Our study included only individual cases of HCL, PLL, and MDS

Our results indicate that the LFA-1 molecule is expressed heterogeneously on the surface of malignant cells without a clear correlation with the stage of cell maturation, and varies considerably between individual tumors of a given group Expression of ICAM 1 was confined to tumors with a phenotype of mature lymphocytes, and it was in many cases low or even missing The same low or missing expression of CD58, CD59, and the homing receptor CD44 were observed in many cases of blood malignancies, a fact which suggests a possible correlation between the immunophenotype, the pathogenesis and the pathophysiology of leukemias and lymphomas (Table 4)

Table 2. Expression of adhesion molecules on leukemic cells.

Anti-gens	NonT ALL**	AML	CLL	CML	AMixL	T ALL	HCL	PLL	MDS
CD2	1/17*	1/30	0/7	0/5	0/3	1/1	0/1	0/1	0/1
CD5	2/7	0/30	6/6	0/5	1/3	1/1	0/1	1/1	0/1
CD11a	10/17	25/30	2/7	5/5	2/3	1/1	1/1	1/1	0/1
CD11c	1/17	5/30	0/7	1/5	3/3	0/1	0/1	0/1	0/1
CD18	10/17	25/30	2/7	5/5	2/3	1/1	0/1	1/1	0/1
CD54	2/14	14/30	3/7	1/1	2/3	1/1	0/1	1/1	0/1
CD44	6/15	13/30	3/7	3/5	3/3	0/1	0/1	0/1	0/1
CD58	2/14	9/30	4/7	0/5	0/3	1/1	0/1	0/1	0/1
CD59	2/11	1/24	0/4	1/5	0/3	1/1	0/1	0/1	-
CD31	7/13	20/24	3/4	1/5	3/3	0/1	0/1	0/1	-

NOTE:* = number of positive cases (>30%)/ total number tested ** = NonT ALL- Non T acute lymphoblastic leukemia. T ALL - Acute lymphoblastic leukemia. AML - Acute myeloblastic leukemia. AMixL - Acute mixed lineage leukemia. CLL -Chronic lymphoblastic leukemia. CML - Chronic myeloid leukemia. HCL - Hairy cell leukemia. PLL - Prolymphocytic leukemia. MDS - Myelodysplastic syndrome.

Table 3. Expression of adhesion molecules on leukemic cells from leukemic sub-types.

Anti-gens	NonT ALL I	NonT ALL II	NonT ALL III	NonT ALL IV+V	AML M1	AML M2	AML M3	AML M4+5
CD2	0/7	0/3	0/3	0/4	1/7	0/8	0/3	0/14
CD5	1/4	0/3	0/3	2/3	0/7	0/8	0/3	0/14
CD11a	3/7	2/3	2/3	3/4	4/7	8/8	2/3	13/14
CD11c	0/7	0/3	0/3	0/4	-	-	-	5/7
CD18	3/7	2/3	2/3	3/4	4/7	8/8	2/3	13/14
CD54	2/6	1/3	0/3	3/4	4/7	2/8	1/3	8/14
CD44	2/7	0/3	3/3	2/4	2/7	4/7	1/3	6/14
CD58	2/6	0/3	2/3	0/4	2/6	0/8	1/3	7/14
CD59	0/5	-	-	1/3	1/5	0/8	0/3	0/8
CD31	1/3	-	-	2/3	4/6	6/8	2/3	7/7

NOTE:* = number of positive cases (>30%) /total number tested

Table 4. Expression of adhesion molecules on cell lines.

Name of cell lines	CD2	CD5	CD11a	CD11c	CD18	CD54	CD58	CD44	CD31	CD59
REH	0	0	4	0	NT	4	3	0	1	0
JURKAT	4	3	3	0	3	NT	1	0	4	2
RAJI	0	0	3	0	3	4	0	0	0	0
U937	0	0	3	0	4	4	4	0	4	3
K562	0	0	0	0	0	3	0	4	0	3
HL60	1	1	4	0	4	3	0	0	4	0
PS1	0	0	4	0	4	3	0	0	4	0
HEL	0	0	4	0	3	3	0	3	4	0
UHKT2	0	0	4	0	3	4	0	0	3	4
KG1	0	0	3	1	4	4	4	4	3	0

NOTE: 0= 0 -10%; 1= 10-20%; 2= 20-40% ; 3= 40-60% ; 4= >60% NT= not tested.

REFERENCES

1. E. Horst, T. Radaszkiewicz, O. Hooftman-den Otter, R. Pieters, J. M. van Dongen, J. L.M. Meijer, and S. T. Pals, *Leukemia* 5:848 (1991).
2. T. A. Springer, *Nature* 346:425 (1990).
3. K. Koubek, Atlas of human leukocyte antigens, p.199, J+J, Prague (1992).
4. L. Osborn, *Cell Vol.* 62:3 (1990).
5. A. R. de Fougerolles, and T. A. Springer, *J. Exp. Med.* 175:185 (1992).

ERGOT ALKALOID-INDUCED CELL PROLIFERATION, CYTO-TOXICITY, AND LYMPHOKINE PRODUCTION

A. Fiserová,[1] G. Trinchieri,[3] S. Chan,[3] K. Bezouska,[2]
M. Flieger.[1] and M. Pospisil[1]

[1]Institure of Microbiology, Czech Academy of Sciences
[2]Institute of Biotechnology, Prague, Czech Republic, and
[3]The Wistar Institute, Philadelphia, USA

INTRODUCTION

On the basis of the structural similarities among ergot alkaloids (EA)[1] and important mediators such as noradrenaline, serotonin and dopamine, the interaction of EA with receptors for these mediators and wide range of their biological effects could be explained.[2] From the functional point of view these effects are mediated by Q-adrenergic[3,4], dopamine[5] or serotonin receptors.[6] Lymphoid cells possess membrane receptors for a variety of hormones that affect DNA synthesis and proliferation via the ubiquitous cAMP (cGMP) system.[7,8] The process of hormone action often goes through cell receptors and membrane adenylcyclase to the cyclic nucleotides that function as transmitters of the hormonal message to the genome, thereby resulting in a change in the physiological status of the cell in question. Neurotransmitters may modulate immune functions, e.g., T cell differentiation, NK cell function[6], T and B cell response to lectins, and macrophage functions.[9] Corresponding receptors to neurotransmitters have been described on mature and developing lymphocytes.[10,11] Therefore, we have focused on determining the direct effects of EAs on some immune functions. We tested lymphocyte proliferative responses, cell-mediated cytotoxicity, and capability to produce cytokines.

MATERIALS AND METHODS

Cytokines and Reagents

rIL-2 (107 U/mg) was purchased from Takeda Chemical Industry, Inc. (Osaka, Japan), PHA-M from Wellcome Diagnostic (Dartford, England) and MAbs anti-IFN-γ (B133.1, B133.3) were supplied by Wistar Institute (Philadelphia, USA). EAs were prepared by Galena (Opava, Czech Republic) and were used for immunomodulation studies as described. Cell Preparations: PBMC separated on Ficoll-Verografin density gradient (from heparinized peripheral blood of healthy donors), purified as T cells (upon a column) were used as effector cells.

Long-term cultures and all experiments were performed in RPMI-1640 medium enriched with L-glutamine, antibiotics (PNC/STM/Fungizone) and supplemented with 5%-10% neonatal calf serum. Incubation was carried out at 37°C (5% CO_2). NK-cell activity was measured by 4 h ^{51}Cr-release assay as reported previously[12] against NK-sensitive K562 and resistant Raji target cells

Lymphokine Release Induction and Assays

The PBMC preparation (2×10^6 cells/ml) were cultured for 5 days with indicated stimulators To detect IFN-γ release the radioimmunoassay (RIA) has been applied [13] IL-2 production by lymphocytes generated in MLC in the presence of EAs was measured by sandwich ELISA method using microtitration plates coated with Tamm-Horsfall glycoprotein, secondary biotinylated goat anti-human IL-2 polyclonal antibody (Wistar Institute), and Extravidin-alkaline phosphatase (Sigma, USA)

RESULTS

Screening a wide panel of natural as well as semi-synthetic EAs of *Claviceps purpurea* revealed new immunomodulatory properties besides their action on neural and endocrine systems It is generally accepted that lymphoid cells require more than one signal for their activation

Proliferative Response and Cell-Mediated Cytotoxicity

In our experiments, PHA was selected as a second signal inducing proliferation of lymphocytes stimulated by EAs As shown in Table 1, EA increased in a dose-dependent manner both cell proliferation and cytotoxic activity predominantly against NK-resistant target cells

Table 1. Effect of EAs (A-agroclavine, EM-ergometrine) on proliferative response of lymphocytes ($2 \times 10^5/200\mu l$) and cell-mediated cytotoxicity (E T=20 1) of human PBMC

Inducer		^{3}H-TdR (SI)		% specific lysis	
		Control	PHA	K562	Raji
None		1 0	1 0	16	5
A	100ng/ml	1 5	1 5	40	30
	1 ng/ml	4 2	1 5	14	34
	10 pg/ml	8 0	1 6	17	18
	100 fg/ml	1 5	1 5	20	14
EM	100 ng/ml	1 8	1 2	10	10
	1 ng/ml	1 0	1 5	16	18
	10 pg/ml	1 4	1 4	19	20
	100 fg/ml	1 2	1 5	20	28

Lymphocytes were cultured 6 days in the presence of EA or EA with PHA (1%) ^{3}H-TdR incorporation was determined after 6-h pulse before the end of cultivation The ^{3}H-TdR incorporation of control cell proliferation = 860 cpm (100%), PHA induced cell proliferation = 30,600 cpm (100%) The results represent 3 experiments performed

Reactivity of PBMC in MLC in the Presence of EA

Responder PBMC were cultivated with gamma-irradiated (45 Gy) stimulator PBMC from unrelated allogeneic donors for 5 days and assayed for production of IL-2 The cell-free supernatants were tested by ELISA for IL-2 release (Fig. 1). No significant changes were detected in the proliferative response of EA stimulated PBMC in MLC.

EAs have also shown a stimulating effect on specific cytotoxic T lymphocytes (CTLs) generated in one-way MLC. EAs have also an enhancing effect on NK-like activity of MLC-derived lymphocytes, since K562 were used as target cells.[14]

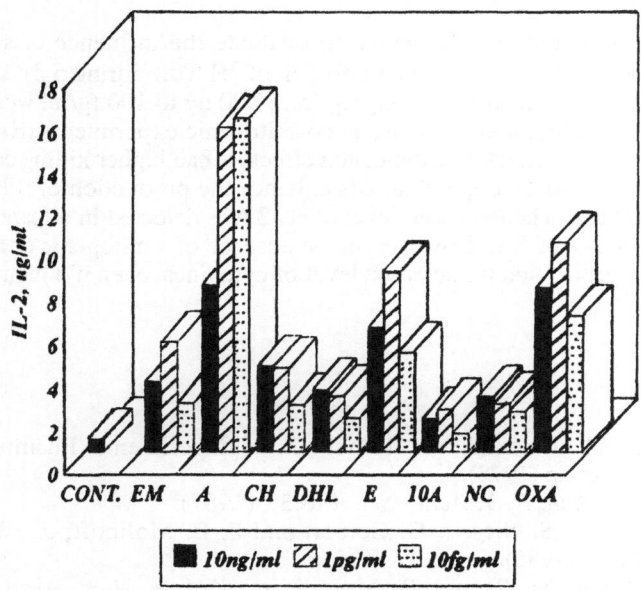

Figure 1. Production of IL-2 by human PBMC stimulated with EA in MLC. 5×10^5 responder cells were stimulated with 1×10^6 of allogeneic irradiated (45Gy) stimulator cells for 5 days. Abbreviations A-agroclavine, CH-chanoclavine, DHL- dihydrolysergol, E-elymoclavine, 10A-10α-methoxy-DHL, NC-nicergoline, OxA-Ox-agroclavine. All experiments were done in triplicates. The results are representative of 4 experiments performed.

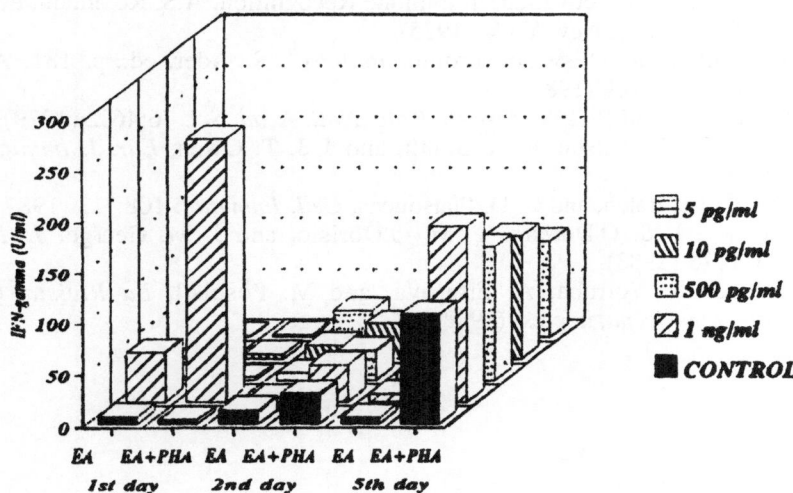

Figure 2. IFN-γ production by PBMC induced with various doses of agroclavine in the absence or presence of PHA assayed 1st, 2nd and 5th days of cultivation. IFN-γ production was detected by ^{125}I-labeled MAbs against human IFN-γ (B133.3). The results represent three experiments performed.

Induction of IFN-γ Production of EA-Stimulated PBMC

EA synergize with mitogens in the production of IFN-γ by freshly isolated lymphocytes. Cell-free supernatants were assayed after 24, 72 and 120 h of cultivation with agroclavine in the presence or absence of PHA (Fig. 2)

CONCLUSIONS

In the present study we attempted to evaluate the influence of selected EAs on lymphocyte functions: a) increased incorporation of ^{3}H-TdR during 5 days of cultures was detected when EAs in concentrations ranging from 100 ng to 100 fg/ml were added; b) we identified different proliferative response in co-mitogenic experiments (EA+PHA); c) we noted the important fact that the EA stimulated effectors had higher killing capability against NK-resistant targets; and d) ergot alkaloids enhance the production of IFN-γ by mitogen-activated lymphocytes. The increased level of IL-2 was detected in alloantigen-stimulated cultures in the presence of EA. However, in the absence of a mitogenic or antigenic signal, EA did not induce production of increased level of cytokines, even if a high killing capacity was retained.

REFERENCES

1. B. Berde and H. O. Schild, *in:* "Handbook of Experimental Pharmacology", p. 49, Springer Verlag (1978).
2. J. W. Hadden, *Ann. N. Y. Acad. Sci.* 496:39 (1987).
3. R. D. Aarons, A. S. Nies, J. G. Gerber, and P. B. Molinoff, *J. Pharmacol. Exp. Ther.* 224:1 (1982).
4. B. G. W. Arnas, M. Brown, R. Maselli, *et al.*, *Ann. N. Y. Acad. Sci.*, 540:585 (1988).
5. W. Kehr, *J. Pharmacol.* 97:111 (1984).
6. K. Hellstrand and S. Hermodsson, *J. Immunol.* 139:869 (1987).
7. E. Chelmicka-Schorr and G. W. Arnason, *in:* "Immunologic Mechanisms in Neurologic and Psychiatric Disease", B.H. Waksman, ed., p. 67 (1990).
8. H. J. Wedner, F. E. Bloom, and C. W. Parker, *in:* "The Role of Cyclic Nucleotides in Lymphocyte Activation, in Immune Recognition, A.S. Rosenthal, ed., p. 337, Academic Press, New York, (1975).
9. A. A. Monjan, *in:* "Psychoneuroimmunology", R. Ader, ed., p. 181, Academic Press, New York (1981).
10. D. P. Richman and B.G.W. Arnaso, *Proc. Natl. Acad. Sci.* 76:4632 (1979).
11. U. Singh, D. S. Millson, P. A. Smith, and J. J. T. Owen, *Eur. J. Immunol.* 9:31 (1979).
12. K. Itoh, C. M. Balch, and C. D. Platsoucas, *Cell. Immunol.* 108:313 (1987).
13. A. Danek, M. S. O'Dorisio, T. M. O'Dorisio, and J. M. George, *J. Immunol.* 131:1173 (1983).
14. M. Flieger, J. Votruba, A. Fiserová, and M. Pospisil, *La Rivista Quichica Latinoamericana*, in press (1992).

EFFECT OF CORTICOSTEROID ON LYMPHOCYTE ADHESION

D. Pearson and P. Sheldon

Department of Microbiology, Medical
Sciences Building, PO Box 138, University of
Leicester, LE1 9HN, United Kingdom

INTRODUCTION

Corticosteroids, such as methylprednisolone are used in the treatment of a number of ailments, including active rheumatoid arthritis.[1] It is generally noted that corticosteroids have a suppressive effect on the immune system both *in vivo* and *in vitro*. In particular, it has been noted that there is a depletion of lymphocytes from areas of the gut associated with mucosal immunity, and the suggestion has been made that this may be due to corticosteroid induced redistribution of lymphocytes.[2] *In vivo* work in male rats has also shown that a large proportion of methylprednisolone partitions to the intestine upon *intravenous* injection.[3] Previous work has demonstrated a link between mucosal lymphocytes and rheumatoid arthritis.[4] Using an *in vitro* cytoadherence assay[5], adapted to a porcine model[4], we have investigated the effect of methylprednisolone on lymphocyte adhesion, known to involve adhesion molecules which effect the trafficking of gut mucosal lymphocytes.[6]

MATERIALS AND METHODS

General Method

A fresh porcine peripheral blood lymphocyte suspension in Iscove's culture medium (ICM) was prepared by erythrocyte sedimentation and density centrifugation. The suspension was divided into two or more aliquots. The cells were pelleted and resuspended in ICM only, for the control, and ICM with added methylprednisolone (as sodium succinate - Solu-Medrone™ Upjohn) for the test suspension(s).

Sections of porcine Peyer's patch tissue (PPT) were prepared by embedding small (1cm^2) pieces in Tissue Tek® OCT embedding medium and using iso-pentane dipped in liquid Nitrogen to cool the samples. Sections were then cut a minimum of 2 h later, when the PPT had equilibrated to freezer temperature (-20°C).

The cell suspensions were left overnight at the appropriate temperature on a rotating table at 400 rpm. After syringe dispersal of clumping, the cells were layered onto 15μm sections of PPT and left for 30 min at 4-6°C on a rotating table at 70 rpm.

The sections were then fixed in formal saline, washed in phosphate-buffered saline and air-dried. Haematoxylin and Eosin staining was used to visualise the tissue and adherent lymphocytes. Equal areas (2mm^2 per section) of lamina propria and lymphatic nodule (follicular area) were used and the results of the latter were subtracted from the former to account for background adhesion and expressed as adherent lymphocytes per mm^2 (n=number of tissue sections examined).

Preliminary Experiments:

Experiment 1. A crossover study was performed to determine likely effects. Both tissue and lymphocytes were treated at 4°C with 0.46mM methylprednisolone. Controls were treated with ICM alone.

Since it appeared that clumping of cells may have affected the results, extra care was taken in subsequent experiments to disperse treated and control cells, to ensure that equal numbers were used.

Experiment 2. Lymphocytes only treated at room temperature. Methyl-prednisolone concentrations were as follows: 0mM (control), 3.15mM, 6.3mM and 12.6mM.

In order to eliminate the possibility of observer bias, in the third experiment counting was performed "blind".

Experiment 3. Lymphocytes treated at 4°C. Methylprednisolone concentrations were as follows: 0mM (control), 0.46mM (pharmacological), 3.15mM, 6.3mM and 12.6mM.

RESULTS AND DISCUSSION

Note

Where mean values are used, error bars are assigned which represent Standard error of the mean.

Viability Testing

Viability tests were performed on cells, prior to the cytoadherence assay, using Trypan Blue. Although cell death was apparent at higher methylprednisolone concentrations, at 4°C cell death was not significant at concentrations less than 6.3mM.

At room temperature, however, cell viability was worse overall.

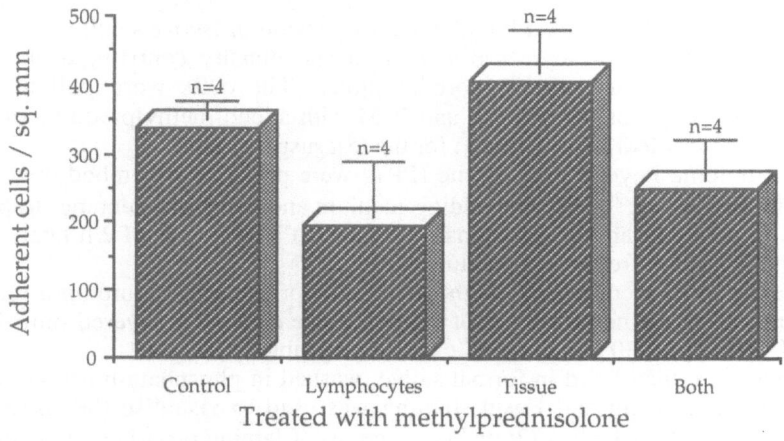

Fig. 1

Experiment 1

As can be seen from Fig. 1, methylprednisolone markedly decreased the ability of lymphocytes to bind to Peyer's patch lamina propria endothelium at a pharmacologically relevant concentration (0.46mM). Incubation of the PPT with methylprednisolone increased its ability to bind lymphocytes, though this was not statistically significant.

Experiment 2

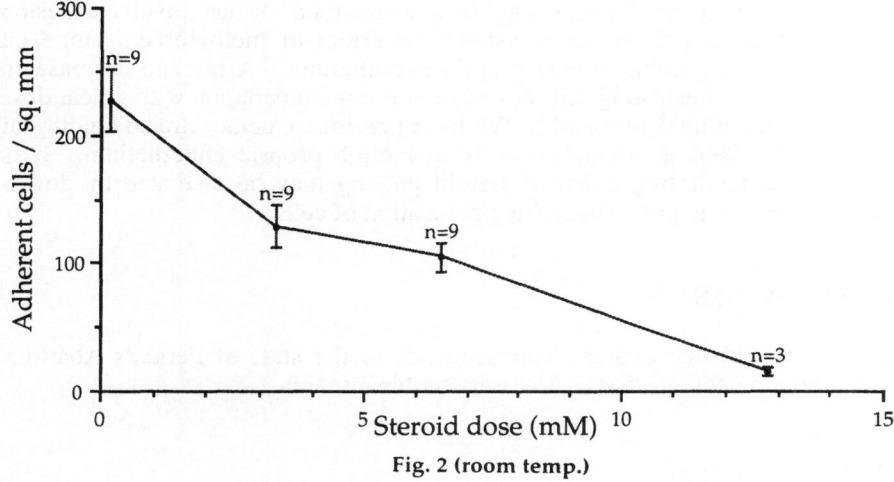

Fig. 2 (room temp.)

The lymphocytes exhibited a clear dose-related reduction in adhesion after pre-treatment with methylprednisolone (Fig. 2) at room temperature.

Experiment 3

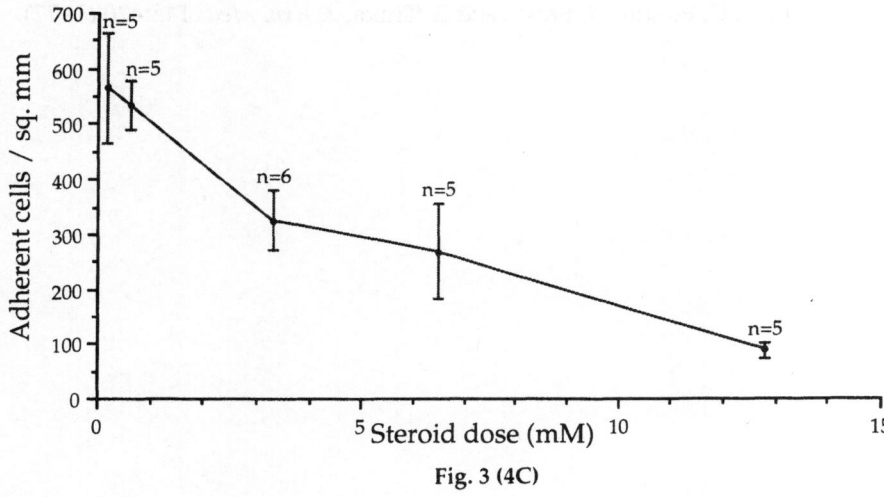

Fig. 3 (4C)

After pre-treatment of lymphocytes at 4°C, an identical effect was observed (Fig. 3). It is likely that the higher numbers recorded overall were due to better cell survival at 4°C.

Methylprednisolone treatment of porcine peripheral blood lymphocytes *in vitro* was shown to down regulate their adherence to Peyer's patch lamina propria endothelium. Since adhesion molecules involved in this process may also be involved pathogenically in some disease processes such as rheumatoid arthritis, it is suggested that the ameliorating effect of pulse steroid therapy may be mediated by the down regulation of adhesion molecules involved in such processes of inflammation.

SUMMARY

Pulse steroid therapy is utilised in the treatment of vasculitis and active rheumatoid arthritis. Decreased entry of lymphocytes to sites of cell-mediated immune reactions is well recognised, though the mechanism is not fully understood. It may involve adhesion to endothelium. We have therefore measured the effect of methylprednisolone on lymphocyte adhesion to porcine lamina propria endothelium. A marked decrease in adhesion was found after incubating cells at 4°C or at room temperature, with a clear dose response effect, from 0.46mM to 6.3mM. We have previously demonstrated binding of rheumatoid synovial fluid mononuclear cells to lamina propria endothelium. It is suggested that the ameliorating effect of steroid pulsing may be mediated by down-regulating adhesin expression of a gut-seeking population of cells.

ACKNOWLEDGMENTS

The authors would like express their gratitude to the staff of Parker's Abattoir, Leicester, for supplying porcine tissues necessary to this research.

REFERENCES

1.   M. A. Byron and A. G. Mowat, *Brit. J. Rheumatol.*, 24:164 (1985).
2.   M. J. Roy and T. J. Walsh TJ, *Lab. Invest.* 64 (4):437 (1992).
3.   H. Kitigawa, Y. Esumi , T. Ohtsuki T *et al:*, *Pharmacometrics* 13 (2): 235 (1977).
4.   A. Kadioglu and P. Sheldon, *Ann. Rheum. Dis.* 51 (1):126 (1991).
5.   H. B. Stamperand and J. J. Woodruff, *J . Exp. Med.* 144:823 (1976).
6.   R. N. P. Cahill, D. C. Poskitt, H. Frost , and Z. Trnka, *J. Exp. Med.* 145:420 (1977).

ADHERENCE OF RHEUMATOID LYMPHOCYTES TO ENDOTHELIUM OF LAMINA PROPRIA

Aras Kadioglu and Peter Sheldon

Department of Microbiology, Medical Sciences Building
University of Leicester, PO Box 138, LE1 9HN, England

INTRODUCTION

It is well established that there are at least two pathways of organ-specific lymphocyte migration, one of which is selective for Mucosa-Associated Lymphoid Tissue (MALT) and the other for peripheral lymph node tissue.[1-3]

It has been suggested that rheumatoid arthritis may involve the pathological homing of such recirculating mucosal lymphocytes to joints, instead of their specific destinations within MALT.[4] Chronic inflammation is characterised by extra-vascular collections of mononuclear cells including lymphocytes, macrophages, and plasma cells. This emigration of mononuclear cells is achieved by means of complementary receptors found on both the recirculating lymphocytes, called "homing receptors" and on the tissue itself; "vascular addressins".[5]

The aim of this study was to see whether RA lymphocytes express adhesion molecules for MALT endothelium, and if so, to compare the adherence of these cells obtained from peripheral blood and synovial fluid.

MATERIALS AND METHODS

Eighteen patients (11 female and 7 male) who satisfied the modified criteria for the diagnosis of rheumatoid arthritis were studied.[6] All had active disease at the time of sampling (at least one painful swollen joint, a raised plasma viscosity, and an increased C-reactive protein).

Their average age was 64 years and their disease duration ranged from 5 months to 12 years. Twelve patients (8 male and 4 female, mean age 47 years) admitted to hospital on account of back pain due to soft tissue damage, and suffering from rheumatoid arthritis were used as controls. Porcine Peyer's patch (PP) lymphocytes were used as a MALT control.

Preparation of Lymphocyte Suspensions

Mononuclear cells were obtained by density gradient centrifugation from paired peripheral blood and synovial fluid samples of 18 patients and from peripheral blood of 12 control patients. This procedure gave yields greater than 90%.

Porcine Peyer's Patch Lymphocytes

PP were identified as raised, dome-like areas on the outer epithelium of the small intestine. Whole lengths of porcine small intestine were collected from the abattoir, the segments of PP excised from the rest of the tissue and cut into small pieces (1-2 cm^2 in size). A total of 20 to 25 PP sections were cut. In order to obtain the lymphocytes the freshly cut sections were washed in phosphate-buffered saline twice and added to 100 ml of Hank's balanced salt solution (Gibco), which also contained 150 mg of protease (Dispase Neutral protease from *Bacillus polymyxa* grade 2, Boehringer), and then stirred at 37°C for 45 min. The supernatant, containing the dissociated lymphocytes was aspirated, washed twice and resuspended in supplemented Iscove's culture medium (sICM from Flow) at a concentration of 5 x 10^6. The same procedure was used for lymph node.

In Vitro Adherence Assay

PP tissue was embedded in Tissue Tek OCT compound (Miles), snap-frozen in liquid nitrogen and stored at -80°C. Before sectioning the tissues were kept at -20°C overnight. 10 μm sections were cut from frozen tissue, using a cryostat and the adherence assay[7] was carried out. A known concentration (5x10^6) and volume (50 μL) of lymphocytes was layered onto the tissue section, encircled by a wax pen, agitated and incubated at 70 rpm for 30 min at 7°C .

After incubation the non-adherent cells were decanted from the tissue by gentle tapping and with the aid of a Q-tip applicator. The tissue sections were then fixed in 2% glutaraldehyde (Sigma) for 10 min and repeatedly washed in PBS. Once dried they were stained with Haematoxylin and Eosin. Lymphocytes appeared blue while the tissue was pink/red. This enabled easy visual quantification of adherent lymphocytes. After staining, the sections were fixed with DPX mountant (BDH) for permanent storage.

Quantification

Quantification was carried out by bright field microscopy, x 250 magnification, by one observer enumerating lymphocytes overlying 1 mm^2 of lamina propria (using a graticule) and subtracting the number adherent over similar areas elsewhere e.g., follicular area of PP.

RESULTS AND DISCUSSION

The adherence of peripheral blood (PB) and synovial fluid (SF) mononuclear cells to endothelial cells within lamina propria of MALT was compared to non-lamina propria. This adherence was expressed as cells adhered per mm^2. The observations were performed on control peripheral blood and porcine PP lymphocytes.

Our previous studies had shown that adherence in this system was metabolically active (complete inhibition by sodium azide), and calcium-dependant (70% inhibition by EDTA), and involved endothelium (80% inhibition by mannose 6-phosphate). The results are shown below.

Table 1. Cell adherence to lamina propria/mm^2.

Source of lymphocytes	n	Mean	SEM
SF lymphocytes	18	296	27
PB lymphocytes	18	148	17
Control PB lymphocytes	12	160	19
Porcine PP lymphocytes	12	498	54

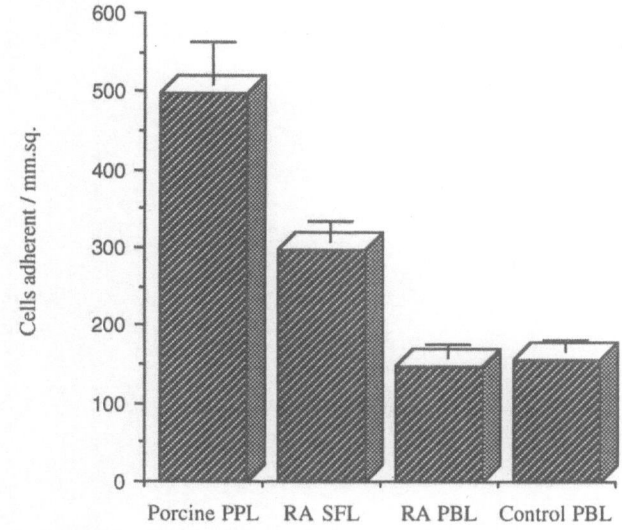

Figure 1. Adherence of Procine, Rheumatoid and Control Lymphocytes to Lamina Propria of Peyer's Patch Rich Tissue.

The adherence of SF lymphocytes was significantly greater than that of PB lymphocytes (p< 0.02, Mann-Whitney). This would imply that SF lymphocytes have a greater proportion of cells which have homing receptors complementary to the vascular addressins of mucosal tissue than do PB lymphocytes.

It further suggests that they share the property of the MALT population and implies a possible mucosal origin of SF lymphocytes in RA. This is in keeping with the hypothesis that in rheumatoid arthritis at least a proportion of the mononuclear cells found within the joints share adherence characteristics with lymphocytes of the gut, from where they might have originated. If so then events occurring at the mucosal level could be pivotal in subsequent homing behaviour of lymphocytes of mucosal origin, and possibly in the pathogenesis of rheumatoid arthritis.

CONCLUSION

Mononuclear cells isolated from paired peripheral blood and synovial fluid (SF) of 18 patients with rheumatoid arthritis (RA), peripheral blood (PB) from 12 control patients and porcine Peyer's patch (PP) were layered onto frozen sections of porcine small intestine. Using this technique a significantly greater proportion of paired synovial fluid lymphocytes were found to adhere to lamina propria. This property was shared by porcine PP lymphocytes, by definition, of mucosal origin. This is in keeping with the suggestions that in RA, the lymphocytes that migrate to the joints may originate from intestinal mucosa.

REFERENCES

1. E. C. Butcher, R. G. Scollay, and I. L. Weissmann, *Eur. J. Immunol.* 10:556 (1980).
2. P. R. Streeter, E. L. Berg, B. T. N. Rouse, R. F. Bargatze, and E. C. Butcher, *Nature* 331:41 (1988).
3. S. Jalkanen, *Springer Semin. Immunopathol.* 11:187 (1989).
4. P. Sheldon, *Ann. Rheum. Dis.*, 47:697 (1988).
5. B. Holzmann, *Cell* 56:37 (1989).
6. F. C. Brown, F. M. Edworthy, and D. A. Bloch, *Arthr. Rheumat.* 31:315 (1988).
7. H. B. Stamper and J. J. Woodruff, *J. Exp. Med.* 144:828 (1976).

EXPRESSION OF SC, IL-6 AND TGF-β1 IN EPITHELIAL CELL LINES

Itaru Moro,[1] Tomihisa Takahashi,[1] Takashi Iwase,[1]
Masatake Asano,[1] Nobuko Takenouchi,[1] Satoshi
Nishimura,[1] Itsuro Kudo,[1] Peter Krajci,[2] Per
Brandtzaeg,[2] Zina Moldoveanu,[3] and Jiri Mestecky[3]

[1]Nihon University, Tokyo, Japan
[2]University of Oslo, Oslo, Norway
[3]University of Alabama at Birmingham, USA

INTRODUCTION

It has been well documented that various kinds of cytokines secreted from immunocytes in the lamina propria of glandular tissues play an important role in the secretory immune system.[1,2] The presence of polymeric Ig receptor, SC in glandular tissues, B cell terminal differentiation factor, IL-6 in keratinocytes and carcinoma cells, and a potential IgA switching factor, TGF-β in various epithelial cells has been reported[3,4,5], but little is known on the role of cytokines secreted from epithelial cells in the secretory immune system. It has been described that the synthesis of SC protein in adenocarcinoma cell line, HT-29 is up-regulated by IFN-γ, TNF-α, and IL-4.[6,7,8] Our previous study also indicated that mRNA expression for SC and IL-6 was detected in HT-29 and the squamous cell carcinoma cell line, Ca9-22.[9]

The purpose of this study is to examine mRNA and protein expression for SC, IL-6 and TGF-β1 in HT-29 and Ca9-22 cultured for various intervals in the presence and absence of IFN-γ or TNF-α by use of the polymerase chain reaction (PCR), Northern blot, ELISA, and immunohistochemistry.

MATERIALS AND METHODS

Cell Lines and Processing

HT-29 cells and Ca9-22 cells (obtained from Japanese Cancer Research Resources Bank) were maintained in RPMI-1640 medium supplemented with 10% FCS and antibiotics. When confluent, cells were transferred to culture medium without FCS, incubated overnight to allow adherence, and 100 u/ml of recombinant IFN-γ (Collaborative Research Inc., Bedford, MA) or 10 ng/ml of recombinant TNF-α (Genzyme Corp., Boston, MA) was added. Cells were harvested at various intervals and subjected to RNA extraction by an acid guanidine thiocyanate phenol chloroform method. Ten ng of RNA was reverse-transcribed for PCR, and 10 μg of RNA was used for Northern blot analysis. Culture supernatants were also collected at various intervals, concentrated 20 times by ultrafiltration and applied to ELISA.

Advances in Mucosal Immunology, Edited by
J. Mestecky *et al.*, Plenum Press, New York, 1995

PCR

cDNA was synthesized by use of murine leukemia virus reverse transcriptase as described by Ferre et al.[10] Samples were added to the reverse transcriptase reaction buffer in a final volume of 20 μl [1 x reverse transcription buffer, 2 units of RNase inhibitor (Takara, Kyoto, Japan)], 1.0 mM each of dNTP's, 10 pM of the 3' PCR primer, 200 units of reverse transcriptase (BRL, Gaithersburg, MD)]. The reaction mixture was incubated at 37°C for 45 min, cooled and diluted with 80 μl of PCR buffer, followed by addition of 40 pM of the 3'PCR primer, 50 pM of the 5'PCR primer and 2.5 units of Taq polymerase (Cetus, Emeryville, CA). The reaction was started by heat-denaturing the hybrid for 20 sec. at 95°C, annealing for 25 sec. at 55°C and then extending for 60 sec. at 72°C. This cycle was repeated 35 times by using a DNA-thermal cycler (Perkin-Elmer Cetus, Norwalk, CT). After amplification, one-tenth volumes of samples were applied to 1.7% agarose gel-electrophoresis and visualized by UV fluorescence after ethidium bromide staining.

Southern Blotting

PCR products electrophoresed in 1.7 % gel were transferred to a nylon membrane and analysed by the Southern blot method with a ^{32}P-labeled oligonucleotide probe. The membrane was washed twice with 2xSSE at room temperature. Following prehybridization, the membrane was hybridized with the ^{32}P-labeled internal probe at 42°C for 3 h. The filter was washed three times with 6xSSPE and subjected to autoradiography. The intensity of the signal was determined by densitometric tracing.

Primers and Probes for PCR

Primers and probes for SC, IL-6, and TGF-β were synthesized on a DNA synthesizer (Model 380, Applied Biosystems, Inc., Foster City, CA) based on the published DNA sequences. The sequences of primers and probes are listed in Table 1.

Northern Blot Analysis

To detect SC, IL-6, and TGF-β1 mRNAs, cDNA fragments were used as probes for northern blot analysis and labelled with ^{32}P-dCTP using a random priming labeling kit from Amersham (Buckinghamshire, UK). The cDNA for SC (position 789-1455) was described previously,[11] and that for IL-6 (position 185-623)[12] was generously provided by Toshio Hirano (University of Osaka, Osaka, Japan). The cDNA for TGF-β1 (position 1679-2014) was obtained from British Bio-technology Ltd. (Oxon, UK).

ELISA

ELISA kits for IL-6 and TGF-β were purchased from Amersham (Buckinghamshire, UK) and for sIgA from Medical & Biological Laboratories Co. Ltd. (Nagoya, Japan). ELISA for SC was developed in our laboratory. In brief, monoclonal anti-human SC (40 μg/ml) was coated onto microplates overnight at 4°C. After washing, each well was incubated with BSA-PBS to block non-specific binding, and standard solution and samples were applied to wells. After extensive washing by PBS containing 0.05% tween 20, peroxidase labeled polyclonal anti-human SC was incubated for 30 min, and 0.1M citrate-phosphate buffer containing 1mg/ml of o-phenylene-diamine and 0.3 μl/ml H_2O_2 was added to wells. The reaction was stopped by the addition of 2M H_2SO_4 and optical density was measured at H492nm.

Immunohistochemistry

After cell culture at a concentration of 1x10^4 in Lab-Tek Chamber slide (Nunc,

Roskilde, Denmark) for various intervals, cells were fixed in cold acetone for 10 min and immersed in methanol containing 0.3% H_2O_2 for 20 min, then washed with PBS. Washed cells were incubated with normal horse serum and subsequently with monoclonal anti-human SC, IL-6, TGF-β1, or with polyclonal anti-human IgA, respectively. After washing, biotinylated horse anti-mouse IgG (for monoclonal antibodies) or biotinylated goat anti-rabbit IgG (for polyclonal antibody) was applied to cells, then avidin-biotin-peroxidase complex and 0.05% 3, 3'-diamnobenzidine tetrahydrochloride containing 0.05% H_2O_2 substrate was added for 10 min. Cells were washed, counterstained with hematoxylin, and mounted.

RESULTS AND DISCUSSION

The Southern blot analysis after PCR indicated that expression of the IL-6 gene was detected in HT-29 cells cultured for various intervals regardless of the presence or absence of IFN-γ or TNF-α (Fig. 1). The addition of IFN-γ or TNF-α in culture resulted in no change in density. Similarly, IL-6 mRNA was detected in Ca9-22 cells at various culture intervals in either the presence or absence of IFN-γ or TNF-α (Fig. 2). mRNA from peripheral blood mononuclear cells stimulated with Con A as positive controls showed a positive band for IL-6 in Southern blot analysis. Even though a clear β-actin band was seen on 1.7% agarose gel, this was not seen after Southern hybridization with an IL-6 probe.

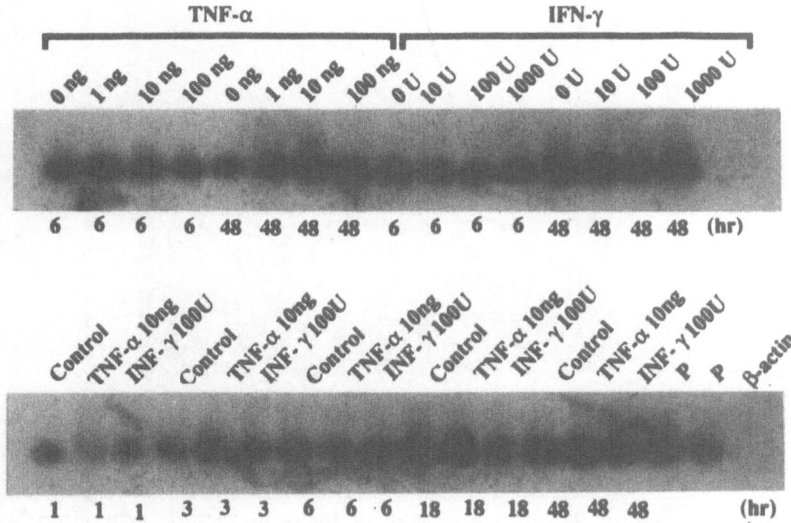

Figure 1. IL-6 expression in HT-29, shown by PCR-Southern blot. P: RNA from peripheral blood mononuclear cell(PBMC) stimulated with Con A.

Since it has been clearly shown that 100 u/ml of IFN-γ or 10 ng/ml of TNF-α were enough to induce up-regulation of SC in HT-29 cells[6,7,8], the same amounts of these cytokines were used for the following experiments. PCR analysis of SC in HT-29 revealed that mRNA expression of SC was detected at all time intervals in the presence or absence of IFN-γ or TNF-α (Fig. 3). When 100 u/ml of IFN-γ or 10 ng/ml of TNF-α was added to cell culture, the SC signal increased in a time-dependent manner. It seems likely that SC

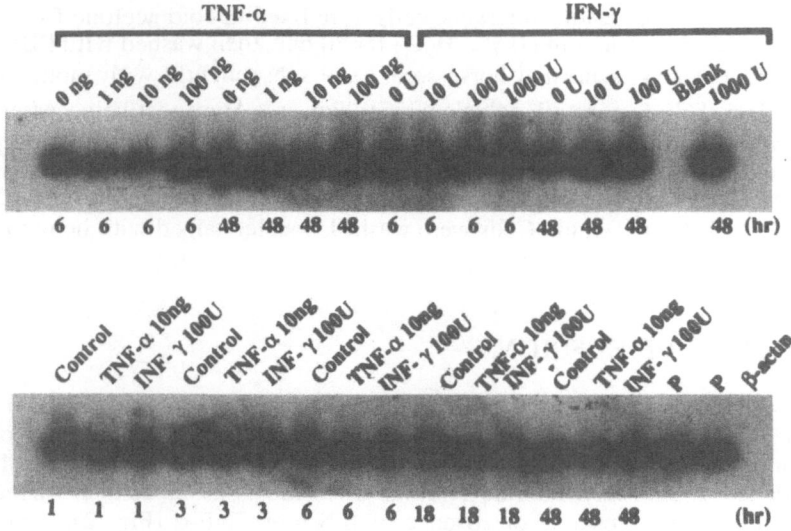

Figure 2. IL-6 expression in CA9-22, shown by PCR-Southern blot. P: RNA from PBMC stimulated with Con A.

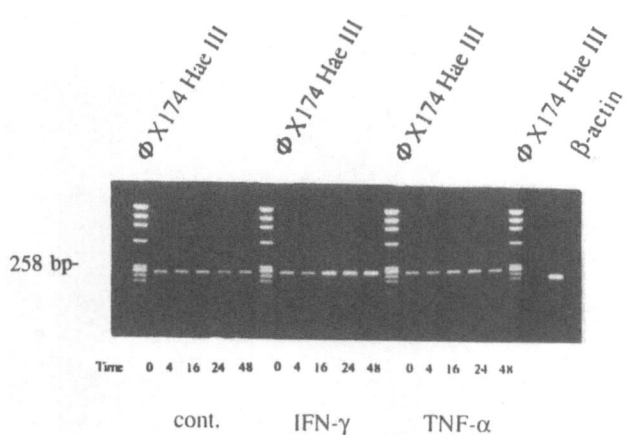

258 bp-

Figure 3. SC expression in HT-29 with or without IFN-γ or TNF-α. PCR (1.7% agarose gel electrophoresis).

mRNA expression is up-regulated by IFN-γ or TNF-α. Although SC is not found in squamous epithelia such as skin by immunological and immunohistochemical methods, SC mRNA was detected in Ca9-22 cells with or without IFN-γ or TNF-α by PCR (Fig. 4). Very weak signals for of TGF-β1 in HT-29 and Ca9-22 cells were also found.

Although PCR is a very sensitive method to detect a small amount of DNA, this technique may not be quantitative. Northern blots were used to detect expression of mRNA for IL-6, SC, and TGF-β1 in HT-29 and Ca9-22 cells. Northern blot analysis for IL-6 in both HT-29 and Ca9-22 cells revealed that mRNA for IL-6 was expressed in both cell lines with or without stimulation by IFN-γ or TNF-α. Figure 5 indicates the expression of IL-6

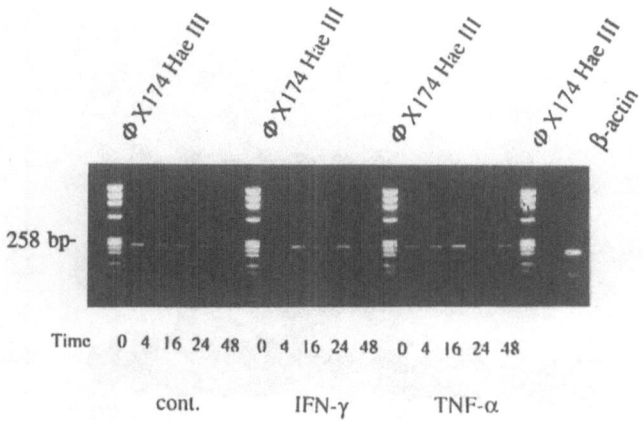

258 bp-

Time 0 4 16 24 48 0 4 16 24 48 0 4 16 24 48

cont.   IFN-γ   TNF-α

Figure 4. SC expression in Ca9-22 with or without IFN-γ or TNF-α. PCR (1.7% agarose gel electrophoresis).

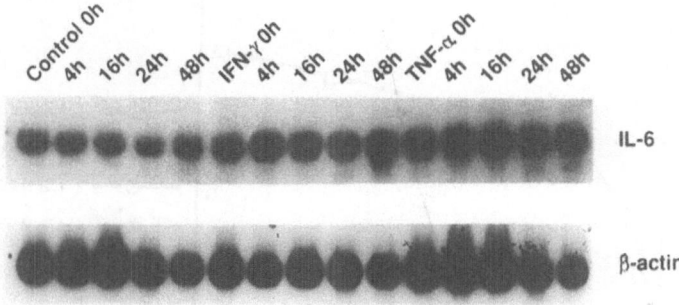

IL-6

β-actin

Figure 5. IL-6 expression in HT-29 with or without IFN-γ or TNF-α. Northern blot analysis.

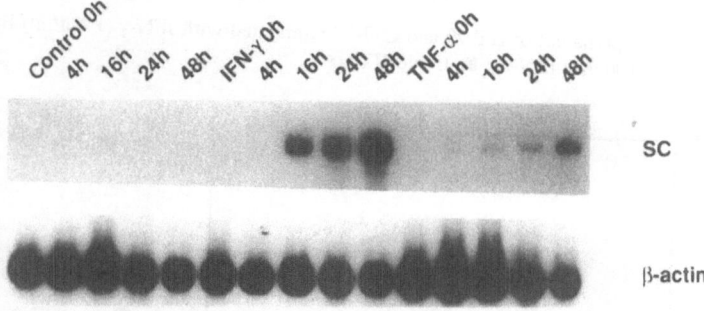

SC

β-actin

Figure 6. SC expression in HT-29 with or without IFN-γ or TNF-α. Northern blot analysis.

in HT-29 cells. Expression of mRNA for SC examined by Northern blot showed very weak signals without stimulation of HT-29 cells by IFN-γ or TNF-α. However, IFN-γ or TNF-α stimulation of HT-29 cells resulted in a dramatic increase of SC mRNA in a time-dependent manner from 16h to 48 h (Fig. 6). The stimulation of HT-29 cells by IFN-γ was more effective than that of TNF-α in the up-regulation of SC. A very weak TGF-β1 signal was detected in HT-29 (Fig. 7) and Ca9-22 cells in the presence or absence of IFN-γ or TNF-α but with little variation according to incubation time.

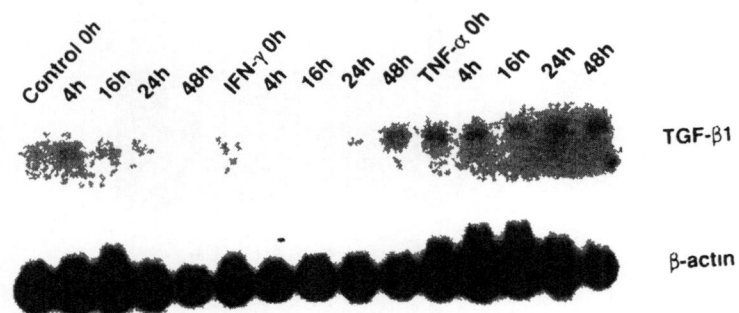

Figure 7. TGF-β1 expression in HT-29 with or without IFN-γ or TNF-α Northern blot analysis

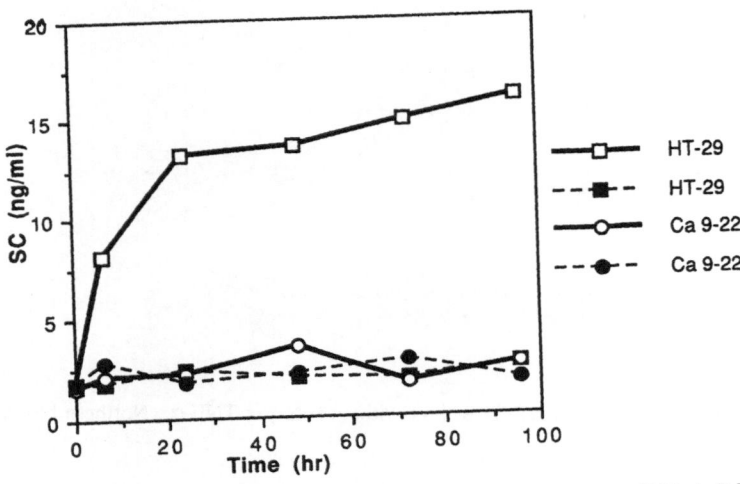

Figure 8. SC in supernatant of HT 29 and Ca9-22 stimulated with IFN-γ (100u/ml) ELISA
◻, 0   100u/ml IFN-α, ■, 0 No IFN-α

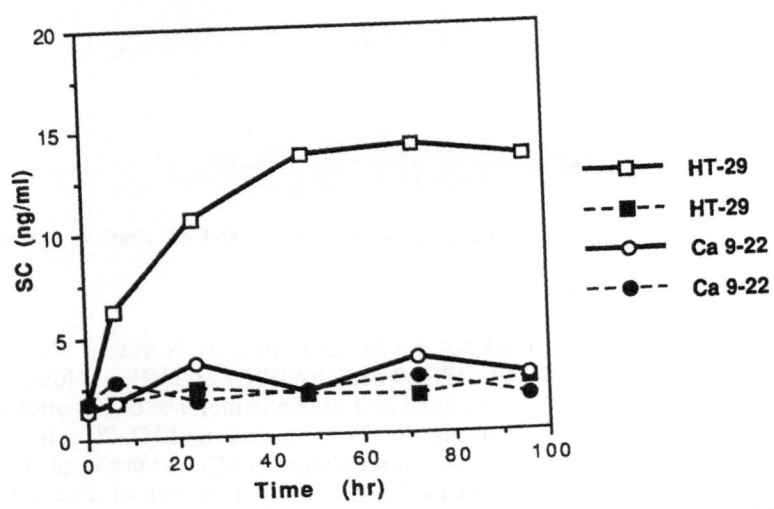

Figure 9. SC in supernatant of HT 29 and Ca9-22 stimulated with TFN-α (10ng/ml) ELISA
◻, 0   10ng/ml TNF- α, ■, 0 TNF α

Since expression of mRNA for IL-6, SC, and TGF-β1 was detected in HT-29 and Ca9-22 cells with or without stimulation by IFN-γ or TNF-α by PCR and Northern blot analysis, the amounts of these proteins in culture supernatants were measured by ELISA. The stimulation of HT-29 and Ca9-22 cells with IFN-γ or TNF-α resulted in the increase of IL-6 in culture supernatants compared to controls. The stimulation of HT-29 cells with IFN-γ or TNF-α increased the level of SC in culture supernatants while little or no SC was detected in culture supernatants of Ca9-22 cells (Fig. 8,9). A small increase of TGF-β1 was detected in HT-29 and Ca9-22 supernatants after stimulation by IFN-γ or TNF-α, and more TGF-β1 was produced by HT-29 cells compared to Ca9-22 regardless of the presence or absence of these cytokines.

The localization of IL-6 in epidermal keratinocytes, SC in various glandular tissues, and of TGF-β1 in intestinal mucosal epithelium, bronchial epithelial cells, biliary ductal cells, and skin has been described.[3,4,5] Since HT-29 and Ca9-22 cells were originally established from colon and squamous cell carcinomas, respectively, we have examined whether these cell lines express IL-6, SC, and TGF-β1 by immunohistochemistry. A small increase in IL-6 positive cells was seen in both HT-29 and Ca9-22 cultures after stimulation with IFN-γ or TNF-α compared to controls. When these cells formed an epithelial island, IL-6 positivity became prominent. A few TGF-β1 positive cells were also detected, but the number of positive cells remained unchanged through whole culture period in either the presence or absence of IFN-γ or TNF-α. Philips et al.[8] reported that approximately 9% of HT-29 E10 cells, a subclone of HT-29, displayed membrane SC expression while 48 h preincubation with 100 u/ml of IFN-γ increased SC positivity to approximately 24%. Similarly, a few SC-positive cells were detected before stimulation by IFN-γ or TNF-α and an increased number of SC-positive cells was seen after stimulation. This result indicated that the production of SC protein in HT-29 cells is up-regulated by IFN-γ or TNF-α. A negligible number of SC-positive cells was also detected in Ca9-22. The detection of SC mRNA in Ca9-22 cells by PCR may reflect the presence of a small number of SC-positive cells because PCR is a highly sensitive method.

We next examined in vitro the production of secretory IgA (sIgA) by ELISA in HT-29 and Ca9-22 cells with or without stimulation by IFN-γ or TNF-α. When confluent, cells were transferred to RPMI-1640 medium containing 10% FCS, and polymeric IgA1 or IgA2 preparations (generously provided by Kunihiko Kobayashi, Hokkaido University, Japan) at a concentration of 10 μg/ml were added to the culture. ELISA results indicated that an increased level of sIgA was detected in HT-29 cultures at day 6 in the presence of IFN-γ, and lesser amounts in the presence of TNF-α. There was no significant difference in amount between sIgA1 and sIgA2. Addition of only polymeric IgA1 or IgA2 to HT-29 as well as Ca9-22 cells, without IFN-γ or TNF-α, also produced a small amount of sIgA1 or sIgA2.

It has been well described that free SC can bind in vitro to polymeric IgA.[13] There is a possibility that sIgA detected in culture supernatants may be formed extracellularly in vitro. Therefore, localization of IgA in HT-29 cells after incubation with polymeric IgA1 or IgA2 in the presence or absence of IFN-γ or TNF-α was examined. There was no IgA in HT-29 cells without incubation with polymeric IgA, while there were many positive cells in HT-29 after incubation with polymeric IgA1 or IgA2 in the presence of IFN-γ. This result indicates that some amount of sIgA is produced intracellularly and secreted into culture supernatants in vitro.

CONCLUSIONS

1. Gene expression of SC, IL-6, and TGF-β1 was detected in HT-29 and Ca9-22 cells with or without stimulation of IFN-γ or TNF-α.

2. SC mRNA was up-regulated by the presence of IFN-γ or TNF-α at the transcriptional level.

3.   Immunohistochemical examination revealed that increased numbers of SC-positive cells were found in HT-29 cultures after stimulation with IFN-γ or TNF-α.

4.   In the presence of IFN-γ or TNF-α, incubation of polymeric IgA with HT-29 cells resulted in an increase of secretory IgA.

5.   SC, IL-6, and TGF-β1 derived from epithelial tissues may play an important role in the production of sIgA.

ACKNOWLEDGMENTS

This work was supported by a Grant-in-Aid for Scientific Research from the Ministry of Education (No. 03404052), the Norwegian Research Council for Science and the Humanities, and the Norwegian Cancer Society.

REFERENCES

1.   J. Mestecky and J. R. McGhee, *Adv. Immunol.* 40:154 (1987).
2.   K. W. Beagley, J. H. Eldridge, F. D. Lee, H. Kiyono, M. P. Everson, W. J. Koopman, T. Kishimoto, and J.R. McGhee, *J. Exp. Med.* 169:2133 (1989).
3.   P. Brandtzaeg, *Scand. J. Immunol.* 22:11 (1985).
4.   K. Yoshizaki, N. Nishimoto, K. Matsumoto, H. Tohgo, T. Taga, Y. Deguchi, T. Kuritani, T. Hirano, K. Hashimoto, N. Okada, and T. Kishimoto, *Cytokine* 2:381 (1990).
5.   N. C. Tompson, K. C. Flanders, J. M. Smith, L. R. Ellingworth, A. B. Roberts, and M. B. Sporn, *J. Cell Biol.* 108:661 (1987).
6.   L. M. Sollid, D. Kvale, P. Brandtzaeg, and D. Lovhaung, *J. Immunol.* 138:4363 (1987).
7.   D. Kvale, P. Brandtzaeg, and D. Lovhaung, Scand. *J. Immunol.* 28:351 (1988).
8.   J. O. Philips, M. P. Everson, Z. Moldoveanu, C. Lue, and J. Mestecky, *J. Immunol.* 145:1740 (1990).
9.   M. Asano, I. Saito, P. Krajci, P. Brandtzaeg, and I. Moro, *in*: "Frontiers of Mucosal Immunology", M. Tsuchiya, H. Nagura, N. Hibi, and I. Moro, eds., Vol. 1., p. 85 (1991).
10.   F. Ferre and F. Garduno, *Nuc. Acid Res.*, 17:2141 (1989).
11.   P. Krajci, R. Solberg, M. Sandberg, O. Øyen, T. Jahnsen, and P. Brandtzaeg, *Biochem. Biophys. Res. Commun.* 158:783 (1989).
12.   T. Hirano, Y. Yasukawa, H. Harada, T. Taga, Y. Watanabe, T. Matsuda, S. Kashiwamura, K. Nakajima, K. Koyama, A. Iwamura, S. Tsunasawa, F. Sakiyama, H. Matsui, Y. Takahara, T. Taniguchi, and T. Kishimoto, *Nature* 324:73 (1986).
13.   J. Radl, H. R. E. Schuit, J. Mestecky, and W. Hijmans, *Adv. Exp. Med. Biol.* 45:57 (1974).

BUTYRATE DIFFERENTIALLY AFFECTS CONSTITUTIVE AND CYTOKINE-INDUCED EXPRESSION OF HLA MOLECULES, SECRETORY COMPONENT (SC), AND ICAM-1 IN A COLONIC EPITHELIAL CELL LINE (HT-29, clone m3)

Dag Kvale[1,2] and Per Brandtzaeg[2]

[1]Medical Dept. A
[2]Laboratory for Immunohistochemistry and Immunopathology (LIIPAT), Institute of Pathology, University
of Oslo, The National Hospital, Rikshospitalet
N- 0027 Oslo, Norway

INTRODUCTION

Short chain fatty acids (SCFAs) are produced in the human body only by anaerobic bacterial fermentation of non-absorbed carbohydrates in the colon. SCFAs (mainly acetic, propionic, and butyric acids) are the predominant aqueous solutes of normal stool, amounting to a total concentration of 100 to 240 mM. Butyric acid, which accounts for about 17% of the SCFAs, is perhaps the most important energy source for normal colonic epithelial cells.[1] Diversion colitis, which may resemble inflammatory bowel disease both clinically and histologically, can be effectively treated by local application of SCFAs.[2] Patients with ulcerative colitis (UC) apparently have decreased ability to metabolize butyrate[3] and also decreased fecal butyrate concentrations.[4] Case reports even suggest that active UC may be dampened by "fecal" enema from healthy donors,[5] possibly through reconstituting a normal bacterial flora.

Butyrate concentrations from 1 to 10 mM apparently have profound effects on the phenotype and proliferation of cultured human cells; transformed cells usually develop differentiation characteristics resembling the tissue of origin.[6] These effects of butyrate have been extensively studied *in vitro* and also *in vivo* over the last few years.

Epithelial cells of the gut exert more or less well-defined immunological functions.[7] We have previously studied how various peptides may modulate such functions in colonic adenocarcinoma cell lines[8-10] because primary epithelial cultures are difficult to establish. The aim of this work was to study butyrate in relation to expression and regulation of immunological characteristics of colonic epithelial cells in an *in vitro* model.

MATERIALS AND METHODS

Cell Lines

The human colonic adenocarcinoma cell line HT-29m3 was used. These cells were subcloned from the HT-29m2 cell line previously selected for high expression of secretory component (SC).[11] The cells were grown in microplates under standard culture conditions as described elsewhere.[10]

Antibodies

The following monoclonal (mAb) and polyclonal antibody reagents were used: (a) anti-human secretory component (SC) mAb, ascites at 1/1800 dilution;[11] (b) anti-human HLA class I mAb, clone W6/32, ascites at 1/3000 (Sera-lab, Sussex, England); (c) anti-human HLA-DR mAb, clone L243, at 1/1600 (Becton Dickinson, Sunnyvale, CA); and (d) anti-human intercellular adhesion molecule-1 (ICAM-1) mAb, clone 84H10, purified IgG 0.25 μg/ml (Serotec, Oxford, U.K.); (e) rabbit anti-mouse IgG at 1/800 dilution (Dakopatts, Glostrup, Denmark); and (f) peroxidase-conjugated swine anti-rabbit IgG at 1/3000 dilution (Dakopatts). A control mAb to an *Aspergillus* enzyme (Dakopatts) was used as purified IgG at 1 μg/ml as control for non-specific binding of mAbs.

Reagents

Recombinant human interferon-γ (IFN), tumor necrosis factor-α (TNF), interleukin-1β (IL-1), and IL-4 were purchased from Genzyme (Boston, MA), and their activities defined as IU according to standard procedures performed by the manufacturers. Sodium butyrate and o-phenylenediamine were purchased from Sigma (St. Louis, MO).

Quantitative Cellular Enzyme-Linked Immunosorbent Assay (CELISA) for HLA Class I, HLA Class II, SC, and ICAM-1 Expression

CELISA with solid phase of fixed HT-29m3 monolayers was performed as described in detail elsewhere.[10] This assay was made quantitative by including reference cells on each microplate, consisting of cytokine-stimulated standard HT-29m3 cells (IFN, 50 IU/ml and TNF, 50 IU/ml; 24 h) grown at graded densities on the same microplate. Immunofluorescence of monolayers grown in Leighton tubes was used to confirm that the expression of the various markers was fairly equal at different sub-confluent cellular densities.

The standard curve from the reference wells was constructed by plotting the CELISA $OD_{492 \, nm}$ values against the actual standard cell counts which were expressed as the crystal violet content measured at $OD_{550 \, nm}$.[10] The linear correlation coefficient r in the log-log standard plot was usually above 0.85. The total cellular expression of the relevant marker protein was measured in CELISA by plotting $OD_{492 \, nm}$ of the test sample wells against the standard curve. This result was then divided by the actual cell number in the well as measured by the crystal violet method to estimate marker expression per cell unit. Standard cells were given an arbitrary value of 1000 units.

Results of experiments are for each test point presented as medians of quadruplicate measurements. The CELISA technique for surface membrane expression of the various markers on viable cells has been described elsewhere.[10]

RESULTS

Effects of Butyrate on Cell Growth

Butyrate inhibited cell growth in a dose-dependent manner. Growth was further inhibited by about 15% when IFN, TNF, and IL-1 was co-incubated at 100 IU/ml with butyrate, and even more so when combinations of these cytokines were used (data not shown).

General Observations Based on CELISA

CELISA with fixed cells, measuring the marker expression per cell unit, showed results almost parallel to those obtained with CELISA on viable cells for determination of surface membrane expression (data not shown). Only CELISA results obtained with fixed cells will be presented below. CELISA values obtained with control mAb were always similar to background OD appearing after omitting the primary mAbs.

Butyrate Affects Basal and Cytokine-Induced HLA Class I, HLA Class II, SC, and ICAM-1 Expression Differentially and in a Dose-Dependent Manner

When HT-29m3 cells were incubated for 96 h with various concentrations of butyrate, the constitutive expression of SC and HLA class I molecules increased 2-4 times above the control in a dose-dependent manner. HLA class II was undetectable in controls and in cells incubated with butyrate < 4 mM, but *de novo* DR expression took place after incubation with butyrate at 8 mM. Constitutive ICAM-1 expression was very low and was not enhanced by the presence of butyrate (Fig. 1).

When cells preincubated with butyrate were co-stimulated with the combination of TNF and IFN, all four markers were increased compared to controls. However, the dose-response relationships were different: (i) SC expression was further enhanced by butyrate; (ii) HLA class II and ICAM-1 were maximally enhanced by butyrate at 4 mM; and (iii) HLA class I expression was reduced by butyrate in a dose-dependent manner (Fig. 1).

Kinetics of Butyrate and Cytokine-Induced HLA Class I, HLA Class II, SC, and ICAM-1 Expression

When HT-29m3 cells were treated with 3 mM butyrate for up to 5 days, a small increase in HLA class I and SC expression was observed, whereas the HLA class II and ICAM-1 levels remained negative (data not shown).

When IFN and TNF at 50 IU/ml were co-incubated with butyrate for the last 24 h of these experiments, three different response patterns evolved: (i) SC expression increased slightly from day 1; (ii) the cytokine-induced enhancement of HLA class II and ICAM-1 showed an initial decrease over the first 48 h, but was higher than cytokine-treated control cells from day 3; and (iii) the HLA class I response to cytokines decreased markedly the first 48 h similar to HLA class II and ICAM-1, but class I expression remained lower than cytokine-stimulated controls when butyrate was present (data not shown).

Effects of IL-1, IL-4, IFN, and TNF on HLA class I, HLA Class II, SC, and ICAM-1 Expression are Differentially Modulated by Butyrate

The overall results of these experiments are shown in Fig. 2. In general, 60 h pretreatment with 3 mM butyrate followed by 36 h co-incubation of butyrate and various combinations of cytokines, tended to further enhance the cytokine-induced increase of HLA class I, HLA class II, SC, and ICAM-1, but with two exceptions: (i) Co-incubation with butyrate selectively decreased the TNF-induced enhancement of HLA class I expression even in the presence of other cytokines; and (ii) the IL-4 induced enhancement of SC expression decreased in the presence of butyrate. These data indicated that the unexpected response patterns described above for HLA class I molecules after co-incubation of butyrate, IFN, and TNF were mainly caused by the presence of TNF.

DISCUSSION

Butyrate is a fermentation product of anaerobic bacteria in the normal human colon which has profound effects on cellular biology in several human tissues. Butyrate regularly induces or enhances cellular differentiation and decreases the growth rate of

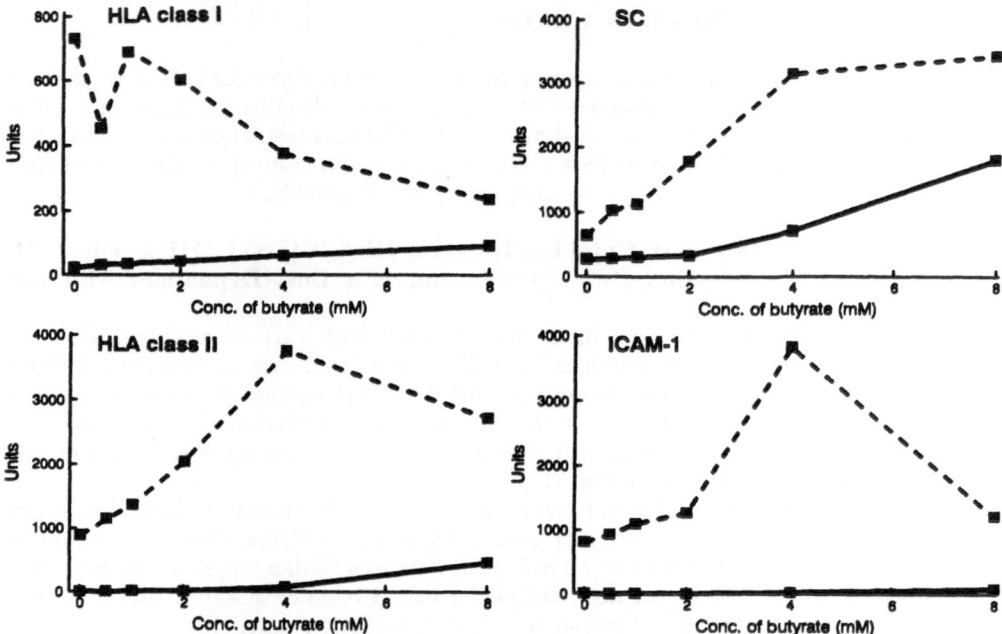

Figure 1. Quantitative CELISA results for HLA class I, HLA class II, SC, and ICAM-1 after incubation (96 h) with butyrate at various concentrations (solid lines) and after coincubation with IFN (50 IU/ml) and TNF (50 IU/ml) for the last 24 h (ICAM-1) or 48 h (HLA molecules and SC) (broken lines). Each data point represents the median cellular expression (quadruplicate wells) in relation to reference cells assigned an arbitrary value of 1000 U. The dip of HLA class I expression obtained with 0.5 mM butyrate was consistent in three different experiments.

transformed cells.[6] Butyrate and its analogues are therefore used in clinical trials for treatment of cancer.[12] The mechanisms by which butyrate exerts its effects are still unclear. Because butyrate inhibits histone deacetylase, it was initially suggested that alterations in chromatin structure and butyrate-associated gene activation were caused by hyperacetylation of nuclear histones.[6] However, a recent reports suggests that specific 5'-flanking sequences mediate butyrate-dependent gene regulation.[13]

Only the large bowel tissue is substantially influenced by butyrate *in vivo*; normal colonic and rectal epithelial cells may even depend on butyrate for normal metabolism and function. We have been particularly interested in the immunological functions of intestinal epithelial cells and wanted to examine whether their immunological features might be modulated by butyrate in our *in vitro* model. It is noteworthy that butyrate has been reported to induce other characteristics of differentiation in the HT-29 cell line.[14]

Butyrate was shown to increase the constitutive expression of HLA class I and SC molecules in the HT-29m3 cells employed in our model system. Interestingly, *de novo* synthesis of HLA class II molecules took place at higher butyrate doses. We have previously found that under normal culture conditions, IFN appears to be necessary for expression of HLA class II, and that TNF[15] and IL-1 (unpublished) may potentiate this effect. Epithelial HLA-DR expression in normal colonic mucosa remains controversial, but the balance of evidence suggests that it is mainly negative.[7] One may speculate whether butyrate under certain conditions might induce a variable and marginal constitutive epithelial expression of HLA-DR in the normal colon. Moreover, the possibility exists that butyrate facilitates cytokine-induced class II upregulation during inflammation.[7] By contrast,

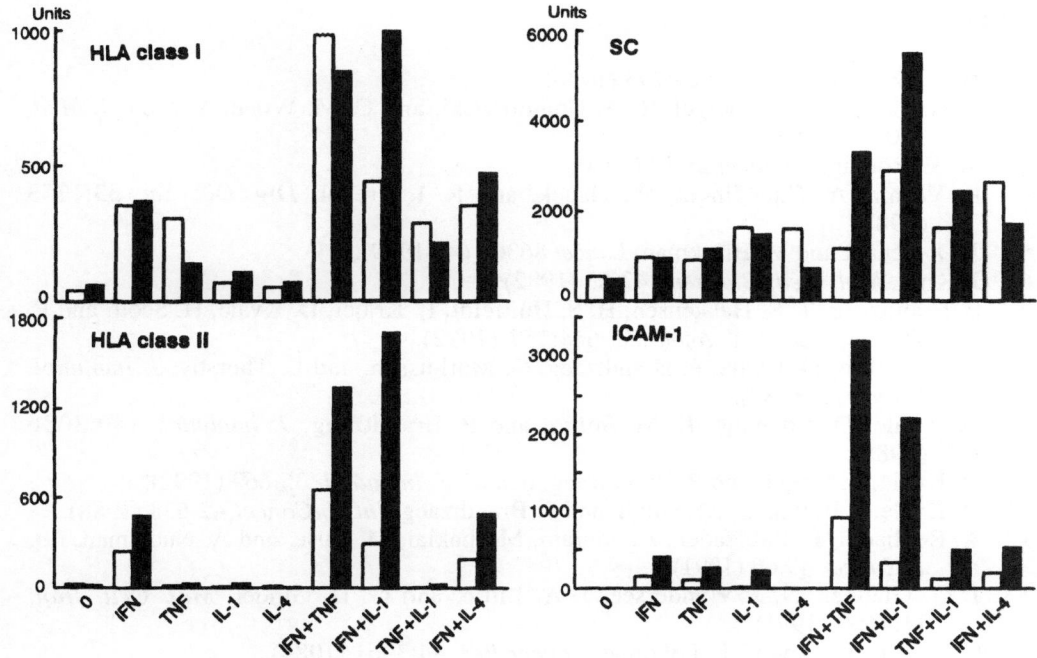

Figure 2. Quantitative CELISA results for HLA class I, HLA class II, SC, and ICAM-1 after incubation (36 h) with IFN (100 IU/ml), TNF (100 IU/ml), IL-1 (200 IU/ml), IL-4 (200 IU/ml), or various combinations of these cytokines in the absence (open columns) or presence (filled columns) of 3 mM butyrate (84 h). Each column represents the median cellular expression (quadruplicate wells) in relation to reference cells assigned an arbitrary value of 1000 U.

ICAM-1, which also becomes considerable upregulated by cytokines in HT-29m3 cells,[10] was not stimulated at all by butyrate alone.

Butyrate enhanced most of the cytokine-induced responses of HLA class I, class II, SC, and ICAM-1 molecules, perhaps because of a variable additive effect of butyrate itself or increased expression of cytokine receptors or molecules involved in intracellular signalling. However, butyrate reduced consistently the effect of TNF on HLA class I expression. The possibility exists that TNF-specific regulatory elements on the HLA class I gene are sensitive to butyrate. The effect of IL-4 on SC expression was likewise reduced by butyrate.

SUMMARY

In conclusion, we have found that physiological concentrations of butyrate increase the constitutive levels of HLA class I and SC molecules in HT-29m3 cells. Moreover, butyrate at high concentrations induces *de novo* synthesis of HLA-DR but not ICAM-1 molecules. Our data further showed that butyrate generally facilitates the cytokine-induced expression of immunological molecules in these cells but, interestingly, it specifically reduces the stimulatory effects of TNF and IL-4 on HLA class I and SC, respectively. We are currently studying how butyrate affects transcriptional regulation of the genes encoding these molecules. Further knowledge about the effects of butyrate in relation to gene regulation is required. It is possible that butyrate might interfere with specific, cytokine-dependent regulatory elements in certain genes which, in turn, could have implications for immune regulation of colonic epithelial cells *in vivo*.

REFERENCES

1. W. E. W. Roediger, *Gut* 21:793 (1980).
2. J. M. Harig, K. H. Soergel, R. A. Komorowski, and C. M. Wood, *N. Engl. J. Med.* 320:23 (1989).
3. W. E. Roediger, *Lancet* 2:712 (1980).
4. P. Vernia, A. Gnaedinger, W. Hauck, and R. I. Breuer, *Dig. Dis. Sci.* 33:1353 (1988).
5. J. D. Bennet and M. Brinkman, *Lancet* 8630:164 (1989).
6. J. Kruh, *Mol. Cell. Biochem.* 42:65 (1982).
7. P. Brandzaeg, T. S. Halstensen, H. S. Huitfeldt, P. Krajci, D. Kvale, H. Scott, and P. Thrane, *Ann. N.Y. Acad. Sc.* 664:157 (1992).
8. L. M. Sollid, D. Kvale, P. Brandtzaeg, G. Markussen, and E. Thorsby, *J. Immunol.* 138:4303 (1987).
9. D. Kvale, D. Løvhaug, L. M. Sollid, and P. Brandtzaeg, *J. Immunol.* 140:3086 (1988).
10. D. Kvale, P. Krajci, and P. Brandtzaeg, *Scand. J. Immunol.* 35:669 (1992).
11. D. Kvale, J. Bartek, L. M. Sollid, and P. Brandtzaeg, *Int. J. Cancer* 42:638 (1988).
12. A. Rephaelii, E. Rabizadeh, A. Aviram, M Shaklai, M. Ruse, and A. Nudelman, *Int. J. Cancer* 49:66 (1991).
13. J. G. Glauber, N. J. Wandersee, J. A. Little, and G. D. Ginder, *Mol. Cell. Biol.* 11:4690 (1991).
14. C. A. Augeron and C. L. Laboisse, *Cancer Res.* 44:3961 (1984).
15. D. Kvale, P. Brandtzaeg, and D. Løvhaug, *Scand. J. Immunol.* 28:351 (1988).

CYTOKINES INDUCE AN EPITHELIAL CELL CYTOKINE RESPONSE

Spencer Hedges, Majlis Svensson, William Agace
and Catharina Svanborg

University of Lund
Department of Medical Microbiology
Clinical Immunology Slvegatan 23
Lund S223-46, Sweden

INTRODUCTION

Intravesical inoculation of gram negative bacteria or isolated bacterial products into mice induces an IL-6 response which can be measured within miutes of the infection.[1,2] In such infections, IL-6 is initially detected in the urine and subsequently in the serum. Similar IL-6 responses were found in human patients deliberately colonized in the urinary tract with *E. coli* Hu7343. IL-6 was secreted intermittantly into the urine in response to continuous bacterial infection in humans, but was not detected in serum. The rapid secretion of IL-6 into urine after bacterial stimulation, and the separation of local from systemic secretion in both humans and mice suggested that IL-6 was produced at the site of infection. Since epithelial cells dominate the naive urinary tract mucosal surface, we suggested that epithelial cells are one source of mucosally produced cytokines.[4] Epithelial cell lines of urinary tract origin secrete IL-6 and IL-8, and can produce other cytokines after bacterial stimulation.[5-7] In this paper we demonstrate that epithelial cell lines of urinary tract origin can respond to exogenous cytokines. This response includes the up-regulation of a variety of cytokine mRNA species as well as secretion of IL-6 and IL-8. Furthermore, this response is specific to the stimulus used, with different cytokine mRNA species induced by different cytokine stimuli.

RESULTS

Epithelial Cell Lines Produce Cytokines Constitutively

The A-498 (kidney) and J82 (bladder) epithelial cell lines were tested for the production of cytokine mRNA and cytokine secretion after stimulation with fresh medium. The supernatants were removed at zero, two, six, and 24 hours post stimulation, the cells were lysed, and the mRNA was extracted. Total mRNA was reverse transcribed and aliquots of cDNA were used to detect specific mRNA species with PCR.[8] The concentrations of secreted cytokines in the supernatents were determined by bioassay (IL-6 (B99), TNF (WEHI10)) or ELISA (IL-1α, IL-1β, IL-8). Both cell lines produced constitutive levels of mRNA and secreted protein for different cytokines (Table 1). PCR amplification products of the correct size were detectable from the bladder mRNA for IL-1α, IL-1β, IL-6, and IL-8. In comparison, only IL-6 and IL-8 amplification products were detectable from the kidney cell line. Both cell lines secreted IL-6 constitutively, however constitutive IL-8 was only detected in the kidney cell line supernatants.

Advances in Mucosal Immunology, Edited by
J. Mestecky *et al.,* Plenum Press, New York, 1995

Table 1. Constitutive cytokine mRNA production by epithelial cell lines and secreted protein levels in supernatants after 24 hours in culture.

| | CELL LINE | | | |
| | A-498 (kidney) | | J82 (bladder) | |
	mRNA	protein[1]	mRNA	protein[1]
actin	+		+	
IL-1α	-	-	+	-
IL-1β	-	-	+	-
IL-6	+	180	+	400
IL-8	+	300	+	<200
TNFα	-	-	-	-

Notes: - not detected, + detected during the 24 hour test period.
  [1]Protein measurements are in pg/ml

Epithelial Cell Cytokine Secretion In Response To Cytokine Stimulation

We tested whether the human kidney (A-498) and bladder (J82) epithelial cell lines could secrete cytokines in response to IL-1α or TNFα stimulation. The cells were exposed to either IL-1α (1ng/ml) or TNFα (10 ng/ml) and the cytokine concentrations in the supernatants were determined at different times after stimulation. IL-1α and TNFα stimulated epithelial cytokine secretion above constitutive levels. Maximum levels of secreted IL-6 were detected at 24 hours post stimulation for both stimulants (Fig. 1A). However, IL-1α induced a rapid IL-6 response with a peak rate of secretion between two and six hours in both cell lines. Maximum levels of secreted IL-8 were also detected at 24 hours post stimulation for both stimulants and cell lines (Fig. 1B). IL-1β was detectable above background in the bladder cell line supernatants after cytokine stimulation, but the levels were not measurable (data not presented). TNFα and IL-1α were not detected in the supernatants.

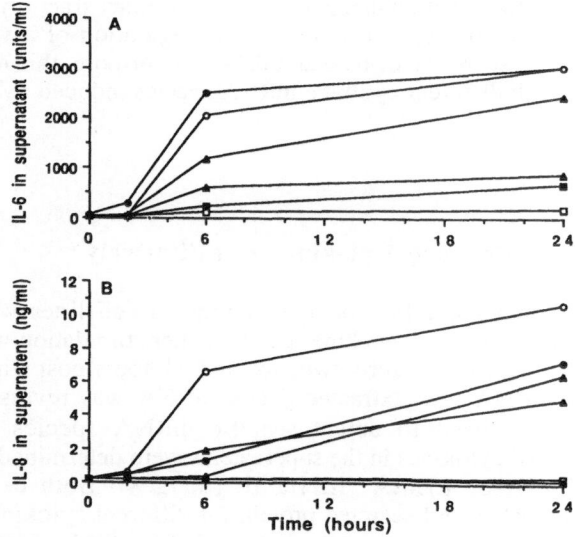

Figure 1. Epithelial cell line IL-6(A) and IL-8 (B) secretion after stimulation with either media, IL-1α, or TNFα. ——■—— J82 media stimulated, ——△—— J82 TNFα stimulated, ——●—— J82 IL-1α stimulated, ——□—— A-498 media stimulated, ——△—— A-498 TNFα stimulated, ——○—— A-498 IL-1α stimulated.

Epithelial Cell Cytokine mRNA production In Response To Cytokine Stimulation

We tested whether cytokine stimulation of epithelial cell lines up-regulated different cytokine mRNA species. The profile of cytokine mRNA species which could be detected over the 24 hour period was compared to the levels of secreted cytokines in the 24 hour supernatants (Table 2). IL-1α stimulation induced up-regulation of IL-1α, IL-1β, IL-6, and IL-8 mRNA in both cell lines compared to the levels at time zero. TNFα stimulation up-regulated IL-1α, IL-1β, IL-6, IL-8, and TNFα mRNA in the bladder cell line, but only IL-6 and IL-8 in the kidney cell line. A comparison of supernatant cytokine levels with detectable mRNA after cytokine stimulation demonstrated three patterns: high levels of mRNA and high levels of secreted protein (IL-6, IL-8), high levels of mRNA and no or low levels of secreted protein (IL-1α, IL-1β), low or no levels of mRNA and no detection of secreted protein (TNFα).

Table 2. Epithelial cell line cytokine mRNA detection compared to secreted cytokine levels at 24 hours post stimulation in response to either TNFα or IL-1α stimulation.

	STIMULUS			
	TNFα		IL-1α	
A-498 (kidney)	mRNA	protein[1]	mRNA	protein[1]
actin	+		+	
IL-1α	-	-	++	ND
IL-1β	-	-	++	-
IL-6	++	760	++	2880
IL-8	++	6400	++	10600
TNFα	-	ND	-	-
J82 (bladder)	mRNA	protein[1]	mRNA	protein[1]
actin	+		+	
IL-1α	++	-	++	ND
IL-1β	++	<7.8	++	<7.8
IL-6	++	2715	++	2885
IL-8	++	5000	++	7200
TNFα	++	ND	+	-

Notes: - not detected, + detected during the 24 hour test period,
++ detected with increased band density compared to 0 hour levels, ND not done.

[1]Protein measurements are in pg/ml

DISCUSSION

Epithelial cell lines of urinary tract origin secrete IL-6 in response to bacterial stimulation.[4,5] Recently it has been shown that these epithelial cell lines can also produce IL-8 and IL-1α as well as IL-6 in response to bacterial stimulation.[6,7] In this study, we have shown that epithelial cell lines of urinary tract origin respond to stimulation with exogenous cytokines. IL-1α and TNFα stimulated IL-6 and IL-8 secretion and induced mRNA production for a variety of cytokines. Furthermore, the types of mRNA produced were found to depend on the cytokine used for stimulation. The ability of the epithelial cell lines to respond to cytokines, as well as to secrete cytokines, suggests that these cells may be capable of functioning within a mucosal cytokine network. Epithelial cytokines may either stimulate or inhibit secondary cytokine production at the site of infection. Target cells for epithelial cytokines could be other epithelial cells, another local cytokine responsive cell type, or influxing lymphocytes. These lymphocytes may in turn produce more cytokines to modify the epithelial cytokine production. Such a mucosal cytokine network may be responsible for some of the observations relating to *in vivo* IL-6 secretion such as intermittant IL-6 secretion, and the IL-6 response to trauma in the presence of bacterial infection.[3,11]

The regulation of cytokine secretion by cytokines has mainly been examined with respect to T cells and monocytes.[12-15] In those cells cytokine induced cytokine production can be stimulated via both autocrine or paracrine pathways, depending on the cytokine stimulus, the type of cytokine produced, and the cell type being stimulated. Growing evidence suggests that many non-lymphoid cell types, including epithelial cells of non-urinary origin, can also produce cytokines in response to different stimuli.[16-22] These non-lymphocyte derived cytokines may play a direct role in the hosts immune response. Epithelial cytokines such as IL-8 may direct the migration of lymphocytes to the site of infection. At the mucosal epithelium, cytokines may control lymphocyte responses such as PMN activation,[23] as well as having direct effects on the epithelium itself such as up-regulation of adhesion molecules.[22] Host interactions with bacteria, viruses, or parasites require a variety of specific responses such as the secretion of different cytokines at different times. The urinary tract epithelial cell lines are capable of flexible cytokine production; different stimuli induce different cytokine mRNA profiles. The complexity of the epithelial cells cytokine response, including their ability to respond within a cytokine network may be crucial to the maintenance of mucosal immunity. The activities of mucosal epithelial cells suggests that they should be viewed as mucosal sentry cells, capable of defense, and early warning to the immune system.

ACKNOWLEDGMENTS

We would like to thank Ann Catrin Simonsson and Carl Borrebaeck for their help in the developement of the PCR technique. This study was supported by grants from The Swedish Medical Research Council (grant no.7934-01), The Medical Faculty at the University of Lund, Sweden, The Tesdorpf, Sterlund and Crawford Foundations.

REFERENCES

1. P. de Man, C. Van Kooten, L. Aarden, I. Engberg, and C. Svanborg-Eden, *Infect. Immun.* 57:3383 (1989).
2. H. Linder, I. Engberg, H. Hoschtzky, I. Mattsby Baltzer, and C. Svanborg-Eden. *Infect. Immun.* 59:4357 (1991).
3. S. Hedges, P. Anderson, G. Lidin-Janson, and C. Svanborg-Eden, *Infect. Immun.* 59:421 (1991).
4. S. Hedges, P. de Man, H. Linder, C. Van Kooten, and C. Svanborg-Eden. *in*: "Advances in Mucosal Immunology, 5th International Conference of Mucosal Immunity", T. Macdonald, ed., p.144, Kluwer, London (1990).
5. S. Hedges, M. Svensson and C. Svanborg, *Infect. Immun.* 60:1295 (1992).
6. W. Agace, S. Hedges, U. Andersson, J. Andersson, and C. Svanborg-Eden, Submitted.
7. W. Agace, S. Hedges, M. Ceska, and C. Svanborg-Eden, Submitted.
8. C. Brenner, *et al.,, BioTechniques* 7:1096 (1989).
9. L. Aarden, E. de Groot, O. Schaap, and P. Lansdorp, *Eur. J. Immunol.* 17:1411 (1987).
10. T. Espevik and J. Nissen-Meyer, *J. Immunol. Methods* 95:99 (1986).
11. S. Hedges, H. Linder, P. De Man, and C. Svanborg-Eden, *Scand. J. Immun.* 31:335 (1990).
12. J. Le, *et al., Proc. Natl.Acad. Sci. USA,* 85:8643 (1988).
13. S. Navarro, N. Debili, J. Bernaudin, W. Vainchenker, and J. Doly, *J. Immunol.* 142:4339 (1989).
14. D.Cheung, P. Hart, G. Vitti, G. Whitty, and J. Hamilton, *Immunology* 71:70 (1990).

15. C. van Kooten, I. Rensink, D. Pascual-Salcedo, R. van Oers, and L. Aarden, *J Immunol.* 146:2654 (1991).
16. H. Grace, W. Wong and D.V. Goeddel, *Nature* 323:819 (1986).
17. P. Defilippi, P. Poupart, J. Tavernier, W. Fiers, and J. Content, *Proc. Natl. Acad. Sci. USA* 84:4557 (1987).
18. P. Sehgal, Z. Walther, and I. Tamm, *Proc. Natl. Acad. Sci. USA* 84:3663 (1987).
19. F. Jirik, *et al.*, *J. Immunol.* 142:144 (1989).
20. P. Guerne, D. Carson, and M. Lotz, *J. Immunol.* 144:499 (1990).
21. K. Shirota, L. LeDuy, S. Yuan and S. Jothy, *Virchows Archiv B* 58:303 (1990).
22. A. Galy and H. Spits, *J. Immunol.* 147:3823 (1991).
23. J. Brom and W. Knig, *Immunology* 75:281 (1992).

MODULATION OF THE MHC CLASS I AND II MOLECULES BY BACTERIAL PRODUCTS ON INTESTINAL EPITHELIAL CELLS

Eduardo J. Schiffrin, Yves Borel and Anne Donnet-Hughes

Nestlé Research Centre, Vers-Chez-les-Blanc
1000 Lausanne 26, Switzerland

INTRODUCTION

Communicating signals between different cells of the immune system involve both direct cell-cell interactions and soluble factors. In the intraepithelial compartment, enterocytes and lymphocytes are closely associated and can interact through both mechanisms. An example of such interactions is the ability of epithelial cells to present antigen to specific T lymphocytes *in vitro*.[1,2] Class I and Class II major histocompatibility complex molecules displaying antigenic peptides are recognized by specific T cell subsets.[3] MHC class II molecules on antigen presenting cells interact with CD4+ helper T cells while MHC class I molecules interact with CD8+ suppressor cytotoxic lymphocytes. Enterocytes constitutively express Class I and II molecules whose levels are modulated by environmental factors. Since epithelial cells of the gut are in permanent contact with food and bacterial products we examined the modulation of MHC molecules expression on these cells by bacterial products.

METHODS

The human colon adenocarcinoma cell line HT-29 displays features of foetal enterocytes and differentiates, after confluency, when cultured in glucose-free media. Cells grown in the presence of glucose remain undifferentiated. HT-29 cells were thus grown under the two conditions. Upon confluency, the cells were exposed for 48 hours to various bacterial products and the expression of Class I and Class II MHC molecules was examined by FACScan analysis. Immunofluorescence staining was performed using monoclonal antibodies W6/32 and L243. Cholera toxin (CT), staphyloccocal enterotoxin B (SEB), and the bacterial chemoattractant f-MLP (f-MLP) were tested at doses of 10 and 100 ng/ml. E. coli LPS was used at 10 and 100 µg/ml.

RESULTS

Basal expression of Class II was minimal on both cell types and was unaffected by any of the treatments (data not shown). However a constitutive expression of Class I, higher on differentiated cells than on undifferentiated cells, was observed.

SEB and f-MLP down-regulated the expression of Class I molecules on differentiated cells (Fig. 1B & C). No modification of the basal expression was detected on the undifferentiated HT-29 cells with either of these two treatments. CT produced a

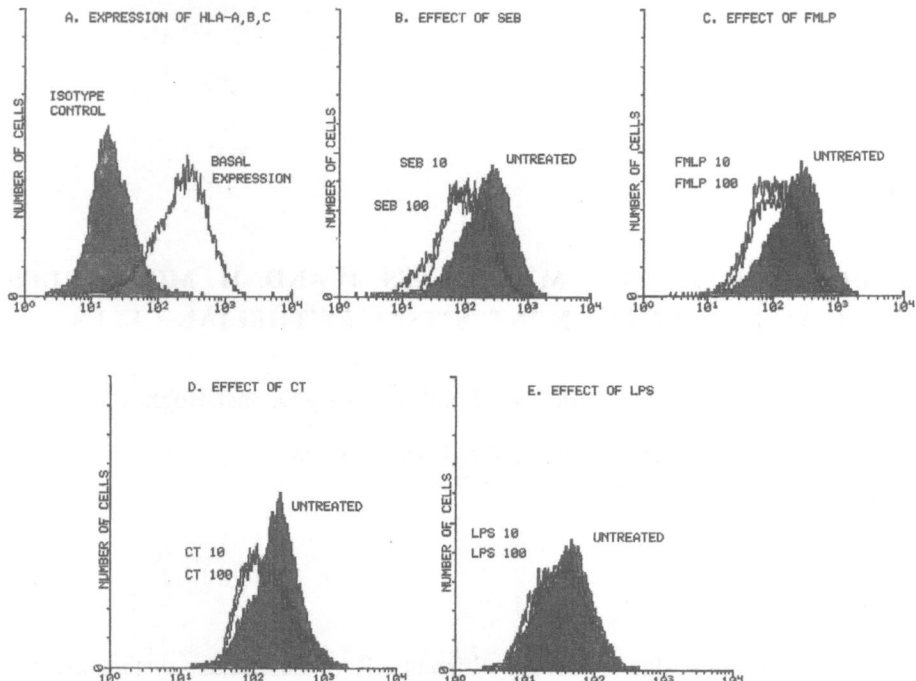

Figure 1. Expression of MHC Class I molecules by HT-29 cells exposed to CT , SEB, f-MLP (10 and 100 ng/ml) or LPS (10 and 100 ug/ml). Basal expression of HLA-ABC antigens on differentiated HT-29 cells (A), treated with SEB (B) f-MLP (C), CT (D) and LPS (E).

diminished expression of MHC Class I on both states of maturation (Fig. 1D). Only a slight down-regulation of Class I was detected on differentiated HT-29 with LPS (Figure 1E), while no effect was observed on undifferentiated cells.

CONCLUSIONS

The bacterial products studied affected Class I but not Class II molecule expression on HT-29 intestinal epithelial cell line. Furthermore, all of the treatments showed a similar trend and the modulation was more apparent for differentiated cells. An interesting finding was that SEB, a superantigen that binds to Class II MHC histocompatibility proteins and activates T lymphocytes bearing particular Vß sequences of the alpha-beta T cell receptor,[4,5] affected the expression of Class I MHC antigens. It is important to note that there is little, if any, Class II expression by the epithelial cells we examined, in consequence, the modification of class I expression may depend on the interaction of SEB with a low amount of class II molecules or with another ligand on the epithelial cell. The ligands involved in epithelial cell and bacterial product interactions are not fully characterized. We are therefore unable to conclude whether our observations result from specific interactions between ligand and bacterial product or from a non-specific mechanism of the bacterial products. However we are able to speculate that these products may influence antigen presentation at the mucosal level.

REFERENCES

1.   D. Kaiserlian, K. Vidal, and J.-P. Revillard, *Eur. J. Immunol.* 19:1513 (1989).
2.   L. Mayer, and R. Schein, *J. Exp. Med.* 166:1471 (1987)
3.   R. N. Germain, and L. Hendrix, *Nature* 353:134 (1991).
4.   J. A. Mollick, R. G. Cook, and R. R. Rich, *Science* 244:817 (1989).
5.   J. D. Fraser, *Nature* 339:221 (1989).

THE EFFECT OF MEDIUM DERIVED FROM ACTIVATED PERIPHERAL BLOOD MONONUCLEAR CELLS ON TWO INTESTINAL CELL LINES

Ulla Kärnström, Ove Norén, and Hans Sjöström

Department of Biochemistry C
The Panum Institute
University of Copenhagen
Copenhagen, Denmark

INTRODUCTION

In celiac disease both immunological reactions and gliadins are supposed to contribute to epithelial damage with villus atrophy, crypt hyperplasia and lymphocyte infiltration.[1,2] A graft versus host reaction in the intestine has similar histology.[3] The same kind of reaction can also be evoked experimentally in human embryonal small intestine by activation of mucosal T-lymphocytes with pokeweed mitogen (PWM).[4] To further study this phenomenon, we have investigated the effect of conditioned medium from PWM stimulated peripheral blood mononuclear cells (PBMC) on the intestinal cell lines HT-29 and IEC-18.

METHODS

PBMC

Human PBMC were isolated by use of Lymphoprep®, according to the guidelines of the manufacterer, and cultured with and without PWM (15 µg/ml) in Dulbecco's modified Eagle medium with 4.5 g/L glucose (DMEM) supplemented with heat-inactivated fetal bovine serum (FBS) 10%, non-essential amino acids (NEAA), penicillin 50 U/ml and streptomycin 50 µg/ml (penstrep). The activation was determined microscopically by estimation of rosette formation and by blast cell count on May-Grünwald Giemsa stained cytocentrifuge slides. The supernatants were filtered through a 0.22 µm filter and added to the intestinal cells after one and two days of PWM stimulation.

Intestinal Cell Lines

The human colon carcinoma cell line HT-29 (ATCC HTB 38) was cultured in DMEM with NEAA, FBS 10% and penstrep. IEC-18 (ATCC CRL 1589), a rat small intestinal crypt cell line, was cultured in DMEM, FBS 5%, bovine insulin 0.1 U/ml and penstrep.

The HT-29 and IEC-18 cells were grown to confluence. They were then cultured in the constant presence (1 µCi/ml) of ^{3}H-methyl thymidine (^{3}H-TdR) with 50% conditioned medium from the PBMC culture. The control cultures with medium from unstimulated PBMC were supplied with PWM. The medium was changed after 1 day. After 2 days the cells were rinsed with cold saline followed by methanol fixation. Cytosolic thymidine was removed with trichloroacetic acid, DNA incorporated ^{3}H-TdR was extracted with perchloric acid and analysed with liquid scintillation.[5] Protein was determined according to Lowry.[6]

The HT-29 cells were cultured in the same manner with conditioned medium from tuberculine (PPD, 10 µg/ml) stimulated PBMC.

RESULTS

PWM-stimulated lymphocytes formed numerous large rosettes, whereas the unstimulated cells had very few and small rosettes. After 2 days about 30% of the stimulated cells were lymphoblasts. No blasts were found among the unstimulated lymphocytes.

Phase contrast microscopy of the HT-29 cells showed a continuous, asynchronous loss of adhesions between cells and to the bottom, only one third of the cells remained attached as compared to the control. In addition there were many cells with membrane blebs and some with karyorrhexis. The morphological characteristics fit well with apoptosis.[7] The IEC-18 cells were morphologically unaffected.

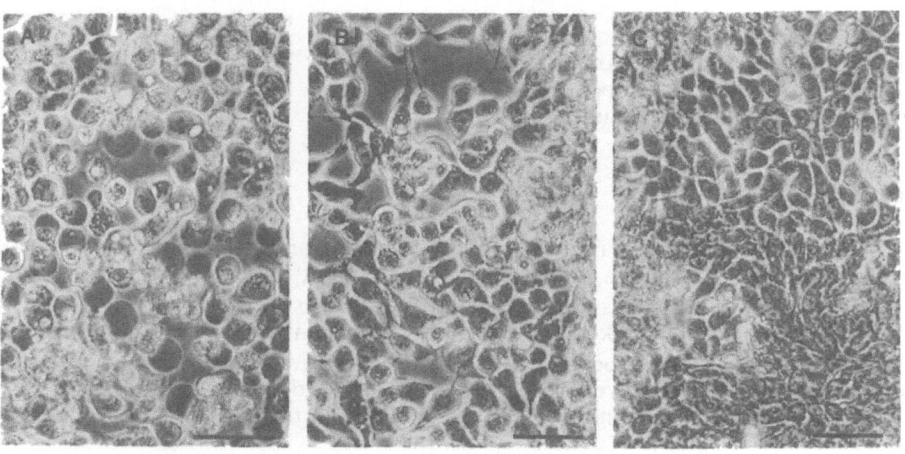

Figure 1. HT-29 cells after 2 days treatment with 50% lymphocyte conditioned medium. A and B supernatants from PWM stimulated PBMC, C control. A before, B and C after rinsing with saline. Bars 50 µm.

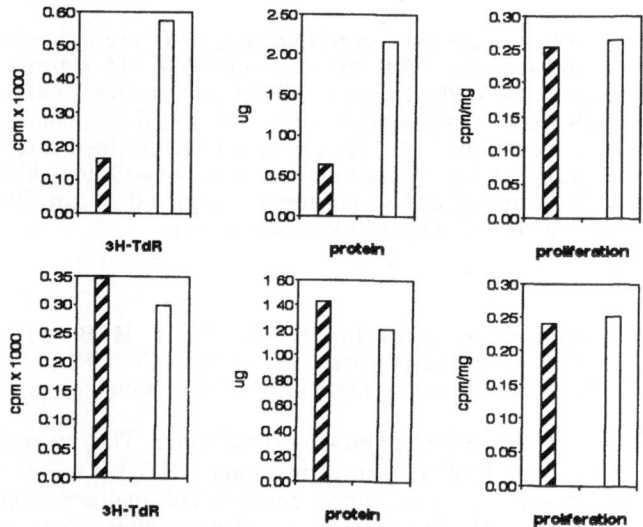

Figure 2. Results from experiments with the same blood donor. Total ^{3}H-TdR, protein and proliferation, in cells cultured for 2 days with 50% PBMC conditioned medium. Upper panel HT-29, lower panel IEC-18. Hatched bars medium from PWM-stimulated PBMC, empty bars control. Mean of triplicates.

The results of ^{3}H-TdR and protein determinations correlate well with the amount of cells remaining at the bottom. Because of the cell loss we have chosen to express proliferation as the ratio of DNA incorporated ^{3}H-TdR and the total protein content. No change in proliferation was noticed within 2 days (Fig. 2). Stimulation with PPD gives similar but less pronounced results with the HT-29 cells.

DISCUSSION

Medium from PWM-activated human PBMC is toxic to confluent cultures of a human, but not a rat, intestinal epithelial cell line and does not affect the proliferation. PBMC activation with the mitogen PWM and the antigen tuberculin gives similar results with the HT-29 cells. We propose that cytokines, mainly from activated T-cells, are responsible for the apoptosis-like cell damage. The difference in sensitivity between the HT-29 and IEC-18 cell lines indicates a specific mechanism, which may be explained by cytokine species differences or lack of appropriate cytokine receptors on the IEC-18 cells. Amomg possible cytokine candidates lymphotoxin and TNF are known to cause apoptsis[7], and human IFN-γ is species-specific.[8] The results indicate that cytokines might play an important role for the mucosal damage in celiac disease, however, we are aware that cancer cells can be a better target for lymphokines then untransformed small intestinal cells.

REFERENCES

1. A. G. F. Davidson and M. A. Bridges, *Clin. Chim. Acta* 163:1 (1987).
2. S. Auricchio, L. Greco, and R. Troncone, *Pediat. Clin. North. Am.* 35:157 (1988).
3. M. N. Marsh and D. E. Loft, *Dig. Dis.* 6:216 (1988).
4. T. T. MacDonald and J. Spencer, *J. Exp. Med.* 167:1341 (1988).
5. R. I. Freshney, *in:* "Culture of Animal Cells" p. 236, Alan R. Liss, Inc., New York (1987).
6. O. H. Lowry, N. J. Rosebrough, A. L. Farr, and R. J. Randall, *J. Biol. Chem.* 193:265 (1951).
7. L. E. Gerschenson and R. J. Rotello, *FASEB J.* 6:2450 (1992).
8. J. Klein, *in:* "Immunology", p. 237, Blackwell Scientific Publications, Inc., Cambridge, Massachusetts (1990).

THE LEUCOCYTE PROTEIN L1 (CALPROTECTIN): A PUTATIVE NONSPECIFIC DEFENCE FACTOR AT EPITHELIAL SURFACES

Per Brandtzaeg,[1] Tor-Øivind Gabrielsen,[1,2] Inge Dale,[1] Fredrik Müller,[1,3] Martin Steinbakk,[4] and Magne K. Fagerhol[4]

[1]Laboratory for Immunohistochemistry and Immunopathology (LIIPAT), Institute of Pathology, [2]Department of Dermatology and [3]Institute of Bacteriology, The National Hospital, Rikshospitalet, and [4]Departments of Microbiology and Immunology, Ullevål Hospital, University of Oslo, Oslo, Norway

INTRODUCTION

The leucocyte antigen L1 is a calcium-binding, highly immunogenic molecule which predominantly consists of different polypeptide chains adding up to an $M_r \approx 36$ K noncovalently stabilized complex.[1] Peptides I and II are virtually identical ($L1_H$ chain) whereas the smaller peptide III is structurally and antigenically in the main different ($L1_L$ chain).[2] L1 was discovered more than 10 years ago as a major cytosol protein fraction (50-60%) of neutrophilic granulocytes.[3] In addition to being expressed by most circulating (and emigrated) neutrophils and monocytes[4], it is also found in a subset of reactive tissue macrophages and many tissue eosinophils.[5] Its abundant occurrence *in vivo* as an $M_r \approx 36$ K peptide complex both in neutrophils and monocytes has recently been confirmed[6,7].

NOMENCLATURE OF THE L1 COMPLEX

In 1987 Odink *et al.*[8] described two macrophage proteins, based on cloning of cDNA, which they called macrophage migration inhibitory factor (MIF)-related proteins 8 and 14 (MRP-8 and MRP-14). The same year Dorin *et al.*[9] had cloned cDNA for a polypeptide that was thought to be the complete cystic fibrosis antigen (CFA), originally isolated from chronic myeloid leukaemia cells and present in large quantities in serum of patients with cystic fibrosis. The deduced sequences of MRP-8 and the putative CFA were nearly the same, and subsequently we showed that MRP-8 and MRP-14 were identical to the $L1_L$ and $L1_H$ chain, respectively.[10] Until now, we have been using the original provisional name L1 for the protein complex. However, in recent *in vitro* experiments the purified composite molecule was shown to exhibit striking antimicrobial properties.[11,12] Therefore we proposed the functional name calprotectin[11], which also has been adopted by others[13], although we still prefer L1 for short.

Advances in Mucosal Immunology, Edited by
J. Mestecky *et al.*, Plenum Press, New York, 1995

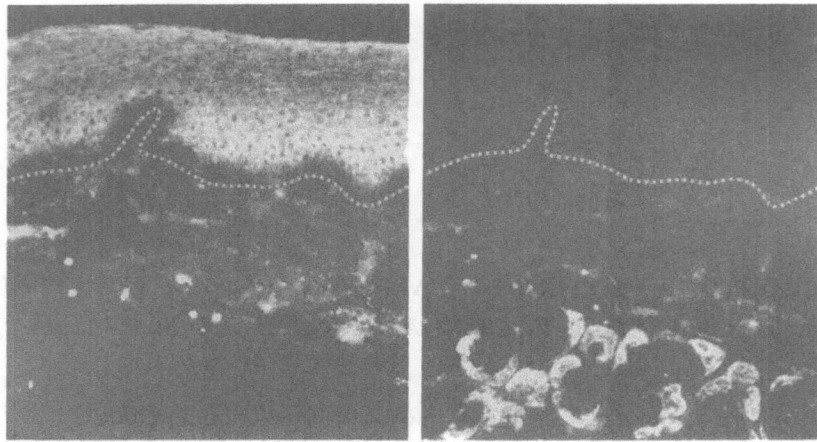

Figure 1. Two-colour immunofluorescence staining for L1 antigen (<u>left</u>, fluorescein) and lysozyme (<u>right</u>, rhodamine) in same field from section of ethanol-fixed normal human labial mucosa. Squamous epithelium except for basal cell layer (above *dashed line*) is selectively L1-positive, whereas serous demilunes of the small salivary gland (at the bottom) are selectively lysozyme-positive. These two myelomonocytic proteins thus show strikingly different epithelial expression patterns, whereas both appear in a few scattered leucocytes and at low levels diffusely in the connective tissue.

We have identified an epithelial protein that is antigenically and physicochemically identical to the myelomonocytic L1 protein[14]; it is expressed normally by squamous epithelia of mucous membranes but by epidermis only in a "reactive state", except for the pilosebaceous units which are often positive.[14,15] Its distribution is similarly visualized by our original rabbit antiserum and the murine monoclonal antibody Mac 387.[16] In other studies of calcium-binding proteins present in keratinocytes[17], the molecule detected by Mac 387 has been termed calgranulin, which is a proposed substitute for CFA, or in fact actually the L1 complex.

L1 IS RELATED TO S-100 PROTEIN

The stoichiometric ratio of $L1_H$ to $L1_L$ chains in the major $M_r \approx 36$ K protein complex remains controversial; 2:1, 1:1 and 1:2 have been proposed in the literature.[2,6,7] Methodological problems may explain these discrepancies, both the striking Ca^{2+} dependency of noncovalent complex formation and artifactual covalent bonding *in vitro*. The fact that the two chains are found to be associated in both neutrophils and monocytes[6,7] contrasts the somewhat incongruent reports on their separate expression by macrophages in acute inflammation[8,20] and perhaps T-cell induced granulomas[20,21], but not in chronic "nonspecific" inflammation[8,20] and "nonimmune" granulomas.[21]

L1 belongs to the steadily growing S-100 protein family of relatively small Ca^{2+}-binding proteins, such as S-100α, S-100β ICaBP, calpactin I, and calcyclin.[3,22] The $L1_L$ chain is most similar to S-100α, while the $L1_H$ chain is more closely related to S-100β. Although the functions of the S-100 proteins are largely unknown, their role in signal transduction is in the focus of interest. Some of the proteins have been suggested as modulators of cell cycle progression, cell differentiation, and cytoskeletal-membrane interactions.[22] It is interesting in this context that L1 has been claimed to be associated with the cytoskeleton of keratinocytes in a calcium-dependent manner.[23]

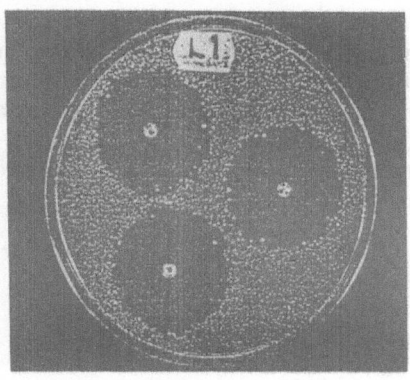

Figure 2. Typical inhibition zones obtained for *Candida albicans* on YNB agar around paper disks containing 10 μg L1 protein.

PUTATIVE BIOLOGICAL FUNCTION OF L1

The abundant expression of L1 in neutrophils, monocytes, certain reactive macrophages, and squamous mucosal epithelia (Fig. 1) may reflect an innate defence function of this protein. It was therefore of considerable interest when we were able to document an *in vitro* antimicrobial effect of L1 tested at biological levels.[11] This effect was particularly striking for *Candida albicans* (Fig. 2). Minimum inhibitory concentrations were 4-32 mg/l for various *Candida* spp. (Fig. 3), 64 mg/l for *Staphylococcus aureus,* 64-256 mg/l for *S. epidermidis*, and 256 mg/ml for *Escherichia coli* and *Klebsiella* spp. Killing was observed at 2-4 times higher concentrations.[11] Similar results were subsequently reported by Sohnle *et al.*[12] Moreover, the C-terminal sequence of the $L1_H$ chain is identical to the N-terminus of peptides known to have neutrophil immobilizing factor (NIF) activity.[24] It has been suggested that generation of NIF activity depends on phosphorylation of the L_H chain.[25] This event could be important for the accumulation of vital granulocytes, while L1 released from dead granulocytes might exert antimicrobial activity by depriving microorganisms of zinc.[26] Perhaps this explains lack of bacterial proliferation in abscess fluid.[13] Alternatively or additionally, L1 might be protective intracellularly against microorganisms that, because of poor opsonization, evade the phagosomes and gain access to the cytosol of phagocytes. In serum, however, there is an as yet unknown factor inhibiting the antimicrobial effect of L1.[11]

L1 expression by normal mucosal squamous epithelia (Fig. 1) and "reactive" epidermis might likewise be related to the possible antimicrobial properties of this protein in primary defence. Interestingly, we have recently found in patients with HIV infection that those who developed oral candidiasis had significantly lower levels of L1 in parotid saliva than those who did not (67 μg/l *vs.* 216 μg/l).[27] This observation, along with reduced parotid output of secretory IgA[28], is probably at least part of the explanation for the strikingly defective salivary anticandidal activities reported in patients with AIDS.[29] The parotid L1 level most likely reflects an individual's release capacity of this antimicrobial factor into the oral cavity, thus being a useful indicator of L1 production by squamous epithelia in this region.

Involvement of L1 in regulation of epithelial cell proliferation and differentiation is another possibility as mentioned above. Murao *et al.*[30] reported that the proliferation of several types of transformed and nontransformed cell lines was inhibited by L1 and proposed that this effect might be explained by inhibition of casein kinase II. However, the high keratinocyte expression of L1 in proliferative conditions such as psoriasis[15] appears ineffective if its upregulation is an immunologically induced phenomenon aiming at epidermal preservation. Our recent observations suggest that although proinflammatory cytokines may enhance epidermal L1 expression, its modulation appears in some way to be linked to keratinocyte differentiation (Gabrielsen *et al.*, unpublished observations). Further work is clearly needed to understand the *in vivo* regulation and function of this intriguing protein.

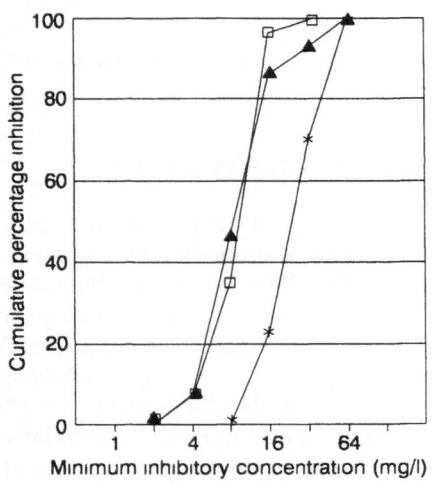

Figure 3. Inhibition of various *Candida* spp. by L1 protein after 20 h incubation (open squares, *C. albicans;* stars, *C. glabrata* and *C. tropicalis;* and triangles, other *Candida* spp).

SUMMARY

The L1 protein occurs at high concentrations in neutrophils, monocytes, certain reactive tissue macrophages, squamous mucosal epithelia, and reactive epidermis. It constitutes in fact about 60% of the neutrophilic cytosol protein fraction. The two L1 chains ($L1_H$ and $L1_L$) are referred to by a bewildering collection of names, various authors having different preferences (MRP-8 and MRP-14; CFA or calgranulin A and B). The most recent proposal is calprotectin because of its calcium-binding properties and antimicrobial effect shown *in vitro*. L1 belongs to the S-100 protein family and may be involved in the regulation of keratinocyte proliferation and differentiation. It exists at high levels in blood and interstitial tissue fluid in several infectious, inflammatory, and malignant disorders, and it is released abundantly in foci of granulocytes and macrophages. The C-terminal sequence of the $L1_H$ chain has been shown to be identical to the N-terminus of peptides known as neutrophil immobilizing factors. Such an activity of L1 could be important for the accumulation of vital granulocytes, while L1 released from neutrophils, macrophages and epithelial cells might exert antimicrobial activity, perhaps by depriving microorganisms of zinc. The minimum inhibitory concentrations of L1 *in vitro* were found to be 4-32 mg/l for *Candida albicans*, 64 mg/l for *Staphylococcus aureus*, 64-256 mg/l for *S. epidermidis*, and 256 mg/ml for *Escherichia coli* and *Klebsiella spp.* Killing was observed at 2-4 times higher concentrations. In patients with HIV infection, those who developed oral candidiasis had significantly lower parotid L1 levels than those who did not (67 µg/l *vs.* 216 µg/l).

ACKNOWLEDGEMENTS

Studies in the authors' laboratories are supported by the Norwegian Cancer Society and by the Norwegian Research Council for Science and the Humanities.

REFERENCES

1. I. Dale, M. K. Fagerhol, and I. Naesgaard, *Eur. J. Biochem.* 134:1 (1983).
2. H. B. Berntzen and M. K. Fagerhol, *Scand. J. Clin. Lab. Invest.* 50:769 (1990).
3. M. K. Fagerhol, K. B. Andersson, C.-F. Naess-Andresen, P. Brandtzaeg, and I. Dale, *in*: "Stimulus Response Coupling: The Role of Intracellular Calcium-Binding Proteins", N. L. Smith, and T. R. Dedman, eds, pp. 187, CRC Press, Boca Raton, Florida (1990).
4. I. Dale, P. Brandtzaeg, M. K. Fagerhol, and H. Scott, *Am. J. Clin. Path.* 84:24 (1985).
5. P. Brandtzaeg, I. Dale, and M. K. Fagerhol, *Am. J. Clin. Path.* 87:681 (1987).
6. S. Teigelkamp, R. S. Bhardwaj, J. Roth, G. Meinardus-Hager, M. Karast, and C. Sorg, *J. Biol. Chem.* 266:13462 (1991).
7. J. Edgeworth, M. Gorman, R. Bennett, P. Freemont, and N. Hogg, *J. Biol. Chem.* 266:7706 (1991).
8. K. Odink, N. Cerletti, J. Brüggen, R. G. Clerc, L. Tarcsay, G. Zwadlo, G. Gerhards, R. Schlegel, and C. Sorg, *Nature* 330:80 (1987).
9. J. R. Dorin, M. Novak, R. E. Hill, D. J. H. Brock, D. S. Secher, and V. van Heyningen, *Nature* 326:614 (1987).
10. K. B. Andersson, K. Sletten, H. B. Berntzen, I. Dale, P. Brandtzaeg, E. Jellum, and M. K. Fagerhol, *Scand. J. Immunol.* 28:241 (1988).
11. M. Steinbakk, C.-F. Naess-Andresen, E. Lingaas, I. Dale, P. Brandtzaeg, and M. K. Fagerhol, *Lancet* 336:763-765 (1990).
12. P. G. Sohnle, C. Collins-Lech, and J. H. Weissner, *J. Infect. Dis.* 163:187-192 (1991).
13. Editorial, *Lancet* 338:855 (1991).
14. P. Brandtzaeg, I. Dale, and M. K. Fagerhol, *Am. J. Clin. Path.* 87:700 (1987).
15. T.-Ø. Gabrielsen, I. Dale, P. Brandtzaeg, P. S. Hoel, M. K. Fagerhol, T. Eeg Larsen, and P. O. Thune, *J. Am. Acad. Dermatol.* 15:173 (1986).
16. P. Brandtzaeg, D. B. Jones, D. J. Flavell, and M. K. Fagerhol, *J. Clin. Pathol.* 41:963 (1988).
17. S. E. Kelly, D. B. Jones, and S. Fleming. *J. Pathol.* 159:17 (1989).
18. M. M. Wilkinson, A. Busuttil, C. Hayward, D. J. H. Brock, J. R. Dorin, and V. van Heyningen, *J. Cell Sci.* 91:221 (1988).
19. C. Barthe, C. Figarella, J. Carrére, and O. Guy-Crotte, *Biochim. Biophys. Acta* 1096:175 (1991).
20. G. Zwadlo, J. Brüggen, G. Gerhards, R. Schlegel, and C. Sorg, *Clin. Exp. Immunol.* 72:510 (1988).
21. J. Delabie, C. De Wolf-Peeters, J. J. van den Oord, and V. J. Desmet, *Clin. Exp. Immunol.* 81:123 (1990).
22. D. Kligman and D. C. Hilt, *TIBS* 13:437 (1988).
23. S. E. Kelly, J. A. A. Hunter, D. B. Jones, B. R. Clark, and S. Fleming, *Br. J. Dermatol.* 124:403 (1991).
24. P. Freemont, N. Hogg, and J. Edgeworth, *Nature* 339:516 (1989).
25. J. Edgeworth, P. Freemont, and N. Hogg, *Nature* 342:189 (1989).
26. P. G. Sohnle, C. Collins-Lech, and J. H. Wiessner, *J. Infect. Dis.* 164:137 (1991).
27. F. Müller, S. S. Frøland, P. Brandtzaeg, and M. Fagerhol, *Clin. Infect. Dis.,* 16:301 (1993).
28. F. Müller, S. S. Frøland, M. Hvatum, J. Radl, and P. Brandtzaeg, *Clin. Exp. Immunol.* 83:203 (1991).

29. J. J. Pollock, R. P. Santarpia, H. M. Heller, L. Xu, K. Lal, J. Fuhrer, H. W. Kaufman, and R. T. Steigbigel, *J. Acquir. Immune Defic. Syndr.* 5:610 (1992).
30. S. Murao, F. R. Collart, and E. Huberman, *Cell Growth Differentiation* 1:447 (1990).

DIFFERENTIAL EXPRESSION OF LEUCOCYTE PROTEIN L1 (CALPROTECTIN) BY MONOCYTES AND INTESTINAL MACROPHAGES

Jarle Rugtveit, Helge Scott, Trond S. Halstensen,
Olav Fausa, and Per Brandtzaeg

LIIPAT, Institute of Pathology and Medical
Department A, University of Oslo, The National
Hospital, Rikshospitalet N-0027 Oslo, Norway

INTRODUCTION

L1 is a major cytosol protein complex present in virtually all circulating monocytes and neutrophilic granulocytes[1], but expressed by only a subset of tissue macrophages.[2] It consists of two subunits (L1$_L$ and L1$_H$ chains) which are identical to the proteins called MRP-8 and MRP-14.[3] Its abundant occurrence *in vivo* as heterodimers (and perhaps tetramers) in neutrophils and monocytes has recently been confirmed.[4,5] The cDNA sequence of MRP-8 was reported to be nearly identical to that of the cystic fibrosis antigen.[6,7] L1 has calcium-binding properties and the alternative name calprotectin reflects its antimicrobial properties [8] Antiproliferative properties against tumour cell lines have also been found.[9] L1 has been reported to be associated with the cytoskeleton of keratinocytes[10,11] and monocytes[1,2] in a calcium-dependent manner; this could possibly be of importance for motility and recruitment of the latter cells. Thus, the abundant expression of this protein by inflammatory leucocytes, including reactive macrophages,[2] is supposedly of significant biological importance in local defence.

In a recent study of mucosal macrophages in inflammatory bowel disease (IBD), we found a large number of L1+ macrophages, becoming more predominant with increasing degree of inflammation (Rugtveit *et al.*, manuscript in preparation). We wanted to study whether this feature represents an upregulation of L1 in resident macrophages or recruitment of monocytes from blood.

MATERIALS AND METHODS

Cell Culture

Monocytes from four healthy blood donors were isolated by separation on Lymphoprep (Nycomed, Oslo, Norway) and subsequent adherence to plastic cell culture wells (Costar, Cambridge, MA) coated with 0.1% gelatin (Sigma, St. Louis, MO) and 50% human plasma.[1,3] Briefly, the blood mononuclar cells (3-4 x10^6 cells/ml medium) were incubated for 1 hour in RPMI-1640 (Whittaker Bioproducts, Walkersville, MD) with 5% human AB serum in the wells before non-adherent cells were removed by three gentle washings with prewarmed medium. The adherent cells were then cultured in the medium with 10% AB serum for 9 days, the last 2 days with 200 U/ml IFN-γ (Genzyme Corp., Cambridge, MA) with or without addition of 1 μg/ml lipopolysaccharide (LPS from *E. coli* Serotype 026 B:6; Sigma, St. Louis, MO) and with LPS alone. Penicillum (100 U/ml), streptomycin (100 μg/ml), gentamycin (40 μg/ml), and L-glutamine (2 mmol/l) were added to all media.

Macrophages were isolated from normal small intestinal mucosa of three necro-organ donors by EDTA/collagenase treatment, using a modification of the technique described by Bull and Bookman[1,4] followed by adherence separation. Briefly, mucosal tissue was transported from the operating room in 0.9% saline at 4°C and washed extensively in Ca- and Mg-free phosphate-buffered saline, pH 7.4 (CMF-PBS), before incubation for 15 min at 20 °C (with shaking) in CMF-PBS with 1 mM dithiothreitol. This was followed by four incubations (with shaking) at 37 °C, 30 min each, in CMF-PBS with 5 mM EDTA to remove epithelial cells. The lamina propria was then dissected and cut into small fragments which were incubated for 3-5 hours at 37°C in RPMI-1640 with collagenase A (from *Clostridium perfringens*; Boehringer Mannheim Biochemica, Mannheim, Germany) at 1 mg/ml in 25 mM HEPES buffer. The cell suspension was filtered through a 200 μm nylon mesh and centrifuged on Lymfoprep (Nycomed A/S, Oslo, Norway) to obtain mononuclear cells. The suspension was then washed three times with CMF-PBS before adherent cells were obtained as described above for blood monocytes. The adherent cells were cultured for 2-3 days with 1000 U/ml granulocyte/macrophage colony-stimulating factor (GM-CSF) (kindly provided by Schering-Plough Research, Bloomfield, NJ) with or without 200 U/ml IFN-γ. Adherent cells from one of the donors were also stimulated with IFN-γ and LPS in combination.

The endotoxin content in the cell suspension during collagenase treatment was 180-320 ng/l (cromogenic limulus amebocyte lysate assay; KabiVitrum AB, Stockholm, Sweden); we thus had to consider all the isolated intestinal macrophages as LPS-stimulated.

Cells for cytospins were obtained at the desired time intervals by incubation with 5 mM EDTA in CMF-PBS at 4°C for 15-30 min and using a "rubber policeman". The cytospins were acetone-fixed for 5 min at room temperature and stored at -70°C until used.

Two-colour Immunofluorescence Staining for CD68 and L1

Cytospins fixed for 10 min in 10% formalin at 4°C were incubated for 1 hour with a 1:50 dilution of rhodamine-conjugated (IgG, 0.035 g/l) anti-L1[1,5] and the monoclonal antibody (mAb) KP1 (courtesy Dr. D. Mason, Oxford, UK) which recognizes the myelomonocytic marker CD68.[16] This was followed by fixation in periodate-lysine-(2%) paraformaldehyde at 4°C for 10 min before incubation with biotinylated horse anti-mouse IgG antibody (IgG 0.025 g/l) (Vector Laboratories Inc., Burlingame, CA) for 2 hours, and fluorescein isothiocyanate (FITC) -labelled streptavidin (Boehringer Mannheim Biochemica, Mannheim, Germany) at 0.02 g/l for 30 min.

Two-colour Sequential Staining for ICAM-1 (CD54) and CD68

Acetone-fixed cytospins were incubated with mAb KP1 (anti-CD68) at 1:50 dilution for 1 hour, followed by FITC-conjugated rabbit anti-mouse IgG (Zymed Laboratories Inc., San Francisco, CA) at 1:10 dilution for 30 min. This was followed by incubation with mAb to ICAM-1 (Serotec Ltd., Oxford, UK) at 1:400 for 1 hour, then with biotinylated horse-anti mouse IgG (0.025 g/l) (Vector) for 2 hours and Streptavidin Texas Red (Bethesda Research Laboratories, Life Technologies Inc., Gaithersburg, MD) at 0.0034 g/l for 30 min. As a control, this paired staining was also done in the opposite sequence, starting with anti-ICAM-1.

RESULTS

Evaluation of L1 expression in cytospins of monocytes and isolated intestinal macrophages by paired immunofluorescence staining in relation to CD68 (mAb KP1) demonstrated a marked decrease in the monocytes after 3 days in culture (Fig. 1). There was less than 5% L1$^+$ intestinal macrophages and no increase appeared after the stimulations described above either in monocytes or macrophages. By contrast, paired sequential staining for ICAM-1 (CD54) and CD68 demonstrated upregulation of ICAM-1 in 15% of monocytes (range, 15-20%) after incubation with IFN-γ and LPS for 2 days, and in 15% (range, 10-30%) of intestinal macrophages cultured with IFN-γ for 2-3 days.

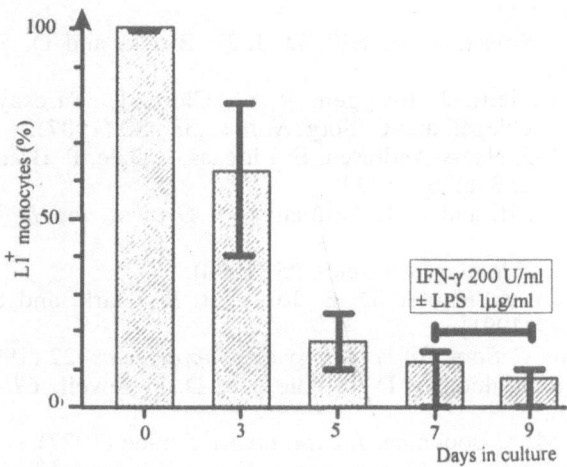

Figure 1. Fraction of CD68$^+$cells expressing L1 in cultures of isolated blood monocytes stimulated with IFN-γ (200 U/ml) with or without addition of LPS (1 µg/ml) from day seven to nine of the culture period (cultures from four healthy donors). Values (%) are expressed as median and observed range (n=8).

DISCUSSION

Downregulation of L1 expression in monocytes during the first week in culture has also been found by others and reported to be accompanied by upregulation of the maturation antigen 25E10.[1,7] Our results indicated that L1 synthesis cannot be induced in mature macrophages by cytokines such as IFN-γ in the abscence or precence of LPS, although a fraction of the monocytes/macrophages demonstrated the ability to upregulate ICAM-1 under these conditions. This suggested that the appearance of L1$^+$ macrophages in inflamed tissue is most likely a result of an increased influx of monocytes from peripheral blood and not caused by an upregulation of L1 in resident macrophages.

Differential regulation of the two L1 subunits has been described in the reactive macrophages in inflammatory tissue, differing between various chronic granulomatous conditions[1,8], with downregulation of the L1$_L$ chain in reactive macrophages in acute, but usually not in chronic inflammatory conditions.[1,9] It was therefore speculated that one of the failures in chronic inflammation is dysregulation of monocyte-to-macrophage differentiation.[1,9] Further studies of such putative differential regulation of cellular L1$_L$ and L1$_H$ expression, comparing isolated macrophages from normal and diseased intestinal mucosa, are certainly needed to evaluate this interesting hypothesis in relation to the pathogenesis of IBD.

REFERENCES

1. I. Dale, P. Brandtzaeg, M. K. Fagerhol, and H. Scott, *Am. J. Clin. Pathol.* 84:2 (1985).
2. P. Brandtzaeg, I. Dale, and M. K. Fagerhol, *Am. J. Clin. Pathol.* 87:681 (1987).
3. K. B. Andersson, K. Sletten, H. B. Berntzen, M. K. Fagerhol, I. Dale, and P. Brandtzaeg, *Nature* 332 688 (1988).
4. S. Teigelkamp, R. S. Bhardwaj, J. Roth, G. Meinardus-Hager, M. Karas, and C. Sorg, *J. Biol. Chem.* 266:13462 (1991).
5. J. Edgeworth, M. Gorman, R. Bennett, P. Freemont, and N. Hogg, *J. Biol. Chem.* 266:7706 (1991).
6. J. R. Dorin, M. Novak, R. E. Hill, D. J. H. Brock, and D. S. Secher, *Nature* 326:614 (1987).
7. K. Odink, N. Cerletti, J. Brüggen, R. G. Clerc, L. Tarcsay, G. Zwadlo, G. Gerhards, R. Schlegel, and C. Sorg, *Nature.* 330:80 (1987).
8. M. Steinbakk, C.-F. Naess-Andresen, E. Lingaas, I. Dale, P. Brandtzaeg, and M. K. Fagerhol, *Lancet* 336:763 (1990).
9. S. Murao, F. Collart, and E. Huberman, *Cell Growth and Differentiation* 1:447 (1990).
10. B. K. Clark and S. Fleming, *J. Pathol.* 25 (1990).
11. S. E. Kelly, J. A. A. Hunter, D. B. Jones, B. R. Clark, and S. Fleming, *Br. J. Dermatol.* 403 (1991)
12. K. B. R. Mahnke, C. Sorg, *Pathobiology 60 (Supplement)* :22 (1992).
13. K. C. Wu, Y. R. Mahida, J. D. Priddle, and D. P. Jewell, *Clin. Exp. Immunol.* 79:35 (1990).
14. D. M. Bull and M. A. Bookman, *J. Clin. Invest.* 59:966 (1977).
15. I. Dale, M. K. Fagerhol, and I. Naesgaard, *Eur. J. Biochem.* 134:1 (1983).
16. K. A. F. Pulford, E. M. Rigney, K. J. Micklem, M. Jones, W. P. Stross, K. C. Gatter, and D. Y. Mason, *J. Clin. Pathol.* 42:414 (1989).
17. G. Zwadlo, R. Schlegel, and C. Sorg, *J. Immunol.* 137:512 (1986).
18. J. Delabie, C. DeWolf-Peeters, J. J. Van Den Oord, and V. J. Desmet, *Clin. Exp. Immunol.* 81:123 (1990).
19. G. Zwadlo, J. Bruggen, G. Gerhards, R. Schlegel, and C. Sorg, *Clin. Exp. Immunol.* 72:510 (1988).

COMPLEMENT COMPONENT C3 PRODUCTION AND ITS CYTOKINE REGULATION BY GASTROINTESTINAL EPITHELIAL CELLS

Akira Andoh,[1] Yoshihide Fujiyama,[1] Tadao Bamba,[1] Shiro Hosoda,[1] and William R. Brown[2]

[1]Department of Internal Medicine, Shiga University of Medical Science, Otsu, Japan; [2]Division of Gastroenterology, Veterans Affairs Medical Center and University of Colorado School of Medicine, Denver, CO 80220

INTRODUCTION

In the gastrointestinal tract, several studies have focused on evidence of local complement activation as one of the factors concerned with the pathogenesis of gastrointestinal disorders. Recently, evidence of local complement component C3 production in intestinal tract were demonstrated *in vivo* by Ahrenstedt *et al.*,[1] but its cellular origin has not been identified. In this study, we examined the possibility of C3 production by intestinal epithelial cells using colonic adenocarcinoma cell lines; Caco-2, HT-29, and HT-29N$_2$ cells. The *de novo* synthesis and secretion of C3 by all three cells were observed, and its production was up-regulated by certain cytokines such as interleukin (IL)-1β, IL-6 and tumor necrosis factor (TNF)α.

MATERIALS AND METHOD

Cytokines

Cytokines were kindly provided or purchased as follows: recombinant (r) human IL-1β (2×10^7 U/mg, defined by murine thymocyte co-stimulation assay), Otsuka pharmaceutical Co., Ltd. (Tokushima, Japan); r-human IL-6 (2×10^4 U/mg, by T-1165 cell proliferation assay), Genzyme Co. (Cambridge, MA); r-human TNFα (2.5×10^6 U/mg, by cytotoxic assay against murine L-M cells), Dainippon Pharmaceutical Co., Ltd. (Osaka, Japan).

Cell Culture

Caco-2 cells were maintained in Dulbecco's modified Eagle medium (DMEM) supplemented with 20% FCS, and 1% non-essential amino acids in a 24-well tissue culture plate. HT-29 and HT-29N$_2$ cells were maintained in RPMI-1640 supplemented with 10% FCS in a 12-well tissue culture plate. Cells were grown in a humidified atmosphere of 5% CO$_2$/95% air. All experiments were started on 4 days after confluence.

Advances in Mucosal Immunology, Edited by
J. Mestecky *et al.*, Plenum Press, New York, 1995

Sandwich ELISA for the Quantification of Immuno-Reactive C3

Wells of 96-well flat-bottomed microtiter plates were coated with goat IgG anti-human C3 (Cappel, Cochranville, PA) in a 50 mM carbonate-bicarbonate buffer (pH 9.6) overnight at 4°C. After washing and blocking of non-specific protein binding, samples were incubated for 2h at 37C°. The wells were then washed and and incubated with rabbit IgG anti-human C3 (Sigma, St. Louis, MO). Finally, peroxidase-conjugated goat IgG anti-rabbit IgG (Zymed, San Francisco, CA) was used. The lower limit of this ELISA was 500 pg/ml of C3.

Metabolic Labeling and Immunoprecipitation

Caco-2 cells were incubated in methionine-free DMEM supplemented with 20% dialyzed FCS and [^{35}S]-L- methionine (18.5MBq/ml) for a predetermined period. HT-29 cells were also incubated in methionine-free RPMI-1640 supplemented with 10% dialyzed FBS and [^{35}S]-L-methionine in the presence of TNFα (500U/ml), because the amount of C3 production by HT-29 cells was minimal in the absence of stimulators (see results). At the end of the incubation time, supernatants were harvested and cells were lysed in 20 nM Tris-HCl buffer (pH 7.4) containing 10mM EDTA, 100mM NaCl, 0.5% NP-40 and 2mM PMSF. Aliquots of supernatants and lysates were incubated overnight at 4°C with excess rabbit anti-human C3. A suspension of protein A-Sepharose CL-4B was then added and incubated for 2 h at 4°C. After extensive washing, antigen-antibody complexes were released by boiling in sample buffer and applied to 10% SDS-PAGE under reducing conditions. After electrophoresis, gels were impregnated with 2,5-dipenyloxazole and dried for fluorography.

RESULTS

Intra and Extracellular C3 Molecules Synthesized by Caco-2 and HT-29 Cells

In metabolic labeling experiments on Caco-2 cells, the intracellular single-chain precursor of C3 (pro C3; 180kD) and disulphide-linked double-chains of native C3 (115kDa α chain and 70kDa β chain) were detected (Fig. 1). In the same way, intracellular pro-C3 and extra-cellular C3, synthesized by HT-29 cells cultured in the presence of 500 U/ml of TNFα, were also detected (Fig. 1). These C3 molecules were compatible with prevously reported C3 forms.[3]

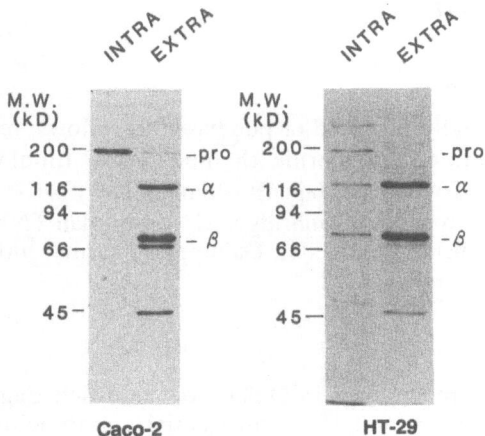

Figure 1. Biosynthesis of C3 by Caco-2 cells and HT-29 cells. Intracellular pro-C3 and extracellular C3 were immunoprecipitated and analyzed by SDS-PAGE under reducing conditions

The Up-Regulation of C3 Production by IL-1β and TNFα

Because it was previously reported that IL-1β and TNFα induce C3 production hepatocytes[4], we examined the effects of these cytokines on C3 production by intestinal epithelial cells. The C3 production of Caco-2 cells was enhanced by the addition of IL-1β or TNFα in a dose-dependent manner (Fig. 2). Similarly, C3 production by HT-29 and HT-29N2 cells was also enhanced by the addition of IL-1β of TNFα in a dose-dependent manner (Fig. 3). In the absence of cytokines, the C3 levels produced by HT-29 and HT-29N2 cells were under the lower limit of ELISA, but Caco-2 cells was constitutively produced large amounts of C3.

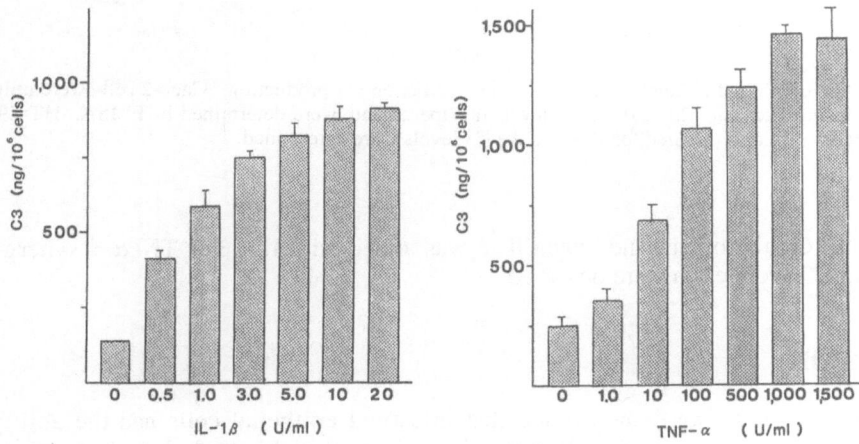

Figure 2. C3 production by Caco-2 cells in response to IL-1β and TNFα. The supernatants of Caco-2 cells, cultured in various concentrations of cytokines for 72h, were collected and C3 levels were determined by ELISA

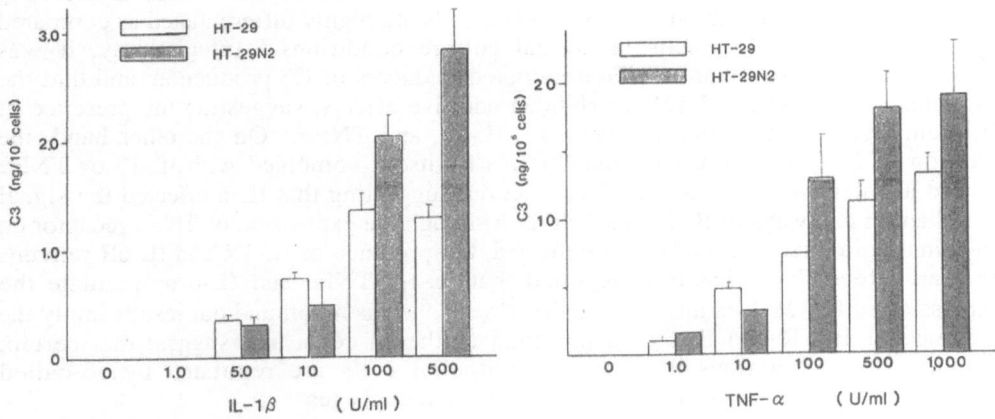

Figure 3. C3 production by HT-29 and HT-29N cells. The supernatants of cells, cultured in various concentration of cytokines for 48h, were collected and C3 levels were determine by ELISA

Combined Effects of IL-1β, IL-6 and TNFα on C3 Production

The simultaneous addition of IL-1β and TNFα markedly enhanced C3 production by Caco-2, HT-29, and HT-29N2 cells (Fig. 4). IL-6 is also known as a cytokine that modulates inflammatory events[5], but it exhibited only a weak enhancement effect on C3

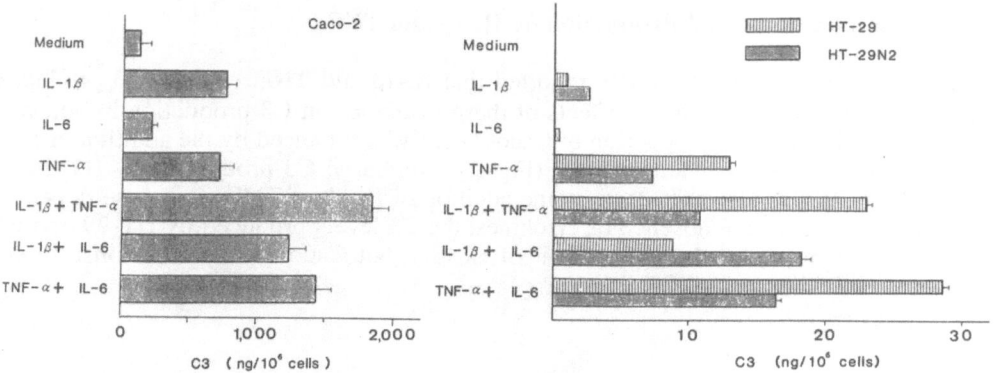

Figure 4. The combined effects of IL-1β, IL-6 and TNFα on C3 production. Caco-2 cells were cultured with various cytokines for 72h, and the C3 levels in supernatants were determined by ELISA. HT-29 and HT-29N2 cells were also cultured for 48h, and the C3 levels were determined.

production. On the other hand, when IL-6 was added with IL-1β or TNFα, a synergistic increase in C3 production were observed.

DISCUSSION

In this study, we demonstrated that intestinal epithelial cells had the ability to synthesize and secrete native C3. Because the cells used in this study have been regarded as good experimental models of normal epithelial cells by morphological and functional criteria[2], our results suggest that the complement components in intestinal fluid are locally synthesized by intestinal epithelial cells *in vivo*. The differences between the C3 production levels of Caco-2 cells and those of HT-29 or HT-29N2 cells might be related to their degree of differentiation, because Caco-2 cells are highly differentiated as compared to HT-29 or HT-29N2 cells in normal culture conditions. Interestingly, it was demonstrated that IL-1β and TNFα were potent inducers of C3 production, and that the combination of IL-1β and TNFα exhibited additive effects, suggesting the presence of different signal transduction pathways for IL-1β, and TNFα. On the other hand, the addition of IL-6, which had minimal effects by itself, combined with IL-1β or TNFα caused a synergistic increase of C3 production, suggesting that IL-6 affected the signal transduction pathways of IL-1β and TNFα. Although the expression of TNFα-receptor on intestinal epithelial cells has been confirmed, the presence of IL-1R and IL-6R remains unclear. Recently, it has been reported that IL-1β, TNFα and IL-6 up-regulate the expression of ICAM-1 on intestinal epithelial cells.[6] This report and our results imply the presence of IL-1R and IL-6R on intestinal epithelial cells, and suggest that certain immunological functions of intestinal epithelial cells are regulated by so-called inflammatory cytokines derived from monocytes/macrophages.

CONCLUSION

Intestinal epithelial cells are considered to be a local site of C3 production. We have demonstrated a novel immunological function of intestinal epithelial cells as producers of C3.

REFERENCES

1. O. Ahrenstedt, L. Knotson, B. Nilsson, K. Nilsson-Ekdahl, B. Odlind, and R. Hallgren, *N. Engl. J. Med.* 322:1345 (1990).
2. A. Zweibaum, M. Laburthe, E. Grasset, and D. Louvard, *in*: "Handbook of Physiology - The gastrointestinal system", B.B. Rauner, ed., Volume IV, p.223, American Physiological Society, Maryland (1991).
3. H. S. Auerbach, R. Burger, A. Dodds, and H. R. Colten, *J. Clin. Invest.* 86:96 (1990).
4. D. H. Perlmutters, C. A. Dinarello, P. I. Punsal, H. R. Colten, *J. Clin. Invest.* 78:1349 (1986).
5. P. C. Heinrich, J. V. Castell, and T. Andus, *Biochem. J.* 265:621 (1991).
6. D. Kvale, P. Krajci, and P. Brandtzaeg, *Scand. J. Immunol.* 35:669 (1992).

THE POLYMERIC IMMUNOGLOBULIN RECEPTOR: SIGNALS FOR POLARIZED EXPRESSION

O. Poulain-Godefroy,[1] R.P. Hirt,[2] N. Fasel,[2] and J.P. Kraehenbühl[2]

[1]Centre d'Immunologie et Biologie Parasitaire, Institute Pasteur, Lille, France; and [2]Institute of Biochemistry, University of Lausanne, Epalinges, Switzerland

INTRODUCTION

Cells such as epithelial or neuronal cells, when fully differentiated, exhibit a clear functional asymmetry related to morphological asymmetry. Differences in plasma membrane composition are maintained by the presence of intercellular tight junctions which prevent lateral diffusion of membrane components,[1] and by cytoskeletal elements that interact with plasma membrane proteins and restrict their movement.[2,3] The rules that govern intracellular sorting and polarized delivery of membrane proteins to the cell surface is complex and still poorly understood.

MDCK cells, a canine kidney cell line, display a polarized phenotype and have been extensively studied as a model for the polarized expression of membrane proteins. In MDCK cells, membrane proteins for the two cell surfaces are sorted in the trans Golgi network and they enter distinct transport vesicles for direct delivery either to the apical or to the basolateral surface.[4] Recent studies using MDCK cells transfected with cDNAs for plasma membrane proteins indicate that their cytoplasmic domain contains signals for cell surface targeting.[5] The polymeric immunoglobulin receptor (pIgR) is synthesized as a transmembrane glycoprotein, initially sorted to the basolateral pathway, and then reendocytosed and sorted to the apical membrane in transcytotic vesicles.[6,7] During transepithelial transport or at the apical surface, the receptor is cleaved and secretory component (SC) is released into the apical medium.[8,9] Signals for basolateral delivery have been localized on the cytoplasmic tail of pIgR.[10] In contrast, membrane proteins anchored to the plasma membrane via a glycosylphosphatidyl (GPI) anchor are directly sorted to the apical cell surface.[11,12] A chimeric protein composed of the ectoplasmic domain of the basolateral G protein of vesicular stomatitis virus and the sequence necessary for the GPI anchor was shown to be targeted to the apical domain in transfected MDCK cells.[11] All these results suggest that sorting signals are in the cytoplasmic or GPI tail and not in the ectoplasmic domain. To test this hypothesis, we fused the cDNA coding for a single Ig-like domain of Thy-1 (a glycolipid anchored protein), to the cDNA coding for the transmembrane segment and cytoplasmic tail of the pIgR, and transfected it into MDCK cells.

CONSTRUCTION AND TRANSFECTION OF cDNAS CODING FOR THY-1, pIgR OR THY-1/pIgR CHIMERA

MDCK cells were transfected with cDNAs inserted into the pLK-neo glucocorticoid-inducible vector.[13] This inducible expression system has been designed for MDCK cells, because it allows uninduced cells to proliferate and differentiate, before accumulation of the heterologous protein, when high amounts of this alter the phenotype of transfected cells.

The cDNAs encoding rabbit pIgR[6] and murine plasma membrane Thy-1 antigen[14] were used. A 0.35 kb *SacI* fragment encoding the ectoplasmic domain of Thy-1 was fused with the *SalI-BamHI* 0.88 kb fragment from rabbit pIgR cDNA corresponding to a portion of the 3' untranslated region, and encoding the cytoplasmic tail, the transmembrane segment, and a small ectoplasmic stretch including the receptor cleavage sites. After insertion in the pLK-neo vector, these clones were transfected according to a modified polybrene protocol.[15] The selection of the transfected cells was based on their capacity to maintain a polarized phenotype including high electrical resistance and polarized distribution of cell surface markers, and on the level of foreign protein expression following dexamethasone induction.

THY-1, pIgR AND THY-1/pIgR STEADY STATE EXPRESSION IN MDCK CLONES

Protein accumulation was analyzed by western blot using a polyclonal antibody specific for Thy-1 and a monoclonal antibody specific for the pIgR cytoplasmic tail.[16] Thy-1 was detected as a single 25 kDa band in Thy-1 transfected clones, and pIgR was resolved with anti-pIgR tail antibody as a doublet of 112-115 kDa in pIgR clones. The upper band corresponded to a major phosphorylated and glycosylated polypeptide, and the lower band to a minor core glycosylated form. Both antibodies were used to identify in transfected cells the chimeric protein which appeared as a 40-42 kDa doublet. Protein accumulation rate was similar in the three clones following dexamethasone induction for 24 h, but appearance of the upper band of chimeric protein was delayed by 2 h when compared to the lower band.

To determine polarized membrane expression of these three proteins, transfected clones were grown on Transwell filters (Costar, Cambridge, MA, USA). After reaching confluency (resistance higher than 1000 Ohm.cm^2), protein expression was induced by addition of dexamethasone for 16 hours. Cell surface iodination was selectively performed from the apical or basolateral compartment, and after cell lysis, proteins were immunoprecipitated and analyzed on SDS-PAGE. Thy-1 was exclusively recovered from the apical membrane, while pIgR was immunoprecipitated from both cell surface domains. Neither polyclonal anti-Thy-1 antibody nor monoclonal anti-pIgR tail antibody was able to immunoprecipitate any protein from the chimeric clone.

Further analysis by confocal microscopy on saponin-permeabilized MDCK clones was performed. Thy-1 was detected exclusively at the apical cell surface and was almost totally absent from intracellular vesicles. pIgR was detected on the apical and basolateral membrane and also in intracellular vesicles. When analyzed with anti-pIgR tail antibody, the chimeric protein displayed a distribution pattern similar to pIgR. We were not able to detect any labeling of the chimeric protein with polyclonal anti-Thy-1 antibody. These results together with the radioiodination experiment are consistent with a conformational change in the Thy-1 moiety of the chimeric protein resulting from its fusion to the pIgR cytoplasmic tail.

KINETIC ANALYSIS OF THY-1/pIgR TRAFFICKING

Pulse-chase experiments followed by selective cell surface biotinylation were performed. Confluent monolayers were labeled for 30 min with ^{35}S-methionine and ^{35}S-cysteine, chased for different time intervals, and biotinylated at 4°C from one side or the other. After a first immunoprecipitation with anti-pIgR tail antibody, membrane biotinylated proteins were recovered by immunoprecipitation with streptavidin-agarose, and intracellular proteins present in the supernatant were precipitated again with anti-pIgR tail antibody for analysis on SDS-PAGE. Newly synthesized chimeric protein appeared first at the apical surface 30 min after the pulse, followed by a second wave of membrane insertion at the basolateral membrane 30 min later.

The lower Mr polypeptide was processed differently from the upper band, and appeared exclusively at the apical cell surface. A pool of the upper band appeared rapidly at the apical surface, and a second pool reached the basolateral surface 60 min after the pulse. No Thy-1 was recovered in the apical medium suggesting that cleavage did not occur despite the presence of receptor cleavage sites in the chimeric protein.

To analyze differential N-linked glycosylation which could be responsible for the heterogeneity of the chimeric protein, cells were treated with tunicamycin during induction with dexamethasone. Analysis by western blot of cell extracts, showed that the rate of protein synthesis was reduced by more than 50%, and that the chimeric protein was still resolved as a doublet. The observed shift corresponded to the loss of the two glycosylated side chains in Thy-1 after its fusion to the pIgR tail.

DISCUSSION

Thy-1 and pIgR have distinct cell surface expression patterns in transfected MDCK cells; pIgR is initially delivered to the basolateral membrane and then redirected by transcytosis to the apical membrane, while Thy-1 directly reaches the apical membrane. Several studies have shown that the pIgR cytoplasmic tail contains signals for sorting, cell surface targeting, and endocytosis.[10,17,18] In this report, we show that by fusing the cytoplasmic tail of pIgR to Thy-1, we were able to redirect part of Thy-1 from its apical membrane localization to the basolateral cell surface. The cytoplasmic tail of pIgR drives only part of Thy-1 from the apical to the basolateral membrane, suggesting that the ectoplasmic domain of the chimeric protein induces a conformational change on pIgR tail that could partially interfere with basolateral sorting signals. In the case of Thy-1, it has recently been shown that apical sorting information is associated both with the GPI-anchor and the ectoplasmic domain.[19] An apical sorting signal in the ectoplasmic domain might confuse the sorting machinery of the chimeric protein, and raises the problem of dominance between such sorting signals.

Finally, two conformations of the same fusion protein may coexist in the transfected MDCK cells, which are not due to differences in glycosylation. Since wild type pIgR also appears as a doublet when synthesized in the presence of tunicamycin[10,20], it is likely that differences in molecular weight of the chimeric protein are due to differences in the degree of phosphorylation. In the chimera, loss of recognition of the Thy-1 moiety by anti-Thy-1 antibody, its acquired inaccessibility to radioiodination enzymes and probably of cleavage sites may be due to a conformational change in the Thy-1 ectoplasmic domain.

Tail-less pIgR deletion mutants are preferentially delivered to the apical surface, suggesting that the pIgR tail acts as the dominant signal.[10] The chimeric protein only partly behaves as the native receptor, suggesting that the basolateral signal may be modulated by factors including the protein's ectodomain, as reported for an EGF/mannose-6-phosphate receptor chimera. [21]

REFERENCES

1. B. Gumbiner and D. Louvard, *Trends Biochem. Sci.* 10:453 (1985).
2. W. J. Nelson and P. J. Veshnok, *J. Cell Biol.* 103:1751 (1986).
3. E. Rodriguez-Boulan and W. J. Nelson, *Science* 245:718 (1989).

4.   M. P. Lisanti, A. Le Bivic, M. Sargiacomo, and E. Rodriguez-Boulan, *J. Cell Biol.* 109:2117 (1989).

5.   W. Hunziker and I. Mellman, *J. Cell Biol.* 109:3291 (1989).

6.   E. Schaerer, F. Verrey, L. Racine, C. Tallichet, M. Reinhardt, and J.-P. Kraehenbühl, *J. Cell Biol.* 110:987 (1990).

7.   K. E. Mostov and D. L. Deitcher, *Cell* 46:613 (1986).

8.   L. S. Musil and J. U. Baenziger, *J. Cell Biol.* 104:1725 (1987).

9.   R. Solari, E. Schaerer, C. Tallichet, L. T. Braiterman, A. L. Hubbard, and J.-P. Kraehenbühl, *Biochem. J.* 257:759 (1989).

10. J. E. Casanova, G. Apocada, and K. E. Mostov, *Cell* 66: 65 (1991).

11. D. A. Brown, B. Crise, and J. K. Rose, *Science* 245:1499 (1989).

12. M. P. Lisanti, M. Sargiacomo, L. Graeve, A. R. Saltiel, and E. Rodriguez-Boulan, *Proc. Natl. Acad. Sci. USA* 85:9557 (1988).

13. R. Hirt, O. Poulain-Godefroy, J. Billotte, J.-P. Kraehenbühl, and N. Fasel, *Gene* 111:199 (1992).

14. J. M. Wilson, N. Fasel, and J.-P. Kraehenbühl, *J. Cell Sci.* 96:143 (1990).

15. H. P. Wessels, G. H. Hansen, C. Fuhrer, A. T. Look, H. Sjöström, O. Norén, and M. Spiess, *J. Cell Biol.* 111:2923 (1990).

16. R. Solari, L. Kühn, and J.-P. Kraehenbühl, *J. Biol. Chem.* 260:1141 (1985).

17. K. E. Mostov, A. De Bruynkops, and D. L. Deitcher, *Cell* 47: 359 (1986).

18. P. P. Breitfeld, J. E. Casanova, W. C. McKinnon, and K. E. Mostov, *J. Biol. Chem.* 265:13750 (1990).

19. S. K. Powell, M. P. Lisanti, and E. Rodriguez-Boulan, *Am. J. Physiol.* 260: C715 (1991).

20. M. A. Bakos, A. Kurosky, C. S. Woodard, R. M. Denney, and R. M. Goldblum, *J. Immunol.* 146: 162 (1991).

21. S. M. Dintzis and S. R. Pfeffer, *EMBO J.* 9: 77(1990).

NEURAL, ENDOCRINE AND IMMUNE REGULATION OF SECRETORY COMPONENT PRODUCTION BY LACRIMAL GLAND ACINAR CELLS

Ross W. Lambert, Jianping Gao, Robin S. Kelleher,
L. Alexandra Wickham, and David A. Sullivan

Department of Ophthalmology, Harvard Medical School
Immunology Unit, Schepens Eye Research Institute
20 Staniford Street, Boston, MA, USA 02114

INTRODUCTION

Our previous research has shown that the endocrine, nervous and immune systems regulate the production of secretory component (SC), the polymeric immunoglobulin receptor, by acinar (epithelial) cells from the rat lacrimal gland.[1] Thus, acinar cell exposure in vitro to androgens (e.g., dihydrotestosterone [DHT]), vasoactive intestinal peptide (VIP), the adrenergic agonist, isoproterenol, cyclic AMP analogues (e.g., 8-bromoadenosine 3':5'-cyclic monophosphate [bcAMP]), cyclic AMP inducers (e.g., cholera toxin, PGE_2), phosphodiesterase inhibitors (e.g., 3-isobutyl-1-methylxanthine), IL-1α, IL-1β, or TNF-α results in a significant increase in SC output.[2-4] Conversely, cellular treatment with the cholinergic agent, carbachol, causes a significant suppression of both androgen-induced and basal SC release by acinar cells.[3,4]

To extend these findings, the current investigation endeavored to: (1) examine the temporal association between the acinar cell reception of stimulatory signals and the consequent SC response; (2) assess whether the androgen and cholinergic control of SC production is receptor-mediated; (3) evaluate whether the androgen modulation of SC synthesis involves the induction of gene transcription; and (4) determine whether the hormonal, neural and lymphokine influence on lacrimal SC is unique to the eye or similar to regulatory processes controlling SC production in salivary and intestinal epithelial cells.

MATERIALS AND METHODS

Intact or castrated male Sprague-Dawley rats (6 to 8 weeks old; Zivic-Miller Laboratories) were provided food and water ad libitum and maintained in constant temperature rooms with light/dark cycles of 12 hours duration.

Methods for the culture of acinar cells from intact rat lacrimal and submandibular glands, as well as epithelial cells from the rat intestine (IEC-6; ATCC), followed previously reported protocols.[2-5] Briefly, lacrimal or submandibular glands were excised, minced and processed through a series of 37°C incubations in EDTA (Gibco), or collagenase (Calbiochem-Behring), hyaluronidase (Calbiochem-Behring), and DNase I (Boehringer Mannheim) in HBSS- or DMEM (Gibco)-based buffers. The resulting digest was filtered through 500 μm and 25 μm Nitex mesh (Tetko Inc.), then spun through a Ficoll 400 (Pharmacia) step gradient (2-4%). Pelleted acinar cells were plated in DMEM at a typical

Advances in Mucosal Immunology, Edited by
J. Mestecky *et al.*, Plenum Press, New York, 1995

density of 2×10^6 cells/well on a reconstituted basement membrane (Matrigel; Collaborative Research) in 35 mm Primaria culture dishes (Falcon). After overnight culture (37°C, 95% air/5% CO_2), attached cells were incubated in supplemented serum-free modified Oliver's media (SFMOM) in the presence of vehicle or specified compounds for designated periods of time. With regard to rat IEC-6 cells, these cells were grown to confluence in 35 mm Primaria culture dishes containing DMEM, 5% FCS (Hyclone) and 1% penicillin/streptomycin (Gibco) and then, as with acinar cells, exposed to various treatment regimens involving one or more of the following compounds: DHT, carbachol, cyproterone acetate, atropine sulfate, actinomycin D, bcAMP, 8-bromoguanosine 3':5'-cyclic monophosphate (bcGMP) (Sigma); VIP, human recombinant IL-1α, IL-1β (Boehringer Mannheim); cholera toxin (Calbiochem-Behring); or rat recombinant IFN-γ (Amgen Biologicals). At the termination of culture experiments, media were centrifuged at 10,000 x g for 4 minutes and resulting supernatants were stored at -20°C until analysis of SC levels by RIA.[6] In addition, attached cells were harvested, examined for viabilty and counted, as previously described.[5]

Procedures for the measurement of SC mRNA content in rat lacrimal tissue, including the isolation of total RNA, preparation and radiolabeling of a 5' ~950 bp cDNA probe (note: this fragment contains part of the SC coding region and was obtained from the full length rat SC cDNA, generously provided by Dr. George Banting, United Kingdom), hybridization and autoradiographic protocols with Northern blots, and densitometry, have been explained in detail.[7] Lacrimal tissues originated from castrated male rats treated with placebo compounds or physiological amounts of testosterone (Innovative Research of America).[7]

Statistical analysis was performed by using the unpaired, two-tailed Student's t test.

RESULTS

Temporal Association Between the Lacrimal Gland Acinar Cell Reception of Stimulatory Signals and the Eventual SC Response

A continuous, 4 day exposure of lacrimal gland acinar cells to DHT or cholera toxin *in vitro* results in a significant increase in SC production.[2-5] This cellular response may first be detected after 48 h of treatment and then rises in magnitude during the ensuing 2 days of culture.[2-5] However, considering the rapid association of androgens or cholera toxin with intracellular or surface receptors, respectively, it may be that acinar cells receive the necessary, regulatory signals from these compounds within the first few hours of exposure, thereby triggering a complete SC response. To test this hypothesis, we compared the influence of acute (i.e. 3 hours) versus chronic (i.e. 96 hours) administration of DHT (10^{-6} M), cholera toxin (10 μg/ml) or vehicle on the extent of acinar cell SC output following a 4 day culture period (n = 5 wells/group). In these studies[5], all experimental media were removed after the first 3 h, and then, depending upon the treatment group, appropriate control or compound-supplemented media were added to the culture dishes. Analysis of SC levels in media after 4 days demonstrated that the magnitude of stimulated SC production was almost equivalent in both DHT or both cholera toxin groups, irrespective of whether cells were exposed to specified agents for an abbreviated interval or throughout the entire culture time course. These findings indicate that a significant delay exists between the initial acinar cell reception of regulatory signals and the eventual SC response.

Receptor Involvement in the Androgen and Cholinergic Control of Acinar Cell SC Production

To determine whether the DHT stimulation of SC production is receptor-mediated, cultured acinar cells (n = 9 to 10 wells/group; 4 day culture) were pretreated with cyproterone acetate (10^{-6} M), an androgen antagonist, for 30 min prior to the addition of

DHT (10^{-8} M) to the incubation media. Our results[5] showed that cyproterone acetate, which competitively binds to androgen receptors, abolished the DHT-induced increase in acinar cell SC synthesis. To examine whether the cholinergic suppression of androgen-stimulated SC output is mediated through muscarinic receptors, acinar cells (n = 9 to 10 wells/group; 4 day culture) were administered DHT (10^{-6} M) and atropine (10^{-5} M) for 30 min before the addition of carbachol (10^{-4} M) to the culture media. In these studies, atropine, which specifically binds muscarinic sites, blocked carbachol's inhibition of DHT-stimulated SC production.

Molecular Biological Mechanisms Involved in the Androgen Control of Acinar Cell SC Synthesis

To assess whether the androgen control of lacrimal SC synthesis involves alterations in acinar cell gene expression, cells (n = 10 wells/ group) were cultured for 2 days in the presence of vehicle, DHT (10^{-6} M) and/or the transcriptional inhibitor, actinomycin D (0.1 μg/ml). Our results[5,7] demonstrated that actinomycin D significantly (p < 0.0001) decreased the ability of DHT to enhance SC production. Given the mechanism of action of actinomycin D, this finding would suggest that the androgen regulation of acinar cell SC may involve an increase in gene transcription and the generation of SC mRNA. In support of this hypothesis, testosterone administration for 7 days to castrated rats (n = 5/group) induced a significant (p < 0.0001) rise in the SC mRNA content of lacrimal glands, compared to levels in lacrimal tissues of placebo-treated controls.

Site-Specificity of the Endocrine, Neural and Immune Regulation of SC Production

To evaluate whether the endocrine, neural and immune control of SC production in lacrimal gland acinar cells is site-specific, various compounds, known to either stimulate (e.g., DHT, cholera toxin, bcAMP, VIP, IL-1α, or IL-1β), suppress (carbachol), or have no effect (e.g., bcGMP, IFN-γ) on lacrimal SC synthesis, were tested for their influence on submandibular acinar, or IEC-6, cell SC output. Studies involved the culture of salivary or intestinal cells (n = 5 to 25 wells/group) for 4 days in the presence of the following agent concentrations: vehicle (control), DHT (1 μM), cholera toxin (10 μg/ml), bcAMP (1 mM), bcGMP (1 mM), VIP (1 μM), carbachol (100 μM), IL-1α (2 ng/ml), IL-1β (2 ng/ml) or IFN-γ (1000 or 3000 U/ml). Analysis of media SC levels after the 4 day period demonstrated[5] that: (1) cholera toxin, bcAMP, bcGMP and IL-1α significantly increased SC production by submandibular acinar cells, whereas carbachol occasionally stimulated SC output (2/4 studies) and DHT had no effect; and (2) IL-1α, IL-1β, and IFN-γ significantly enhanced media SC content in IEC-6 cultures. In contrast, cholera toxin exposure led to augmented SC concentrations in 3 of 5 experiments, while DHT, VIP, bcAMP, bcGMP or carbachol treatment elicited no consistent impact on media SC levels in IEC-6 cultures.

DISCUSSION

The present studies explored the endocrine, neural, and immune regulation of SC production in rat mucosal epithelial cells, with a particular focus upon acinar cells from the lacrimal gland. Our findings demonstrated that: (1) striking, temporal differences exist between regulatory signal reception and the lacrimal SC response: acinar cell processing of androgen or cholera toxin stimulatory signals requires only 3 h of exposure to these agents, but associated SC secretory responses are delayed at least 48 h. Moreover, the extent of acinar cell SC output following 4 days of culture is similar if cells are treated acutely or chronically with androgens or cAMP-inducing secretagogues; (2) androgen stimulation and cholinergic suppression of acinar SC are mediated through interactions with specific cellular receptors; (3) androgen control of acinar SC synthesis appears to involve modulation of SC mRNA levels; and (4) the nature of the neuroendocrinimmune regulation of SC output by lacrimal gland acinar cells appears to be unique, when compared to factors controlling submandibular acinar and intestinal epithelial cell SC production. Overall, our results indicate that the hormonal, neural and lymphokine modulation of SC dynamics by lacrimal acinar cells is site-selective and may be mediated through the regulation of genomic processes.

ACKNOWLEDGMENTS

This research was supported by NIH grants EY-07074 and EY-05612 and a grant from the Massachusetts Lions Research Fund.

REFERENCES

1. D. A. Sullivan, *in:* "Mucosal Immunology", P.L. Ogra, J. Mestecky, M.E. Lamm, W. Strober, J. McGhee, and J. Bienenstock, eds., Academic Press, Orlando, FL, in press (1992).
2. D. A. Sullivan, R. S. Kelleher, J. P. Vaerman, and L. E. Hann, *J. Immunol.* 145:4238 (1990).
3. L. E. Hann, R. S. Kelleher, and D. A. Sullivan, *Invest. Ophthalmol. Vis. Sci.* 32:2610 (1991).
4. R. S. Kelleher, L. E. Hann, J. A. Edwards, and D. A. Sullivan, *J. Immunol.* 146:3405 (1991).
5. R. W. Lambert, R. S. Kelleher, L. A. Wickham, J. P. Vaerman, and D. A. Sullivan, submitted (1992).
6. D. A. Sullivan and C. R. Wira, *J. Immunol.* 130:1330 (1983).
7. J. Gao, R. W. Lambert, and D. A. Sullivan, Submitted (1992).

ANTIGEN PRESENTATION BY A MOUSE DUODENAL EPITHELIAL CELL LINE (MODE-K)

Karine Vidal, Isabelle Grosjean, and Dominique Kaiserlian

Unitéd Immunologie et de Stratégie Vaccinale,
Institut Pasteur, Avenue Tony-Garnier
69365 Lyon Cedex 07, France

INTRODUCTION

Studies on the immunological aspects of lymphoepithelial interactions in the intestine have been hampered by the difficulty to maintain enterocytes viable *in vitro* after isolation from the mucosa We have recently established[1] a mouse duodenal epithelial cell line, MODE-K, immortalized by the SV40 large T antigen, which retained phenotypical characteristics of normal enterocytes We have examined the immunologic function of MODE-K cells in antigen presentation to a specific class II-restricted T cell hybridoma

MATERIALS AND METHODS

Monoclonal Antibodies, Antigen, and Cell Lines

The specific antibodies used in this study are listed in Table 1

Table 1. Antibodies used in this study

	Specificity	References
Villin	Villin	Immunotech France
CK 18	Cytokeratin 18	Sigma
166	Secretory component	Dr Kraehenbuhl
H 2K^k	MHC class I antigens K^k	Becton Dickinson
CD311	MHC class II antigens I-E^k	Dr Glasebrook
10 2 16	MHC class II antigens I-A^k	ATCC
GK1 5	CD4 molecule	ATCC
SN1	LFA- 1 molecule	Dr Pierres
YN1/1 74	ICAM-1 molecule[2]	Dr Patarroyo

The MODE-K (mouse duodenal epithelial) cell line established[1] from young (2 week-old) axenic C3H mouse duodenum by retrovirus-mediated transfer of the immortalizing SV40 large T gene, displays morphologic and phenotypic features of

Advances in Mucosal Immunology, Edited by
J Mestecky *et al* , Plenum Press, New York, 1995

epithelial cells. IFN-γ treated MODE-K cells were produced by culturing the cells for 7 days in the presence of 100 IU/ml of IFN-γ. The hen egg-white lysozyme (HEL) (46-61)-specific I-A^k-restricted T-cell hybridoma 3A9 (kindly provided by Dr. P. Allen, Washington University School of Medicine, St. Louis, MO) was produced as previously described.[3]

Native HEL was purchased from Sigma, and HEL tryptic peptides were kindly provided by Dr. D. Gerlier (ENS, Lyon, France).

Cell Surface and Intracytoplasmic Labeling of MODE-K Cells

Cell surface labeling was performed by incubating the cells (10^6) for 30 min at 4°C with the specific Ab. After two washes in PBS, the staining was revealed using either an FITC-conjugated rabbit F(ab')$_2$ fragment specific for mouse IgG (H+L) (Zymed, France), or an FITC-conjugated goat anti-rat IgG (H+L) (Jackson Laboratory, Immunotech, France). For intracytoplasmic labeling of cell suspension, the cells were first permeabilized by 15 min incubation at 4°C with 0.1% saponin in PBS-1% BSA. Specific immunofluorescence staining was analyzed using a FACScan (Becton Dickinson). For immunocytological staining, cells, grown on glass coverslips, were fixed for 10 min in cold methanol, and rehydrated in PBS before incubation for 30 min with the specific Ab. The specific staining was revealed by an immunoperoxidase detection system using streptavidin-biotin amplification and AEC substrate (Peroxydase LSAB kit, DAKO, France), according the manufacturer's instructions.

Antigen Presentation Assay

IFN-γ treated or untreated MODE-K cells (10^5 cells/well) were cultured with the T-cell hybridoma (10^5 cells/well) in the presence of native HEL (final dose of 300 μg/ml). T-cell activation was quantified by the IL-2 released in 24 h culture supernatant, using the IL-2 dependent CTLL-2 cells. Briefly, 2 x 10^3 CTLL-2 cells per well were incubated 48h at 37°C in 5% CO_2, with log dilution of supernatant. One μCi per well of (^{3}H) thymidine was added during the last 4 h of culture. Results were expressed as IL-2 U/ml. In antibody blocking experiments, various doses of specific mAbs, or irrelevant Ab were added at the initiation of cultures.

RESULTS

The MODE-K cell line express phenotypical characteristics of intestinal epithelial cells, including expression of cytokeratin, villin, and secretory component (Table 2). In addition, MODE-K cells constitutively express surface MHC class I antigens, and only intracytoplasmic MHC class II molecules which appear concentrated in the perinuclear area of the cells (Fig. 1).

Table 2. Phenotypical characterisation of MODE-K cells.

Antibodies to	Gut epithelial cells	MODE-K cells
Cytokeratin 18	++	++
Villin	++	++
Secretory component	++	++
MHC class I molecules	++	++
MHC class II molecules	++	++ (cytoplasmic)

FACS analysis
Immunocytochemical analysis of MHC class II

Intracellular   Cell surface

Class I

Class II

Cell Number

Mean Fluorescence Intensity

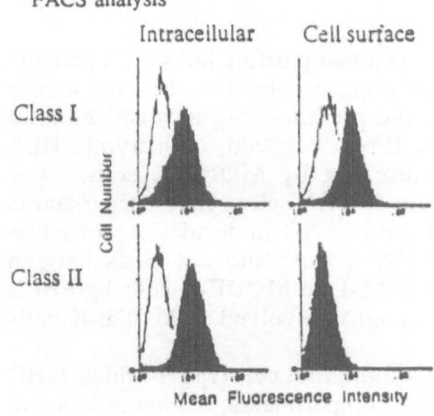
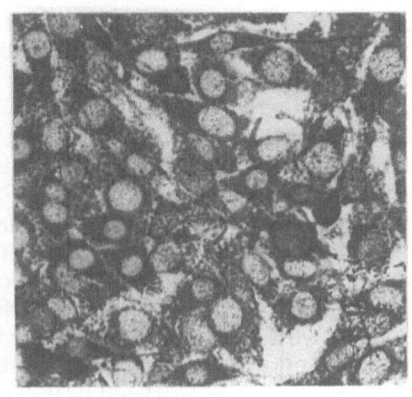

Figure 1. MHC molecules expression on MODE-K cells.

IFN-γ treatment (100 IU/Ml) of MODE-K cells for 7 days resulted in a maximal induction of cell surface MHC class II molecules expression[1], and was associated with the ability of these cells to present native or tryptic peptides of HEL to the specific and I-A-restricted T cell hybridoma 3A9 (Table 3).

Table 3. Ability of MODE-K cells to present HEL to T-cell hybridoma 3A9.

| | IL-2 production (U/ml) by 3A9 cells in presence of HEL | |
	native (300 µg/ml)	tryptic peptides (100 µg/ml)
Untreated MODE-K cells	0	0
IFN-γ-treated MODE-K	100	50

IL-2 production by 3A9 following presentation of native HEL by MODE-K cells is inhibited by anti-I-A^k, but not anti-I-E^k (Fig. 2), indicating a class II-restricted mechanism of antigen presentation. Moreover, addition of anti-CD4, and anti-LFA-1 mAbs at the initiation of culture strongly inhibited T-cell activation whereas mAb against ICAM-1 had no effect.

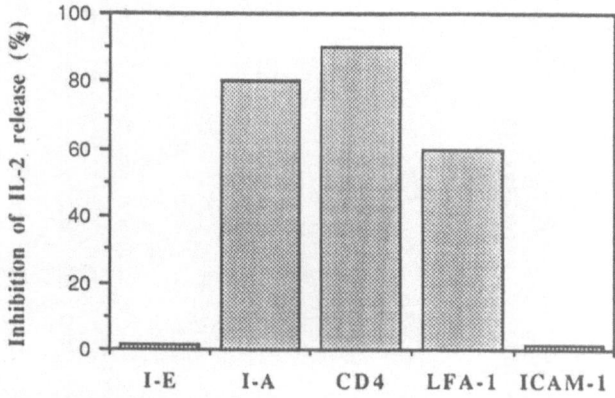

Figure 2. Inhibition of antigen presentation by MODE-K cells.

DISCUSSION

Studies in rats[4], in humans[5] and in mice[6] showed that purified intestinal epithelial cells could present antigen to specific T cells. The present data show that the mouse duodenal epithelial cell line, MODE-K can process and present exogenous antigen after cell surface MHC class II molecules induction by IFN-γ. Indeed, both tryptic HEL peptides and native HEL could be efficiently presented by MODE-K cells. The mechanism of antigen presentation by MODE-K cells is MHC class II restricted and is strengthened by accessory molecules, including CD4 and LFA-1 molecules expressed on the T-cell hybridoma 3A9. However, the anti-ICAM-1 mAb did not block antigen presentation and was unable to detect cell surface ICAM-1 on MODE-K cells by FACS analysis (data not shown). Whether the ICAM-2 ligand is involved in MODE-K cells interaction with LFA-1 on T cells is not known.

It should be emphasized that MODE-K cells are the first cell type in which MHC class II expression is restricted to the intracytoplasmic compartments, but not detected at the cell surface. Class II-restricted presentation of exogenous antigen requires processing of the antigen into peptides which encounter class II molecules in the endosomes. Since at the basal state MODE-K cells are unable to present antigen and peptides, this raises the possibilities that in these epithelial cells, class II molecules are not stabilized by peptides in the endosomes, but may be stable in other compartments such as the endoplasmic reticulum.

This cell line represents a useful model for analyzing the mechanisms of antigen processing and presentation by intestinal epithelial cells, which represent a non-conventional type of APC, and especially for studying the role of the lymphoepithelial interactions in mucosal immunity.

REFERENCES

1. K. Vidal, I. Grosjean, C. Gespach, J. P. Revillard, and D. Kaiserlian, (submitted).
2. J. Prieto, F. Takei, R. Gendelman, B. Christenson, P. Biberfeld, and M. Patarroyo, *Eur. J. Immunol.* 19:1551 (1989).
3. P. M. Allen, D.J. McKean, B. N. Beck, J. Sheffield, L. H. and Glimcher, *J. Exp. Med.* 162:1264 (1985).
4. P. W. Bland and L. G. Warren, *Immunology* 58:1 (1986).
5. L. Mayer and R. Shlien, *J. Exp. Med.* 166:1471 (1987).
6. D. Kaiserlian, K. Vidal, and J. P. Revillard, *Eur. J. Immunol.* 19:1513 (1989).

THE REGULATION OF IL-6 SECRETION FROM IEC-6 INTESTINAL EPITHELIAL CELLS BY CYTOKINES AND MUCOSALLY IMPORTANT ANTIGENS

Dennis W. McGee[1], Kenneth W. Beagley[2],
Wilhelm K. Aicher[3], and Jerry R. McGhee[1]

Departments of Microbiology[1] and Medicine,[2]
University of Alabama at Birmingham,
Medical Center, Birmingham, Alabama, USA
and The Max Planck Institute for Immunobiology,[3]
KFR, University of Freiburg, Freiburg, Germany

INTRODUCTION

Intestinal epithelial cells (IEC) are known to secrete a variety of important cytokines including the inflammatory cytokine interleukin-6 (IL-6).[1, 2] Along with its role in the inflammatory response, IL-6 has also been shown to act as a co-stimulator for T cell proliferative responses[3] and is known to induce Peyer's patch and appendix B cells to secrete high levels of IgA.[4, 5] Therefore, the IEC, through its secretion of IL-6, has the potential to be a very important factor in inflammation and IgA immune responses at the intestinal mucosa.

We have recently begun to study the mechanisms which regulate IL-6 secretion by IEC using the non-transformed rat crypt-like small intestinal epithelial cell line, IEC-6. Initially, transforming growth factor-β (TGF-β) was shown to enhance IL-6 secretion by the IEC-6 cells.[2] This finding was significant because of the known role of TGF-β in inflammatory responses, such as the induction of IL-1 and IL-6 secretion by certain cell types, monocyte chemotaxis, and wound healing,[6] and TGF-β has been shown to enhance IgA secretion while inhibiting IgG and IgM secretion by B cells.[7, 8] In the present study, we have extended these experiments to include the regulatory effects of the inflammatory cytokines IL-1β and tumor necrosis factor-α (TNF-α) on IL-6 secretion by the IEC-6 cells.

Finally, it is now well established that cholera toxin (CT), the enterotoxin from the bacterium *Vibrio cholerae*, can induce a robust immune response to itself after oral administration.[9] Also, CT given orally with an unrelated antigen can act as a mucosal adjuvant to potentiate secretory IgA and systemic IgG immune response to the unrelated antigen which normally by itself would not induce a mucosal immune response.[10,11] Since the majority of orally administered or naturally produced CT would bind to the IEC, and IL-6 secreted by the IEC could have an important effect on the mucosal inflammatory and IgA responses, we have determined the effect of CT on IL-6 secretion by the IEC-6 cells.

MATERIALS AND METHODS

The IEC-6 cells (ATCC #CRL 1592) were cultured in Dulbecco's Modified Eagle's medium with 4.5% glucose, 5% fetal calf serum, 0.1 IU/ml bovine insulin, and antibiotics. All IEC-6 cultures contained 2×10^5 cells/well in 12 well tissue culture plates. After two days, the culture supernatants were removed and replaced with medium containing 1% fetal calf serum and either CT (Sigma Chemical Co., St. Louis, MO, or List Biological

Laboratories, Campbell, CA), recombinant human (rh) IL-1β, rhTNF-α or porcine TGF-β1, (R & D Systems, Minneapolis, MN). At the appropriate interval, the culture supernatants were collected and the adherent cells were removed with trypsin and EDTA for determination of total cell numbers.

The IL-6 content in the culture supernatants was determined by a proliferative bioassay using the IL-6 dependent 7TD1 mouse hybridoma.[12] Proliferation of the 7TD1 cells was determined by the MTT colorimetric assay.[13] One unit of IL-6 represents the reciprocal of the dilution yielding half-maximal proliferation of the 7TD1 cells.

RESULTS AND DISCUSSION

Significantly enhanced levels of IL-6 are known to be present in inflamed mucosal tissue associated with inflammatory bowel disease and gastritis.[14, 15] IEC are known to be capable of producing IL-6[1, 2], yet little is known of the mechanisms which regulate IL-6 secretion by these cells. In a previous report, we have found that TGF-β could enhance IL-6 secretion by the IEC-6 cells.[2] Yet, inflamed mucosal tissues can also contain elevated levels of the inflammatory cytokine IL-1[16, 17] which can induce IL-6 secretion by other cell types.[3] Therefore, the effect of IL-1β, as well as TNF-α which is also known to induce IL-6 secretion,[3] on IL-6 secretion by the IEC-6 cells was determined. Culture of the IEC-6 cells for four days with IL-1β or TNF-α enhanced IL-6 secretion, but to a greater level (Table 1). These results indicate that the IEC may be a very important source of IL-6 during mucosal inflammatory responses.

TGF-β has been shown to induce the immature IEC-6 cell line to show more mature characteristics such as the expression of sucrase and the inhibition of proliferation.[18] Therefore, the effect of TGF-β along with either IL-1β or TNF-α was next determined. Culture of the IEC-6 cells with both TGF-β and either IL-1β or TNF-α resulted in greatly enhanced levels of IL-6 secretion by the IEC-6 cells (Table 1). In an attempt to dissect the mechanism by which TGF-β acted in synergy with IL-1β, the IEC-6 cells were stained for the binding of phycoerythrin-conjugated rhIL-1β. Pretreatment of the IEC-6 cells with TGF-β1 resulted in an enhanced binding of IL-1β (data not shown) suggesting that TGF-β enhanced the expression of IL-1 receptors by the cells. Therefore, the presence of TGF-β in an inflammatory response at the mucosa could induce immature crypt IEC to undergo a differentiation to a more immunologically mature state accompanied by an enhanced expression of IL-1 receptors by the cells and the ability to secrete greatly enhanced levels of IL-6.

Table 1. TGF-β1, IL-1β, and TNF-α induce the IEC-6 cells to secrete enhanced levels of IL-6.

Condition[1]	Units of IL-6 Secreted/10^5 Cells
Control	9 ± 2
TGF-β1	36 ± 4
IL-1β	141 ± 32
TNF-α	124 ± 25
TGF-β1 + IL-1β	3,071 ± 625
TGF-β1 + TNF-α	6,029 ± 732

[1]The IEC-6 cells were cultured for 4 days in the presence or absence of pTGF-β1 (2 ng/ml), rhIL-1β (1 ng/ml), or rhTNF-α (50 ng/ml). Culture supernatants were collected and assayed for IL-6 content.

In an attempt to determine if the IEC could respond to a mucosally important antigen by secreting an enhanced level of IL-6, the IEC-6 cells were cultured for 24 hours in the presence of CT. CT was found to enhance IL-6 secretion (Table 2) suggesting that the secretion of IL-6 by IEC may be important in the mucosal immune response to CT. This IL-6 may potentiate the immune response to CT, therefore providing a partial explanation for why CT can produce an excellent mucosal immune response[9]. CT has also been shown to act as an mucosal adjuvant to enhance the IgA immune response to unrelated antigens[10, 11]. An enhanced secretion of IL-6 by IEC upon stimulation by CT could also contribute to this adjuvant effect.

Interestingly, several reports have found that the B subunit of CT (CT-B), which mediates the binding of CT to cells, did not show the adjuvant effect when co-administrated with an unrelated antigen.[11, 19] When the IEC-6 cells were cultured with CT or CT-B, only the intact CT induced IL-6 secretion (Table 2). This finding not only suggests that the mere binding of CT to the cells is insufficient to induce IL-6 secretion, but lends some support to the hypothesis that the enhanced secretion of IL-6 by IEC may partially account for the adjuvant effect of CT.

Table 2. Cholera toxin, but not the B subunit alone, enhances IL-6 secretion by the IEC-6 cells.

Condition[1]	Units of IL-6 Secreted/10^5 Cells
Control	5 ± 1
CT	129 ± 23
CT-B	11 ± 1

[1]The IEC-6 cells were cultured for one day with or without 1 µg/ml of CT or CT-B. Culture supernatants were then collected and IL-6 content was determined.

In the final set of experiments, the effect of CT in the presence of cytokines known to enhance IL-6 secretion by the IEC-6 cells was determined. Culture of the IEC-6 cells with both CT and IL-1β for three days induced a 130 fold increase in IL-6 secretion, as compared to cultures with IL-1β alone, and a 2000 fold increase in IL-6 secretion when compared to the unstimulated controls (Table 3). Culture of the cells with both CT and TNF-α also greatly enhanced IL-6 secretion as much as 27 and 276 fold over the levels seen with the unstimulated controls or TNF-α treated cultures, respectively. Since IL-1β and/or TNF-α may be present in the mucosal tissue during a cholera infection or oral administration of CT, these results indicate that the IEC may be a very important source of IL-6 in response to CT. As for the adjuvant effect of CT, this greatly enhanced levels of IL-6 would non-specifically enhance immune responses, especially the production of IgA, in the area where CT would bind to the IEC.

Because of its anatomical position, the IEC represents the first line of defense against noxious or infectious agents. The results presented in this study suggest that the IEC may have a more active role in mucosal immune responses than simply acting as a physical barrier. The production of IL-6 may initiate, alter, or enhance a mucosal inflammatory or immune response making this cell type a potentially important factor in these responses. The IEC may also be able to produce other cytokines, such as IL-1, during an inflammatory response or in response to CT which may further affect the mucosal immune response and we are now begining experiments to address this issue.

Table 3. Cholera toxin enhances cytokine induced IL-6 secretion by the IEC-6 cells.

Condition[1]	Units of IL-6 Secreted/10^5 Cells
Control	11 ± 2
CT	45 ± 4
TGF-β1	41 ± 4
IL-1β	161 ± 6
TNF-α	107 ± 16
CT + TGF-β1	354 ± 96
CT + IL-1β	20,991 ± 3,868
CT + TNF-α	2,922 ± 368

1 The IEC-6 cells were stimulated with either CT (1 µg/ml) and/or pTGF-1β (2 ng/ml), rhIL-1β (1 ng/ml), or rhTNF-α (50 ng/ml) for 3 days. The culture supernatants were then collected and IL-6 content was determined.

ACKNOWLEDGMENTS

This work was supported by U.S. Public Health Services Contract AI 15128 and a grant from DRFZ, Berlin, Germany.

REFERENCES

1. L. Mayer, E. Siden, S. Becker and D. Eisenhardt, Antigen handling in the intestine mediated by normal enterocytes, *In:* "Advances in Mucosal Immunology: Proceedings of the 5th International Congress of Mucosal Immunology," T. MacDonald, S. J. Challacombe, P. W. Bland, C. R. Stokes, R. V. Heatley, and A. McI Mowat, eds., Kluwer Academic Publishers, Boston, MA (1990).
2. D. W. McGee, K. W. Beagley, W. K. Aicher and J. R. McGhee, *Immunology.* 76:7 (1992).
3. S. Akira, T. Hirano, T. Taga and T. Kishimoto, *FASEB J.* 4:2860 (1990).
4. K. W. Beagley, J. H. Eldridge, F. Lee, H. Kiyono, M. P. Everson, W. J. Koopman, T. Hirano, T. Kishimoto and J. R. McGhee, *J. Exp. Med.* 169:2133 (1989).
5. K. Fujihashi, J. R. McGhee, C. Lue, K. W. Beagley, T. Taga, T. Hirano, T. Kishimoto, J. Mestecky and H. Kiyono, *J. Clin. Invest.* 88:248 (1991).
6. S. M. Wahl, *J. Clin. Immunol.* 12:61 (1992).
7. R. L. Coffman, D. A. Lebman and B. Shrader, *J. Exp. Med.* 170:1039 (1989).
8. E. Sonoda, R. Matsumoto, Y. Hitoshi, T. Ishii, M. Sugimoto, S. Araki, A. Tominaga, N. Yamaguchi and K. Takatsu, *J. Exp. Med.* 170:1415 (1989).
9. N. Pierce and J. Gowans, *J. Exp. Med.* 142:1550 (1975).
10. C. O. Elson and W. Ealding, *J. Immunol.* 132:2736 (1984).
11. N. Lycke and J. Holmgren, *Immunology.* 59:301 (1986).
12. J. Van Snick, S. Cayphas, A. Vink, C. Uyttenhove, P. G. Coulie, M. R. Rubira and R. J. Simpson, *Proc. Natl. Acad. Sci. USA.* 83:9679 (1986).
13. T. Mossman, *J. Immunol. Methods.* 65:55 (1983).
14. K. Mitsuyama, E. Sasaki, A. Toyonaga, H. Ikeda, O. Tsuruta, A. Irie, N. Arima, T. Oriishi, K. Harada, K. Fujisaki, M. Sata and K. Tanikawa, *Digestion.* 50:104 (1991).
15. J. E. Crabtree, T. M. Shallcross, R. V. Heatley and J. I. Wyatt, *Gut.* 32: 1473 (1991).
16. D. Rachmilewitz, P. L. Simon, L. W. Schwartz, D. E. Griswold, J. D. Fondacaro and M. A. Wasserman, *Gastroenterology.* 97: 326 (1989).
17. M. Ligumsky, P. L. Simon, F. Karmeli and D. Rachmilewitz, *Gut.* 31: 686 (1990).
18. M. Kurokowa, K. Lynch and D. K. Podolsky, *Biochem. Biophys. Res. Comm.* 142:775 (1987).
19. S. J. McKenzie and J. F. Halsey, *J. Immunol.* 133:1818 (1984).

EPITHELIAL CELLS IN SOW MAMMARY SECRETIONS

Christian Le Jan

Laboratoire de Pathologie Infectieuse
et d'Immunologie
I.N.R.A.
37380 Nouzilly, France

INTRODUCTION

Sow milk contains 1 - 2.5 x 10^6 cells per ml: the predominant population is epithelial cells, in addition to lymphocytes, polymorphonuclear leukocytes, and macrophages.[1,2] This is at variance with most other mammalian species, where macrophages predominate. These observations lead to a possible contribution of epithelial cells from sow milk to lactogenic immunity, and more specifically IgA. S-IgA results from active processing of dimeric IgA by epithelial cells. The aim was to establish if epithelial cells exfoliated from alveoli to milk contain IgA. Therefore, we have looked for secretory component (SC) and concomitant IgA expression in epithelial cells from milk.

METHODS

Sows were milked manually, after 10-20 I.U. oxytocin administration. Cells were prepared by centrifugation of milk and washings of cell pellets in PBS with 5% of SVF, and cellular composition was evaluated after May Grünwald Giemsa staining. Monoclonal antibodies used for labelling of cells were: - K60 IFI (mouse IgG1, Bourne) directed against porcine secretory component; - K61 IB4 (mouse IgG1, Bourne) directed against porcine IgA; - 74.22.15 (mouse IgG1, Pescovitz) directed against porcine macrophages. Indirect membrane immunofluorescence[4] was performed on live cells in suspension, using monoclonal antibodies K60 IFI and swine Ig anti-mouse Ig conjugated to fluorescein isothiocyanate (SwAM/FITC, Nordic Immunology); preparations were examined under an ultraviolet microscope (Leitz Orthoplan, ocular 12,5x, objective 40/1.30, immersion). Fixed cells were labelled by alkaline phosphatase-anti alkaline phosphatase method (APAAP).[5] Fixation of monoclonal antibodies 77.22.15, K60 IFI and K61 IB4 was visualized by use of complexes of mouse monoclonal antibodies and alkaline phosphatase (Dako APAAP, lot n° 091), and coloration substrate containing Naphtol AS-MX phosphate-free acid (SIGMA).

RESULTS

Identification of Epithelial Cells from Milk

68% (± 14% n=11) of milk cells are epithelial cells, as ascertained by morphologic criteria and no labelling with monoclonal antibody directed against pig macrophages. These large cells (15-40 μm) are strongly vacuolated. 53% (±11%) of them are mononucleated, 10% (± 9%) binucleated, and 35% (± 12%) have lost their nuclei (n=11).

Advances in Mucosal Immunology, Edited by
J. Mestecky *et al.*, Plenum Press, New York, 1995

Detection of Secretory Component in Epithelial Cells from Milk

Most epithelial cells express high density of membrane SC, whatever the number of nuclei. All milk epithelial cells contain intracytoplasmic SC, and 66% contain intracytoplasmic IgA. However, incubation of epithelial cells one hour at 37°C reduces the proportion of cells expressing intracytoplasmic IgA, which indicates that IgA diffuses, passively or by active cellular processing.

DISCUSSION

Epithelial cells from milk contain intracytoplasmic IgA. Piglets ingest daily $5 - 7 \times 10^8$ maternal cells from milk, and intracellular IgA could represent a significant proportion of total IgA afforded by milk. Processing of dimeric IgA by epithelial cells is well established.[6] Since all milk epithelial cells express SC, intracellular IgA is probably the result of linkage of dimeric IgA to poly-Ig receptor and intracellular internalization. As immunoglobulin characteristic of lactogenic protection is S-IgA, excreted from epithelial cells of the mammary gland,[7] intracellular IgA represents a potential contribution to neonate intestinal mucosa. However, the mechanism of liberation of IgA by milk epithelial cells needs further investigation, as lysis of cells in the neonatal digestive tract before complete processing of IgA could result in liberation of IgA linked to poly-Ig receptor, which could have different properties than S-IgA.

REFERENCES

1. A. Schollenberg, T. Frymus, A. Degorski, and A. Schollenberg, *J. Vet. Med.* 33:31 (1986).
2. U. Magnusson, H. Rodriguez-Martinez, and S. Einarsson, *Vet. Rec.* 30:485 (1991).
3. F. Klobasa, E. Werhahn, and J.E. Butler, *J. Anim. Sci.* 64:1458 (1989).
4. B. Kaeffer, E. Bottreau, L. Phan Thanh, M. Olivier, and H. Salmon, *Int. J. Cancer* 46:481 (1990).
5. J. L. Cordell, B. Falini, W. N. Erber, A. K. Ghosh, Z. Abdulaziz, S. MacDonald, K. A. F. Pulford, H. Stein, and D. Y. Mason, *J. Histochem. Cytochem.* 32:219 (1984).
6. E. Schaerer, F. Verrey, L. Racine, C. Tallichet, M. Reinhardt, and J. P. Kraehenbuhl, *J. Cell Biol.* 110:987 (1990).
7. R. A. Goldblum, *J. Clin. Immunol.* 10:64 (1990).

MICRODISSECTED DOMES FROM GUT-ASSOCIATED LYMPHOID TISSUES: A MODEL OF M CELL TRANSEPITHELIAL TRANSPORT *IN VITRO*

Zita Kabok, Thomas H. Ermak, and Jacques Pappo

Vaccine Delivery Research, OraVax Inc.
Cambridge, MA 02139

INTRODUCTION

Specialized M cells in gut-associated lymphoid tissues (GALT) comprise 10-50% of the epithelial cell population overlying GALT domes,[1,2] and exhibit short microvilli, numerous apical vesicles, and a basolateral pocket infiltrated by clusters of lymphoid cells.[3,4] Functionally, M cells endocytose protein molecules,[5] viruses,[6] and Gram⁻ microorganisms.[7] Particle binding to the M cell apical membrane results in rapid (2 μm/min) internalization and shuttling to pocket domains and to underlying dome lymphoid tissue.[8] While it has been shown that lymphocytes which infiltrate M cell pockets and localize to subepithelial GALT domes are MHC class II⁺ activated T cells,[4,9] and that mucosal immunization with structures which are taken up by M cells into GALT may result in the development of IgA responses,[10] the relationship between M cell-mediated antigen uptake and the evolution of IgA antibody has not been examined directly. A model using microdissected domes from GALT was developed to enable the *in vitro* screening of a large repertoire of antigens for M cell tropism, and to study the development of IgA antibody responses to antigens recognized by M cells.

MATERIALS AND METHODS

Isolation and Culture of Lymphoepithelial Domes

Peyer's patches (PP) and appendix were excised from 2.0 kg NZW rabbits, rinsed with RPMI 1640 medium,[2,4] and domes microdissected. Isolated domes were placed in organ culture dishes containing 0.45 μm HA membranes in RPMI 1640 medium supplemented with 10% fetal bovine serum in the presence or absence of 1-1000 μg/ml cationized ferritin (CF; Sigma). After 10 min to 18 h, the domes were rinsed extensively with RPMI 1640 medium at 4°C to avoid carryover of unbound CF into the cultures, and transferred to wells of 24 well plates for analysis of immunoglobulin (Ig) and antibody production *in vitro*, or fixed and processed for transmission electron microscopy.[2] In some experiments, microdissected domes were cultured with varying concentrations of pokeweed mitogen (PWM), PHA, or ConA. For experiments involving *in vivo* exposure to CF, intestinal loops containing PP[8] were instilled with 350 μl of CF in phosphate-buffered saline (PBS). After 1 h, the PP were removed, the domes microdissected, and subsequently cultured in the absence of CF.

Advances in Mucosal Immunology, Edited by
J. Mestecky *et al.*, Plenum Press, New York, 1995

Analysis of Ig and Antibody Production

Culture supernatants from GALT dome explants were assayed at sequential time periods for IgA, IgG or IgM concentration using a biotin-avidin isotype-specific ELISA. Briefly, 96-well plates were incubated with a monoclonal antibody against rabbit Ig, or with CF, followed by dome culture supernatants. The wells were then serially incubated with the appropriate dilution of biotinylated goat anti-rabbit IgA, IgG or IgM, avidin-alkaline phosphatase and phosphatase substrate. The sensitivity of the ELISA was ~1.5 ng/ml for rabbit IgA and IgG, and 62 ng/ml for rabbit IgM. Purified rabbit milk IgA, or serum IgG and IgM were used to generate standard curves to estimate the Ig concentration by linear regression analyses ($R^2 = >0.95$).

RESULTS AND DISCUSSION

M cells populating the follicle epithelium in GALT were examined for their ability to bind and translocate antigen *in vitro*. Microdissected lymphoepithelial domes derived from rabbit PP and appendix harbored an intact sheet of follicle epithelium and M cells overlying a dome and one submucosal lymphoid follicle (Fig. 1a). CF administered to dome explants in organ culture bound to follicle epithelial apical cell membranes. M cells *in vitro* exhibited numerous endosomes containing CF, which trafficked vectorially across the M cell cytoplasm, fused with M cell pocket membranes, and delivered CF to lymphocytes within the M cell pocket (Fig. 1b). The kinetics and magnitude of M cell uptake and transepithelial transport of CF *in vitro* resembled those previously observed *in vivo*.[5,11]

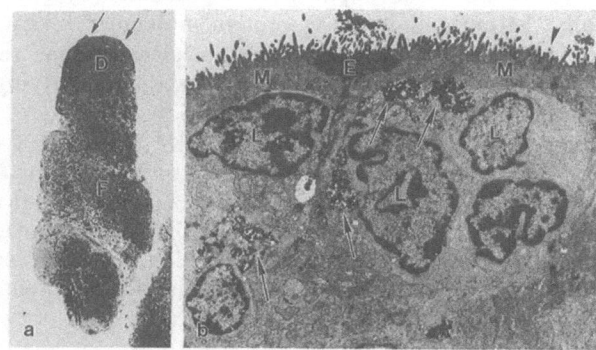

Figure 1. (a) Cross-section of microdissected dome from GALT showing intact M cell-containing follicle epithelium (arrows) overlying dome (D), and mucosal lymphoid follicle (F). (b) After a 90 min pulse *in vitro*, most bound CF has been internalized by M cells (M), and transported to the pocket domain (arrows) containing lymphocytes (L). While only small aggregates of CF remain at the M cell apical membrane (arrowhead), clusters of CF remain at the enterocyte (E) apical surface without detectable uptake.

Studies of follicle epithelial cells in isolated domes cultured from 1-6 h showed that M cells remained attached to the underlying basal lamina, to adjacent enterocytes via tight junctions, and maintained the associations with lymphocytes infiltrating the pocket. Thus, microdissected GALT domes contained structurally intact M cells during short-term organ culture.

Lymphoepithelial domes from PP and appendix synthesized substantially greater levels of IgA than IgM or IgG (Table 1). Higher concentrations of IgM were measured in appendiceal dome supernatants than in PP dome cultures during the same time intervals. Studies on the kinetics of Ig secretion showed that 50-60% of the total Ig was detected after 1 day, approximately 30% during days 2-4, and 10% from 5 to 7 days after culture. Incubation of appendiceal domes with mitogens, particularly with PWM (10 μg/ml) resulted in greater levels of IgA production (1057 vs 870 ng/ml) but significantly lower IgM secretion (40 vs 131 ng/ml). Taken together, these observations suggested active Ig synthesis by GALT domes *in vitro*.

Table 1. Ig secretion *in vitro* by microdissected domes from GALT.

		ng / ml / dome explant[a]		
		IgA	IgG	IgM
Peyer's patch				
	Exp 1	288	50	<5
	2	1194	37	24
	3	400	<5	<5
	4	322	<5	29
Appendix				
	Exp 1	255	56	175
	2	477	24	215
	3	223	<5	22
	4	269	<5	37

[a]Cumulative total in supernatants from days 1-7 of culture.

Table 2. IgA levels after *in vivo* and *in vitro* M cell antigen uptake.

		ng / culture[a]	
		IgA	anti-ferritin IgA
	Treatment[b]		
in vivo	PBS	3746 ± 576	-
	CF + CT	3742 ± 444	33 ± 9
in vitro	Medium	2959 ± 473	-
	CF + CT	2581 ± 370	42

[a]Summary data from 4 experiments showing the cumulative total in supernatants from days 5-7 of culture.
[b]Peyer's patch M cells were exposed to CF (3.5 µg/ml) and cholera toxin (CT; 35 µg/ml) in intestinal loops or in organ culture for 1 hr. Microdissected domes were subsequently incubated, and the supernatnants assayed for total IgA, and IgA antibody against CF.

The development of IgA antibody by microdissected domes was next examined. Isolated domes from PP and appendix were pulsed in organ culture with CF, or with CF and cholera toxin (CT) for 1 h. Total IgA concentrations were somewhat lower in PP domes incubated *in vitro* than in domes microdissected after *in vivo* treatment (Table 2). Analyses of culture supernatants revealed the presence of IgA anti-ferritin antibody in antigen-pulsed domes (Table 2). Exposure of follicle epithelial M cell populations to antigen in ligated loops *in vivo*, followed by culture of dome explants was also found to result in secretion of anti-CF IgA after 5-7 days in culture (Table 2). While cultures containing mesenteric lymph node explants secreted primarily IgA *in vitro* (1170±121 ng/culture at day 5), no anti-CF IgA antibody was detected in supernatants of lymph node explants pulsed in organ culture with equivalent concentrations of CF, or with CF and CT. These findings showed that isolated GALT domes responded to primary stimulation with antigen, and suggested the development of an IgA antibody response to antigen which is bound and translocated efficiently by M cells *in vivo* or *in vitro*.

ACKNOWLEDGMENT

This work was supported by NIH grants DK 36563 and DK 38550.

REFERENCES

1. M. W. Smith and M. A. Peacock, *Am. J. Anat.* 159:167 (1980).
2. J. Pappo, H. J.Steger, and R. L. Owen, *Lab. Invest.* 58:692 (1988).
3   W. A. Bye, C. H. Allan, and J. S. Trier, *Gastroenterology* 86:789 (1984).
4. T. H. Ermak, H. J. Steger, and J. Pappo, *Immunology* 71:530 (1990).
5. D. E. Bockman and M. D. Cooper, *Am. J. Anat.* 136:455 (1973).
6. J. L. Wolf, D. H. Rubin, R. Finberg, R. S. Kauffman, A. H. Sharpe, J. S. Trier, and B. N. Fields, *Science* 212:471 (1981).
7. R. L. Owen, N. F. Pierce, R. T. Apple, and W. C. Cray, *J. Infect. Dis.* 153:1108 (1986).
8. J. Pappo and T. H. Ermak, *Clin. Exp. Immunol.* 76:144 (1989).
9. T. H. Ermak, H. J. Steger, and J. Pappo, *FASEB J.* 5:A1693 (1991).
10 S. D. London, D. H. Rubin, and J. J. Cebra, *J. Exp. Med.* 165:830 (1987).
11. M. R. Neutra, T. L. Phillips, E. L. Mayer, and D. J. Fishkind, *Cell Tissue Res.* 247:537 (1987).

EVIDENCE THAT MEMBRANOUS (M) CELL GENESIS IS IMMUNO-REGULATED

T. C. Savidge and M. W. Smith

Department of Cell Biology, AFRC IAPGR
Babraham, Cambridge CB2 4AT
England

INTRODUCTION

Small intestinal membraneous (M) cells are specialised epithelial antigen-transporting cells, confined to follicle-associated epithelia (FAE). Recent findings in germfree BALB/c mice demonstrated that the interaction of *Salmonella typhimurium aroA⁻* with FAE of small intestinal Peyer's patch (PP) tissue caused a rapid expansion of the M cell population that could be measured within this tissue.[1] In addition, M cells were shown to transcytose *S. typhimurium aroA⁻* to juxtaposed lymphocytes.[2] It could be suggested that M cell genesis may feasibly result as a consequence of direct cross-talk between *Salmonella* and FAE. To address this question further, a severe combined immunodeficient (*scid*) mouse model was established with the aim to induce M cell numbers in the absence of an invasive bacterial vector.

METHODS

Age and sex matched (8 week old female) BALB/c and C.B-17 *scid* mice were bred and maintained under isolator conditions throughout the experimental procedures. Leaky *scid* mice were omitted from the investigation by preliminary screening for serum IgM as described by Bosma et al.,[3] BALB/c mice were primed *in vivo* by intra-gastric injection of 5×10^9 live *S. typhimurium aroA⁻* in 0.5 ml brain heart infusion media (BHI). Control unprimed BALB/c mice received BHI alone. Lymphocytes were harvested from spleen, mesenteric lymph node (MLN) and PP tissue two weeks after infection. Application of antibiotics (100 u/ml penicillin & 100 µg/ml streptomycin) to the cellular transfer ensured that no *Salmonella* were transferred to *scid* recipients. This was confirmed by an absence of bacterial growth on Columbia blood and MacConkey agar plates. 1×10^8 unprimed, 1 or 5 $\times 10^8$ *Salmonella* primed lymphocytes were transferred intra-peritoneally into 8 recipient *scid* mice and small intestinal PP tissue was processed for quantitative M cell analysis 1 week after immunoreconstitution as described by Savidge *et al.,*[1] Serum anti- *S. typhimurium aroA⁻* antibodies were also measured using a modification of the enzyme-linked immunosorbant assay (ELISA) described by Maskell *et al.*[4]

Advances in Mucosal Immunology, Edited by
J. Mestecky *et al.*, Plenum Press, New York, 1995

RESULTS

Successful immunoreconstitution of *scid* mice was demonstrated in all cases. This was evident from the formation of *de novo* PP tissue which was consistently absent in unreconstituted *scid* mice. Also, the use of rabbit anti-mouse Ig^+ and anti-$CD3^+$ (145-2C11) antibodies demonstrated the presence of both B and T lymphocytes in spleen and MLN of immunoreconstituted mice. Further application of these antibodies demonstrated that novel PP tissue harboured conventional B and T cell compartments. No germinal centres were evident within these structures, however. ELISA demonstrated the presence of *S. typhimurium aroA⁻* specific antibodies in *scid* mice that received *Salmonella* primed lymphocytes. Serum antibody titres were directly dependent on the numbers of cells transferred but were consistently lower than those detected in BALB/c mice infected orally with *S. typhimurium aroA⁻* for a similar period of time. Titres were, however, significantly higher than those measured in mice that received an equal number of unprimed cells, therefore demonstrating the presence of a preferentially activated immune system in such recipients (Average ELISA OD's for unreconstituted and immunoreconstituted *scid* mice receiving 1×10^8 unprimed, 1×10^8 or 5×10^8 *Salmonella* primed lymphocytes were 0.10 ± 0.02, 0.18 ± 0.04, 0.39 ± 0.05 and 0.62 ± 0.07 respectively. Unprimed and *Salmonella* primed BALB/c sera demonstrated an OD of 0.22 ± 0.05 and 1.29 ± 0.30 respectively; $\pm$ SEM; sera diluted 1:100 in PBS and applied to sonicated *S. typhimurium aroA⁻* antigen at 2.5 μg protein/ml PBS in ELISA wells).

Novel PP tissue harbored M cells that were amenable to quantitative analysis. FAE of immunoreconstituted *scid* mice characteristically possessed an M cell representation and an FAE surface area as is indicated in Table 1.

Table 1. FAE characteristics in immunoreconstituted *scid* mice.

Experimental Group	% of FAE area		Av. follicle SA $(x10^5\mu m^2)$	Av. # follicles/PP
	% M cell	% G cell		
1×10^8 unprimed cells	3.05±0.44(8)	1.83±0.21(8)	1.83±0.15(8)	3.4±0.2(8)
1×10^8 *Salmonella* primed cells	6.09±1.19(8)*	2.21±0.24(8)^{NS}	2.11±0.13(8)^{NS}	3.2±0.3(8)^{NS}
5×10^8 *Salmonella* primed cells	7.43±0.80(8)**	1.86±0.15(8)^{NS}	2.59±0.36(8)*	3.8±0.3(8)^{NS}

M cell and goblet (G) cell representation within FAE was measured as a percentage (%) of FAE area in the uppermost region of the dome. The average follicle surface area (Av. follicle S.A.) in immunoreconstituted *scid* mice was too small to make peripheral M cell measurements; Av. # follicles/PP = average number of follicles per PP; * and **; $P<0.05$ and $P<0.001$ respectively compared with values for unprimed cells (Welch test); NS = no significant difference; $\pm$ SEM; () = number of observations per group.

In addition, *Salmonella* primed lymphocytes significantly altered the distal ileal morphology as compared to unprimed cells. An increased enterocyte migration rate and, hence an elevation of crypt cell proliferation, was also recorded within this mucosal location. This is shown in Table 2. and indicated that increased cell mediated immune responses were operating within this small intestinal location.

Table 2. Effects of full immunoreconstitution on *scid* intestinal morphology and enterocyte migration rate (EMR).

Experimental parameter			
Small intestinal location 75% SI (distal ileum)	1×10^8 unprimed cells	1×10^8 *Salmonella* primed cells	5×10^8 *Salmonella* primed cells
Villus height (μm)	232±10(8)	214±12(8)^{NS}	227±11(8)^{NS}
Crypt depth (μm)	98±3(8)	114±4(8)*	122±7(8)*
EMR (μmh^{-1})	3.4±0.2(8)	4.1±0.3(8)*	4.3±0.4(8)*

*; P<0.05 (Welch test); NS = no significant difference; ± SEM; () = number of animals in each group.

DISCUSSION

Immunoreconstitution of *scid* mice resulted in the formation of novel PP tissue which harbored a *de novo* M cell population. The level of M cell genesis, which was reflected by the abundance of this cell type within FAE, was demonstrated to be highly dependent upon both the nature and the size of the cellular transfer. Increased lymphocyte activation, induced by the priming of such cells with *S. typhimurium aroA*⁻, caused an approximate doubling in M cell genesis. This was extended further by increasing the number of cells transferred. The level of M cell genesis within FAE is, therefore, regulated by the immunological characteristics imparted to the *scid* recipients. In this model such characteristics are assumed to constitute increased lymphocyte activation within the follicle micro-environment and, also, preferential homing of *Salmonella* primed cells to PP tissue. These results propose that the M cell abundance within FAE is largely dictated by the level of immunological activity within the PP. This is in agreement with previously published data demonstrating that M cell genesis may be down regulated by means of immuno-suppressive therapy using cyclosporin A.[5]

REFERENCES

1. T. C. Savidge, M. W. Smith, P. S. James, and P. Aldred, *Am. J. Pathol.,* 139:177 (1991).
2. T. C. Savidge, M. A. Jepson, P. S. James, N. L. Simmons, and B. H. Hirst, *Proc. Roy. Micro. Soc.,* 25: 5 (1990).
3. G. C. Bosma, M. Fried, R. P. Custer, A. Carroll, D. M. Gibson, and M. J. Bosma, *J. Exp. Med.,* 167:1016 (1988).
4. D. J. Maskell, K .J. Sweeney, D. O'Callaghan, C. E. Hormaeche, F. Y. Liew, and G. Dougan, *Micro. Path.,* 2:211 (1987).
5. T. C. Savidge, and M. W. Smith, *J. Physiology* 422:84 (1990).

A CONFOCAL MICROSCOPICAL ANALYSIS OF PEYER'S PATCH MEMBRANOUS (M) CELL AND LYMPHOCYTE INTERACTIONS IN THE *SCID* MOUSE

T. C. Savidge, M. W. Smith and P. S. James

Department of Cell Biology, AFRC IAPGR
Babraham, Cambridge CB2 4AT, England

INTRODUCTION

Small intestinal membraneous (M) cells have been labelled functionally as epithelial antigen transporting cells.[1] The unique ability of these cells to bring in enteric antigen and/or immune complex into the follicular immune environment plays a fundamental role in eliciting and maintaining mucosal immune responses. The absence of afferent lymphatics the PP tissue and the impervious nature of PP endothelium to antigen and immune complex implicates M cell mediated transport of antigen as an important rate limiting step in the initiation of immune reactions idiosyncratic to PP tissue.[2] Investigation of the trancytotic capabilities of M cells and their potential to act as antigen presenting cells is crucial in providing a better understanding of how this cell type elicits mucosal immune responses.

We have previously demonstrated the ability of M cells to transport a range of oral vaccines and/or drug vehicles to juxtaposed leucocytes situated within M cell intra-cellular pockets.[3,4] These represent the first cell types to encounter trancytosed enteric material and, as such, are likely to play a fundamental role in deciding the fate of ensuing immune reactions. The purpose of the present study was to develop novel methodology able to identify the nature of such cellular interactions in whole PP tissue. The methodology specifically applied confocal laser scanning microscopy to *de novo* PP tissue generated in severe combined immunodeficient (*scid*) mice. The potential of such a model to provide critical physiological appraisal of M cell-lymphocyte interactions and relate these to the transport of antigenic matter in living tissue is discussed.

METHODS

Age and sex matched (8 week old female) Balb/c and C.B.-17 *scid* mice were bred and maintained under isolator conditions throughout the experimental procedures. Leaky *scid* mice were identified as possessing serum IgM levels in excess of 0.05 mg/ml and such mice were omitted from further study. Congeneic lymphocytes used for the immunoreconstitution of *scid* mice were harvested from Balb/c mouse PP and mesenteric lymph node tissues. The plasma membrane of these cells was labelled specifically with the fluorescent lipophilic probe PKH26-GL (Zynaxis Cell Science Ltd.), as is specified in the instructions accompanying the kit (5 μM PKH26-GL for 1-2 minutes at room temperature). 1×10^8 flourescently labelled lymphocytes were adoptively transferred into age and sex matched *scid* recipients by intra-peritoneal injection. Small intestinal PP tissue was harvested 7 days after the cellular transfer and was either fixed in 2% paraformaldehyde for 30 minutes at room temperature, snap frozen in liquid nitrogen cooled isopentane or

Advances in Mucosal Immunology, Edited by
J. Mestecky *et al.*, Plenum Press, New York, 1995

maintained at 37° C in oxygenated RPMI 1640 + 10 % heat inactivated fetal calf serum in a specifically designed superfusion chamber that fits onto an MRC 500/600 Bio-rad Confocal Microscope. Fixed PP tissue was subsequently reacted with alkaline phosphatase to identify M cells as described by Savidge *et al* .[5]

RESULTS

The fluorescent membrane probe PKH26-GL proved non-toxic to lymphocytes at the concentrations specified and was exceedingly stable *in vivo* for the length of the experimental duration. It did not interfere with the ability of lymphocytes to home *in vivo*, as was observed by the formation of *de novo* PP tissue which was consistently absent in control unreconstituted *scid* mice. Alkaline phosphatase stained PP tissue demonstrated the presence of M cells within this tissue. A quantitative study demonstrating the temporal appearance of M cells within FAE following the immunoreconstitution of *scid* mice is shown in Figure 1. A sharp rise in the number of M cells detected was recorded 3-4 days following the adoptive transfer of cells and this corresponded approximately to the time of arrival of large numbers of B lymphocytes into subepithelial and FAE compartments of PP tissue. This was verified by the application of rabbit anti-mouse Ig[+] and anti-CD3[+] (145-2C11) antibodies to 8 µm thick frozen cryostat sections.

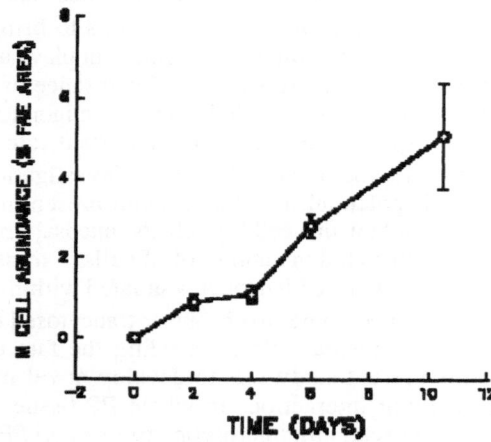

Figure 1. Temporal relationship of M cell reconstitution following adoptive transfer.

Confocal microscopical analysis of whole PP tissue demonstrated the existence of large numbers of lymphocytes within FAE and subepithelial immune compartments that retained a strong fluorescent signal characteristic of PKH26-GL. Viewing PP tissue maintained in a specifically designed superfusion chamber under physiological conditions, demonstrated the ability of this technique to view lymphocyte interactions and movement in whole living PP tissue.

A quantitative confocal study performed on alkaline phosphatase reacted PP tissue demonstrated a good correlation between M cells and the presence of lymphocytes within M cell intra-cellular pockets, as has been previously described using transmission electron microscopy.[6] This correlation was, however, highly dependent upon the follicle dome location from where measurements were taken. As is recorded in Table 1., the highest incidence of M cell-lymphocyte associations was evident in peripheral regions of the dome. Such cellular interactions were the least apparent in regions approaching the dome apex. Applying anti-CD3[+] (145-2C11), rabbit anti-mouse Ig[+] and anti-common leukocyte antigen

(T200) antibodies demonstrated no obvious replacement of PKH26-GL positive lymphocytes for unlabelled cells. Also, lymphocytes were shown to be the predominant leukocyte cell type to be present within FAE.

Table 1. M cell-lymphocyte associations in *scid* mice.[1]

Dome region	Distance from dome apex (μm)	Lymphocytes/M cell[2]
Apical dome	<20	0.23 ± 0.04(32)
Middle dome	20-50	0.54 ± 0.11(36)[3]
Peripheral dome	>50	0.85 ± 0.14(59)[4]

[1]M cells located <20, 20-50 or >50 μm radially from the dome apex are referred to as apical, middle and peripheral dome regions respectively.
[2]The number (#) of fluorescent lymphocytes found to associate with M cells was determined from 9 follicles obtained from 4 PPs (Mean small intestinal site = 38 ±12% SI) from 4 immunoreconstituted *scid* mice. Goblet cells were eliminated from the study by post-staining with 1% alcian blue;
[3] and
[4];P<0.05 and P<0.001 (Welch test); ± SEM; () = number of M cells measured.

DISCUSSION

The present study demonstrated the successful immunoreconstitution of *scid* mice using congeneic lymphocytes isolated from age and sex matched Balb/c mice. When donor cells were harvested from PP and mesenteric lymph node tissues, rapid homing and formation of novel PP tissue in the small intestine of these animals was recorded. In control unreconstituted *scid* mice, PP tissue was consistently absent. The novel PP tissue proved amenable to quantitative M cell analysis and substantial numbers of such cells could be demonstrated. Confocal microscopy revealed a large population of PKH26-GL positive lymphocytes to be present within the FAE of this tissue and a substantial proportion of these were shown to associate positionally with M cells. This specific cellular interaction was, however, highly dependent upon the dome location. M cells showed a reduced tendency to associate with lymphocytes as they matured and migrated towards the follicle dome apex. The physiological significance of this previously undescribed finding remains unclear. Studies of M cell transport characteristics in our laboratory have demonstrated that the majority of bound and trancytosed material occurs within the dome periphery, where there is an abundance of M cells. There may, as a consequence, be a reduced tendency for lymphocytes to remain in more apical dome regions where there is a reduced antigenic presence.

REFERENCES

1. M. R. Neutra, and J. -P. Kraehenbuhl, *Trends in Cell Biology* 2:134 (1992).
2. C. H. Allan, and J. S. Trier, *Gastroenterol.* 100:1172 (1991).
3. T. C. Savidge, and M. A. Jepson, *J. Cell. Biol.* 111:462a (1990).
4. T. C. Savidge, M. A. Jepson, P. S. James, N. L. Simmons, and B. H. Hirst, *Proc. Roy. Micro. Soc.* 25:5 (1990).
5. T. C. Savidge, M. W. Smith, P. S. James, and P. Aldred, *Am. J. Pathol.* 139:177 (1991).
6. R. L. Owen, and A. L. Jones, *Gastroenterol.* 66:189 (1974).

THE MODULATION OF CLASS I AND CLASS II MHC MOLECULES ON INTESTINAL EPITHELIAL CELLS AT DIFFERENT STAGES OF DIFFERENTIATION

Anne Donnet-Hughes, Yves Borel, and
Eduardo, J. Schiffrin

Nestlé Research Centre, Vers-chez-les-Blanc
1000 Lausanne 26, Switzerland

INTRODUCTION

Two classes of effector T cells exist. The first is normally CD4+, has a TCR associated with CD3 and interacts with foreign peptides bound to MHC Class II antigens on antigen presenting cells. The second, is CD8+, has TCR-CD3 and interacts with foreign peptides associated with Class I MHC molecules on target cells. However, little is known of the interaction between these differenteffector cells and intestinal epithelial cells. For this reason, the regulation of MHC molecules by two T cell products, IFN-γ and IL-4, was examined. Cytokines such as the IFN-γ produced by Th1 cells, induce MHC Class II expression on intestinl epithelial cells (IEC) both "in vitro"[1] and "in vivo"[2] but the effect on MHC Class I, and on cells at different stages of differentiation has received essattention. IL-4, a product of Th2 cells, is known to up-regulate Class II expression on B cells and monocytes[3] but its effect on that of IEC has not been studied. We,therefore, used FACScan analysis to compare the expression of MHC antigens on human intestinal epithelial cell lines and the subsequent effect of a 48hr exposure to IFN-γ or IL-4. Furthermore, since IEC "in vivo" are present at different stages of differentiation depending on their location within the intestinal villus, two epithelial cell lines, representing different maturation states were compared.

MATERIALS AND METHODS

The human colon adenoncarcinoma cell lines HT-29 and T84 were used in this study. The HT-29 cell line displays characteristics of foetal enterocytes and can be grown in an undifferentiated state when cultured in the presence of glucose.[4] T84, on the other hand, exhibits some of the features of crypt epithelial cells and represents a more differentiated cell line. When the cell monolayers reached 70-80% confluence, they were exposed over a 48hr period to IFN-γ or IL-4 at 10, 100, or 1000 U/ml. FACScan analysis of MHC Class I and Class II expression on detached cells was then carried out after immunofluorescence labelling with monoclonalantibodies W6/32 and L243 respectively.

Advances in Mucosal Immunology, Edited by
J. Mestecky *et al.*, Plenum Press, New York, 1995

RESULTS

Expression of MHC Molecules by Undifferentiated HT-29 Cells

All cells expressed moderate levels of HLA-A,B,C (Fig. 1a), and little detectable HLA-DR (Fig.1c). However, IFN-γ at all the doses tested was capable of inducing expression of both antigens (Figs. 1b and 1c). This inductionwas dose-dependent. In contrast, IL-4 had no apparent effect on HLA-DR, and the effect, if any on Class I was variable (Results not shown).

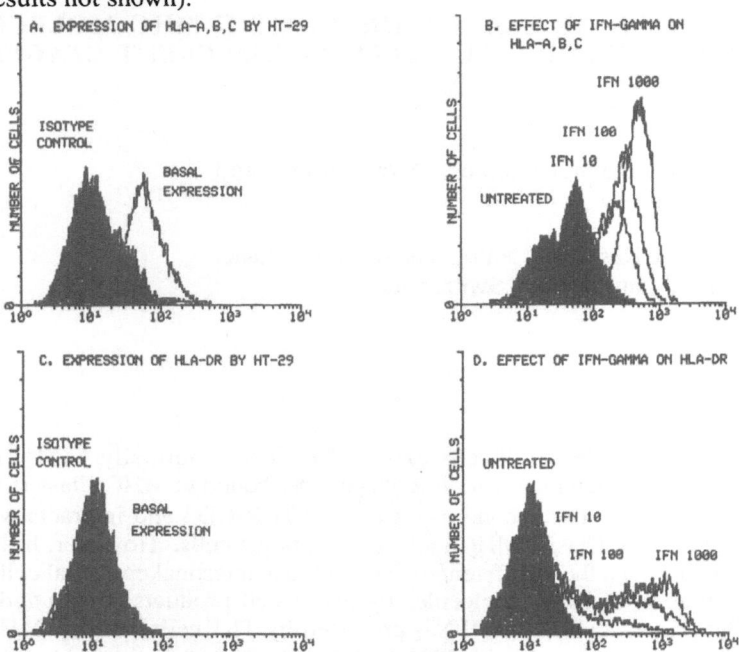

Figure 1. FAC^{Scan} analysis of MHC antigen expression by HT-29 cells. Shaded curves in Figures 1a and 1c represent the appropriate antibody control, whilst unshaded curves represent basal expression of Class I (Fig. 1a) and Class II (Fig. 1c) antigens. Shaded curves in Figs. 1b and 1d represent basal expression of Class I and Class II respectively, whilst unshaded curves show the effect of IFN-γ treatment.

Expression of MHC Molecules by T84 Cells

As for HT-29 cells, the T84 cells constitutively expressed moderate levels of Class I molecules (Fig. 2a) but no detectable Class II antigens (Fig. 2d). Once again, exposure to IFN-γ led to increased expression of HLA-DR in a dose-dependent manner (Fig. 2e). However, in this case it was not accompanied by a similar increase in HLA-A,B,C expression. On the contrary, a slight decrease in Class I expression was often seen (Fig. 2b). As before, IL-4 had no obvious effect on HLA-DR expression but HLA-A,B,C expression was decreased (Fig. 2c).

DISCUSSION

Intestinal epithelial cells of adult tissue express MHC Class I and Class II antigens[3,5] and are thought to be involved in antigen processing and presentation.[6,7] The extent of Class II expression is dependent on both the location of the cell within the villus and the intestine itself.[2] Immature cells in the crypt region only express Class II in distal intestine whilst mature enterocytes show a patchy distribution within the villus of duodenal tissue and a profound expression covering the entire villus in distal intestine. The epithelial cell lines used in this study were similar to IEC from normal duodenal tissue in that they

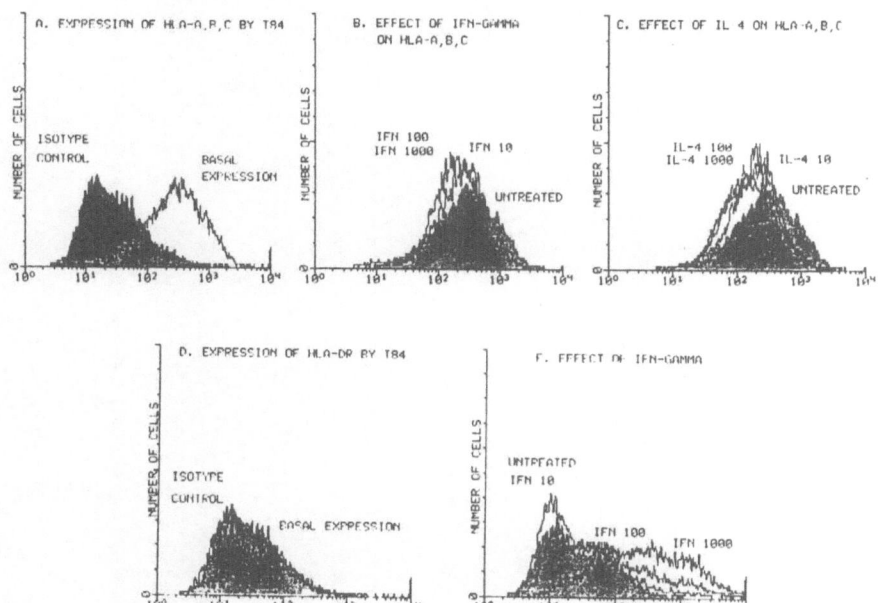

Figure 2. FACScan analysis of MHC antigen expression by T84 cells. In Figs. 2a and 2d, shaded curves represent antibody controls and the unshaded curves, the basal expression of Class I and Class II antigens respectively. In the remaining graphs, the shaded curve represents the basal expression of MHC antigen and the unshaded curve, the effect of IFN-γ or IL-4.

expressed little, if any, Class II molecules but could be induced to do so following exposure to IFN-γ. Moreover, less differentiated cells, in keeping with 'in vivo' studies[2] are more readily inducible. However, differential regulation of Class I molecules was observed. Whilst IFN-γ increased expression of HLA-A,B,C on the more immature HT-29 cells, it often caused a slight decrease in this expression on T84 cells. IL-4, a product of Th2 cells, has previously been reported to augment Class II expression on B cells and monocytes.[3] However, our results suggest that it has no significant effect on Class II expression by IEC. IL-4 did, however, influence expression of Class I but only on the more differentiated T84 cells. In conclusion, individual T cell products 'in vivo' may have different capacities to modulate the ratio of MHC molecules expressedby IEC. The concentration of these lymphokines, the resultant changes in MHC antigen expression and the location of the epithelial cells affected, will determine the type and the extent of the interaction with the different effector T cells in the gut mucosa.

REFERENCES

1. P. Hoang, B. Crotty, H. R. Dalton, and D. P. Jewell, *Gut* 33:1089 (1992).
2. A. Hughes, K. J. Bloch, A. K. Bhan, D. Gillen, V. C. Giovino, and P. R. Harmatz, *Immunology* 72:491 (1991).
3. T. L. Gerrard, D. R. Dyer, and H. S. Mostowski, *J. Immunol.* 144:4670 (1990).
4. A. Zweibaum, M. Pinto, G. Chevalier, E. Dussaulx, N. Triadou, B. Lacroix, K. Haffen, J. L. Brun, and M. Rousset, *J. Cell. Physiol.* 122:21 (1985).
5. W. S. Selby, G. Janossy, G. Goldstein, and D. P. Jewell, *Clin. Exp. Immunol.* 44:453 (1981).
6. P. W. Bland and C. V. Whiting, *Immunology* 68:497 (1989).
7. L. Mayer and R. Shlien, *J. Exp. Med.* 166:1471 (1987).

CRYPTDINS: ENDOGENOUS ANTIBIOTIC PEPTIDES OF SMALL INTESTINAL PANETH CELLS

Sylvia S L Harwig, Patricia B Eisenhauer,
Nancy P Chen, and Robert I Lehrer

Department of Medicine
UCLA-Center for the Health Sciences
Los Angeles, CA 90024-1678

INTRODUCTION

Despite its nutrient-laden content, only a sparse microbial flora persists within the small intestine's lumen How the small intestine controls microbial populations at its critical sites of food absorption and epithelial cell renewal is poorly understood, but multiple factors are likely to contribute Traditionally, these are believed to include the effects of gastric acidity on organisms that enter it via the oral cavity, the combined mechanical effects of peristalsis and epithelial cell shedding, and possibly the antimicrobial effects of various detergent-like bile salts, fatty acids and lysolipids A recent report that murine small intestinal Paneth cells contained mRNA encoding a peptide, "cryptdin",[1] that was homologous to the antimicrobial defensins[2] found in many mammalian phagocytes, raised the possibility that endogenous antibiotic peptides might also mediate innate small-intestinal resistance to colonization and infection Because these intestinal defensins had not previously been isolated and tested for antimicrobial activity, we undertook the studies described below

MATERIAL AND METHODS

Defensins were purified from the small intestines of female adult Swiss Webster mice (Harlan Sprague Dawley, Indianapolis, IN) Briefly, after the small intestines had been excised and rinsed, the organ was sectioned into quarters transversely The segments were inverted onto thin glass pipettes and immersed for 60 min in a saline solution that contained EDTA and dithiothreitol At intervals, the exfoliated cells were collected, washed and resuspended on 0 34M sucrose The "granule"-enriched postnuclear fraction, prepared after homogenization and centrifugation at 27,000 x g, was extracted with acetic acid, and the intestinal defensins were purified from it by sequential gel permeation chromatography on a SynChropak GPC100 column, and several modes of reversed phase HPLC on a Vydac C-18 column Three purified peptides were obtained and subjected to quantitative amino acid analysis and gas phase sequencing Two of them were also tested for antimicrobial activity A detailed description of these procedures has been provided elsewhere [3]

The primary amino acid sequences were confirmed by performing gas phase sequencing and amino acid analyses on peptide fragments prepared after endoproteinase (Lys-C) digestion Briefly, approximately 100-400 pmoles of each purified cryptdin was carboxymethylated (CM) and suspended in 10 µL of a 25 mM Tris buffer (pH 8.5) that also contained 1 mM EDTA, 0.5M urea and 20 mM methylamine. One part, by weight, of freshly prepared Lys-C enzyme (Boehringer Mannheim, Indianapolis, IN) was added per 10 parts of peptide and digestion was allowed to proceed for 16h. The reaction was stopped by adding glacial acetic acid and the digest's contents were resolved on a narrow bore (2.1 x 150 mm) Vydac C18 column (The Separations Group, Hesperia, CA).

Additional structural confirmation was obtained by subjecting CM-Cryptdins 1 and 2 to electrospray mass spectroscopy, performed by Dr. Kristine Swiderek of the Protein Core Facility of the Beckman Research Institute, City of Hope, Duarte, CA. Mass spectra were recorded in the positive ion mode using a TSQ-700 triple quadrupole instrument (Finnigan-MAT, San Jose, CA) equipped with an electrospray ion source. The observed masses were: for CM-cryptdin-1, 4601 (expected 4601.4), and for CM-cryptdin-2, 4485.0 (expected 4484.3).

Antimicrobial activity was tested in thin agarose gels by an ultrasensitive radial diffusion assay that has been described elsewhere.[4] The test organisms were *Listeria monocytogenes,* Strain EGD, *Escherichia coli* ML-35, and an isogenic pair of *Salmonella typhimurium* strains that were provided by Dr. Fred Heffron of the University of Oregon: 14028S (wild type, mouse virulent) and 7953S (*phoP*, mouse-avirulent).

RESULTS AND DISCUSSION

The three defensins (cryptdins) purified from the mouse small intestine differed in their net cationicity, allowing their resolution by acid-urea (AU)-PAGE (Fig. 1). Although all were less cationic than rabbit defensin NP-1, which was used as a control, each was more cationic than the four human neutrophil defensins (data not shown).

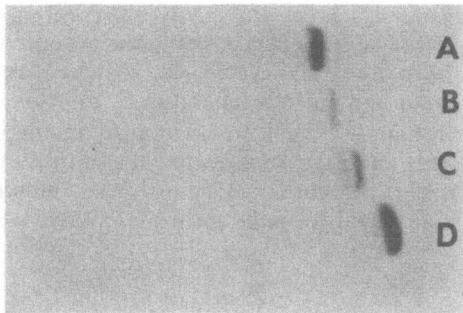

Figure 1. Acid-urea polyacrylamide gel of purified defensins Lane A contained Cryptdin 2, Lane B contained Cryptdin 1, Lane C contained Cryptdin 1α (accompanied by a more slowly migrating peptide that was not further characterized) and Lane D contained rabbit defensin NP-1

Fig. 2 shows the amino acid sequences of our three mouse intestinal defensins and of the peptide fragments purified from their Lys C digests. Each cryptdin species contained 35 amino acid residues. Cryptdins 1 and 2 were 88.6% identical, differing only at residues 10, 15, 29, and 31, where Cryptdin 1 contained Thr_{10}, Arg_{15}, Met_{29} and Thr_{31} and Cryptdin 2 contained Ala_{10}, Gly_{15}, Leu_{29} and Met_{31}. The extra, positively-charged arginine in Cryptdin 1 would account for its more cathodal migration in AU-PAGE, relative to Cryptdin 2 (Fig. 1). Cryptdin 1α and cryptdin 1 were identical except for residue 10, a lysine in Cryptdin 1α and a threonine in Cryptdin 1. This additional lysine explained why Cryptdin 1α migrated more cathodally than Cryptdin-1 (Fig. 1).

Ouellette *et al.*[5] previously localized the cryptdin (*Defcr*) gene to the proximal region of mouse chromosome 8 and, by analyzing restriction fragment length polymorphisms, inferred the existence of at least three potential *Defcr* alleles in mice.[5] The human defensin genes were found on a region of human chromosome 8 (8p23) that is homologous (syntenic) to the region of mouse chromosome containing the *Defcr* gene(s).[6]

Since cryptdin 1 and cryptdin 2 differed in 4/35 residues, they are more likely the products of reduplicated rather than of allelic genes. In contrast, while the single residue difference between cryptdin 1 and cryptdin 1α does not preclude an origin from reduplicated genes, their respective genes are more likely to be allelic. Evidence for reduplicated and allelic defensin genes has been provided in other mammalian systems.[2]

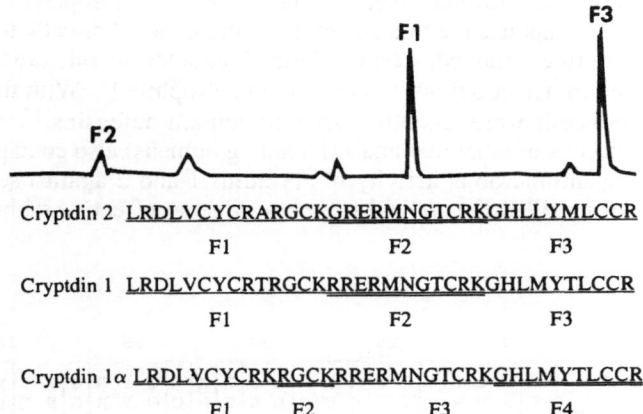

Figure 2. Endoproteinase (LysC) digestion of murine cryptdins. The complete primary amino acid sequences of Cryptdins 1, 1α and 2 are shown. The LysC fragments that we recovered and sequenced have been underlined (F1, F2, etc.). The RP-HPLC chromatogram for the LysC digest of carboxy-methylated cryptdin-2 is also shown.

Both myeloid and intestinal defensins are produced by post-translational proteolysis of 93-95 amino acid precursors whose respective defensin sequences reside in their extreme carboxy-terminal portion.[2] Each of the intestinal defensins we characterized contained five amino terminal residues preceding their initial cysteine. This elongated amino-terminus contrasts with all of the previously described myeloid defensins, two which contain 0 to 2 residues in this location. Since it was recently noted that murine neutrophil 32-D cells processed the aminoterminal portion of transfected human HNP-1 pre-prodefensins to form HNP molecules with 0-1 residues amino-terminal to the first cysteine,[7] the processing machinery of murine myeloid and Paneth cells evidently differs. It remains to be determined if additional N-terminal proteolysis after Paneth cell cryptdins are secreted to the intestinal lumen truncates them to the usual size of fully processed myeloid defensins.

Although none of our three intestinal defensins corresponded precisely to the cryptdin domain of the pre-procryptdin cDNA initially reported by Ouellette *et al.*,[1] Cryptdin 2 differed from it at only two sites (residues 10 and 31) and Cryptdins 1 and 1α differed at only three (residues 10, 15, and 29). Because Swiss Webster mice are outbred and we did not use the same breeder used by Ouellette *et al.* (Charles River, North Wilmington, MA). Consequently, these discrepancies probably reflect the complexity and polymorphisms of the *Defcr* locus in mice. Indeed, Selsted *et al.*[8] recently purified and characterized five distinct intestinal defensins obtained from outbred Swiss Webster mice purchased from the same breeding colony initially used by Ouellette. The sequence of their most abundant intestinal defensin (unfortunately, also named "Cryptdin-1" in their publication) agreed precisely with the previously reported cDNA sequence, reaffirming one's faith in the genetic code. Their next most abundant peptide (designated "Cryptdin-2") was identical to the peptide called Cryptdin 1 in this report, and their "Cryptdin-3" was identical to our Cryptdin 1α. Fig. 3 combines their findings with ours, to shows the six structurally distinct murine intestinal cryptdin species discovered to date. In an attempt to mitigate the potentially confusing numbering system for these peptides, we have added the letter "S" to the designations used by Selsted et al.[8]

Fig. 3 also shows the primary structures of the four rat neutrophil defensins previously purified by us.[9] Sixteen of the 31 amino acid residues (51.6%) in rat neutrophil defensins (RatNPs) 1 and 2 were identical to those in mouse intestinal cryptdins 1, 1α, and 2. These included the principal residues that govern the three-dimensional structure of defensins, chiefly its cysteines, and certain glycine, glutamic acid and arginine residues.[10] Residue 10 (numbering from the amino terminus of Cryptdin 1) was especially mutable, differing in 5 of the 6 mouse intestinal defensins and in all 4 rat neutrophil defensins.

We cannot yet extrapolate the findings in this murine model directly to the biology of the human small intestine. Indeed, despite their abundance in rat, rabbit, and human neutrophils, we failed to detect defensins in murine neutrophils.[11] With the exception of rabbits, whose Paneth cells were recently shown to contain defensins,[12] it remains to be shown that the Paneth cells of other mammals (including humans) also contain defensins.

We tested the antimicrobial activity of cryptdins 1 and 2 against several bacteria. Both intestinal defensins killed *E. coli* and *L. monocytogenes* effective rabbit defensin NP-

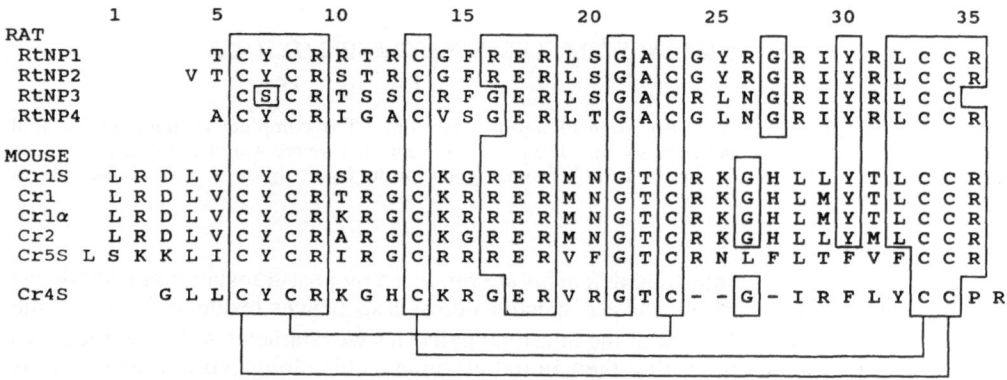

Figure 3. Primary structures of rat and mouse defensins. Conserved residues have been boxed and the cysteine connectivity was inferred from prior studies with rabbit and human defensins.[2,10]

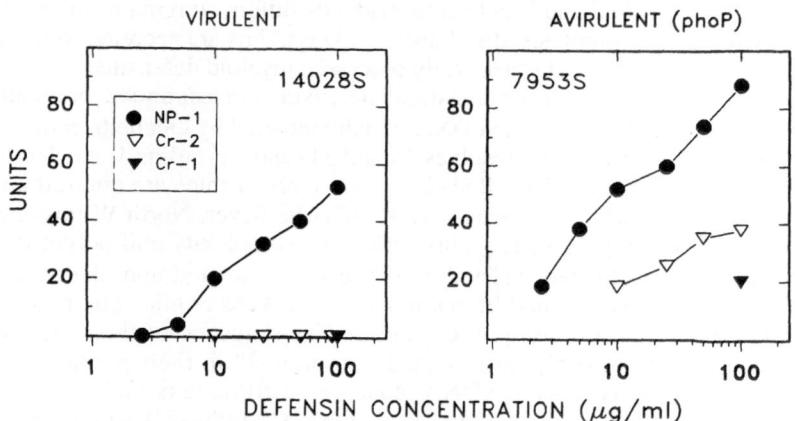

Figure 4. Bactericidal activity of cryptdins. Radial diffusion assays were performed against two smooth (S) isogenic strains of *S. typhimurium*: 7953 (mouse-avirulent, *phoP*) and 14028 (mouse-virulent, wild-type). Modified from reference 3, with permission.

1 and the less potent human defensin HNP-1.[3] Figure 4 shows the effects of rabbit NP-1 and the murine intestinal defensins Cryptdin-1 and Cryptdin-2 against *S. typhimurium*.

Whereas both Cryptdin 1 and Cryptdin 2 readily killed the avirulent smooth 7953 strain (*phoP*), they were ineffective against its isogenic wild-type parent, Strain 14028S. We have yet tested the antimicrobial potency of Cryptdin 1α, but based on the pattern seen with other defensins it would not be surprising if its greater cationicity endowed it with enhanced potency, relative to Cryptdin 1. The nature and significance of the *phoP* mutation in *S. typhimurium* has received intensive study by others, whose work should be consulted for details.[13,14]

SUMMARY

We purified three peptides ("cryptdins") from the small intestines of mice, established their primary amino acid sequences and examined their antimicrobial activity. Their primary sequences revealed approximately 50% identity to a group of antimicrobial defensins that we had previously isolated from the granules of rat neutrophils. In addition to their ability to kill Gram-positive (*L. monocytogenes*) and Gram-negative bacteria (*E. coli* and *S. typhimurium*) *in vitro*, the peptides were much more active against an avirulent (*phoP*) *S. typhimurium* strain than against its isogenic, mouse-virulent progenitor. Overall, these data suggest that endogenous antimicrobial peptides produced by Paneth cells may protect small intestinal crypts, which are critical sites of epithelial cell renewal, from invasion by autochthonous flora or by perorally acquired potential pathogens, such as *Listeria* and *Salmonella*.

ACKNOWLEDGMENTS

Our work on the presence of defensins in murine intestinal and myeloid cells was supported by grants from the National Institutes of Health: AI 29595 and AI 29839.

REFERENCES

1. A. J. Ouellette, R. M. Greco, M. James, D. Frederick, J. Naftilan, and J. T. Fallon, *J. Cell Biol.* 108:1687 (1989).
2. R. I. Lehrer, A. K. Lichtenstein, and T. Ganz, *Annu. Rev. Immunol.* 11:105 (1993).
3. P. B. Eisenhauer, S. S. L. Harwig, and R. I. Lehrer, *Infect. Immun.* 60:3556 (1992).
4. R. I. Lehrer, M. Rosenman, S. S. L. Harwig, R. Jackson, and P. B. Eisenhauer, *J. Immunol. Meth.* 137:167 (1991).
5. A. J. Ouellette, D. Pravtcheva, F. H. Ruddle, and M. James, *Genomics* 5:233 (1989).
6. R. S. Sparkes, M. Kronenberg, C. Heinzmann, K. A. Daher, I. Klisak, T. Ganz, and T. Mohandas, *Genomics* 5:240 (1989).
7. L. Liu, E. V. Valore, A. Oren, and T. Ganz, *Blood* 82:641 (1992).
8. M. E. Selsted, S. I. Miller, A. H. Henschen, and A. J. Ouellette, *J. Cell. Biol.* 118:929 (1992).
9. P. B. Eisenhauer, S. S. L. Harwig, D. Szklarek, T. Ganz, and R. I. Lehrer, *Infect. Immun.* 58:3899 (1990).
10. C. P. Hill, J. Yee, M. E. Selsted, and D. Eisenberg, *Science* 251:1481 (1991).
11. P. B. Eisenhauer and R. I. Lehrer, *Infect. Immun.* 60:3446 (1992).
12. T. Tominaga, J. Fukata, Y. Hayashi, Y. Satoh, N. Fuse, H. Segawa, O. Ebisui, Y. Nakai, Y. Osamura, and H. Imura, *Endocrinology* 130:1593 (1992).
13. E. A. Groisman, E. Chiao, C. J. Lipps, and F. Heffron, *Proc. Natl. Acad. Sci. USA* 86:7077 (1989).
14. S. I. Miller, *Mol. Microbiol.* 5:2073 (1991).

PHYSICAL INTERACTION BETWEEN LUNG EPITHELIAL CELLS AND T LYMPHOCYTES

René Lutter,[1] Ben Bruinier,[1,2] Bernard E.A. Hol,[1]
Frans H. Krouwels,[1] Theo A. Out[2,3] and Henk M. Jansen[1]

[1]Dept. of Pulmonology, F4-206, [2]Clinical Immunology
Laboratory B1-236, Academic Medical Centre,
University of Amsterdam, [3]Lab. Clinical Experimental Immunology
CLB, Amsterdam, The Netherlands

INTRODUCTION

The proximal airways, including the bronchi, are exposed to the external environment and thus frequently challenged by allergens and micro-organisms. In fact, these represent the insults that cause bronchial responses in patients with allergic asthma or chronic obstructive pulmonary disease.[1] Several lines of defence in the airways exist against these challenges. The epithelial cell layer covering the upper airways contributes, by various means, significantly to the protection of the airways. In healthy individuals, the epithelium forms a tight, impermeable cell layer. This barrier function is maintained largely by apically positioned tight-junction complexes as well as the capacity of the epithelium to migrate and regenerate swiftly upon mechanical damage.[2,3] The mucus layer on top of the epithelium is another mechanical barrier. Specialized epithelial cells produce mucus components whereas other epithelial cells provide for the clearance of mucus and can modulate the rheological properties of the mucus, for example by altering the ionic composition of the periciliar fluid.[4] Epithelial cells also play a key role in the protection of the airways by transepithelial transport and secretion of microbicidal proteins. In this respect, the transport of dimeric IgA and pentameric IgM by the polymeric immunoglobulin receptor and the production of lysozyme and lactoferrin are well-studied functions of epithelial cells.[5]

Over the last three years it has been recognized that epithelial cells may influence activities of both immunocompetent and inflammatory cells. From their topology, epithelial cells are in a key position to modulate inflammatory and immunological processes. In addition, epithelial cells produce a limited repertoire of mediators, including the cytokines IL-1, IL-6, IL-8, GM-CSF, TNF-α and arachidonic acid metabolites, that can modulate activities of other cells.[6,7] And further, epithelial cells express receptors for cytokines and various cell adhesion molecules that allows them to respond and to bind to other cells.[8,9] Our research into the pathogenesis of allergic asthma is focusing on the role of bronchial epithelium in modulating inflammatory and immunological responses.

Recent studies have implicated T cells in the pathogenesis of asthma. In both bronchial biopsies as well as bronchoalveolar lavages derived from asthmatics, increased numbers of lymphocytes and of activated lymphocytes, as evident from an increased expression of the receptor of IL-2 (IL2-R)[10,6,] have been described. Here we report that epithelial cells can activate T cells in vitro in an apparent antigen-independent manner. We propose that at mucosal surfaces such as the bronchi, epithelial cells can function as accessory cells in T cell activation.

Advances in Mucosal Immunology, Edited by
J. Mestecky *et al.*, Plenum Press, New York, 1995

METHODS

Cells

Throughout the present studies we used the human lung derived mucoepidermal cell line, H292 (ATCC CRL 1848, Rockville, Maryland, USA) serving as a highly reproducible model of airway epithelium. These cells express a cytokeratin profile that resembles that of differentiated airway epithelium. And further, these cells can be grown on permeable filters in tight monolayers with characteristics indicative of a polarized cell (Lutter *et al.*, in preparation). The cell line was maintained in monolayers in plastic culture flasks in RPMI 1640 (Gibco) supplemented with 10 % heat-inactivated foetal calf serum (Gibco), penicillin (100 U/ml, Gist-Brocades, Delft, The Netherlands), streptomycine (100 µg/ml, Gibco) and 0.5 mM glutamic acid (Merck). Cell suspensions from these monolayers were obtained by a standard trypsin treatment.

Cells from two different human lung derived T cell clones, which were prepared from cells in the broncho-alveolar lavage fluid from healthy individuals, were used.[11] Both clones were CD4+ and expressed the memory T cell-associated antigen, CD45RO, but differed with respect to the expression of the fibronectin receptor, VLA4. Cells from the clone 7B8 were VLA4+ whereas cells from clone 2B1 were VLA4-. Stimulation of the cells from each clone with immobilized anti-CD3 induced the production of both IL-4 and IFN-γ and thus these clones were denoted as Th0-like.

Clonal T cells grown at standard conditions, were in an activated state as supported by an increased expression of the IL-2 receptor (CD25). For our adhesion studies, stored T cells were thawed and stimulated with phytohaemagglutinin (PHA; 2 µg/ml, Wellcome, Dartford, UK) and irradiated (4000 Rad) autologous blood mononuclear cells (FC) for one week. Then, T cells were stimulated again for another three days with immobilized anti-CD3 (mouse IgG2a, CLB T3/3, 0.5 µg/ml; Central Laboratory of the Netherlands Red Cross Blood Transfusion Service (CLB), Amsterdam, The Netherlands). We also employed clonal T cells that had been depleted of stimuli by washing the cells four days after the initial stimulation with PHA/FC and cultured for another six days (i.e., resting).

In our initial studies we employed CD2+ T lymphocytes isolated from peripheral blood by E-rosetting. After thawing, these CD2+ cells were used either directly (i.e. resting) or after stimulation with PHA for three days (i.e. activated).

Cell Adhesion Assay

Confluent monolayers of epithelial cells were grown in 96-well microtitre plates. Epithelial cells were exposed to human recombinant IFN-γ (100 U/ml, Genentech) for different periods, as indicated in the text. Before adding T cells, IFN-γ was removed in three washes with medium. Clonal T cells (5.10^6) were labelled with ^{51}Cr by incubation for 1 hour at 37° C with 200 µCi of Na^{51}CrO4 (350-600 µCi/mg, Amersham). Free label was removed in four washes with warm medium. Labelled cells (2×10^5) in Earle's medium were added to the well and incubated at 37° C for 1 hour. Non-adherent cells were removed by gently washing the wells four times with warm medium. The adherent cells were subsequently lysed by the addition of 200 µl of 1% (w/v) Triton X-100 (Sigma) in distilled water. The contents of the wells were collected on cotton swabs and radioactivity was determined in a gamma-counter. The spontaneous release of ^{51}Cr amounted less than 6 % and was not affected by epithelial cells. Epithelial cells did not acquire ^{51}Cr released by T cells. Assays were carried out in quadruplicate and repeated at least twice. Adherence was expressed as percentage of bound counts over added counts, and corrected for spontaneous release. Standard deviations were all less than 10 % of the mean.

For some experiments, T cells were stained green with sulfofluoresceine diacetate (SFDA; 5 µg/ml, Molecular Probes, Junction City, OR) or red with hydroethidine (HE; 4 ng/ml, Polysciences, Warrington, PA) by incubation with the dye for 1 hour at 37° C.

Antibodies

The hybridoma (TS2/9), that produces mAbs against the lymphocyte functional antigen-3, LFA-3 (CD58), was obtained from the ATCC and cultured in RPMI-1640 supplemented with 10 % foetal calf serum, 0.5 mM sodium-pyruvate (Merck), penicillin and streptomycin (for concentrations see above). Culture supernatants were concentrated by ultrafiltration and used without further purification. mAbs (clone MEM 111: IgG2a) against the intracellular adhesion molecule-1, ICAM-1, were purchased from Monosan (Uden, The Netherlands) and used in a 1 in 10 dilution. Anti-CD 71 was bought from Becton Dickinson. The mAbs against HLA-DR (CLB-HLA-class II) and HLA-ABC (CLB-HLA-class I) were obtained from the CLB. Antibodies against CD3 (CLB T3/3) were a gift from Dr. R.A.W. van Lier (CLB, Amsterdam, The Netherlands).

FACS Analysis

Single cell suspensions of monolayers of epithelial cells were obtained by incubation with 0.1 g/l of trypsin and 0.04 g/l EDTA in phosphate-buffered saline, PBS (pH 7.2-7.4) for 60 min at room temperature. This treatment did not affect the expression of the molecules of interest. Further, wells were inspected visually for the removal of all cells.

T cells and loosened epithelial cells were incubated on ice for 30 min with 100 μl of monoclonal antibodies, at the indicated dilution in PBS, supplemented with 0.5 % BSA (PBA) and 1% normal goat serum. The cells were washed three times with cold PBA to remove unbound antibody, and then resuspended in 100 μl of phycoerythrine-labelled goat anti-mouse immunoglobulin (Southern Biotechnology, Birmingham, UK) diluted in PBA and incubated for 30 min on ice. After three washes, the percentage and intensity of stained cells was determined using a FACSScan (Becton Dickinson, Mountain View, CA). To determine the aspecific binding, subclass specific irrelevant mouse antibodies were used.

Table 1. Adherence of CD2[+] peripheral blood T lymphocytes to epithelial cells.

| | | % binding of T cells to epithelial cells | | | |
	state of T cells	0h	4h	24h	48h
1st experiment	activated	26	-[1]	35	-
2nd experiment	activated	-	63	-	54
	resting	-	8	-	8
3d experiment	resting	5	-	14	-

[1]-; not done

RESULTS AND DISCUSSION

T cells Adhere to Airway Epithelial Cells

Epithelial cells release various mediators such as interleukin-6 and prostaglandin E2 that may affect proximate T cells. We addressed the issue whether epithelial cells communicate directly with T cells, in particular by binding T cells. To establish the cell adhesion assay we initially used ^{51}Cr-labelled cells from one pool of peripheral blood CD2[+] T lymphocytes (Table 1). As shown in Table 1, there was substantial binding of peripheral T cells to untreated epithelial cells, i.e. at t_{0h}. Markedly, peripheral T cells stimulated by phytohaemagglutinin (activated T cells) show a higher percentage of binding to the epithelium than do the non-activated T cells. After treatment of epithelial cells with IFN-γ the percentage of bound peripheral T cells increased, and again activated peripheral T cells showed a higher percentage of binding than non-activated T cells.

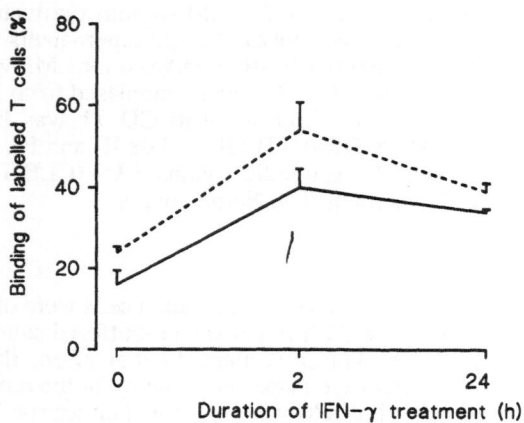

Figure 1. Adhesion of clonal lung T cells (clone 7B8) to IFN-γ treated epithelial cells. The continuous and broken curves are of resting and activated cells, respectively.

Since our interest was to see whether airway epithelium interact with lung T cells we used T cells cloned out of bronchoalveolar lavage fluid from healthy individuals. Clonal lung T cells differ to some extent both phenotypically and functionally from clonal T cells derived from blood as reported elsewhere.[11,12] Figure 1 shows the adherence of clonal lung T cells with the phenotype CD4+, CD45Ro and VLA4+ to epithelial cells, before and after treatment of the epithelial cells with IFN-γ. Identical results were obtained with cells that were CD4+, CD45Ro and VLA4- (not shown). Alike the results with peripheral CD2+ T-lymphocytes, there is a significant binding of clonal lung T cells to untreated epithelial cells. And, at any time point, more activated T cells than resting T cells bound to epithelial cells. Further, both maximally stimulated T cells and resting T cells showed an increased binding when exposed to epithelial cells that had been treated with IFN-γ for two hours prior to the assay. The binding to epithelial cells fell off again to lower levels when T cells were added to epithelium treated with IFN-γ for 24h.

The Effect of Adherence on Clonal Lung T Cells

After activation and prior to proliferation, T cells express characteristic antigens on their cell surface like CD71 and the receptor for interleukin-2 (IL-2R, i.e. CD25). Resting clonal T cells (see Figure 1), that were exposed for one hour only to either non-activated epithelium or to epithelial cells treated with IFN-γ for two hours (see Material and Methods), were cultured further. The expression of CD25 and CD71 on T cells was determined by FACS analysis (Table 2). No changes in the level of expression of CD25 and CD71 were seen on T cells cultured for an additional 4 or 8 hours after exposure to either activated or non-activated epithelium (not shown). However, T cells co-cultured with activated epithelium and subsequently cultured for 16-20 hours showed an increase in the expression of both CD25 and CD71. This expression of CD25 was transient as T cells that were activated by epithelial cells and cultured for 40 hours express little CD25 (not shown). T cells exposed to non-activated epithelium and cultured for 16-20 hours did not show an increase in the expression of these markers.

Epithelial cells, when activated, can secrete various mediators that could influence T cell activation. Therefore, parallel to the experiments mentioned above, T cells were allowed to adhere to epithelial cells and left for an additional 20 hours with the epithelial cells. The expression of CD25 on these T cells was not different from T cells that were exposed to activated epithelium and cultured in the absence of epithelium for 20 hours. The expression of CD71 was not determined in these experiments.

Table 2. Changes in the expression of IL2-R (CD25) and CD71 by T cells after exposure to resting or activated epithelial cells.

T cells	relative mean fluorescence intensity		
	isotype control	CD25[1]	CD71[2]
at rest	100	316	250
exposure to resting epithelium	100	333	300
exposure to activated epithelium	100	1666	1050

[1]measured after 20 hours
[2]measured after 16 hours

It is clear from these results that activation of T cells by epithelial cells is determined solely by signals exchanged during the one-hour period of increased adherence. Both soluble components and cell adhesion molecules may be implicated in T cell activation by epithelial cells during that one hour period.

The Process and Mechanism of Adherence

From the previous it follows that a proportion of the T cells that are non-adherent at the time of recovery is committed by epithelial cells to activation. This raises the question whether or not these T cells bound to epithelial cells during the one hour co-culture period. To gain insight in this process, T cells were labelled with fluorescent dyes. Red-coloured (hydroethidine) cells that were allowed to adhere to epithelial cells for one hour were subsequently chased off in approximately 30-60 min by green-labelled (fluorodiacetate) cells. This observation indicated that binding of T cells to epithelial cells is a rapid and dynamic process and thus non-adherent cells could have been bound by epithelial cells prior to their recovery.

Several molecules are implicated in the binding of T cells to epithelial cells. In initial experiments we determined the level of expression of various of these molecules on epithelial cells following activation and over time. Whereas the level of expression of HLA-class I, HLA-class II and ICAM-1 gradually became higher over 24-48 hours, the number of LFA3 molecules on epithelial cells increased significantly within 60 minutes after stimulation (Figure 2). The increased binding of T cells paralleled the increased expression of LFA3 by epithelial cells. LFA3 is the counterstructure of the common T cell antigen, CD2.[13] At 48 hours after stimulation, the expression of LFA3 returned to basal level (not shown). Although suggestive, these results do not prove the involvement of LFA3 in the binding of T cells via CD2 to epithelial cells or indeed the lack of involvement of other adhesion molecules. Studies with inhibitory antibodies should provide further insight into the adhesion molecules that are involved.

In addition, other, yet unknown factors determine the binding of T cells to epithelial cells based on results of identical cell adhesion experiments in which varying numbers of T cells were used. Exposure of activated epithelial cells, that were activated with IFN-γ for two hours prior to the adhesion assay, to either $0.5.10^5$ T cells or 2.10^5 T cells resulted in very similar percentages of binding, 36% and 30% respectively. It can be argued that only a subpopulation of the T cells bound to epithelial cells. Therefore, non-adherent T cells taken from an adhesion experiment were pelleted to restore the original cell density and were subsequently used in an identical adhesion experiment. The percentage of binding was similar (47% and 44%, respectively), indicating that within the population of T cells there is not a subgroup of cells that adhere. Why, at lower T cell densities, not all available binding sites are occupied by the remaining non-adherent cells, that are occupied at higher T cell densities, remains to be studied.

SUMMARY

The present results support a role for epithelial cells in the activation of T cells in an apparent antigen-independent manner. The transient expression of CD25 indicates a short acting T cells activation. Possibly, this event primes T cells to respond swiftly upon antigen-specific stimulation or to synthesize mediators that affect the local milieu.

The molecular mechanism of interaction, although not well defined possibly involves LFA3-CD2 interactions. In T cell activation, via LFA3-CD2 interaction, the density of presented LFA3 molecules is critical[14]. With the increase in the level of expression of LFA3 by epithelial cells this critical density may have been reached. However, based on what is known about T cell activation and CD25 expression in particular it is likely that additional signals such as soluble mediators are required for T cell activation by epithelial cells.[13]

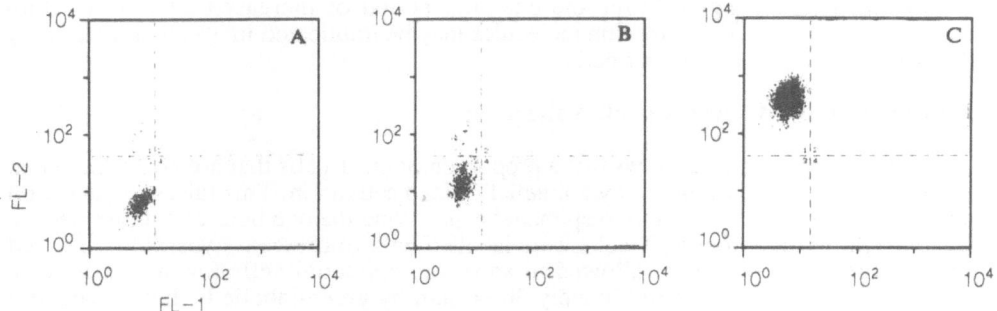

Figure 2. FACS analysis of LFA3 by confluent epithelial cells after 60 min in the absence and the presence of IFN-γ. The horizontal axes shows the fluorescence intensity of phycoerythrine-labelled goat anti mouse immunoglobulin. (A): Level of fluorescence with an irrelevant antibody on non-activated epithelial cells. There is no increase in this level of fluorescence 60 min after activation of epithelial cells. (B): anti-LFA3 on non-activated epithelial cells. (C) anti-LFA3 on activated epithelial cells.

Whether this mode of activation occurs in vivo remains to be established by studying *ex vivo* and in situ material. Not much is known about the expression of LFA3 by epithelial cells *in vivo*, nor about the stimuli that induce the upregulation of LFA3. In preliminary experiments with fluorescence microscopy we found that neither TNF-α nor IL-1β induce LFA3 in the same fashion as IFN-γ.

In conclusion, T cell activation by epithelial cells could be an important feature in inflammatory and immunological processes in mucosal systems such as the bronchi and deserves further research.

ACKNOWLEDGEMENTS

Frans H. Krouwels is supported by Glaxo Research Laboratory (The Netherlands). The authors thank Mr. R. M. R. Reijneke for providing CD2+ T lymphocytes.

REFERENCES

1. American Thoracic Society, *Am. Rev. Resp. Dis.* 136:225 (1987).
2. S. G. Milton and V. P. Knutson, *J. Cell Physiol.* 144:498 (1990).
3. C. Basbaum and B. Jany, *Am. J. Physiol.* 259:L38 (1990).
4. P. Verdugo, M. Aitke, L. Langley, and M. J. Villalon, *Biorheology* 24:625 (1987).
5. P. P. Breitfeld, J .E. Casanova, N. E. Simiter, S. A. Ross, W. C. McKinnon and K. E. Mostov, *Am. J. Respir. Cell Mol. Biol.* 1:257 (1989).
6. S. Mattoli, V. L. Mattoso, M. Soloperto, L. Allegra, and A. Fasoli, *J. Allergy Clin. Immunol.* 87:794 (1991).
7. Y. Ohkawara, K. Yamauchi, Y. Tanno, G. Tamura, H. Ohtani, H. Nagura, K. Ohkuda and T. Takishima, *Am. J. Respir. Cell Mol. Biol.* 7: 385 (1992).
8. M. F. Tosi, J. M. Stark, C. W. Smith, A. Hamedani, D. C. Gruenert and M. D. Infeld, *Am. J. Respir. Cell Mol. Biol.* 7:214 (1992).
9. D. Kvale, P. Krajci and P. Brandtzaeg, *Scand. J. Immunol.* 35:669 (1992).
10. M. Azzawi, B. Bradley, P. K. Jeffery, A. J. Frew, A. J. Wardlaw, G. Knowles, B. Assoufi, J. V. Collins, S. Durham and A. B. Kay, *Am. Rev. Resp. Dis.* 142:1407 (1990).
11. B. E. A. Hol, F. H. Krouwels, B. Bruinier, R. M. R. Reijneke, H. J. J. Mengelers, L. Koenderman, H. M. Jansen and T. A. Out, *Am. J. Respir. Cell Mol. Biol.* 7:523 (1992).
12. F. H. Krouwels, B. E. A. Hol, B. Bruinier, H. J. J. Mengelers, R. Lutter, H. M. Jansen and T. A. Out, (submitted for publication).
13. T. A. Springer, *Annu. Rev. Cell Biol.* 6:359 (1990).
14. S. M. Denning, M. L. Dustin, T. A. Springer, K. H. Singer and B. F. Haynes, *J. Immunol.* 141:2980 (1988).

THE ROLE OF LYMPHO-EPITHELIAL INTERACTIONS IN THE REGULATION OF SMALL INTESTINAL EPITHELIUM PROLIFERATION

Andrej N. Shmakov, Natalija G. Panteleeva,
Andrej V. Fedjanov, and Valerkij A. Trufakin

Siberian Branch of the Russian Academy of
Medical Sciences, Institute of Clinical and
Experimental Lymphology,
Novosibirsk 630117, Russia

INTRODUCTION

There is evidence that immune system participates in the regulation of intestinal epithelial proliferation and differentiation. It has been proposed[1] that activated T lymphocytes affect crypt cell turnover through release of soluble products, i.e., lymphokines. This point of view has more recently received additional experimental confirmation. It was shown[2] that Con-A-stimulated intraepithelial lymphocytes and spleen cells suppress growth and modulate Ia expression by intestinal epithelial cell line. The stimulation of T lymphocytes with pokeweed mitogen or anti-CD3 monoclonal antibodies applied to intestinal organ culture enhanced the rate of crypt cell proliferation.[3] Lymphocyte culture supernatants modulate $[^3H]$-thymidine ($[^3H]$TdR) incorporation into isolated small intestinal crypt cells.[4] There are also data suggesting that lymphocytes induce formation of M cells[5] and regulate Goblet cell development.[6] The involvment of immune factors in regulation of intestinal epithelium functions is supported by data on Interleukin-6 receptor expression in intestinal epithelial cells.[7]

At the same time, it is important for studies of regulation to understand how feedback is functioning. One may speculate that if immune system actually participates in the regulation of intestinal epithelium renewal, then 'information" from epithelium should be taken by lymphocytes and modulate their influence on epithelium. We have obtained data indicating that T-lymphocytes transfered to syngeneic mice affect in dependence on the rate of recipients crypt cell proliferation (number of cells in cycle): suppress normal proliferation level and stimulate depressed slates.[8,9] We postulated that crypt cells and/or villus epithelial cells provide signal(s) modulating lymphocytes activity. However, the evidence is indirect and need further experimental verification. The present study was carried out to explore the direct effect of epithelial cells on lymphocyte ability to modulate crypt cell proliferation.

METHODS

Male (CBA/Ca x C57BL/6J) F1 mice, aged 10-12 weeks were used. Mice were killed by cervical dislocation.

Design of Experiments

Lymphocytes and small intestinal epithelial cells were incubated separately or

Advances in Mucosal Immunology, Edited by
J. Mestecky *et al.*, Plenum Press, New York 1995

together in tissue culture flasks (Nunclon, Denmark) for 2 hr at 37° C in 5% CO_2. The initial concentration of 5 x 10^6 cells per ml in RPMI 1640 medium (Flow Laboratories, Scotland) + 10 % heat-inactivated fetal calf serum was employed. Cells were incubated in different ratios in total volume of 12 ml: 1) 1 portion (2 ml) of one cell type + 5 portions (10 ml) of other cell type and vice versa, or 2) cells of individual type constituted 1 or 5 portions and the rest volume added was medium without cells. Thymocytes, nonfractionated spleen lymphocytes (spleen cells), spleen T lymphocytes, small intestinal epithelial crypt and villus cells were used for incubation. After incubation periods, supernatants were collected and its actions on [3H]TdR incorporation into isolated small intestinal crypt cells (SICC) were assesed by method described by Speekenbrink & Parrott.[4] Crypt cell number were adjusted to 1 x 10^7 million per ml in cold medium, and 0.74 MBq of [3H]thymidine (Izotop, Russia) were added to each ml of suspension. Supernatants were dispensed into wells of 96-well microtiter plate, 0.05 ml per well, and 0.05 ml of SICC was added to each well. The experiments were performed in quadruplicate if not otherwise indicated. The plates were incubated for 1 hr at 37° C in 5% CO_2. Cells were collected using a cell harvester and [3H]TdR uptake was done.

Cell Preparations

Segments of small intestine were excised, washed and everted. After rinsing with calcium and magnesium free Hanks balanced salt solution (HBSS) the segments were placed in flasks containing HBSS + 1.5 mM EDTA and incubated at 37° under gentle agitation. Initial 15-min incubation was used to remove bacteria, debris and mucus. The villus cells were isolated after 35-min incubation and crypt cells - after 60 min incubation. The released cells were filtered and centrifuged twice at 100 g for 5 min at 4° C. Separation of villus and crypt cells was confirmed by microscopic observation of histological sections of intestinal fragments retrieved during isolation procedures and in separate experiments by assessing radioactivity of cell fractions obtained after *in vivo* injection of [3H]TdR (data not shown). Spleen T-lymphocytes were isolated by passage through nylon wool columns.[10] Cells viability was > 90% as assessed by trypan blue.

Statistics

Results are expressed as the mean + SEM, and were compared by Wilcoxon-Mann-Whitney test.

RESULTS

The effects of supernatants from lymphocytes and small intestinal epithelial cells incubations on [3H]TdR incorporation into SICC are shown in Fig. 1.A & B. When crypt or villus epithelial cells constituted 5 portions of incubated volume, the resultant supernatants suppressed [3H]TdR incorporation into SICC. Supernatants of 1 portion of crypt cells had no significant effect. The inhibitory effect of supernatants from 1 portion of villus cells was slightly less marked than such effect of supernatants from 5 portions of villus cells (Fig. 1A) or was not significant (Fig. 1B). No significant effect was seen when supernatants of thymocytes or nonfractionated spleen cells were used. Supernatant from 5 portions of spleen T lymphocytes stimulated incorporation of label by SICC. No additional incorporation of [3H]TdR was seen when supernatant from 1 portion of spleen T-lymphocytes was used. measured as cpm in a liquid scintillation -system (Tm Analytic, USA).

Data on the effects of supernatants from co-incubations of lymphocytes and epithelial cells could be divided into two sets of results. The first one: co-incubation had no marked influence on the ability of cells to release into medium inhibitors or stimulators of [3H]TdR incorporation into SICC. Supernatants from villus cells and thymocytes

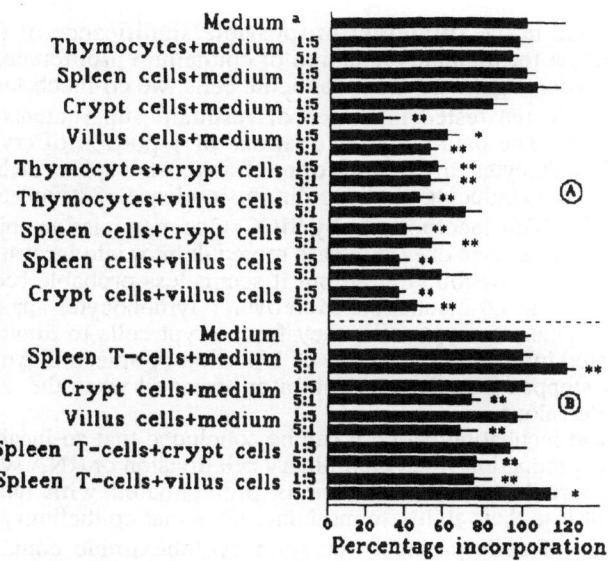

Figure 1 A & B. Influence of supernatants from lymphocytes and small intestinal crypt and villus epithelial cells on [³H]TdR incorporation by isolated small intestinal crypt cells. Results are shown as a percentage of incorporation obtained in medium without supernatants. Experiments were performed in quadruplicate if not otherwise indicated. Data are expressed as mean + SEM. A: 100 % incorporation = 1227+225 cpm; as performed in 8 separate experiments. B: 100% incorporation = 1381 ± 38 cpm. * P<0.05, ** P<0.01.

co-incubated in the volume ratios 1:5 or 5:1 suppressed incorporation of label into SICC like supernatants of villus cells incubated alone did. Similar situation was observed when spleen cells were co-incubated with villus cells. Supernatant from spleen T-lymphocytes and villus cells coinhcubated in ratio 5:1 exerted the same stimulatory effect like supernatants of 5 portions of spleen Tlymphocytes did. (Supernatant from 1 portion of villus cells had no effect in this experiments, Fig. 1B). The addition of 1 portion of thymocytes or spleen cells to 5 portions of crypt cells did not remove apparently the ability of crypt cells to release inhibitor(s) - the supernatants from such co-incubations suppressed incorporation of label by SICC. Supernatants from villus cells co-incubated with crypt cells inhibited [³H]TdR incorporation into SICC. The second set of these data: co-incubation had an influence on cells ability to release factors modulating [³H]TdR incorporation by SICC. Supernatants from 5 portions of thymocytes or spleen cells co-incubated with 1 portion of crypt cells inhibited incorporation of label by SICC as contrasted with supernatants from separate incubations of the same numbers of thymocytes, spleen or crypt cells, that had no marked influence on the incorporation of label. The effect of coincubation was even more visible when 5 portions of spleen T-lymphocytes were combined with 1 portion of crypt cells. The resultant supernatant suppressed [³H]TdR incorporation into SICC in contrast with stimulatory influence of supernatants from 5 portions of spleen T-cells incubated alone. The supernatant from 1 portion of spleen T-cells coincubated with 5 portion of crypt cells had no effect on the incorporation of label contrasted with suppressive effect of supernatants from 5 portions of crypt cells.

DISCUSSION

It was shown that isolated murine SICC may have the potential to provide a useful *in vitro* model for crypt cell responses to stimuli.[4] In this study we have observed that supernatants of crypt and villus epithelial cells suppressed [³H]TdR incorporation into SICC. These data confirm that may be a feedback both from functional villus cells and from crypt cell population to proliferating crypt cells.[11]

We have stressed in the **Introduction** possible significance of feedback from epithelium to lymphocyte for immune regulation of epithelium proliferation. To reveal if lymphocytes actually receive signal(s) from epithelial cells, we co-incubated lymphocytes and crypt or villus cells and tested influence of resultant supernatants on [³H]TdR incorporation into SICC. The data on co-incubations of 1 portion of crypt cells and 5 portions of spleen T lymphocytes (thymocytes or nonfractionated spleen cells) indicate that crypt cells provide signal(s) inducing Tcells to diminish release of stimulator(s) and/or to release inhibitor(s) of [³H]TdR incorporation by SICC. One may raise an objection that not crypt cells, but T-cells induce crypt ones to release more inhibitor into incabation medium. It is difficult to ignore such a possibility at all but it seems less probable because of small numbers of crypt cells in this co-incubation. Moreover T-lymphocytes apparently provide completely different signals to crypt cells: they force crypt cells to diminish release of inhibitor. It is suggested by data that addition of 1 portion of spleen T-lymphocytes to 5 portion of crypt cells stopped the release of inhibitor observed when the same number of crypt cells was incubated alone.

Because of short incubation period it can be concluded that co-incubations change release of factors or its production when preliminary cell division or RNA synthesis are not needed. This is supported by data that T-cells preincubated with mitomycin C or actinomycin D did not lose their ability to modulate intestinal epithelium proliferation *in vivo*.[8] On the contrary T-cells preincubated with cycloheximide could not suppress [3H]thymidine labelling index in the jejunal crypts when iniected to syngeneic mice.[8]

Summarizing the results of our experiments and taking into consideration available literature data we proposed speculative model of lympho-epithelial interactions in the regulation of small intestinal epithelium proliferation (Fig. 2). According to this scheme T-lymphocytes are involved in the regulation of cellular output from crypts through release of stimulator(s) and inhibitor(s) of crypt cell proliferation. The additional mechanism stimulating epithelium proliferation consist in T-lymphocyte ability to suppress release of inhibitor by crypt cells. The capacity of crypt cells for inducing T-cells to diminish release of stimulator and/or to produce inhibitor of crypt cell proliferation is an important element of this regulation. Together with negative feedback from villus and crypt cell populations to crypt cells it ensue control of crypt cell population growth and epithelium steady state maintenance.

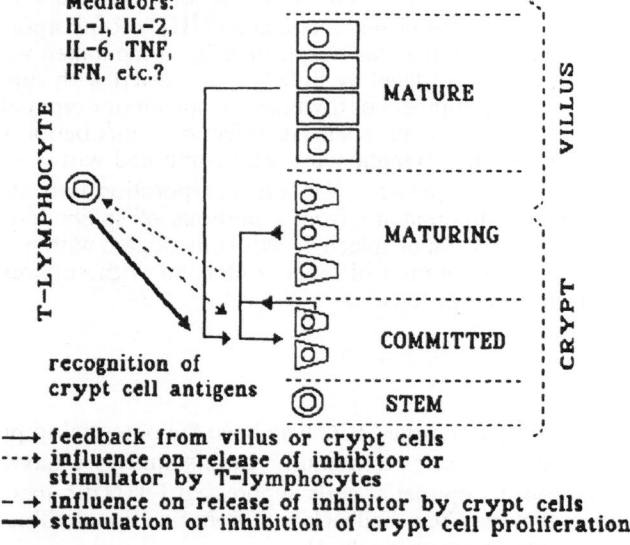

Figure 2. Schematic representation of lympho-epithelial interactions in the regulation of small intestinal epithelium proliferation.

It may be supposed that different mechanisms are involved in signalling from epithelial cells to lymphocytes and vice versa. Firstly, it is cell communications via cytokines. Such possibility is supported by data that intestinal epithelial cells produce factor with characteristics of IL-1[12] and express Interleukin-6 and its receptor.[7] In addition, tumor necrosis factor mRNA localized to Paneth cells.[13] Another possibility can be based on lymphocyte capacity for self-recognition. It is not unlikely that differentiation antigens of epithelial cells may be recognized by lymphocytes. It may be speculated that some increase in concentration of fetal antigens retained as shown[14] in the adults crypt cells may trigger T-cell to produce inhibitor of crypt cell proliferation that will reduce this concentration as a consequence of resultant decrease in proliferating crypt cells.

In spite of this report does not deals with intraepithelial lymphocytes, it is very attractive to propose that IEL may be induced to release inhibitor of crypt cell reproduction when proliferative cells appear onto villi. In this case local immune system pretend to participate in control over the maintanance of optimum numbers of specific cells in a given location.

However, the last possibility and the exact roles of different cytokines, T-lymphocyte subpopulations, nervous and hormonal influences, etc. in bidirectional lymphoepithelial interactions in normal and pathophysiological conditions remains to be clarified.

ACKNOWLEDGMENTS

We thank Dr. D.S. Mirsayafov and Mr. I.B. Belan for help in the preparation of the manuscript.

REFERENCES

1. A. M. Mowat and A. Ferguson, *Gastroenterology* 83:417 (1982).
2. N. Cerf-Bensussan, A. Quaroni, J. T. Kurnick, and A. K.Bhan, *J. Immunol.* 132:2244 (1984).
3. T. T. MacDonald, and J. Spencer, *J. Exp. Med.* 167:1341 (1988).
4. A. B. J. Speekenbrink, and D. M. V. Parrott, *Cell Tissue Kinet.* 20:135 (1987).
5. M. W. Smith, and M. A. Peacock, *Am. J. Anat.* 159:167 (1980).
6. S. Ahlstedt ,and I. Enander, *Int. Arch. Allergy Appl. Immunol.* 82:357 (1987).
7. K. Shirota, L. LeDuy, S. Yuan, and S. Jothy, *Archiv. B. Cell. Pathol.* 58:303 (1990).
8. V. A. Trufakin and A. N. Shmakov, *Archiv. Anat. Histol. Embriol.(RS).* 98:73 (1990).
9. A. N. Shmakov and V. A. Trufakin, *Tsitologiya (RS).*31:1074 (1989).
10. M. H. Julius, E. Simpson, and L. A. Herzenberg, *Eur. J. Immunol.* 3:645 (1973).
11. N. A. Wright and A. Al-Nafussi, *Cell Tissue Kinet.* 15:611 (1982).
12. P. W. Bland and L. G. Warren, *Immunology* 58:1 (1986).
13. S. Keshav, L. Lawson, P. Chung, M. Stein, V. H. Perry, and S. Gordon, *J. Exp. Med.* 171:327 (1990).
14. A. Quaroni, *J. Cell Biol.* 100:1601 (1985).

DENDRITIC CELLS *"IN VIVO"*: THEIR ROLE IN THE INITIATION OF INTESTINAL IMMUNE RESPONSES

Liming Liu and G. Gordon MacPherson

Sir William Dunn School of Pathology
South Parks Road Oxford OX1 3RE
England

INTRODUCTION

There is general agreement that the CD4[+] T lymphocyte is responsible for initiating and regulating the immune response to most antigens. The activation of CD4[+] T cells requires recognition of antigen-derived peptides associated with MHC class II. In rodents, whereas many cells expressing MHC II can present antigen to activated T cells, there is good evidence that resting Tcells are only activated efficiently by a specialised cell type, the lymphoid-dendritic cell.[1] There exist two distinct, and very probably unrelated kinds of dendritic cells (DC) which are involved in immune responses. The follicular DC is present B cell areas of secondary lymphoid tissues, is probably not bone-marrow derived, is long lived, retains immune complexes on its surface for presenting to B cells and will not be considered further here. The other, the lymphoid DC is specialised for the presentation of antigen to T cells, and is probably of particular importance in primary responses.

The T cell-associated dendritic cell is bone-marrow derived, usually short-lived and is a migratory cell involved in the presentation of antigen to T cells. A primary function of this DC is to acquire antigen in peripheral tissues and to transport it to lymph nodes where the antigen is presented as peptides to T lymphocytes. Research into the DC system has demonstrated many features which underly this specialised *"in vivo"* function in antigen presentation. This paper will review recent research into *"in vivo"* DC properties and function, and will include recent data obtained in the model which we have been investigating for several years in which DC derived from the small intestine can be collected under "near physiological" conditions.

DC can be viewed as a lineage of cells specialised to monitor tissues for the presence of antigen and to maximise the chances of antigen-specific T cells being able to recognize and react to the antigen. A current description of DC functionis that they are bone-marrow derived, have a blood-borne precursor which is now being characterised and that this precursor enters tissues in a relatively immature state. At this time it is capable of endocytosing and processing antigen for MHC class II association. After a period of residence in the periphery, the DC enters peripheral lymphatics and migrates to the draining node where it enters the T-dependent paracortical area and is able to present peptides to recirculating CD4[+] T cells. After a few days in the node it dies. DC are extremely efficient accessory cells; we can routinely detect the presence of < 50 DC in allogeneic MLR's or oxidative mitogenesis assays.

Recent studies have largely focused on the *"in vitro"* properties and functions of DC. It is however, essential to point out that in addition to the exploration of DC cell and molecular biology, the complete understanding of DC function demands knowledge of the evolutionary role of DC in the induction of protective immune responses to pathogens *"in *vivo"*, and we are still a long wayfrom this position. We are, however, accumulating data which enable us to generate some testable hypotheses.

In this review we will first describe experiments showing that DC can acquire antigen in the periphery. We will then discuss the regulation of DC migration and finally will examine the role of DC in activating naive T cells. We will also point out some critical areas where knowledge is lacking and will try to suggest potential approaches to these problems. We will concentrate on experiments carried out in the rat but will relate our results to those obtained in other systems. In view of the current interest in the development of vaccines suitable for oral administration, it is critical that their handling and uptake by antigen-presenting cells be understood. At present such studies are in their in fancy but we hope that the model we have developed will enable us to gain understanding of the *"in vivo"* uptake and presentation of intestinal antigens.

INTESTINALLY-DERIVED DC IN THE RAT

DC have been extracted from both Peyer's patches[2,3] PP and lamina propria of mice[4] and rats (Liu and MacPherson, in preparation) and there are claims that PP DC may be involved in IgA switching.[5] DC in peripheral lymph (veiled cells) are normally filtered out in the draining lymph node. Following mesenteric lymphadenectomy, peripheral and central lymphatics join and DC derived from the small intestine can be collected in thoracic duct lymph.[6] These DC are a heterogenous, rapidly turning-over population and are potent stimulators of the MLR and other T-dependent responses.[6] In contrast to DC from solid tissues, they can be concentrated up to 60-80 % by a single centrifugation over Metrizamide and can be collected from TDL for up to 4 days in as "close to physiological" condition as is at present possible. We do not yet know what proportion of lymph DC are derived from PP and lamina propria.

ACQUISITION OF ANTIGENS BY DC IN PERIPHERAL TISSUES

One of the first functional properties of DC to be described was their ability to stimulate T cells, both in the allogeneic MLR and by the presentation of protein antigens to sensitized cells.[1] It is only more recently that it has been shown that DC can acquire antigen in the periphery. Thus Knight's group has shown that after skin painting with contact-sensitizing agents, DC extracted from draining lymph nodes can stimulate sensitized T cells,[7] and Bujdoso *et al.,*[8] showed that after the subcutaneous injection of soluble antigen in sheep, lymph draining the site of injection contained DC bearing antigen which could be presented to T cells. We have recently shown that for a period of about 24h following the injection of ovalbumin (OVA) or horse-radish peroxidase into the intestinal lumen of rats, DC in lymph draining the intestine carry antigen and can present it to sensitized T cells in an antigen-specific, CD4+ and MHC II-dependent manner.[9] We also showed that DC were the only cells on this lymph able to presentantigen, and that B cells were inert in this assay.

It is very clear that DC in the periphery can acquire contact-sensitizing agents and soluble protein antigens but it is not at all clear how this relates to their role in defence. Most pathogens are particulate and apart from the isolatedcase of allogeneic lymphocytes,[10] mature DC are not, or only very weakly phagocytic. DC in lymph draining the intestine do, however contain inclusions, some of which contain DNA and others peroxidatic material, and electronmicroscopy shows that there is cellular debris in some inclusions+.[10] Reis and Sousa (personal communication) has shown that Langerhans cells are phagocytic,but much less so than macrophages. Mayrhofer et al.,[11] have shown that following a Salmonella infection, rat DC derived from the intestine carry Salmonella antigens but the functional status of these cells was not examined.[1]

Do DC Lose the Ability to Acquire Antigens? Fresh LC are very good presenters of soluble proteins but after *"in vitro"* culture, this property is lost, although cultured LC retain the ability to present peptides and are potent stimulators of the MLR.[12] The implications of these observations are that once DC have acquired antigen in the periphery, peptides derived from this antigen may be retained for presentation for long periods. It has not yet been shown whether this maturation also occurs *"in vivo"*. Thus there remain several important questions concerning antigenacquisition by DC: (1) What are the "physiological" antigens acquired by DC. Can they phagocytose bacteria, can they endocytose or be infected by viruses during natural infections ? (2) What receptors are involved in antigen uptake ? (3) Does antigen acquisition signal changes in DC properties e.g., stimulate their release into lymph ? (4) Is the loss of the ability to process and present whole protein antigens a programmed stage of DC differentiation ?

MIGRATION OF DC AND ITS REGULATION

Normal Migration. All mammalian peripheral lymph so far examined contains migrating DC (veiled cells) although the actual numbers vary according to the tissue of origin. Under normal conditions the output of DC into lymph remains relatively constant. It is presumed that these migratory DC migrate into lymph nodes and become interdigitating cells (IDC). Fossum[13] has shown that rat intestinally-derived veiled cells injected into rat foot pads develop into IDC in the draining node and he and Austyn[14] have shown that DC injected IV migrate into T cell areas of the spleen. Thus a migratory pathway has been defined but many questions remain : (1) How is the release of DC determined? LC may spend many weeks in the epidermis before exiting whereas intestinal DC spend only 2-4 days. Is release a programmed maturation step or is it determined by local microenvironmental changes? What changes in the microenviron mentstimulate release ? Can changes in surface phenotype be correlated withrelease? (2) How do DC recognise lymphatic endothelium in tissues? (3) What directs DC to appropriate locations in secondary lymphoid tissue? (4) DC spend only a few days in secondary lymphoid tissue and are presumed to die *in situ*. LC spend a long time in epidermis but die rapidly in culture and *"in vivo"*. Is this death programmed? Is it due to lack of a survival factor present only in peripheral tissues? Is DC death apoptopic?

Stimulated Migration. Kelly[15] described a marked increase in the output of veiled (frilly) cells in peripheral lymph draining the site of alum-precipitated diphtheria toxoid or bovine RBC injection, detectable at 2 days and peaking at 4 to 5 days. Other groups have shown an increase in the numbers of DC extractable fromn odes draining the sites of skin painting with contact sensitizers[7] and also from non-draining nodes.[16] We have shown that the output of DC into intestinal lymph is markedly increased after intravenous endotoxin injection[17] andAustyn's group (Austyn, J.M. Personal communication), has shown that after endotoxin administration to mice, the majority of DC leave hearts and kidneys. The actual mediators of DC migration are not yet defined but it has been shown that following intradermal injection of TNF-α increased numbers of DC can be extracted from draining nodes[18] and we have preliminary evidence that an anti-TNF-α monoclonal antibody can block the endotoxin stimulated release of DC into lymph (MacPherson, in preparation). The DC released by endotoxin in rats appear functionally similar those normally found in lymph in their capacity to stimulate a MLR and to present antigen to sensitized T cells. Thus it is clear that DC release from peripheral tissues can be modulated,and that TNF-α may be involved at some stage in this release, but many questions remain to be answered. Whether TNF is the only mediator and whether TNF acts directly on DC are unknown, as are the changes in DC phenotype and function which may accompany stimulated release.

"IN VIVO" FUNCTIONS OF ANTIGEN-BEARING DC

In rodents, DC stimulate the allogeneic MLR much more efficiently than do other cells expressing MHC II, suggesting that DC may have a particular role in initiating primary immune responses. Knight's group has shown that DC extracted from nodes draining the site of skin painting with FITC can adoptively sensitize naive recipients.[7] Steinman's group has shown that following the IV injection of antigen, the only cells present in spleen that can present antigen are DC.[19] Similarly, we have shown that following antigen injection into the intestine, or after "in vitro" pulsing, DC but not B cells or macrophages are able to present antigen to primed T cells[9] and only DC can prime T cells after injection into naive rats (Liu & MacPherson, in preparation). As few as 500 antigen-pulsed DC are sufficient to prime a naive animal. In these models it is difficult to exclude a role for host antigen-presenting cells in processing and presenting antigens delivered by the injected DC. To overcome this, Inaba et al.,[20] used parental strain antigen-pulsed DC to prime an F1 mouse. T cells from the draining node could only be restimulated efficientlyby antigen-presenting cells from the priming parental strain, showing that the original priming was restricted to the MHC of the priming DC, and excluding a significant role for host processing and presentation. We have performed similar experiments using DC from animals which had been injected intra-intestinally with OVA and shown a similar restriction of priming (Liu & MacPherson, in preparation). We have recently been examining the acquisition by DC of antigens given by gastric intubation. It is clear that OVA given in this way is acquired by DC which migrate into lymph, and that as little as 1mg can load DC sufficiently for them to be able to sensitize naive T cells "in vivo" (Liu and MacPherson, inpreparation). Thus, we can conclude that DC which have acquired antigen "in vivo" in a relatively natural manner from the small intestine can prime T cells in naive recipients. Any explanation of oral tolerance must take into account the obseravations that following oral administration of antigen, potent immunostimulatory DC migrate from the intestine to the mesenteric nodes.

ACKNOWLEDGEMENTS

We are extremely grateful for the expert technical assistance of Chris Jenkins.

REFERENCES

1. R. M. Steinman, *Annu. Rev. Immunol.* 9:271 (1991).
2. T. T. MacDonald and P. B. Carter, *Immunology* 45:769 (1982).
3. D. M. Spalding, W. J. Koopman, J. H. Eldridge J. R. McGhee, and R. M. Steinman, *J. Exp. Med.* 157:1646 (1983).
4. P. Pavli, C. E. Woodhams W. F. Doe, and D. A. Hume, *Immunology* 70:40 (1990).
5. D. M. Spalding, S. I. Williamson W. J. Koopman, and J. R. McGhee, *J. Exp. Med.* 160:941 (1984).
6. C. W. Pugh, G. G. MacPherson, and H. W. Steer, *J. Exp. Med.* 157:1758 (1983).
7. S. E. Macatonia, A. J. Edwards, and S. Knight, *Immunology* 59:509 (1986).
8. R. Bujdoso, J. Hopkins, B. M. Dutia, P. Young, and I. McConnell, *J. Exp. Med.* 170:1285 (1989).
9. L. M. Liu and G. G. MacPherson, *Immunology* 73:281 (1991).
10. S. Fossum and B. Rolstad, *Eur.J. Immunol.* 16:440 (1976).
11. G. Mayrhofe, P. G. Holt, and J. M. Papadimitriou, *Immunology* 58:379 (1986).
12. N. Romani, S. Koide, M. Crowley, M. Witmer-Pack, A. M. Livingstone, C. G. Fathman, K. Inaba, and R. M. Steinman, *J. Exp. Med.* 169:1169 (1989).
13. S. Fossum, *Scand. J. Immunol.* 27:97 (1988).
14. J. M. Austyn, *Res. Immunol.* 140:898, discussion p. 918 (1989).
15. R. H. Kelly, *Nature* 227:510 (1970).
16. S. Hill, A. J. Edwards, I. Kimber, and S. C Knight, *Immunology* 71:277 (1990) .
17. G. G. MacPherson, *Immunology* 68:108 (1989).
18. M. Cumberbatch, and I. Kimber, *Immunology* 75:257 (1992).
19. M. Crowley, K. Inaba, and R. M. Steinman, *J. Exp. Med.* 172:383 (1990).
20. K. Inaba, J. P. Metlay, M. T. Crowley, and R. M. Steinman, *J. Exp. Med.* 172:631 (1990).

DIRECT MODULATION OF ENTEROCYTE GROWTH BY ACTIVATED MACROPHAGES

Angela K. Hutton and Allan McI. Mowat

Department of Immunology
Western Infirmary
Glasgow, Scotland, United Kingdom

INTRODUCTION

Several small intestinal disorders, including coeliac disease, are associated with an enteropathy characterised by crypt hyperplasia and villus atrophy.[1] There is indirect evidence that the pathology is caused by a local cell-mediated immune response in which cytokines act directly on crypt stem cells.[2] Much attention has focused on the role of T lymphocytes and their cytokine products; however, little attention has been paid to the role of macrophages. Here, we have examined the effects of macrophages on the growth of crypt intestinal epithelial cells *in vitro*.

MATERIALS AND METHODS

The non-transformed rat intestinal epithelial (RIE) cell line was used (from passage 6-20) as a model of dividing enterocytes.[3] Its growth *in vitro* was measured using an MTT colorimetric assay. All assays were carried out in quadruplicate. Rat macrophages were obtained by peritoneal lavage of a Wistar rat and then stimulated overnight with 500U/ml IFNγ and indomethacin as required, before being co-cultured at a ratio of 1:1 with 10^4 RIE cells in culture medium. Those wells requiring LPS, and L-NMMA had them added at this point. The cultures were then incubated at 37°C in 5% CO_2 and harvested daily.

RESULTS

In the first experiments, we examined the effects of resting or activated macrophages on the growth of RIE cells *in vitro*. As seen in Fig. 1, non-activated macrophages had a cytostatic effect on the RIE cells. Similar levels of growth inhibition were seen when the macrophages had been stimulated with LPS. Macrophages stimulated with IFNγ, or both IFNγ and LPS had an even greater inhibitory effect on the growth of the RIE cells.

Next we investigated whether soluble mediators could account for the inhibitory effects of macrophages on RIE cells. Fig. 2 shows that addition of the prostaglandin synthetase inhibitor, indomethacin, at a concentration of 10^{-4}M, had no effect on the ability of resting or activated macrophages to inhibit the growth of the RIE cells.

We next examined the role of nitric oxide (NO), another product of activated macrophages which has recently been implicated in enteropathy *in vivo*.[4] Fig. 3 shows that addition of the specific inhibitor of macrophage-dependent NO synthesis, L-NG-

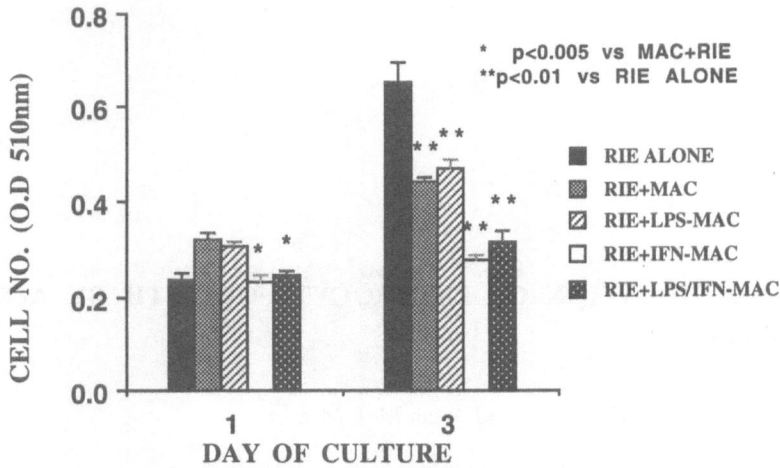

Figure 1. The effect of macrophages on the growth of RIE cells.

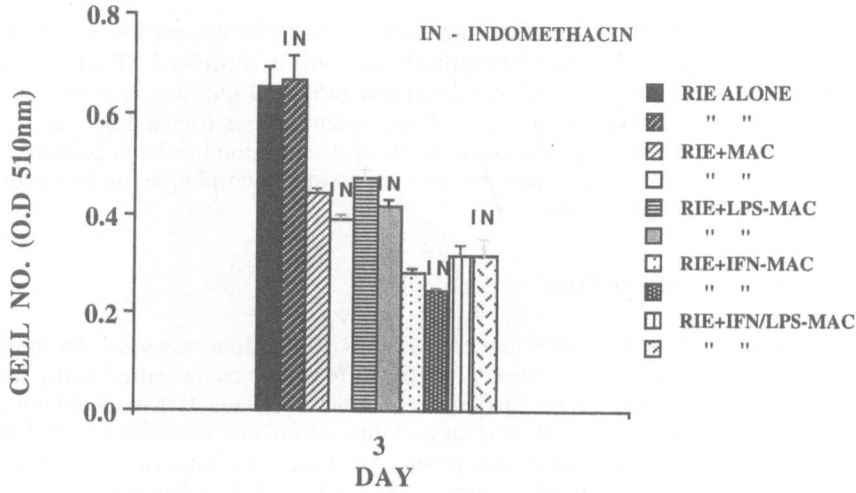

Figure 2. The effect of indomethacin on cytostatic activity of macrophages.

monomethyl arginine (L-NMMA), partially but significantly reversed the inhibitory effects of macrophages which had been activated with both IFNγ and LPS. L-NMMA had no effect on the inhibition caused by resting macrophages or LPS stimulated macrophages.

We then examined whether NO itself could inhibit the growth of the RIE cells. Fig. 4 shows that addition of S nitroso-N-acetyl-DL-penicillamine (SNAP), a compound which releases NO in solution also had a cytostatic effect on the RIE cells and this effect was dose dependent.

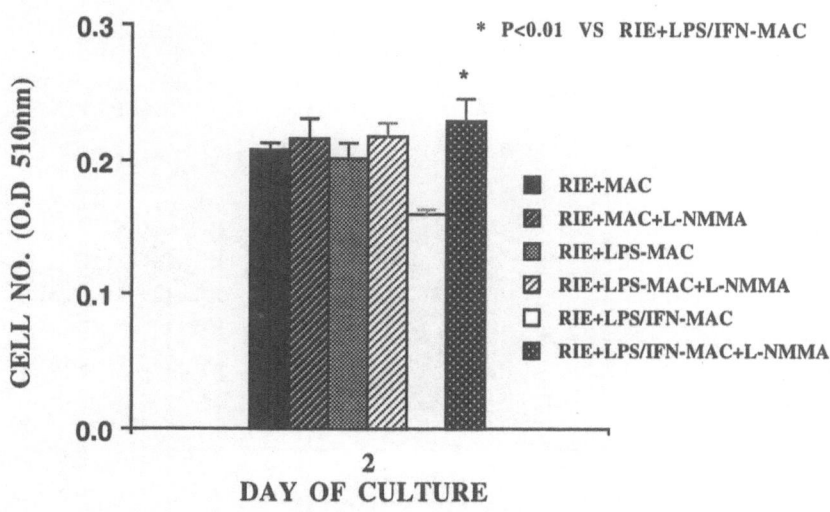

Figure 3. Effect of L-NMMA on cytostatic activity of macrophages.

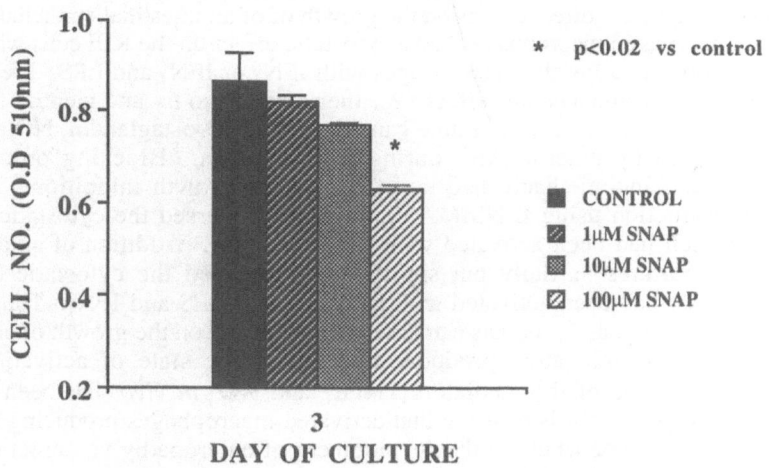

Figure 4. Effect of Nitric Oxide on growth of RIE cells.

Finally, we examined whether TNFα, a cytokine which can cause enteropathy *in vivo* might be important in the macrophage growth inhibition of the RIE cells. Fig. 5 shows that the cytostatic effect of macrophages activated by LPS and also IFNγ and LPS could be partially but significantly reversed by a polyclonal anti-TNFα antibody.

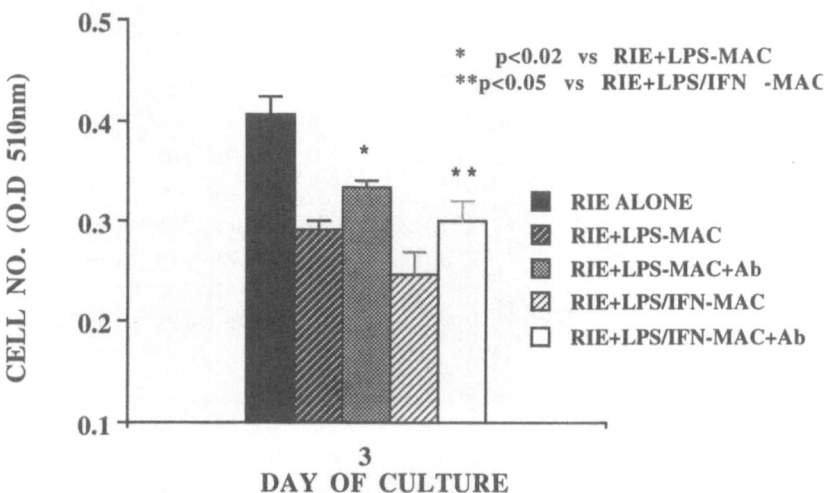

Figure 5. Effect of anti-TNF α on cytostatic effect of macrophages.

DISCUSSION

It is known that cell-mediated immune responses play an important role in enteropathy *in vivo* , possibly via effects on crypt epithelial cells.[2] Here we have shown that macrophages can have direct effects on the growth of of an intestinal epithelial crypt cell line *in vitro*. Unactivated macrophages had a cytostatic effect on the RIE cells which could be enhanced by preactivating the macrophages with IFNγ or IFNγ and LPS. Preactivation with LPS alone did not enhance this effect. We then went on to try and identify the factors involved in this cytostatic effect. Possible candidates were prostaglandin, NO and TNFα which are produced by macrophages during inflammation. Blocking production of prostaglandin using indomethacin had no effect on the growth inhibition. However, blocking NO production using L-NMMA significantly reversed the cytostatic effect of macrophages which had been activated with LPS and IFNγ. Addition of an anti TNFα antibody to the cultures partially but significantly reversed the cytostatic effects of macrophages which had been activated with LPS and also LPS and IFNγ. Thus it seems that a number of macrophage products can have a direct effect on the growth of enterocytes *in vitro*, and that the mediators produced depend on the state of activation of the macrophages. Blocking of the mediators TNFα[5] and NO[4] *in vivo* has been shown to prevent enteropathy and so it is possible that activated macrophages producing mediators such as NO and TNFα contribute to the development of enteropathy via direct effects on dividing crypt epithelial cells.

REFERENCES

1. A. Ferguson and A. McI. Mowat, *in*: "Recent Advances in Gastrointestinal Pathology", R. Wright, ed., p. 93, Saunders, Piladelphia, Pennsylvania (1980).
2. T. MacDonald and J. Spencer, *J. Exp. Med*. 167:1341 (1988).
3. R. Blay and K. Brown, *Cell. Biol. Int. Rep*. 8:551 (1984).
4. P. Garside, A. Hutton, A. Severn, F. Liew, and A. McI. Mowat, *Eur. J. Immunol.* 22:2141 (1992).
5. P. Piguet, G. Grau, B. Allet, and P. Vassalli, *J. Exp. Med.* 166:1280 (1987).

PENETRATION OF FLUORESCENT NEUTROPHILS THROUGH CULTURED EPITHELIUM STUDIED BY CONFOCAL MICROSCOPY

Kajsa Holmgren Peterson, Birgitta Johansson,
Maria Johansson and Karl-Eric Magnusson

Dept. of Medical Microbiology
University of Linköping
S-581 85 Linköping, Sweden

INTRODUCTION

The migration of leukocytes is an important part of the inflammatory response. To reach the site of infection and/or inflammation, leukocytes must be able to cross the endothelium lining blood vessels, and also, in case of intestinal infections, the epithelium.

The normal intestinal mucosa demonstrates physiologic state of inflammation in the lamina propria with numerous neutrophils, macrophages, plasma cells and lymphocytes which serve to maintain the integrity of the mucosa.[1] Leukocytes are, unfortunately, also able to damage the intestinal barrier by production of inflammatory mediators. In addition, they can be stimulated to migrate across an epithelium by bacterial products such as the chemotactic peptide N-formyl-methionyl-leucyl-phenyl-alanine (fMLF).

Neutrophils have been shown to migrate between the epithelial cells[2] and this process can cause a temporary fall in epithelial electrical resistance.[2] The fall in resistance is due to altered tight junction permeability[3]. During the transmigration intimate focal contacts are made between neutrophils and epithelial cells which enable the neutrophils to manipulate the junctional barrier[3]. Following neutrophil migration the tight junctions appear to reform.[4]

The aims of this work were a) to study the effect of neutrophil penetration on epithelial integrity and b) to develop methods for studies of neutrophil-epithelial interaction using double-fluorescence labelling and confocal microscopy. For this, we have used an *in vitro* model of inflammation in the intestine. Migration of fluorescently labelled human neutrophils across a monolayer of cultured MDCK-II cells[5,6] in response to fMLF was studied with confocal laser scanning microscopy.[7] We have also studied the effects of neutrophil migration on epithelial permeability to fluorescent macromolecules.

MATERIALS AND METHODS

Chemicals

The fluorescent membrane label PKH2-PCL was obtained from Zynaxis Cell Science Inc. (Malvern, PA); rhodaminated phalloidin (rhod-phalloidin), which labels F-actin, from Molecular Probes Inc. (Eugene, OR); N-formyl-methionyl-phenylalanine (fMLF) from Sigma (St Louis, MO); and Na-fluorescein (MW 330 Da) from Merck (Darmstadt, Germany). Krebs-Ringer glucose phosphate buffer (10 mM glucose) with 1

Advances in Mucosal Immunology, Edited by
J. Mestecky *et al.*, Plenum Press, New York, 1995

mM Ca^{2+} and Mg^{2+}, pH 7.3 (KRG), and phosphate buffered saline, pH 7.3 (PBS), were used.

Cells

Neutrophils were isolated from heparinized venous blood from healthy blood donors by dextran sedimentation and centrifugation at 4° C on a Ficoll-Paque (Pharmacia LKB Biotechnology, Uppsala, Sweden) gradient.[8] Madine-Darby canine kidney II (MDCK-II) cells were used as a model of intestinal epithelium. The cells were grown to confluency (4 days) on 6.5 mm Transwell nucleopore polycarbonate filters, pore size 3.0 µm (Costar, Cambridge, MA), in Dulbeccos modified Eagle medium (DMEM) with 5 % fetal calf serum, 10 mM HEPES, 4 mM L-glutamine and standard concentrations of penicillin and streptomycin.

Purified neutrophils (2×10^7 cells/ml) were labelled with PKH2-PCL according to the protocol provided by Zynaxis Cell Science Inc., suspended in KRG and kept on ice. The filters with MDCK-II were gently washed, equilibrated to 37° C and tranferred to plastic wells with 700 µl of 10 µM fMLF in KRG. 150 µl of labelled neutrophils (2×10^5-2×10^7 cells/ml in KRG), equilibrated to 37° C, was added to the apical side of the epithelium and incubated for 10-90 min at 37° C. KRG was used as control. After gentle washing, the filters were fixed in 4 % (w/w) paraformaldehyde at 4° C for 60 min and labelled from the apical side with 150 ul of rhod-phalloidin (1 µg/ml, 30 min). The filters were washed, cut out and mounted in PBS on microscope slides. In the case of neutrophil migration from the basal to the apical side of the epithelium, filter wells were turned upside down and filled with fMLF (10 µM). Neutrophils were added at the bottom of the filter and the set-up was incubated, fixed and counterstained as above.

Confocal Microscopy

A Phoibos 1000 confocal microscope with a two-wavelength detection system (Molecular Dynamics, Sunnyvale, CA) and a 40x oil immersion objective (N.A. 1.0) was used, yielding lateral and vertical resolutions of 0.25 and 1.2 µm, respectively.

Permeability Measurement

Non-labelled neutrophils (2×10^5-2×10^7 cells/ml in KRG) were added to filters with MDCK-II and incubated for 60 min at 37° C as above. After gentle washing, the filters were transferred to 4° C. 100 µl of of the permeability marker Na-fluorescein (50 µg/ml in KRG) was added to the apical side of the epithelium. Samples (10 µl) were drawn at regular intervals from the surrounding buffer (750 µl KRG) and analyzed in a fluorescence spectrometer (Perkin-Elmer LS3B).

RESULTS

PKH2-PLC did not significantly impair the neutrophils ability to move on an albumin-coated surface at 37° C. Confocal microscopy showed that the probe labelled the cell membrane as well as intracellular structures. PKH2-PLC-labelled neutrophils were easily detected both in horizontal and vertical confocal scans of epithelial monolayers, counter-stained with rhod-phalloidin.

Clusters of labelled neutrophils, adhering to the epithelial cells, were observed after administration of neutrophils to the apical side of the epithelium. The neutrophils migrating through the epithelium were followed in a series of horizontal sections from the confocal microscope (Fig. 1a). Migrating neutrophils could also be viewed in vertical sections (Fig. 1b). Neutrophils that had penetrated the epithelium were observed on the basal side of the epithelium (not shown).

When neutrophils were placed on the basal side of the epithelium, fewer clusters of neutrophils adhered to the epithelium. Migrating neutrophils could, however, be found and visualized in vertical sections (not shown).

Measurements of Na-fluorescein permeability indicated a neutrophil concentration-dependent, increased transepithelial leakage as a result of the interaction between neutrophils and epithelium (Fig. 2).

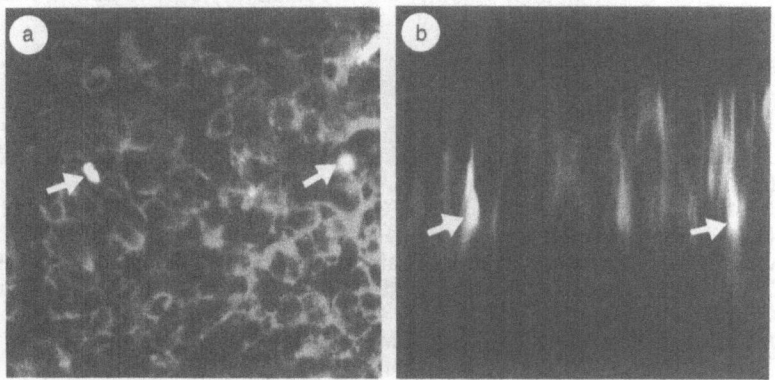

Figure 1. Dual detector confocal micrographs (a: horizontal section; b: vertical section) showing the passage at 37° C of PKH2-PLC-labelled human neutrophils (2×10^7/ml, arrows) through a monolayer of MDCK-II cells grown on filter support. A chemotactic gradient was obtained by adding 10 μM fMLF to the basal side of the epithelium. The filter was counter-stained with rhod-phalloidin (1 μg/ml) after fixation. Images from the two detectors have been superimposed in the pictures. Bar=10 μm.

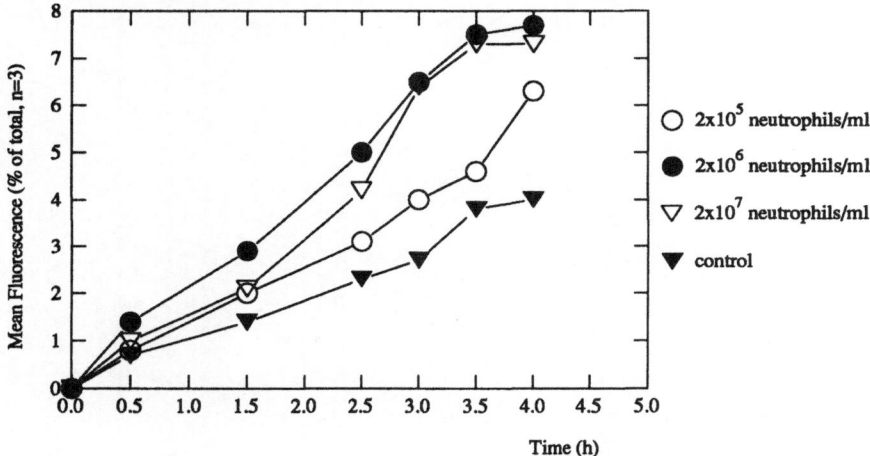

Figure 2. Effect of neutrophil passage on the permeability of Na-fluorescein through monolayers of MDCK-II grown on filter supports. The graph shows the permeability of the Na-fluorescein at 4° C after incubation of unlabelled neutrophils and epithelial cells for 60 min at 37° C. KRG was used as control.

DISCUSSION

This study shows that neutrophils, fluorescently labelled with PKH2-PCL, can be stimulated to penetrate a monolayer of cultured MDCK-II cells in response to the chemotactic peptide fMLF. The system can be used as a model of intestinal inflammation and the penetration of neutrophils through the epithelium can be studied with confocal laser scanning microscopy.

Neutrophils have been shown to migrate between epithelial cells[2] which can also be seen in our pictures (Fig. 1). During their transmigration, the neutrophils seem to transfer some of their fluorescence to the surrounding epithelial cells, possibly because of intimate contacts established between the membranes of the two cell types.[3] As the fluorescently labelled neutrophils maintain their viability and are relatively easy to detect, transmigration may also be followed in non-fixed material.

REFERENCES

1. G. L. Mandell, R. G. Douglas, and J. E. Bennett, *Prin. Pract. Infect. Dis.* (Third Edition) (1990).
2. L. C. Milks, G. P. Conyers, and E. B. Cramer, *J. Cell Biol.*, 103: 2729 (1986).
3. S. Nash, J. Stafford, and J. L. Madara, *Lab. Invest.* 4: 531 (1988).
4. G. Migliorisi, E. Folkes, and E. B. Cramer, *J. Leukocyte Biol.*, 44:485 (1988).
5. J. C. W. Richardson, V. Scalera, and N. L. Simmons, *Biochim. Biophys. Acta* 673:26 (1981).
6. S. D. Fuller and K. Simons, *J. Cell Biol.* 103:1767 (1986).
7. K. Carlsson and N. Åslund, *Appl. Optics* 26:3232 (1987).
8. A. Böyum, *Scand. J. Clin. Lab. Invest.* 97:77 (1968).

IMMUNOSTAINING WITH MONOCLONAL ANTIBODIES TO EOSINOPHIL CATIONIC PROTEIN (EG1 AND EG2) DOES NOT DISTINGUISH BETWEEN RESTING AND ACTIVATED EOSINOPHILS IN FORMALIN-FIXED TISSUE SPECIMENS

Frode Jahnsen, Trond S. Halstensen, and Per Brandtzaeg

LIIPAT, Institute for Pathology, University of Oslo, The
National Hospital, Rikshospitalet, N-0027 Oslo, Norway

INTRODUCTION

Activated eosinophils release highly toxic cationic proteins which are important in the defence against parasites. However, activated eosinophils may also induce tissue damage and have been reported to be associated with lesions of various allergic and inflammatory disorders. Tai et al.[1] raised two monoclonal antibodies (mA) against eosinophil cationic protein (ECP); mA EG1 recognized both the stored and secreted forms of ECP, whereas mA EG2 was claimed to identify only the secretory product. mA EG2 has therefore been used in several immunohistochemical studies to identify activated eosinophils in formalin-fixed tissue specimens.[1-10] The aim of our investigation was to examine whether this application of EG2 is valid.

MATERIAL AND METHODS

Peripheral Blood Eosinophils

Eosinophils were isolated by centrifugation on discontinuous Percoll density gradients and activated with C3b-coated zymosan particles. Cytospins of stimulated and unstimulated cells were either used unfixed (air-dried only) or fixed in different ways: with acetone alone (10 min, 20°C); acetone followed by 0.5% paraformaldehyde-lysine-periodate (PLP)(10 min, 4°C); 2% PLP (10 min, 4°C); 10% formalin (10 min, 4°C); 96% ethanol (10 sec, 1 min or 10 min, 20°C); or 96% methanol (10 sec, 1 min or 10 min, 20°C).

Tissue Specimens

Routinely formalin-fixed and paraffin-embedded tissue specimens were obtained from different sites of non-inflamed gastrointestinal mucosa from 17 patients: stomach (n=3); small intestine (n=8); and large bowel (n=6).

Staining Procedures

Dewaxed tissue sections, treated with trypsin (1 mg/ml, 10 min, 37°C), and cytospins prepared in various ways (see above), were incubated with mA EG1 or EG2 in an alkaline phosphatase-antialkaline phosphatase (APAAP) staining method [11] Sequential two-colour immunofluorescence staining with mA EG1 and EG2 was performed to examine whether all eosinophils reacted with both EG1 and EG2 or not

RESULTS

MAbs EG1 and EG2 Applied on Peripheral Blood Eosinophils

The mAb EG1 reacted with the vast majority of both unstimulated and stimulated eosinophils The mAb EG2 reacted selectively with eosinophils containing zymosan particles in unfixed cytospins, and also in preparations fixed with acetone or acetone/0 5% PLP However, most stimulated as well as unstimulated eosinophils reacted with EG2 after fixation in 2% PLP or 10% formalin Methanol and ethanol fixation afforded weak and variable staining (Table 1)

Table 1. APAAP staining of peripheral blood eosinophils prepared in different ways and stained with EG1 or EG2

	Stimulated eosinophils*		Unstimulated eosinophils*	
	EG1	EG2	EG1	EG2
Unfixed	+	++ §	+	−
Acetone	ND #	++ §	ND #	−
Acetone/0 5% PLP	ND #	++ §	ND #	−
2% PLP	++	++	++	++
10% Formalin	++	+++	++	+++
96% Methanol	+/−	+/−	+/−	+/−
96% Ethanol	ND #	+/−	ND #	−

* Staining was graded from negative ( ) or inconsistent (+/) to definitely positive with increasing intensity (+ to +++)
§ EG2 reactivity with a variable number of eosinophils containing zymosan particles
# ND = Not done

MAbs EG1 and EG2 Applied on Formalin-Fixed Tissue Specimens

Two-colour immunofluorescence staining on sections of formalin-fixed tissue specimens from normal gut mucosa revealed that all EG1-positive cells also reacted with EG2 (Fig 1) The same pattern was seen in adjacent sections stained first with EG1 and then with EG2 (not shown) In addition, EG1 and EG2 decorated a comparable number of cells in adjacent tissue sections when applied in the APAAP method

DISCUSSION

It has been shown by Tai et al [1] that normal blood eosinophils react with mAb EG1 but not mAb EG2 However, after in vitro activation, many eosinophils react also with the latter antibody suggesting that it selectively identifies activated eosinophils secreting (or ready to secrete) ECP On the basis of these results, several investigators have applied EG2 on formalin-fixed tissue specimens, the high numbers of EG2-positive cells found in a variety of tissue lesions have been taken to imply important pathogenic roles for eosinophils in different disorders [1 10]

In our hands, however, the epitope for mAb EG2 was found to be remarkably fixative-dependent, EG2 did in fact stain both stimulated and unstimulated formalin-fixed

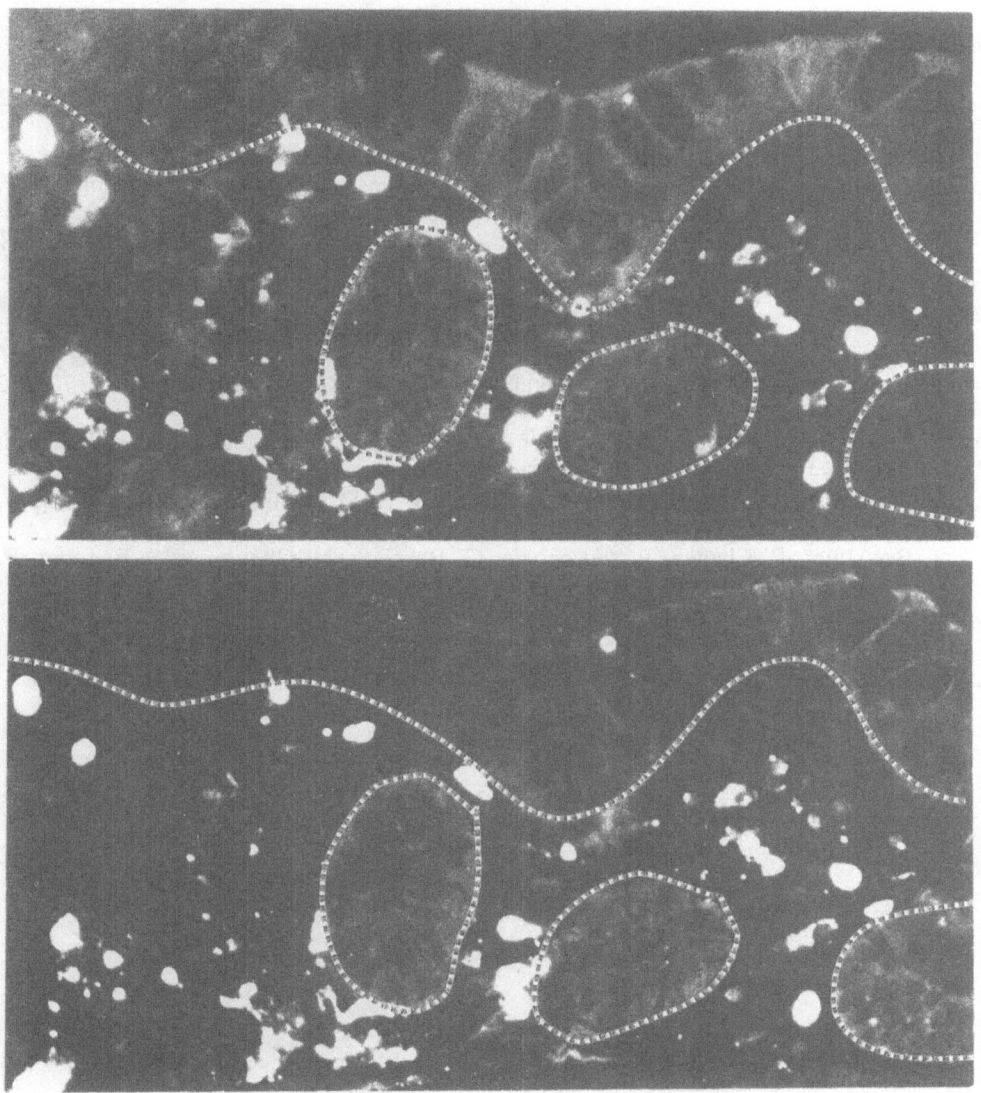

Figure 1. Two-colour sequential immunofluorescence staining in the same field from a section of normal colonic mucosa showing EG2- (upper panel) and EG1- (lower panel) positive cells. No additional positivity was revealed by EG1. Broken line indicates epithelial basement membrane zone. Original magnification X 320.

peripheral blood eosinophils. Moreover, all EG1-positive cells also reacted with EG2 in formalin-fixed, paraffin-embedded tissue specimens from normal gastrointestinal mucosa.

We conclude, therefore, that EG2-reactivity in formalin-fixed tissues only reflects the number of eosinophils present, without distinguishing between resting and stimulated eosinophils. Thus, this staining procedure does not afford results that can be validly interpreted in terms of local eosinophil activation. Our results do suggest that formalin demasks the epitope for EG2. Further studies are needed to see whether this antibody may be used to detect eosinophil activation when applied on tissue specimens prepared in alternative ways.

REFERENCES

1. P. C. Tai, C. J. F. Spry, C. Peterson, P. Venge, and I. Olsson, *Nature* 309:182 (1984).
2. K. Fredens, H. Dybdahl, R. Dahl, and U. Baandrup, *APMIS* 96:711 (1988).
3. R. Hällgren, J. F. Colombel, R. Dahl, K. Fredens, A. Kruse, N. O. Jacobsen, P. Venge, and J. C. Rambaud, *Am. J. Med.* 86:56 (1989).
4. R. A. Lundin, K. Fredens, G. Michaëlsson, and P. Venge, *Br. J. Dermatol.* 122:181 (1990).
5. R. Hällgren, S. O. Bohman, and K. Fredens, *Nephron* 59:266 (1991).
6. P. C. Tai, S. J. Ackerman, C. J. F. Spry, S. Dunnette, E. G. J. Olsen, and G. J. Gleich, *Lancet* 1:643 (1987).
7. C. J. F. Spry, P. C. Tai, and J. Barkans, *Int. Archs Allergy Appl. Immunol.* 77:252 (1985).
8. A. Tøttrup, K. Fredens, J. P. Funch, S. Aggestrup, and R. Dahl, *Dig. Dis. Sci.* 34:1894 (1989).
9. P.-C. Tai, M. E. Holt, P. Denny, A. R. Gibbs, B. D. Williams, and C. J. F. Spry, *Br. Med. J.* 289:400 (1984).
10. J. Bousquet, P. Chanez, J. Y. Locoste, G. Barneon, N. Ghavanian, I. Enander, P. Venge, S. Ahlstedt, J. Simony-Lafontaine, P. Godard, and F.-B. Michel, *N. Engl. J. Med.* 323:1033 (1990).
11. D. Mason, *in*: "Techniques in Immunocytochemistry", G.R. Bullock and P. Petrusz, eds., p. 25, Academic Press (1985).

THE ROLE OF MAST CELLS IN INTESTINAL IMMUNOPHYSIOLOGY

Lu Wang, Sheila Savedia, Michelle Benjamin, and Mary H. Perdue

Intestinal Disease Research Unit
McMaster University
Hamilton, Ont L8N 3Z5
Canada

INTRODUCTION

Immunophysiology of the gut is the study of the regulation/modulation of its function by immune cells and factors. Fluid secretion is a key function of the intestine which aids digestion by maintaining the lumenal contents in a liquid state, promoting mixing of nutrients with digestive enzymes which are then presented to the absorptive epithelial cells. Fluid secretion is also a host defence mechanism of mucosa that serves to wash away noxious material from the epithelial surface. In the gut, active transport of negatively charged Cl^- ions into the lumen by epithelial cells has been clearly shown to be the driving force for fluid secretion, while positively charged ions such as Na^+ follow in response to the electrical gradient. Cl^- channels open in response to increased concentrations of intracellular second messengers such as cAMP and Ca^{2+}, resulting in ion and fluid movement across the gut wall.[1]

The regulation of ion and fluid secretion is a complex process involving neural and hormonal components. Over the last few years, it has become increasingly obvious that ion secretion is also modulated by immune cells and their mediators/cytokines. In this paper, we will concentrate on mast cells (MC), which are examples of immune cells. Based on their locations and respective histochemical, biochemical, pharmacological, and functional characteristics, rat MC have been classified as connective tissue (isolated from the peritoneal cavity) or mucosal (isolated from intestinal mucosal tissues).[2] Lately, the role of MC in regulation of intestinal ion secretion has been studied extensively.[3] In the gastrointestinal tract, MC are present in the lamina propria, and have been found in increasing numbers during inflammatory conditions.[4,5] Activation of MC by antigen in hypersensitivity reactions, or by other stimuli (such as complement fragments, cytokines, neuropeptides) in inflammatory reactions, causes the release of mediators such as histamine, 5HT, arachidonic acid metabolites, proteases, reactive oxygen species, etc., which may result in pathophysiology altering epithelial functions including ion and water transport.[6,7,8,9,10] However, such evidence does not constitute proof that MC are solely responsible for these transport changes. A variety of cell types such as basophils, monocytes/macrophages, platelets, and eosinophils possess similar or identical mediators as those found in MC,[11,12] and can also bind IgE or aggregates of IgE.[11]

To further investigate the role of MC in intestinal physiology, two different mutant strains of mast cell-deficient mice, W/W^v and Sl/Sl^d have proven to be good models.[13,14] It is known that MC are derived upon stimulation by appropriate growth factors from their precursor cells which originate in bone marrow.[2] If one or more of these precursors or growth factors is defective or deficient, the animal may lack a mature MC population.

Advances in Mucosal Immunology, Edited by
J. Mestecky *et al.*, Plenum Press, New York, 1995

W/W^v and Sl/Sl^d mice are devoid of MC due to mutations on two completely different chromosomes.[15,16] The W/W^v have mutations on chromosome 5, resulting in an abnormal tyrosine kinase receptor for stem cell growth factor (SCF), so that the precursor cells are unable to differentiate into functional MC. The W/W^v have less than 0.3% MC in the skin and none in the gut. In addition to the absence of MC, other abnormalities such as anemia, lack of skin pigment and sterility have also been found in W/W^v. However, MC populations can be reconstituted in W/W^v mice by the injection of precursor cells from the bone marrow of congenic control mice.[13] In contrast, the Sl/Sl^d mice lack MC due to mutations on chromosome 10 resulting in the defective production of SCF.[16]

These studies examined, by using MC-deficient mice (W/W^v and Sl/Sl^d) and their congenic normal (+/+) controls, the role of MC in ion transport in intestinal responses to three known secretagogues: egg albumin antigen (EA) in sensitized mice; and cholera toxin (CT) and substance P (SP), a proinflammatory neuropeptide, in untreated mice.

METHODS

Animals

MC-deficient (W/W^v, Sl/Sl^d) and the congenic normal (+/+) mice were used in our studies. For some studies, MC populations in W/W^v mice were restored by intravenous injection with 2×10^7 bone marrow cells from +/+ congenic controls[13,17] and experiments were conducted 10 weeks later. These mice are designated as W/W^vR mice. The mice used in these experiments were either untreated or sensitized.

Sensitization

In the experimental model of hypersensitivity, mice were sensitized with 0.1 mg chicken egg albumin (EA) in alum solution, and pertussis vaccine was used as adjuvant for IgE. The mice were studied 12-15 days after sensitization.[18]

Ussing Chambers

Mice were killed by cervical dislocation. Segments of mid-small intestine were removed and placed in oxygenated Krebs buffer. The segments were opened along the mesenteric border and cut into sheets, 2 cm in length. At least four sheets from each animal were mounted in Ussing-type flux chambers that had been modified to contain Ag/AgCl stimulating electrodes on opposite sides of the tissue.[19] Agar-salt bridges were used to monitor potential difference (PD in mV) and inject an appropriate amount of current to maintain a PD of zero across the tissue. The tissues were clamped at zero volts using an automatic voltage clamp and the short-circuit current (Isc in $\mu A/cm^2$) was recorded continuously. Electrical transmural stimulation (TS) was carried out by passing rectangular current pulses (10 mA, 10 Hz, 0.5 m) for a total time of 5s across the tissue in a perpendicular direction. The change in Isc (_Isc) was calculated as the difference between basal Isc and maximal Isc seen in response to TS. In Ussing chambers, Cl^- ion secretion is indicated by an increase in Isc.[20,21]

RESULTS AND DISCUSSION

Responses to Antigen

It is generally accepted that the IgE antibody has a high affinity for MC and basophil granulocytes.[22] The bridging of cell-bound IgE antibody by antigen triggers the release of a variety of chemical mediators from MC.[11,18] In our first study, we sensitized mice to EA using alum and pertussis adjuvants to stimulate IgE production. In Ussing chambers, addition of serosal EA (100 µg/ml) caused increases in Isc in intestinal preparations from sensitized mice, both +/+ and W/W^v. The maximal response to EA in W/W^v was significantly less (~70%) than that in normal control mice and was completely restored in intestine from W/W^vR mice (Fig. 1).

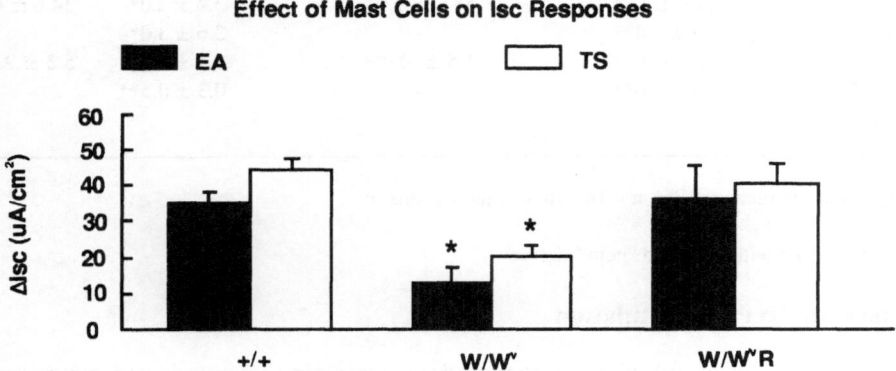

Figure 1. Values represent the means ± SEM ; n = 43, 26, and 14 for +/+, W/W^v, and W/W^vR, respectively. *P M 0.05 compared with +/+ controls.

Preincubation with either the H_1 anti-histamine, diphenhydramine (DPH, $10^{-5}M$), the $5HT_2$ serotonin antagonist, ketanserin (KET, $10^{-5}M$), or the cyclooxygenase inhibitor, piroxicam (PIR, $10^{-5}M$), alone or in combination significantly reduced secretory responses in +/+, but only PIR or the combination resulted in a significant inhibition in W/W^v (Table 1). The neurotoxin, tetrodotoxin (TTX, $10^{-6}M$), inhibited the antigen response in +/+ but not in W/W^v.[18] Antigen-induced secretion was also significantly reduced in Sl/Sl^d mice compared to +/+. These results provide convincing evidence that MC play an important role in stimulating ion secretion during intestinal anaphylaxis, even though other cells may be involved in this reaction. However, the increase in ion secretion is thought to be a result of MC mediators acting directly on the epithelium and/or indirectly via intestinal nerves.[18] Besides the MC mediators, histamine and serotonin, products of arachidonic acid metabolism such as prostaglandins (PG) or thromboxanes may influence ion transport. Although some MC can themselves release PG upon appropriate stimulation,[23,24] the ability of mouse intestinal MC to produce cylooxygenase products has not yet been determined. In our studies, PIR treatment of W/W^v intestine resulted in inhibition of the antigen-induced increase in Isc, suggesting that MC are presumably not the only source of these arachidonic acid metabolites in normal mice. In addition, blocking nerves by TTX reduced the antigen response in +/+, while the response to antigen in W/W^v mice was not affected by TTX, suggesting that there is an interaction between nerves and MC in normal mice. This interaction is not required for the function of the "non-mast cell" involved in anaphylactic responses, in MC-deficient mice.

Table 1. Effect of Various Pharmacological Agents on Isc Responses.

| | Isc (μA/cm2) | | | |
| | +/+ | | W/Wv | |
	EA	TS	EA	TS
None	43.1 ± 4.2	67.8 ± 4.7	11.9 ± 2.0	41.6 ± 3.6
DPH	27.2 ± 4.4+	52.4 ± 4.0*	11.9 ± 2.2	39.2 ± 5.7
KET	20.5 ± 4.5*	38.8 ± 5.0*	11.5 ± 3.1	39.0 ± 5.7
PIR	24.9 ± 2.7*	58.2 ± 9.9	3.4 ± 2.0*	34.6 ± 4.1
DPH + KET + PIR	5.6 ± 2.4**	-	2.6 ± 3.0*	-
TTX	12.2 ± 3.5**	1.5 ± .09**	8.4 ± 1.6	5.3 ± 2.1**
DPH + KET + PIR + TTX	4.4 ± 1.0**	-	0.3 ± 0.5**	-

Values represent the mean ± SEM. n = 16 - 18 for each treatment;
*P <0.05
**P< 0.01 compared with untreated "none" tissues

Responses to Nerve Stimulation

Enteric nerves in full thickness preparations containing myenteric and submucosal plexuses were stimulated via TS. Gut from MC-deficient mice responded abnormally to TS compared to MC-replete mice. As shown in Figure 1, the TS response in intestine from W/W^v mice was ~ 50% of that in intestine from +/+ mice but was restored in intestine from W/W^vR mice. In +/+, the TS-induced response was abolished by TTX and reduced by MC mediator antagonists, DPH and KET, but not PIR. In W/W^v, only TTX inhibited the TS response[18] (Table 1). Bienenstock et al.,[25] have reviewed the evidence to suggest that MC and nerves may form specific and selective associations together as is demonstrated in our data. The interaction between the nervous system and MC may result in various regulatory effects. One possible explanation is that TS affects MC indirectly via release of neurotransmitters from enteric nerves which then induced chloride secretion. Although a gross neuroanatomical defect in enteric nerves was not detected in W/W^v mice, our data do not eliminate the possibility that W/W^v mice may exhibit abnormalities in intestinal nerve function. These results demonstrate the importance of MC-nerve interactions in regulation of ion secretion.

Responses to Cholera Toxin (CT)

To examine the role of MC in the regulation of ion secretion in unsensitized mice, CT was chosen as a stimulus. In the small intestine, CT produced by the organism *Vibrio cholerae* can cause severe dehydrating diarrhea. The diarrhea is thought to be due to electrolyte and fluid secretion in the intestine by both direct and indirect actions of CT on the epithelium. The former action results in increased activity of adenylate cyclase which elevates enterocyte cAMP leading to a sustained opening of Cl^- channels in the brush border membrane of crypt epithelium.[26] The latter may involve enteric nerves and other cellular mediators, because CT-induced secretion can be inhibited by blocking receptors for neurotransmitters,[27] $5HT_2$ and $5HT_3$, on nerves. The source of the 5HT was postulated to be from enteroendocrine cells.[28] We propose that CT-induced secretion may also be regulated in part by MC because they are a major source of 5HT. In this study, CT caused a dose dependent increase in Isc in both normal and W/W^v mice. In W/W^v intestine, the response began later and reached maximum Isc values which were only approximately 50% of those in controls. The response to CT in intestine from controls was significantly reduced 59%, 64%, and 57% by TTX, DPH, and the combination of DPH and KET, respectively. These inhibitors had no effects on the response to CT in intestine from W/W^v. Response to CT was restored to control values in successfully reconstituted W/W^v using bone marrow from +/+. These results support the hypothesis that CT-induced secretion in mice small intestine is mediated by direct and indirect activation of MC.

To determine if the defects in ion transport in the W/W^v mice were due to their MC deficiency rather than to some other consequence of their W mutations, we also examined responses in Sl/Sl^d mice. The same reduced responses to EA, TS and CT were observed in these mice compared to the respective congenic controls.

Responses to Substance P

Recently, we have been focusing on the role of MC in the neuropeptide, SP-induced ion secretion. SP is an 11 amino acid peptide which produces effects via its C-terminal binding to a specific receptor (NK_1).[29] Evidence suggests that during inflammatory reactions, both MC and SP are important components and that MC and SP nerves may interact with each other to regulate intestinal pathophysiology.[30] SP-containing nerves are found in close apposition to MC.[2,30] SP can activate intestinal mucosal MC resulting in the release of its mediator, histamine.[31] Our preliminary data have shown that the secretory response (_Isc) to SP $(5x10^{-6}M)$ is smaller in W/W^v mice $(45.7 \pm 5.5 \ \mu A/cm^2)$ than in +/+ $(70.6 \pm 2.9 \ \mu A/cm^2)$. Although precise mechanisms of this effect are unclear, we propose that the effect of SP in regulation of ion secretion may be, at least in part, mediated via MC.

CONCLUSIONS

In conclusion, MC-deficient and normal mice do respond differently to EA antigen, CT and SP in the absence or presence of various inhibitors. Even though the precise mechanism of responses is uncertain, our studies strongly support the concept of a MC-nerve unit which is important functionally in the regulation of intestinal secretion due to immunological and non-immunological stimuli. In addition to MC playing an important role in the regulation of intestinal ion transport, they may also be essential in other pathological situations and perhaps even in homeostasis of the normal gut.

ACKNOWLEDGEMENTS

The work described in this paper was supported by grants from the Medical Research Council (Canada), the National Institutes of Health, and the Canadian Foundation for Ileitis and Colitis.

REFERENCES

1. D. R.Halm and R. A. Frizzell, *In*: "Textbook of Secretory Diarrhea" E. Lebenthal and M. Duffey (eds.). Raven Press, Ltd., New York. 47 (1990).
2. R. H. Stead, M. H. Perdue, *et al., In::* The Neuroendocrine-Immune Network, S. Freier (ed.) Boca Raton, FL. CRC Press 9 (1990).
3. M. H. Perdue, S. E. Crowe, *et al., In*: Proceedings of Sixth International Symposium on Neural of Bodily Function", Hosre & Huber Publishers, Toronto, 1 (1990).
4. A. M. Dvorak, *Pathol. Annu.,* 18 (Pt 1):181 (1983).
5. M. H. Perdue, J. K. Ramage, *et al., Dig. Dis. Sci.* 35:724 (1989).
6. L. B. Schwartz, K. F. Austen, *et al., Prog. Allergy* 34:271 (1984).

7. L. Enerback, *In*: Mast Cell Differentiation and Heterogeneity, A. D. Befus, J. Bienenstock, *et al.,* (eds.) Raven Press, New York 1 (1986).

8. J. R. Gordon, P. R. Burd, *et al., Immunol. Today* 11:458 (1990).

9. G. A. Castro, *In*: Textbook of Secretory Diarrhea, E. Lebenthal and M. Duffey (eds.) Raven Press, Ltd., New York. 31 (1990).

10. M. H. Perdue and J. Bienenstock, *Curr. Opinion Gastroenterol.,* 7 421 (1991).

11. A. Capron, J. P. Dessaint, et al., *Immunol. Today* 7:15 (1986).

12. S. J. Galli, Fed. Proc. 46:1906 (1987).

13. S. J. Galli and Y. Kitamura, *Am. J. Pathol.* 127:191 (1987).

14. B. K. Wershil and S. J. Galli, *Gastroenterol. Clin. North Amer.* 20:613 (1991).

15. K. Nocka, S. Majumder, *et al.,* Genes Dev. 3:816 (1989).

16. O. Witte, Cell 63:5 (1990).

17. Y. Kitamura, S. Go, *et al., Blood* 52:447 (1978).

18. M. H. Perdue, S. Masson, *et al., J. Clin. Invest.* 87:687 (1991).

19. M. H. Perdue, J. S. Davison, *et al., Am. J. Physiol.* 254:G444 (1988).

20. S. Crowe, P. Sestini, *et al., Gastroenterology* 99:74 (1990).

21. M. Perdue and D. G. Gall, *Am. J. Physiol.* 250:G427 (1986).

22. T. Ishizaka and K. Ishizaka, *Prog. Allergy* 34:188 (1984).

23. S. I. Wasserman and D. L. Marquardt, *In*: Allergy: Principles and Practice. 3rd ed. Middleton E, Jr, *et al.,* eds. Mostby Press, St Louis. 1365 (1988).

24. D. J. Heavey, P. B. Ernst, *et al., J. Immunol.* 140:1953 (1988).

25. J. Bienenstock, M. Blennerhassett, *et al., In*: Mast Cell and Basophil Differentiation and Function in Health and Disease, S. Galli and K. F.Austen, (eds.) Raven Press, New York. 275 (1989).

26. M. Donowitz and M. J. Welsh, *In*: Physiology of the Gastrointestinal Tract, L. R.Johnson (eds.) Raven Press, New York, 1351 (1987).

27. J. Cassuto, M. Jodal, *et al., Scand. J. Gastroerol.* 16:377 (1981).

28. E. Beubler and G. Horina, *Gastroenterology* 99:83 (1990).

29. M. Mousli, J. -L. Bueb *et al., Trends Pharm. Sci.* 11:358 (1990).

30. J. Bienenstock, G. MaCqueen, *et al., Am. Rev. Respir. Dis.* 143:S55 (1991).

31. F. Shanahan, J. A. Denbury, *et al., J. Immunol.* 135:1331 (1985).

MODULATION OF TUMOUR NECROSIS FACTOR-ALPHA mRNA LEVELS BY INTERFERONS IN DIFFERENT POPULATIONS OF MAST CELLS

J. Antonio Enciso, Elyse Y. Bissonnette, and A. Dean Befus

Department of Microbiology and Infectious Diseases
University of Calgary, Calgary, Canada

INTRODUCTION

Tumour necrosis factor alpha (TNF-α) is a multipotent cytokine[1] which can be produced by mast cells[2] (MC). Given the regulatory roles of interferons (IFN) and the ability of IFN to potentiate TNF-α activity of macrophages[3] we previously studied the effects of IFN on MC TNF-α activity.[4] In contrast with effects on macrophages, IFN-α/ß and IFN-γ inhibited TNF-α activity of mast cells. Thus, we analyzed the effects of IFN-α/ß and IFN-γ on MC mRNA expression for TNF-α, high affinity immunoglobulin E receptor-alpha chain (FcERI-α chain), 2'5' oligo adenylate synthetase (2'5'OAS) and ß-actin.

METHODS AND RESULTS

Rat tissue-cultured mast cells (RCMC) and rat hybrid mast cell lines (HRMC) kindly provided by Dr. A. Froese, The University of Manitoba, Winnipeg, Manitoba, were used.[5] Total RNA from mast cell lines treated with different doses (100, 200, 400 and 800 U/ml) of IFN-α/ß or IFN-γ for 24 h was isolated according the procedure of Chomczynski and Sacchi[6], with some modifications. RNA samples were resolved on 1.3 % agarose-formaldehyde, blotted onto nylon membranes[7] and the messages for TNF-α, ß-actin, FcERI-α chain and 2'5' OAS, were detected as 1.7, 1.9, 1.2, and 1.6 kb mRNA by hybridization of the filters with these probes labelled with ^{32}P. Band intensities were quantified using a laser scan densitometry and compared with the band intensities obtained from sham treated cells.

Analysis of the effects of different doses of IFN on MC showed that mRNA levels for TNF-α decreased following most treatments. However, at lower doses of IFN-γ, there was a slight increase in mRNA levels for TNF-α (Table 1). IFN-α/ß also inhibited the TNF-α mediated cytotoxicity of these MC lines by up to 67 % (data not shown). Interestingly, mRNA levels for FcERI-α chain and 2'5'OAS were enhanced by IFN treatment in a dose-dependent manner. Levels of mRNA for ß-actin were not modified by IFN treatment.

Table 1. Dose dependency of IFN-γ and IFNα/ß on mRNA levels for TNF-α, FcERI-α chain, 2'5'OAS and ß-actin in RCMC and HRMC.

	HRMC					RCMC			
					U/ml IFN-γ				
	100	200	400	800		100	200	400	800
Probe									
TNF-α	+	+	***	***		+	+	***	***
FcERI-α chain	+	+	++	+++		+	++	+++	+++
2'5'OAS	+	+	++	+++		+	++	+++	+++
ß-actin	-	-	-	-		-	-	-	-
					U/ml IFN-α/ß				
TNF-α	+	*	*	**		*	*	*	*
FcERI-α chain	+	+	++	+++		+	+	++	+++
2'5' OAS	+	+	+	++		+	+	+	++
ß-actin	-	-	-	-		-	-	-	-

Table 2. Time course analysis of IFN-γ or IFN-α/ß on mRNA levels for TNF-α, FcERI-α chain and ß-actin in HRMC and RCMC.

	HRMC				RCMC			
					IFN-γ			
	Time In Hours				Time In Hours			
	1	4	16	24	1	4	16	24
Probe								
TNF-α	*	*	**	***	*	*	**	**
FcERI-α chain	+	++	+++	+++	++	+++	+++	+++
ß-actin	-	-	-	-	-	-	-	-
				IFN-α/ß				
TNF-α	**	**	**	**	**	**	**	**
FcERI-α chain	+	+	+	+	+	+	+++	+++
ß-actin	-	-	-	-	-	-	-	-

Analysis of the effects on mRNA levels between control cells and IFN-treated cells estimated on base of the values of band intensities are considered as follows:
No effect of IFN (-)
High increase +++; medium increase ++; low increase +
High decrease ***; medium decrease **; low decrease *

To analyze the effects of duration of IFN pre-treatment on mRNA levels, MC were pre-treated with 400 U/ml of IFN-γ or IFN α/ß for 1, 4, 16 and 24 h. A time-dependent decrease in mRNA levels of TNF-α occurred in both MC lines (Table II). There was a time dependent increase in mRNA levels for FcERI-a chain; this effect of IFN-γ was more pronounced than that of IFN-α/ß. No significant change occurred in the levels of mRNA for ß-actin.

DISCUSSION

Pre-treatment with IFN-α/ß and IFN-γ depressed the mRNA and protein levels for TNF-α in MC. The effect of IFN at 400 and 800 U/ml was a marked decrease in the TNF-α mRNA levels. This effect was also time dependent, being maximal after 24 h. Moreover, given that mRNA levels for ß-actin were stable following treatment with interferons, we conclude that the decrease in the mRNA levels of TNF-α was not the result of a general mRNA breakdown. TNF-α-dependent cytotoxicity assays showed a similar dose-dependent inhibition by IFN. Studies previous in our laboratory indicate that this phenomena is also observed in rat peritoneal MC.[4] This inhibition is not observed with monocytes/macrophages and T cells, where IFN-γ can induce TNF-α expression.[3] Our results support the hypothesis that there are different mechanisms of action of IFN in MC than in macrophages.

Given that TNF-γ mediates citotoxicity and has many other roles,[4,7] the inhibitory effect of IFN on this MC cytokine might be an important component in the regulation of inflammatory reactions. Because MC are able to produce several cytokines similar to those from T_H2 cells[8] and in this way have been implicated in the regulation of IgE-dependent immediate hypersensitivity reactions, our observations with IFN regulation of MC may have broad implications.

In contrast to the reduction in TNF-α levels, we observed that the mRNA levels of FcERI-α chain were increased in a dose- and time dependent manner. Interestingly, recent reports have shown that IFN-γ and IFN-α induce enhanced protein and mRNA expression for receptors for the Fc portion of IgA (Fc-αR) in polymorphonuclear leucocytes and macrophages. [9,10] Moreover, Bolt-Nitulescu and colleagues[11] established that IFN-γ enhanced the expression of receptors for IgE on rat bone marrow-derived macrophages.

Why TNF-α levels should be depressed and FcERI enhanced in MC by the same IFN treatment is unknown. Nevertheless our observations suggest that IgE-dependent and TNF-α functions of MC are regulated in different ways. Perhaps this implies that these functions of MC are segregated and not necessarily active concurrently.

The mechanisms of IFN action are complex. In other systems IFN induces the synthesis of many new proteins[12,13] that act in numerous ways, including in gene activation or depression. Whether IFN induced transcriptional regulatory factors and the gene expression they modulate are the same in MC as in other cell types remains to be investigated. Knowledge of these issues will contribute to our understanding of the cytokine network and the regulation of MC function in inflammation and allergy.

ACKNOWLEDGMENTS

This work was supported by the Medical Research Council of Canada. EYB and ADB are funded by the Alberta Heritage Foundation for Medical Research.

REFERENCES

1. P. Scheurich, B. Thomas, U. Ucer, and K. Pfizenmaier, *J. Immunol.* 138:1786 (1987).
2. J. D. Young, C. C.Lice, G. Butler, Z. A. Cohn, and S. J. Galli, *Proc. Natl. Acad. Sci. USA* 84:9175 (1987).
3. L. M. Pelus, O. G. Ottomann, and K. H. Nocka, *J. Immunol.* 140:479 (1988).
4. E. Y. Bissonnette and A. D. Befus, *J. Immunol.* 145(10):3385 (1990).
5. Y. Zheng, B. M. C. Chan, E. S. Rector, I. Istvan, and A. Froese, *Exp. Cell Res.* 194:301 (1991).
6. P. Chomczynski and N. Sacchi, *Anal. Biochem.* 162:156 (1987).
7. J. Sambrook, E. F. Fritsch, and T. Maniatis, Molecular Cloning: A Laboratory Manual. Second edition, Cold Spring Harbor Laboratory Press, New York, (1989).
8. P. R. Burd, H. W. Rogers, J. R. Gordon, C. A. Martin, J. Sundararajan, S. D. Wilson, A. M. Dvorak, S. Galli, and M. E. Dorf, *J. Exp. Med.* 170:245 (1989).
9. M. A. Cassatella, F. Bazzoni, F. Calzetti, I. Guasperri, F. Rossi, and G. Trinchieri, *J. Biol. Chem.* 266:22079 (1991).
10. M. J. Fultz and S. N. Vogel, *J. Leuk. Biol.* 51:300 (1992).
11. G. Boltz-Nitulescu, C. Wiltschk, K. Langer, H. Nemet, C. Holzinger, A. Gessl, O. Forster, and E. Penner, *Immunology* 63:529 (1988).
12. S. L. Gupta, B. Y. Rubin, and S. L. Holmes, *Proc. Natl. Acad. Sci. USA* 76:4817 (1979).
13. J. Weil, C. J. Epstein, L. B. Epstein, J. J. Sedmak, J. L. Sabra, and S. E. Grosberg, *Nature* 301:437 (1983).

AGGREGATION OF THY-1 GLYCOPROTEIN INDUCES TYROSINE PHOSPHORYLATION OF DIFFERENT PROTEINS IN ISOLATED RAT MAST CELLS AND RAT BASOPHILIC LEUKEMIA CELLS

Lubica Dráberová,[1,2] and Petr Dráber[2]

[1]Department of Immunology, Institute of Microbiology
[2]Department of Mammalian Gene Expression,
 Institute of Molecular Genetics, Czech Academy
 of Sciences, Prague 4, Czech Republic

INTRODUCTION

Cross-linking of cell-bound immunoglobulin E (IgE) on mast cells, basophils, and related cultured cell lines induces activation of various membrane-associated enzymes, which results in cell degranulation. The early biochemical signals initiated by aggregation of the high-affinity IgE receptor (FcεRI) include activation of phospholipase C leading to hydrolysis of phosphatidylinositol to inositol 1,4,5-triphosphate (IP$_3$) and diacylglycerol (DAG), mobilization of intracellular Ca^{2+} from internal stores by IP$_3$, increased Ca^{2+} influx into the cell, activation of protein kinase C by DAG, membrane depolarization and others. It has also been shown that FcεRI aggregation induces phosphorylation of several proteins on serine/threonine and tyrosine residues. Phosphorylation products include not only components of FcεRI[15] but also phospholipase Cγ1[16], pp72[2,18], and others.[3]

Mast cells and basophils express on their surfaces several glycoproteins (gp) that modulate FcεRI-induced activation or may play a role as an alternative pathway of cell activation. Basciano et al. (1986)[1] described a monoclonal antibody (mAb) that bound to ganglioside G$_{D1b}$ and inhibited FcεRI- or ionophore A23187-induced release of histamine from rat basophilic leukemia cells, RBL-2H3.[9] Several other antibodies have been described that inhibit FcεRI-induced activation.[14]

Recently we have found that rat peritoneal and pleural mast cells express on their surfaces Thy-1 glycoprotein (gp) as detected by binding of MRCOX7 mAb.[5] Scatchard analysis of the binding data revealed surface expression of at least one million of Thy-1 molecules per cell. Interestingly, incubation of the isolated mast cells with various concentrations of MRCOX7 mAb induced a rapid and concentration-dependent increase in the intracellular free calcium concentration ([Ca^{2+}]$_i$), which was followed by histamine release. In an attempt to analyze the mechanism of Thy-1-mediated mast cell activation we studied expression of the Thy-1 gp in cultured cells and found that RBL-2H3 are Thy-1 positive and that aggregation of Thy-1 induced their activation.[7] Availability of cloned Thy-1 genes allowed us to prepare stable transfectants expressing both the endogenous Thy-1.1 and transfected mouse Thy-1.2 genes. The transfected cells could be activated not only by MRCOX7 mAb but also by 1aG4 mAb that recognizes Thy-1.2 antigen.[7]

Thy-1 gp displays strong homology with the variable domain of Ig and may correspond to the putative primordial domain of the Ig superfamily.[23] In several cell types Thy-1 gp was found to be anchored to the membrane through a glycosylphosphatidylinositol (GPI) tail.[12,21] To determine the mechanism of Thy-1-mediated activation of mast cells, we analyzed sensitivity of the Thy-1 gp to phosphatidylinositol-specific phospholipase C (PI-PLC), and changes in protein tyrosine phosphorylation induced by Thy-1-specific antibodies in isolated peritoneal and pleural mast cells, and in RBL-2H3 cells.

MATERIALS ND METHODS

Cells

The origin of RBL-2H3 cells and culture conditions have been described previously.[7] Peritoneal and pleural mast cells were recovered by lavage of the cavities as described.[6] The suspensions contained >95% mast cells as determined by staining with toluidine blue.

PI-PLC Treatment and Radioantibody Binding Assay

Cells (5×10^6) in 500 µl of culture medium (MEM) diluted 1:1 with phosphate-buffered saline, pH 7.2 (PBS), and supplemented with 0.2% bovine serum albumin (BSA) were incubated at 37°C with 0.6 U PI-PLC (Boehringer Mannheim). At various time intervals, 100-µl aliquots were removed, centrifuged and resuspended in 120 µl of MEM/PBS/BSA. Thirty-µl aliquots, containing 0.25×10^6 cells, were removed and assayed for MRCOX7 mAb binding by a direct radioantibody binding assay as described.[7]

Cell Activation

RBL-2H3 cells grown as monolayers were detached by incubation for 5 min with PBS containing 0.02% EDTA and 0.01% trypsin. Trypsin activity was stopped by addition of complete culture medium. The cells were centrifuged, washed in Tyrode's solution with 0.1% BSA, and counted. The cells (0.5×10^6) in 250 µl of Tyrode's solution with 0.1% BSA were activated by MRCOX7 mAb at 37°C.[13] At various time intervals, the reaction was stopped by transfer of the tubes on ice and brief centrifugation to pellet the cells.

Identification of Tyrosine-Phosphorylated Proteins

Pelleted cells were solubilized for 30 min at 4°C in lysis buffer containing 1% Triton X-100, 0.1% SDS, 0.5% sodium deoxycholate, 10 mM sodium phosphate (pH 7.2), 50 mM NaCl, 50 mM NaF 1 mM Na_3VO_4, 1 mM PMSF, 20 µg/ml leupeptin, 0.33 U/ml aprotinin, and 0.1% NaN_3. Insoluble material was removed by centrifugation at 12,000 x g for 15 min at 4°C. Protein concentration of the supernatants was determined by an assay according to Bradford (1976).[4] Supernatant was mixed with 2x SDS polyacrylamide gel electrophoresis sample buffer[10], boiled, and run on a 10% polyacrylamide gel. The separated proteins were electrotransferred onto nitrocellulose membrane, and free binding sites were blocked by incubating the membrane overnight at 4°C with PY 20 buffer (10 mM Tris, pH 7.4, 75 mM NaCl, 0.05% Tween 20) containing 5% BSA. After washing, the membrane was incubated for 2 h at room temperature with anti-phosphotyrosine mAb PY 20 coupled to horseradish peroxidase (ICN ImmunoBiologicals, Costa Mesa, CA) in PY 20 buffer with 0.5% BSA. After washing (3 x 15 min in 50 ml PY-20 buffer), the PY 20 mAb-reactive proteins were visualized with a chemiluminescent procedure using ECL Western blotting detection reagent (Amersham) according to the manufacturer's recommendations.

RESULTS

Sensitivity of Surface Thy-1 gp to PI-PLC Treatment

To verify that Thy-1 gp in RBL-2H3 cells is anchored via GPI linkage, we analyzed the binding of ^{125}I-labeled MRCOX7 mAb to the cells pretreated for various time intervals with PI-PLC (Fig. 1). As expected, PI-PLC treatment drastically decreased the surface expression of Thy-1 gp; cells treated with PI-PLC for 40 min bound less than 0.5% of ^{125}I-MRCOX7 whereas control cells bound 13% of the mAb. These results were confirmed by flow cytofluorometry and extended to isolated rat peritoneal and pleural mast cells (not shown).

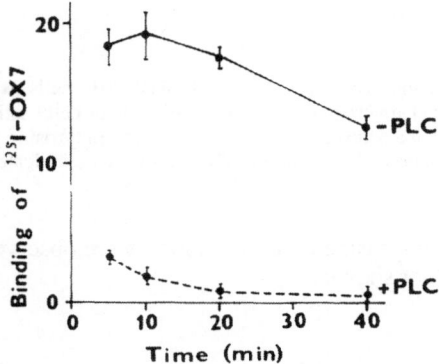

Figure 1. Sensitivity of the surface Thy-1 gp to PI-PLC treatment. RBL-2H3 cells were incubated for various time intervals without (solid line) or with (dashed line) PI-PLC. At various time intervals their ability to bind ^{125}I-MRCOX7 (OX7) mAb was determined in a direct radioantibody binding assay. Points are the mean values + S.D. of triplicate determinations.

Changes in Protein Tyrosine Phosphorylation Induced in RBL-2H3 Cells by Thy-1 Glycoprotein Aggregation

Recent data have suggested that there are multiple tyrosine signalling pathways in RBL-2H3 cells and that their selective utilization may be determined by the nature of the stimulus.[24,18] Therefore, we analyzed cellular proteins that are tyrosine phosphorylated as a result of Thy-1-induced activation of RBL-2H3 cells. The cells were activated by exposure to MRCOX7 mAb and cell lysates were analyzed by immunoblotting employing peroxidase-labeled PY 20 mAb (Fig. 2a). In lysates from unstimulated cells several proteins phosphorylated on tyrosine were observed; the bands of 53, 55, 58, and 96 kDa were prominent. When the cells were exposed to MRCOX7 mAb, an increase in the intensity of several bands, including 53, 55, 58, and 96 kDa, was observed. However, no tyrosine-phosphorylated protein of 72 kDa was observed in MRCOX7-activated RBL-2H3 cells.

Changes in Protein Tyrosine Phosphorylation Induced in Isolated Rat Mast Cells by Thy-1 Glycoprotein Aggregation

Rat peritoneal and pleural mast cells were used as an alternative system to study Thy-1-induced protein tyrosine phosphorylation. In lysates from unstimulated cells the PY 20 mAb reacted with a major gp of 42 kDa and with two minor gps of 28 and 52 kDa (Fig. 2b). Cross-linking of Thy-1 gp induced in mast cells a rapid increase in phosphorylation of proteins of approximately 32, 52, and 60 kDa which declined after 30 min incubation.

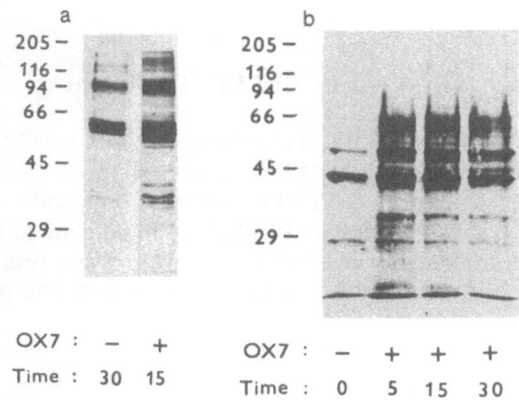

Figure 2. Protein tyrosine phosphorylation induced in RBL-2H3 cells (a) and isolated rat peritoneal and pleural mast cells (b) by Thy-1 gp cross-linking. Control cells or cells activated by exposure to MRCOX7 mAb (100 μg/ml) were lysed in a detergent lysis buffer and protein tyrosine phosphorylation was analyzed as described in Materials and Methods. The positions of M_r markers are indicated.

Similar changes in protein tyrosine phosphorylation were observed in mast cells activated by cross-linking of FcεRI (not shown).

DISCUSSION

Antigen-specific (FcεRI-mediated) and Thy-1 gp-mediated activation of mast cells and RBL-2H3 cells results in the release of histamine and other mediators of allergic reactions. These pathways, therefore, converge on a common route leading to cell degranulation. The present study was designed to investigate whether the convergence occurs immediately after Thy-1/FcεRI cross-linking or whether it could involve separate intracellular biochemical events before merging into a common pathway. The data presented in this paper indicate that Thy-1 gp-mediated activation of RBL-2H3 cells induces increased phosphorylation of several proteins, namely of 53, 55, 58, and 96 kDa. There was no tyrosine phosphorylation of pp72 which is found in antigen-activated cells.[2] Thus, the early activation pathways induced by Thy-1 gp and FcεRI cross-linking are different.

The mechanism by which Thy-1 gp mediates cell activation is not known. We have found that Fab' fragments of MRCOX7 mAb, in contrast to F(ab')$_2$ fragments, are unable to induce RBL-2H3 cell activation (unpublished); this suggests that Thy-1 gp aggregation plays a decisive role in this proces. Thy-1 is anchored to the cell surface by GPI linkage and, therefore, its transmembrane signalling capacity is probably mediated via its association with other transmembrane surface molecules. In T cells, the Thy-1 gp was found to interact with several surface molecules such as CD45 tyrosine phosphatase[22], p100 transmembrane protein[11], and protein tyrosine kinases p56[lck][19] and p60[fyn].[20] RBL-2H3 cells express *src* protein-related tyrosine kinases p53/p56[lyn] and pp60[c-src], and their activity is increased in antigen-stimulated cells.[8] It is likely that they are also involved in the observed Thy-1-mediated protein tyrosine phosphorylation. In fact, we have recently found physical association of p53/p56[lyn], but not pp60[c-src], with Thy-1 in Nonidet P40-solubilized cells (unpublished).

Compared to RBL-2H3 cells, activation of isolated rat peritoneal and pleural mast cells with MRCOX7 mAb resulted in tyrosine phosphorylation of different sets of proteins. This may reflect a different origin of these cells; RBL cells are similar to mucosal mast cells, whereas peritoneal and pleural mast cells belong to connective tissue mast cells.[17] Both cell types differ in many properties including histology and secretory granule biochemistry. The data presented here suggest that they may also differ in proteins phosphorylated on tyrosine residues. Alternatively, the observed difference may reflect the tumor origin of RBL-2H3 cells This underlines the necessity to analyze mast cell activation not only on cultured cell lines but also on isolated cells before any generalization about the activation pathway is made It should also be noted that activation of isolated mast cells via Thy-1 gp and FcεRI produces almost identical patterns of proteins phosphorylated on tyrosine This could reflect engagement of the FcεRI or IgG receptors in Thy-1-mediated activation of mast cells and/or a different mechanism of Thy-1-mediated activation in both cell types These problems are addressed in our recent studies

ACKNOWLEDGMENTS

This work was supported by a collaborative grant from the Fogarty International Center's Central and Eastern European Initiative and by grant No 52 034 from the Czech Academy of Sciences

REFERENCES

1 L K Basciano, E H Berenstein, L Kmak, and R P Siraganian, *J Biol Chem* 261 11823 (1986)
2 M Benhamou, J S Gutkind, K C Robbins, and R P Siraganian, *Proc Natl Acad Sci USA* 87 5327 (1990)
3 M Benhamou and R P Siraganian, *Immunol Today* 13 195 (1992)
4 M M Bradford, *Anal Biochem* 72 248 (1976)
5 L Dráberová, *Eur J Immunol* 19 1715 (1989)
6 L Dráberová, *Eur J Immunol* 20 1469 (1990)
7 L Dráberová, and P Dráber, *Eur J Immunol* 21 1583 (1991)
8 E Eiseman and J B Bolen, *Nature* 355 78 (1992)
9 N Guo, G Her, V N Reinhold, M J Brennan, R P Siraganian, and V Ginsburg, *J Biol Chem* 264 13267 (1989)
10 U K Laemmli, *Nature* 227 680 (1970)
11 A Lehuen, R C Monteiro, and J F Kearney, *Eur J Immunol* 22 2373 (1992)
12 M G Low and P W Kincade, *Nature* 318 62 (1985)
13 D W Mason and A F Williams, *Biochem J* 187 1 (1980)
14 E Ortega Soto and I Pecht, *J Immunol* 141 4324 (1988)
15 R Paolini, M -H Jouvin, and J -P Kinet, *Nature* 353 855 (1991)
16 D J Park, H K Min, and S G Rhee, *J Biol Chem* 266 24237 (1991)
17 D C Seldin, S Adelman, K F Austen, R L Stevens, A Hein, J P Caulfield, and R G Woodbury, *Proc Natl Acad Sci USA* 82 3871 (1985)
18 V Stephan, M Benhamou, J S Gutkind, K C Robbins, and R P Siraganian, *J Biol Chem* 267 5434 (1992)
19 I Stefanová, V Horejsí, I J Ansotegui, W Knapp, H Stockinger, *Science* 254 1016 (1991)
20 P M Thomas and L E Samelson, *J Biol Chem* 267 12317 (1992)
21 A G D Tse, A N Barclay, A Watts, and A F Williams, *Science* 230 1003 (1985)
22 S Volarevic, C M Burns, J J Sussman, and J D Ashwell, *Proc Natl Acad Sci USA* 87 7085 (1990)
23 A F Williams and A N Barclay, *Annu Rev Immunol* 6 381 (1988)
24 K -T Yu, R Lyall, N Jariwala, A Zilberstein, and J Haimovich, *J Biol Chem* 266 22564 (1991)

EFFECTS OF SUBDIAPHRAGMATIC VAGOTOMY ON MUCOSAL MAST CELL DENSITIES IN STOMACH AND JEJUNUM OF RATS

Thomas P. Gottwald,[1] Sarka Lhotak[2] and Ron H. Stead[2]

[1]Department of General Surgery, Eberhard-Karls-University, 7400 Tübingen, Germany; [2]Intestinal Disease Research Unit and Department of Pathology, McMaster University, Hamilton, Ontario, Canada

INTRODUCTION

There is increasing evidence for the innervation of intestinal mucosal mast cells (IMMC), including both microanatomical and functional data.[1] Some of this work suggests peptidergic or cholinergic nerve-induced activation of mast cells. For example, Bani-Sacchi *et al.*, reported a decrease in the stainability of IMMC in the ileum after field stimulation of isolated intestines.[2] This was accompanied by an increased release of acetylcholine and histamine, and was depressed by atropine and tetrodotoxin.

Studies in the dura mater have shown decreased densities of connective tissue mast cells (CTMC) after electrical stimulation of the trigeminal ganglion, and an increase in this cell type after capsaicin pretreatment.[3] Conversely, stimulation of the cervical sympathetic ganglion appears to increase the serotonin content of dural CTMC and sympathetic ganglionectomy reduces this.[4] These data can be explained by the hypothesis that mast cells may be tonically activated by sensory afferent nerves and stabilized by the sympathetic system.

Two earlier publications (and one abstract) by Ganguly and colleagues[5,6] have shown decreased mast cell densities in the gastric mucosa following pylorus ligation; and significantly greater mast cell numbers in paired animals subjected to sub-diaphragmatic vagotomies one week previously. Concurrent pylorus ligation and sub-diaphragmatic vagotomy was further found to prevent the loss of gastric tissue histamine induced by pylorus ligation alone. However, this work was presented in an incomplete manner, in that non-stressed groups were not included in the histological study. Therefore, we decided to document the effect of vagotomy *per se* (without specific stress stimuli) on gastric mucosal mast cell densities. Furthermore, since it is known that the vagus impinges on the enteric nervous system throughout the gastrointestinal tract, we wished to determine the effect of vagotomy on mast cells in the jejunum, where extensive mucosal mast cell nerve associations have heen recorded.[7,8]

METHODS

Male Lewis rats (150-200 g) were given food and water *ad libitum*, except for 24 hours prior to surgery when chow was withheld. All procedures were carried out under

Advances in Mucosal Immunology, Edited by
J. Mestecky *et al.*, Plenum Press, New York, 1995

anaesthesia with sodium pentobarbitol (50 mg/kg). Three groups of animals were employed: vagotomy with pyloroplasty (to prevent motility problems), pyloroplasty only and laparotomy only. Animals were sacrificed 1 and 21 days after surgery and the stomachs and segments of jejunum (beginning 15 cm after the pylorus) removed for histological examination. For sacrifice, animals were etherized and exanguinated by cardiac puncture, and the serum used for analysis of gastrin levels, as an indicator of vagotomy. The stomachs were opened along the greater curvature and any contents gently removed before pinning on cork boards and immersion in acetic acid/ethanol (1:9; AA). Jejunums were opened along the antimesenteric border, again cleaned and fixed by immersion in AA. After 6 - 8 hours, tissues were transferred to absolute alcohol and processed to paraffin. Sections were cut at 6,um and stained with Alcian blue in 0.7 N HC1 for 2 hours. Mast cell densities were determined using a Quantimet 520D image analysis system. For analysis, 40x magnification was used and the entire mucosa was assessed for both stomach and jejunum (total area per slide was c.15.0mm^2). Serum samples were frozen and stored at -70° C before performing radioimmunoassay to determine gastrin levels, using rabbit anti-human gastrin (courtesy of Prof. Becker, Tuebingen) and ^{125}I-labelled gastrin (NEN Research Products). Standards were obtained from Sigma Chemical Co. and control sera for low, middle and high gastrin ranges from DRG Instruments GmbH. Data were analyzed using one way analysis of variance (ANOVA) and two sample t-tests.

RESULTS

Mast cell densities in the stomach and jejunal mucosae at days 1 and 21 after vagotomy plus pyloroplasty, pyloroplasty only or laparotomy are shown in Table 1. In the stomachs, one way analysis of variance revealed no difference between the day 1 groups of animals (F = 0.04, p = 0.966); however, the day 21 stomach mucosae had significantly higher mast cell densities in the vagotomy group (F = 4.91, p = 0.018). In addition, the differences in mast cell densities between day 1 and day 21 were significant in all groups p < 0.05)

In the jejunums, one way analysis of variance revealed statistically significant differences at both day 1 (F = 4.01, p = 0.032) and at day 21 (F = 5.84, p = 0.011). At day 1, both vagotomy (t = 2.18, p = 0.044) and pyloroplasty (t = 2.42, p = 0.029) groups had significantly lower mast cell densities than the laparotomy animals. However, at day 21, only the vagotomy (t = 2.98, p = 0.013) but not pyloroplasty (t = 1.55, p = 0.16) group had significantly fewer mast cells than the laparotomized rats.

To ensure that the reduced mast cell densities in the jejunums were not due to increased lamina propria areas, sections from 6 animals in each of the 1 day laparotomy (maximum mast cell density) and 21 day vagotomy (minimum mast cell density) groups were further analyzed to determine the proportion of villus lamina propria in the measured fields. This was found to be 10.8 % and 10.2 %, respectively and was not significantly different (P = 0.315).

Table 1. Mast cell densities in stomach and jejunal mucosae.

Group	Stomach[1]		Jejunum[1]	
	Day 1	Day 21	Day 1	Day 21
Laparotomy	54 ± 7 (9)	91 ±12 (8)	140 ± 10 (9)	125 ± 9 (7)
Pyloroplasty	57 ± 8 (9)	93 ± 8 (8)	109 ± 8 (8)	110 ± 4 (6)
Vagotomy	55 ± 6 (9)	148 ± 20 (8)	115 ± 6 (10)	94 ± 6 (8)
ANOVA (F,p)	0.04, 0.966	4.91, 0.018	4.01, 0.032	5.84, 0.011

[1]Mast cell densities (mm-^{2}); mean ± S.E. (n)

Gastrin levels were significantly increased in the sera of day 21 vagotomized animals, indicating successful vagotomy (Table 2). The gastrin assays were done in two groups. In the first set, gastrin controls were 16 pg/mL (low), 148 pg/mL (medium) and 297 pg/mL (high); and in the second set, controls were 7 pg/mL (low), 19 pg/mL (medium) and 90 pg/mL (high). The stomach mucosal mast cell densities were found to correlate with gastrin levels at 21 days post-vagotomy ($r = 0.60$ and $r = 0.83$ for the first and second groups, respectively).

Table 2. Gastrin levels at 21 days post-surgery.

Group		Gastrin Levels *	ANOVA
Expt. 1	Laparotomy	69.0 ± 27.4 (2)	
	Pyloroplasty	107.7 ± 27.1 (3)	$F = 2.21$
	Vagotomy	415.5 ± 165.3 (3)	$p = 0.191$
Expt. 2	Laparotomy	55.0 ± 23.2 (6)	
	Pyloroplasty	24.5 ± 13.2 (4)	$F = 11.16$
	Vagotomy	109.4 ± 38.5 (5)	$p = 0.002$

[1]Gastrin levels presented as pg/mL; mean $\pm$ S.D. (n)

DISCUSSION

Total subdiaphragmatic vagotomy causes an increase in the density of mucosal mast cells in the stomach 21 days post surgery. Conversely, there is a decrease in the mucosal mast cell density in the jejunums of animals both 1 and 21 days after surgery. Vagotomy was successfully performed, as indicated by gastrin levels being significantly increased at 21 days post vagotomy. The mechanisms through which these changes occur are not known but appear to be different for stomach and jejunum.

Several studies on the effects of sustained hypergastrinaemia have shown a trophic effect of gastrin on the oxyntic mucosa of the stomach exclusively (no trophic effects have been observed in other regions of the gastrointestinal tract).[9,10] The enterochromaffin-like (ECL) cells which produce and store histamine[9] are known to be influenced by this trophic effect of gastrin.[11] Perhaps mast cells are similarly affected. Loss of tonic nerve stimulation, resulting in mast cell stabilization, is an alternative explanation in the stomach. However, this explanation could not apply to the jejunum where mast cell densities were found to decrease following vagotomy, unless the net excitatory/inhibitory action of vagal nerves is different in these two regions of the gastrointestinal tract. It is possible that the decreased mast cell densities in the jejunums reflect initial activation during the surgical procedure (day 1) and subsequent loss of vagal or second order nerves impinging on jejunal mucosal mast cells. The latter change should result in decreased mast cell densities if IMMC in normal rats are innervated, since nerve/target tissues generally exhibit trophic interdependence. We are currently examining the intestinal innervation in vagotomized animals to further address this possibility.

ACKNOWLEDGEMENTS

Assunta Beltrano and Marianne Hannon are gratefully acknowledged for their expert assistance. The Medical Research Council of Canada supported this work.

REFERENCES

1. R. H. Stead, M. H. Perdue, M. G. Blennerhassett, *et al. In*: "The Neuroendocrine-Immune Network" S. Freier, ed., p. 19, CRC Press Inc., Boca Raton (1990).
2. T. Bani-Sacchi, M. Barattini, S. Bianchi, *et al., J. Physiol.* 371:29 (1986).
3. V. Dimitriadou, M. G. Buzzi, M. A. Moskowitz, *et al., Neuroscience* 44:97 (1991).
4. F. Ferrante, A. Ricci, L. Felici, *et al., Acta Histochem. Cytochem.* 23:637 (1990).
5. A. K. Ganguly, P. Gopinath, *Quart. J. Exp. Physiol.* 64:1 (1979).
6. A. K. Ganguly, S. S. Sathiamoorthy, O. P Bhatnagar, *Quart. J. Exp. Physiol.* 63:89 (1978).
7. R. H. Stead, M. Tomioka, G. Quinonez, *et al., Proc. Natl. Acad. Sci. USA* 84:2975 (1987).
8. R. H. Stead, M. F. Dixon, N. H. Bramwell, *et al., Gastroenterology* 97:575 (1989).
9. H. Larsson, E. Carlsson, H. Mattson, *et al., Gastroenterology* 90:391 (1986).
10. F. Sundler, R. Hakanson, E. Carlsson, *et al., Digestion* 35:56 (1986).
11. R. Hakanson, G. Bottcher, F. Sundler, *et al., Digestion* 35:23 (1986).

MECHANISMS OF NK RECOGNITION AND ACTIVATION BASED ON LECTIN-SACCHARIDE INTERACTIONS

M. Pospisil,[1] K. Bezouska,[2] M. Campa,[3] A. Fiserova,[1]
J. Kubrycht,[1] N. Huan,[1] S. Chan,[4] and G. Trinchieri[5]

[1]Department of Immunology, Institute Microbiology,
[2]Institute of Biotechnology, Praha, Czech Republic
[3]Dipartemento di Biomedicina, Universitá di Pisa, Italy
[4]CNRS, Institut de Chimie et de Biologie, Strasbourg, France
[5]The Wistar Institute, Philadelphia, PA, USA

INTRODUCTION

Cell-mediated cytotoxicity is one of the most important effector function of the immune system. Activation of both antigen-specific T lymphocytes (TCR/CD3+) and non-MHC-restricted lymphocytes (predominantly NK cells-TCR/CD3⁻) occurs in response to microorganisms, tumor cells or allografts. Critical events in the cytolytic cascade are the abilities to recognize target antigens and to transmit intracellular signals from the cell surface receptors. The nature of the target structures and NK cell receptors remains unknown.

The endogenous ligand activating the NK cells is largely unknown, although evidence has been accumulating to indicate the importance of altered oligosaccharide structures on the surface of target cells as specific triggering molecules. Simple sugars and oligosacharides are potent inhibitors of NK cytotoxicity[1,2] and β-D-galactose-terminated oligosaccharides (OS) have highly specific binding to NK cells.[3] Recently, several molecules supposed to play a role in NK cell activation have been described, including leukocyte common antigens[4] and C-type lectins.[5] Here we report the identification of sugar ligands, and association with intracellular protein tyrosine kinases for group V of C-type NK cell receptors.

MATERIALS AND METHODS

Cell Preparation

NK and other cells were isolated from pig peripheral blood leukocytes using Ficoll-Verografin centrifugation (Ficoll 400, Pharmacia, Uppsala, Sweden; and Verografin Leciva, Prague, Czech Republic). Mononuclear cells of different phenotypes were separate as described elsewhere.[3] Targets were maintained in complete medium used for lymphocytes, e.g., RPMI 1640 (Inst. Sera and Vaccines, Prague) supplemented with 10% (v/v) heat-inactivated FCS, 2mM L-glutamine and 50 µg/ml gentamycin (GIBCO). A standard 4-h chromium release assay was used for the measurement of cell-mediated cytotoxicity.

Advances in Mucosal Immunology, Edited by
J. Mestecky *et al.*, Plenum Press, New York, 1995

Saccharides

All monosaccharides were from Sigma (St. Louis, USA) and oligosaccharides were prepared from serum and egg-white glycoproteins.[5]

Lectin Preparations

The preparation of porcine spleen integral[2] and peripheral[3] membrane lectins, and the 205 kDa porcine NK lectin[4] have been described. The soluble extra-cellular portion of NKR-P1[6] was expressed in *E. coli* as a HinfI - Bg1II fragment inserted between the EcoRI and HindIII sites of the pINIIIompA3 expression vector. The induction and purification of NKR-P1 protein was performed according to Taylor *et al.*[7]

Binding Studies

Interaction of lectins with carbohydrates was measured by equilibrium dialysis[5], Eastern ligand blotting, or plate assay.[7]

Immunopurification and Immunoblotting

Polyclonal rabbit antiserum against porcine spleen 30 kDa lectin ($3\mu l$)[2] was used for immunopurification of C-type lectins from NK cell and lymphocyte lysates, and analyzed by immunoblotting with primary rabbit antisera against p56[lck] and p59[fyn].[8]

RESULTS AND DISCUSSION

Lectin-like recognition in NK cytotoxicity was first suggested after observing significant inhibition of the process by simple sugars. From results shown on Fig. 1 we could differentiate between systems specific for D-galactose, D-mannose and L-fucose, respectively. Moreover, the participation of acidic sugars in NK cytotoxicity was suggested.[9]

Table 1 represents a summary of our current knowledge on the lectin molecules important in NK cell activation and cytotoxicity. They include various representatives of the sub group V of C-type lectins[5], most notably several representatives of the NK cell receptor multigene family, and porcine 205 kDa NK cell lectin, identified as a member of leukocyte common antigens (CD45-ref. 4). As a part of our preliminary studies aimed at better understanding of the biological relevance of these molecules, we identified the carbohydrate ligands of recombinant NKR-P1 protein, expressed as a 42 kDa disulfide-linked dimer. Ligand blotting and plate assays identified several acidic sugars as the best ligands for these lectins (Fig. 2). Moreover, we identified a specific temporary association of C-type lectins with the key regulatory protein tyrosine kinases p56[lck] and p59[fyn] that are induced upon interation of these molecules with macromolecular or particulate, but not with soluble, carbohydrate ligands (Fig. 3). Similarly, strong NK cell activation was achieved by particular carbohydrate ligands bound to CD45. Furthermore, a peripheral form of the LAK-1 molecule has been identified as the soluble targeting receptor for tumor necrosis factor.[10]

Thus, the present results argue for an important role of NK cell surface lectins in their activation and cytotoxicity. Current experiments are aimed at the identification of cell-surface saccharides as the possible target structures.

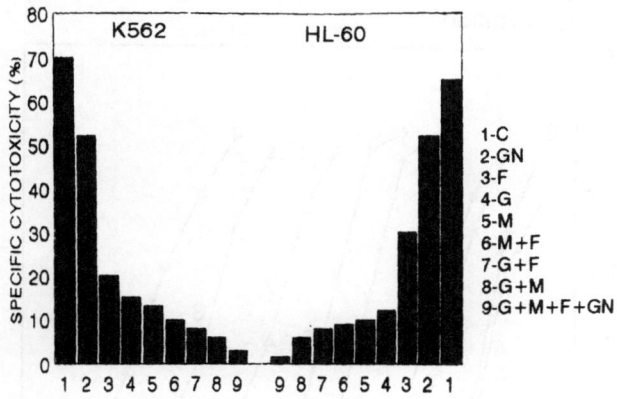

Figure 1. Inhibition of NK cell activity by monosaccharides. Experiments were performed at the effector: target cell ratio, 300 : 1. Abbreviations: C, control experiments, F, L-fucose, G, D-galactose, GN, N-acetyl-D-glucosamine, M, D-mannose.

Table 1. NK cell surface lectins in NK cytotoxicity.

Family	Lectin	Mr(kDa)	Structure	Ligands	Function	Ref.
C-type lectins	Pig spleen	45	Integral extr. CRD	Galactose, Galoligos	Conjugate formation	1
Inte-grins	Mannose lectin	180 +95	Integral membrane	Mannose mann.olig.	Conjugate formation	3
C-type lectine, group v	Pig NK cell	30	Integral extr. CRD	Sulf. carb. Sialoolig.	NK cell activation	1
	NK recep. multig. family	30	Integral extr. CRD	Fucose sulf. carb.	NK cell activation	this work
LCA (CD45)	Pig NK cell	205	Integral phosphatase	Triantenn deglyc.OS	NK cell triggering	4
?	Periph. galactose	65	Peripheral, Man anchor	Galactose Lactose	Opsonization of targets	3
C-type lectins	Periph. galactose	180	Peripheral, soluble	Mannose mann. olig.	Opsonization of targets	5

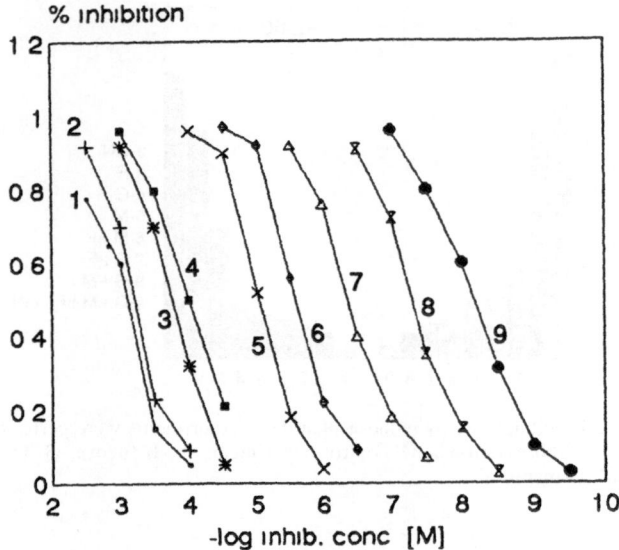

Figure 2. Binding of recombinant NKR-P1 to neutral and acidic carbohydrates determined by the plate assay inhibition method with ^{125}I-Fuc32-BSA as ligand Inhibitors 1, glucose, 2, fucose, 3, glucose-6-phosphate, 4, glucose-6-sulfate, 5, glucose-1-phosphate, 6, fucose-1-phosphate, 7, sialooligosaccharides from orosomucoid, 8, sulfated OS from ovalbumin, 9, fucoidan

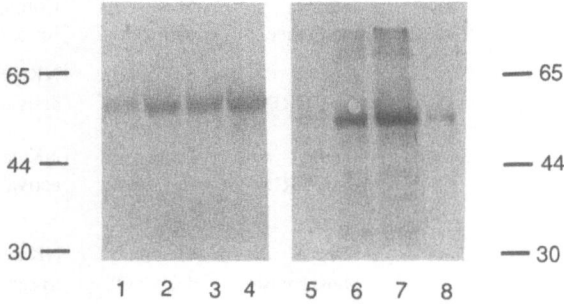

Figure 3. Association of NK receptors with cytoplasmic tyrosine kinase p56[lck] C type NK cell receptors were immunopurified from 10[7] digitonin-lysed human NK cells with immobilized rabbit antiserum against pig spleen lectin, and the associated tyrosine kinase detected by immunoblot with rabbit anti-peptide antiserum against p59[fyn] (lane 1-4) or p56[lck] (lane 5 8), (these antisera were kindly provided by Drs D Davidson and A Veilette, McGill Cancer Center, Montreal, Canada) swine anti-rabbit antibodies conjugated with horseradish peroxidase, and diaminobenzidine - Ni^{2+} Lane 1, 5 stimulation with soluble sialooligosaccharides (4), lane 2,6 stimulation with particle-bound sialooligosaccharides, lane 3, 7 costimulation with soluble sialooligosaccharide and polyclonal anti-lectin antibodies, lane 4, 8 stimulation with the polyclonal antibodies alone Mr of the protein markers is given in kDa

REFERENCES

1. K. Bezouska, M. Pospisil, J. Kubrycht, Z. Holan, D. Lukesova, and J. Kocourek, *Lectin Rev.* 1:81 (1991).
2. M. Pospisil, J. Kubrycht, K. Bezouska, O. Taborsky, M. Novak, and J. Kocourek, *Immunol. Lett.* 12:83 (1986).
3. M. Pospisil, J. Kubrycht, K. Bezouska, O. Taborsky, and J. Kocourek, *Lectins* 6:145 (1988).
4. K. Bezouska, A. Krajhanzl, M. Pospisil, J. Kubrycht, J. Stajner, K. Felsberg, and J. Kocourek, *Eur. J. Biochem.* 213:1303 (1993).
5. K. Bezouska, V. E. Piskarev, G. J. van Dam, J. Kubrycht, M. Pospisil, and J. Kocourek, *Mol. Immunol.* 29:1437 (1992)..
6. R. Giorda, W. A. Rudert, C. Vavasori, W. H. Chambers, J. C. Hiserodt, and M. Trucco, *Science* 249:1298 (1990).
7. M. E. Taylor, K. Bezouska, and K. Drickamer, *J. Biol. Chem.* 267:1719 (1992).
8. K. Toyoshima, Y. Yamanashi, K. Inoue, K. Semba, T. Yamamoto, and T. Akiyama, *Ciba Found. Symp.* 164:240 (1992).
9. W. W. Young, M. Durdik, D. Urdal, S. Hakomori, and C. S. Henney, *J. Immunol.* 136:1 (1981).
10. J. Kubrycht, P. Malikova, A. Fiserova, K. Bezouska, P. Kruzik, N. H. Huan, and M. Pospisil, *Eur. J. Immunol.* (submitted).

DETECTION OF CELLS WITH NK ACTIVITY FROM HISTOLOGICALLY NORMAL MUCOSA IN RELATION TO DISEASE

E.A.F. van Tol, H.W. Verspaget, A.S. Peña, and C. B. H. W. Lamers

Department of Gastroenterology and Hepatology
University Hospital Leiden
P.O. Box 9600, 2300 RC Leiden
The Netherlands

INTRODUCTION

Cells exerting natural killer (NK) activity are sparsely present in the intestinal mucosa and were therefore found to be hardly detectable by assays determining their activity.[1-3] Earlier studies have been rather disappointing because of unsuccessful immunohistologic detection and/or measurement of NK activity in mucosal mononuclear cell preparations obtained by enzymatic isolation procedures from tissue sections.[2,4,5] Identification of these cells with specific monoclonal antibodies (mAb) against NK cell surface markers, however, has greatly improved the assessment of these cells in the intestinal mucosa.

The low distribution of these cells in the intestine still intrigues many researchers to what extent these cells can have functional implications for mucosal immune reactivity. Histological studies in particular describing alterations in the presence of NK cells with respect to intestinal disease are increasing in the literature, supporting the putative involvement of these cells in intestinal disease. The low presence of cells with NK markers in the intestinal lamina propria and their functional activity in both diseased and normal gut mucosa deserves further study using more specific mAb. In the present studies we evaluated the functional contribution of cells with NK surface markers isolated from histologically normal mucosa by depletion studies and their activity in relation to intestinal disease. In most studies reported so far, NK activity of isolated lamina propria mononuclear cells (LPMC) was assessed by introducing pre-culturing conditions, high effector-to-target ratios, and through enrichment of NK cells by density gradient centrifugation, elutriation centrifugation or panning techniques. We assessed the phenotype of NK cells indirectly by complement-mediated lysis using specific mAb recognizing the NK cell surface markers CD16 and CD56. Peripheral blood mononuclear cells (PBMC) exerting NK activity have been found to express almost exclusively both CD16 and CD56 surface markers, but there is a small population of $CD3^+CD16^-CD56^+$ (non-MHC-restricted) T lymphocytes involved in this type of spontaneous cell-mediated cytotoxicity.[6,7]

In general, studies on non-MHC-restricted cytotoxicity or NK activity in LPMC preparations from inflammatory bowel disease (IBD) patients have been discouraging. NK activity levels were found to be absent or low which was accompanied by the inability to detect NK cells immunohistochemically or only at very low proportions.[1,4,8-10] Moreover, studies on mucosal NK activity in IBD frequently dealt with cells isolated from inflamed specimens compared with control LPMC isolated from normal mucosa of patients with carcinoma.

In our studies we used histologically normal intestinal mucosa distant from the involved region obtained from patients with IBD, colorectal carcinoma or a variety of other diseases, including diverticular disease, familial adenomatous polyposis coli, Hirschsprung's disease or volvulus. NK activity was determined in a 4 and 18 h ^{51}Cr-release assay with E/T ratios of 50:1 (PBMC) or 500:1 (LPMC). The sensitivity of the colonic target cell line (Caco-2 colon adenocarcinoma cells) for NK activity by LPMC was also tested in addition to the ubiquitously used standard NK sensitive K-562 (erythroleukemia) cell line.

RESULTS AND DISCUSSION

We demonstrated considerable NK cell activity against both K-562 and Caco-2 target cells by LPMC isolated from histologically normal mucosa of nearly all patients by using the higher E/T ratio. From the functional depletion experiments with anti-CD16 and anti-CD56 it became clear that cells exerting NK activity or non-MHC-restricted cytotoxicity in the intestinal lamina propria are of a different phenotype, i.e. all CD56$^+$ but only partly CD16$^+$, compared to their peripheral blood counterparts which are mainly both CD16$^+$ and CD56$^+$.[6] By functional depletion experiments removing CD16$^+$ or CD56$^+$ LPMC we found almost complete inhibition of the NK activity in the 4 and 18 h assay by removal of the CD56$^+$ cell population, whereas depletion of CD16$^+$ cells resulted in 30-35% reduction of the NK activity in both assays.[11] These data from experiments with LPMC markedly contrast with depletion of NK cells from PBMC where we found depletion of both CD16$^+$ as well as CD56$^+$ cells to almost completely abrogate NK activity in both assays (Table 1). Moreover, the relative contribution of CD16$^+$ or CD56$^+$ cells to NK activity by LPMC was not found to be disease dependent since we found similar percentages of inhibition of NK activity among the various disease groups. CD56$^+$ cell depletion also resulted in abrogation of NK activity by LPMC in the prolonged assay indicating that the lytic activity in this extended assay is also mediated by cells expressing the CD56 adhesion molecule. The importance of intestinal CD56$^+$ cells was previously demonstrated by Shanahan et al.[12], using panning techniques, however, they were not able to assess the relative contribution of CD16$^+$ or CD56$^+$ cells to cytotoxicity because CD16$^+$ cells were found to be absent by immunohistochemistry in their studies.

Table 1. Mean cytotoxicity levels found against the NK sensitive K-562 erythroleukemia target cells before and after depletion of CD16- or CD56-positive cells from mononuclear cell preparations from peripheral blood (PBMC) of healthy individuals or from the intestinal lamina propria (LPMC) from histologically normal mucosa of patients with intestinal disease. Mononuclear cells were also incubated with baby rabbit serum to determine the effect of this low toxic complement source (BRC) on cell activity. [a]n=17, [*] p<0.005, [**] p<0.001.

	PBMC (n=11)		LPMC (n=20)	
Assay	K-562 4h	K-562 18h	K-562 4h	K-562 18h
Control MNC	40±6	67±7	20±3	44±5
CD16 depleted	4±0.4[**]	9±2[**]	12±3[a][*]	29±6[a][**]
CD56 depleted	3±0.8[*]	6±2[**]	2±1[**]	-3±3[**]
MNC and BRC	34±6	60±6	17±3	42±4

Taken together, the CD16 marker seems to be less adequate than the CD56 marker for detecting NK cells in LPMC preparations. The small subset of T cells with CD3$^+$CD16$^-$ CD56$^+$ phenotype described by Lanier et al.[6] which are capable of mediating non-MHC-restricted cytotoxicity in peripheral blood do not seem to contribute to this type of cellular cytotoxicity in the intestinal lamina propria since we found depletion of CD3$^+$ T cells not to change NK activity in the 4 and 18 h assay.

Cytotoxicity levels by LPMC were reported to be low in IBD when using low E/T ratios as shown by Fiocchi et al.[8] and MacDermott et al.[2] while no differences were found between Crohn's disease (CD) and ulcerative colitis (UC) patients. By using a low E/T ratio but a prolonged 24 h assay, Beeken et al.[9] did find significant levels of cytotoxicity in LPMC preparations of both normal and inflamed mucosa from CD patients. By combining the high E/T ratio and a prolonged 24 h assay Gibson et al.[10] could measure significant levels of NK activity in LPMC preparations but they concluded the activity levels to be independent of the underlying disease, drug therapy or anatomical origin of the effector cells.

We observed important differences between CD and UC patients regarding the NK activity levels of LPMC isolated from histologically normal mucosa. Previous studies on NK activity in IBD usually comprised inflamed mucosa, and LPMC from unaffected normal mucosa from carcinoma patients or from patients of a miscellaneous group served as controls. Our data from experiments with LPMC isolated from normal mucosa distant from the tumour region in carcinoma patients, as well as normal mucosa obtained from the miscellaneous disease group indicated that these specimens provide appropriate control LPMC populations to compare with normal IBD mucosa. Intriguingly, we found that histologically normal IBD mucosa concealed alterations in NK activity; i.e. high levels in CD versus low levels in patients with UC (Table 2). These observed abnormalities were strengthened by the findings that NK activity of CD patients or UC patients were also higher or lower, respectively, compared to the control carcinoma patients. It needs to be emphasized here, however, that despite interesting differences of NK activity between the IBD subgroups, the biological relevance of such a sparsely distributed cell in the intestinal mucosa still remains to be clarified, in particular with respect to their role in the pathogenesis of IBD. Nevertheless the observed differences in NK activity between UC and CD patients may support the concept that we are dealing with different disease processes.

Table 2. Mean NK activity of LPMC preparations from histologically normal mucosa of patients with a variety of intestinal diseases including colorectal carcinoma (carc.), inflammatory bowel disease (IBD, CD=Crohn's disease, UC=ulcerative colitis), and a miscellaneous (misc.) group. * p=0.005, ** p=0.001 when comparing NK activity levels between CD and UC patients.

	Total (n=55/54)	Carc. (n=26/25)	IBD (n=17)	Misc. (n=12)	CD (n=9)	UC (n=8)
K-562 4h	13±2	13±2	14±3	13±4	21±4*	7±2
K-562 18h	28±3	28±4	34±6	23±6	49±7**	16±3

The origin of the tissue from which the LPMC have been isolated might contribute to the observed differences of NK activity between CD and UC. However, the following important observations caused us to consider the differences indeed to be disease related, namely *colonic* cells from two patients with Crohn's colitis also revealed markedly higher NK activity levels than those from control colonic tissue of the carcinoma or miscellaneous diseases group, and even much higher activity than those of the UC group, whereas cells isolated from *small* bowel mucosa of an UC patient revealed much lower NK activity than the LPMC isolated from normal ileum of CD patients.

Autologous epithelial cells from the colon can be used as "physiological targets" to study spontaneous or antibody-armed cytotoxicity but colon target cells are not routine in use for peripheral blood or mucosal NK activity studies. Despite the lower susceptibility of Caco-2 cells for NK activity as detected in our studies, we found significant correlations between the activities against both K-562 and Caco-2 target cell lines in both 4 and 18 h assays (0.54<R<0.82; 0.0001<p<0.02). Other groups have also reported colon derived tumor target cells to be sensitive targets for detecting mucosal NK activity,[9,13] which in particular may be of interest in mucosal cytotoxicity studies, because these cells originate from colonic epithelium, a putative target for local NK activity. Hence, our observation as well as those made by others indicate that: a) cells exerting NK activity in the intestinal lamina propria are of a different phenotype compared to their peripheral blood counterparts; b) the mucosal NK cell has a different profile of target cell recognition compared to its peripheral blood counterpart; and c) NK activity may already have undergone important alterations in histologically normal intestinal mucosa adjacent to the inflamed region in IBD patients.

The implications of these findings for the pathogenesis of CD or UC, however, still remain to be elucidated. Future studies should focus on the distribution and migratory patterns of NK cells within the mucosa since there are indications from several studies that the presence of NK cells changes in IBD as well as in coeliac disease.[14-16]

REFERENCES

1. R. P. MacDermott, G. O. Franklin, K. M. Jenkins, I. J. Kodner, G. S. Nash, and I. J. Weinreib, *Gastroenterology* 78: 47 (1980).
2. R. P. MacDermott, M. J. Bragdon, I. J. Kodner, and M. J. Bertovich, *Gastroenterology* 90: 6 (1986).
3. P. R. Gibson, E. L. Dow, W. S. Selby, R. G. Strickland, and D. P. Jewell, *Clin. Exp. Immunol.* 56: 438 (1984).
4. I. Hirata, G. Berrebi, L. L. Austin, D. F. Keren, and W. O. Dobbins,. *Dig. Dis. Sci.* 31: 593 (1986).
5. C. Fiocchi, R. R. Tubbs, and K. R. Youngman, *Gastroenterology* 88: 625 (1985).
6. L. L. Lanier, A. M. Le, C. I. Civin, M. R. Loken, and J. H. Phillips, *J. Immunol.* 136: 4480 (1986).
7. L. L. Lanier and J. H. Phillips, *Immunol. Today* 7: 132 (1986).
8. C. Fiocchi, K. R. Youngman, B. Yen-Lieberman, and R. R. Tubbs, *Dig. Dis. Sci.* 33: 1305 (1988).
9. W. L. Beeken, R. M. Gundel, S. St. Andre-Ukena, and T. McAuliffe, *Cancer* 55: 1024 (1985).
10. P. R. Gibson and D. P. Jewell, *Gastroenterology* 90: 12 (1986)
11. E. A. F. van Tol, H. W. Verspaget, A.S. Peña, C. V. Elzo Kraemer, and C. B. H. W. Lamers, *Eur. J. Immunol.* 22: 23 (1992).
12. F. Shanahan, M. Brogan, and S. Targan, *Gastroenterology* 92: 1951 (1987).
13. J. Taunk, A. I. Roberts, and E. Ebert, *Gastroenterology* 102: 69 (1992).
14. F. Hadziselimovics, L. R. Emmons, and U. Schaub, *Can. J. Gastroenterol.* 4: 303 (1990).
15. J. Haruta, K. Kusugami, and A. Kuroiwa, *Am. J. Gastroenterol.* 87: 448 (1992).
16. F. Hadziselimovics, L. R. Emmons, U. Schaub, E. Signer, A. Bürgin-Wolff, and R. Krstic, *Gut* 33: 767 (1992).

EFFECTS OF MESALAZINE ON LAMINA PROPRIA WHITE CELL FUNCTIONS

Hein W. Verspaget

Department of Gastroenterology and Hepatology,
University Hospital Leiden, Building I C4-P, P.O.
Box 9600, 2300 RC Leiden, The Netherlands

INTRODUCTION

Mesalazine (5-aminosalicylic acid, 5-ASA) is known to be effective in the treatment of inflammatory bowel disease for more than a decade. Originally 5-ASA was developed as the salicylate component, linked to sulfapyridine (SP) through on azo bond, of sulfasalazine (SASP) about 50 years ago. After proving its efficacy in rheumatoid arthritis, it was also used for the treatment of ulcerative colitis and Crohn's disease.[1] The mode of action of SASP and its active moiety 5-ASA, however, has remained obscure up till now. As discussed in the symposium "Mechanism of Action of Mesalazine" during the Mucosal Immunology Meeting in Prague and presented by the speakers in these proceedings, 5-ASA and SASP are able to interfere with cellular functions at multiple levels and in numerous systems. Examples highlighted were intestinal epithelial cell function, cytokine production, leukocyte-endothelium interactions, arachidonic acid metabolism, and reactive oxygen metabolite scavenging. Recently, several reviews appeared on the actions of these compounds on the mentioned and other cellular activities.[2-5] In the present short and selective summary, some aspects of lamina propria white cell functions will be addressed which have gained little attention over the years, i.e., immunoglobulin synthesis, macrophage activity, mast cell function, and cellular cytotoxicity or natural killer cell (NK) activity.

IMMUNOGLOBULIN SYNTHESIS

Intestinal inflammation in ulcerative colitis and Crohn's disease is associated with an increase in the number of lamina propria white cells. Recently, Zaitoun et al.[6] showed in a double blind trial that 5-ASA and corticosteroids significantly reduced the number of these intestinal cells based on computerized morphometric analyses. A major component of the mucosal white cells pool consists of immunoglobulin synthesizing and secreting cells, and the major class of immunoglobulins produced at the intestinal level is IgA. Early reports in rheumatoid arthritis indicated that SASP was able to induce a selective IgA deficiency in some of these patients by an unknown mechanism.[7] A more recent study by MacDermott et al.[8] was fully dedicated to the influence of 5-ASA and SASP on the antibody secretion by mononuclear cells. Pokeweed mitogen-induced IgA, IgG, and IgM secretion by peripheral blood cells was greatly inhibited by SASP and to a lesser extent by 5-ASA, whereas SP had little effect. Similar results were obtained regarding the spontaneous IgA secretion by intestinal mononuclear cells. To elucidate the mechanism of inhibition of IgA secretion, cell viability and proliferation studies were performed which revealed that SASP at higher concentrations ($\geq$ 0.5 mM) had a direct toxic effect as opposed to 5-ASA which neither influenced cell viability nor proliferation. Therefore, they concluded that 5-ASA directly interferes with antibody secretion, whereas SASP exerts its effect by decreasing cell viability and proliferation.

MACROPHAGE AND MAST CELL ACTIVITY

Lymphocytes are not the only type of white cells which account for the increased cell numbers in the mucosa of patients with inflammatory bowel disease. Phagocytes, like macrophages and neutrophils, are also prominent cell types in the inflamed tissue. One of the mechanisms by which these chemotactic cells are attracted to the lamina propria is by the increased local production of inflammatory mediators.[2,4] The most potent 5-lipoxygenase metabolite known to induce chemotaxis is leukotriene B4 (LTB$_4$). The influence of 5-ASA containing drugs on the LTB$_4$ induced chemotaxis of purified intestinal macrophages *in vitro* was analysed by Nielsen *et al.*[9] Although both SASP and olsalazine (azo-disalicylic acid, ADS) were able to inhibit the LTB4 induced macrophage chemotaxis, 5-ASA was found to be the most potent inhibitory drug, whereas acetylated-5-ASA had no effect at all.

Mast cells are able to produce many inflammatory mediators themselves and, though present in low numbers in the normal intestine, can be impressively increased in inflammatory bowel disease. Stimulation of isolated intestinal mast cells with anti-IgE-induced histamine release and prostaglandin D$_2$ production was dose-dependently inhibited by 5-ASA in a manner similar to that of peripheral blood basophils.[10] In contrast, SASP induced release of histamine by these intestinal mast cells in dose-dependent fashion which might explain the allergic reaction sometimes observed after treatment with this drug in some patients. This histamine release could be initiated by the direct toxic effect of SASP on the cells as reported by MacDermott *et al.*,[8] and Gibson and Jewell.[11]

NATURAL KILLER CELL ACTIVITY

Cytotoxicity of drugs is a completely different phenomenon from spontaneous cell mediated cytotoxicity. This latter function is primarily exerted by a specific cell type, sparsely present in both blood and intestine, the NK cell. Its role in the pathogenesis of inflammatory bowel disease is unknown, but nevertheless a few studies have been performed on the effects of 5-ASA drugs on the function of these cells. Peripheral blood NK cells were found to be dose-dependently inhibited in their *in vitro* cytotoxic activity by ADS and SP, but most prominently by SASP.[11] In contrast, 5-ASA did not affect NK activity at all. Although SASP caused cell death of up to 50% of the mononuclear cells by a direct drug toxicity, preincubation experiments showed that the inhibition of NK activity was readily reversible. This obervation was confirmed by Aparicio *et al.*[12] showing complete recovery of NK suppression after SASP treatment *in vivo*. The inhibition of NK activity by SASP was also observed with intestinal NK cells. MacDermott *et al.*[13] found not only spontaneous but also the interleukin-2 (IL-2)-induced cellular cytotoxicity to be decreased by SASP. In this system 5-ASA and SP were not found to affect intestinal NK activity neither spontaneous nor IL-2 induced. Recent studies, however, indicated that 5-ASA may not influence the response to IL-2, but it markedly inhibits the production of IL-2 by lamina propria mononuclear cells.[14]

CONCLUSIONS

Mesalazine influences multiple lamina propria white cell functions, particularly regarding their activity in the context of the immune response. The interference by 5-ASA on these immunoregulatory functions is, however, by no means specific for this drug, i.e., SASP and ADS, as well as corticosteroids,[6,13,14] indomethacin,[8,11] and cyclosporin,[14] can have similar effects, with some exceptions, e.g., histamine release by intestinal mast cells. The 50% inhibitory dose of 5-ASA and SASP is usually within the 0.1 to 2.0 mM range (table) and the direct toxic effect of these drugs, expecially SASP, at low dose is readily reversible.[11,12,15]

Table 1. Fifty-percent inhibitory dose (ID_{50}) of 5-aminosalicylic acid containing drugs on mononuclear cell function.

Cell function	ID_{50} (mM)		
	SASP	5-ASA	ADS
IgA synthesis[8]	0.1 - 0.2	1.4 - 2.0	n.d.
Macrophage chemotaxis[9]	0.4	0.2	0.4
Mast cell mediator release[10]	+	0.5	n.d.
Natural killer cell activity[11,13]	0.7 - 1.0	-	4.0

SASP = Sulfasalazine; 5-ASA = mesalazine; ADS = olsalazine
n.d. = not done; ± = increase; - = outside range.

The mechanisms by which these drugs interfere with cellular activity still remains unknown. Recent studies speculate that this can be on the level of transcription and translation,[16] intracellular transport of molecules,[10,17,18] and cytokine production and receptor expression.[19,20]

REFERENCES

1. Sulphasalazine, Drugs 32, suppl 1 (1986).
2. D. J. Fretland, S. W. Djuric, and T. S. Gaginella, *Prostagland. Leuk. Essent. Fatty Acids* 41:215 (1990).
3. A. Ireland and D. P. Jewell, *Clin. Sci.* 78:119 (1990).
4. I. Ahnfelt-Rønne, *Dan. Med. Bull.* 38:291 (1991).
5. T. S. Gaginella and R. E. Walsh, *Dig. Dis. Sci.* 37:801 (1992).
6. A. M. Zaitoun, I. Cobden, H. Al Mardini, and C. O. Record, *Gut* 32:183 (1991).
7. J. P. Delamere, M. Farr, and K. A. Grindulis, *Brit. Med. J.* 286:1547 (1983).
8. R. P. MacDermott, S. R. Schloemann, M. J. Bertovich, G. S. Nash, M. Peters, and W. F. Stenson, *Gastroenterology* 96:442 (1989).
9. O. H. Nielsen, H. W. Verspaget, and J. Elmgreen, *Aliment. Pharmacol. Therap.* 2:203 (1988).
10. C. C. Fox, W. C. Moore, and L. M. Lichtenstein, *Dig. Dis. Sci.* 2:36 (1991).
11. P. R. Gibson and D. P. Jewell, *Clin. Sci.* 69:177 (1985).
12. M. N. Aparicio-Pagés, H. W. Verspaget, J. C. M. Hafkenscheid, G. E. Crama-Bohbouth, A. S. Peña, I. T. Weterman, and C. B. H. W. Lamers, *Gut* 31:1030 (1990).
13. R. P. MacDermott, M. G. Kane, L. L. Steele, and W. F. Stenson, *Immunopharmacology* II:101 (1986).
14. W. E. Pullman and W. F. Doe, *Clin. Exp. Immunol.* 88:132 (1992).

15. H. W. Verspaget, M. N. Aparicio-Pagés, S. Verver, P. M. Edelbroek, J. C. M. Hafkenscheid, G. E. Crama-Bohbouth, A. S. Peña, I. T. Weterman, and C. B. H. W. Lamers, *Scand. J. Gastroenterol.* 26:779 (1991).
16. Y. R. Mahida, C. E. D. Lamming, A. Gallagher, A. B. Hawthorne, and C. J. Hawkey, *Gut* 32:50 (1991).
17. B. Crotty, P. Hoang, H. R. Dalton, and D. P. Jewell, *Gut* 33:59 (1991).
18. J. H. Wandall, *Pharmacol. Therap.* 5:609 (1991).
19. W. E. Pullman, S. Elsbury, M. Kobayashi, A. J. Hapel, and W. F. Doe, *Gastroenterology* 102:529 (1992).
20. S. Schreiber, R. P. MacDermott, A. Raedler, R. Pinnau, M.J. Bertovich, and G.S. *Gastroenterology* 101:1020 (1991).

THE SIGNIFICANCE OF CULTIVATING CELLS AND HEMOPOIETIC TISSUE FROM TUNICATES

Edwin L. Cooper[1] and David A. Raftos[2]

[1]Department of Anatomy and Cell Biology
UCLA School of Medicine
University of California at Los Angeles
10833 Le Conte Avenue
Los Angeles, CA 90024-1763

[2]School of Biological and Biomedical Sciences
University of Technology, Sydney
P.O. Box 123
Broadway NSW 2007
Australia

INTRODUCTION

The Solitary Tunicate Model: *In vivo* evidence for Immunocompetent Cells

Solitary tunicates are excellent models for unraveling the intricacies of the immune system[1] and where its development might fit into immunoevolution.[2] At the cellular level, they possess numerous leukocytes that have been identified by electron microscopy and classified.[3,4] Moreover, putative lymphocyte-like phenotypes which may emerge as homologs of vertebrate equivalents have also been observed, including Thy1[5,6] and Lyt 1,2.[7,8,9] Lymphocyte-like cells (LLCs) proliferate in response to allogeneic stimuli and after antigenic challenges *in vivo* and *in vitro,* hemocytes revealed significantly greater proliferative activity after allogeneic immunization than autogeneically primed and naive recipients.[10] Proliferative responses derived from *in vivo* activity of LLCs prompted the development of *in vitro* methods[11] and confirmation (Sawada, Zhang and Cooper, in preparation) with wide application to problems of cellular communication by means of putative cytokine-like molecules.[12-17]

THE *IN VITRO* ASSAY AND CONFIRMATION OF VIABILITY AND MULTIPLICATION

The pharynx, an epithelial-mesenchymal organ, represents an early mucosal site in evolution, and these results represent the first time that the main source of hemopoiesis has been cultivated. Explants of pharyngeal tissue from the tunicate *Styela clava* are amenable to culture, since survival has been observed for up to 72 days *in vitro*. Tissue morphology remained unaltered except for the migration of hemocytes from explants into culture supernatants. Since the pharynx is composed of three basic components, ciliary activity and muscular contraction were observed for extended periods, and hemocytes which lodge in the sinuses incorporated the thymidine analog, BrDu. This was in marked contrast to the absence of specific BrDu uptake in controls in which irradiation curtailed DNA replication or in which BrDu assimilation was competitively inhibited by using thymidine. These results strongly support the view that BrDu incorporation represents active DNA replication associated with cell division. Proliferation in pharyngeal sinuses contributes substantially to the pool of hemocytes in culture and seems to mimic the *in vivo* situation where there is rapid turnover.[18] *In vivo*, circulatory hemocytes survive for only several weeks[19] and we found it difficult to cultivate them, *in vitro* since their viability declines rapidly. Yet the hemocyte pool of pharyngeal cultures expands, despite this short life span. During 61 days of culture, approximately 3.5×10^6 cells migrated into supernatants while serial reconstructions suggest that hemocytes' numbers within explants declined by only 1.6×10^5 cells (40%). Such a disparity indicates that proliferation renews the hemocyte pool in pharyngeal cultures. To substantiate renewal capacity, we irradiated cultures to inhibit cell division and found that the afflux of hemocytes from explants was inhibited as revealed by decreased cell proliferation. At doses of 5000 rads, proliferation ceased and there was a decline (80%) of hemocytes entering the culture supernatants. Since hemocyte viability was unaffected by irradiation, we concluded that live cells were not the source of this depressed migration.

EVIDENCE OF PROLIFERATIVE ACTIVITY

Analyses by microscopy and autoradiography have confirmed that the pharynx is a major hematopoietic site in tunicates.[18,19] Therefore, proliferative activity, which we have demonstrated, is consistent with hematopoiesis in the pharynx. *In vivo*, lymphocyte-like cells and granular amoebocytes divide within proliferative nodes of pharyngeal sinuses and other somatic tissues.[10,19] Stem cells divide and then undergo differentiation, producing morphologically diverse, circulatory hemocytes. In our view, all hemocytes in *Styela clava* may represent varied differentiation states of a single cell lineage, i.e., from lymphocyte-like cells (LLCs) the progenitors of this lineage.[20] This is supported by the view that LLCs differentiate rapidly, producing all other forms of hemocytes. This differentiation process, as with proliferative activity, is apparently restricted primarily to the hematopoietic nodes of somatic organs. According to Ermak[18] and Wright,[20] most hemocytes *in vivo* enter the circulating hemolymph as end products of a regulated differentiation pathway. We have observed a similar process of maturation occurring *in vitro* as measured by proliferative activity in cultured explants restricted to LLCs and granular amoebocytes. Moreover, we believe that the proportion of precursor LLCs contributing to the proliferative population increases over time. This strengthens the view that they function as primary stem cells. However, all hemocyte types, including non-proliferative phagocytic amoebocytes and vesiculated hemocytes, were observed in culture supernatants, and their frequencies remained constant for 44 days. In summary, this maintenance of a diverse hemocyte population derived from only two proliferative precursor types is indicative of regulated differentiation similar to that evident *in vivo*.

DIFFERENTIATION OF HEMOCYTES FROM HEMOPOIETIC SITES

There may be several explanations for the lack of vitality in circulatory hemocytes. First, the differentiation state of hemocytes migrating from fixed hematopoietic sites may explain the failure of circulatory hemocytes to survive in culture. Cell suspensions from the hemolymph remain viable for only short periods. Since circulatory hemocytes are incapable of proliferation *in vitro*, according to one possible explanation, circulatory cells may be committed to advanced differentiation states. Moreover, hematopoiesis could be restricted totally to somatic tissues; thus only committed cells which are incapable of further division are released into the circulation. However, this interpretation conflicts with previous analyses that identified at least limited proliferation among circulatory cells.[19] Second, an alternative explanation proposes that absence of proliferation when hemocytes are cultured, is due to inappropriate culture conditions. Although the medium was the same as that which maintained viability and proliferation in pharyngeal explants, additional factors, both physical and chemical, may be required for the successful propagation of single cell suspensions, or such suspensions may require purification and enrichment of those cells capable of division. Another point considers the substratum provided by tissue culture plates, which may not have been amenable to division. Although a variety of seeding densities were tested, appropriate cellular interactions could also have been absent. For this reason, we have begun cultivating hemocyte pellets formed by centrifuging hemocyte suspensions (Sawada and Cooper, in preparation). Moreover, stimulatory factors that may be produced by explanted tissue would not have been available to circulatory hemocytes.

There is evidence that cytokine-like molecules are active in invertebrates.[12-17] For example, a protein from echinoderms induces proliferation of mammalian thymocytes and fibroblasts,[12] and tunicates possess molecules that stimulate the division of mammalian thymocytes and cytotoxic T-cells similar to that of interleukin- 1.[13] Moreover, we showed that tunicate plasma was essential for maintaining proliferation *in vitro*, and it declined rapidly in media that excluded plasma even though nutrients were supplied to explants by adding RPMI-1640. Serum supplements as used in mammalian tissue cultures may be similar, essential as factors in tunicate plasma which facilitate cell division. We have been investigating this hypothesis by testing the effect of vertebrate cytokines and mitogens on tunicate hemocyte proliferation.

EVIDENCE FOR CYTOKINE-LIKE ACTIVITY

Cell Proliferation Stimulated by an Interleukin 1-like Molecule and Its Release from Hemocytes Stimulated with Zymosan

According to Raftos *et al.,*[14] little is known about the regulation or control of growth, differentiation and function of LLCs and granular amoebocytes that comprise the pharynx. In recent experiments, a tunicate interleukin 1 (IL-1)-like fraction has been shown to stimulate the proliferation of cells *in vitro*. This fraction, designated tunicate IL-1β, was isolated from tunicate hemolymph by gel filtration and chromatofocusing chromatography. Mitogenic responses to tunicate IL-1β were dose dependent and could be eliminated rapidly by removing tunicate IL-1β from the culture medium. In addition, a second tunicate hemolymph fraction exerted no effect on tunicate cell proliferation even though it exhibited IL-1-like activity in a mouse thymocyte proliferation assay. The mitogen, phytohemagglutinin, did not act synergistically with either of the fractions. Conditioned media and cell extracts from tunicate hemocytes that had been cultured with various antigens were tested for interleukin 1 (IL)-like activity.[16] Results revealed that media conditioned by hemocytes that had been stimulated with zymosan significantly increased the proliferative and phagocytic activities of tunicate hemocytes. As indicated by SDS-PAGE, proliferation and phagocytosis were associated with the adaptive release of tunicate IL 1-like (tunIL1) molecules by stimulated hemocytes. From these results, we suggest that tunIL1 molecules are expressed in response to selected antigenic stimuli. Since there is controversy regarding mechanisms, we propose that cytokine-like activity involves mechanisms that form the basis for non-clonal, adaptive immune responses among phylogenetically primitive animals.

Interleukin-2 and Phytohaemagglutinin Stimulate the Proliferation of Tunicate Cells

To extend assays concerned with interleukin-like molecules, especially another, IL-2, proliferative responses of cells in tunicate pharyngeal explants to human interleukins and mitogenic lectins have been tested.[15] Among pharyngeal cells incubated with recombinant human interleukin-2 (IL-2), and phytohaemagglutinin-P (PHA-P), increased tritiated-thymidine ([^{3}H]-TdR) uptake has been detected. Responses to IL-2 were dose-dependent and LLCs were sensitive. Furthermore, enhanced proliferation was stimulated by IL-2 without co-stimulants and was not synergized by co-incubation with human interleukin-1 (IL-1) or PHA-P. In addition, anti-IL-2 polyclonal antibody was found to inhibit the stimulatory activity of recombinant human interleukin-2 (rhIL-2). Only PHA-P proved to be mitogenic when compared with the effects of two other lectins (concanavalin-A [Con-A], pokeweed mitogen [PWM] and PHA-P). In contrast, Con-A and PWM did not significantly increase proliferative activity even though both lectins were able to bind pharyngeal cells when observed by flow cytometry and fluorescence microscopy. Finally, human IL-1 had no effect on [^{3}H]-TdR uptake either alone or in combination with IL-2 and PHA-P. In summary, the functions of certain interleukin-like cytokines have been conserved during evolution, and it seems that the pharynx may serve as an excellent source of their synthesis by cells which are contained within it.

PERSPECTIVES

There is a need for invertebrate immunologists to actively develop *in vitro* assays, not only of whole organs that are clearly hemopoietic sites, the source of stem cells or progenitors but suspensions of cell types. Moreover, we should improve the existing *in vitro* technology to allow the successful cultivation and demonstrated replication of certain cell types within the entire hemocyte population through methods of purification and enrichment. One promising approach has been to provide conditions in which circulatory hemocytes are forced together in a compact situation that may mimic their arrangement within the whole organ. In this way we have created aggregates of circulatory hemocytes by centrifuging them and cultivating the resulting pellets (Sawada and Cooper, in preparation). Promising results from either of these approaches must then be coupled with the use of specific antisera raised against purified cell fractions with the aim of identifying specific antigens. Only then can we move toward isolation of membrane components that can be subjected to biochemical and later molecular analyses.[21]

ACKNOWLEDGEMENTS

This research was partially supported by grants from the National Science Foundation (DCB 8519848 and 9005061). David A. Raftos was a Fulbright Postdoctoral Fellow and recipient of a Frederick B. Bang Scholarship in Marine Invertebrate Immunology, administered by the American Association of Immunologists.

REFERENCES

1. E. L. Cooper, D. A. Raftos, and K. L. Kelly, *Boll. Zool.* 59:175 (1992).
2. E. L. Cooper, *Boll. Zool.* 59:119 (1992).
3. H. Zhang, T. Sawada, E. L. Cooper, and S. Tomonaga, *Zoo. Sci.* 9:551 (1992).
4. T. Sawada, J. Zhang, and E. L. Cooper, *Biol. Bull.* In Press (1992).
5. M. H. Mansour, R. DeLange, and E. L. Cooper, *J. Biol. Chem.* 260:2681 (1985).
6. M. H. Mansour, and E. L. Cooper, *Eur. J. Immunol.* 14:1031 (1984).
7. H. I. Negm, M. H. Mansour, and E. L. Cooper, *Biol. Cell* 72:249 (1991a).
8. H. I. Negm, M. H. Mansour, and E. L. Cooper, *Comp. Biochem. Physiol.* 99B:741 (1991b).
9. H. I. Negm, M. H. Mansour, and E. L. Cooper, *Comp. Biochem. Physiol.* 101B:55 (1992).
10. D. A. Raftos and E. L. Cooper, *J. Exp. Zool.* 260:391 (1991).
11. D. A. Raftos, D. L. Stillman, and E. L. Cooper, *In Vitro* 26:962 (1990).
12. G. Beck and G. S. Habicht, *Proc. Natl. Acad. Sci.* 83:7429 (1986).

13. G. Beck, G. R. Vasta, J. J. Marchalonis, G. S. Habicht, *Comp. Biochem. Physiol.* 92B:93 (1989).

14. D. A. Raftos, E. L. Cooper, G. S. Habicht, and G. Beck, *Proc. Natl. Acad. Sci. USA* 88:9518 (1991).

15. D. A. Raftos, D. L. Stillman, and E. L. Cooper, *Immunol.* and *Cell Biol.* 69:225 (1991).

16. D. A. Raftos, E. L. Cooper, D. L. Stillman, G. S. Habicht, and G. Beck, *Lymphokine* and *Cytokine Research* (in press) (1992).

17. G. Beck and G. S. Habicht, *Mol. Immunol.* 28:577 (1991).

18. T. H. Ermak, *Amer. Zool.* 22:795 (1982).

19. T. H. Ermak, *Experientia* 31:837 (1975).

20. R. K. Wright, *in* "Invertebrate Blood Cells" N.A. Ratcliffe, A.F. Rowley, eds., p 565, Academic Press, London, (1981).

21. E. L. Cooper, B. Rinkevich, G. Uhlenbruck, and P. Valembois, *Scand. J. Immunol.* 35:247 (1992).

PERIENTERAL CHLORAGOGEN TISSUE AND ITS ROLE IN DEFENSE IN LUMBRICID WORMS

Petr Sima,[1] Martin Bilej,[1] and Jaroslav Slipka[2]

[1]Department of Immunology Institute of Microbiology
Czech Academy of Sciences Prague, Czech Republic
[2]Faculty of Medicine, Charles University, Pilsen,
Czech Republic

INTRODUCTION

Annelids are able to express a type of adaptive immunity. The main work in that respect has been done on lumbricids in which the perienteral chloragogenic tissue seems to be the only precursor organ that produces free defense cells and humoral defense factors. In this work we demonstrate morphologically a perivascular localization of phagocytozed radioactively-labeled particles inside the typhlosole, an enteric organ considered as an analog of the gut-associated lymphoid structures of vertebrates. The inflammation-like reactions in the typhlosole after challenge by endotoxin are documented histologically.

The annelids are continuously exposed to antigens in a similar way to vertebrates. They do not possess an immunomorphological structure comparable to that of vertebrates; nevertheless, they are able to react to antigenic substances in a similar way. Their defense reactions are based on self-nonself recognition followed by an elimination of antigens by means of humoral lytic molecules of coelomic fluid[3,4], phagocytosis[2], encapsulation, and expulsion of foreign particles from the body via dorsal pores. Moreover, these animals are able to reject allo- and xenografts with elements demonstrating immunological memory. All these immunological phenomena are mediated by free coelomocytes which originate from epithelial linings of the coelomic cavity.[1] The aim of our study was to morphologically demonstrate the defense function of predilected immunocompetent structures (epithelia of the coelom, chloragogenic perienteral tissue, and the typhlosole).

PHAGOCYTOSIS

Recently a series of papers has been published dealing with phagocytosis by free coelomocytes. We focused on the capture of methacrylate co-polymer injected into the coelomic cavity. Fig. 1 shows the distribution of labeled particles particularly in somatic and enteric epithelia of the coelom, inside the nephridia, within chloragogenic tissue adjacent to the dorsal vein, and in the center of the typhlosole. The elimination of the phagocytozed superfluous material within the free coelomic cells through the dorsal pores is seen on Fig. 2.

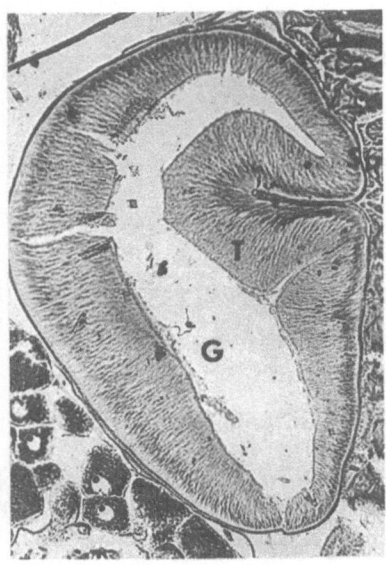

Figure 1. Distribution of ^{125}I-labeled particles inside the coelom and typhlosole. T-typhlosole; G-Gut; C-coelom

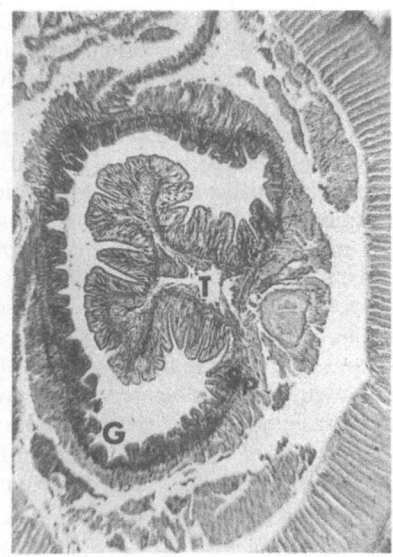

Fiugre 2. Elimination of ^{125}I-labeled particles through the dorsal pore. D.V. - dorsal vein; D. P. - dorsal pore

POIETIC FUNCTION

It is generally accepted that the prime source of coelomocytes in earthworms is the lining of the coelomic cavity. The relatively low cell renewal in the coelomic epithelia can be increased by antigenic stimulation. The results are detailed in our previous paper[1] and in another article in these proceedings.

INFLAMMATION-LIKE RESPONSES

We observed inflammatory-like changes 3 days after intra-coelomic injection of endotoxin (lipopolysaccharide) Comparison between stimulated and unstimulated animals is seen in Fig 3 a and b The inflammatory-like changes occur predominantly in the typhlosole and perienteral chloragogenic tissue, whereas the somatic coelomic lining remains inactive

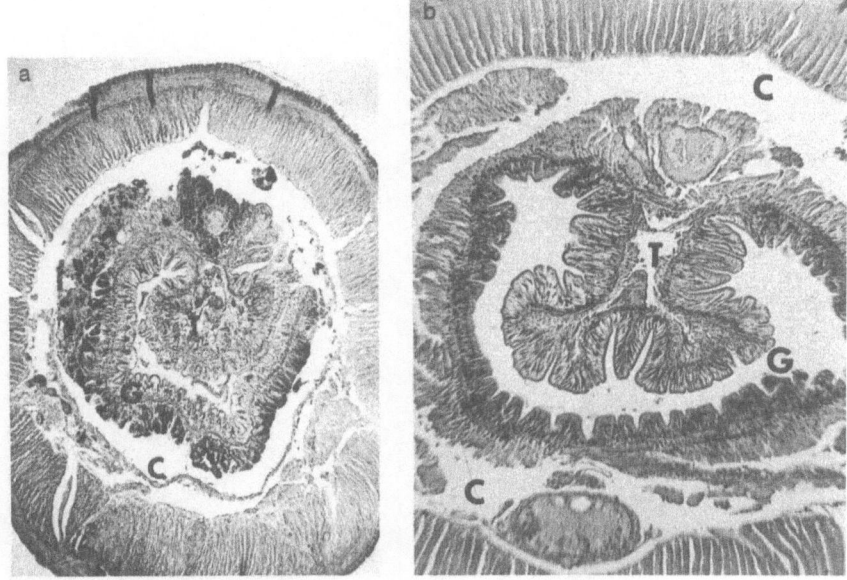

Figure 3a. Inflammatory-like changes in the gut and typhlosole after endotoxin stimulation T typhlosole, G - gut, C- coelom, **3b.** Unstimulated animal intact tissue T - typhlosole, G - gut, C coelom

REFERENCES

1   M Bilej, P Sima, and J Slipka, *Immunol Lett* 32 181 (1992)
2   E Stein, R R Avtalion, and E L Cooper, *J Morph* 153 467 (1977)
3   L Tuckova, J Rejnek, P Sima, and R Ondrejova, *Develop Compar Immunol* 10 181 (1986)
4   P Valembois, P Roch, and M Lassegues, *in* 'Immunity in Invertebrates", M Brehelin, ed , Springer-Verlag, Berlin (1986)

References

THE SPLEEN AND ITS COELOMIC AND ENTERIC HISTORY

P. Sima,[1] and J. Slipka[2]

[1]Department of Immunology, Institute of Microbiology
Czech Academy of Sciences, Prague, Czech Republic
[2]Institute of Histology and Embryology, Faculty of
Medicine, Charles University, Pilsen, Czech Republic

The spleen is a typical vertebrate organ whose homolog is said to be found in cyclostomes and chondrichthian in the form of a spiral fold. The typhlosole of some invertebrates is considered to be its analog. Here, the authors describe a typhlosole-like structure also in the cephalochordates. Both the spiral fold and typhlosole are derived from the lining of the coelomic cavity and appear as a fold of the intestinal wall with invaded perienteric coelomic tissue of splanchnopleuric origin. Similarly the spleen develops in the close vicinity of the digestive tube in the dorsal mesentery. Its ability to phagocytose resembles the function of the typhlosole of annelids. The hemopoietic function is secondary, and it evolved gradually in close association with other important immune organs, especially the thymus.

The spleen may be regarded as a mysterious organ from other points of view. For many years there have been doubts about its significance because it has been more or less accepted that after splenectomy, or in case of congenital asplenia, the immune potential of an organism is not substantially affected[1], even if the spleen is one place where the interaction of immunocytes with antigens occurs. The same is true in the case of evolutionarily lower vertebrates such as sharks, where the immune and hemopoietic functions after splenectomy are promptly transferred to other lymphoid organs, predominantly the Leydig's organ.[2]

The spleen is a typical organ of vertebrates and fulfills two main functions: hemopoietic and immune. Contrary to lymph nodes, the spleen is situated directly in the blood circulation and has a major role as a blood filter. This is done by destroying worn-out blood corpuscles and foreign materials which may appear in the blood. The direct relation of the spleen to the coelomic epithelium as well as to the alimentary tract suggests its importance in gut immunity.

Elucidation of the functional dichotomy united in such a distinct organ requires an approach from various angles. We have tried, therefore, to find a partial explanation of the "spleen dilemma" by comparing the origin of the spleen in ontogeny and phylogeny.

The spleen primordium appears in human ontogeny at the end of the fourth week as a swelling of the coelomic epithelium which covers the dorsal mesentery (for reviews see 3,4). The mesenchyme of this gut suspension becomes gradually vascularized. At the end of the second month, the primordium separates and remains connected with the mesentery only by its ligaments. Hemopoiesis reaches its peak at the fifth month, but this activity declines and finally stops at the sixth month. At the seventh month the first signs of Malpighian corpuscles appear. Full differentiation into white and red pulp appears after delivery.[5,6]

Similar development occurs in other vertebrates but spleen hemopoiesis starts in various animals at different times.[7] In species in which the gestation period is short, hemopoiesis usually continues also in the postnatal period. In ontogeny, the spleen develops primarily as a hemopoietic organ which, contrary to the thymus, acquires its immune function much later and not before first contact with environmental antigenic stimuli.[2]

In the evolutionary line, the first spleen emerges in chondrichthian, the sharks, as an elongated, anatomically highly organized organ situated in the dorsal mesentery, and supplied by arteria lienalis which brings blood from the aorta next to the arteria mesenterica superior, in the shallow groove of the spiral fold.[9] In these animals, teleostean fish, amphibians, and other tetrapods immune activity in the spleen is initiated much later than the thymus.

In the evolutionarily lowest vertebrates, the jawless agnathans, there is a spiral fold in the mid-gut wall consisting of lymphoid tissue around the submucous vessels. This structure corresponds functionally to the spleen and bone marrow of higher vertebrates (for review see 10).

We could not find any previous references to a similar organ in the vast literature on acranians.[11] Therefore we studied *Amphioxus* in serial sections from the Oxford embryonic collections. In transverse sections of this young cephalochordate, the pharyngeal cavity has branchial clefts on the sides. Next to it can be seen a diverticulum of the mid-gut (often described as the hepatic diverticulum) as a blind evagination of the mid-gut turned to the anterior direction.

This picture is well known from all the textbooks, but in our sections we could discern an interesting typhlosole-like invagination (T) of the gut wall on the medial side (Fig. 1). The surrounding splanchnopleura (Sp) becomes invaginated in this fold and attaches the blind gut to the pharyngeal structures. By this arrangement, contact of the external antigenic environment with this mesentery and gut wall is realized.

There is no doubt that the invaginated fold structure of *Amphioxus* may be analogous to a morphologically and developmentally similar structure in ammocoetes, and to the spiral valve of adult agnathans and selachians. Recently it has been shown that the typhlosole and spiral valve could be functionally similar, in that they play the same role as organs of immunity, and fulfill the function of both important immune organs of endothermic tetrapods, the spleen and bone marrow.[12]

If homology is accepted between the amphioxus typhlosole-like gut structure and the spiral valve of agnathans, some chondrichthians, and primitive neopterygian fish[13,14], the question arises whether such an organ appears in representatives of some invertebrate taxa. Among annelids, a phylum which forms an evolutionarily important and very successful animal assemblage, such a structure in the mid-gut exists. The annelid typhlosole is seen as a dorsal invagination of chloragogenic tissue into the gut. This tissue is located both in the center of the typhlosole and around the dorsal vessel forming direct contact with the coelomic cavity (Fig. 2). Recent reports have documented the importance of this organ in the annelid internal defense, especially in the selective localization, capturing and destruction of antigenic substances.[15] Unfortunately, and in contrast to the above-mentioned morphofunctional homology of the typhlosole, spiral valve and spleen within the chordate-vertebrate lineage, it is impossible to formulate any idea concerning the evolutionary kinship of the annelid typhlosole with these organs.

From our observations as well as from studies of our predecessors, it can be concluded that the spleen has a long evolutionary history (Fig. 3) which begins in parallel with the emergence of vertebrates. Undoubtedly, the structural and functional history of the spleen were originally different, and have changed during the ages. It began as a structure with phagocytic activity, then it became an organ of hemopoiesis and a cemetery of blood cells, and finally it acquired the function of an immune organ. Hence, even if the spleen history equals approximately that of the thymus, its immune history is quite modern and can be compared to the late appearance of lymph nodes. Thus the phylogeny of the spleen is a recapitulation of its ontogeny, and it acquires its immune function late, not metamorphosis or delivery.

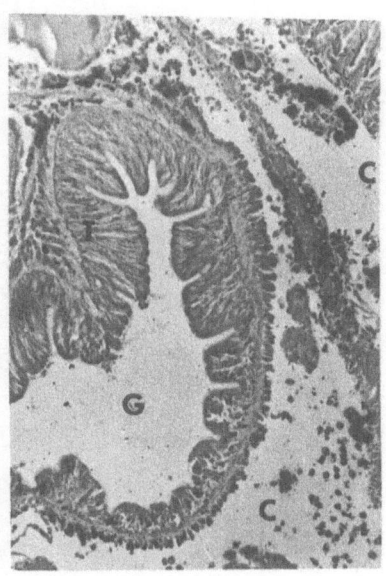

Figure 1. *Amphioxus* t typhlosole like invagination, Sp splanchnopleura, G gut

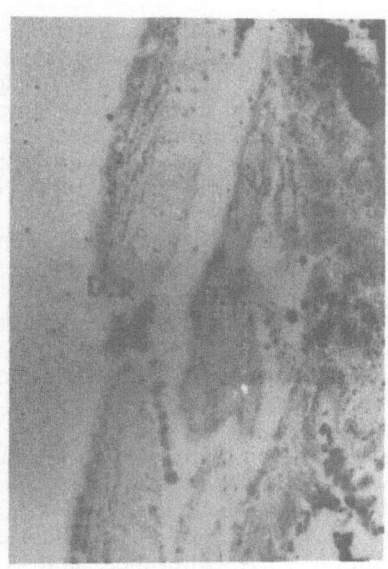

Figure 2. *Lumbricus* T - typhlosole, Sp - splanchnopleura, G gut

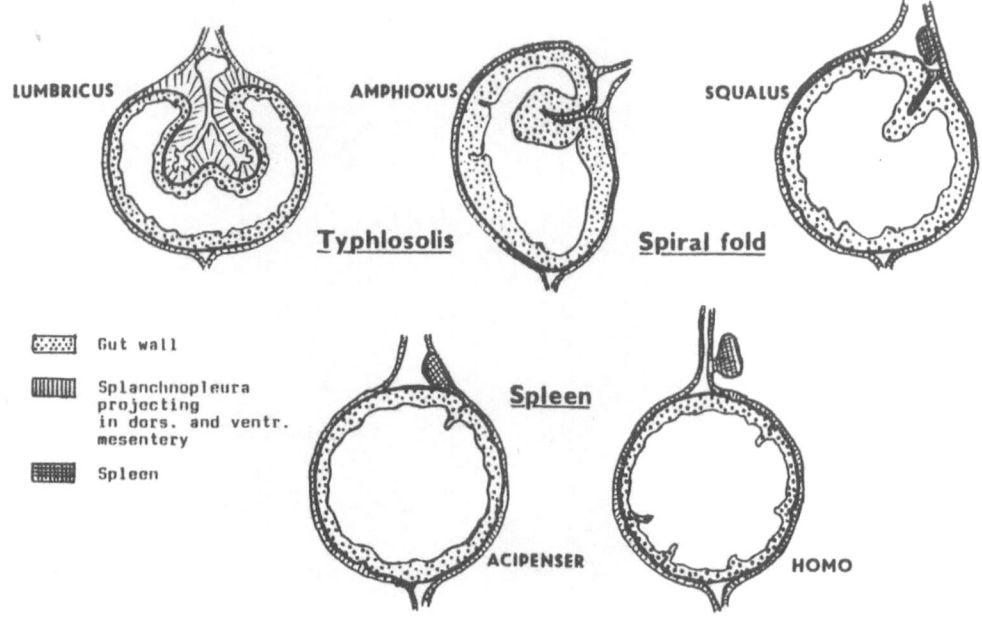

Figure 3. Evolution of the spleen

REFERENCES

1. S. W. Gray and J. E. Skandalis, *in*: "Embryology for Surgeons" Saunders Co., Philadelphia (1972).
2. R. Fange and M. L. Johansson-Sjobeck, *Comp. Biochem. Physiol.* 52A:577 (1975).
3. F. Tischenforf, *in*: "Handbuch der mikroskopischen Anatomie des Menschen", W. Mollendorf and W. V. Bargmann, eds., Springer Verlag, Berlin (1969).
4. O. F. Kampmeier, *in*: "Evolution and Comparative Morphology of the Lymphatic System", Charles C. Thomas, ed, Springfield (1969).
5. M. Caar, D. Stites, and H. Fudenberg, *Transplantation* 20:410 (1975).
6. J. Sterzl, *in*: "Vyvoj a indukce imunitni odpovedi" (in Czech), Academia, Praha (1989).
7. H. Z. Movat, G. A. van Erkel, and N. V. P. Fernando, *Fed. Proc.* 22:600 (1963).
8. F. Kovaru, R. Stepankova, L. Mandel, J. Kruml, and E. Kenig, *Folia Microbiol.* 24:32 (1979).
9. W. Marinelli and A. Strenger, *in*: "Vergleichene Anatomie und Morphologie der Wirbeltiere", Deuticke, Wien (1959).
10. A. G. Zapata and E. L. Cooper, *in*: "The Immune System: Comparative Histophysiology", Wiley, New York (1990).
11. E. Conklin, *J. Morphol.* 54:69 (1932).
12. P. Sima and V. Vetvicka, *in:* "Evolution of Immune Reactions", CRC Press, Boca Raton (1990).
13. R. Fange, *Vet. Immunol. Imunopathol.* 12:153 (1986).
14. R. A. Good, J. Finstad, B. Pollara, and A. E. Gabrielsen, *in*: "Phylogeny of Immunity", R.T. Smith, P. A. Miescher, and R. A. Good, eds., p.149, University of Florida Press, Gainesville (1966).
15. J. Rejnek, L. Tuckova, P. Sima, and M. Bilej, *Immunol. Lett.* 36:131 (1993).

THE FATE OF PROTEIN ANTIGEN IN ANNELIDS - *IN VIVO* AND *IN VITRO* STUDIES

L. Tucková, M. Bilej, and J. Rejnek

Institute of Microbiology CSAS
Prague, Czech Republic

INTRODUCTION

Administration of protein antigens into the coelomic cavity of earthworms *Lumbricus terrestris* (*L.t.*) and *Eisenia foetida* (*E.f.*) leads to a marked increase of the coelomic fluid protein concentration (up to 3 times of original level) and to the formation of a protein that binds the antigen used for stimulation, i.e., antigen-binding protein (ABP).

The specificity of the binding is much lower than that of vertebrate antibodies or T cell receptors since the earthworms ABP binds the antigen used for stimulation, as well as other protein antigens, although the most effective binding was always found with the antigen used for stimulation.[4,5]

The ABP after stimulation was found not only in the coelomic fluid (CF) but also on the surface of some coelomocytes. Maximal binding activity and the highest number of ABP-bearing cells in the CF was shown to occur between days 4 and 8 after primary stimulation, while the response after secondary challenge was faster and more pronounced.[1,2]

Since the response to antigenic stimulation in vertebrate species depends on antigen processing and since a powerful proteolytic system capable of digesting foreign, but not self proteins, exists in CF and coelomocytes of earthworms we consider it of intrest to find out whether the adaptive-like response of earthworms also requires antigen processing.[3]

RESULTS AND DISCUSSION

The administration of the labelled antigen (^{125}I- ARS-HSA) into the coelomic cavity showed that 50% of the radioactivity in *E.f.* CF disapeared during the first 48 h and the rest was eliminated after 5 days when the radioactivity reached background level. The decreased radioactivity in *L.t.* CF was somewhat slower, although almost 50% of the radioactivity was again eliminated in 48 h (Fig. 1).

The TCA precipitation of *L.t.* (similar to *E.f.*) CF samples collected 2, 4, 8 and 24 h after antigen administration demonstrated that as early as after 2 h more than 40% of antigen was present in the form of fragments non-precipitable with TCA and after 24 h it was more than 60% non-precipitable (Fig. 2a). When the samples obtained after 2 and 4 h were analyzed by SDS-PAGE and autoradiography, it was found that small amounts of intact (non-digested) antigen was detectable only in the 2 h samples, which confirms that the TCA

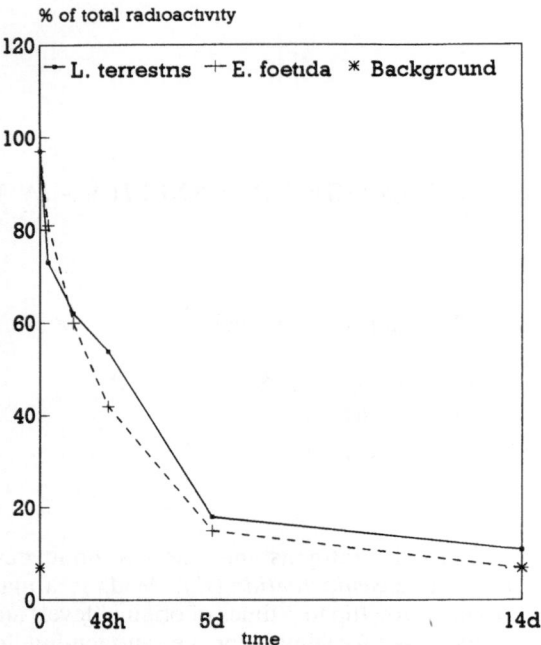

Figure 1 The kinetics of ^{125}I- ARS-HSA elimination in *L terrestris* (--) and *E foetida* (- -) earthworms

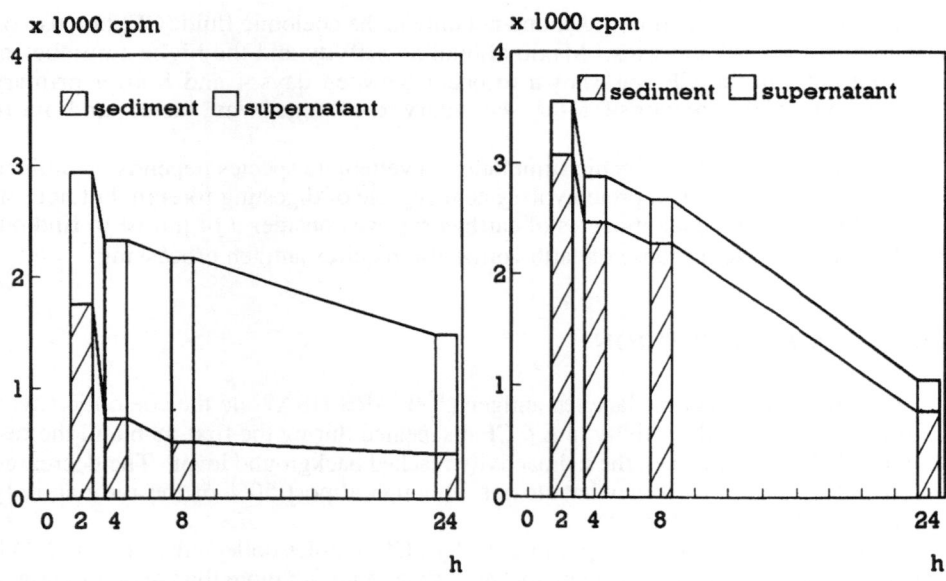

Figure 2 Proteolytic activity in coelomic *L t* fluids (a) and coelomocytes (b), indicated as relative distribution of radioactivity of *in vivo* administered ^{125}I- ARS-HSA in sediments and supernatants after TCA precipitation

336

precipitates were mostly composed of larger TCA precipitable fragments and that all the administered antigen is split during the first 4 h.

The degradation of labelled antigen internalized into coelomocytes was similar, about 60% of radioactivity was eliminated in 24 h, but the ratio of TCA precipitable and non-precipitable radioactivity was different. While in TCA precipitates of CF the amount of radioactivity mounted to 30-50%, in coelomocyte lyzates it was 70-90%. One plausible explanation for this phenomenon is that the small proteolytic fragments are rapidly released from the cell into the CF (Fig. 2b).

The fate of antigen was also followed in *in vitro* experiments. The mixture of cold and labelled antigen was added to CF and the radioactivity checked in TCA sediments and supernatants after 2, 4, 8, 24 and 48 h incubation at room temperature. About half the antigen (56% in *E.f.* and 42% in *L.t.* CF) was found in the TCA precipitates after 24 h of incubation. After 48 h, it was 84% (*E.f.*) and 71% (*L.t.*).

The antigen processing in *E.f.* and *L.t.* coelomocytes was followed by cultivation of separated cells again with the mixture of cold and ^{125}I- ARS-HSA. After the same time intervals as mentioned above, the cultivation medium was separated and the cells lyzed and centrifuged. Radioactivity was measured in the insoluble fraction (cell debris) and TCA supernatants and precipitates of cell lyzates and cultivation media. It was shown that during first 2 h almost 20% of the antigen entered *L.t.* coelomocytes (10% in the case of *E.f.* coelomocytes) and 35 and 30% was internalized by 24 h. Between 24 and 48 h no significant internalization of antigen was observed (Table 1). This could be caused either by continuous cleavage of antigen which enters the cells and the subsequent release of small fragments into the medium or by inability of antigen already fragmented in medium to enter the cells. The antigen seems to be more efficiently cleaved in *L.t.* than in *E.f.* coelomocytes, since in the latter, less than 20% of internalized antigen was present in the form of TCA non-precipitable fragments after 24 h while in the case of *L.t.* coelomocytes it was more than 50%.

Table 1. The kinetics of proteolytic activity in cell cultures measured as relative distribution of radioactivity (%).

Lumbricus terrestris	2 h	4 h	8 h	24 h	48 h
Culture medium TCA-supernatant	5.7	10.6	10.9	20.4	52.7
" TCA-precipitate	65.4	65.1	66.5	41.6	11.3
Cell lysate TCA-supernatant	1.1	1.4	1.4	12.9	12.7
" TCA-precipitate	11.6	11.5	10.7	11.3	12.3
Insoluble remnants of cells	6.1	7.5	8.6	13.9	11.0
Eisenia foetida					
Culture medium TCA-supernatant	11.9	11.7	11.7	15.5	32.4
" TCA-precipitate	79.0	75.1	76.1	56.3	40.2
Cell lysate TCA-supernatant	0.9	1.2	1.1	2.9	4.2
" TCA-precipitate	5.0	7.0	5.5	18.1	15.8
Insoluble remnants of cells	3.1	5.1	5.4	7.3	7.4

The proteolytic enzymes produced by coelomocytes were released into the culture medium, because in the cell free medium antigen is very efficiently digested. The comparison of media obtained from coelomocyte culture cultivated with or without antigen showed that proteolytic activity of those cultivated in the absence of antigen was significantly lower. This indicates that the release (and probably also formation) of proteolytic enzymes by coelomocytes was induced by antigenic stimulation. The results did not show whether the same cells that internalize and digest antigen were also responsible for the release of enzymes, but the release of enzymes by cells containing substrate (internalized antigen) seems to be less probable.

The fact that as early as 4 h after administration no intact antigen was detected in *E.f.* and *L.t.* coelomic fluids, whearas all antigenic material was present in the form of proteolytic fragments. Together with the fact that the ABP formation in CFs and on the surface of coelomocytes was detectable as late as 4 days later indicates that the ABP response was induced by proteolytic fragments, i.e., requires antigen processing.

In order to obtain some evidence regarding antigen processing, we have modified the technique of cultivating *L.t.* tissue explants, as described by Janda and Bohuslav almost 60 years ago, and used it for testing ABP responses *in vitro*. Small tissue fragments of *L.t.* gut walls were cultured either alone or in the presence of a mixture of cold and ^{125}I- ARS-HSA or its small (<10 kD) or large (>10 kD) proteolytic fragments obtained after incubation of the mixture with *E. f.* CF. The ABP response was tested in culture medium after one week by ELISA using mAbs to ABP prepared in our laboratory.

The results confirmed our anticipation that antigen processing is needed in ABP response and showed that the stimulation of *L.t.* tissue culture with small, TCA non-precipitable fragments led to a similar ABP response (only 7% lower) as when the intact antigen was used. However, to our surprise, the larger fragments had very little, if any, stimulatory effect since the values obtained were comparable to those (or even lower) where the cultivation was performed in the absence of antigen (Fig. 3).

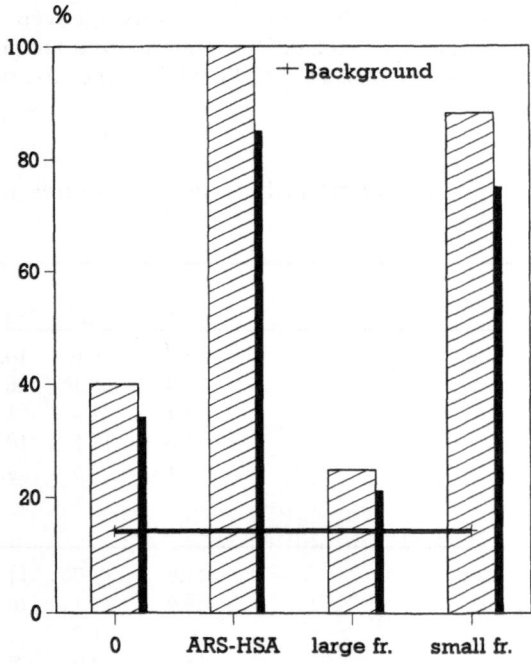

Figure 3. ABP formation in *L. terrestris* tissue cultures of non-stimulated (O) and stimulated with ARS-HSA or its large and small proteolytic fragments followed by ELISA, indicated as relative values of response to ARS-HSA (100 %).

We can only guess that the molecules on the coelomocyte surface that are responsible for binding of small antigenic fragments and subsequently for stimulation of ABP formation carry a binding site which can accomodate only peptides of certain size similarly as MHC molecules of vertebrates.

REFERENCES

1. M. Bilej, L. Tucková, J. Rejnek, and V. Vetvicka, *Immunol. Letters* 26:183 (1990).
2. M. Bilej, P. Rossmann, T. VandenDriessche, J.P. Scheerlinck, P. De Baestselier, L. Tucková, V. Vetvicka, and J. Rejnek, *Immunol. Letters* 29:241 (1991).
3. L Tucková, J. Rejnek, P. Sima, and R. Ondrejová, *Dev. Comp. Immunol.* 10:181 (1986).
4. L. Tucková, J. Rejnek, and P. Sima, *Dev. Comp. Immunol.* 12:287 (1988).
5. L. Tucková, J. Rejnek, M. Bilej, and R. Pospisil, *Dev. Comp. Immunol.* 15:263 (1991).

HEMOLYTIC FUNCTION OF OPSONIN-LIKE MOLECULES IN COELOMIC FLUID OF EARTHWORMS

Marek Sinkora, Martin Bilej, Karel Drbal, and
Ludmila Tucková

Department of Immunology, Institute of Microbiology Czech Academy of Sciences, Prague,
Czech Republic

Factors with hemolytic, agglutinating, and bacteriostatic activities have been detected in coelomic fluid of annelids. Protein factors found in the coelomic fluid in normal conditions are expressed constitutively, but higher levels can be induced by stimulation (immunization). In the hemolytic system of *Eisenia foetida* several proteins possess strong lytic activity against vertebrate red blood cells (RBC) and this can be enhanced after stimulation.[4] Two glycoproteins with mol. wt. 40 kDa and 45 kDa have proved to be involved in RBC lysis[8] but it is not yet clear whether hemolysis is caused by direct action of the glycoproteins, or if associated factors are responsible for the RBC destruction. Some authors argue that hemolytic factors are not proteins but are thermostable[3], while others supose that hemolytic factors are associated with antibacterial protein molecules that are able to oligomerize and thus produce of transmembrane channels. The transmembrane complex, however, is not analogous to that of C9 or perforin, and no homology with C3 has been found.[10]

Two proteins involved in the hemolytic processes are encoded by two distinct genes, monoallelic and polyallelic (4 alleles) with different isoelectric points. They are supposed to be secreted by chloragocytes and are heat-sensitive. Eleven different hemolysin gene combinations have been described: 4 homozygous, 6 heterozygous, and 1 not elucidated. Their frequency of occurrence is extremely variable and 10 out of 11 allelic combinations (one of which is the most frequent) display high inhibitory activities. On the other hand no difference can be observed when testing individual coelomic fluids for hemolysis.[9,11]

In addition, a poorly agglutinating, hemolytic protein with mol. wt. 20 kDa was found.[9,12] Some low mol. wt. hemolytic molecules have been detected also in the coelomic fluid of *Eisenia foetida*. A 15 kDa lysozyme-like protein and a small 2.6 kDa flavin-like non-polypeptide molecule play a significant roles in hemolysis.[5,6]

The effect of opsonization of synthetic hydroxyethylmethacrylate copolymer microparticles (HEMA) was tested by Bilej and co-workers.[1,2] Both the phagocytic activity and phagocytic index were increased after preincubation of HEMA particles with coelomic fluid. Futhermore, the proteins adsorbed on the surface of the particles were isolated and characterized as two proteins: 62 and 68 kDa.

Function of hemolytic opsonin-like molecules was tested by compensation assays as shown in Fig. 1. Hemolytic activity of the incomplete coelomic fluid (after adsorption of opsonizing proteins on the surface of HEMA particles) was decreased and could be renewed almost to the control level by the addition of opsonizing proteins isolated from the surface of HEMA particles. The decrease of the hemolytic activity was quadratically

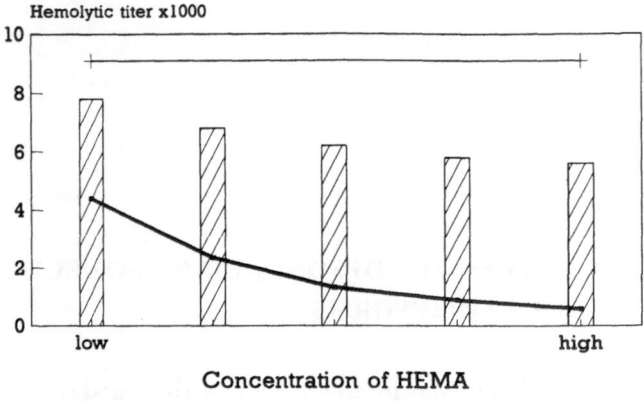

Concentration of HEMA

--- After adsorption   + Basal hemolysis   ▨ +opsonizing protein

Figure 1. Hemolytic activity: Coelomic fluid of *E. foetida*

dependent on the concentration of HEMA particles (i.e. on the surface area of the particles). On the other hand the isolated opsonizing proteins were not hemolytic at any concentration tested, and could probably be involved in a cascade of hemolytic proteins.

HPLC studies indicated that the mol. wt. of isolated opsonizing proteins is between 11-17 kDa. After immunization, elution profiles of complete coelomic fluid changed in the range 110 kDa, 30 kDa, and 10 kDa. The concentration of hemolytic factors with opsonizing function was also enhanced after immunization, but their mol. wt. and numbers remained identical. Nevertheless, it is necessary to realize that mol. wt. estimated by HPLC could be different from those estimated by SDS-PAGE. Further characterization of the opsonizing proteins is a future topic of investigation in our laboratory.

REFERENCES

1.   M. Bilej, V. Vetvicka, L. Tuckova, and I. Trebichavsky, *Folia Biol.* 36:273 (1990).
2.   M. Bilej, J. P. Scheerlinck, T. Van den Driessche, P. DeBaetselier, V. Vetvicka, *Cell Biol. Int. Rep.* 14:831 (1990).
3.   P. Cenine, *Dev. Comp. Immunol.* 7:637 (1983).
4.   P. Chateraureynaud-Duprat, *C. R. Acad. Sci.* 273D:1647 (1971).
5.   F. Lassalle, P. Roch, and M. Lassegues, *Dev. Comp. Immunol.* 10:645 (1986).
6.   F. Lassalle, M. Lassegues, and P. Roch, *Comp. Biochem. Physiol* (1B:187 (1988).
7.   M. Lassegues, P. Roch, M. A. Cadoret, and P. Valembois, *C. R. Acad. Sci.* 299:691 (1984).
8.   P. Roch, P. Valembois, N. Davant, and M. Lassegues, *Comp. Biochem. Physiol.* 69B:329 (1981).
9.   P. Roch, P. Valembois, and M. Lassegues, *Prog. Clin. Biol. Res.* 233:91 (1987).
10. P. Roch, C. Canicatti, and P. Valembois, *Biochim. Biophys. Acta* 983:193 (1989).
11. P. Roch, M. Lassegues, P. Valembois, *Dev. Comp. Immunol.* 15:27 (1991).
12. J. Vailler, M. A. Cadoret, P. Roch, and P. Valembois, *Dev. Comp. Immunol.* 9:11 (1985).

ROLE OF LECTINS (C-REACTIVE PROTEIN) IN DEFENSE OF MARINE BIVALVES AGAINST BACTERIA

Jan A. Olafsen

Department of Marine Biochemistry
University of Tromsø, The Norwegian College of Fishery Science
N-9037 Tromsø, Norway

INTRODUCTION

Lectins, sugar-binding proteins or glycoproteins of non-immune origin that agglutinate cells or precipitate glycoconjugates, are found in species of all taxa from viruses and bacteria to vertebrates. Their presence in the body-fluids of invertebrates is well documented, but their biological role(s) are not yet firmly established. Invertebrates phylogenetically below the *Agnatha* lack immunoglobulins, T-cell receptors or lymphoid cells, and thus have no protection due to clonal selection and immune memory as we have observed from studies with vertebrates. However, these primitive invertebrates have survived with success in environments where they are intimately exposed to bacteria, and yet we have relatively scarce information about their humoral defense reactions.

Lectins in the body-fluids (hemolymph) of invertebrates are believed to take part in defense by reacting with bacteria and augmenting their phagocytic uptake.[1-4] The function of lectins in such recognition phenomena would require different recognition determinants for non-self and the phagocytic cells (self), and also the ability to recognize different bacterial cell-surface glycoconjugates. Ideally opsonization or binding to self (hemocytes) should not take place until there is positive recognition of the antigen. This would require heterogeneity of binding-sites, diversity in sugar specificity and a degree of regulation beyond what is normally observed in lectins. Even though lectins with heterogeneic binding-sites exist in invertebrates, their biological function(s) or regulation are not well described.[2,5]

Lectins are identified by their ability to agglutinate mammalian erythrocytes (RBC), but agglutinins for bacteria have been described from invertebrates[6-9], suggesting their involvement in defense reactions. The activity of hemolymph lectins may be augmented by challenge with bacteria[9,10] and by trauma[11], and they may act as opsonins by increasing uptake of bacteria by hemocytes.[10,12] Lectins have also been found on the plasma membrane of circulating phagocytic cells[12,13].

Recent evidence suggests a structural and functional relationship between invertebrate lectins and vertebrate recognition molecules in immune-regulation and communication since some lectins are related to C-Reactive Proteins (CRP). CRP is a member of the pentraxin family of acute-phase proteins, an early indicator of various infections or inflammatory diseases in mammals. This group includes lectins from scorpions, tunicates, and horseshoe crabs.[5] Recently we have found this property in lectins from molluscs.[14] These marine filter-feeders accumulate bacteria, and may also have bacteria in their tissues and circulatory fluid.[15,16] In contrast to most mammals, their humoral lectins (CRPs) are constitutive, but the activity may be augmented by challenge with bacteria.[9] The possible role(s) of these lectins in defense reactions against bacteria will be briefly discussed.

CHARACTERISTICS OF OYSTER LECTINS

Crassostrea gigas hemolymph contain lectins that agglutinate human RBC (gigalin H) and horse RBC in presence of Ca^{++} (gigalin E).[17] The gigalins agglutinate a variety of RBC and other cells, but from cross-adsorption the agglutination of horse- and human-RBC appeared to be caused by separate entities.[9,18] *In vivo* challenge of *C. gigas* by bacteria in seawater resulted in an increased hemolymph lectin titer compared to unchallenged animals (Table 2).[9,10] We also demonstrated that the gigalins were opsonins in that hemolymph stimulated *in vitro* uptake of bacteria by oyster hemocytes.[10] Incubation of hemocytes and bacteria with affinity-purified gigalins in saline did not stimulate uptake in hemocytes, whereas preincubation of bacteria with gigalins did. This suggested that affinity purified lectins had high hemocyte-directed affinity, and/or that binding to bacteria increased the self-affinity.

The agglutination of human RBC was inhibited by a sialic acid (NANA; N-acetyl-neuraminic acid) and by bovine submaxillary mucin (BSM; a glycoprotein rich in terminal NANA). The inhibiton by BSM was dramatically greater than of free NANA. Essentially all of the gigalin activities were found in cell-free hemolymph, and no activity was associated with isolated hemocytes. The oyster lectins were purified by affinity chromatography on BSM coupled to Sepharose 4B. Further characteristics of the gigalins are listed in Tables 1 and 2.

Table 1. Properties of oyster (*Crassostrea gigas*) lectins (gigalins).

Biological properties of the gigalins:[2,9,10,18]

- Act as opsonins by stimulating in vitro phagocytosis of bacteria by oyster hemocytes
- Activity in hemolymph augmented following in vivo challenge to bacteria.
  - mean ratio (titer) of challenged:control, Gigalin E 6.3; Gigalin H 4.6

Biochemical characteristics of native gigalins:[2,18,19]

- Exist as distinct aggregates with M_r from 0.5×10^6 to 2×10^6
- Composed of subunits (8 M urea) of M_r 21,000; 22,500 and 33,000
- RBC agglutinating specificity determined by subunit composition
- Contains (heterogeneic) binding sites for different cell-surface ligands
- Binding sites may be hidden (cryptic), and not reactive with cell-surface ligands
- Hidden binding sites may be re-exposed by ligand binding
- Ca^{++} and ligand binding affect gigalin conformation
- Reacts as CRP by reacting with phosphorylcholine in presence of Ca^{++}

Gigalin H was quantitatively adsorbed to BSM, but also Gigalin E was partially (77%) bound to this ligand even though agglutination of horse RBC was unaffected by NANA or sialidase. In contrast, the sialic acid-specific lectin limulin from the horseshoe crab (*Limulus polyphemus*) agglutinates horse RBC.[20] Affinity purification of oyster

Table 2. Biochemical characteristics of oyster (*Crassostrea gigas*) lectins.[2,9,18]

	Gigalin E	Gigalin H
Agglutination of erythrocytes (RBC)	Agglutinates horse RBC in presence of Ca^{++}	Agglutinates human (group 0) RBC
Activity (titer) in hemolymph (95% confidence interval)	117-2163	65-147
Sensitive to neuraminidase	-	+
Inhibited by BSM	-	+
Inhibited by NANA	-	+
Adsorbed to BSM-Sepharose	Partially	Completely
Recovered activity	30%	40%
Specific activity in % of activity in hemolymph	108%	120%
Composition (protein:carbohydrate:lipid in % of protein)	100:19:22	100:15:21
Effect of metals on lectin activity:	+	-
RBC agglutination inhibited by EDTA >1mM	+	-
Ca^{++} needed for binding to BSM	0.1 µM	n.d.
Ca^{++} needed for full RBC agglutination	0.1 mM	n.d.
Inhibition of RBC agglutination by Zn^{++}	+	-
Protein in hemolymph (mg per ml)	4.2 ± 1.2	
Ca^{++} in hemolymph (mg per ml)	0.55	
Zn^{++} in hemolymph (µg per ml)	8.4	

lectins resulted in preparations with mixed activities for human- and horse-RBC that constituted about 2% (50 µg per ml) of hemolymph protein.[18]

Invertebrate lectins may be heterogeneic polymers of subunits with different binding sites, which would allow flexibility in structure and specificity. The oyster lectins (gigalins) exist as distinct macromolecular aggregates with M_r in the range 0.5-2 million.[2,18] The aggregates were dissociated in 8M urea to subunits α_1(21,000 d), α_2(22,500 d) and β(33,000 d). A combination of adsorption by erythrocyte ghosts followed by BSM-Sepharose revealed that gigalin H activity was dominated by the subunits $\alpha_1\alpha_2$, whereas gigalin E activity was $\alpha_1\beta$.[2,18] Thus different compositions of gigalin subunits result in distinct agglutinating specificity. Polymeric aggregates of subunits would allow multipoint ligand-binding, and explain why the glycoprotein BSM was a more potent inhibitor of agglutination than free NANA.

Most invertebrate lectins need divalent cations, in particular Ca^{++}, for agglutination of some erythrocytes, but may agglutinate other RBC without such requirements. Thus agglutination of human RBC by the eastern oyster (*C. virginica*) lectin was dependent on Ca^{++}, whereas agglutination of sheep and rabbit RBC was not.[21] In contrast gigalin required Ca^{++} for agglutination of horse RBC, 0.1µM was sufficient to restore full activity, whereas agglutination of human RBC took place in 1mM EDTA. The agglutination of horse RBC was completely inhibited by Zn^{++} at about the concentration normally found in hemolymph (0.1mM), but the effect could be reversed by excess Ca^{++}. Zinc also increased the Ca^{++} concentration required for gigalin E activity from 0.1µM to 1mM. The ratio of

Ca^{++}:Zn^{++} in hemolymph (100:1) prevents *in vivo* inhibition (Table 2). Marine bivalves, however, may concentrate Zn^{++} into granular amoebocytes.[22] Exposure of *C. virginica* hemocytes to an ionophore and *Escherichia coli* resulted in rapid exocytosis of Zn^{++}.[23] It is thus feasible that Zn^{++} could affect the activity or specificity of hemocyte lectins during bacterial challenge, and that the opsonic properties of *C. gigas* lectins might be influenced by the intra/extra-cellular ratio of divalent cations.

ROLE OF LECTINS/CRP IN INVERTEBRATE HUMORAL DEFENSE

Protein/carbohydrate interactions constitute basic phenomena common to recognition events in all organisms, and lectins may play important roles in such immune-regulation and communication. Cells of the immune-system may be activated or depressed through lectin-like receptors, and various signalling-molecules have lectin-like properties.[24] The relationships between invertebrate lectins and vertebrate recognition molecules include the observation that lectins from some invertebrates are related to C-reactive proteins. The exact biological function of CRP is not known, but it is assumed to modify the behaviour of the immune system, opsonize foreign invaders and activate complement.[25] We have recently demonstrated that sialic acid-specific lectins from the oyster *C. gigas* (gigalins) and the horse-mussel *Modiolus modiolus* are CRP[14], since they precipitate C-polysaccharide from *Streptococcus pneumoniae* and may be affinity-purified with phosphorylcholine as ligand in presence of Ca^{++}. Bivalve lectins have a number of biological and functional characteristics related to mammalian CRP[5], but also some differences, since oyster and horseshoe crab lectins are glycoproteins[2,26,27], whereas mammalian CRP is not.

CRP is secreted by hepatocytes, and in monkeys, dogs, and rabbits it shows high levels during the acute phase response, whereas in other animals (cow, goat, and rat) levels are normally high and increases further during the acute phase.[25] Invertebrate humoral lectins are constitutive, and their activities are augmented by challenge with bacteria.[9] Marine filter-feeders accumulate large numbers of bacteria that may also enter body-fluids and tissues.[15] We observed that healthy oysters and horse-mussels contain bacteria in hemolymph at temperatures < 8°C (CFU, colony forming units, 10^2-10^3) and soft tissues (CFU 10^5). Exposure to psychrophilic, fish-pathogenic bacteria (*Vibrio salmonicida*) in the seawater boosted hemolymph and tissue CFU by a factor of about 100.[16] The two species were obtained from different habitats. Oysters were kept in sand-filtered seawater, whereas horse-mussels were collected from stones in free-flowing seawater at 1-8°C. The dominating bacteria were for both animals pseudomonas-, vibrio- and aeromonas-types. Thus, in contrast to mammals, hemolymph of healthy invertebrates may contain bacteria, and also constitutive lectins/CRP. It is not certain whether hemocytes release intact lectins or lectin subunits as a result of such challenge, even though lectins may be found on the plasma membrane of hemocytes from some invertebrates.[12,13]

Oyster lectins (gigalins) precipitated easily and affinity-purified lectins could not be concentrated above 100 µg per ml without precipitation. The native lectin concentration in oyster hemolymph is about 10-50 µg per ml. At this concentration the gigalins precipitate well with C-polysaccharide from *S. pneumoniae*, whereas precipitation of limulin (Sigma) was not higher than the control (BSA).[14] Precipitation by limulin was obtained at 1 mg per ml, which is similar to the concentration of limulin in horseshoe crab plasma.[27] This makes possible the hypothesis that reaction with non-self would induce precipitation-reaction which would in turn facilitate opsonisation and phagocytosis. Lipid associated with the gigalins results in their hydrophobicity, and possibly takes part in their regulation.

The molecular configuration of gigalins appears to support the above contention. Cryptic or hidden binding sites were demonstrated by rechromatography of fractions following adsorption by RBC and BSM. This procedure revealed emergence of binding sites that were depleted, and suggested that the gigalins contained masked binding sites.[2,18]

In this context it is of interest that Ca^{++} affects the binding specificity of the pentameric ring-like structures of CRP such that reaction with galactans and phosphorylcholine is facilitated in the presence of Ca^{++}, whereas in its absence it binds various polycations. The two different conformations can be distinguished by specific antibodies. Following denaturing procedures and attachment to polystyrene-plates, free CRP subunits (neoCRP) are formed. NeoCRP also occurs naturally as on various macrophages and natural killer (NK)-cells, where it can function as a galactose-specific receptor. Attachment of CRP to fibronectin or LDL (low density lipoprotein) requires aggregation or dense spatial packing. When this requirement is fulfilled, complement is activated through the classical pathway (reviewed in ref. 25).

In comparison, native gigalins have been proposed to exist as polymeric aggregates of ring-shaped structures.[2] These aggregates were dissociated in 8M urea to subunits that could not agglutinate cells, but did inhibit agglutination of human RBC. By measuring the effect of metal substitution on intrinsic protein fluorescence we observed that Ca^{++} affects protein conformation, but is not directly involved in ligand binding.[19] Changes in structure and biological activity may occur by intramolecular subunit rearrangement facilitated by ligand binding or by the ratio of Ca^{++}/Zn^{++}.

Thus invertebrates have a constant level of lectins in the circulatory fluid that may be mobilized to react with invading bacteria. In addition to structural similarities[5], these lectins are also functionally alike to C-reactive protein. A predominant feature of opsonization in invertebrates is precipitation with antigen that results in changes in molecular configuration and augments the phagocytic uptake of bacteria by hemocytes.

ACKNOWLEDGMENTS

The author acknowledges financial support from the Norwegian Research Council for Science and the Humanities and the Norwegian Fisheries Research Council. The technical assistance of Helene V. Mikkelsen in experiments with CRP is greatly appreciated.

REFERENCES

1. L. Renwrantz, *Symp. Zool. Soc. London.* 56:81 (1986).
2. J. A. Olafsen, *in:* "Immunity in Invertebrates", M. Brehelin, ed., p. 94, Springer-Verlag, Berlin (1986).
3. J. A. Olafsen, *Am. Fish. Soc. Special Publ.* 18:189 (1988).
4. G. R. Vasta and J. J. Marchalonis, *in:* "Invertebrate Models. Cell Receptors and Cell Communication", A. H. Greenberg, ed., p. 104, Karger, Basel (1987).
5. G. R. Vasta, *in:* "Defense Molecules", J. J. Marchalonis and C. L. Reinisch, eds., p. 183, Wiley-Liss, New York (1990).
6. T. G. Pistole, *J. Invertebr. Pathol.* 28:153 (1976).
7. D. Zipris, N. Gilboa-Garber, and A. J. Süsswein, *Microbios.* 46:193 (1986).
8. E. A. Stein, S. Younai, and E. L. Cooper, *in:* "Developmental and Comparative Immunology", E. L. Cooper, C. Langlet, and J. Bierne, eds., p. 79, Alan R. Liss, New York (1987).
9. J. A. Olafsen, T. C. Fletcher, and P. T. Grant, *Dev. Comp. Immunol.* 16:123 (1992).
10. S. W. Hardy, T. C. Fletcher, and J. A. Olafsen, *in:* "Developmental Immunobiology", J. B. Solomon and J. D. Horton, eds., p 767, Elsevier/North Holland Biomedical Press, Amsterdam (1977).
11. H. Komano, D. Mizuno, and S. Natori, *J. Biol. Chem.* 256:7087 (1981).
12. L. Renwrantz and A. Stahmer, *J. Comp. Physiol.* 149:535 (1983).

13. G. R. Vasta, J. T. Sullivan, T. C. Cheng, J. J. Marchalonis, and G. W. Warr, *J. Invertebr. Pathol.* 40:367 (1982).
14. J. A. Olafsen, Manuscript in preparation.
15. C. A. Farley, *Ann. N.Y. Acad. Sci.* 298:225 (1977).
16. J. A. Olafsen, H. V. Mikkelsen, H. Giæver, and G. H. Hansen, *Appl. Environ. Microbiol.* 59:1848 (1993).
17. S. W. Hardy, P. T. Grant, and T. C. Fletcher, *Experientia (Basel).* 33:767 (1977).
18. J. A. Olafsen, T. C. Fletcher, and P. T. Grant, submitted.
19. R. Hovik and J. A. Olafsen, submitted.
20. J. J. Marchalonis and G. M. Edelman, *J. Mol. Biol.* 32:453 (1968).
21. J. E. McDade and M. R. Tripp, *J. Invertebr. Pathol.* 9:523 (1967).
22. S. G. George, B. J. S. Pirie, A. R. Cheyne, T. L. Coombs, and P. T. Grant, *Marine Biol.* 45:147 (1978).
23. T. C. Cheng, *J. Invertebr. Pathol.* 59:308 (1992).
24. V. Kéry, *Int. J. Biochem.* 23:631 (1991).
25. V. Kolb-Bachofen, *Immunobiol.* 183:133 (1991).
26. R. T. Acton, J. C. Bennett, E. E. Evans, and R. E. Schrohenloher, *J. Biol. Chem.* 244:4128 (1969).
27. A. C. Roche and M. Monsigny, *Biochem. Biophys. Acta.* 371:242 (1974).

BACTERIAL ANTIGEN PRIMING OF MARINE FISH LARVAE

Jan A. Olafsen

Department of Marine Biochemistry
University of Tromsø
The Norwegian College of Fishery Science
N9037 Tromsø, Norway

The production of large quantities of larvae is a prerequisite for the successful aquaculture of marine fish. Fish eggs are incubated at high density in incubators with a microflora that differs in numbers and characteristics from that in the sea. The eggs become heavily overgrown with bacteria.[1] Marine fish larvae ingest bacteria by drinking, and thus bacteria enter the digestive tract of fish larvae before active feeding commences and may possibly affect antigen priming.[2,3] Even though an intimate relationship exist between fish larvae and bacteria, little is known about the early onset of mucosal immunity or tolerance in fish. This is notable in a species whose entire interface against the environment is guarded by the mucosal epithelium. Some aspects of bacterial colonization and uptake of bacterial antigens by marine fish larvae are discussed below.

The numbers of bacteria in incubators for aquaculture typically increase from 10^3 to 10^6 per ml during hatching, and eggs become overgrown by bacteria after a few days.[1] Some adherent bacteria erode the chorion by exo-enzymatic activity.[1,4] This could affect the quality of eggs and subsequently the microflora and health of larvae. Fish eggs may contain substances such as immunoglobulins (IgM) and lectins that possibly could affect bacterial invasion,[5-7] and inhibit the growth of bacterial pathogens.[8] The exact functions of these substances in protection of eggs against pathogens are not known. The microflora adhering to the egg surface was heterogeneous and did not appear to be regulated. Incubation of germ-free eggs with various antibiotic-producing and probiotic bacteria of marine origin had no lasting effect on the microflora.[1] It has been argued that the indigenous microflora of fish reflects bacteria in the feed or habitat.[9] However, several reports describe bacteria firmly attached to the gut mucosa of fish, and it is now accepted that fish contain a resident intestinal microflora of aerobic, facultative anaerobic, and obligate anaerobic bacteria.[10,11] The composition of this microflora may change following feeding,[11] starvation[12] or antibiotic treatment.[13] Most studies have been concerned with adults, and only few reports exist on the intestinal microflora of fish larvae.[11,13,14] Pathogenic vibrios may colonize the intestinal epithelium of cod larvae and result in extensive damage to microvilli and the brush border, observed by immunohistochemistry and electron microscopy (Hansen and Olafsen, unpublished results). The gut microflora of wild captured cod juveniles consisted of 50% vibrios, of which 30% cross-reacted with monoclonal antibodies against the fish pathogen *Vibrio salmonicida* and were also related to this pathogen in biochemical traits.[11] The gut vibrios disappeared from the intestinal microflora after feeding a commercial diet, and could not be detected following an outbreak of coldwater vibriosis caused by *V. salmonicida*.

Advances in Mucosal Immunology, Edited by
J. Mestecky *et al.*, Plenum Press, New York, 1995

Thus changes in the food regimen may affect the equilibrium on the gut mucosa between commensal gut vibrios and their pathogenic relatives occupying the same microenvironments.

Fish are among the most primitive animals capable of mounting an anticipatory immune response, and we would expect their intimate relationship with bacteria to be effectively regulated. The epithelium with its mucus layer forms a barrier against an external environment that harbours a multitude of potentially harmful microorganisms. A continuous secretion and shedding of mucus produced by goblet cells in skin, gills, and mucosa of the gastrointestinal tract may prevent microbial colonization. Defense factors in mucus such as immunoglobulins, complement, lysozyme, and lectins may bestow protection,[15,16] and possibly explain observations that body surface mucus of healthy fish normally do not contain bacteria.[17] However, large numbers of *Vibrio anguillarum* may be found in mucus during advanced infections.[18] Following stress or infection, mucus secretion may be increased.[19] Increase in the number of bacteria in the seawater results in an increase of epidermal mucus production of 6-week old halibut larvae (Ottesen and Olafsen, unpublished results). However, as most bacterial infections begin on mucosal surfaces, increased production of mucus may not be unequivocally beneficial.

In adult fish evidence suggests that local mucosal and secretory immunity is important in protection against bacterial infections.[20] Teleosts possess intraepithelial lymphoid tissue, although less organized than in mammals. Macrophages, lympoid cells and secretory Ig-forming cells infiltrate the intestinal epithelium.[21,22] Uptake of intact protein antigens by intestinal epithelial cells of adult fish has been extensively reported. Fish apparently lack specialized intraepithelial M cells for the uptake of antigens, and it appears that enterocytes serves a similar purpose of antigen sampling and presentation to lymphoid cells.[21,23] Cells capable of sequestering intact antigens are found particularly in the hindgut of teleosts, and the induction of mucosal immune response against particular antigens can possibly be obtained by the hindgut. The absorptive cells may act to present antigens to the immune system, and possibly result in secretory immunoglobulins.[21] Secretion of interleukin-like factors suggests that epithelial cells may have an immunomodulatory function.[24] However, local secretory mucosal antibody similar to the IgA class of mammals has not been detected in fish, and it has been suggested that its mucosal and serum antibodies are identical.[16]

Relatively little is known about ontogeny of immunity in fish, but it has been inferred that the "immune capacity" of fish larvae is not fully developed until they are several months old.[25] Until then they probably rely on non-specific immunity, active macrophages, or in some species immunity acquired from the mother by ingestion of mucus.[7] The lymphoid functions develop asynchronously. At 4 days post hatching, and before lymphocytes mature, functional macrophages are present in skin, gills and kidney, and thereafter cellular immunity develop rapidly.[26] Ig positive cells are detected after a few weeks, and from then the thymus resembles that of adult fish. Antibody responses and memory have been induced in 2-month old fish, whereas the ability to respond to some T cell-dependent antigens develops later.[27] Premature injection (before 3 months) of an antigen may result in tolerance, in particular to soluble protein antigens. In contrast, a response to bacteria has been observed 3 weeks post-hatching.[28] Fish of 2-3 weeks have been successfully vaccinated by injection,[29] and some protection was obtained in young fish following immersion. It appeared that the attainment of immune competence is correlated with the involvement of a critical number of immune cells rather than with age.[30] Oral administration of vaccine against various fish pathogenic bacteria usually yields less protection than injections, but high levels of protection have been obtained with antigens that reach the posterior part of the intestine unaltered.[31] Vaccination by the oral route may prevent colonization of *V. anguillarum* resulting from anti-*Vibrio* agglutinins in skin mucus.[32] However, this did not result in antibodies in serum or mucosa, nor in decreased mortality following injection of the bacterium.

Fish larvae ingest bacteria that are propelled by peristaltic movement and ciliated cells towards the posterior gut segment. In larvae with a straight gut like herring, the hindgut may be totally occluded with bacteria.[13] Bacteria are endocytosed by epithelial cells in the hindgut of immature larvae,[3,13] probably by nonspecific uptake. However, since bacteria were engulfed whereas erythrocytes were not, some specificity may be involved.[2] Antigens sequestered in the hindgut may possibly be affected by digestive enzymes. However, very young cod larvae also take up bacteria or intact bacterial antigens in columnar epithelial cells in the fore gut. These antigens were intact, as demonstrated by immunohistochemistry, and penetrated beyond the gut epithelium.[3] The active cells demonstrated some preference in the uptake of different bacterial antigens, and did not ingest latex beads.[3] However, production of specific antibodies has not been observed in fish at this age, and it is believed that the immature larvae rely mostly on non-specific defense mechanisms and phagocytosis.[15]

In mammals pre-feeding with protein antigens causes reduction in the subsequent systemic immune response. This is known as oral tolerance, a complex immunological process influenced by the nature of the antigen. It is most easily elicited with T cell-dependent antigens and facilitated by T suppressor cells. Oral tolerance to various antigens has been observed also in fish. Peroral administration of soluble antigens may result in tolerizing fragments entering the blood stream, whereas bacterial antigens do not.[21,33] Thus, acquired tolerance appears to affect only the gut-associated immunity in fish. It is not known whether sequestering of intact bacterial antigens would result in immunity or tolerance. Vibrios are common members of the commensal microflora of marine fish, and wild captured fish may have antibodies to pathogenic *V. anguillarum* strainss.[34] Bacterial colonization of marine fish eggs and larvae has been previously reviewed[35], and we have recently demonstrated that marine bivalves could serve as hosts for fish-pathogenic vibrios.[36] Vibrios appear to dominate the intestinal microflora of marine fish, and active uptake of bacterial antigens by fish larvae could result in tolerance to such commensals. Changes in the commensal "gut microflora" by domestication could facilitate invasion of related pathogenic vibrios in the gut mucosa facilitated through acquired tolerance to shared antigens.

AKNOWLEDGMENTS

Supported by grants from the Norwegian Research Council for Science and the Humanities and the Norwegian Fisheries Research Council.

REFERENCES

1. G. H. Hansen and J. A. Olafsen, *Appl. Environ. Microbiol.* 55:1435 (1989).
2. J. A. Olafsen, *in*: "The Propagation of Cod, *Gadus morhua* L.", E. Dahl, D. S. Danielsen, E. Moksness, and P. Solemdal, eds., Institute of Marine Research, Bergen (1984).
3. J. A. Olafsen and G. H. Hansen, *J. Fish Biol.* 40:141 (1992).
4. G. H. Hansen, Ø. Bergh, J. Michaelsen, and D. Knapskog, *Int. J. Syst. Bacteriol.* 42:451 (1992).
5. H. Fuda, A. Hara, F. Yamazaki, and K. Kobayashi, *Dev. Comp. Immunol.* 16:415 (1992).
6. A. Mor and R. R. Avtalion, *Bamidgeh.* 40:22 (1988).
7. A. Mor and R. R. Avtalion, *J. Fish Biol.* 37:249 (1990).
8. E. W.J. Voss, J. Fryer, and G. Banowetz, *Arch. Biochem. Biophys.* 186:25 (1978).
9. R. W. Horsley, *J. Appl. Bacteriol.* 36:377 (1973).
10. M. M. Cahill, *Microb. Ecol.* 19:21 (1990).

11   E Strøm and J A Olafsen, The indigenous microflora of wild captured juvenile cod in net pen rearing *in* "Microbiology in Poecilotherms", R Lésel, ed , Elsevier Science Publishers B V , Amsterdam (1990)

12   P L Conway, J Maki, R Mitchell, and S Kjelleberg, *FEMS Microbiol Ecol* 38 187 (1986)

13   G H Hansen, E Strøm, and J A Olafsen, *Appl Environ Microbiol* 58 461 (1992)

14   K Muroga, M Higashi, and H Keitoku, *Aquaculture* 65 79 (1987)

15   T C Fletcher, *Dev Comp Immunol* Suppl 2 123 (1982)

16   L W Harrell, H M Etlinger, and H O Hodgins, *Aquaculture* 7 363 (1976)

17   R A Crouse-Eisnor, D K Cone, and P H Odense, *J Fish Biol* 27 395 (1985)

18   T Kanno, T Nakai, and K Muroga, *Dis Aquat Org* 8 72 (1990)

19   N Blackstock and A D Pickering, *J Zool* 197 463 (1982)

20   S Hart, A B Wrathmell, J E Harris, and T H Grayson, *Dev Comp Immunol* 12 453 (1988)

21   J H W M Rombout and A A van den Berg, *J Fish Biol* 35 13 (1989)

22   S Tomonaga, K Kobayashi, K Hagiwara, K Sasaki, and K Sezaki, *Dev Comp Immunol* 9 617 (1985)

23   J H W M Rombout, A A van den Berg, C T G A van den Berg, P Witte, and E Egberts, *J Fish Biol* 35 179 (1989)

24   M M Sigel, B A Hamby, and E M Huggins Jr , *Vet Immunol Immunobiol* 12 47 (1986)

25   C Chantanachookhin, T Seikai, and M Tanaka, *Aquaculture* 99 143 (1991)

26   M J Manning, M F Grace, and C J Secombes, *in* "Microbial Diseases of Fish", R J Roberts, ed , Academic Press, London (1982)

27   C H J Lamers, Ph D, Agricultural University, Wagen ingen, The Netherlands (1985)

28   M S Mughal and M J Manning, *in* "Fish Immunology", M J Manning and M F Tatner, eds , Academic Press, London (1985)

29   K A Khalifa and G Post, *Prog Fish Cult* 38 66 (1976)

30   M F Tatner and M T Horne, *Dev Comp Immunol* 7 465 (1983)

31   K A Johnson and D F Amend, *J Fish Dis* 6 473 (1983)

32   K Kawai, R Kusuda, and T Itami, *Fish Pathol* 15 257 (1981)

33   E McLean and R Ash, *J Fish Biol* 31, Suppl A 219 (1987)

34   J A Olafsen, M Christie, and J Raa, *System Appl Microbiol* 2 339 (1981)

35   J A Olafsen, *in* "Salmon Aquaculture An Overview of Recent Research", K Heen, R L Monahan, and F Utter, eds , p 166, Blackwell, Oxford (1993)

36   J A Olafsen, H V Mikkelsen, H M Giaever, and G H Hansen, *Appl Environ Microbiol* 59 1848 (1993)

PHYLOGENY OF THE IMMUNOGLOBLIN JOINING (J) CHAIN

Tomihisa Takahashi,[1] Takashi Iwase,[1] Kunihiko Kobayashi,[2] Jaroslav Rejneck,[3] Jiri Mestecky,[4] and Itaru Moro[1]

[1]Department of Pathology, Nihon University School of Dentistry, Tokyo, Japan; [2]Department of Laboratory Medicine, Hokkaido University, Sapporo, Japan; [3]Institute of Microbiology, Czechoslova Academy of Sciences, Prague, Czechoslovakia, and [4]Department of Microbiology, University of Alabama at Birmingham, USA

INTRODUCTION

J chain was first fractionated electrophoretically as a component of human colostral IgA that migrated ahead of bands typical of light chains.[1] J chain is an acidic 15 KDa polypeptide found in polymeric immunogloblins, covalently linked to dimeric IgA and pentameric IgM molecules.[2,3] J chain, which is synthesized by plasma cell, appears to be disulphide-linked to half-cysteine residues of two α chains in IgA and of at least one μ chain in IgM.[4,5] J chain was found to be associated with secreted polymeric immunogloblins, but its biological functions are not at all clear.

Previously, J chain expression has been detected by various techniques in many vertebrate species including fish, amphibians, reptiles, birds, and mammals. In mammalian species, nucleotide sequences encoding J chain protein have been identified in human and mouse using DNA cloning technology.[6,7,8] The amino acid sequences of rabbit and bullfrog J chain protein have also been determined.[9,10] We have become interested in the presence of J chain in lower animal species which are known to lack immunoglobulins.

In this report, we examined the phylogeny of J chain expression in various lower animal species by use of the PCR technique, and determined the partial sequence of earthworm J chain derived from nucleotide sequence analysis of cloned cDNA amplified by PCR. Earthworm J chain DNA and amino acid sequence data obtained in this study were compared to those of human, rabbit, mouse, and bullfrog.

MATERIALS AND METHODS

Animal and Cell Line

Earthworm, clam, slug, silkworm, ascidia, lamprey, spotted garpike, African clawed frog, and newt were chosen in this study. For earthworm, clam, slug, silkworm, and ascidia, 1.0 g of whole body tissue was homogenized with 5.0 ml of 4 M guanidine thiocyanate solution and stored at -80 °C until RNA extraction. For lamprey, spotted garpike, African clawed frog, and newt, 1.0 g of the liver was excised and homogenized in 5.0 ml of the same reagent for RNA isolation. Furthermore, we used Epstein-Barr virus-infected human B lymphoma (Daudi) cells as a positive control, becouse this cell line was reported to be J chain-positive. 1 x 10[6] Daudi cells were collected for the extraction of RNA.

RNA Analysis

Total RNA was isolated by the guanidine thiocyanate procedure, and 10 μg per lane was electrophoresed on a 1.0 % agarose-formaldehyde gel, then transferred to Hybond[TM]-N nylon membrane (Amersham Corp. Buckinghamshire, England). Membranes were hybridized with ^{32}P-labelled human J chain cDNA probe and were exposed to Kodak X-OMAT AR film for 16 hr at -80 °C.

Plasmid Preparation

Two human J chain cDNA-containing plasmids were transformed in our laboratory into JM109 cells and digested by Bam HI restriction enzyme to obtain the insert for hybridization. The purified inserts were labelled with ^{32}P by random hexanucleotide priming (random primer DNA labelling kit, Takara Shuzou Co., Ltd., Tokyo, Japan).

RT-PCR Procedure

The reverse transcriptase PCR was performed as follows. Random hexamer-primed single stranded cDNA was synthesized in a final volume of 10μl with 2 units of Rous sarcoma-associated virus 2 reverse transcriptase (Takara Shuzo) and 1 μg of total RNA. An aliquot (5μl) of the reaction mixture was diluted with 50 μl of the PCR buffer, and the PCR reaction was performed using 2.5 units of Taq DNA polymerase (Perkin Elmer Cetus). The conditions for the PCR were; 60 sec. denaturing at 95°C, 120 sec. annealing at 60°C, and 180 sec extension at 72°C, repeated for 35 cycles. The oligonucleotides used as primers were GAGGACATTGTGGAGAGAAA (5' primer; UJ-1) and GTCAGGATCAGCA -GGCAT (3' primer ; IS4-2), which were synthesized according to published DNA and amino acid sequences common to human, mouse, rabbit, and bullfrog J chains. The amplified DNA fragments were electrophoresed on 1.5% agarose gel, then visualized on a UV illuminator and photographed with type 55 positive/negative film (Polaroid, Cambridge, MA.). For Southern blot analysis, after the electrophoresis, the amplified DNA fragment was transferred to Hybond-N and the membrane was hybridized with a ^{32}P-labelled cDNA probe, then exposed to the film for 16 hr at -80°C.

DNA Sequencing

Nucleotide sequencing was perfomed using the TA cloning system (Invitrogen Corp.). In brief, the template cDNA was purified on an Ultrafree centrifuge tube (Millipore Corp., Bedford, Mass), and the purified insert DNA (10 ng of template DNA) was ligated into PCR TM[1000] vector. These recombinants were then transfected into E. coli INVαF competent cells. After restriction analysis by means of Hph I, 6 positive clones were selected. Double-stranded DNA sequencing was performed by the dideoxynucleotide chain-termination procedure using sequenase Ver 2.0 (United States Biochemical) as recommended by the manufacturer.

RESULTS AND DISCUSSION

In this study, we chose PCR to identify J chain expression at the transcription level in various lower animals, because this technique makes it possible to detect and amplify a small amount of gene by a combination of primer sets. We could find the expression of a J chain gene with approximately 300 base pairs in 8 lower animal species (earthworm, clam, slug, silkworm, ascidia, lamprey, African clawed frog, and newt), except for the spotted garpike (Figs. 1 and 2). The data suggested that lower animals, which do not produce immunoglobulin at all, expressed a J chain gene. To ascertain whether these J chain genes in lower animals have high homology to those of mammals, we performed sequencing on the PCR-amplified product of the earthworm J chain gene as a representative. We used 2 types of human J chain cDNA (59 J and 60 J) as probes, which comprise 1.4 Kb encoding exons 2 and 3, or 3 and 4 respectively. Southern blot analysis revealed that both types of human J chain cDNA hybridized with the PCR amplified product of earthworm J chain giving 310 bp bands. A similar 310 bp band was identified from human Daudi cell RNA, which is known to be J chain positive, using the same J chain probes. Furthermore, Northern blot analysis of earthworm as well as Daudi cell RNA indicated J chain transcription. These results suggested that the earthworm may have a J chain similar to mammals.

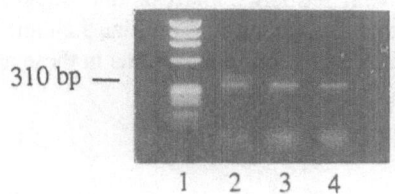

310 bp —

1 2 3 4

Figure 1. J chain gene detection in earthworm by PCR. After 35 cycles of amplification, 10 µl of samples were electrophoresed on 1.5 % agarose gel. DNA size marker φX174/Hae III (lane 1), earthworm (lane 2), mouse intestine (lane 3), Daudi cell (lane 4).

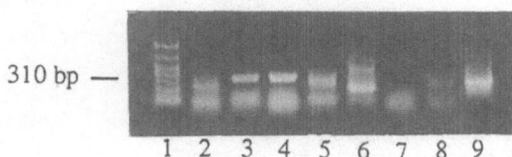

310 bp —

1 2 3 4 5 6 7 8 9

Figure 2. J chain gene detection in various lower animals by PCR. After 35 cycles of amplification, 10ul of samples were electrophoresed on 1.5 % agarose gel. DNA size marker φX174/Hae III (lane 1), clam (lane 2), slug (lane 3), silkworm (lane 4), ascidia (lane 5), lamprey (lane 6), spotted garpike (lane 7), newt (lane 8), African clawed frog (lane 9).

Six positive clones with an insert length of 300 bp were selected for sequence analysis. The results revealed a 312 base sequence from 5' to 3' primers which represented 104 amino acids (Fig. 3). The amino acid sequence deduced from the amplified cDNA of earthworm had high homology to the amino acid sequence of J chain from human, rabbit, mouse, and bullfrog. Interestingly, the earthworm sequence had 67 % homology to human, 58 % to rabbit, 68 % to mouse and 55 % to bullfrog J chain. The results of sequencing therefore supported the concept that lower animals, especially earthworm, have a J chain gene similar to higher animals.

CONCLUSION

J chain gene expression was found in various lower animal species including earthworm, clam, slug, silkworm, ascidia, which do not produce immunoglobulins and also

```
earthworm: E D IVQRYI RI N VPLK NRGN ISDP T S P I R NQ F VTH L S NS FRR CDP
   human : E D IVERN I RI L VPLN NREN ISDP T S P LR TR F VYH L S DLCKK CDP
  rabbit : E D IVE RNI RI I VPLN TREN ISDP T S P L R TE F KYN L ANLCKK CDP
   mouse : E D IVE RNI RI V VPL N NREN ISDP T S P LR RN F VYH L S DVCKK CDP
 bullfrog: E I LERNIQ IT I PTS S RMX ISDP Y S P LR TQ F V YN L WDI CQK CDP


YED E VV ——— T A T Q TN IC TPDQGVPQYCRDYDRNK CYT VL V PLGTTGE
T EV E LDNQ IV T A T Q SN IC DEDSAT− ETCYTYDRNK CYT AV V PLVYGGE
T EI E LDNQVF T A S Q SN IC PDDDYS−ETCYMYDRNK CYT TL V PITH RGV
VEV E LEDQVV T A T Q SN IC NEDDGVPETCYMYDRNK CYT TM V PLRYHGE
VQL E I GG I PV L A S Q PXXS KPDd E————————— CYT TE V NFK———


TKMVQNA LTPD N CY PD
TKMVET A LTPD A CY PD
TRMVK AT LTPD S CY PD
TKMVQA A LTPD S CY PD
———KKVP LTPD S CY EY
```

Figure 3. Comparison of the deduced amino acid sequence of the PCR-amplified earthworm J chain gene with the partial J chain sequence of 4 other animals. Boxed areas show amino acids to all 5 species.

in lamprey, African clawed frog, and newt. Partial sequence analysis data suggested that earthworm J chain gene has high homology to mammalian and amphibian J chain, and that the function of J chain may not be the polymerization of immunoglobulins in these animals.

ACKNOWLEDGMENTS

This work was supported by a Grant-in-Aid for Scientific Research from the Ministry of Education, Science and Culture (No. 03404652).

REFERENCES

1. J. Rejnek, J. Kostka, and O. Kotynek, *Nature* 329:926 (1966).
2. M. E. Koshland, *Annu. Rev. Immunol.* 3:425 (1985).
3. J. Mestecky, J. Zikan, and W. T. Butler, *Science* 171:1163 (1971).
4. A. Garcia-Pardo, M. E. Lamm, A. G. Plaut, and B. Frangione, *J. Biol. Chem.* 256:11734 (1981).
5. J. Mestecky and R. E. Schrohenloher, *Nature* 249:650 (1974).
6. E. M. Max and S. Korsmeyer, *J. Exp. Med.* 161:832 (1985).
7. G. M. Cann, A. Zartisky, and M. E. Koshland, *Proc. Natl. Acad. Sci. USA* 79:6656 (1982).
8. L. Matsuuchi, G. M. Cann, and M. E. Koshland, *Proc. Natl. Acad. Sci. USA* 83:456 (1986).
9. G. J. Hughes, S. Frutiger, N. Paquet, and J. C. Jaton, *Biochem. J.* 271:64 (1990).
10. C. A. Mikoryak, M. N. Margolies, and L. A. Steiner, *Proc. Natl. Acad. Sci. USA* 140:4279 (1988).

GENITAL-ASSOCIATED LYMPHOID TISSUE IN FEMALE NON-HUMAN PRIMATES

Thomas Lehner,[1] Christina Panagiotidi,[1] Lesley A. Bergmeier,[1]
Louisa Tao,[1] Roger Brookes,[1] Andy Gearing,[2] and Sally Adams[2]

[1]Department of Immunology
UMDS of Guy's and St. Thomas' Hospital
London Bridge, London SE1 9RT
[2]British Biotechnology Ltd.
Watlington Road
Cowley, Oxford OX4 5LY
United Kingdom

INTRODUCTION

The gut-associated[1] and bronchial-associated lymphoid tissues[2] are well recognized as part of the mucosal associated lymphoid tissue in which IgA isotype of B cells circulate and home to a variety of glandular tissues.[3-7] However, genital-associated lymphoid tissue has received limited attention. Cervico-vaginal immunization with inactivated poliovirus in human subjects resulted in anti-polio IgA antibodies in the vaginal washings;[8] intra-muscular immunization did not induce vaginal IgA antibodies. Adoptive transfer experiments with radiolabelled lymphocytes suggests that the origin of IgA-producing B cells in murine genital tissue appears to be the mesenteric lymph nodes.[6] Direct administration of SRBC into Peyer's patches of Sprague-Dowley rats induced vaginal IgA and IgG antibodies;[9,10] this suggested that homing of sensitized B cells has taken place from GALT to the genital tract. Intra-uterine immunization elicited vaginal IgA, as well as serum IgG antibodies. In another series of experiments, vaginal application of horse ferritin to mice induced specific vaginal IgA antibodies.[11] They suggested that the paucity of T and B cells and lymphoid follicles in the cervico-vaginal mucosa and stimulation of vaginal IgA antibodies by immunization of the pelvic lymph nodes, argues in favour of the iliac lymph nodes being involved in generating IgA antibodies in the genital tract.[12,13]

T cell responses may play an essential part, not only as CD4 helper cells in antibody synthesis, but also as CD8 cytotoxic cells. Indeed, protection from lethal vaginal infection with HSV-2 was transferred to non-immune mice with the para-aortic lymph node cells but not with other lymphoid cells or with serum.[14,15] Furthermore, after vaginal inoculation with HSV-2, adoptive transfer of genital lymph node T cells but not B cells home preferentially to HSV-2 challenged vaginal mucosa and some of these cells may be cytotoxic T cells.[16] However, rhesus monkeys immunized with the T4-coliphage by the vaginal route elicited only a low proliferative response of the circulating lymphocytes to the T4 antigen and in only one third of the macaques.[17]

A series of investigations have recently been pursued in the non-human primate, to compare the oral, vaginal, and systemic routes of immunization, using simian immunodeficiency virus (SIV) antigens.[18] The experiments were carried out predominantly in rhesus and to a lesser extent in cynomolgus macaques, as SIV infection of macaques appears to be the most important non-human primate model for studying vaccines, routes of immunization, the immune responses, and the protective mechanisms against AIDS.[19-22]

The objectives were to investigate the immune responses in the genital lymph nodes, spleen and other lymphoid tissues after genital immunization was augmented by oral immunization with the hybrid virus-like particles (VLP), containing SIV p27 fused to the yeast retrotransposon Ty and linked to cholera toxin B subunit (CTB) in macaques.[23] This mode of immunization elicited specific CD4[+] T cell proliferative responses and helper function in B cell IgA antibody synthesis in the genital lymphoid tissue but not in other lymph nodes. Splenic CD4 proliferative and helper T cells were also observed but the helper function was directed predominantly towards B cells producing the IgG class of antibodies. Thus, augmented vaginal by oral immunization with p27:Ty-VLP/CTB stimulates specific regional and central CD4 and B cell responses which may play an important role in preventing vaginal infection by viral and other microbial agents.

MATERIALS AND METHODS

Vaccine

The construction of hybrid virus-like particles containing the SIV p27 sequence of SIVmac251 fused to the p1 protein of Ty has been described previously.[24,25] The SIV gag p27 gene was derived from the clone pNIBSCI and the SIV p27:Ty-VLP and control Ty-VLP were purified from yeast extracts.[26] Non-particulate p27 was prepared by cleavage from the p27:Ty-VLP and further purified by ion-exchange chromatography. The absence of any Ty protein in the p27 preparation was confirmed by Western blotting. The recombinant antigens were covalently linked to CT-B, (Sigma Chemical Co.) at a ratio of 1:1, using SPDP (N-succinimidyl-3-2 pyridyl dithiopropionate).[27]

Immunization

Six rhesus macaques received 2 oral followed by 3 vaginal immunizations (O-V) of p27:Ty-VLP/CTB (n=3), p27:Ty-VLP without CTB (n=1), p27/CTB without Ty-VLP (n=1), or p27 (n=1) at monthly intervals. Three rhesus and 1 cynomolgus macaque received 2 vaginal followed by 3 oral (V-O) monthly immunizations of p27:Ty-VLP/CTB and 1 control rhesus macaque received Ty-VLP/CTB. Topical vaginal administration of 200 mg of p27:Ty-VLP/CTB (or Ty-VLP/CTB) was carried out using soft lubricated pediatric naso-gastric tubes. Oral administration was performed by intra-gastric intubation of gelatin-coated capsules, containing 500 mg of the vaccine and 400 ml cholera vibrio (Institut Merieux, Lyon, France), in the presence of sodium bicarbonate. One month after the last mucosal administration, all macaques were challenged by intramuscular immunization with 200 mg of the preparation used to immunize that animal and mixed with aluminium hydroxide (AluGel, Uniscience Ltd, London), except for the p27/CTB immunized animal which received 130 mg of p27/CTB in AluGel. Blood was collected from the femoral vessels and the serum was separated. All procedures in the macaques were carried out by prior sedation with IM Ketamine hydrochloride (10 mg/kg; Parke-Davis Veterinary, U.K.).

Mononuclear Cell Separation

Mononuclear cells were separated from defibrinated blood by lymphoprep (NYCOMED, Oslo) density gradient centrifugation by established procedures. The spleen, genital (obturator), iliac and para-aortic, mesenteric, bronchial, and axillary lymph nodes were removed at autopsy from 4 macaques immunized by the vagino-oral (n=2) or oro-vaginal (n=2) route and boosted by IM immunization. The cells were separated after breaking up the tissues by established procedures. Enriched monocytes were isolated by incubating the cells in plastic plates in RPMI (containing 10% FCS), for 1 hr at 37° C, with 5% CO_2 (Fig. 1). Non-adherent cells were removed and adherent cells were incubated with RPMI overnight at 37° C and recovered by washing the plates. The non-adherent cells were separated into 2-aminoethylisothiouronium-treated sheep red blood cells binding (T+ cells) and nonbinding (T-; enriched B cells) by established procedures. The T cells were further separated by panning 5 x 10^6 cells with a pre-determined optimal amount of monoclonal anti-T4 culture supernatant (100 ml per 10^6 cells) in Hanks solution (with 10% FCS), overnight at 4°C. After washing, 15 x 10^6 cells were added to petri dishes which had been coated with affinity-purified goat anti-mouse IgG (at 5 mg/ml, in 0.5M Tris-HCl pH 9.5), for 70 min, at 4°C. The non-adherent cells consisted of enriched CD8 cells and the adherent cells were enriched CD4 cells).

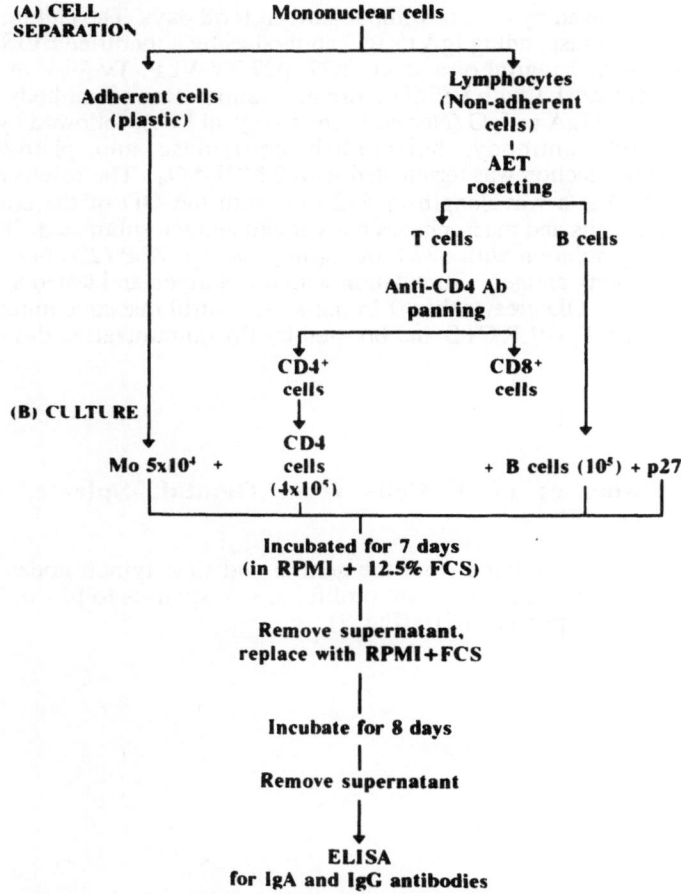

Figure 1 *In vitro* antibody synthesis

Lymphoproliferative Assay

The enriched T-cell subsets (10^5) were reconstituted with 10% monocytes and cultured without antigen and with 1, 10 and 20 mg/ml of p27, p27:Ty-VLP, Ty-VLP, CTB, R20 and concanavalin A (Con A) in 96 well round bottomed plates (Costar, Cambridge, MA), containing RPMI 1640 (GIBCO), supplemented with penicillin (100 mg/ml; Sigma) and streptomycin (100 mg/ml), 2 mmol/1-L glutamine (Sigma) and 10% autologous serum, for 4 days and then pulsed with 0.5 m Ci ^3H-thymidine for 4 h. The cells were then harvested on filter paper discs and the ^3H-thymidine uptake determined by scintillation counting. The results were expressed as stimulation indices (ratio of counts with and without antigen), and as cpm for cultures stimulated with 10mg/ml of p27, those stimulated with p27 Ty-VLP gave similar results All cultures yielded high stimulation indices and counts with concanavalin A, and no significant counts were found with CTB or R20 The mucosal route of immunization failed to elicit a rise in ^{3}H-thymidine uptake when the cells were stimulated with Ty-VLP However, after IM administration of the immunogens, moderate responses were elicited by stimulation with Ty-VLP

CD4 Helper Function in B Cell Antibody Synthesis

In vitro IgA and IgG antibody synthesis was determined with reconstituted B cells, CD4 cells and monocytes (Figure 1B) [28] Enriched B cells (10^5), CD4 cells (4×10^5) and monocytes (5×10^4) were reconstituted and stimulated with p27, Ty-VLP or TT (200 or 20 ng/ml) for 7 days, followed by culture without antigen for 8 days The culture supernatants were assayed for the corresponding IgA or IgG antibodies by a modified ELISA Microtiter plates were coated with 1 mg/ml of antigen (p27, p27 Ty-VLP, Ty-VLP or TT) Culture supernatants were diluted 1 1 with RPMI before incubation Bound antibody was detected using goat anti-monkey IgA or IgG (Nordic Immunological Lab), followed by biotinylated rabbit anti-goat IgG antibody, horseradish peroxidase and phenylenediamine dihydrochloride The reaction was terminated with 2 M H_2SO_4 The results are expressed as mean ($\pm$ sem) OD at a wavelength of 492 nm, with the OD of the control culture, consisting of CD4, B cells and macrophages but without antigen subtracted The results are presented only for stimulation with p27 (200 ng/ml) and Ty-VLP (200 ng/ml) and tested against the corresponding antigen Stimulation with one antigen and tested against another antigen did not show an OD greater than 0 15 units A control macaque immunized by the vagino-oral route with Ty-VLP/CTB and boosted by IM immunization did not yield p27 antibodies

RESULTS

Proliferative Responses of T Cells from Genital, Splenic, and Other Lymphoid Tissues

Mononuclear cells isolated from the genital and iliac lymph nodes draining the cervico-vaginal tissues showed significant proliferative responses to p27 or Ty-VLP (but not to CTB or the random peptide, R20) (Fig 2)

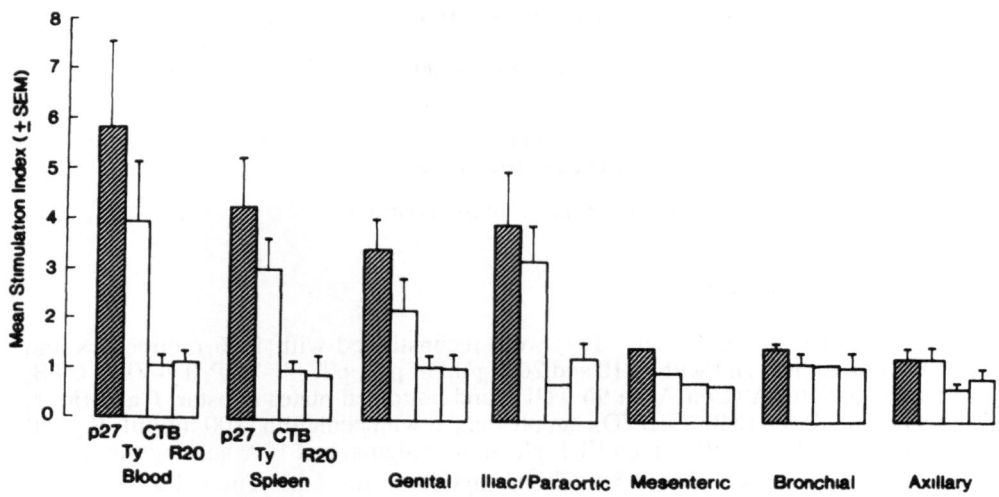

Figure 2. Stimulation indices of the mononuclear cells from 4 macaques of 5 regional lymph nodes, spleen and blood stimulated with p27, Ty VLP, CT B or T20

However, the superior mesenteric, bronchial or axillary lymph node cells failed to respond to these antigens, although they were readily stimulated with Con A. This suggests that augmented vaginal immunization does not seem to involve the entire mucosa-associated lymphoid tissue, since neither the superior mesenteric nor the bronchial lymph nodes yield specific lymphoproliferative responses. Splenic and circulating mononuclear cells also showed significant T-cell responses. Comparative T-cell proliferative assays of the enriched CD4 and CD8 subsets, reconstituted with 10% monocytes and stimulated with p27 showed that the responding cells belong to the CD4 subset. Genital, iliac, splenic, and circulating CD4-enriched cells yielded SI of 3.8 to 6.3, as compared with SI < 2.0 with CD8 enriched cells (Fig. 3). Intra-muscular without prior mucosal immunization elicited circulating (SI 8.8; 6012 cpm) and splenic (SI 8.4; 3633 cpm) T-cell responses but the genital and other lymph node cells did not respond to any of the antigens (SI < 2.0).

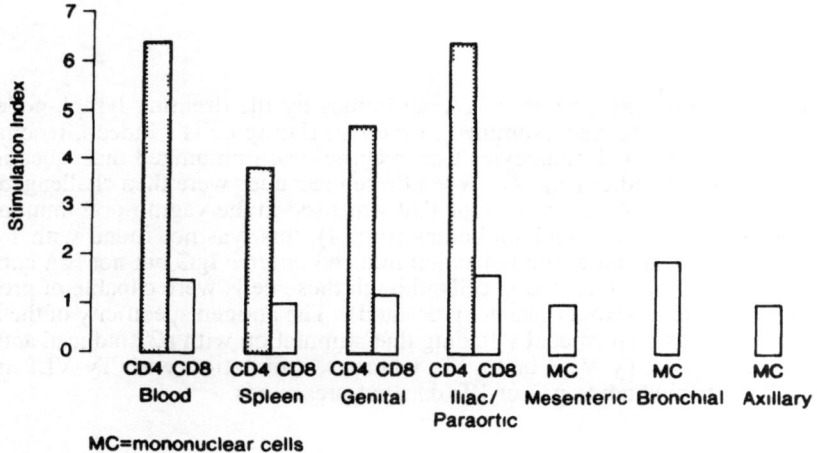

Figure 3. *In vitro* antibody synthesis.

CD4 Helper Function in B Cell Antibody Synthesis

In order to determine if augmented vaginal immunization induces CD4+ helper T cells capable of helping B cells to produce specific antibodies we carried out *in vitro* reconstitution experiments.[28] Enriched B cells, CD4 cells and macrophages were separated from splenic and genital cells and cultured without and with p27. Anti-p27 IgA or IgG antibodies were determined by ELISA. As can be seen from Table 1 no IgA or IgG antibodies were elicited by B cells alone, B cells reconstituted with enriched CD4 cells or with macrophages. Significant anti-p27 antibodies were only elicited in the presence of B cells, CD4 cells, macrophages and p27 antigen. Indeed, substitution of the CD4 cells for CD8 cells abrogated the antibody response. A comparison of genital with splenic CD4, B cells and macrophages showed that the genital cells produced more anti-p27 IgA than IgG antibodies and the reverse was found with the splenic cells (Table 1).

Table 1. *In vitro* antibody synthesis by reconstituted B cells, CD4 cells and macrophages with P27 antigen from the genital lymph nodes or spleen of macaques immunized by the oro-vaginal route with P27 Ty-VLp/CTB

		Antibodies to p27				
Cell		Spleen			Genital	
Reconstition	p27	IgA	IgG	IgA	IgG	
B		0 02	0 11	0 04	0 03	
B	+	0 13	0 12	0 03	0 02	
B + CD4		0 11	0 14	0 02	0 0	
B + CD4	+	ND	ND	0 02	0 01	
B + M		0 10	0 11	0 0	0	
B + M	+	0 13	0 13	03	0 0	
B + CD4 + M		0 12	0 13	0 02	0 02	
B + CD4 + M	+	0 28	0 51	0 26	0 18	
B + CD8 + M		0 11	0 15	0 0	0 02	
B + CD8 + M	+	0 13	0 15	0 0	0 0	

The results are expressed in OD at 405 nm

The preferential expression of IgA antibodies by the draining lymph nodes after vaginal immunization was also examined in the circulating cells   Indeed, reconstituted circulating CD4, B cells and monocytes from vagino-oral immunized macaques induced predominantly IgA antibodies (Fig 4)   When these macaques were then challenged by the intra-muscular route with the same vaccine that was used in the vagino-oral immunization there was a shift from IgA to IgG antibodies (Fig 4), this was not found with Ty-VLP Systemic without prior mucosal immunization induced *in vitro* IgG but not IgA antibodies to p27 by the reconstituted circulating cells, though these cells were capable of producing IgA antibodies to tetanus toxoid (data not presented)   The antigen specificity of the *in vitro* antibody synthesis was confirmed, by finding that stimulation with p27 induced antibodies only to p27 and not to Ty-VLP or to TT, whereas stimulation with Ty-VLP induced antibodies to Ty-VLP but not to p27 or TT (data not presented)

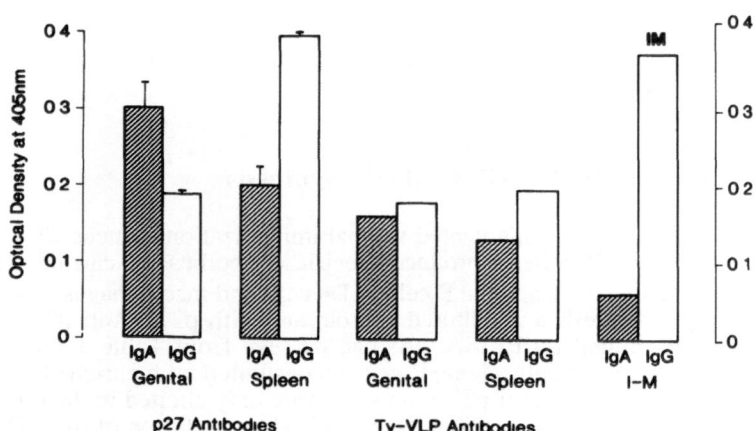

Figure 4. *In vitro* IgA and IgG anti p27 or Ty VLP antibodiees reconstituted CD4 cells B cells and macrophages comparing genital lymph node with spleen cells in macaques immunized by the vagino oral followed by the IM route (n=3)

DISCUSSION

Vaginal augmented by oral immunization with a particulate p27 Ty-VLP vaccine covalently linked to CTB elicited secretory IgA and IgG antibodies in vaginal fluid and specific serum IgA and IgG antibodies in non-human primates [23] [29] These macaques also developed specifically sensitized CD4 cells in the circulation which were stimulated *in vitro* by p27 to undergo proliferative responses and to help B cells in producing anti-p27 antibodies [23] [30] Hence, local mucosal and central systemic immunity were induced by the augmented vaginal immunization In addition, however a third level of immunity was induced in the regional lymph nodes (genital and iliac), as well as in the spleen CD4[+] T cells from the genital lymph nodes or the spleen were specifically stimulated with p27 to undergo proliferation and the CD4[+] T cells were capable of helping B cells to produce anti-p27 antibodies However, the reconstituted CD4 cells, B cells, and macrophages from the genital lymph nodes induced more IgA than IgG antibodies, whereas the corresponding splenic cells induced more IgG than IgA antibodies Finding specifically primed T and B cells in the genital but not in the superior mesenteric, bronchial or axillary lymph nodes after vaginal immunization in non-human primates is consistent with the concept of genital-associated lymphoid tissue developed in mice [14] [15] Whether the T cells home to the genital mucosa in primates, as has been demonstrated by adoptive transfer of genital lymph node T cells primed to HSV-2 in mice[16] needs to be investigated

It was surprising to find that T cells from the superior mesenteric lymph nodes failed to respond to p27, although the vaccine was administered orally before or after vaginal immunization Peyer's patches are difficult to isolate and culture in the macaque, but recently we have found that these cells can be primed to p27 [31] Furthermore, T cells from the inferior mesenteric lymph nodes respond to p27, so that part of the GALT is primed Whereas genital immunization alone or oral immunization failed to elicit vaginal antibodies,[18] it seems that oral immunization augments genital immunity Although the mechanism is unknown, it is envisaged that homing of B and T cells from GALT to genital-associated lymphoid tissue (GENALT) is facilitated within the mechanism of MALT, and this might expand the specific B- and T-cell populations required for secretory IgA production Alternatively, specific regulatory T cells generated in the GALT may home to GENALT and trigger off secretory IgA production

We suggest that this pre-clinical primate model of non-invasive mucosal immunization may provide a basis for immunization against a variety of sexually transmitted microbial diseases In this primate model recombinant particulate SIV p27 Ty-VLP vaccine linked to CTB elicited 3 levels of immunity in the genital mucosa, in the genital lymph nodes, and in the circulating lymphocytes and serum Cervico-vaginal secretory IgA antibodies may function as a first line of defence, preventing epithelial attachment and transmission of SIV through the epithelium Recent evidence suggests 3 possible mechanisms for the function of secretory IgA extracellular prevention of mucosal attachment, intracellular interference with viral replication by IgA antibodies binding to newly synthesized viral proteins,[32] and elimination of any viral peptide-antibody complexes found in the tissues by epithelial transcytosis [33] If this barrier were however, to be breached and the SIV (or HIV) is carried by Langerhans cells or macrophages to the draining genital lymph nodes, the primed T and B cells might prevent infection A failure of both the mucosal and lymphoid barriers still leaves the circulating IgA and IgG antibodies and specifically sensitized T and B cells to prevent SIV infection

SUMMARY

We investigated genital-associated lymphoid tissue (GENALT) in non-human primates (macaques), by augmenting vaginal with oral immunization The vaccine was a recombinant particulate SIV antigen (p27 Ty-VLP), linked to CT-B, and administered into the vagina by a paediatric naso-gastric tube and into the stomach by a gastric tube Oro-vaginal or vagino-oral sequence of immunization elicited specific CD4[+] T cell proliferative responses to p27 antigen in the genital lymph nodes and the spleen but not in unrelated lymph nodes CD4[+] T cells reconstituted with B cells and macrophages from the genital lymph nodes induced specific IgA and to a lesser extent IgG anti-p27 antibodies. However, the corresponding splenic cells induced greater IgG than IgA antibody synthesis. Intramuscular immunization primed splenic but not genital lymph node cells, and induced CD4$^+$ T cell proliferative responses and predominantly B cell IgG antibody synthesis. Finding primed B and T cells in the genital lymph nodes after augmenting vaginal by oral immunization provides experimental evidence for GENALT in non-human primates. This primate model of vaginal immunization suggests 3 levels of specific immunity: (1) secretory IgA (and IgG) in the cervico-vaginal mucosal epithelium; (2) primed CD4$^+$ T cells and B cells in the genital lymph nodes and the spleen; and (3) circulating CD4$^+$ T cells, B cells and IgG and IgA antibodies specific to the immunizing antigen.

REFERENCES

1. C. W. Craig and J. J. Cebra, *J. Exp. Med.* 134:188 (1971).
2. O. Rudzik, R. L. Clancy, D. Y. E. Perey, R. P. Day, and J. Bienenstock. *J. Immunol.* 114:1599 (1975).
3. P. C. Montgomery, J. Cohn, and E. T. Lally, *Adv. Exp. Med. Biol.* 45:453 (1974).
4. P. C. Montgomery, I. Lemaitre-Coelho, and E. T. Lally, *Ric. Clin. Lab. 6 (Suppl. 3)* 93:93 (1976).
5. A. J. Husband and J. L. Gowans, *J. Exp. Med.* 148:1146 (1978).
6. M. R. McDermott and J. Bienenstock, *J. Immunol.* 122:1892 (1979).

7. P. Weisz-Carrington, M. E. Roux, M. McWilliams, J. M. Phillips-Quagliatta, and M. E. Lamm, *J. Immunol.* 123:1705 (1979).
8. P. L. Ogra and S. S. Ogra, *J. Immunol.* 110:1307 (1973).
9. C. R. Wira and C. P. Sandoe, *J. Immunol.* 138:4159 (1987).
10. C. R. Wira and C. P. Sandoe, *Immunology* 68:24 (1989).
11. E. L. Parr, M. B. Parr, and M. A. Thapar, *J. Reprod. Immunol.* 14:165 (1988).
12. M.B. Parr and E.L. Parr. *Biol.Reprod.* 44:491-498 (1991).
13. M. A. Thapar, E. L. Parr, and M. B. Parr, *Immunology* 70:121 (1990).
14. M. R. McDermott, J. R. Smiley, P. Leslie, J. Brais, H. E. Rudzroga, and J. Bienenstock, *J. Virol.* 51:747 (1984).
15. M. R. McDermott, P. L. J. Brais, G. C. Goettsche, M. J. Evelagh, and C. H. Goldsmith, *Arch. Virol.* 93:51 (1987).
16. M. R. McDermott, C. H. Goldsmith, K. L. Rosenthal, and L. J. Brais, *J. Infect. Dis.* 159: 460 (1989).
17. S. L. Yang, and G. F. B. Schumacher, *Fertil. Steril.* 32:588 (1979).
18. T. Lehner, C. Panagiotidi, L. A. Bergmeier, T. Ping, R. Brookes, and S. E. Adams, *Vaccine Res.* 1:319 (1992).
19. N. L. Letvin, M. D. Daniel, P. K. Sehgal, R. C. Desrosiers, R. D. Hunt, L. M. Waldron, J. J. Mackey, D. K. Schmidt, L. V. Chalifoux, and N. W. King. *Science* 230: 71 (1985).
20. M. D. Daniel, L. N. Letvin, N. W. King, M. Kannagi, P. K. Sehgal, R. D. Hunt, P. J. Kanki, and R. C. Desrosiers, *Science* 228:1201 (1985).
21. M. Murphey-Corb, L. N. Martin, S. R. S. Rangan, B. J. Gormus, R. H. Wolf, W. A. Andes, M. West, and R. C. Montelaro, *Nature* 321:435 (1986).
22. E. J. Stott, W. L. Chan, and K. H. G. Mills, *Lancet.* 336:1538 (1990).
23. T. Lehner, L. A. Bergmeier, C. Panagiotidi, L. Tao, R. Brookes, L. Klavinskis, P. Walker, R. G. Ward, L. Hussain, A. J. H. Gearing, and S. E. Adams. *Science* (In press) (1992).
24. N. R. Burns, S. Craig, S. R. Lee, S. M. H. Richardson, N. Sterner, S. E. Adams, S. M. Kingsman, and A. J. Kingsman, *J. Mol. Biol.* 216:207 (1990).
25. N. R. Burns, J. E. M. Gilmour, S. M. Kingsman, A. J. Kingsman, and S. E. Adams, *Methods in Molecular Biology* 8, (ed.) M. Collins) p. 277 (1991).
26. N. Almond, M. Page, K. Mills, A. Jenkins, C. Ling, R. Thorpe, and P. Kitchin. *J. Virol. Methods* 28:301 (1990).
27. C. Czerkinsky, M. W. Russell, N. Lycke, M. Lindblad, and J. Holmgren, *Infect. Immun.* 57:1072 (1989).
28. R. Fellowes and T. Lehner, *J. Immunol. Methods* 132:165 (1990).

29. L. A. Bergmeier, L. Tao, A. J. M. Gearing, S. Adams, and T. Lehner, *Advances in Experimental Medicine and Biology* (this volume).

30. C. Panagiotidi, L. A. Bergmeier, A. J. M. Gearing, S. E. Adams, and T. Lehner, *Advances in Experimental Medicine and Biology* (this volume).

31. R. Brookes, L. A. Bergmeier, L. Klavinskis, and T. Lehner, (In preparation) (1992).

32. M. B. Mazanec, C. S. Kaetzel, M. E. Lamm, D. Fletcher, and J. G. Nedrud, *Proc. Natl. Acad. Sci. USA* 89:6901 (1992).

33. C. S. Kaetzel, J. K. Robinson, K. R. Chintalachuvu, J. P. Vaerman, and M. E. Lamm, *Proc. Natl. Acad. Sci. USA* 75:8796 (1991).

IMMUNOMORPHOLOGIC STUDIES OF HUMAN DECIDUA-ASSOCIATED LYMPHOID CELLS IN NORMAL EARLY PREGNANCY

Lucia Mincheva-Nilsson,[1] Vladimir Baranov,[2] Moorix Mo-Wai Yeung,[1] Sten Hammarstrom[1] and Marie-Louise Hammarstrom[1]

[1]Department of Immunology, University of Umea
90185 Umea, Sweden 2All Union Cancer Research Center
Academy of Medical Sciences, Moscow, Russia

INTRODUCTION

The pregnant uterus is an immunologically privileged site. A semiallogeneic fetus is allowed to implant in the mucosa, survive, develop and grow, while the same embryonic tissue, transplanted elsewhere in the body will be rejected.[1] The mechanisms underlying immunological acceptance of the fetus are not understood. One factor believed to be of importance is the strict regulation of MHC antigen expression on trophoblasts - fetal cells of the placenta that form the outermost border toward the mother. In human trophoblasts polymorphic MHC class I and II antigens are absent. However, non-polymorphic HLA-G antigens are expressed at certain stages of placental development.[2] Another factor believed to be of importance is local immunosuppression. Immunosuppresive factors, notably TGFβ[3] and suppressor T cells, have been demonstrated in mice. The pregnant uterus mucosa - decidua - consists of epithelial cells, lining the metrial glands, stromal cells, lymphoid cells and the endothelium of the vessels. Lymphoid cells are abundant in decidua. Already in 1921 it was reported[4] that decidua contained granulated cells now identified as CD56+ large granular lymphocytes (LGL). We have shown[5] that decidua lymphocytes are comprised of 4 major cell populations of similar sizes. They are: TCRγδ+/CD56- cells, TCRγδ+/CD56+ cells, TCRγδ+/CD56+ cells and TCRαβ+/CD8+ cells. Minor populations are TCRαβ+/CD4+ cells and B-cells. The γδ cells and the CD56+ cells express activation markers, e g CD45RO, KP43 and the mucosa associated integrin/activation marker HML-1.

In this report we show that human decidual lymphocytes are to a large extend organized in lymphoid cell clusters (LCC) that comprise different population of activated cells. We also show that two morphologically different populations of γδT-cells are present in decidua.

METHODS

Decidua Specimens and Isolation of Decidual Mononuclear Cells (DMC)

Vaccuum extraced decidual tissue was donated by healthy women undergoing elective termination at 6-10 weeks of gestation. DMC were isolated by mechanical disruption of thoroughly washed tissue followed by gradient centrifugation as described.[5]

Advances in Mucosal Immunology, Edited by
J. Mestecky et al., Plenum Press, New York, 1995

Monoclonal Antibodies

The following mAbs were used: anti-CD45 (PD7/26 and 2B11), anti-CD45RO (UCHL-1), anti-CD3 (T3-4B5), anti-CD4 (T3-10), anti-CD8 (DK25), anti-CD5 DK 23), anti-CD7 (DK24) all from Dakopatts, Glostrup, Denmark; anti-TCRX (CyM1 and bV1) from T cell Sciences, Cambrige, MA; anti-TCRoc,δ(WT31), antiCD14 (Mf-P9), anti-CD56 (MY-31) and anti-HLA-DP (B7/21) all from Beckton-Dickinson, Mountain View, Ca; anti HLA-DR(BL2), anti-HLA-DQ (SPVL3) and anti-HML-1 (2G5.1) all from Immunotech, France and anti-Kp43 (HP3B 1), a kind gift from Dr. M. Lopez-Botet, Seccion de Immunologia, Hospital de la Princessa, Madrid, Spain.

Immunohistochemistry

Pieces of decidua were snapfrozen in isopenthane precooled in liquid nitrogen. The tissue was cut into 5 µm sections in a cryostatic microtom, fixed in acetone for 10 minutes at 4° C and subjected to immunoperoxidase staining as described.[5]

Immunoelectron Microscopy

Small pieces of decidua were fixed in 4 % paraformaldehyde in 0.1 M cacodylic buffer for 2 hrs at 4° C. After washing the fixed tissue was frozen and cut into 15 µm thick sections. Floating sections were stained using the immunoperoxidase staining procedure.[5] The sections were again fixed with 1.33 % OSO_4, dehydrated in acetone and flat embedded in a mixture of Epon and Araldite. Ultrathin sections were examined in an electron microscope Zeiss EM 109 without additional staining.

Electron Microscopy of Sorted Cells

DMC were treated with anti-TCRγδ mAbs followed by FITC labled anti-mouse Ig and sorted in a fluorescence activated cell sorter (>95 % TCRγδ+ cells). The γδ cells were fixed in 2 % paraformaldehyde and 2.5 % glutaraldehyde in 0.1M cacodylic buffer at 4° C for 1 hour. After washing the cells were fixed in 1 % OSO_4 and embedded in Epon and Araldite. Ultrathin sections were examined before and after staining with lead citrate.

RESULTS

Morphometric analysis[6] revealed that the CD45+ cells comprise 10-15 % of all cells in decidua. The leucocytes in decidua were situated intra- or subepithelially, or in the stroma or around vessels. A large fraction of the leukocytes were present in aggregates, lymphoid cell clusters (LCC), mainly localized in close proximity to metrial glands (Table 1).

LCC were found in all decidual samples examined (6-10 weeks of gestation). All cells were CD45+ . The surface marker profile of the cells within the LCC is shown in Table 1. At least four different cell populations were detected, namely TCRγδ+ cells, CD56+ cells, TCRαβ+/CD8+ cell and TCRαβ+/CD4+ cells. Many of the CD56+ cells are probably TCRγδ+ since such double positive cells have been identified by two color immuno flow cytometry on isolated DMC.[5] The cells in the clusters were activated as revealed by the high frequency of cells with activation markers (Kp43; MHC class II antigens). B cells were not seen.

Immunoelectron microscopy revealed that the dominating CD45+ cell population had long thick processes, microvilli and cytoplasmatic granules (Fig. 1, anti-CD45). The CD45+ cells had close contact with each other in the LCC (Fig. 1, anti-CD45. The intraepithelial lymphocytes had intimate contact with the surrounding epithelial and stromal cells (Fig. 1, ant-Kp43) and sometimes also with other leukocytes (data not shown). Decidual tissues contained lymphocytes showing signs of cellular movement and granula excretion.

Table 1. Characteristics of the lymphoid cell clusters (LCC).

Frequency in the tissue		$0.33/mm^2$
Location	Close to gland	72 %
	Close to vessel	18 %
	Distant from gland or vessel	10 %

Surface marker	Frequency of marker positive cells
TCR γδ	++[1]
TCR αβ	+
CD4	+
CD8	+
CD3	++[2]
CD2, CD5, CD7	++
CD56	+++
CDl9, CD20, CD22	-
CD14	-
Kp43	+++
HML- 1	++
HLA-DR	+++
HLA-DP	++
HLA-DQ	++

[1]+++, 60 100 % marker positive cells
++, 30-60 % marker positive cells
+, 5-30 % marker positive cells
-, <5 % marker positive cells
[2]heterogeneous staining, only a small fraction of the cells were strongly positive

When TCRγδ+ cells were enriched by cell sorting and analyzed by electron microscopy two cell types were observed. One had a plain surface with short microvilli, membrane bound granules and nuclear inclusions consisting of short tubular membranous structures (Fig. 1A, γδ T cell). The other had a rough surface with many microvilli, non-membrane bound electron-dense granules and micro vesicular bodies characteristic of cytotoxic cells (Fig. 1B, γδ cell).

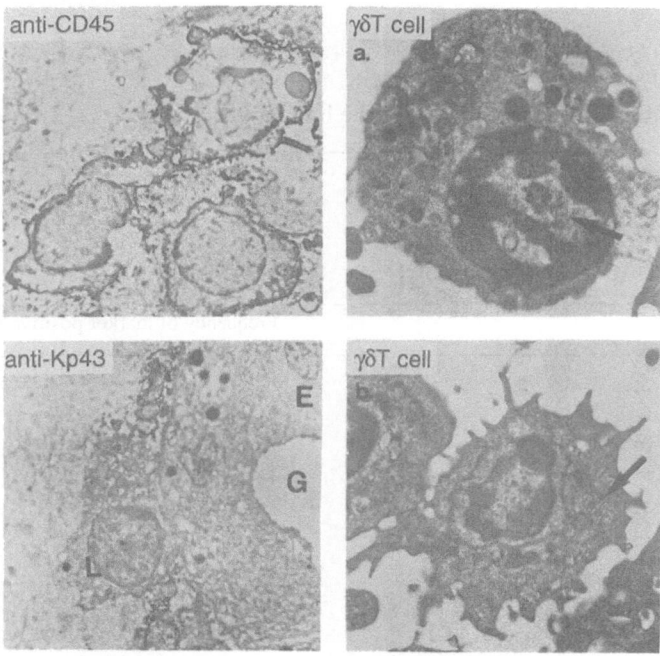

Figure 1. Immunoelectron microscopy of decidual tissue stained with anti-CD45 mAbs (PD7/26 and 2B11) and anti-Kp43 mAb (HP3B1) and electron microscopy of isolated γδ-cells G, glandular lumen, E, glandular epithelium, L, lymphocyte Arrows a) nuclear inclusions, b) microvesicular body

DISCUSSION

Human decidua from early pregnancy is rich in leukocytes, mainly TCRγδ+ cells and CD56+ cells. Decidual leukocytes comprise as much as 10-15 % of all decidual cells. The majority of the lymphocytes are either located intra- or subepithelially in metrial glands or in large aggregates termed the lymphoid cell clusters (LCC). The cells in the LCC are in close contact with each other and express activation markers. This close proximity suggests that the LCCs are sites of local immunoreactivity. Lack of B-cells and presence of different types of T-cells as well as NK cells further suggest that cell mediated immune reactions and/or cytokine production are induced in the LCC. In early pregnancy the immune system of the mother has to meet special demands: 1) the semiallogeneic fetus must be accepted; 2) ingrowth of trophoblast into vessels must be allowed in order to develop the chorionic villi and to insure that the fetus retains the necessary supply of nourishment ; 3) the ingrowth must be controlled so that undue throphoblast invasion is prevented; and 4) the maternofetal unit must be protected from infection.

It is possible that local immunosuppression is achieved in the LCC. We have shown that decidual lymphocytes respond poorly to alloantigen, although their response to mitogenic lectins is comparable to that of peripheral lymphocytes.[5] Another sign of immunosuppression in decidua is the low cytotoxic activity of the CD56[+] cells compared to NK cells in peripheral blood.[7]

To further understand the mechanism of suppression, studies of the cytokine production of decidual lymphocytes are in progress.

REFERENCES

1.   J. W. Streilein and T. G.Wegmann, *Immunol. Today* 8:362 (1987).
2.   S. Kovats, E. K.Main, C. Librach, M. Stublebine, S. J. Fisher, and R. DeMars, *Science* 248:220 (1990).
3.   D. A. Clark, K. Flanders, D. Banwatt, W. Millar-Book, J. Manuel, J. Stedronska-Clark, and B. Rowley, *J. Immunol.* 144:3008 (1990).
4.   P. Weill, *Arch. d~AnatMicrosc.* T.XVII:77 (1921).
5.   L. Mincheva-Nilsson, S. Hammarstrom and M-L Hammarstrom, *J. Immunol* 149:2203 (1992).
6.   E. R. Weibel, "Stereological Methods", Academic Press, London, New York, Toronto, Sydney, San Francisco (1979).
7.   A. King, and Y. W. Loke, *Cell. Immunol.* 118:337 (1989).

IMMUNOCYTES AND CELL-MEDIATED IMMUNITY IN THE PATHOLOGY OF REPRODUCTION

K. Nouza, R. Kinsky, M. Petrovská, D. Dimitorv,
R. Sedlák, J. Laitl, and J. Presl

Institute for Mother and Child Care, Praha, Czech
Republic

INTRODUCTION

The reproductive system is from the immunological point of view highly heterogeneous and sophistically regulated. The germinative central organs hold a privileged status: they must guarantee full tolerance of developing and mature male and female germinal cells. This tolerance is secured by anatomical barriers and complex immunoregulatory mechanisms.[1] General ovarian functions, as well as the particular development and functional maturation and regression of the human corpus luteum (hmCL) are under the well-known strict neuroendocrine control. The roles played by the immune system in ovarian control have been hypothetically outlined[2], but they are still incompletely proven. To understand better the roles of immune cells and their products during the whole life span of hmCL we have studied, using the immunoperoxidase method and its evaluation by computer-assisted image analysis, the dynamic distribution of leukocytes within various structures of the hmCL.

Mucosal immunocytes of the reproductive ducts and peritoneal environment represent a special but fully integrated part of the mucosal immune system. They restrict the entry and multiplication of infectious agents and participate in the support of passing germ cells. The leukocytes and their products are regularly found in semen and cervical mucus as well as in the uterine and peritoneal cavity. Recent studies further indicate dramatically increased numbers and unusual profiles of leukocytes in male and female genital tracts in some infertile patients and in the peritoneal fluid of women with endometriosis. [3,4] These often activated immunocytes and their products (lymphokines and monokines) may have detrimental effects on germinal cells, fertilization, the trophoblast, and embryo. [5] These and other data weaken the validity of the prevailing opinion, that humoral factors, especially antibodies against sperm, ovarian, endometrial, trophoblastic and embryonic antigens play decisive roles in the immunopathology of reproduction, and support the idea that pathologic cell-mediated immunity (CMI) may be involved in reproductive failures. This situation led us to optimize one of the classical tests of CMI, the one-step agarose indirect LMIF assay, and to adapt it for the exact detection of women and men hypersensitive to spermatozoa (and other reproductive targets). The use of a computer-assisted image analysis system for the estimation of leukocyte migration ensures precise quantitative measurement and avoids subjective factors affecting other methods.

The peritoneal fluid is an important contributor to the environment in which fertilization and early embryonic development take place. While the peritoneal fluid of healthy women has been reported to contain less than 1×10^6 leukocytes per milliliter, their numbers are significantly increased in women with endometriosis and unexplained

Advances in Mucosal Immunology, Edited by
J. Mestecky *et al.,* Plenum Press, New York, 1995

infertility. Also elevated levels of interleukin (IL)-1 and tumor necrosis factor (TNF)-α have been identified in the peritoneal fluid of infertile women with endometriosis and some other peritoneal diseases: these cytokines are supposed to have unfavorable effects on reproductive performance. For these reasons we have decided to determine the levels of IL-1, TNFα and IFNγ in the peritoneal fluids of our patients with endometriosis, infertility of known etiology, unexplained infertility and fertile controls, and compare them with cellular profiles characterized by a panel of monoclonal antibodies.

RESULTS

The hmCL study was performed on 10 hmCL specimens obtained from women of reproductive age hysterectomized for non-ovarian gynecological diseases. Dating was based both on the patient's information and on Corner's morphologic criteria. Individual structures were defined according to Nelson. Detailed description of the materials and methods is given in Pertovská *et al.*[6]

The most frequent immunocytes in young hmCL (1-6 days) were monocytes/macrophages (MM) and MHC-Class II+ cells (MHCII+). The former were most abundant within fibrous and theca-lutein trabeculae, and in the stroma, the latter in the granulosa-luteal layer. During the period of maximum hmCL activity (day 7-9, Fig. 1), the number of MHC II+ increased by a factor of 3 to 4, MM increased only by a factor ~2. At this stage, total T lymphocytes and Tc/s and Th/i cells first appeared in theca-lutein trabeculae, but their numbers were relatively low. In the regressing hmCL (over 10 days) MM still maintained the same distribution pattern, but neither of the remaining structures was dominant for HMC II+. Numbers of Tc/s cells decreased to one half in theca-luteal trabeculae. Th/i cells were detected in negligible numbers only in the stroma.

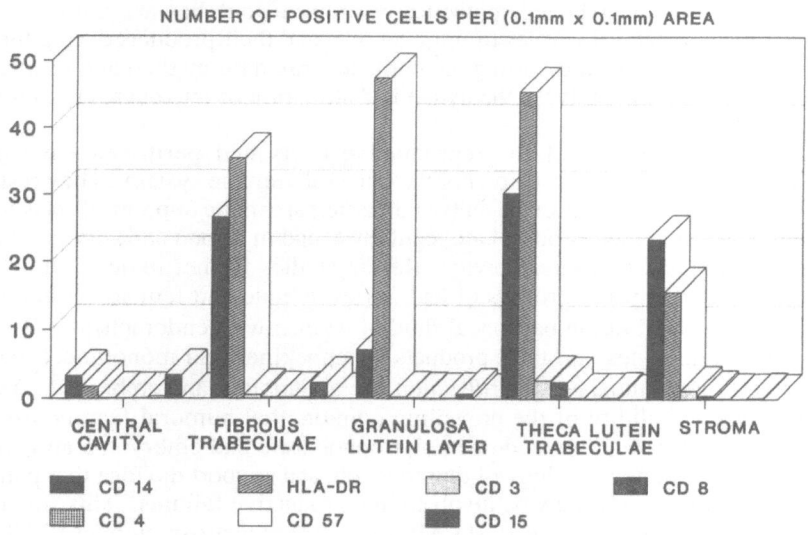

Figure 1. The peak of endocrine CL activity (day 7-9)

Our results suggest that MM and MHC II+ cells are the most prominent elements through the whole life span of hmCL. While the latter cells are probably of local origin, the former might be largely of blood origin. At least a subpopulation of MM probably supports the differentiation and functional activity of luteal cells. We failed to identify significant numbers of T lymphocytes in the hmCL; moreover, their presence in thecal trabeculae may be at least partially attributed to the vascularization of this structure. The idea that T lymphocytes may attract and activate macrophages therefore seems doubtful.

A detailed description of our method for determining antisperm CMI (asCMI) is given in Dimitrov et al.[7] The clinical trials included a total of 190 patients (103 females and 87 males), aged 21 to 40 years. All women were divided after thorough examination into 4 groups (fertile-24, unexplained infertility-25, infertility of known etiology-39, and recurrent abortions-15). Men were divided into 3 groups (normospermic-37, asthenospermic-40 and teratospermic-10). Patients with antisperm antibodies were omitted. All fertile women fell into the negative range of asCMI. Of women with unexplained infertility 80% revealed positive indexes. Among infertile women with known etiology only 7.6% reacted positively, while 60% of recurrently aborting women showed reactivity (Fig. 2). Among normospermic fertile men 4 revealed a positive reaction (10.8%). A high proportion of infertile men with asthenospermia revealed positive indexes (57.5%), while in 10 patients with teratospermia only 1 (10%) reacted positively (Fig. 3).

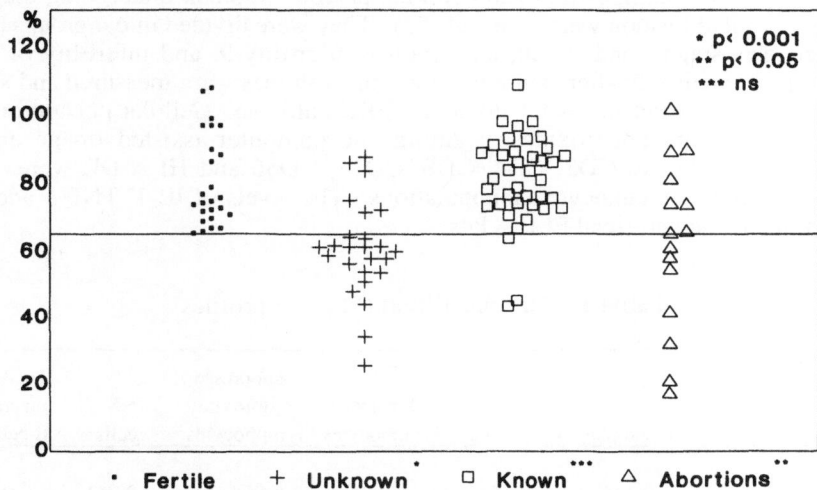

Figure 2. Migration indices of women with infertility and recurrent abortions

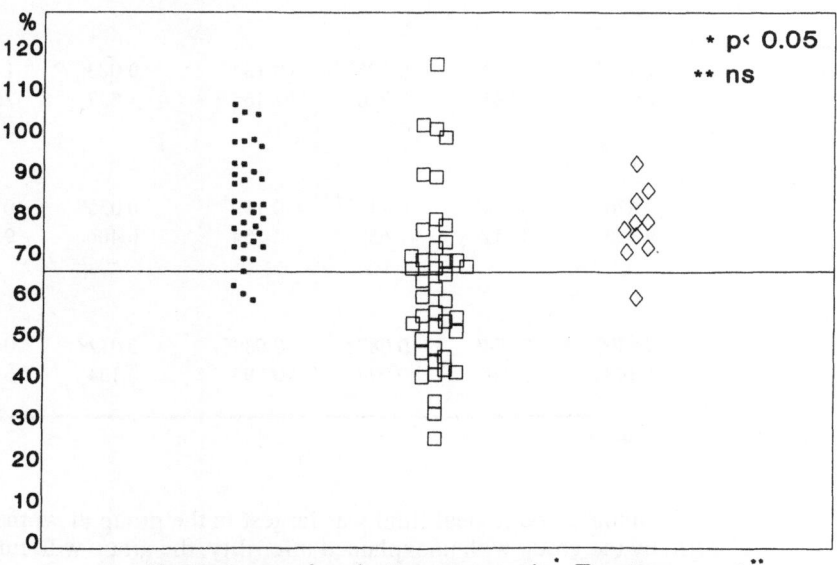

Figure 3. Migration indices of normo, astheno, and teratospermic men

The most important finding in our studies is a highly significant increase of positive asCMI in females with unexplained idiopathic infertility It therefore appears important to evaluate precisely asCMI in all cases where anatomical, infectious, and hormonal causes of infertility can be excluded Also increased was the incidence of asCMI in a group of recurrently aborting women, this favors the opinion concerning the existence of cell-mediated cross-reactivity against sperm antigens and epitopes expressed by elements of the fetoplacental unit Very interesting is the finding of pathologic asCMI in more than half of the infertile men with asthenospermia This phenomenon deserves further study

The incidence, albeit low, of positive asCMI infertile women of known etiology and normospermic men may indicate the presence of specific concomitant reactivity below the level of clinical manifestation or the existence of nonspecific (e g , inflammatory) processes affecting regulatory mechanisms The respective individuals will be further observed

The purpose of studying the cellular and humoral composition of peritoneal fluid was to determine whether various forms of female infertility are associated with a characteristic cellular composition and cytokine profile Women undergoing diagnostic laparoscopy or tubal ligation were included (52) They were divided into 4 groups (fertile-5, endometriosis stage I and II -20, unexplained infertility-9, and infertility of known etiology-19) Immediately after aspiration, the fluid volumes were measured and samples centrifuged The supernatants were stored at -20°C until use Cellular phenotyping was performed using immunoperoxidase staining and computer-assisted image analysis Monoclonal antibodies to CD3, CD4, CD8, CD14, CD56 and HLA-DR were used to differentiate individual leukocyte subpopulations The levels of IL-1, TNF-α and INF-γ were determined by commercial ELISA kits

Table 1. Peritoneal fluid (PF) cell profiles

Groups of patients	All cells	Macrophages	All T cells	T helper lymphocytes	T-suppressor/ cytotoxic lymphocytes	NK cells	HLA-DR+ (macrophages, B cells)
Fertile n=5							
per 1 ml PF	0 574	0 424	0 134	0 062	0 064	0 011	0 445
	4 707	3 476	1 098	0 508	0 524	0 090	3 649
Total mean vol 8 2 ml							
Endometriosis n=20							
per 1 ml PF	1 709*	1 181*	0 324*	0 175*	0 134*	0 083*	1 365*
	31 44	21 73	5 961	3 220	2 465	1 527	25 11
Total mean vol 18 4 ml							
Unexplained infertility n=9							
per 1 ml PF	1 022*	0 676[a]	0 274[a]	0 112**	0 159*	0 030[a]	0 713[a]
	13 84	9 159	3 712	1 653	2 154	0 406	9 661
Total mean vol 13 5 ml							
Known infertility n=19							
per ml PF	0 798[a]	0 549[a]	0 174[a]	0 087[a]	0 089[a]	0 012[a]	0 547[a]
	8 937	6 148	1 948	0 974	0 996	0 134	6 126
Total mean vol 11 2 ml							

*$p<0 01$, **$p<0 05$, [a]not significant

The mean total volume of peritoneal fluid was largest in the group of women with endometriosis followed by the group with unexplained infertility, the group with infertility of known etiology, and the fertile group The cellularity per milliliter of the fluid and per total fluid volume followed the same order The increase in patients with endometriosis and in patients with unexplained infertility was significant as compared with fertile controls (p <0.01 or <0.05). The dominating type of cells in any peritoneal fluid was macrophages and HLA-DR+ cells. Also both subpopulations of T cells and NK cells were present. Surprisingly, the proportions of individual cell types were quite similar in all groups of women. The only one difference characteristic for patients with endometriosis appeared to be an increase of NK cells (Table 1).

The levels of IL-1, TNF-α and IFN-γ were increased in all patients with endometriosis and in none of the fertile women. In the group of patients with infertility of known etiology, the levels of IL-1 were increased in 4, of TNF-α in 7, and of IFN-γ in 1 of 18 cases. In the group with unexplained infertility the levels of IL-1 were increased in 4, of TNF-α in 5, and of IFN-γ in 3 of 9 cases. In some but not all patients, the levels of at least two cytokines were correlated. A good correlation was also found in most cases between the concentration of cells and the levels of cytokines. It cannot be excluded that some positive findings in non-endometric patients may signal a hidden or microscopic endometriotic process. Other factors leading to an increase of cellularity and cytokine levels may also exist. However, the most specific indicator of endometriosis may be IFN-γ (Fig. 4).

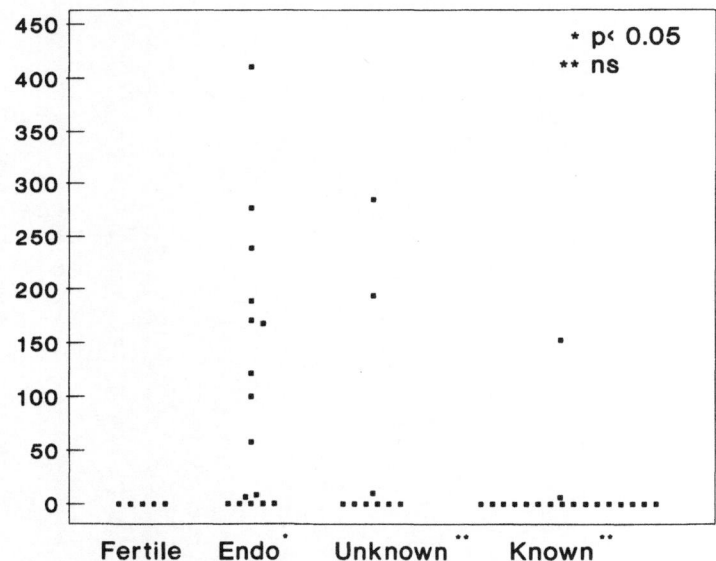

Figure 4. Peritoneal fluid INF-γ concentrations

REFERENCES

1. K. Nouza, R. Kinsky, and D. G. Dimitrov, *Folia Biol.* (Praha) 38:170 (1992).
2. A. Bukovsky and J. Presl, *Med. Hypoth.* 7:1 (1981).
3. N. J. Alexander and D. J. Anderson, *Fertil. Steril.* 47:192 (1987).
4. J. A. Hill, H. M. P. Faris, I. Schiff, and D. J. Anderson, *Fertil. Steril.* 50:216 (1988).
5. J. A. Hill, *Colloque INSERM* 212:269 (1991).
6. M. Petrovská, R. Sedlák, K. Nouza, J. Presl, and R. Kinsky, *Am. J. Reprod. Immunol.* 28:77 (1992).
7. D. G. Dimitrov, R. Sedlák, K. Nouza, and R. Kinsky, *J. Immunol. Methods* 154:147 (1992).

INFLUENCE OF ESTROUS CYCLE AND ESTRADIOL ON MITOGENIC RESPONSES OF SPLENIC T- AND B- LYMPHOCYTES

Rao H. Prabhala,[1] and Charles R. Wira[2]

[1]Department of Microbiology, Chicago College of
Osteopathic Medicine, Downers Grove, IL 60515
[2]Department of Physiology, Dartmouth Medical
School, Lebanon, NH 03756

INTRODUCTION

Reproductive tract tissues are known to undergo continuous cycles of cell proliferation, differentiation, and shedding. These events are precisely regulated by sex hormones during the normal menstrual cycle.[1] In earlier studies, we demonstrated that IgA-positive cells enter uterine tissues of ovariectomized rats in response to estradiol.[2] When uteri from saline- or estradiol-treated rats were analyzed, estradiol treatment increased the movement of T-lymphocytes and B-lymphocytes into uterine tissues[2,3]. The recognition that sex hormones influence lymphoid-cell movement to selected mucosal surfaces, directly affect B-lymphocyte responses to mitogens[4], inhibit T-suppressor cells[5,6], influence the maturation T lymphocytes[7,8], and alter macrophage function[9], provides overwhelming evidence that sex hormones are involved at the cellular level in immune regulation.

To investigate further the role of sex hormones in the regulation of the immune system under *in vivo* conditions, the present studies were undertaken to 1) examine the influence of the estrous cycle on the response of spleen cells to mitogens, and 2) determine whether estradiol affects splenic T- and B- lymphocyte responses to mitogens.

ENDOCRINE EFFECTS ON CELL MITOGENESIS

To examine the effect of the estrous cycle on the response of spleen cells to mitogens, animals were selected by daily vaginal smears after they had gone through at least one normal (4 day) estrous cycle. Spleen cells were prepared and incubated with Con A (1 mg/ml), PHA (5 mg/ml), or LPS (10 mg/ml) in 96-well microtiter plates for 3 days as described previously.[10] Cell proliferation was evaluated by measuring the incorporation of ^{3}H-thymidine added to incubation wells for the last 24 h. As shown in Fig. 1, stage of the estrous cycle had a marked effect on spleen cell mitogenesis. When responses were analyzed, mitogenesis of T and B lymphocytes from proestrous rats in response to Con A, PHA and LPS was significantly greater (2-3 fold) than that seen with spleen cells from animals at estrous and/or diestrous stages of the cycle.

To determine whether changes in spleen cell mitogenesis observed during the estrous cycle are under hormonal control, ovariectomized animals were treated with estradiol (2mg/day) or saline (0.1ml) for three days prior to sacrifice 24 h after the third injection.

Advances in Mucosal Immunology, Edited by
J. Mestecky *et al.*, Plenum Press, New York, 1995

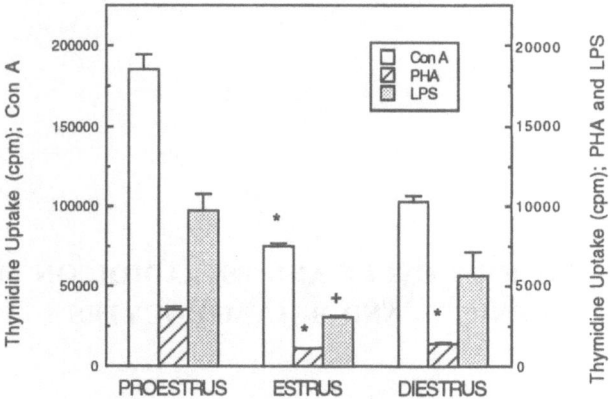

Figure 1. Mitogenic response of isolated spleen cells from intact female rats at various stages of the reproductive cycle. Animals were selected by routine vaginal smears and spleens were recovered during proestrus, estrus, and diestrus (day 2). Splenocytes were prepared as described previously 10, and incubated with ConA (1 mg/ml), PHA (5 mg/ml), and LPS (10 mg/ml) for 3 days. [3]H thymidine was added to culture wells for the last 24 h of each incubation. Each bar represents the mean ± SE minus background (counts per min/well in the absence of mitogen) of splenocytes pooled from (4-6) animals/group. +, Significantly (P<0.05) lower than proestrous values; *, significantly (P<0.01) lower than proestrous values.

As seen in Table 1, estradiol treatment had a stimulatory effect on spleen cell responses to both T and B cell mitogens relative to saline controls. These results indicate mitogenesis increases at the proestrous stage of the estrous cycle, when serum levels of estradiol are known to be elevated[11], and in response to estradiol given *in vivo*. Since sex hormones are known to play a central role in the recruitment of lymphocytes into the female reproductive tract, the magnitude of lymphocyte responses at a particular stage of the estrous cycle and/or following hormone stimulation may reflect either altered responsiveness of spleen cells to mitogen or the number of cells present in the spleen in response to hormonal influences.

Table 1. Effect of estradiol on the mitogenic response of spleen cells.

Groups	Thymidine incorporation (cpm)[a]		
	Con A	PHA	LPS
Control	36,813 ± 3061	3037 ± 436	6430 ± 1218
Estradiol	61,949 ± 10,065*	11,684 ± 869*	14,738 ± 1936*

[a]Results represent mean and SEM of a pool of six animals. *Significantly different (p<0.05) than control.

SUMMARY

Previous studies from our laboratory have shown that the stage of the estrous cycle and hormone treatment regulates the presence of immune cells in the female reproductive tract. The purpose of the present study was to determine whether sex hormones influence spleen cell responses to mitogens. The present study demonstrates that spleen cell responses to mitogens vary with the stage of the estrous cycle in intact rats. Further, our findings indicate that estradiol administered to ovariectomized animals increases the mitogenic response of spleen cells to both B and T lymphocyte mitogens.

ACKNOWLEDGMENTS

We wish to thank Mr. Richard M. Rossoll for his technical assistance and to Edie Coates for typing this manuscript. Supported by research grants AI-13541 and AI-07363 from NIH and CA-23108 from NCI.

REFERENCES

1. A. Ferenczy, G. Bertrand, and M. M. Gelfand, *Am. J. Obstet. Gynecol.* 133:859 (1979).
2. C. R. Wira, E. Hyde, C. P. Sandoe, D. A. Sullivan, and S. Spencer, *J. Steroid Biochem.* 12:451 (1980).
3. C. R. Wira and R. Prabhala, Eleventh Ann Bristol Myers-Squibb/Mead Johnson Symposium (1992) in press.
4. Z. M. Sthoeger, N. Chiorazzi, and R. G. Lahita, *J. Immunol.* 141:91 (1988).
5. T. Paavonen, L. C. Anderson, and H. Adlercreitz, *J. Exp. Med.* 154:1935 (1981).
6. C. J. Grossman, *Science* 227:257 (1985).
7. S. A. Ahmed, M. J. Dauphinee, and N. Talal, *J. Immunol.* 134:204 (1985).
8. Y. Weinstein and Z. Berkovich, *J. Immunol.* 126:998 (1981).
9. A. D. Schreiber, F. M. Nettl, M. C. Sanders, M. King, P. Szabolcs, D. Friedman, and F. Gomez, *J. Immunol.* 141:2959 (1988).
10. R. H. Prabhala and C. R. Wira, *Endocrinology* 129:2915 (1991).
11. A. A. Shaikh, *Biol. Reprod.* 5:297 (1971).

SECRETORY COMPONENT PRODUCTION BY RAT UTERINE EPITHELIAL CELLS IN CULTURE

Jan M. Richardson and Charles R. Wira

Department of Physiology
Dartmouth Medical School
Lebanon, New Hampshire 03756

INTRODUCTION

The female sex hormones have marked effects on the mucosal immune system in the female reproductive tract. Previous studies in this laboratory have shown that IgA, IgG and secretory component levels change in the uterus and the vagina during the reproductive cycle, as well as after hormone treatment.[1] Our laboratory and others[2,3,4] have found that estradiol regulates the mucosal immune system by a) controlling IgA and IgG movement from blood to tissue, b) stimulating the movement of IgA-positive cells into the female genital tract, and c) regulating the transfer of immunoglobulins from tissue to lumen. In the last case, IgA was found to move through uterine epithelial cells bound to its receptor, secretory component (SC), the external domain of the polymeric IgA receptor. SC production is also regulated by estradiol, IFN-γ[5], and IL-6.[6]

The purpose of this study was to define an *in vitro* system that can be used to analyze the mechanisms of hormone and cytokine action on IgA transport and SC synthesis in uterine epithelial cells. Studies were carried out to: 1) determine if luminal epithelial cells from ovariectomized and intact mature rats grow to confluence on Millicell chambers, retain polarity, and synthesize SC *in vitro* and 2) examine whether SC production is different in epithelial cells isolated from conventional and viral free (VAF) raised animals of the same strain.

METHODS

Adult female Sprague-Dawley rats (150-200 g) were obtained from Charles River breeding laboratories (Kingston, NY). Animals were maintained in a constant temperature room with light/dark intervals of 12 h and allowed food and water *ad libitum*. In experiments using cells from castrate rats, ovariectomies were performed 7-10 days before the start of each experiment.

To isolate uterine epithelial cells for culture, uterine horns from mature rats were slit lengthwise and incubated with 0.5% trypsin/2.5% pancreatin for 1 h at 4°C followed by 1 h at 22°C, as previously described.[7] Uteri were removed from enzyme, placed in Hanks buffered saline solution (Gibco, Grand Island, NY) and vortexed to release plaques of cells. Following centrifugation (5 min, 500 x g), and gravity sedimentation of plaques, cells were resuspended in DMEM:Ham's F-12 nutrient mixed 1:1, containing BSA, HEPES, and pen/strep supplemented with 2.5% FBS and 2.5% Nu-Serum. Cells were seeded into wells

(Millicell HA membranes, Millipore Inc) coated with Matrigel (Collaborative Bioresearch), at a density of 1/4-1/2 uterine epithelial cells/well Media was collected from the basolateral and apical chambers and replaced every 48 h Transepithelial resistance, used as an indication of tight junction formation in the epithelial cell monolayer, was monitored using an epithelial voltohmmeter (World Precision Instruments, Inc) SC levels in culture media were measured by radioimmunoassay (RIA) [8]

RESULTS

Analysis of SC production by uterine epithelial cells isolated from adult intact VAF animals at random stages of the estrous cycle and cultured as described above is shown in Fig 1 Transepithelial resistance was measured to monitor cell cultures for formation of tight junctions High transepithelial resistance indicates that epithelial cells from mature animals form tight junctions Apical and basolateral media from day 7 of incubation were assayed for SC production In this experiment, cells released approximately 65 ng of SC/48 h of culture into the apical media, with almost no SC found in the basolateral media, indicating that cells maintain polarity in culture In other studies, we have found that the amounts of SC released from epithelial cells on day 7 vary in different experiments from 10 ng to 65 ng in 48 h A 10 day time course study showed a correlation between high levels of SC in the apical medium and high transepithelial resistance (data not shown)

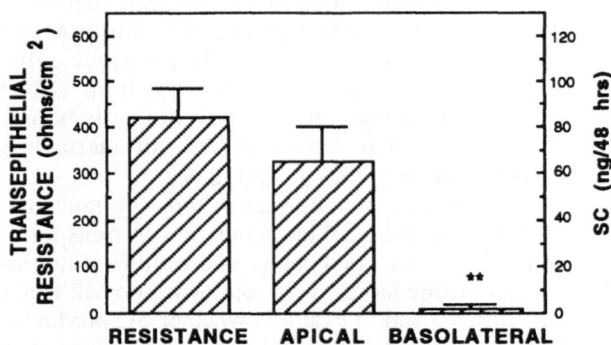

Figure 1. Uterine epithelial cells from intact VAF animals were isolated and cultured on Millicell chambers Apical and basolateral media from day 7 of culture were assayed for SC production and transepithelial resistance was measured ** significantly (P<0 001) less than apical media

Uterine epithelial cells from conventionally raised and viral free ovariectomized animals were isolated and cultured for 12 days and SC levels in the apical media were measured by RIA With ovariectomies, the hormone state of adult animals is similar to immature animals, in that both are hormonally deprived As seen in Fig 2, cells from both VAF and conventional animals grew to confluence and released similar amounts of SC into the apical media Virtually no SC was found in the basolateral media, SC levels in the basolateral media were consistently less than 2 0 ng in all studies (data not shown) both before and after cells grew to confluence and formed tight junctions When compared to previous studies of SC production by immature epithelial cells in culture[9], the results shown in Fig 2 demonstrate that SC production by uterine epithelial cells from ovariectomized rats is similar to that seen with epithelial cells from immature animals

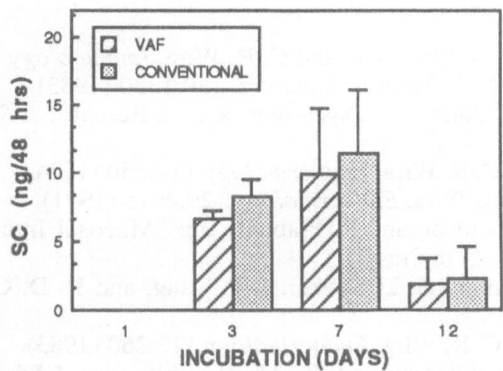

Figure 2. Uterine epithelial cells from conventionally raised and VAF ovariectomized animals were isolated and cultured for 12 days. Cell cultures were fed every 48 h. SC levels in the apical media were measured by RIA and represent ng SC produced/48 h.

DISCUSSION

SC, the external domain of the polymeric IgA receptor, transports dimeric or larger IgA polymers into external secretions. Using the rat model system, studies from this laboratory have demonstrated that SC production and the movement of IgA from uterine tissues into secretions is controlled by sex hormones and cytokines.[5,6] Previous studies have shown that uterine epithelial cells from immature rats grow to confluence, preferentially release SC into apical medium, and remain responsive to estradiol in culture.[9] This study demonstrates that uterine epithelial cells isolated from mature intact rats at all stages of the cycle as well as ovariectomized mature rats can be cultured on Millicell chambers. These polarized uterine epithelial cells produce SC and preferentially release it into the apical medium (analogous to the uterine lumen) of the Millicell chamber and form tight junctions, as measured by electrical resistance. Levels of SC produced by uterine epithelial cells from castrate conventionally raised animals and castrate VAF rats are the same. Additionally, when SC production by confluent epithelial cell cultures from mature intact animals was compared with data from cultures of immature rat epithelial cells[5], both cell populations were found to produce similar amounts of SC within the same time period. The magnitude of transepithelial resistance generated across the cell monolayer appears to be proportional to the amount of SC produced. With higher resistance, generally more SC is found in the apical media. Future studies will utilize this system to study the affects of hormones and cytokines on IgA transport and SC production and secretion. These studies will contribute to understanding the role of hormones and cytokines in mucosal immunity.

ACKNOWLEDGMENTS

We thank Dr. B. Underdown (McMaster University, Hamilton, Ontario) for his gifts of SC and rabbit anti-rat SC antibody and Mr. Richard Rossoll for excellent technical assistance Supported by Research Grant AI-13541 from NIH.

REFERENCES

1. D. A. Sullivan, B. J. Underdown, and C. R. Wira, *Immunology* 49:379 (1983).
2. D. A. Sullivan and C. R. Wira, *J. Immunol.* 130:1330 (1983).
3. F. Rauchman, V. Casimira, A. Psychoyos, and O. Berhard, *J. Reprod. Fertil.* 69:17 (1983).
4. D. A. Sullivan and C. R. Wira, *Endocrinology* 114:650 (1984).
5. R. Prabhala and C. R. Wira, *Endocrinology* 129:2915 (1991).
6. C. R. Wira, J. Richardson and R. Prabhala, *in*: "Mucosal Immunology: Mucosal Diseases", Vol. II, in Press.
7. S. R. Glasser, J. Julian, G. L. Decker, J. P. Tang, and D. D. Carson, *J. Cell Biol.* 107:2409 (1988).
8. D. A. Sullivan and C. R. Wira, *Endocrinology* 112:260 (1983).
9. C. R. Wira, S. R. Glasser and R. M. Rossoll, *in:* "Advances in Mucosal Immunology", T.T. MacDonald, ed., p. 599, Kluwer Academic Publishers, Dordrecht (1990).

HUMAN DECIDUAL CD7+ LYMPHOCYTES DISPLAY A UNIQUE ANTIGENIC PHENOTYPE

Igor I. Slukvin, Victor P. Chernyshov, and Genady I. Bondarenko

Ukrainian Institute of Pediatrics, Obstetrics and Gynecology, 252050 Kiev, Ukraine

INTRODUCTION

After implantation of the embryo, the human endometrium goes through profound changes to form the decidua. Chorionic villi grow into the decidua basalis and form the fetal part of the placenta. The inability to show any significant alterations in maternal systemic immunity in normal pregnancy has led to the suggestion that local intrauterine immunoregulatory mechanisms may be decisive for survival of the semi-allogenic fetus. Therefore, phenotypic and functional characterization of the lymphocytes in human decidua will improve understanding of the mechanisms of feto-maternal interactions.

Immunohistochemical analysis of human decidua in the first trimester of pregnancy identified three principal leukocyte populations: macrophages, large granulated lymphocytes with an unusual antigenic phenotype (CD2+, CD7+, CD38+, CD56+, CD3-, CD16-, CD57-, CD5-) and T lymphocytes as a minor component.[1] The role of decidual leukocytes is uncertain.

The aim of this study was to delineate the phenotype of decidual CD7+ lymphocytes using two-color flow cytometry and to demonstrate the use of a mechanical dispersal technique for the preparation of decidual cell suspensions for flow cytometry.

MATERIALS AND METHODS

Decidua was obtained from healthy women undergoing legal abortion of normal pregnancy at 8-12 weeks of gestation. In order to exclude damage to surface markers, a mechanical dispersal technique for the preparation of decidual cell suspensions was chosen. Suspensions enriched for decidual lymphocytes were obtained by two-step density gradient centrifugation (Percoll 35%/75%).

Monoclonal antibodies labeled with FITC or phycoerythrin (Becton Dickinson) were used for direct immunofluorescence. Cell fluorescence was studied on a FACScan flow cytometer (Becton Dickinson). The phenotype of decidual lymphocytes was analyzed after gating the cell subset that included ~90% of CD45+ cells or >75% of all decidual CD45+ cells. The viability of cells as determined by propidium iodide staining was usually >95%. Preliminary experiments showed a high level of non-specific staining of decidual cells with the Simultest control (mouse IgG1 FITC + IgG2a PE, Becton Dickinson). The problem of non-specific staining of decidual lymphocytes was resolved by pretreatment of cells with aggregated rabbit Ig for the remaining experiments.

Advances in Mucosal Immunology, Edited by
J. Mestecky *et al.*, Plenum Press, New York, 1995

RESULTS AND DISCUSSION

As determined by flow cytometry, the most abundant decidual leukocytes expressed CD7 (73.9±4.1%), CD38 (81.8±4.0), CD56 (73.4±5.0), and CD2 (63.6±5.3) markers; relatively small proportions of CD3 (10.8±0.8), CD8 (17.8±1.9), CD4 (5.4±0.7), CD16 (7.4±3.2), CD45RA (24.8±3.8), CD11b (16.2±4.3), CD14 (10.3±1.2), and Leu-8 (7.2±1.8) cells were also present. Among decidual lymphocytes we did not find in significant numbers B-cells (CD19[+]), CD57[+], CD10[+] or cells which expressed IL-2R-α. Proportions of CD38, CD56, CD3, CD8 and CD16 decidual lymphocytes were rather similar to those obtained by Bulmer *et al*[2] using immunohistochemistry. However, mechanical disaggregation gives a low yield of macrophages. Less than 10% of CD45[+] cells expressed the CD14 marker in our experiments, whereas in cryostat sections ~38% of CD45[+] cells were CD14[+]. It seems that mechanical dispersal techniques are more suitable for the study of decidual large granular lymphocytes and T cells than macrophages.

Like peripheral blood lymphocytes, the majority of decidual lymphocytes expressed the CD7 antigen. The function of the CD7 molecule still remains obscure. It is known that CD7 antigen is acquired by T cell precursors outside the thymic mocroenvironment. CD7 has been demonstrated within bone marrow and fetal lymphocytes before colonization of the thymus.[3] CD7 antigen is present on 85-90% of peripheral blood T cells and thymus-independent NK cells.[4] CD7 is expressed before TCR-β gene rearrangement and surface expression of CD1, CD2, and CD3.[3,5] CD7 antibodies inhibit alloproliferation of naive but not memory lymphocytes, and does not significantly affect NK or LAK cytotoxicity.[6] The CD7 molecule is a member of the Ig superfamily and is capable of transmembrane signaling.[7] It has been suggested that the CD7 molecule contains functional epitopes, which could be a binding site for its physiological ligand similar to other receptors of this multigene family, i.e., TCR, Ig, MHC class I, class II, CD4, and CD8.[8] It is possible that surface CD7 on decidual cells could interact with its physiological ligand to play a major role in activation and differentiation of decidual lymphocytes.

Surface phenotype of decidual CD7 cells is shown in Fig. 1. The vast majority of them expressed CD2, CD56, and CD38 antigens but lacked CD3, CD4, and CD8 antigens. Hence it seems likely, that decidual lymphocytes are of bone marrow origin and the majority of them are thymus-independent.

Comparison of our data with the analysis of CD7 distribution on peripheral blood and intraepithelial lymphocytes showed that decidual CD7[+] cells have a unique antigenic phenotype (Table 1). Contrary to peripheral blood and gut intraepithelial lymphocytes, CD7[+]CD3[-] cells constituted one of the major components of decidual lymphocytes. The majority of decidual CD7[+]CD3[-] cells bore CD38, CD56, and CD2 antigens, but did not express CD16, CD57, or CD11b. However, in all the experiments we did not find a complete overlap between CD7 and other markers such as CD2 and CD56. It is possible to

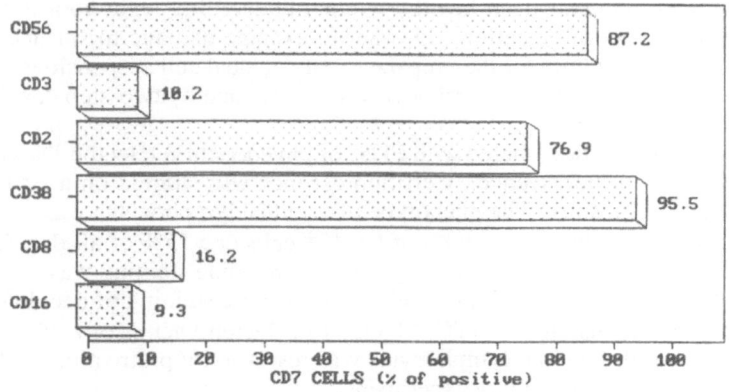

Figure 1. Summary of the phenotype of decidual CD7+ lymphocytes.

suggest the presence, among decidual lymphocytes, of a subset of cells which express the CD7 marker and lack other T lymphocyte and NK cell markers. It has been reported that intrathymic prothymocytes with $CD7^+CD1^-CD2^-CD3^-CD4^-CD8^-$ phenotype were capable of differentiating *in vitro* into $CD3^+$, TCR-1[+,] or TCR-2[+] cells.[12] Extrathymic fetal tissue contained $CD7^+CD3^-$ precursors that differentiate into $CD3^+$ lymphocytes in the presence of IL-2.[3] Purified and cloned $CD7^+CD3^-$ cells from the peripheral blood differentiate into $CD3^+$ cells in the presence of IL-2, PHA and irradiated feeder cells.[10] $CD7^+$ cells, which did not express CD3, CD2, CD8, CD4, CD5 or killer cell markers, but contained conspicuous granules, were found among gut intraepithelial lymphocytes.[11] It has been suggested that these cells represent a subset of T cell precursors capable of differentiation in the intestinal microenvironment. Therefore, it is possible to suggest the presence, among decidual and endometrial lymphocytes, of T cell precursors that do not express any NK or T cell markers except CD7, and that could be induced to differentiate into $CD3^+$ cells outside the thymus.

Table 1. Comparison of the phenotype of $CD7^+$ peripheral blood, gut intraepithelial, and decidual lymphocytes.

Peripheral blood lymphocytes[9,10]	Gut intraepithelial lymphocytes[11]	Decidual lymphocytes
~75% expressed CD7	~85% expressed CD7	~70% expressed CD7
Majority of $CD7^+$ are $CD3^+,CD2^+,CD5^+$	Majority of $CD7^+$ are $CD3^+,CD2^+, CD8^+$	Majority of $CD7^+$ are $CD3^-,CD5^-,CD8^-,CD4^-$, $CD2^+,CD56^+,CD38^+$
$CD7^+CD3^-$ ~13%	$CD7^+CD3^-$ ~ 9%	$CD7^+CD3^-$ ~ 65%
$CD7^+CD3^-$ are $CD16^+$, $CD11b^+$	$CD7^+CD3^-$ are $CD16^-$, $CD2^-,CD5^-,CD4^-$	Majority of $CD7^+CD3^-$ are $CD16^-,CD11b^-$, $CD56^+,CD2^+,CD38^+$
All $CD8^+$ cells bear CD7 antigen		$CD8^+CD7^-$ were found

A small subset of $CD7^+$ lymphocytes expressed the CD3 antigen. These $CD3^+$ cells reacted weakly with anti-TCR-2 monoclonal antibodies and were not stained with anti-TCR-1 antibodies (clone 11F2), which react with a conformational determinant of the TCR-1 molecule thought to be present on all TCR-1[+] cells. Hence, intradecidual $CD3^+$ lymphocytes had detectable surface TCR-2 but not TCR-1 molecules. The low density of TCR-2 on intradecidual lymphocytes may be a result of down-regulation of TCR-2 expression *in situ*. The suppression of TCR-2 and the lack of TCR-1 on decidual lymphocytes could be one of the decisive factors in local tolerance of maternal T cells towards the semi-allogenic fetus.

Thus, we showed that intradecidual $CD7^+$ lymphocytes express a unique antigenic phenotype, but the function of these cells *in vitro* and *in vivo* remains to be established.

REFERENCES

1. J. Bulmer, L. Morrison, M. Longfellow, and A. Ritson, *in*: "Cellular and Molecular Biology of the Materno-Fetal Relationship," G. Chaouat and G. Mowbray eds., p. 189, John Libbey Eurotext, Paris (1991).

2. J. Bulmer, L. Morrison, M. Longfellow, A. Ritson, and D. Pace, *Human Reprod.* 6:791 (1991).

3. B. Haynes, M. Martin, H. Kay, and J. Krutzberg, *J. Exp. Med.* 168:1061 (1988).

4. T. Palker, R. Scearce, L. Hensley, W. Ho, and B. Haynes, *in*: "Leukocyte Typing II Human T Lymphocytes," E. Reichert, B. Haynes, L. Nadler and I. Bernstein eds., p. 303, Springer-Verlag, New York (1985).

5. S. Pittaluga, M. Raffeld, E. Lipfor, and J. Cossman, *Blood* 68:134 (1986).

6. A. Akbar, P. Amlot, C. Hawkins, W. Newsholme, S. Delaney, N. Borthwick, and G. Janossy, *Transplantation* 52:325 (1991).

7. A. Carrera, M. Rincon, F. Sanchez-Madrid, M. Lopez-Botet, and M. De Landzuri, *J. Immunol.* 141:1919 (1988).

8. S. Carrel, S. Salvi, F. Rafti, M. Favort, C. Rapin, and R. Sekaly, *Eur. J. Immunol.* 21:1195 (1989).

9. M. Link, R. Warnke, J. Finlay, M. Amylon, R. Miller, J. Dilley, and R. Levy, *Blood* 62:722 (1983).

10. F. Preffer, C. Kim, K. Fischer, E. Sabga, R. Kradin, and R. Colvin, *J. Exp. Med.* 170:177 (1989).

11. A. Jarry, N. Cerf-Bensussan, N. Brousse, F. Selz, and D. Guy-Grand, *Eur. J. Immunol.* 20:1097 (1990).

12. M. Toribio, A. De La Hera, J. Borst, M. Marcos, C. Marquez, J. Alonso, A. Barcena, and A. Martinez, *J. Exp. Med.* 168:2231 (1988).

IMMUNE CELLS ON INTRAUTERINE CONTRACEPTIVE DEVICES

Ilja Trebichavsky,[1] Otakar Nyklícek,[2] and Miloslav Pospísil[1]

[1]Institute of Microbiology, Czech Academy of Sciences,
Departments of Immunology and Gnotobiology, Prague,
Czech Republic; and [2]Hospital Náchod, Departments of
Gynaecology and Obstetrics, Náchod, Czech Republic

INTRODUCTION

Intrauterine contraceptive devices (IUDs) elicit endometrial erosion and inflammation[1] associated with the failure of pregnancy. The immunological events which are connected with this mechanism remain incompletely defined. It is known that inflammatory cells are potentially hazardous for embryos. IUDs are frequently associated with an increased number of anaerobic bacteria and an increased risk of pelvic inflammatory disease. Bacterial substances lead frequently to failure of the conceptus. Lipopolysaccharide applied *in utero* to pregnant mice causes abortion. It has been proposed that TNF-α, IFN-γ, NK cells, and activated macrophages are responsible for the destruction of the conceptus. Cytotoxic mechanisms leading to death will be however more complicated and delicate since TNF is a physiological placental and embryonic constituent[2], and IFN-γ and IL-6 are produced by the trophoblast. Cytokine message for IL-1, IL-4, TNF and TGF-β was also found in human trophoblast cells.[3] NK cells are regularly present in human, murine and porcine decidua[4], and macrophages form a physiological endometrial defence and react even after each coitus. Endometriosis with frequent abortions and infertility is associated with an elevated number of NK cells, T cells and macrophages in the peritoneal fluid.[5] The cross-talk of various cytokines also increases; many of these pleiotropic factors can elicit different responses according to their concentration and target tissue. TNF, for example, is a differentiation factor at low concentration, a cytotoxic factor at medium concentration, and a shock agent at high concentration.

IUDs cause mechanical injury of the endometrium, haemorrhage, and resulting epithelial signals. This signalling is probably similar to that observed in other epithelial that are structurally threatened, involving the secretion of IL-1, IL-6, IL-8, TNF, phospholipase and PDGF (platelet derived growth factor). Neutrophils and macrophages participate in the initial repair, the former cells mainly in the presence of bacteria and sperm cells, the latter in relatively sterile conditions.

RESULTS AND DISCUSSION

Cell populations were recovered from the surface of Dana IUD plastic loops which were removed for medical reasons. All 70 devices were obtained from healthy women 19 - 45 years old. Leukocyte differentiation antigens were detected with monoclonal antibodies and FITC-labeled swine anti-mouse Ig. Cytokines were detected in cells fixed with 0.5%

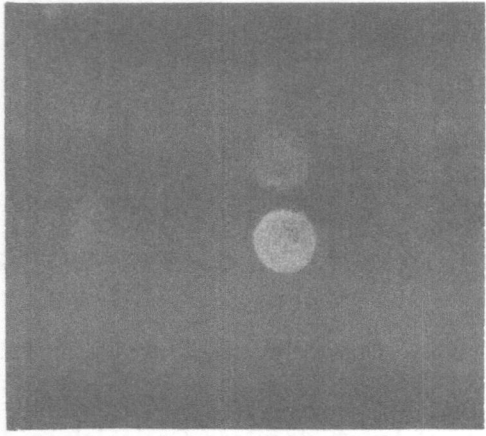

Figure 1. Positive and negative IUD lymphocyte/1500x/.

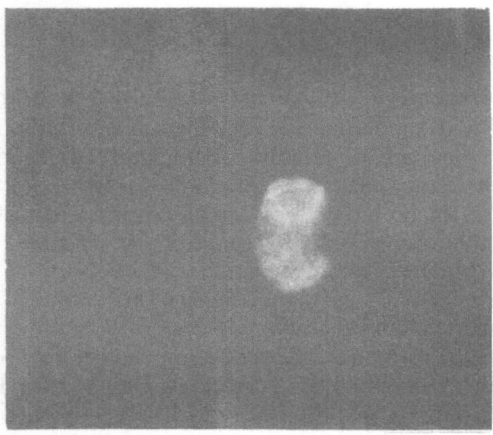

Figure 2. Macrophage with bright perinuclear and weak cytoplasmic IL-6 immunofluorescent positivity/1500x/.

paraformaldehyde in PBS and incubated with monoclonal antibodies in 0.1% saponin (Figs. 1 and 2).

Macrophages were the prevailing cells on IUDs and they covered the plastic loops with a continuous layer beginning on the second day after insertion. These large cells with foamy vacuolized cytoplasm and many phagolysosomes were characterized by activation antigens and molecules typical of mature cells: CD11c, CD14, CD15, CD45, CD71, and HLA-DR. Their cytoplasm contained IL-1, IL-6, and TNF. Cytoplasmic TNF-α was found in $14.3 \pm 8.8\%$ of nucleated IUD cells. TNF-positive cells were either macrophages or NK cells (expressing also CD56 and CD8 membrane antigens).

B cells positive for α heavy and J chains (cells producing polymeric IgA) were the second most frequent cell type. IgA antibodies were detected also on cellular detritus and sperm cell. Cells producing IgG and IgM were less numerous (not exceeding 1%); IgD[+] and IgE[+] cells were not found.

Endometrial mucosa contained frequent T cells and also B cells producing IgG and IgA. CD3[+] T cells were not found on IUDs. T cells as producers of IFN-γ could play also an important role in pregnancy failure. It seems that activated macrophages and NK cells are the most important cells causing destruction of the conceptus in the vicinity of IUDs. Biochemical changes of the endometrium after injury and under the influence of inflammatory cells could play a secondary role in contraceptive mechanisms elicited by IUDs.

REFERENCES

1. I. Trebichavsky and O. Nyklicek, *Acta Cytol.* 23:366 (1979).
2. S. S. Witkin, H. C. Liu, O. K. Davis, and Z. Rosenwaks, *J. Reprod. Immunol.* 19:85 (1991).
3. M. K. Haynes, R. Tuan, L. G. Jackson, and J. R. Smith, *in*: "8th Int. Congress Immunol.", Budapest, p.366 (1992).
4. B. A. Croy, A. Waterfield, W. Wood, and G. J. King, *Cell. Immunol.* 115:471 (1988).
5. D. J. Anderson and J. A. Hill, *Curr. Opinion Immunol.* 1:1119 (1989).

CELLULAR AND HUMORAL IMMUNITY TO SPERM IN OVULATORY CERVICAL MUCUS FROM INFERTILE WOMEN

Zdenka Ulcová-Gallová,[1] and D. Sedláček[2]

[1]Department of Obstetrics and Gynecology
[2]Department of Infectious Diseases
Charles University, Pilsen, Czech Republic

INTRODUCTION

Infertility caused by immunological cervical factors directed against sperm antigens has been identified in 20% of infertile women[1,2] with the diagnosis of unexplained infertility. The complex view of this problem includes humoral, cell-mediated immunity, and soluble products of granulocytes, lymphocytes and macrophages which reduce sperm motility and sperm penetration through the zona pellucida.

In addition to sperm antibodies, we began to characterize the cells in ovulatory cervical mucus in a group of infertile women with and without sperm agglutinating antibodies in the cervical mucus. Our preliminary investigations suggest that the presence of immunocompetent cell may be important in infertility.

PATIENTS AND METHODS

Thirty-one infertile women aged 22-39 years were chosen according to these criteria that exclude known causes of infertility: normal spermiograms of their husbands, normal gynecological and endocrinological examinations, normal sex hormone levels and oncological cytology and histology of endometrium curettage on day 26 of the cycle, normal bacteriological vaginal and cervical examination, normal laparoscopical and hysterosalpingographic findings, repeated negative post-coital tests before and after coitus with condom, and the absence of serum antibodies to sperm to exclude the possibility of antibody transudation from the blood into cervical ovulatory mucus.

Cervical ovulatory mucus was aspirated by a special syringe (Fortuna) from cervical canal, taken directly into glass capillaries, transferred into phosphate buffered saline (PBS), and centrifuged at 600 x g for 10 min. Cell pellet was applied onto the microscope slide (2 cm^2 area), fixed in methanol and stained by Giemsa. Cervical cells were observed in Meopta microscope (magnification 100 - 600x), identified and counted in area 1 cm^2 (10 - 20 cells).

Local antibodies to sperm were measured in the supernatant of cervical ovulatory mucus by an indirect mixed anti-globulin reaction (MAR) test[3] for IgG, S-IgA, IgM and IgE, by tray agglutination (TAT) test[4], and by sperm penetration into cappillaries containing mucus.[5]

Advances in Mucosal Immunology, Edited by
J. Mestecky *et al.*, Plenum Press, New York, 1995

RESULTS

The presence of epithelial cells, polymorphonuclear leukocytes, lymphocytes, and monocytes is shown in Fig. 1. Subtle differences in types of cells present in ovulatory cervical mucus were seen in infertile women with or without antibodies to sperm. Higher numbers of epithelia were found in infertile women without antibodies to sperm. The numbers of polymorphonuclear leukocytes, lymphocytes, and monocytes in both groups were almost the same.

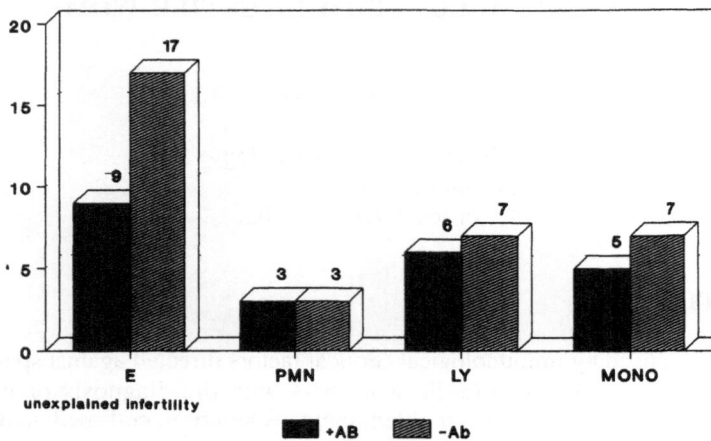

Figure 1. Frequency of epithelia (E), polymorphonuclear leukocytes (PMN), lymphocytes (LY), and monocytes (MONO) in cervical ovulatory mucus in infertile women with (+Ab) and without (-Ab) spermagglutinating antibodies.

No cervical antibodies to sperm were found in 17 infertile women when tested by negative indirect MAR and tray agglutination test, and positive sperm penetration into capillary tubes with mucus more than 5 cm/h. Frequency of local sperm antibodies in the second group of 14 patients is shown in Table 1. S-IgA (7 patients), IgG (2 patients), S-IgA and IgG (1 patient), S-IgA and IgE (1 patient), IgE (1 patient), and cytotoxic reaction in 2 cases were detected.

Table 1. Ig properties of local antibodies to sperm in cervical ovulatory mucus of infertile women with diagnosis of unexplained infertility. Sperm-agglutination was estimated by indirect MAR test. Positivity indicates that more than 41% of motile spermatozoa is involved in mixed agglutinates (spermatozoa incubated in treated cervical ovulatory mucus and sheep erythrocytes coated by corresponding Ig).

	Number of infertile women	
	with	without
Ig	sperm antibodies	
s-IgA	7	17
IgG	2	
s-IgA + IgG	1	
s-IgA + IgE	1	
IgE	1	
cytotoxic reaction	2	

DISCUSSION

We have studied non-specific and specific factors of cervical immunity for several years.[6] Activation of cervical immunocompetent cells by sperm antigens stimulates and creates an immune response that plays an important role directed against normal way of fertilization. Cervical cells involved in such immune responses migrate from the lamina propria the cervical mucus, uterine cavity, oviductus and peritoneal cavity into the cervical mucus.

The immunological response in the cervix as in the entire reproductive tract is a complex process in which many mechanisms participate: plasma cells derived from B-lymphocytes mediate humoral immune responses by producing antibodies, and T-lymphocytes direct cell-mediated immune responses through the cytokine secretions. Other leukocytes such as granulocytes, macrophages, and natural killer cells participate in immunological responses. Immunological causes of infertility are also influenced by hormonal and psychological factors.

Our preliminary results show subtle cell differences in polymorphonuclear leukocytes, lymphocytes, and monocytes between infertile women with and without local sperm-agglutinating antibodies. Currently we characterize immunocompetent cells with monoclonal antibodies, because this approach is likely to facilitate the determination of their phenotype. In addition, we are monitoring of cervical cells and sperm antibodies in infertile women treated by local application of hydrocortisone.[6-8]

REFERENCES

1. J. A. Hill, J. Cohen, and J. D. Anderson, *Am. J. Obstet. Gynecol.* 5:1154 (1989).
2. R. M. Wah, D. J. Anderson, and J. A. Hill, *Fertil. Steril.* 54:445 (1990).
3. H. Meinertz and T. Hjort, *Am. J. Immunol. Micorbiol.* 14:129 (1987).
4. J. Friberg, *Acta Obstet. Gynecol. Scand. (Suppl.)* 36:21 (1974).
5. J. Kremer, *Int. J. Fertil.* 10:209 (1965).
6. Z. Ulcová-Gallová, V. Zavázal, F. Macku, and I. Ulc, *in:* II. Magdeburger Symp. "Die kindrlose Ehe", Abstr. p. 25, (1984).
7. Z. Ulcová-Gallová, L. Mráz, E. Plánicková, F. Macku, and I. Ulc, *Int. J. Fertil.* 33:421 (1988).
8. Z. Ulcová-Gallová, L. Mráz, E. Plánicková, F. Macku, and I. Ulc, *Ztrb. Gynäcol.* 112:867 (1990).

HISTOLOGY OF THE CONTINUOUS PEYER'S PATCHES IN THE TERMINAL ILEUM OF PIGS IN THE PERINATAL PERIOD

R. Halouzka and F. Kováru

Veterinary and Pharmaceutic University Brno
Czech Republic

INTRODUCTION

There are two types of Peyer's patches (PP) in the ileum of sheep,[1] dog, and pig,[2] defined by their different localization, histological structure, and function.[3] Several discrete PP have been described in the upper ileum, and there is a continous PP in the terminal ileum of the pig.[4] Some data suggest that the development and involution of the continous PP in the terminal ileum occurs on the same time scale as in the thymus. Furthermore, their cellular compostion, structure, lymphocyte traffic, and independence of antigenic stimuli suggest that the PP in the ileum of pigs may play a role comparable to that of a primary organ for B lymphocyte development.

The purpose of the present work was to demonstrate the prenatal and postnatal histological structure of the continuous PP in the terminal ileum of germ-free (GF) pigs and its structural similarity to the avian bursa of Fabricius.

MATERIAL AND METHODS

The terminal ileum of experimental piglets of GF Minnesota minipigs from the Novy Hrádek breeding facility, was histologically examined. We used piglets born by caesarean section at the age of 95, 97, and 112 days of intrauterine life, or spontaneously born animals, and piglets at the age of 3, 4, 12, and 28 days after birth, kept under GF conditions with artificial nutrition. The terminal ileum was taken 10 cm before the ileocaecal junction, fixed in 6% neutral formalin solution, processed by routine paraffin technique, and stained with hematoxylin and eosin, and the Van Gieson method.

RESULTS

At the age of 95 and 97 days of intrauterine life, the terminal-ileal continuous PP showed small ovoid follicles with densely packed lymphocytes and small interfollicular areas localized in lamina propria mucosae and separated from the mucosal epithelium by well defined fibres of connective tissue and basal membrane. Formation of mucosal plicae was not observed.

At the age of 112 days of the intrauterine life (Fig. 1), at birth, and at 3 and 4 days after birth (Fig. 2), the gradual development of 2 - 3 plicae of mucosal and submucosal tissue in the area of the continuous PP were invariably observed. The continuous PP showed round follicles separated by more developed interfollicular areas.

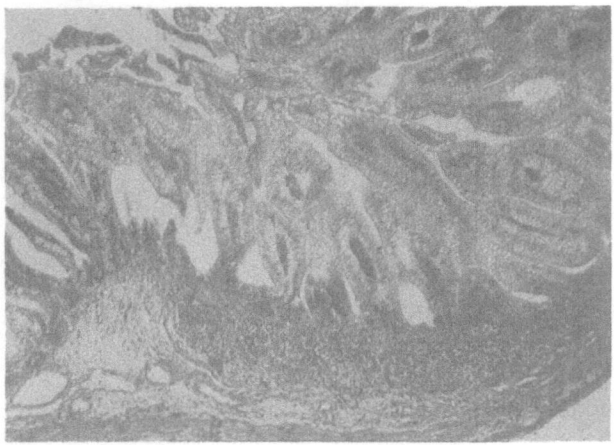

Figure 1. Small round follicle and interfollicular areas separated from the mucosal epithelium on the 112th day of intrauterine life Initiation of mucosal folding H & E, 200x

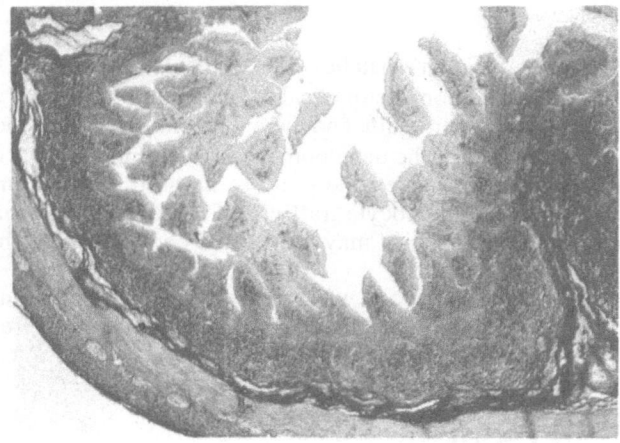

Figure 2. Developing mucosal plicae in the continuous PP on the 4th day after birth VG, 200x

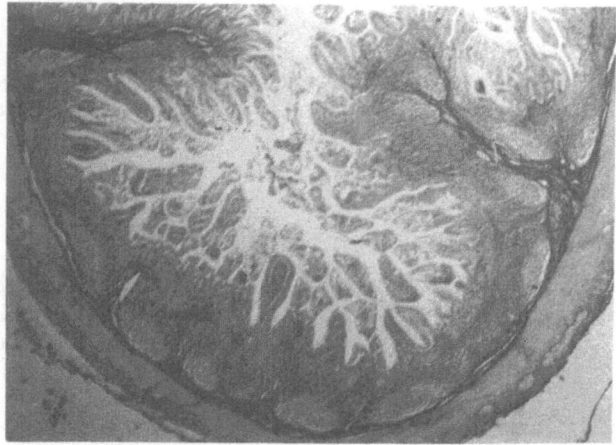

Figure 3. Mucosal plicae with compartments of lymphoid tissue resembling the histological structure of the avian bursa of Fabricius on the 28th day after birth VG, 200x

At 12 and 28 days after birth (Fig. 3), well-developed mucosal plicae were observed in the area of the terminal ileal continuous PP. Their general appearance, oval and elongated follicles, interfollicular areas with prominent collagenous fibres, and close contact with the surface epithelium resemble the histological structure of the plicae of the avian bursa of Fabricius. The continuous patch in the unfolded area of the terminal ileual mucosa showed typical compartmentalization with oval and elongated follicles, interfollicular areas, and domes. The collagenous fibres of the lamina propria mucosae above the lymphoid follicles were reduced, thus allowing close contact of the lymphocytes with the surface epithelium.

DISCUSSION

Morphological differences of the terminal ileal continuous PP during prenatal and postnatal development in GF pigs were demonstrated. In the intrauterine period (up to the 112th day), the terminal continuous PP consist of less developed individual compartments. Small ovoid follicles were clearly separated from the surface epithelium. Folding of the mucosa was not observed.

In newborn and older animals, compartmentalization of the continuous patch was much more developed, and the contact between the lymphoid follicles and the epithelium became closer. This morphological feature of the continuous patch resembles the avian bursal follicles that have close contact with the bursal lumen through the phagocytic follicle-associated epithelium, which has characteristics similar to mammalian M cells.[5,6,7]

At 28 days after birth, the histological structure of well-developed plicae and lymphoid tissue in unfolded areas strongly resembled the structure of the chicken bursa of Fabricius in the post-hatching period.[8,9]

Our findings demonstrate the developmental morphological differences of the terminal ileal continuous PP during the prenatal and postnatal period in pigs in the absence of antigenic stimuli, and its morphological similarity to the avian bursa of Fabricius.

REFERENCES

1. J. D. Reynolds, R. Pabst, and G. Bordmann, *Adv. Exp. Med. Biol.* 186:101 (1985).
2. R. Pabst, *Anat. Embryol.* 176:135 (1987).
3. R. Pabst, M. Geist, H. J. Rothkötter, and F. J. Fritz, *Immunology* 64:539 (1988).
4. H. J. Rothkotter and R. Pabst, *Immunology* 67:103 (1989).
5. D. D. Joel, B. Sordat, M. W. Hess, and H. Cottier, *Experientia* 26:694 (1970).
6. D. E. Bockman and M. D. Cooper, *Fed. Proc.* 30:511 (1971).
7. M. Lupetti, A. Dolfi, F. Giannessi, and S. Michelucci, *J. Anat.* 136:851 (1983).
8. G. Hoffman-Fezer and R. Lade, *Z. Zellforsch. Mikroskop. Anat.* 124:406 (1972).
9. B. Glick, *Int. Rev. Cytol.* 48:345 (1977).

EMBRYONIC LIVER: DIVERSIFICATION SITE OF LYMPHOCYTE LINEAGES

Ija Trebichavsky,[1] Richard Pospisil,[1] Otakar Nyklicek,[2]
Igor Splichal,[1] and Leos Mandel[1]

[1]Institute of Microbiology, Czech Academy of Sciences,
Department of Immunology and Gnotobiology, Prague; and
[2]Hospital Nachod, Department of Gynecology and
Obstetrics, Nachod, Czech Republic

INTRODUCTION

Common haematopoietic and endothelial stem cells originate in the primitive streak and differentiate in various extra-embryonic foci into angioblasts, erythroid cells, macrophage precursors, and megakaryocytes. The most extensive source of these cells is the yolk sac mesenchyme which continues in the embryonic body in the septum transversum, a mesodermal plate separating the yolk sac and pericardial cavities. This mesenchyme is invaded and encircled by endodermal hepatic cords, cells of the midgut diverticulum which projects into the plexus arising from the vitelline and umbilical veins. These three constituents, vessels, endodermal cords, and mesenchymal islands represent the liver anlage.

The liver plays an important role in the development of the immune system as the site of lymphocyte origin. The aim of this study was to demonstrate that the liver is a site of lymphocyte diversification. Porcine and human embryonic liver was studied using a panel of monoclonal antibodies directed against lymphocyte differentiation antigens, and immunofluorescence and flow cytometry techniques. Porcine embryos were recovered from miniature Minnesota sows under Halothane/oxygen anesthesia; human abortuses were obtained for medical reasons with ethical committee approval.

RESULTS AND DISCUSSION

Porcine Liver

Lymphocytes appear in the liver sinuses on the 28th day of gestation. MHC class II antigens and CD2 and CD45 molecules were detected at this developmental stage for the first time. Erythroid cells expelling nuclei, macrophages engulfing them, platelets, and precursor cells that stained with 8/1 and MG-7 antibodies were detected on the 25th day. T cell markers TCR-$\gamma\delta$, CD4, and CD8 molecules) and B cell markers (low μ chain) were found in the 39th day of gestation (Tables 1,2).

Table 1. Monoclonal antibodies directed against porcine molecules.

MoA	Antigen	Cells	Origin
LK-3	SLA-D/MHCII/	lymphocytes macrophages	Koubek 1992
MG-7	-	T cells, precursors	Pospisil 1992
LK-2	CD45	haematopoiet. cells	Pospisil 1992
11 8 1	8/1	lymph precursors T cells	Saalmuller 1987
MSA4	CD2	pre-T, T cells	Hammerbeg, Schurig 1986
74 12 4	CD4	T cells	Pescowitz 1984
11/295/33	CD8	T cells	Schuller 1987
86D	TCR-γδ	T cells	Mackay 1985
L₁G4	μ chain	pre-B, B cells	Dvorák 1987

Table 2. Lymphocyte differentiation antigens in the porcine embryonic liver.

Day of gestation	Antigens detected
25	8/1, MG-7
28, 30, 35	SLA-D, CD45, CD2, 8/1, MG-7
39, 45, 48	SLA-D, CD45, CD1, CD2, 8/1, MG-7, gamma/delta TCR, CD4, CD8, low μ chain
50	the same antigens and high cytoplasmic μ chain

Human Liver

Lymphocytes appeared in the liver sinuses on the 40th day of gestation. MHC class II antigens were detected early on the 34th day on liver macrophages and erythroblasts. Differentiation molecules typical of granulocytes (CD15), NK cells (CD56) and early B cells (CD5) were found on the 40th day of gestation in human embryonic liver (Table 3).

The early finding of κ light chain exclusively in the cytoplasm of phagocytic cells confirmed the early transfer of maternal IgG.

Table 3. Differentiation lymphocyte antigens in human embroynic liver.

Day of gestation	Antigens detected
34	HLA-DR
40	CD5, CD15, CD45, CD56, kappa light chain
57	cytoplasmic μ chain

A comparison of liver ontogeny in the pig and man is shown in Table 4.

The liver represents in the early developmental period a site where all immune cells meet. Cells from the early embryonic liver are capable of phagocytosis, secretion of complement factors and inhibitors, natural cytotoxicity, production of interleukins, TNF and interferons, MHC antigens, and according to this study, several differentiation antigens. The maturation of functional immune responses is connected to the maturation of biochemical phenotype. Recent studies show that this phenotype develops very early in prenatal ontogeny.

Table 4. Early ontogeny of human and porcine immune cells (days of gestation).

	Man (267)	Pig (114)
Yolk sac haematopoiesis	20	16
Liver anlage	25	18
Liver haematopoiesis	32	20
Yolk sac MHC II$^+$ cells	32	24
Liver MHC II$^+$ cells	34	28
Liver macrophages	35	25
Liver granulocytes	40	28
Liver lymphocytes (NK cells)	40	28
Liver T lineage	49	35
Liver gamma/delta TCR	49	39
Liver B lineage	55	39
Liver B cells	57	44

RETICULUM CELLS IN THE ONTOGENY OF NASAL-ASSOCIATED LYMPHOID TISSUE (NALT) IN THE RAT

J. Biewenga, M.N.M. van Poppel, T.K. van den
Berg, E.P. van Rees, and T. Sminia

Dept. of Cell Biology, Medical Faculty
Vrije Universiteit, Amsterdam

INTRODUCTION

Nasal associated lymphoid tissue (NALT), is a paired organ located at the entrance of the nasopharyngeal duct of the rat. It is covered by a specialized epithelium containing M microfold (M) cells.[1,2] NALT development is first seen at the day of birth. T and B cell areas can be distinguished from 10 days after birth.[3] This development requires a tight regulation in which reticulum cells may play a role by the guidance of lymphocyte immigration.[4] Here we describe the ontogeny of reticulum cells in the NALT of Wistar rats.

MATERIALS AND METHODS

Wistar rat embryos were collected at 15, 17, 19, and 21 days after conception. Newborn rats were sacrificed on day of birth and 2, 4, 7, 14, and 21 days after birth. From the youngest animals the heads were collected, either intact or in two halves. The mandibulae was removed from heads of animals aged 7 or 14 days, and the NALT was removed from the heads of 21 day old rats. All tissue was snap-frozen and stored at - 20°C until used for immunohistochemistry. A standard two-step immunoperoxidase staining was performed on 8μm thick cryostat sections.

A panel of six monoclonal antibodies against reticulum cells, ED10-ED15, was used.[4] In addition, the monoclonal antibodies OX4 (anti-Ia)[5], OX19 (anti-T cell)[6,7], HIS14 (anti-B cell)[8], ED1 (macrophage marker)[9], and ED17 (recognizes a macrophage subset and Langerhans cells)[10] were applied.

RESULTS

General Features

In the prenatal stage macrophages and Ia+ cells are present in the nasal cavity, but lymphocytes appear on the day of birth. At that time a small accumulation of lymphoid and non-lymphoid cells appears at a specific site in the nasopharyngeal area. Ten days after birth distinct T and B cell areas are formed.[3] In adult NALT, ED10 stained reticulum cells in the T cell area, whereas ED11 was mostly confined to the B cell area, where it produced a web-like staining. ED12, ED14, and ED15 staining were more pronounced in the T cell than in the B cell areas. ED13 stained blood vessels and high endothelial venules.

Advances in Mucosal Immunology, Edited by
J. Mestecky *et al.*, Plenum Press, New York, 1995

Embryos

On day 15 after conception no staining was observed in the nasal mucosa with ED10, ED11 and ED12. ED13 recognized the basement membrane of the epithelium and a few randomly distributed cells. The antibodies ED14 and ED15 produce weak staining in the whole mucosa. Furthermore, ED17+ cells, having cell processes, were present in the mucosa. No ED1-, OX4-, OX19-, and HIS14-positive cells were detected. On day 17 after conception the first ED1+ cells were found scattered throughout the whole mucosa. Staining with ED13, ED14, and ED15 was stronger than before. ED13 stained the basement membrane and blood vessels. ED14 and ED15 recognized fibre-like structures throughout the mucosa. On day 19 after conception the first OX4+ cells, which had cell processes, were found scattered throughout the mucosa. The antibodies ED10, ED11, OX19, and HIS14 did not yet recognize any cells. At day 21 after conception no significant differences were observed as compared to the staining pattern on day 19.

Neonates

On the day of birth an area was observed in the nasal cavity with a different staining pattern than the surrounding connective tissue. ED10 now recognized fibre-like structures in this specific area. OX4+ cells accumulated in the area where NALT was developing. OX19+ cells were found scattered throughout the whole mucosa. ED11 and HIS14 stainings were negative. From the second day after birth HIS14+ cells were present in the developing NALT. A weak staining of the reticulum with ED10 and ED12 was seen in the NALT, but not in the connective tissue. From day 4 these antibodies also recognized structures in the mucosa, but the staining was always weaker than in the NALT area. At day 14 separate T and B cell areas were recognized. ED10 expression was more confined to the T cell area, whereas web-like ED11 staining was found more in the B cell area (Fig.1). ED17+ cells were observed in the periphery of the NALT and directly under the epithelium.

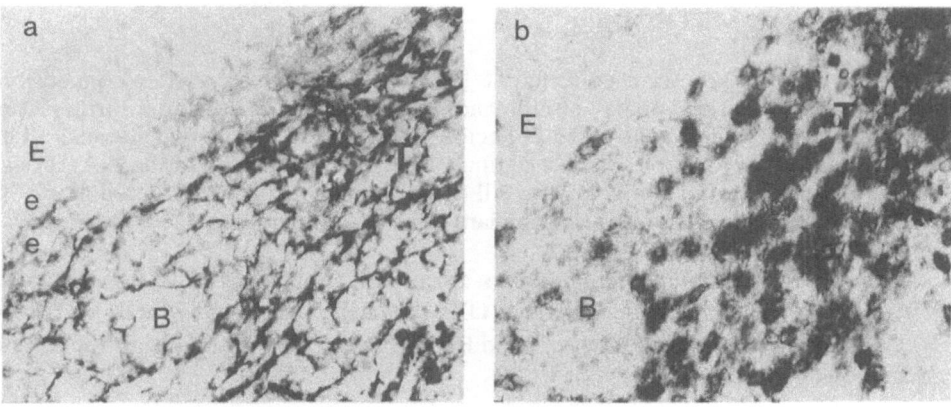

Figure 1. T cells (a) and ED10 expression (b) in rat NALT at the age of 14 days. E= epithelium; T=T cell area; B=B cell area.

DISCUSSION

The various cell types present in adult NALT arise at different stages in ontogeny. Before birth undifferentiated reticulum cells, recognized by ED13, ED14 and ED15, are present in the nasal mucosa. ED17+ cells are present before Ia or ED1 expression is seen. The first sign of specialization is seen on the day of birth, when ED10+ cells recognize a specific area in the nasal mucosa and also T cells are found. Soon after (day 4) ED11 expression and B cells are detected in the developing NALT. In accordance with previous results[3] T and B cell areas had developed by day 14 after birth. As in other lymphoid

Table 1. The development of reticulum cells in the NALT of Wistar rats

Mab	after conception				after birth				
	15	17	19	21	0	2	4	14	21
OX4 (Ia)	–	–	±	±	+	+	++	++	++
OX19 (T cells)	–	–	–	–	±	+	++	++	++
HIS14 (B cells)	–	–	–	–	–	+	++	++	++
ED1 (Mø)	–	±	+	±	+	++	++	++	++
ED10 (RC)	–	–	–	–	+	+	++	++	++
ED11 (RC)	–	–	–	–	–	±	+	+	+
ED12 (RC)	–	–	–	–	–	±	±	+	+
ED13 (RC)	+	+	+	+	+	++	++	++	++
ED14 (RC)*	+	+	+	+	+	++	++	++	++
ED15 (RC)*	±	+	+	+	+	+	++	++	++
ED17 (Mø, LC)	+	+	+	+	+	+	++	++	++

* in connective tissue

tissues[4,11] ED10 is found in the T cell area, whereas most ED11 expression is found in the B cell compartment. At the age of 21 days NALT has the same appearance as in adult rats.

Rather surprising is the finding of ED17[+] cells before Ia or ED1 expression is seen in the nasal mucosa. Elbe et al.[12] described a precursor of Langerhans cells in mice, which is Ia negative, and matures in the epidermis. ED17[+], Ia- cells in the nasal mucosa could represent a comparable population of immature Langerhans cells. Hameleers et al.[3] reported that ED3[+] macrophages in adult NALT were found at the border of NALT and connective tissue, and directly under the epithelium. The ED17[+] cells located between NALT and the connective tissue could represent the same population as the ED3[+] cells in adult NALT, since ED17 and ED3 specificities overlap.[10] ED3 recognizes a subset of macrophages[9] predominantly in spleen and lymph nodes.[13]

Important events in the formation of lymphoid tissue are the influx of lymphocytes and the later development of distinct T and B cell compartments. Our findings demonstrate that the influx of lymphocytes and the expression of ED10 and ED11 in the developing NALT occur simultaneously. Whether expression of the reticulum cell markers is directly responsible for the influx of lymphocytes is not known, but their preferential expression in T and B cell areas, respectively, is now well documented (this study and 4, 11). ED11 was recently shown to recognize complement factor C3, which is associated with follicular dendritic cells.[14] The antigen recognized by ED10 has not been identified.

CONCLUSION

This study showed that in the development of NALT the first appearance of ED10[+] and ED11[+] reticulum cells is around birth when T and B cells are also first seen. These results suggest that the development of the reticulum is important for the positioning of lymphocytes and the compartmentalization of NALT.

REFERENCES

1. B. J. Spit, E. G. J. Hendriksen, J. P. Bruijntjes, and C. F. Kuper, *Cell Tissue Res.* 255:193 (1989).
2. C. F. Kuper, P. J. Koornstra, D. M. H. Hameleers, J. Biewenga, B. J. Spit, A. M. Duijvestijn, P. J. C. Van Breda Vriesman, and T. Sminia, *Immunol. Today* 13:219 (1992).

3. D. M. H. Hameleers, M. Van der Ende, J. Biewenga, and T. Sminia, *Cell Tissue Res.* 259:431 (1989).

4. T. K. Van den Berg, E. A. Döpp, J. J. P. Brevé, G. Kraal, and C. D. Dijkstra, *Eur. J. Immunol.* 19:1747 (1989).

5. W. R. McMaster and A. F. Williams, *Eur. J. Immunol.* 9:426 (1979).

6. M. J. Dallman, D. W. Mason, and M. Webb, *Eur. J. Immunol.* 12:511 (1982).

7. D. W. Mason, R. P. Arthur, M. J. Dallman, J. R. Green, G. P. Spickett, and M. L. Thomas, *Immunol. Rev.* 74: 57 (1983).

8. F. G. M. Kroese, D. Opstelten, A. S. Wubbena, G. J. Deenen, J. Aten, E. H. Schwander, L. de Leij, and P. Nieuwenhuis, *Adv. Exp. Med. Biol.* 186:81 (1985).

9. C. D. Dijkstra, E. A. Döpp, P. Joling, and G. Kraal, *Immunology* 54:589 (1985).

10. J. G. M. C. Damoiseaux, E. A. Döpp, and C. D. Dijkstra. *J. Leukocyte Biol.* 49:434 (1991).

11. M. Soesatyo, T. K. Van den Berg, E. P. Van Rees, J. Biewenga, and T. Sminia, *Regional Immunol.* 4: 46 (1992).

12. A. Elbe, E. Tschachler, G. Steiner, A. Binder, K. Wolff, and G. Stingl, *J. Immunol.* 143:2431 (1989).

13. R. H. J. Beelen, I. L. Eestermans, E. A. Döpp, and C .D. Dijkstra, *Transplant. Proc.*19:3166 (1987).

14. T. K. Van den Berg, E. A. Döpp, M. R. Daha, G. Kraal, and C. D. Dijkstra, *Eur. J. Immunol.* 22:957 (1992).

THE PRODUCTION OF TNF, IL-1 AND IL-6 IN CUTANEOUS TISSUES DURING MATURATION AND AGING

Stanislava S. Stosic-Grujicic,[1] and Miodrag L. Lukic[2]

[1]Institute for Biologica Research
[2]Institute of Microbiology and Immunology
School of Medicine
University of Belgrade
Belgrade, Yugoslavia

INTRODUCTION

Epithelial cells (EC) resident to cutaneous tissues are important producers of various immunostimulatory, proinflammatory and inhibitory cytokines[1,2] Locally produced cytokines in epithelial tissues mediate events relevant to growth, differentiation and function of surrounding cells. Based on their capacity to induce various cell adhesion molecules, as well as "secondary" cytokines,[3] they play an important role in intercellular communications. Since immunocompetence of epithelial tissues may be influenced by maturation and aging,[4-6] the aim of the present study was to obtain more information on the age-related changes of production of relevant cytokines. To examine this, we investigated age-related expression of TNF, IL-1 and IL-6 by comparing secretory potential of EC normally resident to skin of fetal, neonatal and adult rats.

METHODS

Epithelial Cell (EC)-Derived Cytokines

Supernatant containing epithelial cytokines were prepared by using EC resident to skin of inbred Albino Oxford (AO) rats. Epidermal sheets were isolated by day 16 of fetal development. Epidermal cell cultures were derived from back skin of rat fetuses and newborns. To prepare adult EC cultures, ear skin of 3 - 20 month old rats was used. Cell-free supernatant was collected after 24 hr cultivation of freshly prepared cell suspensions (1×10^6 cells/ml) obtained by trypsin digestion of the dermal tissue.[7]

Bioassays to Determine IL-1, IL-6 and TNF Activity

Cytokine activity in cell-free EC-derived supernatants was tested using highly specific bioassays. The assay to detect IL-1 α and β activity using murine T cell subclone D10S has been described in detail.[8] Serial dilutions of purified human rIL-1 β (Biogen) were used as standards.

The IL-6 activity was determined using probit analysis of its ability to stimulate cell proliferation of B9 hybridoma.[9] A unit of IL-6 activity is based on human rIL-6 (Genzyme) as the standard.

TNF-α biological activity was measured in the lytic assay as described[10] by using actinomycin D-treated murine L-929 fibroblast cells. Lysis was normalized to a standard of human rTNF-α (Genzyme) at 2×10^7 units/mg.

RESULTS AND DISCUSSION

Earlier studies on the ontogeny of a diverse spectrum of cells resident to epithelial tissues have demonstrated that age-related changes affect the expression of distinct cytokines.[4,5] Our previous experimental evidence, revealed an early onset of IL-1 production in rat epidermis.[11] Thus, considerable amounts of this cytokine were found at day 16 after conception. From day 16 on, a gradually increasing IL-1 activity was revealed, culminating on the day of birth. Results presented in Fig. 1 show that during postnatal life the constitutive production of IL-1 by EC decreased. The comparison of dose-response curves of IL-1 activity obtained from newborn and adult EC indicated that newborn animals are better producers of IL-1 than adults. With advancing age, the capacity of EC to constitutively produce IL-1 greatly diminished (not shown).

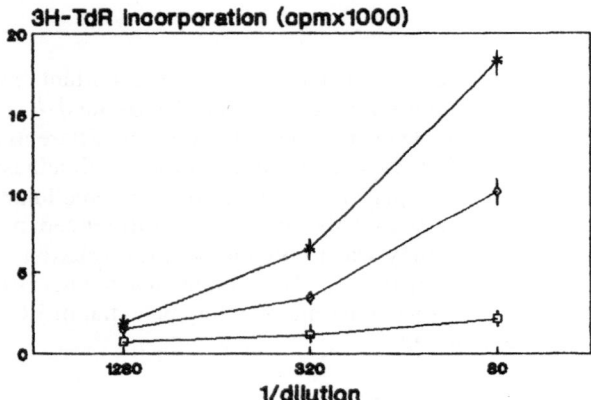

Figure 1. The production of IL-1 by newborn versus adult epithelial cells. Supernatants of freshly isolated rat EC obtained from newborns (◊), or from 12 month old (□), were tested at serial dilutions (v/v) for IL-1 activity in the D10S proliferation assay (mean ± SEM from triplicate cultures). Serial dilutions of human rIL-1 β (*) starting with 2 ng/ml were used as a standard. The spontaneous ^{3}H-TdR uptake of D10S cells was 733 ± 5 cpm.

TNF-α and IL-1 have many common effects and might be expected to be similarly regulated. Therefore, we further examined age-related changes in epithelial TNF-α production during ontogeny and aging. As presented in Fig. 2, significant production of TNF-α was obtained at day 16 of gestation and increased during further fetal development. Maximal TNF-α activity was found in newborn animals, i.e., at the same stage at IL-1 activity reached peak levels. During postnatal life constitutive TNF alpha production paralleled IL-1 secretion.

In order to examine IL-6 producing by EC during ontogeny and aging, the same supernatants were tested in the bioassay specific for IL-6. Results presented in Fig. 3 revealed that the pattern of IL-6 production differed from that of IL-1 and TNF-α. Thus, the IL-6 activity was undetectable in supernatants from day 16 fetal EC. As the age of fetuses advanced, there was a sharp increase in IL-6 production, and IL-6 expression peaked prior to birth. A delay in I-6 appearance may reflect a need for the induction of this

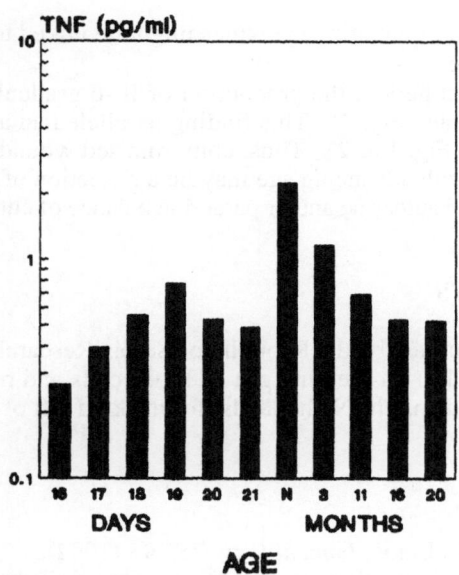

Figure 2. The production of TNF-α by epithelial cells during ontogeny and aging. Supernatants of freshly isolated EC, obtained from 16 to 21 days old fetuses, newborns and 3 to 20 month old adult rats, were tested for TNF activity in the lytic assay using L-929 target cells. Content of TNF represents pg/ml of TNF activity, calculated from triplicate determinations, each point being assayed in four serial 1:2 dilutions of supernatant.

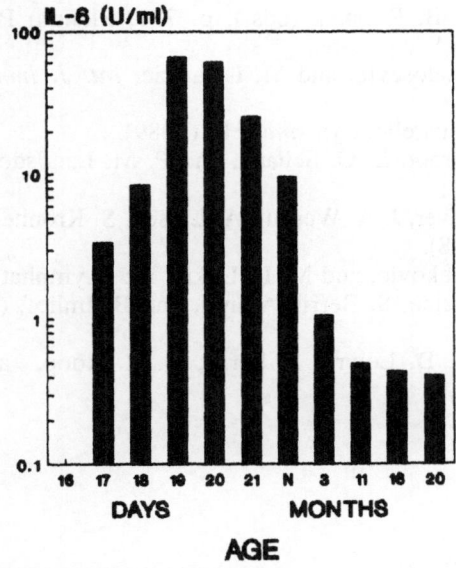

Figure 3. The production of IL-6 by epithelial cells during ontogeny and aging. Supernatants of freshly isolated EC, obtained from 16 to 21 day old fetuses, newborns and 3 to 20 month old adult rats, were tested for IL-6 activity in the B9 proliferation assay. Values represent units/ml of IL-6 activity, calculated from triplicate determinations, each point being assay in eight serial 1:2 dilutions of supernatant.

"secondary" cytokine by IL-1 and TNF as initiating cytokines. Our findings support the concept of IL-1 and TNF-α as especially important "primary" cytokines[3] based on their capacity to induce many other "secondary" cytokines with a diverse spectrum of activities.

413

The constitutive production of initiating cytokines might be linked to EC differentiation in developing cutaneous tissues.[2,12]

During the postnatal period, the production of IL-6 gradually decreased and was very low in 20 month old rats (Fig. 3). This finding paralleled an analogous reduction of IL-1 and TNF-α with age (Fig. 1 & 2). Thus, compromised wound healing, resistance to infections and neoplasms with advancing age may be a reflection of age-associated loss of the production and disturbed autocrine and/or paracrine balance of cutaneous cytokines.

ACKNOWLEDGMENTS

This work was supported by the Republic of Serbia Research Fund. We thank Dr. C. A. Dinarello (Boston, MA) for the kind gift of D10S cells and recombinant cytokines, and Dr. L. Aarden (Amsterdam, The Netherlands) for the kind gift of B9 cells.

REFERENCES

S. Stosic-Grujicic and M. L. Lukic, *Immunology* 75:293 (1992).

C. F. Bigler, D. A. Norris, W. L. Weston, and W. P. Arrend, *J. Invest. Dermatol.* 98:38 (1992).

T. S. Kupper, *J. Clin.. Invest.* 86:1783 (1990).

D. N. Sauder, B. M. Stanulis-Praeger, and B. A. Gilchrest, *Arch. Dermatol. Res.* 280:71 (1988).

D. V. Belsito, S. P. Epstein, J. M. Schultz, R. L. Bear, and G. J. Thorbecke, *J. Immunol.* 143:1530 (1989).

C. D. Dijkstra, E. P. Van Rees, and E. A. Döpp, *In*: "Histophysiology of the Immune System", S. Fossum, and B. Rolstad, (eds.), p. 731, Plenum Press, New York and London (1988).

S. Stosic-Grujicic, D. J. Lalosevic, and M. L. Lukic, *Int. J. Immunopharmac.* 9:577 (1987).

S. F. Orencole and C. A. Dinarello, *Cytokine* 1:14 (1989).

L. A. Aarden, E. R. De Groot, L. O. Schaap, and P. M. Landsdorp, *Eur. J. Immunol.* 17:1411 (1987).

S. K. Burchett, W. M. Weaver, J. A. Westall, A. Larsen, S. Kronheim, and C. B. Wilson, *J. Immunol.* 140:3473 (1988).

S. Stosick-Grujicic, V. Stankovic, and M. L. Lukic, *In*: "Lymphatic Tissue and In vivo Immune Responses", S. Ezine, S. Berrih-Akinin, and B. Imhof, (eds.), p. 321, Marcel Dekker Inc., NY (1991).

J. C. Ansel, T. A. Luger, D. Lowry, P. Perry, D. R. Roop, and J. D. Mounts, *J. Immunol.* 140:2274 (1988).

ONTOGENY OF THE MUCOSAL IMMUNE RESPONSE AGAINST DIFFERENT TYPES OF PNEUMOCOCCAL POLYSACCHARIDE IN RAT

Germie P.J.M. van den Dobbelsteen, Karin Brunekreef,
Hilde Kroes, Taede Sminia, and Emmelien P. van Rees

Department of Cell Biology, Division Histology, Medical
Faculty, Vrije Universiteit, Amsterdam, The Netherlands

INTRODUCTION

Neonates and young children are highly susceptible to infections with encapsulated bacteria, particularly *Streptococcus pneumoniae*. The ability of the host to produce antibodies against capsular polysaccharides plays an important role in the defence against these bacteria.[1] In human development, responsiveness to polysaccharides does not develop until after the first several months of life and does not reach adult levels until after 5 years[2,3]. By contrast, responses to TI-1 antigens and TD antigens can already be evoked at birth.

Until now, studies of the humoral immune response against pneumococcal polysaccharides (PPS) have always been focussed on the IgM or IgG response. Little is known on IgA response against these antigens, although mucosal immune responses play an essential role in the first line of defence against pneumococcal infections.

In this study, we investigated the ontogeny of the mucosal immune response against 3 structurally different pneumococcal polysaccharides. Our studies in adult Wistar rats already showed that both the route of immunization and the structure of these polysaccharides have a profound influence on the immune responses.[4]

MATERIAL AND METHODS

Animals and Antigens

At least 2 Wistar rats of the following ages were used for each PPS studied: 5, 10, 14, 18, 21, 25, 28, 35, and 49 days after birth.

Type 3, 4 and 14 pneumococcal capsular polysaccharides (PPS-3, PPS-4 and PPS-14) were supplied by the American Type Culture Collection (USA). These PPS differ in chemical, structural and immunogenic properties.[5]

Experimental Design

Neonates were immunized intraperitoneally (ip) or intraduodenally (id) with PPS, dissolved in saline without adjuvant. The dose of antigen was increased with age, in a range from 5 to 50 µg (adult dose). Six days after immunization, the rats were killed. Sera were collected and tissues removed for histological examination. Spleen, intestine, PP, mesenteric and parathymic lymph nodes were frozen in liquid nitrogen and stored at -70°C.

Double Immunocytochemical Staining

Double immunohistochemical stainings to demonstrate the localization of antigen and also the specificity and isotype of antibody containing cells (ACC) were performed by means of a combined two- and three-step immuno-enzyme method.[6]

ELISA

Anti-PPS antibodies in serum were detected by a direct-ELISA.[6] All samples were tested in duplicate. Titers are defined as giving an absorption of 0.05 above negative aged-matched control serum. Data are presented as mean of two animals.

RESULTS

Development and Localization of Anti-PPS ACC

In the various tissues of the youngest animals (days 5, 10) a remarkable amount of antigen was still present, especially PPS-3. Most of the antigen was taken up by macrophages. The clearance of the antigen is probably not effective enough at these ages. Anti-PPS ACC could not be found in animals which were immunized before day 14 after birth. The first anti-PPS ACC were found in animals immunized with PPS-14 and these specific ACC were seen in mesenteric lymph nodes, villi and spleen. Depending on their anatomical localization, these specific ACC were either IgM or IgA positive (Table 1).

Table 1. Specific ACC in various tissues six days after ip immunization with PPS types 3, 4, or 14.

Day of immunization		14		21		28		35		49	
		IgM	IgA	IgM	IgA	IgM	IgA	IgM	IgA	IgM	IgA
PPS-3	spleen	-	-	+	+	+	±	+	±	+	+
	PTLN	-	-	+	+	+	+	+	+	+	+
	MLN	-	-	+	+	+	+	+	+	+	+
	PP	-	-	-	±	-	±	-	±	-	±
	villi	-	-	-	±	-	±	-	±	-	±
PPS-4	spleen	-	-	±	±	+	±	+	+	+	+
	PTLN	-	-	±	±	+	+	+	+	+	+
	MLN	-	-	±	±	±	+	+	++	+	++
	PP	-	-	-	±	-	±	-	±	-	±
	villi	-	-	-	±	-	±	-	±	-	±
PPS-14	spleen	±	±	±	±	±	±	±	±	±	±
	PTLN	-	-	-	±	±	±	±	±	±	±
	MLN	±	+	±	+	+	+	+	+	+	++
	PP	-	-	-	±	-	±	-	±	-	±
	villi	-	±	-	±	-	±	-	±	-	±

The symbols represent respectively: - no cells; ± 1-5 cells; + 5-10 cells; ++ 10-50 cells per average organ section. Abbreviations: PTLN: parathymic lymph nodes; MLN: mesenteric lymph nodes; PP: Peyer's patches

The first specific ACC elicited in rats by PPS-3 and PPS-4 appeared one week later. The number of specific ACC increased with age (Table 1). Trapping of immune-complexes in the follicles of spleen, lymph nodes and PP was seen after 3 weeks.

Anti-PPS Antibodies in Serum

Anti-PPS antibodies in serum were detected by ELISA, and are shown in Fig. 1. IgM anti-PPS antibodies were first detected in 16 day old rats, which were immunized on day 10. The specific IgA initially were seen in 20 day old rats. The levels of specific antibodies increased progressively until adult response levels were achieved. Animals immunized with PPS-14 were first to reach adult levels, compared to PPS-3 and PPS-4. Adult levels of IgM were measured at day 21, and for IgA at day 27. For PPS-3 and PPS-4, adult levels of IgA were not reached by the age of 35 days.

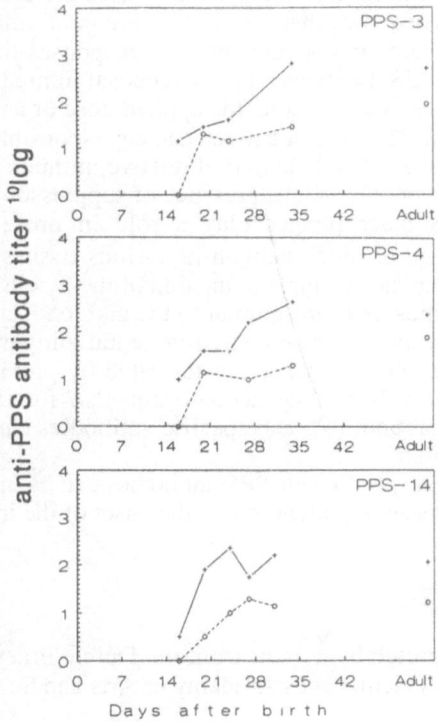

Figure 1. Anti-PPS antibody titers in serum after id immunization.with PPS-3, PPS-4 and PPS-14 Symbols: --+-- IgM and --o-- IgA.

DISCUSSION

The cause of the late development of immune responses to several polysaccharide antigens in healthy human infants is still not known at the present. This late onset is also seen in rodents.[7] Using a rat model, we studied the ontogeny of the mucosal immune response against PPS type 3, 4 and 14. In various lymphoid tissues, IgA specific anti-PPS ACC were found at the same time as IgM specific anti-PPS ACC. In serum, specific IgA antibodies were detected after specific IgM antibodies. Our results are comparable with results shown by Kimura[8], who immunized Fischer rats ip with TNP-Ficoll and TNP-Dextran. Immune responses were first detected in 12-day old rats and had reached 30% of the adult level in 30-day old rats. Another study by Benson and Roberts[9] showed that Sprague-Dawley rats were capable of responding to PPS-3 between the age of 24 and 31 days.

In comparing different ontogeny studies[8-10] and considering the differences in time of onset, several important features have to be taken into account. First, different rat strains were used. The age-related immune response to Ficoll in mice is affected by strain differences.[11] This could also be the case for rats. Second, some of the differences in results may be due to the route of immunization used.[3] The development of different compartments of the immune system does not take place simultaneously. For instance, Peyer's patches, important for taking up antigen from the intestinal lumen, are not developed completely until day 12-14.[12] Even the maturation of different subpopulations of cells can take place at different time intervals. In humans, fetal and neonatal B cell populations differ from the adult splenic B cell population by the lack of $\mu+\partial$- B lymphocytes[13], a subpopulation of B lymphocytes that occur in the marginal zone of the adult spleen, which plays an important role in the immune response against polysaccharide antigens.[13,14]

Third, the structure and dose of the polysaccharide are also important. Regarding the structure, studies[1,2,4] have shown that certain PPS are good immunogens in children of all ages, whereas for other serotypes, serum antibody responses following vaccination are variable. In our rat model, PPS-14 elicits an early mucosal immune response, compared to the other two polysaccharides. Furthermore, the applied dose of antigen may be relevant as this varies in several studies. The dose of antigen can be responsible for inducing tolerance, which is easily induced during the early period of relative immune incompetence. Both the immature macrophage function[15,16] and the presence of suppressor T cells[17] may contribute to this phenomenon. As macrophages play a role in presenting TI-antigen to B lymphocytes[18], the presence of much antigen in various tissues at day 5 and 10 in the present study, is probably due the incomplete function of these cells.

Finally, previous exposure of the animal to the antigen is another important factor, as this may influence the response. Many environmental antigens cross-react with each other, so that a response to one is enhanced or tolerated by previous exposure to related antigens. This counts not only for the neonate but also for the mother, as Lee and coworkers[19] have shown that both PPS and specific antibodies can be transferred through the placenta and mother's milk.

We conclude that IgA specific anti-PPS antibodies can be induced in young rats; but the structure of the PPS plays an important role in the onset of the immune responses.

ACKNOWLEDGMENTS

This work was supported by a grant from the Dutch 'Praeventie Fonds'. E.P. van Rees is a Fellow of the Royal Netherlands Academy of Arts and Sciences.

REFERENCES

1. D. J. Barrett, *Advances Ped.* 18:1067 (1985).
2. H. F. Pabst and H. W. Kreth, *J. Pediatr.* 97:519 (1980).
3. M. J. Cowan, A. J. Ammann, D. W. Wara,V. M. Howie, L. Schultz, N. Doyle, and M. Kaplan, *Pediatrics* 62:721 (1978).
4. G. P. J. M. van den Dobbelsteen, K. Brunekreef, T. Sminia, and E. P. van Rees,*this volume*.
5. J. E. G. Van Dam, A. Fleer, and H. Snippe, *Antonie van Leeuwenhoek* 58:1 (1990).
6. G. P. J. M. Van den Dobbelsteen, N. Van Rooijen, T. Sminia, and E. P. Van Rees, *J. Immunol. Methods* 145:93 (1991).
7. D. E. Mosier and B. Subbarao, *Immunol. Today* 3:217 (1982).
8. S. Kimura, J. H. Eldridge, S. M. Michalek, I. Morisaki, S. Hamada, and J. R. McGhee, *J. Immunol.* 134:2839 (1985).
9. R. W. Benson and D. W. Roberts, *J. Toxicol. Environ. Health* 10:859 (1982).

10. E. P. Van Rees, C. D. Dijkstra, and N. Van Rooijen, *Cell Tissue Res.* 250:695 (1987).
11. T. Hosokawa, A. Aoike, M. Hosono, K. Kawai, and B. Cinada, *Mech. Ageing Dev.* 45:9 (1985).
12. T. Sminia, E. M. Janse, and B. E. C. Plesch, *Anat. Rec.* 207:309 (1983).
13. W. Times, A. Boes, T. Rozeboom-Uiterwijk, and S. Poppema, *J. Immunol.* 143:3200 (1989).
14. P. L. Amlot, D. Grennan, and J. H. Humphrey, *Eur. J. Immunol.* 15:508 (1985).
15. K .E. Schuit, *J. Reticuloendo. Soc.* 30:341 (1981).
16. C. Y. Lu and E. R. Unanue, *Infect. Immun.* 36:169 (1982).
17. H. C. Morse III, B. Prescott, S. S. Cross, P. W. Stashak, and P. J. Baker, *J. Immunol.* 116:279 (1976).
18. H. S. Boswell, S. O. Sharrow, and A. Singer, *J. Immunol.* 124:989 (1980).
19. C.-J. Lee, Y. Takaoka, and T. Saito, *Rev. Infect. Dis.* 9:494 (1987).

INCREASED MANNOSE-SPECIFIC ADHERENCE AND COLONIZING ABILITY OF *ESCHERICHIA COLI* 083 IN BREASTFED INFANTS

Milada Slavíková,[1] Rája Lodinová-Zádníková,[2] Ingegerd Adlerberth,[1] Lars Åke Hanson,[1] Catharina Svanborg[3] and Agnes E. Wold[1]

[1]Department of Clinical Immunology, University of Gothenburg, Sweden; [2]Institute of Care for Mother and Child, Prague, Czech Republic; and [3]Department of Clinical Immunology, University of Lund, Sweden

INTRODUCTION

The large bowel of the newborn is colonized during the first days of life with enterobacteria[1], and in most cases *Escherichia coli* soon becomes the dominant enterobacterial species.[2,3] *E. coli* commonly express type 1 fimbriae that mediate adherence to human colonic epithelial cells.[4] Thus, fimbrial expression may be of importance for the ability of *E. coli* to colonize and persist in the large bowel of the neonate. On the other hand, secretory IgA (S-IgA) carries oligosaccharide chains that function as receptors for type 1 fimbriae.[5] The S-IgA in milk may provide extra binding sites for type 1-fimbriated strains, or block their adherence to colonic epithelial cells. In either case, breast feeding may influence the early enterobacterial colonization of the neonate.

To investigate the role of breast feeding, we studied frequencies and the adhering capacity of type1-fimbriated *E. coli* isolates from breast-fed and non-breast-fed neonates. Half of the babies were deliberately colonized with the non pathogenic *E. coli* strain 083K24H31.[6,7] The other half, who were kept at the same ward, were spontaneously colonized with strains from the environment.

MATERIALS AND METHODS

Infants

Twelve breast-fed and ten bottle-fed infants were given a suspension of 5×10^8 *E. coli* 083K24H31 on day 1, 2 and 3. Thirteen breast-fed and 14 bottle-fed infants in the same ward served as a control group. Both breast-fed and bottle-fed infants had contact with their mothers only during feeding hours.

Bacterial Sampling and Identification

Rectal cultures were obtained on day 1, 3 and 5. Enterobacterial isolates were biotyped using a limited schedule which permitted identification of *E. coli* isolates. These were further characterized with serotyping, using antisera against 69 0- and 17 K-antigens.

Advances in Mucosal Immunology, Edited by
J. Mestecky *et al.*, Plenum Press, New York, 1995

Hemagglutination Pattern and Adherence to Colonic Epithelial Cells

The hemagglutination pattern of the *E. coli* isolates was determined using human, guinea-pig, and chicken erythrocytes, in the presence or absence of mannose. The ability of the *E. coli* isolates to adhere to the human colonic epithelial cell line HT-29 was tested. Cells were incubated with bacteria, in the presence or absence of mannose, washed and fixed. At least 40 cells from each preparation were scored by interference contrast microscopy for the number of adhering bacteria.

RESULTS

E. coli was commonly isolated from days 3-5 in both the colonized and control groups (Table 1).

Table 1. Percent of children with positive rectal cultures for Gram-negative aerobic bacteria and for *E. coli* on their 1st, 3rd and 5th day of life.

	Total Gram-negatives			*E. coli*		
	1	3 (day)	5	1	3 (day)	5
Controls:						
Breast-fed	15%	62%	85%	15%	54%	69%
Bottle-fed	7%	86%	100%	7%	79%	86%
Colonized:						
Breast-fed	33%	92%	100%	25%	83%	100%
Bottle-fed	20%	100%	100%	20%	90%	100%

A majority of the *E. coli* strains, especially in the breast-fed groups, expressed type 1 fimbriae, mediating mannose-sensitive hemagglutination of guinea-pig erythrocytes (Table 2).

Table 2. Hemagglutination pattern of *E. coli* from colonized and control infants.

		Hemagglutination pattern, percent of strains		
		MSHA	MRHA		HAneg
	n		human	chicken	
Control:					
Brest-fed	11	91%		9%	9%
Bottle-fed	17	70%		18%	24%
Colonized:					
Brest-fed	16	100%			
Bottle-fed	13	92%	8%	8%	8%

MSHA=mannose-sensitive hemagglutination pattern; MRHA=mannose-resistant hemagglutination pattern; HAneg=hemagglutination-negative.

The perorally given *E. coli* 083 colonized more efficiently the breast-fed than the bottle-fed infants. Among the breast-fed infants, it replaced other *E. coli* strains in all cases but one (Table 3). In the bottle-fed group, approximately half of the infants acquired *E. coli* 083, and the other half other *E. coli*.

Table 3. Percent of children in the colonized group with positive rectal cultures for *E. coli* 083 and for *E. coli* of other serotypes.

	Breast-fed			Bottle-fed		
	1	3 (day)	5	1	3 (day)	5
E. coli 083	0%	67%	90%	0%	50%	56%
Other *E. coli*	25%	17%	10%	20%	40%	44%

    E. coli 083 expresses type 1 fimbriae and adheres to HT-29 cells by a mannose-dependent mechanism. In the deliberately colonized group, the *E. coli* 083 isolates recovered from breast-fed infants showed a higher mean adherence than those from bottle-fed infants (Fig. 1). Two bottle-fed control infants had acquired *E. coli* 083 spontaneously. These 4 isolates showed a very high adherence to HT-29 cells (Fig. 1).

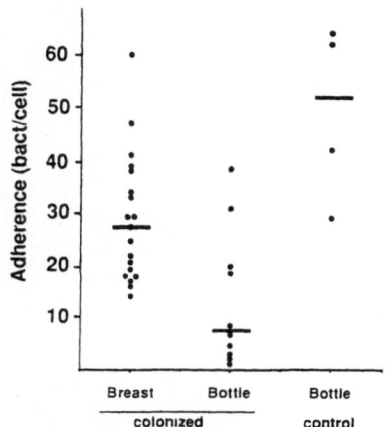

Figure 1. Mannose-sensitive adherence of *E. coli* 083 isolates recovered from breast-fed and bottle-fed infants. Adherence is expressed as the mean number of adhering bacteria per cell. Each dot represents one isolate.

DISCUSSION

    E. coli 083 was strongly favored, relative to other strains, in the breast-fed group. The *E. coli* 083 isolates derived from breast -fed babies displayed a markedly enhanced mannose-specific adherence to colonic epithelial cells compared with those from bottle-fed ones. These findings indicate that receptor analogues in milk influence the colonization. We also found a high expression of mannose-specific binding in the isolates from 2 bottle-fed babies who had acquired *E. coli* 083 spontaneously, compared with babies given the strain perorally. This finding indicates that adherence is a trait which favors colonization in the newborn period.

REFERENCES

1. L. J. Mata and J. J. Urrutia, *Ann. N. Y. Acad. Sci.* 176:93 (1971).
2. K. Tullus, *Acta Paediatr. S cand.* 76, Suppl 334:1 (1987).
3. S. E. Balmer and B. A. Wharton, *Arch. Dis . Child.* 64:1672 (1989).
4. A. E. Wold, M. Thorssen, S. Hull, C. Svanborg-Eden, *Infect. Immun.* 56:2531 (1988).
5. A. E. Wold, J. Mestecky, M. Tomana, A. Kobata, H. Ohbayashi, T. Endo, and C. Svanborg-Eden. *Infect . Immun.* 58:3073 (1990).
6. R. Lodinova, V. Jouja, and A. Lanc, *Z. Immunitätsforsch.* 133:229 (1967).
7. R. Lodinova, V. Jouja, N. Vinsova, J. Vocel, and J. Melkova, *Czech Med.* 3:47 (1980).

THE INFLUENCE OF MILK FACTORS ON THE COURSE OF NEONATAL
MOUSE CRYTOSPORIDIOSIS - A PRELIMINARY REPORT

Jiri Hermanek and Bretislav Koudela

Institute of Parasitology
Czech Academy of Sciences
37005 Ceske Budejovice
Czech Republic

INTRODUCTION

Cryptosporidium parvum (*Coccidia*) is an intracellular protozoan parasite, which causes transient, mostly asymptomatic infections in immunologically competent mammals, while chronic infections associated with profuse diarrhoea occur in immunologically compromised hosts.[1] Both humoral and cell-mediated immunity are involved in anti-cryptosporidial resistance.[2] The most widely used laboratory model for cryptosporidiosis is the neonatal suckling mouse.[1] In two previous studies, mouse pups were not passively protected when allowed to suckle mothers who had prior or immunologically boosted experiences with *Cryptosporidum*.[3,4] We have therefore carried out experiments to examine the influence of nonspecific milk factors on the course of neonatal cryptosporidiosis in cross-feeding experiment using BALB/c and C57B1/6 strains of mice.

MATERIALS AND METHODS

The neonatal mice of BALB/c and C57B1/6 strains originated from pregnant SPF mice (VELAZ, Prague). Each litter of suckling mice was kept individually and remained with their dam during the whole experiment. Cross-exchange of neonatal mice was performed immediately (within 6 h) after birth. Four groups of neonatal mice were used: BALB/c, C57B1/6, BALB/c suckled by C57B1/6 dams and C57B1/6 suckled by BALB/c dams.

Cryptosporidium parvum oocysts were obtained from naturally infected calf and were purified from feces by using discontinous sucrose gradient.[5] All neonatal mice were infected with 10^5 purified oocysts in 100µl of PBS at 7 days of age.

Every second day (starting from day post infection /DPI/3) 3 mice from each

Advances in Mucosal Immunology, Edited by
J. Mestecky *et al.*, Plenum Press, New York, 1995

experimental group were necropsied. Tissue samples harvested from the duodenum, middle jejunum, ileum, cecum, colon, liver, kidney, and pancreas were fixed in formalin, embedded in paraffin, sectioned, and stained with hematoxylin-eosin. Stained sections were examined for *C. parvum* localization and scored 0 to 4+ according to parasite load.[6] Samples from the same animals were also prepared for the examination of the inner intestinal surface by scanning electron microscopy (SEM). On the basis of the SEM findings the degree of infection was classified 0 to 4+ based on *C. parvum* intestinal mucosa colonization.[6]

Spleens aseptically removed from 3 mice of the same experimental group were pooled and suspensions were prepared in RPM-1640 medium (supplemented with 10% of FCS). 2×10^5 splenocytes in 0.2ml were incubated in 96-well plates (Nunc) with PBS-soluble or Triton-extracted membrane fractions of sonicated oocysts (kindly provided by Dr. Ditrich, Institute of Parasitology, Ceské Budejovice). The level of activation of splenocytes was assessed by the modified MTT-assay.[7]

RESULTS AND DISCUSSION

Suckling mice of the C57B1/6 strain were more susceptible to infection with *C. parvum* than BALB/c neonates; the infection process started earlier and persisted longer (Fig. 1- histology showed very closed correlation's with SEM). Similar differences in the susceptibility of these strains to infection with *C. parvum* and another intestinal coccidia (*Eimeria vermiformis*) have been described.[8,9]

The course of cryptosporidial infection was significantly altered in C57B1/6 neonatal mice suckled by BALB/c dams in the cross-feeding experiment. The intensity of infection was lower in all examined parts of intestine and these mice cleared cryptosporidial infection more rapidly in comparison with neonates suckled by their own dams (Fig. 1). The spleen cells of neonatal C57B1/6 mice showed a stronger *in vitro* response to the Triton-extracted membrane antigens prepared from the sonicated oocysts during the second week after infection in comparison with BALB/c neonates (Fig. 2). The response of spleen cells from C57B1/6 neonates suckled by BALB/c dams was remarkably lower in comparison with mice suckled by their own dams, probably due to the limited exposure to cryptosporidial antigens. Similar results were obtained with PBS-soluble antigens (data not shown). In contrast, the course of cryptosporidial infection (Fig. 1) as well as the antigen-induced response of spleen cells (Fig. 2) were only slightly modified by cross-feeding in BALB/c mice.

The results presented demonstrate the possibility of transfer of the "more resistant" phenotype from BALB/c nonimmune dams to C57B1/6 neonates by milk. Because of the SPF status of animals used, the involvement of specific antibodies or immune cells in this transfer appears unlikely. We suppose that nonspecific milk factors produced by more resistant BALB/c dams act rather on the development of resistance of C57B1/6 suckling mice than directly on the parasites. It was shown previously, that the incubation of cryptosporidial oocysts in nonimmune milk (or in immune colostrum) did not alter their infectivity.[1] Our hypothesis is supported by the observation, that milk lymphoid cells invade the suckling neonatal mice.[10]

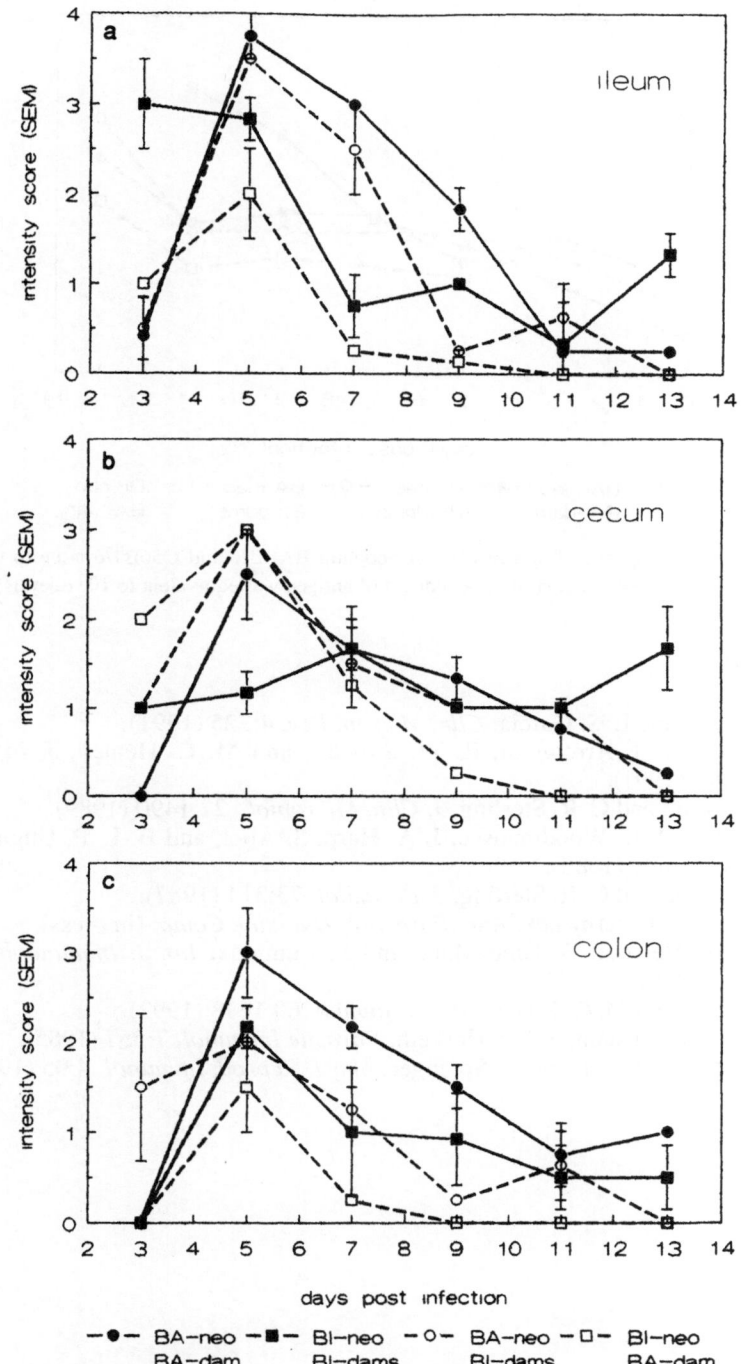

Figure 1. The intensity of cryptosporidial infection in the ileum, cecum, and colon of neonatal mice of BALB/c and C57B1/6 strains after infection with 10^5 oocysts of *C. parvum* on day 7 of life (based on SEM-examination).

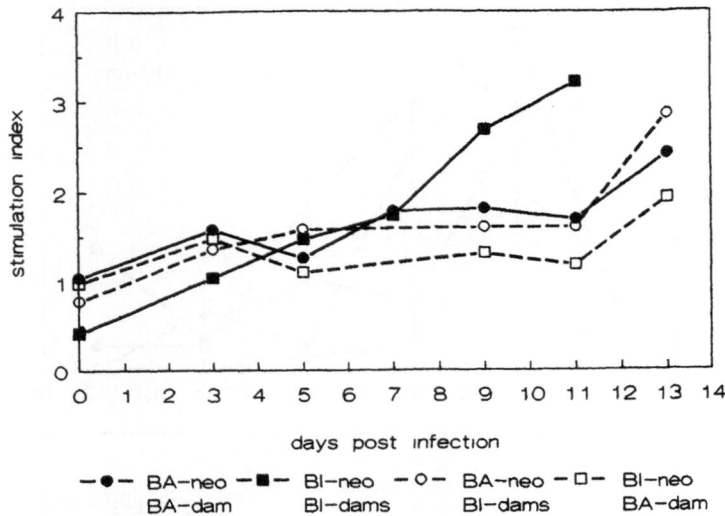

Figure 2. The *in vitro* response of spleen cells of neonatal BALB/c and C56B1/6 mice to the Triton-extracted membrane antigens of *C. parvum* (the amount of antigen was equivalent to 10^4 oocysts).

REFERENCES

1. W. L. Current and L. S. Garcia, *Clin. Microb. Rev.* 4:325 (1991).
2. J. R. Mead, M. J. Arrowood, R. W. Sidwell, and M. C. Healey, *J. Infect. Dis.* 163:1297 (1991).
3. M. J. Arrowood and C. R. Sterling, *J. Clin. Microbiol.* 27:1490 (1989).
4. H. W. Moon, D. B. Woodmansee, J. A. Harp, S. Abel, and B. L. P. Ungar, *Infect. Immun.* 56:649 (1988).
5. M. J. Arrowood and C. R. Sterling, *J. Parasitol.* 73:314 (1987).
6. B. Koudela and J. Hermanek, *Ann. Parasitol. Humaine Comp.* (in press).
7. M. Page, N. Bejaoui, B. Cinq-Mars, and P. Lemieux, *Int. J. Immunopharmacol.* 10:785 (1988).
8. R. Rasmussen and M. C. Healey, *Infect. Immun.* 60:1648 (1992).
9. M. E. Rose, D. Wakelin, and P. Hesketh, *Parasite Immunol.* 7:557 (1985).
10. I. J. Weiler, W. Hickler, and R. Sprenger, *Am. J. Reprod. Immunol.* 4:95 (1983).

CELLULAR AND MOLECULAR BIOLOGIC APPROACHES FOR ANALYZING THE *IN VIVO* DEVELOPMENT AND MAINTENANCE OF GUT MUCOSAL IgA RESPONSES

John J. Cebra,[1] Nico A. Bos,[2] Ethel. R. Cebra,[1] David. R. Kramer,[1] Frans G. M. Kroese,[2] and Carol E. Schrader[1]

[1]Department of Biology, University of Pennsylvania, Philadelphia, PA 19104-6018, and [2]Department of Histology and Cell Biology, University of Groningen, Groningen NL

INTRODUCTION

We will consider three aspects of mucosal immunity we would like to better understand: (1) the ontogeny of preferential switching to IgA expression in Peyer's patches (PP) of neonatal mice and the role(s) of maternal antibodies in modulating this process; (2) the role(s) of dendritic cells in potentiating the expression of IgA by clones derived from both primary and IgA memory B cells; and (3) the contributions of CD5[+] B Cells (B1 B cells) to the population of IgA plasma cells in the gut lamina propria.

Neonatal Mice Are Competent to Develop Preferential IgA Responses in PP Vs. Acute Oral Reovirus Infection

Reoviruses given intraduodenally (i.d.) or orally cause sub-clinical enteric infections in immunocompetent adult mice but are potent stimulators of a mucosal IgA Ab response. Type 1 reovirus has a tissue tropism for M-cells overlying PP and can generate a rapid response in conventionally reared or germ-free (GF) mice as evidenced by PP fragment cultures.[1] These cultures generate specific Ab when established 3-6 days after *in vivo* infection and show peak responses if initiated 6-14 days post infection. We have previously found that neonatal mice show a delay in the development of IgA memory B cells vs. bacterial deter-minants associated with their normal gut flora -- the frequencies of these specific IgA memory cells do not reach adult levels until 10-12 weeks of life.[2] In order to determine whether this delay was due to a delay in the development of fully functional PPs, able to confer preferential switching to IgA expression on locally stimulated B cells, we challenged 10 day old neonatal mice orally with reovirus 1 and compared their responses to 12 week old mice using PP and lamina propria (LP) fragment cultures. Ten day old mice are the youngest that can contain reovirus infections to the gut and forestall often fatal sequelae such as hepatitis, meningoencephalitis, biliary atresia or severe diarrhea. A comparison of the fragment cultures clearly showed the development of a mucosal IgA response, peaking at about 6 days in PP followed by a progressive rise in LP, and that the time course was the same in 10 day or 10-12 week old mice. We next sought to examine the influence of the maternal immune system on the development of humoral mucosal immunity by the pups. Using reciprocal crosses of congenic BALB/c and CB.17 *scid* mice we have generated immunocompetent F_1 pups that are born to and reared by either immunocompetent BALB/c or immuno-incompetent CB.17 *scid* dams. Upon infecting both groups orally with active reovirus 1 we found no differences in the magnitude or kinetics of the developing, reovirus specific IgA Ab responses in PP or LP cultures. PP cultures from both groups produced virus specific IgM when initiated 3 days post infection (p.i.) and began to produce specific IgA if cultured at day 6 p.i. Control, non-challenged littermates of both groups were consistently negative for reovirus-specific Abs. However, the non-challenged F_1 pups of BALB/c immuno-competent dams did not exhibit detectable 'total' or 'non-specific' IgA Ig in either PP or LP cultures until days 19-22 of life, while non-challenged pups born to CB.17 *scid* dams made detectable total IgA as early as day 13 of life. In close correlation with this finding, pups born of *scid* mothers had abundant LP and MLN IgA plasma cells by days 19-22 while pups born of immunocompetent mothers had very few (Table 1).

Table 1. Frequency of IgA and IgM plasma cells in lamina propria and mesenteric lymph nodes of genetically identical neonates born of severe immunodeficient or immunocompetent mothers.

Tissue	Group*/ Age	F_1 neonates from: CB.20♂ x CB.17scid/scid♀ cIgA #/10^5 cells	cIgM	F_1 neonates from: CB.17scid/scid♂ x CB.20♀ cIgA #/10^5 cells	cIgM
L	1-d.18	262	14	12	6
a P	2-d.18	349	12	4	6
m r	3-d.18	672	17	-	-
i o	4-d.18	395	10	-	-
n p	5-d.19	111	9	6	1
a r	6-d.19	93	2	4	3
i	7-d.21	86	6	10	1
a	8-d.21	69	3	23	6
L	1-d.18	38	15	0	0
y N	2-d.18	9	9	0	2
m o	3-d.18	61	30	-	-
p d	4-d.18	6	7	-	-
h e	5-d.19	30	17	1	0
s	6-d.19	3	1	0	1
	7-d.21	23	14	0	2
	8-d.21	3	3	0	3

* each group of cells a pool from three neonatal mice

To determine what aspect of maternal immunocompetence is most significant in mediating the normal delay in expression of IgA after birth, F_1 pups from the same type of reciprocal crosses were exchanged at birth. Thus, some pups from immunocompetent birth mothers were foster-nursed by immunodeficient dams, while some pups from immunodeficient birth mothers were nursed by immunocompetent dams. The outcome was extremely clear. Pups nursed by immunocompetent dams showed a delay of 3-4 weeks after birth before expressing 'natural' IgA in PP and LP cultures, no matter what the immunologic status of their birth mothers. Reciprocally, pups nursed by *scid* dams showed accelerated expression of 'natural' IgA, independent of the immunologic status of their birth mothers. We think it reasonable to assume that maternal secretory IgA Abs in milk, specific for a spectrum of environmental antigens, normally encountered by neonates via the gut route, shield the otherwise competent PP from stimulation. Experiments are underway to test whether enteric reovirus infection can overcome the hypothetical block of maternal Abs vs. particular, commensal enteric bacteria -- perhaps by facilitating uptake of bystander antigens into PPs.

Dendritic Cell Dependent Expression of IgA By Clones in T/B Microcultures is Cyclosporine Resistant

Our original T/B microculture was based on clonal culturing of B cells responsive to thymus-independent Ags, as practiced by the Nossal laboratory using Ag-specific B cells enriched by panning on haptenated gelatin, except that we used cloned, Ag (conalbumin)-specific D10.G.4.1 T_H2 cells and haptenated Ag as stimuli.[3] Small numbers (10-20) of enriched, Ag-specific B cells give Ab secreting clones when placed in microcultures (10 μL) with non-limiting numbers of T_H2 cells (1500-3000) and their specific Ag conjugated to the hapten corresponding to the B cell specificity. These responses are clonal, Ia haplotype restricted, and exhibit requirements for hapten-carrier linkage if the T_H cells are 'rested' prior to use by brief (48 hr.) culture in the absence of Ag, APC, and LKs. The resting period results in a marked decline in cytoplasmic mRNA for IL-4 and IL-5 as detected by *in situ* hybridization. An Ag-independent, haplotype-restricted version of these clonal microcultures has been developed using inputs of 0.5-2 purified F_1, k x b haplotype, B cells and the alloreactivity of D10 cells vs. I-A^b. In either type of microculture, primary B cells can be stimulated to proliferate and generate clones that display isotype switching. However, IgA expression is rare among the clonal Ab products (6 ± 2% of Ab-secreting clones express IgA). Further, B cells shown to include many IgA-memory cells, such as PP B cells enriched on phosphocholine (PC)-gelatin and tested in splenic fragment cultures, failed to generate clones secreting solely IgA in microculture. Even addition of exogenous IL-5 and IL-6 failed to markedly enhance IgA expression in clonal microculture and addition of TGFß markedly reduced the cloning efficiency of B cells without increasing IgA expression.

In an attempt to make T/B microcultures more supportive of IgA expression, we added either peritoneal macrophage, NIH/3T3, BALB/3T3, or dendritic cells (DC) as 'filler' or 'feeder' cells.[4] Although all types of added cells increased the frequency of responding B cells, only DC prepared from either spleen or from PP markedly potentiated IgA expression in both Ag-dependent and allo-stimulated clonal B cell microcultures. We found that as few as 400 DC per culture resulted in a marked increase in the proportion of Ag-dependent clones making anti-PC or anti-N-acetyl glucosaminyl Abs of the IgA isotype (to 30-50%). The IgA Ab-secreting clones included a sizable fraction that secreted IgA Abs exclusively. A similar result was found for allostimulated B cells when DC were added. Neither the source -- spleen or PP -- of either the B cells or the DC affected the clonal expression of IgA when DC were added to the culture. If sIgD$^+$ B cells were prepared from either spleen or PP by fluorescence-activated cell sorting (FACS), then addition of DC to clonal microcultures resulted in a large fraction of clones that expressed IgA as one of several isotypes of Ab or Ig. Thus, addition of DC to these cultures revealed intraclonal isotype switching to IgA.

Using allostimulated PP B cells we were also able to demonstrate that addition of DC to about one B cell and T_H cells allowed the outgrowth of IgA memory cells.[5] These IgA memory cells partitioned into the non-GC subset of PP cells prepared by FACS -- with a surface phenotype of low levels of peanut agglutinin binding (PNA^{low}) and high levels of sIg (kappa chain) (sκ^{high}). They were also markedly enriched by selecting the sIgA$^+$ cell fraction by FACS. When plated into alloreactive T_H-dependent microcultures containing DC these IgA memory cells gave clones that exclusively secreted IgA.

Other features of the DC effect on T-dependent microcultures were that: (1) T_H cells were required for generating Ab-producing clones, but their division during the culture period was not necessary as 2,000 R irradiation did not affect their efficacy; (2) the input of T_H cells could be reduced from about 3,000 without DC to about 375 with DC and comparable frequencies of responding clones were observed; and (3) the presence of DC promoted more vigorous cultures which survived longer and produced more total Ab than in their absence.

DC alone cannot effect IgA expression in T-independent clonal B cell microcultures. PC-specific B cells were enriched on PC-gelatin and were stimulated in clonal microculture with LPS, dextran sulfate (DXS) or both in the presence of DC. Although many highly productive clones secreting IgM and IgG isotypes developed, none secreted IgA. Further, addition of 10% conditioned medium (CM) from ConA-stimulated D10 T_H2 cells also failed to elicit IgA expression. Since the D10 cells themselves were capable of allostimulating a large proportion of clones from PC-specific B cells to express IgA in T/B/DC cultures in the absence of mitogens, we conclude that some sort of physical interactions among the cell types may be required for this effect.

DC plus Ag induces rested D10 T_H2 cells to secrete IL-4 and IL-6 but exogenous LK, D10 CM, or DC/D10 CM cannot replace DC for IgA expression. Co-culturing DC with rested D10 cells results in an Ag dependent output of IL-4 measured using CTLL and CT.4S indicator cells, and IL-6, using B9 cells for bioassay. However, addition of exogenous IL-4, IL-5, IL-6, and combinations of these to T/B microcultures did not replace the effect of DC to promote IgA expression. Further, neither CM from ConA stimulated D10 cells nor 2-10% CM prepared from Ag-stimulated DC/D10 cultures effected IgA expression in T/B microculture. However, all exogenous lymphokines (LK) and CM described above increased the incidence of productive B cell clones -- as did the addition of various macrophage and fibroblast 'filler' cells. Thus the increase in cloning efficiency seems separable from the support of IgA expression. In the course of some of these assays we used cyclosporine A (CyA, 20 ng/ml) to diminish the contributions of endogenous LK made by the D10 cells in T/B microcultures. Indeed, CyA prevented the development of productive B cell clones, but addition of exogenous LK in CM from ConA stimulated D10 cells or in CM from Ag/DC/D10 over-came its effect in T/B microculture, again without effecting IgA expression.

The presence of DC in DC/T/B microcultures counteracts the effect of CyA to prevent development of Ig-producing B cell clones.[4] Remarkably, we found that concentrations of CyA that sharply reduced or eliminated productive clones in T/B micro-cultures, had little effect on their incidence in DC/T/B microcultures. Higher concentrations of CyA did somewhat diminish the total output of Ig secreted by clones developing in the presence of DC, but not the proportion of these that expressed IgA. For instance, 20 ng/ml CyA had no effect on the frequency of responding B cells in DC/T/B microcultures and 87% versus 91% of productive clones expressed IgA in the presence or absence of CyA respectively.

All of our observations are consistent with the requirement that T, B, and DC be physically co-cultured in order to enhance development of B cell clones that secrete IgA. Of course, we cannot rule out sequential interactions of DC with T cells, followed by T with B or of DC with B cells, followed by B with T. The simple propensity of DC to facilitate cell clustering and to nucleate cell aggregates can be observed in microculture and could account for a permissible decrease in T_H2 cell input. However, the remarkable ability of DC to overcome the effects of CyA to prevent the LK-dependent outgrowth of productive B cell clones suggests a more active role for these cells.

The Normal Physiologic Contribution of CD5+ (B1) B Cells in Immuno-Competent Mice to the Pool of LP IgA Plasma Cells and to Protective Mucosal Immunity.

Peritoneal cavity (PeC) B cells (CD5+, Ly1+, or B1 cells) have been found to make significant contributions to the LP IgA plasma cell population in IgM-transgenic mice and upon cell transfer to either x-irradiated, BM reconstituted, adult or to anti-IgM treated neonatal mice.[6] Since the IgM Igs derived from cells of the B1 lineage seem to include a high proportion of Abs cross-reactive with bacterial antigenic determinants as well as autoreactive Abs, it seemed possible that IgA cells derived from this lineage may play a role in mucosal humoral immunity. In the course of using Zynaxis (Zy) fluorescent dye labeled cells to track migration and tissue lodging, we found that PeC B cells were selectively retained in the PeC. Syngeneic adoptive transfer of 6.5 x 10[6], Zy-Red labeled PeC cells intraperitoneally (i.p.) into BALB/c mice resulted in a recovery of donor B cells in PeC of 49, 51, 41, 37, 38, 23, and 24 percent total B cells at days 1, 4, 6, 8, 13, and 22 respectively using FACS analysis -- a decrease of only about 50% in 22 days. At several time points about the same proportion of host and donor PeC B cells incorporate BrdU provided in drinking water (about 25% over 22 days). Exodus of splenic or Peyer's patch B cells from PeC is far more rapid than PeC B cells. If mixtures of PeC and PP cells, contrastingly stained with Zy-Red or Zy-Green, were transferred i.p. into syngeneic/congenic *scid* CB.17 recipients, the PP B cells declined 10-fold more rapidly in the peritoneal cavity than PeC cells over the first 10 days.

The co-transferred PeC and PP B cells also differed in expression of Igh-locus controlled allotype markers. Thus, their plasma cell progeny could be identified and relatively quantitated at times after cell transfer long after Zy-Red or Zy-Green labeled cells could be detected by FACS or fluorescence microscopy. At times from day 10 to day 108 after cell transfer, the plasma cell progeny of the PeC B cells predominated in the spleens of recipients and almost all of these expressed IgM. IgA plasma cells derived from either PP or PeC B cells made a more balanced contribution to the overall pool of IgA cells in the mesenteric lymph nodes and intestinal LP of the immuno-compromised hosts. At earlier times after cell transfer, especially in recipients of higher ratios of PP:PeC B cells, IgA plasma cells derived from the PP source tended to predominate. However, even after long periods (44-108 days), IgA cells from both PP and PeC sources remained rather evenly balanced. This finding is not consistent with the view that B1 cells from PeC are long-lived (self-renewing) while B2 cells (PP) are relatively short-lived and are ordinarily replaced by precursors from bone marrow. The B cells recovered from PeC and spleen of *scid* recipients of mixtures of PeC and PP cells have the characteristics of B1 cells (the PeC source) at 108 days after transfer. Most of the recovered B cells are $IgM^{high} IgD^{low}$, many appear to be Ly1 (CD5)-intermediate, and most bear the IgM allotype of the PeC cell source. In fact, the distribution of lymphocyte subsets in these recipients is very similar to that in PeC and spleen of *scid* mice receiving only PeC cells 125 days previously.

In an attempt to estimate the normal, physiologic contribution of B1 (PeC) B cells to the pool of IgA plasma cells in MLN and LP of immuno-competent mice, we transferred inocula of BALB/c (Igh^a) PeC cells 5×10^6) into congenic CB.17 (Igh^b) recipients. Adult recipients of inocula sufficient to account for an appreciable proportion of PeC B cells in the host displayed, with one exception, only a few percent of MLN or LP IgA plasma cells of donor origin at 13, 23, 50, and 85 days after transfer. Donor PeC B cells could be detected in the PeC of recipients up to 50 days after transfer.

Newborn mice have few IgA or IgM plasma cells in MLN or LP for the first three weeks of life if they are born of immuno-competent mothers (F_1 from CB.20 [*scid/scid*] male x CB.17 [+/+] female). Neonates born of immuno-compromised mothers (F_1 from CB.20 [+/+] male x CB.17 [*scid/scid*] female) exhibit appreciable numbers of IgA plasma cells in MLN and LP by two weeks of life (Table 1). We transferred $1-2 \times 10^6$ BALB/c PeC (Igh^a) cells, into five day old F_1 neonates derived from these reciprocal crosses. Although engraftment was successful, the transferred PeC B cells made little if any contribution to the developing population of MLN or LP IgA plasma cells after 13 or 16 days (by 18-21 days of life).

Congenic transfers of PeC B cells into immuno-competent adult or neonatal (5 day old) recipients suggest a quantitatively minor role for B1 B cells in maintaining the steady state level of IgA plasma cells in the MLN and intestinal lamina propria of immuno-competent mice. A caveat is whether the congenic host treats the Igh allotype-different inocula as 'self' or not.

A final resolution of the relative contributions of B cells stimulated by antigen in mucosal follicles vs. those exposed at other, extramucosal sites -- such as PeC -- to the steady state population of IgA plasma cells in gut lamina propria may depend on a retrospective evaluation of the origin of these end-stage, secretory cells. Adoptive transfer of B cells from a wide variety of lymphoid tissues into immunocompromised recipients always results in the eventual accumulation of some IgA plasma cells in gut lamina propria.[6,7] Presumably these arise by stochastic Ig isotype switching, followed by selective lodging of IgA blasts in mucosal and exocrine tissues.[8] To evaluate the relative contributions of mucosal vs. systemic (including PeC) sources to the IgA plasma cell pool in physiologically normal, immunocompetent hosts we propose a molecular biologic approach. We hypothesize that the preferential isotype switching to IgA expression displayed by B cells responding in PP has as its underlying mechanism direct V/D/J/Cμ to V/D/J/Ca switch recombination at the productive Igh locus. In contrast, switch recombination to IgA expression in B cell clones developing in non-mucosal tissue is likely a stochastic process, frequently involving sequential switching through C_H-genes positioned between Cμ- and Ca-genes in the Igh locus. Evidence that both mechanisms for eventual downstream expression of IgE -- after switch recombination to Cε -- may be operative *in vivo* or *in vitro* has been obtained by DNA sequencing of switch circular DNA or PCR-amplified Sμ/Sε (switch recombination region) junctions.[9,10] Need for a retrospective analysis of switch-recombination pathways for gut IgA plasma cells would preclude use of DNA 'circles' as in the studies of acute switching to IgE. However, we propose to use PCR-amplification and DNA sequencing of switch recombination junctions from IgA hybridomas and IgA plasma cells derived from IgA-committed PP cells and from PeC B cells. Of fundamental interest is whether most examples of IgA plasma cells derived from PP have undergone direct Sμ → Sα recombination. Should the IgA cells derived from non-mucosal sources (PeC) often contain residual segments of Sγ or Sα switch regions between the Sμ/Sα junction, indicating sequential switch recombination, one may be able to use analyses of Sμ/Sαjunctions found in IgA plasma cells from gut lamina propria to roughly estimate the mucosal/non-mucosal origins of these cells. To this end we have developed a method for markedly enriching IgA plasma cells from lamina propria using preparative FACS of cells stained for surface expression of PC-1, a plasma cell specific antigen.[11]

REFERENCES

1.   P. D. Weinstein and J. J. Cebra, *J. Immunol.* 147:4126 (1991).
2.   J. J. Cebra, E. R. Cebra, and R. D. Shahin, *in* "Bacteria and the Host". M. Ryc and J. Franék, eds., p. 303, Avicenum, Prague (1986).
3.   C. E. Schrader, A. George, R. L. Kerlin, and J. J. Cebra, *Int. Immunol.* 2: 563 (1990).
4.   C. E. Schrader and J. J. Cebra, *in* "Dendritic Cells in Fundamental and Clinical Immunology", E. C. M. Hoefsmit, E. W. A. Kamperdijk, and P. Nieuwenhuis, eds., Plenum Press, Amsterdam. (In press).
5.   A. George and J. J. Cebra, *Proc. Natl. Acad. Sci. USA* 88:11, 1991.
6.   F. G. M. Kroese, E. C. Butcher, A. M. Stall, P. A. Lalor, S. Adams, and L. A. Herzenberg, *Int. Immunol.* 1:75, 1989.
7.   J. J. Cebra, P. J. Gearhart, R. Kamat, S. M. Robertson, and J. Tseng, *Cold Spring Harbor Symp. Quant. Biol.* 41:201 (1977).
8.   J. M. Phillips-Quagliata, *in* "Mucosal Immunology", P. L. Ogra, J. Mestecky, M.E. Lamm, W. Strober, J. R. McGhee, and J. Bienenstock, eds., Academic Press, San Diego (1993). (In press)
9.   K. Yoshida, M. Matsuoka, S. Usuda, A. Mori, K. Ishizaka, and H. Sakano, *Proc. Natl. Acad. Sci. USA* 87: 7829 (1990).
10.   S. K. Shapira, H. H. Jabara, C. P. Thienes, D. J. Ahern, D. Vercelli, H. J. Gould, and R. S. Geha, *Proc. Natl. Acad. Sci. USA* 88:7528 (1991).
11.   J. W. Goding and F.-W. Shen, *J. Immunol.* 129: 2636 (1982).

A DUAL ORIGIN FOR IgA PLASMA CELLS IN THE MURINE SMALL INTESTINE

Frans G.M. Kroese,[1] Willem A. Ammerlaan,[1] Gerrit Jan Deenen,[1] Sharon Adams,[2] Leonore A. Herzenberg,[2] and Aaron B. Kantor[2]

[1]Department of Histology and Cell Biology, Immunology Section, University of Groningen, Groningen, The Netherlands; and [2]Department of Genetics, Beckman Center, Stanford University, Medical Center, Stanford, CA, USA

INTRODUCTION

More than two decades ago, Craig and Cebra[1] showed that Peyer's patches are an important source of progenitor cells for intestinal IgA plasma cells. The vast majority of B cells in Peyer's patches are conventional B cells, which are produced throughout the life of the animal and which are responsible for high-affinity antibody responses to a variety of antigens. More recently, we provided evidence that probably also B-1 cells (previously called Ly-1 or CD5 B cells[2]) also contribute significantly to the population of IgA plasma cells in the gut, at least in B lineage chimeras.[3,4] B-1 cells are almost absent from Peyer's patches and are enriched in the peritoneal cavity. These cells are largely self-replenishing and have a selected antibody repertoire with specificities frequently directed towards "natural antigens", autoantigens and bacteria-related antigens.[5,6] In studies presented here we provide additional data, both from transfer studies with sorted B-1 cells and from analysis of μ,κ transgenic mice, to support our hypothesis that B-1 cells can contribute to the IgA response of the gut.

STUDIES WITH B LINEAGE CHIMERAS

In previous experiments B lineage chimeras were constructed by reconstituting lethally irradiated mice with syngeneic bone marrow (BM) and peritoneal cells (PerC) from immunoglobulin allotype congenic donors.[3] Flow-cytometry analysis (FACS) shows that conventional B cells express the BM-donor Ig allotype and B-1 cells express the PerC donor Ig allotype. Since many (40%) IgA plasma cells in the gut express PerC-donor allotype, even up to one year after transfer, these cells most likely belong to the B-1 cell lineage. However, it is possible that long-lived and/or self-replenishing IgA+ memory cells are present among the PerC and expand in the (irradiated) recipient. FACS analysis indicates that some IgA+ cells are present in PerC, although their numbers are very low (approximately 1% of PerC in 10 week old BALB/c mice[3]). Therefore, we have now sorted IgA- B-1 cells (defined by their low levels of sIgD and high levels of sIgM) by three

Advances in Mucosal Immunology, Edited by
J. Mestecky *et al.*, Plenum Press, New York, 1995

color FACS and transferred these cells into Ig congenic lethally irradiated mice, together with syngeneic bone marrow. FACS analysis shows that in these chimeric mice (3 months after transfer), B-1 cells express the Ig allotype of the transferred sorted IgA- B-1 cells. Immunoperoxidase and fluorescence staining of gut sections from these mice demonstrates that high numbers of IgA plasma cells in the lamina propria express the IgA allotype of the sorted B-1 cells. Thus, these experiments show that sorted IgMhighIgDlow peritoneal B-1 cells are able to undergo isotype switching to IgA-expressing cells and to migrate to the gut lamina propria.

IgA PLASMA CELLS IN THE GUT OF μ,κ TRANSGENIC MICE

The introduction of a functionally rearranged μ heavy chain can perturb the function and development of B cells[7-11] and may reflect existing differences between the two B cell lineages.[12-14] In most transgenic mice the vast majority of B cells express transgenic IgM exclusively, as expected by the principle of allelic exclusion. A small population of the B cells, however, is still able to rearrange and express endogenous Ig genes demonstrating that allelic exclusion is not absolute. FACS analysis and transfer studies with M54 transgenic mice, containing a rearranged NP-specific μ heavy chain transgene, have shown that expression of endogenous IgM is largely restricted to the B-1 cell lineage.[12,14] We have recently studied[15] B6-Sp6 mice which carry fully rearranged (BALB/c derived, Igh-C^a allotype) μ heavy chain and κ light chain transgenes, specific for TNP, on a C57Bl background (Igh-Cb).[8] Three criteria demonstrate that endogenous Ig is largely restricted to the B-1 lineage in B6-Sp6 mice. First, FACS analysis shows that the vast majority of B cells in peripheral lymphoid organs and bone marrow of B6-Sp6 mice express transgenic IgM exclusively (Fig. 1). Only a small proportion of B cells (<10%) in these tissues express endogenous IgM, usually simultaneously with the transgenic IgM. Endogenous IgM+ cells in B6-Sp6 μ,κ transgenic mice are, however, clearly enriched in the peritoneal cavity (Fig. 1). Thus, endogenous IgM+ cells have the same anatomical distribution as B-1 cells. Almost half of the peritoneal B cells express endogenous IgM, two-thirds concomitant with trangenic IgM. Second, the peritoneal endogenous IgM$^+$ cells display the B-1 cell phenotype. They express Mac-1 (CD11b) and low levels of IgD (Fig. 1) and many also express Ly-1 (CD5). Third, B6-Sp6 BM poorly reconstitutes endogenous IgM+ cells, just as adult BM poorly reconstitutes B-1 cells. In contrast, B6-Sp6 BM reconstitutes transgene expressing B cells very well, just as normal BM reconstitutes conventional B cells very well. The few endogenous IgM$^+$ cells in the reconstituted mice are predominantly located in the peritoneal cavity and have the phenotype of the CD5- B-1b cells (Ly-1 B sister cells). Likewise, normal adult BM reconstitutes B-1b cells better than B-1a cells.[16] Together, these three criteria (anatomical localization, phenotype and relative BM independency) are sufficient to qualify the large majority of endogenous IgM$^+$ cells in μ,κ transgenic B6-Sp6 mice as B-1 cells.

Immunohistological staining of the small intestine of B6-Sp6 mice shows the presence of high numbers IgA-containing cells (plasmablasts/-cells), and relatively few IgM-containing cells.[15] This IgM is exclusively transgenic IgM. Two-color immunofluorescence staining (IgM and IgA) of cytospins prepared from isolated lamina propria cells demonstrates three subsets of plasmablasts/- cells: a majority population of cells containing only IgA, one-third of the cells containing simultaneously IgA and (transgenic) IgM and a very small population of cells containing only (transgenic) IgM (Fig. 2). This IgA in the gut lamina propria is produced by endogenous Ca immunoglobulin genes, since the transgene in the B6-Sp6 mice encodes only for a μ heavy chain gene. Furthermore, two-color staining of gut sections from these mice with anti-IgA or anti-IgM in combination with a Mab directed to the transgenic idiotype (Id) shows that most IgA containing cells do not express the transgenic Id; only IgM containing cells (with or without IgA) are Id$^+$. Thus, this IgA is the result of the complete, functional VDJ rearrangement,

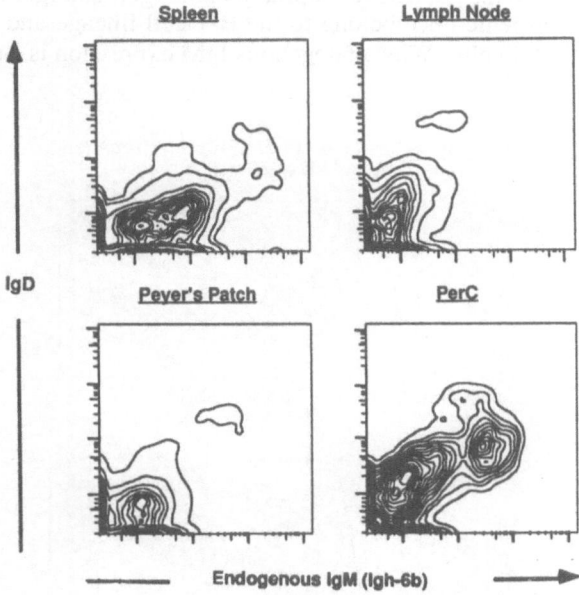

Expression of endogenous IgM in B6-Sp6 μ,κ transgenic mice

Figure 1. Expression of endogenous IgM and IgD on B cells from B6-Sp6 transgenic mice. Cells from various sources were stained and analysed by FACS for endogenous IgM and IgD using fluorescein-labeled anti-Igh-6b (AF6-78.25) and biotinylated anti-Igh-5b (AF6-122.2), respectively. Biotin was revealed by avidin conjugated to Texas red. Only lymphocytes are shown after gating on forward and obtuse scatter and dead cells were excluded by their staining for propidium iodide. Plots display 5% probablity plots.

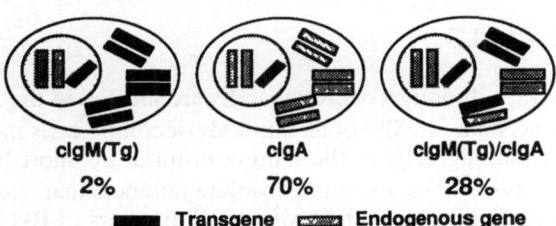

Antibody containing cells in the gut lamina propria of B6-Sp6 μ,κ transgenic mice

Figure 2. Schematic drawing of the types of plasma cells found in the gut lamina propria of B6-Sp6 μ,κ transgenic mice. Cytospins from isolated lamina propria cells were stained by two color immunofluorescence for fluorescein-conjugated anti-IgA (Mab 71.14) and biotinylated anti-IgM (Mab 331), followed by avidin-TRITC. Al least 300 antibody- containing cells were analyzed.

isotype switching and expression of endogenous Ig genes, and is not due to trans-splicing[17] or trans-recombination[18] between endogenous heavy chains and the transgenic VH gene.

Apparently, the relatively few endogenous IgM⁺ cells are responsible for the generation of many IgA-secreting cells in the gut. These data are consistent with findings of Forni[19] and Grandieu *et al.*[20] who observed that a large proportion of the plasmablasts/-cells produce (endogenous) IgG and IgA. Our studies with M54[12,14] and B6-Sp6 transgenic mice[15] demonstrate that endogenous IgM$^+$ expression is almost exclusively confined to the B-1 cell population. As we summarize in Fig. 3, we conclude that the IgA antibody-containing cells in the gut lamina propria (and the IgG- and IgA-containing cells in the spleen) of these transgenic mice belong to the B-1 cell lineage and are derived from isotype switched IgM$^+$ B-1 cells. Why, endogenous IgM expression is largely restricted to

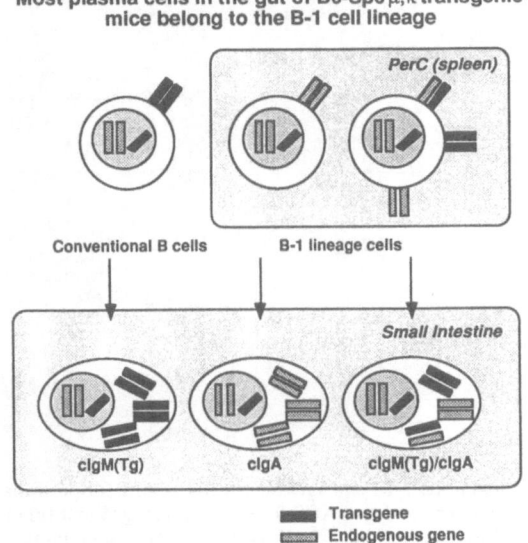

Figure 3. Model of the origin of IgA-containing cells in the lamina propria of the small intestine of B6-Sp6 μ,κ transgenic mice. For explanation see text.

the B-1 cell population is not known, but might be due to differences in development and/or selection mechanisms between B-1 cells and conventional B cells[14].

KINETICS OF B-1 CELLS

Approximately 15 million IgA-secreting cells are located in the murine gut lamina propria, and these cells account for 90% of all antibody-secreting cells in the mouse.[21] The majority of the IgA-containing cells in the lamina propria are short-lived cells with an estimated half-life of 5 days.[22] This means in absolute numbers that about 1.5 million IgA plasma cells are renewed daily. Given the low overall numbers of B-1 cells in the mouse (rouhgly 7-10 million cells6, and the observation that B-1 cells may be responsible for 40% of the IgA cells in the gut, this implies that B-1 cells must expand somewhere in the animal to produce enough IgA precursor cells. To adress this point we have studied the kinetics of peritoneal B-1a cells in mice.[23] Methaphase arrest using vincristine sulphate and S-phase index labelling studies using a single injection of 5'-bromo-2'-deoxyuridine (BrdU) show that B-1a cells do not divide significantly within the peritoneal cavity. Long term oral administration of BrdU in combination with three color immunocytology demonstrates that peritoneal B-1a cells are long-lived cells, and have a renewal rate of only 1% per day (similar to conventional B cells). If our data are correct that B-1 cells account for such a high number of IgA-secreting cells in the gut lamina propria, the total number of (peritoneal) B-1 cells produced daily is clearly not enough to account for the large numbers of IgA precursor cells possibly needed every day. During their differentiation pathway to IgA-secreting cells, B-1 cells must thus divide several times after leaving the peritoneal cavity. Where this expansion (and isotype switching) occurs is currently not known.

CONCLUSIONS

The studies with B lineage chimeras and μ,κ transgenic mice show that B-1 cells potentially can contribute significantly to the generation of IgA plasma cells in the murine intestine. Support for this hypothesis also comes from findings by others.[3,24] For example, several B-1 cell-lines (e.g. CH12) readily switch in vitro from IgM to IgA expresion[25] and transplanting fetal omentum into severe combined immunodeficiency (SCID) mice not only results in the reconstitution of exclusively B-1 cells but also in the development of IgA plasma cells in the small intestine.[26] Finally, studies by Pecquet *et al.*[27] have recently shown that B-1 cells are responsible for a protective mucosal immune response to *Salmonella typhimurium*: only when repopulated with cell sources containing B-1 cells, CBA/N Xid mice (which are normally devoid of B-1 cells[5]) produce serum and mucosal IgA responses after oral immunization with *Salmonella*.

Although the data thus indicate that B-1 cells can generate significant numbers of IgA plasma cells, it is still not known what the relative contribution is of B-1 derived IgA cells in normal, untreated animals. It is also not know whether both B-1a and B-1b cells are involved in the mucosal humoral immune response and transfer studies with sorted B-1 subpopulations are required to resolve this point. The most intriguing questions that also remains to be answerred are of course whether there are differences in the IgA repertoire derived from conventional B cells and of B-1 cells and whether the two IgA's excert different functions.

ACKNOWLEDGMENTS

We like to thank Dr M.C. Lamers for providing us with B6-Sp6 mice and Dr N.A. Bos for discussions. The work described here is supported by NATO collaborative grant CRG 910195, the Netherlands Digestive Diseases Foundation, grant WS 91-25 (to FGMK), the Interuniversity Institute for Radiopathology and Radiation protection I.R.S., grant 7.2.3 (to FGMK), National Institutes of Health grants HD-01287 (to LAH) and DK38707 pilot study 1-04 and National Research Service Award AI07937 (to ABK).

REFERENCES

1. S. W. Craig and J. J. Cebra, *J. Exp. Med.* 134:188 (1971).
2. A. B. Kantor, *Immunol. Today* 12:388 (1991).
3. F. G. M. Kroese, E. C. Butcher, A. M. Stall, P. A. Lalor, S. Adams, and L. A. Herzenberg, *Internat. Immunol.* 1:75 (1989).
4. F. G. M. Kroese, W. A. M. Ammerlaan, and G. J. Deenen, *Ann. N.Y. Acad. Sci.* 651:44 (1992).
5. L. A. Herzenberg, A M. Stall, P. A. Lalor, C. Sidman, W. A. Moore, D. R. Parks, and L. A. Herzenberg, *Immunol. Rev.* 93:81 (1986).
6. A. B. Kantor and L. A. Herzenberg, *Ann. Rev. Immunol.* 11:in press (1993).
7. R. D. Grosschedl, D. Weaver, D. Baltimore, and F. Constatini, *Cell* 38:2263 (1984).
8. S. Rusconi and G. Köhler, *Nature* 314:330. (1985).
9. C. C. Goodnow, J. J. Crosbie, S. Adelstein, T. B. Lavoie, S. J. Smith-Gill, R. A.Brink, H. Pritchard-Briscoe, J. S. Wotherspoon, R. H. Loblay, K. Raphael, R. J. Trent, and S. A. *Nature* 334:676 (1988).
10. D. A. Nemazee and K. Buerki, *Nature* 337:562 (1989).
11. M. C. Lamers, M. Vakil, J. F. Kearney, J. Langhorne, M. H. Julius, H. Mossmann, R. Carsetti, and G. Köhler, *Eur. J. Immunol.* (1989).
12. L. A. Herzenberg, A. M. Stall, J. Braun, D. Weaver, D. Baltimore, L. A. Herzenberg, and R. Grosschedl, *Nature* 329:71 (1987).

13. W. Müller, U. Rüther, P. Vieira, J. Hombach, M. Reth, and K. Rajewsky, *Eur. J. Immunol.* 19:923 (1989).
14. L. A. Herzenberg and A. M. Stall, Cold Spring Harbor Symp. Quant. Biol. 54:219 (1989).
15. F. G .M. Kroese, W. A. Ammerlaan, and A. B. Kantor. submitted. (1992).
16. A. B. Kantor, A. M. Stall, S. Adams, L. A. Herzenberg, and L. A. Herzenberg. *Proc. Natl. Acad. Sci. USA* 89: 3320 (1992).
17. A. Shimizu, M. C. Nussenzweig, H. Han, M. Sanchez, and T. Honjo, *J. Exp. Med.* 173:1385. (1991).
18. R. M. Gerstein, W. N. Frankel, C.-L. Hsieh, J. M. Durdik, S. Rath, J. M. Coffin, A. Nisonoff, and E. Selsing, *Cell* 63:537 (1990).
19. l. Forni, *Eur. J. Immunol.* 20:983 (1990).
20. A. Grandieu, A. Coutinho, and J. Andersson, *Eur. J. Immunol.* 20:991 (1990).
21. P. J. Van der Heijden, W. Stok, and A. T. J. Bianchi, *Immunology* 62:551 (1987).
22. C. Mattioli and T. Tomasi, *J. Exp. Med.* 138:452 (1973).
23. G. J. Deenen and F. G. M. Kroese, *Eur. J. Immunol.*, in press (1992).
24. F. G. M. Kroese, A. B. Kantor, and L. A. Herzenberg, *in*: "Handbook of Mucosal Immunology", P. L. Ogra, J. Mesecky, M. E. Lamm, W. Strober, J. R. McGhee, and J. Bienenstock, eds., Academic Press, Orlando, FL, USA. (1994).
25. L. W. Arnold, T. A. Grdina, A. C. Whitmore, and G. Haughton, *J. Immunol.* 140:4355 (1988).
26. N. Solvason, A. Lehuen, and J. F. Kearney, *Internat. Immunol.* 3:543 (1991).
27. S. S. Pecquet, C. Ehrat, and P. B. Ernst, *Infect. Immun.* 60:503 (1992).

DEVELOPMENT OF MUCOSAL HUMORAL IMMUNE RESPONSES IN GERM-FREE (GF) MICE

Khushroo E. Shroff and John J. Cebra

Department of Biology
University of Pennsylvania, Philadelphia, PA, 19104-6018
USA

INTRODUCTION

GF mice can be effectively mono-associated with gram-negative bacteria *Morganella morganii*. This colonization results in hypertrophy of Peyer's patches (PP), including germinal center reactions (GCR), and the development of specific IgA responses detected *in vitro* in PP fragment cultures. IgA antibody against the phosphocholine (PC) determinant characterizes this response. Although the mucosal response to *M. morganii* develops in 14 days, colonization with another, related gram-negative bacteria-proteus spcs., fails to elicit GCR or IgA antibody responses over a period of weeks. However, oral administration of cholera toxin (CT) during this period results in a prompt mucosal immune response vs. both the CT as well as the commensal bacteria. We are employing a T-B clonal microculture to quantitate and determine the kinetics of development of specific IgA pre-plasmablasts and memory B cells in relation to GCRs.

MATERIALS & METHODS

Mice

Adult C3H germ-free mice were obtained from Taconic and maintained in a sterile environment within the germ-free facility of the Biology department, University of Pennsylvania.

Antigens

M. morganii was kindly provided by Ann Finey, La Jolla. *Proteus rettgerii* used in these studies, was isolated previously in our laboratory. A 1.0 ml aliquot of the organism, stored at -70°C, was thawed and inoculated into a sterile broth of 10% sucrose and brain-heart infusion agar and propagated at 37° C for 16 h. Organisms were harvested and washed. GF mice were inoculated *per os* with a 0.2 ml solution containing 2×10^8 organisms and 5% sodium bicarbonate. Following colonization, the mice were kept in their formerly sterile isolator. CT (List Biological Laboratories, Campbell CA 95008) was used at 10µg/mouse.

PP Fragment Cultures

PP fragment cultures were done as described previously by Logan *et al.*[1]

Advances in Mucosal Immunology, Edited by
J. Mestecky *et al.*, Plenum Press, New York, 1995

Reagents

Conalbumin was purchased from Sigma Chemical Co., (St. Louis. MO) peanut agglutinin (PNA) from Boehringer Manheim Biochemicals (Indianapolis, IN), Cedarlane low-toxicity rabbit complement from Accurate Chemical and Scientific Corp (Westbury, NY) and [methyl-^{3}H]-thymidine (20 Ci/mmol) from NEN Research Products, (Boston, MA). Monoclonal anti-Thy 1.2 (30 H-12) used for T-cell depletion was serum-free culture supernatant from the cell line, purchased from the American Type Culture Collection (ATCC, Rockville, MD). Aphidicholin, recommended by Dr. Richard Schultz, University of Pennsylvania, was obtained from Sigma.

T Cells

The helper T cell clone D.10.G4.1 (D10, ATCC) was maintained by alternate weekly cycles of stimulation and rest in Click's medium (Irvine Scientific, Santa Ana, CA) supplemented with 10% fetal calf serum (FCS) (Gibco, Grand Island, NY), 2mM L-glutamine (Gibco), 5×10^{-2} mM 2-mercaptoethanol (Sigma) and 50 μg/ml of gentamycin (Gibco). Resting cells (1×10^5/ml) were stimulated with 100 /ml of conalbumin, irradiated spleen cells (5×10^5/ml) and 5% rat Con A supernatant (prepared from rat spleens stimulated for 48h with Con A-Sepharose beads). After 7 days, cells were harvested and rested for 48h in medium alone for use in microculture. D10 cells are specific for conalbumin presented in the context of I-A^k and are alloreactive for cells bearing I-A^b [2]

B Cells

PP B cells were isolated as described by Schrader et al.[3] and preparations were found to be >95% pure B cells by staining for surface Ig and I-A (data not shown).

Dendritic Cells (DC)

DC were isolated from the PP of mice by the procedure of Spalding et al.[4] By FACS analysis, the preparations were B220$^-$, Mac-1$^-$, and I-A$^+$ (data not shown).

Fluorescence-Activated Cell Sorting and Analysis

Fluorochrome-labelled PNA was prepared as described.[5] Phycoerythrin-labeled B220 was kindly provided by Dr. Michael Cancro (University of Pennsylvania). PNA coupled to Texas red (Molecular Probes, Eugene, OH) as described[4] was also used. FITC anti-κ was prepared by coupling affinity-purified anti-mouse-κ from hybridoma 187.1 (ATCC) to FITC. PP cells were incubated with the appropriate dilution of fluorochrome-coupled reagent in PBS containing 5% FCS for 30 min on ice. The cells were then washed three times and sorted on the FACS IV Flow Cytometer (Becton Dickinson). Sorted cells were resuspended in RPMI 1640 supplemented with 10% FCS, glutamine, gentamicin, and mercaptoethanol (complete RPMI), as in the preparation of Click's medium, and incubated for 1h at 37° C to allow residual azide to leach from cells before they were put into culture. PNA^{high} sk^{low} populations were 85-90% pure, sIgA$^+$ preparations were 80% pure on reanalysis (data not shown).

Microculture

B cells enriched either for sIgA or for anti-PNA receptor were plated out at 20 cells per well with 3000 D10 cells and 400 dendritic cells in complete RPMI in 60 well Terasaki plates (Miles). The plates were incubated at 37° C in 5% CO_2 in air at >95% humidity for 10 days. Culture supernatants were then collected into 140 μl of RPMI medium and 20 μl was used for detecting hapten-specific antibody by RIA, using plates coated with 1% PC-BSA. RIA was done as described previously.[6] The assay is linear for Ig concentrations in the range of 0.5-10 ng. The frequency of responding clonal precursors was determined as described.[7]

RESULTS & DISCUSSION

The PP of conventionally reared mice are constantly activated and contain chronic GCR presumably due to stimulation by gut antigens. In contrast, the PP of GF mice are quiescent and lack germinal centers. Therefore GF mice are ideally suited to analyze the events leading to IgA commitment in PP GC and to understand the relationship between generation of IgA memory cells and secretory plasma cells. We sought to derive antigen-specific GCR in PP and to determine the kinetics of the specific local antibody response. Formerly GF C3H mice were colonized orally with *M. morganii* which is an occasional gut commensal previously reported to induce an anti-PC response. The most encompassing marker for GC B cells is the capacity to bind relatively high levels of the lectin, PNA (PNA^{high}). PP were removed at various times following inoculation, dispersed and analyzed for PNA binding and B220$^+$ markers by FACS.

In order to determine if a specific local immune response to *M. morganii* was stimulated, PP fragment cultures were set-up and the culture supernatant were examined for anti-PC antibody. Serum from the immunized mice was also examined for systemic anti-PC response.

Although mono-association of germ-free mice with a gram-negative occasional commensal like *M. morganii* results in a germinal center reaction and anti-PC-specific antibody from PP fragment culture supernatant, mono-association with another gram negative, commensal, bacillus, *P. rettgerii* (which is classified along with Morganella in the family *Enterobacteriacae*), fails to result in the generation of GCR.

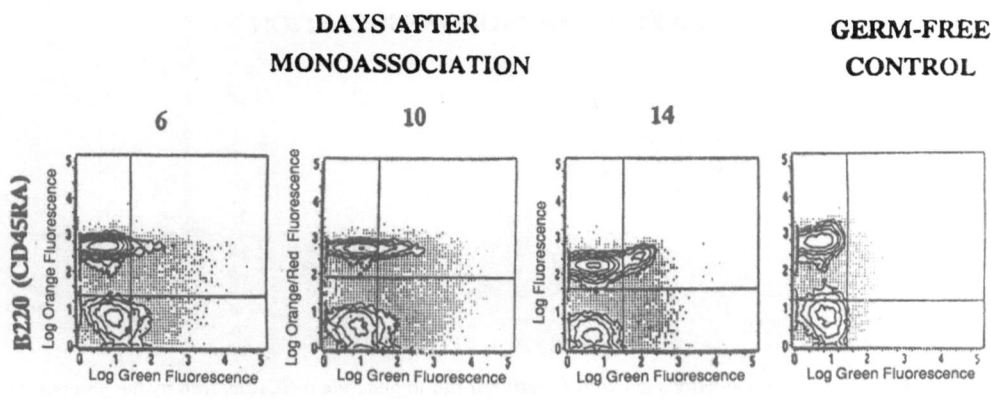

DAYS AFTER MONOASSOCIATION   **GERM-FREE CONTROL**

6   10   14

PNA BINDING

Figure 1 FACS profiles illustrating PP GC development A distinct population of PNA^{high} B220$^+$ cells can be seen to appear at day 6 and peak at day 14 following inoculation with *M morganii*

Having failed to obtain a GCR in the PP after 6 weeks of colonization with *P. rettgerii,* we decided to determine if these mice would respond to oral immunisation with CT, which is an excellent mucosal immunogen. A rapid generation of a GCR indicated by the development of the PNA^{high} Kappa$^+$ cell population 8 days following oral administration of CT was seen in mice that had been mono-associated with *P. rettgerii.* Interestingly,when we analyzed the fragment cultures that were set-up using PP from mice administered CT, we found that the culture supernatant not only contained antibody against CT but also against PC.

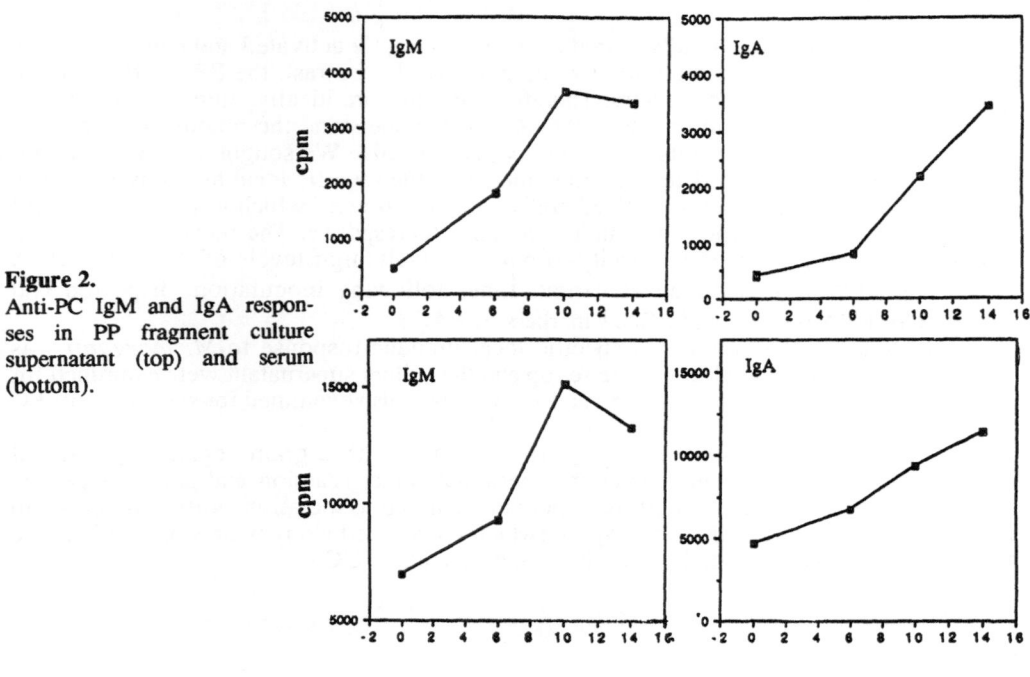

Figure 2.
Anti-PC IgM and IgA responses in PP fragment culture supernatant (top) and serum (bottom).

DAYS AFTER MONOASSOCIATION

Figure 3A. GF mice mono-associated with *P. rettgerii* fail to generate a GCR as seen by the absence of PNA^{high} bearing kappa$^+$ cells at days 14, 28 and 42 after colonization.

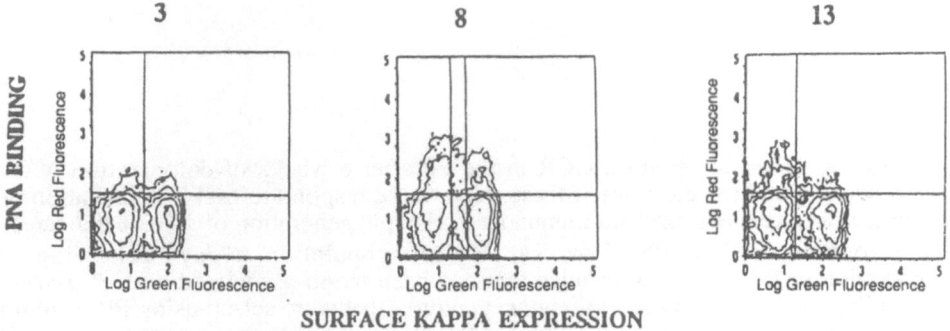

Figure 3B. FACS analysis of PP cell stained for PNA^{high} bearing, Kappa$^+$ cells, at days 3, 8 and 13 after admininistration of CT to mice mono-associated with *P. rettgerri*.

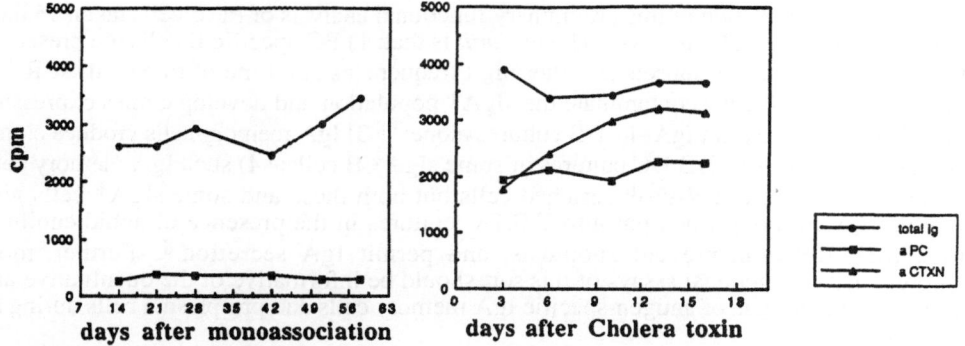

days after monoassociation   **days after Cholera toxin**

Figure 4. Analysis of PP fragment culture supernatant for presence of total antibody as also for anti- PC and anti-CT-specific antibodies. The PPs were from mice that were administered CT following mono-association with *P. rettgerii.*

As no anti-PC antibody response was found in these mice prior to administration of CT we find it reasonable to conclude that oral immunization with CT potentiates' a humoral response against PC-bearing gram-negative bacilli colonizing in the mouse gut.

So, having obtained a specific anti-PC IgA antibody response in fragment cultures of PP of *M.morganii*-immunized mice, we attempted to understand events leading to IgA commitment in PP GC and the relationship between the generation of IgA memory and secretory plasma cells. Roughly 40 % of sIgA+ cells in the PP have elevated mRNA levels for α chain of IgA and as a good percent of these can be recovered from the PNA^high cell population, cells from this subset may become secretory plasma cells. sIgA+ B cells from PP should contain both these GC B cells that have switched to IgA expression and non-GC IgA memory cells. Thus, using flow cytometry we enriched for PNA^high cells and sIgA+ cells from PP of mice mono-associated with *M. morganii.* The enriched cells were then employed in a B cell clonal microculture. The culture results are shown in Table 1.

Table 1. Production of PC-specific Ig isotypes by Peyer's patch B cells in clonal T cell-dependent micrculture.

Cells	Wells # Pos	M	G1	A	M+G1	M+A	G1+A	M+G1+A
			Wells containing anti-PC immunoglobulin* (%):					
DC + T	----	–	–	–	–	–	–	-
PNA^hiB	----	–	–	–	–	–	–	-
PNA^hiB+T	----	–	–	–	–	–	–	-
PNA^hiB+T+DC	14/60	28	8	8	28	14	14	-
PNA^hiB+T+DC+	15/60	33	–	40	7	–	20	-
sIgA+ B+T	7/60	43	–	–	–	14	43	-
sIgA+B+T+DC	23/60	18	4	35	4	26	13	-
sIgA+ B+T+DC+	25/60	12	–	64	–	20	4	-
Unsorted B+T+DC	7/60	14	–	86	–	–	–	-

* % wells positive for Ig expressing the indicated isotypes
† Aphidicholin concentration was 1 μg/ml in culture

445

Our interpretation of this preliminary functional analysis of PP B cells taken 15 days after colonization of GF mice with *M. morganii* is that: 1) PC-specific B cells are present in the PNA[high] and sIgA[+] subsets at rather high frequencies at a time of maximal GCR ; 2) non-GC primary B cells contaminate the sIgA[+] population and develop clones expressing multiple isotypes - except IgA -in T/B cultures alone;[3,8] 3) IgA memory cells produce clones expressing only IgA in T/B/DC cultures of some sIgA[+] B cells;[3] 4) such IgA memory cells are rare in cultures of PNA[high] enriched cells but both these and some sIgA[+] cells give secreted IgA antibody when put into T/B/DC cultures in the presence of aphidicholin to block their division, prevent apoptosis, and permit IgA secretion.[8] Further, more comprehensive, functional assays of this sort should be informative of the quantitative and temporal development of antigen-specific IgA memory cells and pre-plasma cells during *de novo* GCRs.

REFERENCES

1. A. V. Logan, K.-P. N. Chow, A. George, P. D. Weinstein, and J. J. Cebra, *Infect. Immun.* 59: 1024 (1991).
2. J. Kaye, S. Porcelli, J. Tite, B. Jones, and C. C. Janeway, *J. Exp. Med.* 158: 836 (1983).
3. C. E. Schrader, R. L. Kerlin, A. George, and J. J. Cebra, *Int. Immunol.* 2:563 (1990).
4. D. M. Spalding, W. J. Koopman, J. H. Eldridge, J. R. McGhee, and R. M. Steinman, J. *Exp. Med.* 157:1646 (1983).
5. M. L. Rose, M. S. C. Birbeck, V. J. Wallis, J. A. Forrester, and A. J. S. Davies, *Nature* 284:364 (1990).
6. J. L. Hurwitz, V. B. Taggart, P. A. Schweitzer, and J. J. Cebra, *Eur. J. Immunol.* 12:342 (1982).
7   P. A. Schweitzer and J. J. Cebra, *Molec. Immunol.* 25:231 (1988).
8. A. George and J. J. Cebra, *Proc. Natl. Acad. Sci. USA* 88:11 (1991).

INHIBITION OF BACTERIAL TRANSLOCATION FROM THE GASTROINTESTINAL TRACT TO THE MESENTERIC LYMPH NODES IN SPECIFIC PATHOGEN-FREE MICE BUT NOT GNOTOBIOTIC MICE BY NON-SPECIFIC MACROPHAGE ACTIVATION

Rodney D. Berg

Department of Microbiology and Immunology
Louisiana State University Medical Center-Shreveport
Shreveport, LA 71130

INTRODUCTION

Bacterial translocation is defined as the passage of viable indigenous bacteria from the gastrointestinal (GI) tract across the mucosal barrier to extraintestinal organs and sites such as the mesenteric lymph nodes (MLN), liver, spleen, kidney, and blood.[1] In adult, specific pathogen-free (SPF) mice maintained under barrier conditions, indigenous bacteria are continuously translocating from the GI tract to the MLN and other sites, but the host immune defenses are able to kill these low numbers of "spontaneously" translocating bacteria and bacteria are not usually cultured from the MLN and other extraintestinal organs.[1-3] Bacterial translocation from the GI tract is readily detected, however, when: (1) the host immune defenses are compromised, (2) the ecology of the indigenous GI microflora is disrupted to allow intestinal bacterial overgrowth, or (3) the mucosal barrier is physically disrupted[4-11] If only one of these mechanisms promoting bacterial translocation is present in a particular animal model, the indigenous bacteria may translocate only to the MLN and not spread systemically. However, in animal models where more than one promotion mechanism is operating, the indigenous bacteria can spread systemically from the MLN to other organs and sites, and the animal may die of sepsis by its own indigenous GI microflora.

Bacterial translocation is the result of indigenous GI bacteria overcoming the various host defenses, especially immune defenses, to survive in the hostile environment of tissues and body fluids. Consequently, it is important to identify the immune cells, cytokines and effector mechanisms involved in the immune defense against translocation if we are to devise rational strategies for preventing these opportunistic infections originating from the GI tract.

RESULTS AND DISCUSSION

Mice injected once intraperitoneally with immunosuppressive agents, such as cyclophosphamide, prednisolone, methotrexate, 5-fluorouracil, or cytosine arabinoside, exhibit increased bacterial translocation to the MLN, spleen, and liver.[12] The bacterial species isolated from these organs in order of frequency were *Lactobacillus acidophilus*, *Escherichia coli*, *Klebsiella pneumoniae*, *Proteus mirabilis*, *Enterococcus faecalis*, and *Staphylococcus aureus*. Cyclophosphamide and prednisolone were more effective in promoting bacterial translocation than were the various antimetabolites tested. These mice were immunosuppressed as demonstrated by their decreased serum antibody responses to *E. coli* challenge. Serum immunity does not appear effective in preventing the initial translocation of bacteria from the GI tract across the mucosal barrier to the MLN but is

Advances in Mucosal Immunology, Edited by
J. Mestecky *et al.*. Plenum Press. New York. 1995

effective in inhibiting the systemic spread of translocating bacteria from the MLN to other organs and sites.

T cells also appear to be important in the immune defense against bacterial translocation. Athymic nude (nu/nu) mice were tested for "spontaneous" bacterial translocation to the MLN and other organs because they are unique in being congenitally deficient in T cell-dependent immunity. Spontaneous bacterial translocation occurred to 50% of the MLNs, spleens, livers, and kidneys of athymic nu/nu mice compared with only 5% incidence of translocation to these organs in control, euthymic (nu/+) mice.[13,14] *E. coli* and *L. acidophilus* were the predominating bacterial species cultured from these organs. Thymuses from neonatal nu/+ mice (1-2 days old) were grafted to neonatal nu/nu mice (1-2 days old), and the grafted nu/nu mice then tested at 8 wks of age for bacterial translocation from the GI tract. The thymus grafts were successful as demonstrated by increased serum antibody responses of the grafted (nu/nu) mice after vaccination with sheep erythrocytes (T-dependent antigens). The incidence of bacterial translocation to the MLNs, spleens, livers, and kidneys decreased from 50% in the athymic (nu/nu) mice to 8% in the thymus-grafted (nu/nu) mice. This 8% translocation incidence is similar to the 5% incidence of positive organs in control, euthymic nu/+ mice.

Bacterial translocation also was tested in a thymectomized mouse model to insure that the results obtained with nu/nu mice were not due to an unrecognized abnormality of genetically athymic mice unrelated to their lack of a thymus. Neonatally thymectomized +/+ mice tested at 8 wks of age exhibited increased incidences of bacterial translocation to their MLNs, spleens, livers, and kidneys (46%) compared to sham-thymectomized control mice (5%).[15] These results employing both thymectomized and congenitally athymic mouse models demonstrate that T cell-mediated immunity is important in the host defense against bacterial translocation.

It is well known that athymic mice are more susceptible than euthymic mice to a variety of bacterial, mycotic, and viral infections. Consequently, it is not particularly surprising that athymic mice also are more susceptible to the systemic spread of translocating bacteria. However, the specific immunologic mechanisms have not been identified whereby the lack of T cell-mediated immunity allows bacterial translocation in these athymic mice. Because of the lack of T cells, the effector macrophages are not receiving T cell helper function. Also, since secretory immunity is T cell-dependent, these athymic mice lack secretory IgA in their intestinal mucosal secretions. Lack of secretory IgA may allow increased and closer association of indigenous GI bacteria with the intestinal epithelium and thereby allow increased bacterial translocation from the GI tract.

In most bacterial translocation models, the translocating bacteria are cultured from the MLN prior to their appearance in other organs, such as the liver or spleen. Thus, resident macrophages in the MLN appear ideally situated to clear bacteria translocating from the GI tract. Since it cannot be predicted which of the many species of bacteria in the GI tract will translocate in any particular clinical condition, it would be useful to develop an immunologic regimen that would inhibit the translocation of a broad range of bacterial species. Therefore, we tested whether certain immunomodulators known to non-specifically activate macrophages could inhibit bacterial translocation from the GI tract to the MLN.

SPF mice were antibiotic-decontaminated with oral streptomycin and penicillin-G in their drinking water. One group of mice was injected intraperitoneally with muramyl dipeptide, a second group with formalin-killed *Propionibacterium acnes* (formerly classified as *Corynebacterium parvum*), and a third group with particulate yeast glucan. Vaccination with each of these immunomodulators stimulated a lymphoreticular response since the mice demonstrated splenomegaly, a commonly used indicator of macrophage activation. The mice were challenged with viable *E. coli* C25 in their drinking water, sacrificed on day 14, and the MLN cultured for translocating *E. coli* C25. Vaccination with *P. acnes*, but not glucan or muramyl dipeptide, inhibited *E. coli* C25 translocation to the MLN.[16]

P. acnes vaccination of antibiotic-decontaminated mice monoassociated with *E. coli* C25 reduced both the numbers of MLN positive for translocating *E. coli* C25 (translocation incidence) and the numbers of translocating *E. coli* C25 per gram MLN (translocation magnitude). *P. acnes* vaccination decreased bacterial translocation when given at the time of *E. coli* C25 challenge before *E. coli* C25 had begun to translocate or when given after *E. coli* C25 were already translocating to the MLN. *P. acnes* vaccination also decreased translocation to the MLN of other indigenous bacteria, such as *P. mirabilis* and *Enterobacter cloacae,* demonstrating that the *P. acnes* immunostimulation is non-specific.

To further assess the role of macrophages in the defense against bacterial translocation, spleen or MLN cells from *P. acnes*-vaccinated donor mice were adoptively transferred to non-vaccinated recipient mice in an attempt to inhibit *E. coli* C25 translocation. Adoptively transferred MLN or spleen cells from the *P. acnes*-vaccinated donor mice inhibited *E. coli* C25 translocation to the MLN of non-vaccinated recipient mice that had been antibiotic-decontaminated and then challenged orally with *E. coli* C25.[17] Furthermore, *E. coli* C25 translocation to the MLN was reduced by the adoptive transfer of plastic-adherent cells (predominately macrophages) but not adoptively-transferred nonadherent cells (predominately lymphocytes) (Gautreaux and Berg, manuscript in preparation).

Interestingly, *P. acnes* vaccination inhibited *E. coli* C25 translocation in antibiotic-decontaminated SPF mice orally challenged with *E. coli* C25, but did not inhibit either the incidence or magnitude of *E. coli* C25 translocation to the MLN in gnotobiotic (ex-germfree) mice monoassociated with *E. coli* C25. The gnotobiotic mice were immunologically "stimulated" by the *P. acnes* vaccination since they exhibited an even greater splenic index (spleen weight of vaccinated mice/spleen weight of non-vaccinated mice) than the *P. acnes*-vaccinated SPF mice. Also, differences in *E. coli* C25 population levels were not a factor since the cecal levels of *E. coli* C25 were similar in both gnotobiotic mice and antibiotic-decontaminated SPF mice colonized with *E. coli* C25. Thus, it appears that the mouse immune system must be "primed" by antigens of the indigenous GI microflora before a "second stimulation" by *P. acnes* vaccination can activate their macrophages sufficiently to inhibit bacterial translocation to the MLN.

To test this hypothesis, adult germfree mice were colonized with either *E. coli* C25 or the whole cecal microflora for 8 wks prior to *P. acnes* vaccination. Surprisingly, *P. acnes* vaccination did not reduce *E. coli* C25 translocation to the MLN in these adult gnotobiotes monoassociated with *E. coli* C25 or colonized with the whole indigenous microflora for 8 wks.[18] It seems that immunostimulation by antigens of the entire GI microflora for 8 wks would have been sufficient time to "prime" the immune system so that a subsequent *P. acnes* vaccination would be effective in reducing bacterial translocation. However, it may be that this priming effect only occurs if the mouse is exposed to antigens of the indigenous GI microflora very soon after birth before its immune system has fully developed.

Germfree mice were colonized at various times following birth with the entire cecal microflora obtained from SPF mice.[18] The mice then were tested for the effectiveness of *P. acnes* vaccination in reducing *E. coli* C25 translocation to the MLN at 8 wks of age. *P. acnes* vaccination reduced *E. coli* C25 translocation only in the groups of gnotobiotic mice that received the indigenous microflora within 7 days after birth. Other investigators have reported that the immune system of the mouse is not fully developed until 7 days after birth.[19-24] Therefore, it appears that the mouse must be exposed to indigenous microflora antigens soon after birth when the immune system is immature in order for a subsequent *P. acnes* vaccination to inhibit bacterial translocation from the GI tract.

Germfree animals, due to the lack of antigenic stimulation by an indigenous GI microflora, exhibit an underdeveloped immune system compared with conventional animals with a full complement of indigenous GI bacteria.[25] For example, macrophages of germfree animals exhibit phagocytic activities similar to that of conventional animals, but their macrophage bactericidal activities are decreased compared with their conventional counterparts.[26,27] In germfree mice not primed by the indigenous GI microflora, *P. acnes* vaccination may non-specifically stimulate only the phagocytic activities of macrophages, whereas in SPF mice previously primed by the indigenous microflora *P. acnes* vaccination non-specifically activates both phagocytic and bactericidal activities of macrophages. *P. acnes* vaccination of gnotobiotic mice often increases rather than decreases *E. coli* C25 translocation to the MLN. These results are consistent with the hypothesis that *P. acnes* vaccination increases migratory and phagocytic but not bactericidal activities of macrophages of ex-germfree mice. In this case, the increased migratory and phagocytic activities of *P. acnes*-stimulated macrophages promotes ingestion and transport of translocating bacteria from the lamina propria to the MLN. However, the translocating bacteria are not killed even though they are phagocytized because the bactericidal activities of the macrophages have not been activated.

CONCLUSION

There are many questions yet to be answered concerning the immune defense against bacterial translocation. For example, what are the relative contributions of systemic immunity (serum IgG and IgM), T cells (cytokines), and especially mucosal immunity (secretory IgA) in the immune defense? In future studies, the various compartments of the host immune system (systemic, cell-mediated, and secretory immunity) will be compared to determine their relative importance and whether they act in concert to prevent bacterial translocation from the GI tract.

The translocation of indigenous bacteria from the GI tract is not just a laboratory phenomenon. These translocating indigenous bacteria are "classical" opportunistic pathogens and are capable of causing septicemia and death under certain conditions. Disruption by oral antibiotics of the ecologic equilibrium in the GI tract allows intestinal overgrowth by certain antibiotic-resistant indigenous bacteria and is a major mechanism promoting bacterial translocation[2,3,28-30] In fact, there is a direct relationship between the cecal population level of a particular indigenous bacterium, such as *E. coli*, and the numbers of *E. coli* translocating to the MLN.[29] Even though indigenous *E. coli* translocate to the MLN in intestinal overgrowth models, the *E. coli* usually do not spread systemically to other organs or sites in the immunocompetent host. However, if an animal exhibiting *E. coli* intestinal overgrowth also is immunocompromised, the translocating *E. coli* spread systemically to cause life-threatening infection. For example, mice given the combination of oral penicillin plus IP cyclophosphamide die of septicemia within 14 days due to indigenous bacteria translocating from their own GI tract.[31] Oral penicillin disrupts the GI ecology to allow the *Enterobacteriaceae*, i.e. *E. coli*, to overgrow the intestines and translocate to the MLN. Immunosuppression by cyclophosphamide then allows the translocating *E. coli* to spread systemically to other organs, including the peritoneal cavity and bloodstream. Similar results occur with other combinations of antibiotics and immunosuppressive agents, such as oral clindamycin plus prednisolone.[31] Thus, the combination of an oral antibiotic plus an immunosuppressive agent synergistically promotes the translocation and spread of indigenous GI bacteria to cause lethal sepsis.

We have demonstrated bacterial translocation in a variety of animal models including rodents with bacterial intestinal overgrowth due to oral antibiotics,[28,30] streptozotocin-induced diabetes,[8] thermal injury,[32,33] solid tumors,[34] leukemia,[35] endotoxemia,[36,37] hemorrhagic shock,[38,39,40] intestinal obstruction,[41] bile duct ligation,[42] protein malnutrition,[43] or parenteral nutrition.[44] The results from these animal models suggest that bacterial translocation pathogenesis occurs in distinct stages. "Spontaneous" translocation is occurring continuously in the healthy adult animal, but this low rate of translocation is easily controlled by the host immune defenses and bacteria are not cultured from the MLN or other extraintestinal organs. The first stage of bacterial translocation occurs when indigenous bacteria are cultured from nearly 100% of the MLN, as often occurs when animals are given oral antibiotics. The oral antibiotics disrupt the GI ecology and allow intestinal bacterial overgrowth by antibiotic-resistant members of the indigenous GI microflora. Intestinal bacterial overgrowth is a major mechanism promoting bacterial translocation from the GI tract. This low level of bacterial translocation to the MLN can occur for a prolonged period without the animal showing symptoms. This first stage of translocation leads to a more serious infection, however, if the animal's GI tract is colonized by exogenous bacteria more virulent than the indigenous bacteria. The second state of translocation occurs when translocating bacteria spread from the MLN to other organs, such as the liver, spleen, and kidneys. Immunosuppression of the host immune system is the major mechanism promoting this second stage of translocation. The host may still confine the infection depending upon the virulence of the translocating bacteria. The third stage of bacterial translocation occurs when the translocating bacteria spread systemically to cause sepsis, septic shock and even death. The progress of infection during this third stage is dependent upon factors such as the virulence properties of the translocating bacteria, the degree of intestinal mucosal permeability, and the immune status of the host.

Evidence is accumulating that translocation of indigenous bacteria from the GI tract to cause life-threatening infections also occurs in debilitated human patients. A direct relationship has been demonstrated between the bacterial biotype/serotype predominating in fecal surveillance cultures from patients with leukemia or other immunosuppressive diseases and the bacterial biotype/serotype eventually causing septicemia in these patients.[45] Indigenous bacteria also have been cultured from the MLN of patients with bowel obstruction,[46] colorectal cancer,[47] Crohn's disease,[48] or hemorrhagic shock.[49] Studies of the immune defense mechanisms inhibiting bacterial translocation from the GI tract will provide information necessary for controlling opportunistic indigenous infections originating from the GI tracts of debilitated patients, such as those with thermal injury, trauma injury, hemorrhagic shock, endotoxemia, immunosuppressive disorders, and AIDS.

REFERENCES

1. R. D. Berg and A. W. Garlington, *Infect. Immun.* 23:403 (1979).
2   R. D. Berg and W. E. Owens, *Infect. Immun.* 25:820 (1979).
3. R. D. Berg, *Infect. Immun.* 29:1073 (1980).
4. R. D. Berg, *Am. J. Clin. Nutr.* 33:2472 (1980).
5. R. D. Berg, *Microecology Therapy.* 11:27 (1981).
6. R. D. Berg, *in:* "Recent Advances in Germfree Research", S. Sasaki, A. Ozawa and K. Hashimoto, eds., p. 411, Univ. Tokai Univ. Press, Tokyo (1981).
7. R. D. Berg, *in* "The Intestinal Microflora in Health and Disease", D. Hentges, ed., p. 333, Academic Press, New York (1983).
8. R. D. Berg, *Expt. Animals* 34:1 (1985).
9. R. D. Berg, *Microecol. Therapy* 18:43 (1989).
10. R. D. Berg, *in:* "Gut-Derived Infectious-Toxic Shock (GITS). A Major Variant of Septic Shock", H. Cottier and R. Kraft, eds., p. 44, S. Karger AG, Basel (1992).
11. R. D. Berg, *in:* "Scientific Basis of the Probiotic Concept", R. Fuller, ed., p. 55, Chapman and Hall, London (1992).
12. R. D. Berg, *Curr. Microbiol.* 8:285 (1983).
13. W. E. Owens and R. D. Berg, *Infect. Immun.* 27:461 (1980).
14. W. E. Owens and R. D. Berg, *J. Immunol. Methods.* 27:461 (1981).
15. W. E. Owens and R. D. Berg, *Curr. Microbiol.* 7:169 (1982).
16. K. Fuller and R. D. Berg, *in:* "Microflora Control and its Application to the Biomedical Sciences", B.S. Wostmann, ed., p.195, Alan R. Liss, Inc., New York (1985).
17. M. D. Gautreaux and R. D. Berg, *in:* "Proc. 10th Int. Symp. Gnotobiology", P. Heidt, ed., in press. (1992).
18. R. D. Berg and K. Itoh, *Microecology Therapy* 16:131 (1986).
19. C. J. Czuprynski, C. J. and J. F. Brown, *Infect. Immun.* 50:423 (1985).
20. B. Morland and T. Midtvedt, *Infect. Immun.* 44:750 (1984).
21. B. G. Carter and E. S. Rector, *J. Immunol.* 109:1345 (1972).
22. B. J. Hargis and S. Malkiel, *J. Immunol.* 104:942 (1970).
23. G. C. Saunders and D. Swartrzendruber, *J. Exp. Med.* 131:1261 (1970).
24. K. Takeya and K. Nomoto. *J. Immunol.* 99:831 (1967).
25. R. D. Berg, *in:* "The Intestinal Microflora in Health and Disease", D. Hentges, ed., p. 101, Academic Press, New York (1983).
26. M. O. Chiscon and E. S. Golub, *J. Immunol.* 108:1379 (1972).
27. H. Y. Yang and O. K. Skinsnes, *RES J. Reticuloendothel. Soc.* 14:181 (1973).
28. R. D. Berg, *Infect. Immun.* 33:854 (1981).
29. E. K. Steffen and R. D. Berg, *Infect. Immun.* 39:1252 (1983).
30. E. Deitch, K. Maejima and R. D. Berg, *J. Trauma* 25:385 (1985).
31. R. D. Berg, M. E. Wommack and E. Deitch, *Arch. Surg.* 123:1359 (1988).
32. K. Maejima, E. A. Deitch and R. D. Berg, *Arch. Surg.* 119:166 (1984).
33. K. Maejima, E. A. Deitch and R. D. Berg, *Infect. Immun.* 43:6 (1984).
34. R. L. Penn, R. D. Maca and R. D. Berg, *Infect. Immun.* 47:793 (1985).

35. R. L. Penn, R. D. Maca and R. D. Berg, *Microecology Therapy.* 15:85 (1986).
36. E. A. Deitch and R. D. Berg, *J. Trauma* 27:161 (1987).
37. E. A. Deitch, R. D. Berg and R. Specian, *Arch. Surg.* 122:185 (1987).
38. J. W. Baker, E. A. Deitch, R. D. Berg and L. Ma, *Surg. Forum* 37:73 (1987).
39. J. W. Baker, E. A. Deitch, L. Ma and R. D. Berg, *J. Trauma* 28:896 (1988).
40. E. A. Deitch, J. W. Ma, L. Ma, R. Berg, and R. Specian, *Surgery* 106:292 (1989).
41. E. A. Deitch, W. R. Bridges, J. W. Ma, L. Ma, R. D. Berg, and R. D. Specian. *Amer. J. Surg.* 159:394 (1990).
42. E. A. Deitch, K. Sittig, L. Ma, R. Berg, and R. D. Specian, *Amer. J. Surg.* 159:79 (1990).
43. E. A. Deitch, J. Winterton, L. Ma, and R. D. Berg, *Ann. Surg.* 20:8 (1987).
44. G. Spaeth, R. D. Berg, R. Specian, and E. A. Deitch, *Surgery* 108:240 (1990).
45. C. H. Tancrede and A. O. Andremont, *J. Infect. Dis.* 152:99 (1985).
46. E. A. Deitch, Arch. Surg. 124:699 (1989).
47. P. Vincent, J. F. Colombel, D. Lescut, L. Fournier, C. Savage, A. Cortet, P. Quandalle, M. Vankemmel, and H. LeClerc, *J. Infect. Dis.* 158:1395 (1988).
48. M. S. Ambrose, M. Johnson, D. W. Burdon, and M. R. B. Keighley, *Brit. J. Surg.* 71:623 (1984)
49. B. F. Rush, A. J. Sori, T. F. Murphy, S. Smith, J. J. Flanagan, and G. W. Machiedo, *Ann. Surg.* 207:549 (1988).

EXPRESSION OF Thy-1 ANTIGEN IN GERM-FREE AND CONVENTIONAL PIGLETS

Richard Pospísil,[1] Ilja Trebichavsky,[1] Jirí Sinkora,[1] Marie Lipoldová,[2] Leos Mandel,[1] Ludmila Tucková,[1] and Jaroslav Rejnek[1]

[1]Institute of Microbiology, Czech Academy of Sciences
Department of Immunology and Gnotobiology
[2]Institute of Molecular Genetics, Czech Academy of Sciences, Prague, Czech Republic

INTRODUCTION

Thy-1 is considered to be a differentiation-regulated membrane protein, with a variable level of expression on lymphohematopoietic cell types.[3] This structure is rarely expressed on normal human T or B lymphocytes except in their early stages of development.[16] In contrast, the Thy-1 antigen is expressed at high levels on mouse T cells at most stages of differentiation.[14] Rat thymocytes bear high levels of Thy-1, but mature peripheral T cells lack this antigen.[18] Thy-1 analogs have been described also in dogs[11], chickens[17], and frogs.[10]

In humans the Thy-1 antigen is expressed at low levels in early differentiation stages of hematolymphoid cells which do not bear markers associated with mature B, T, or myeloid cells.[13]

In this study we to followed the expression of Thy-1 molecule in germ-free and conventional piglets.

MATERIALS AND METHODS

Animals and Antigens

Fetal pigs of the Minnesota miniature strain at 60 and 90 days of gestation (3 fetuses at each stage, and 3 fetuses of the same litters served as unimmunized controls which were sham-stimulated only with saline), 7 conventional pigs (5 months old) and 4 germ-free piglets (2 weeks old) were injected i.m. in the gluteus muscle with 100 µg of LPS (from *Salmonella typhimurium*) or 50 µg of *Nocardia* delipidated cell mitogen (NDCM),[2] in 1 ml of Al-Span-Oil adjuvant (Velaz, Prague). The whole operation on pregnant sows was accomplished under Halothane-oxygen anaesthesia.[5] Fetuses were removed 7 days later.

Indirect Immunofluorescence

To determine the relative abundance of the Thy-1+ cells in the thymus and lymph nodes, tissues were minced and the cells were obtained by homogenization. Isolated cells

Advances in Mucosal Immunology, Edited by
J. Mestecky *et al.*, Plenum Press. New York. 1995

453

(10^6) were first incubated for 30 min at 4°C with F7D5 monoclonal antibody[7], washed twice and incubated for 30 min at 4°C with FITC-conjugated swine anti-mouse Ig diluted 1:4 by PBS. Thoroughly washed cells were analysed in a fluorescence-activated cell sorter (FACScan, Becton Dickinson)

Immunoprecipitation and Immunoblotting

Lymph node cells were ^{125}I labelled using the lactoperoxidase technique[12] and lysed with 0.75M TRIS-HCl buffer pH 8,2 containing 0.5% Nonidet P-40, 1mM PMSF, 1mM EDTA, and 10μg/ml-1 Aprotinin for 20 min on ice and centrifuged (2000g, 10 min and 10,000g 30 min). 100 μl of lysate was precipitated by a two-step procedure[1] using monoclonal antibody or normal BALB/c serum and corresponding second antibody. Precipitates were washed and resuspended in sample buffer, boiled for 10 min at 90°C and subjected to SDS-PAGE.[6] After separation, gels were dried and exposed to Kodak X-OMAT films (at -70°C).

Northern Blot Analysis

To confirm the Thy-1 gene transcription in the fetal liver we used RNA isolated from the liver of 35 day old fetuses for Northern blot analysis. Total RNA was isolated by the acid guanidine-phenol-chloroform method.[4] The 1350 bp ApaI fragment of the mouse genomic Thy-1 probe (gift from Dr. D. Kioussis, MRC, London, U.K), which hybridizes with both Thy-1.1 and Thy-1.2 mRNA was used. Conditions for hybridization were described in detail elsewhere.[8]

RESULTS AND DISCUSSION

The expression of Thy-1 in pigs differs from all other species studied so far. Thy-1 antigen is expressed in low levels on pig thymocytes (Fig. 1), splenocytes, and peripheral blood T lymphocytes (not shown) but lymph node lymphocytes bear high levels of Thy-1 (Fig. 1).

During the ontogenic development, Thy-1$^+$ cells appeared first in the thymus and later in the embryonic liver and peripheral blood. A substantial increase of Thy-1$^+$ thymocytes was observed after stimulation of fetuses with LPS or NDCM mitogens (Fig. 2).

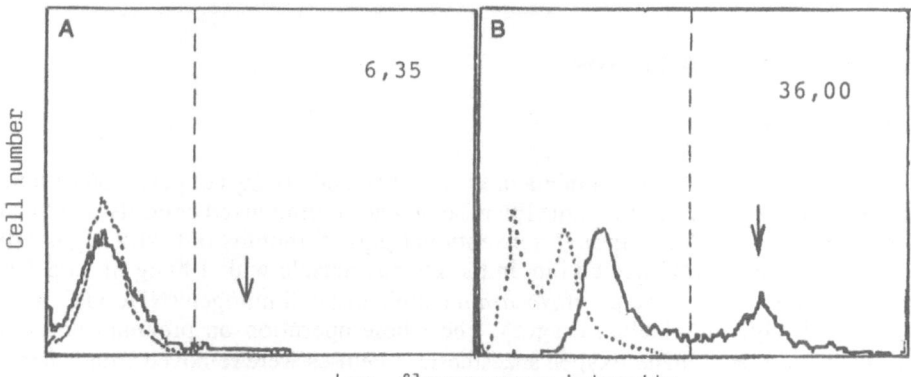

Figure 1. Flow cytometry : Histograms of pig thymocytes (A) and lymph node lymphocytes (B) stained with F7D5 mAb and FITC-conjugated swine anti-mouse Ig (solid line). Labelling with the second antibody alone served as a negative control (dashed line). Histograms represent the number of cells vs. relative fluorescence intensity (logarithmic scale).

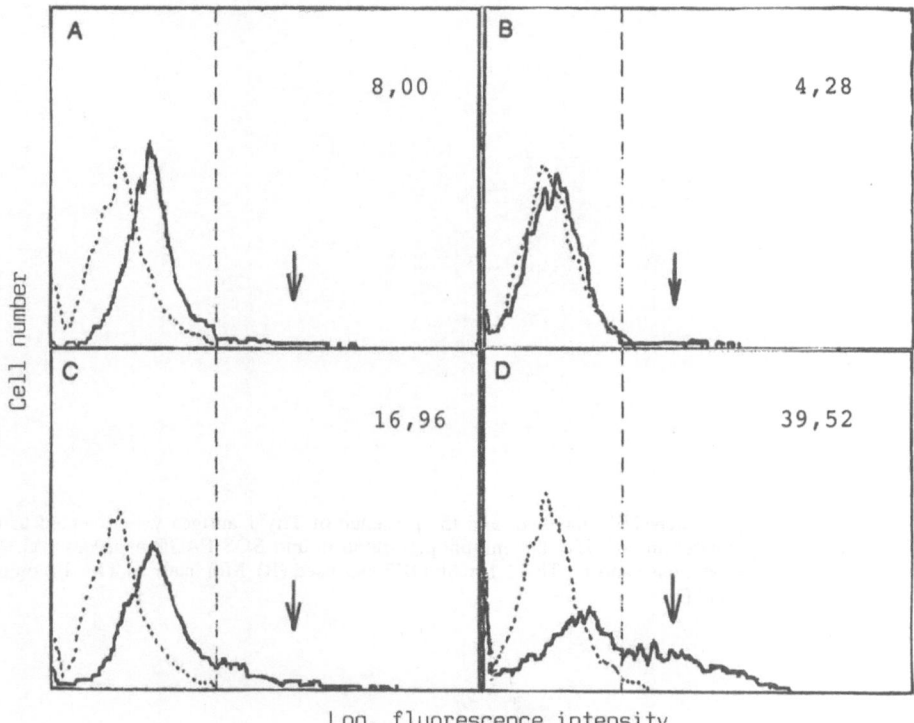

Figure 2. Thy-1$^+$ thymocytes at 60 day of gestation (A,C) and at 90 day of gestation (B,D) in pig fetuses. (C) and (D): fetuses injected with LPS and NDCM, respectively. (A) and (B): fetuses of the same litters which were sham-stimulated only with saline.

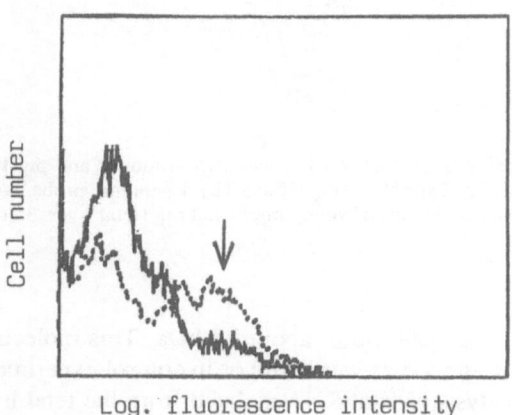

Figure 3. Expression of Thy-1 antigen in germ-free animals (solid line) and in germ-free piglets stimulated with NDCM antigen (dashed line).

Thy-1 antigen was absent from lymph nodes of unstimulated germ-free animals but was highly expressed in germ-free piglets simulated with NDCM antigen (Fig. 3). Thy-1 antigen on pig lymph node cells was characterized by the immunopre-cipitation technique using monoclonal antibody prepared by Lake *et al.*[7] and Manson *et al.*[9] analyses of precipitates by SDS-PAGE showed that mAb to Thy-1.2 antigen reacted

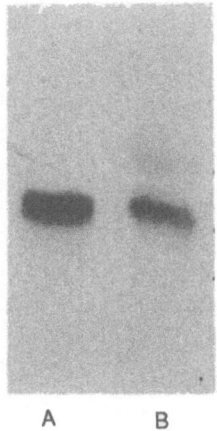

Figure 4 Lymph node cells were I^{125} labelled and the presence of Thy 1 antigen was detected using mouse anti rat Thy 1 2 isolated mAb F7D5 by immunoprecipitation and SDS PAGE analyses (A) No reactivity was observed when mouse anti rat Thy 1 1 mAb OX7 was used (B) Mol mass of Thy 1 2 on pig lymph node cells is about 28 kDa

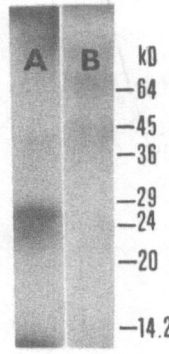

Figure 5 Northern blot analysis of total RNA isolated from mouse and pig tissues Total RNA was analysed for the presence of Thy 1 mRNA using mouse Thy 1 genomic probe (A) and (B) a comparison between probe binding to mouse (thymus of young mice) and pig (fetal liver 35 day of gestation) mRNA respectively

in pig T cell lysates with a molecule of about 28 kDa This molecule was not detectable when either anti-Thy-1 1 antibodies were used or in control experiments (Fig 4)

Northern blot analysis of total RNA isolated from the fetal liver at the 35th day of gestation is shown in Fig 5 The mol wt of porcine mRNA cross-hybridizing with mouse Thy 1 probe and its ability to form hybrids with the probe under stringent conditions suggest that it is an analogue of mouse Thy 1 mRNA

REFERENCES

1   P Baron, F Wernet, F Schunter, and H Wigzell, *Scand J Immunol* 6 3851 (1977)

2   R Barot Ciorbaru, J Brochier, T Miyawaki, J L Preud homme, J F Petit, C Bona, N Taniguchi, and J P Revillard, *J Immunol* 135 3277 (1985)

3. S. H. Feng, C. Woodley-Miller, L. Chao, and A. C. Wang, *Immunol.Lett.* 19:109 (1988).
4. P. Chomczynski and N. Sacchi, *Anal. Biochem.* 162:156 (1986).
5. F. Kovárü and V. Stozicky *in*: "Proceedings of the 9th Congress of the International Pig Veterinary Society, Barcelona" (1986).
6. V. F. Laemmli and M. Faure, *J. Molec. Biol.* 80:575 (1979).
7. P. Lake, E. A. Clark, M. Khoshidi, and G. H. Sunshine, *Eur. J. Immunol.* 9:875 (1979).
8. M. Lipoldov, M. Londei, B. Grubeck-Loebenstein, M. Feldmann, and M. J. Owen, *J. Autoimmun.* 2:1 (1989).
9. W. Manson and A. F. Williams, *Biochem. J.* 187:1 (1980).
10. M. H. Mansour and E. L. Cooper, *J. Immunol.* 132:2515 (1984).
11. J. L. McKenzie and J. W. Fabre, *Transplantation* 31:275 (1981).
12. M. Morrison, *Methods iEnzymol.* 70:214 (1970).
13. C. E. Muller-Sieburg, C. A. Whitlock, and I. L. Weissman, *Cell* 44:653 (1986).
14. A. E. Reif and J. M. V. Allen, *Nature* 200:1332 (1963).
15. A. E. Reif and M. Schlesinger, *in* "Cell Surface Antigen Thy-1," M. Dekker, Inc., New York (1989).
16. M. A. Ritter, C. A. Sauvage, and D. Delia, *Immunology* 49:555 (1983).
17. J. A. P. Rostas, T. A. Shevenan, C. M. Sinclair, and P. L. Jeffrey, *Biochem.J.* 213:143 (1983).

TRANSIENT APPEARANCE OF CIRCULATING INTERLEUKIN-6 AND TUMOR NECROSIS FACTOR IN GERM-FREE C3H/HeJ AND C3H/HeN MICE UPON INTESTINAL EXPOSURE TO *E. COLI*

Ulf I Dahlgren,[1,2] Tore Midtvedt,[3] and Andre Tarkowski[1,4]

[1]Department of Clinical Immunology,
[2]Department of Endodontics/Oral Diagnosis,
University of Goteborg,
[3]Department of Microbial Ecology, Karolinska Institute, Stockholm,
[4]Depattment of Rheumatology, Sahlgrenska hospital, Göteborg, Sweden

INTRODUCTION

The intestinal lumen harbours an enormous number of bacteria. These bacteria carry and release substances, e.g., lipopolysaccharide (LPS), which are potent activators of the immune system and inflammatory reactions. Despite this, the host does not normally show any inflammatory reaction towards the intestinal flora and the blood concentration of inflammatogenic cytokines such as interleukin (IL)-1, IL-6 and TNF are generally below detection levels. In contrast, if other mucosal surfaces, e.g., the urinary tract, are colonized with Gram negative bacteria this results in the appearance of IL-6 in the urine and infiltration with neutrophilic granulocytes in the lamina propria. In this respect it is known that mice which are repeatedly administered LPS become non-responsive in terms of cytokine production.[1] Whether the same type of non-responsiveness exists in response to the inflammatogenic properties of the intestinal flora is at present unknown.

The present study was undertaken in order to investigate the early events, in terms of production of TNF and IL-6, in germ-free mice whose gastrointestinal (GI) tracts were colonized with a Gram negative bacterium. We also wanted to see if we could detected any differences in the cytokine response between LPS sensitive and non-sensitive congenic mouse strains.

METHODS

Animals

Germ-free C3H/HeJ and C3H/HeN mice reared at the Department of Microbial Ecology, Karolinska Institute, Stockholm, Sweden were used.

Colonization and Sample Collection

A suspension was prepared from faeces obtained from isolator reared animals monocolonized with *Escherichia coli* 06. Fifty μl of this suspension was placed in the mouth of the germ-free mice. Blood and tissue samples were taken 8 and 24 hours later. Blood was also obtained from germ-free and conventional mice. Serum was prepared and stored at -70 ° C until analyzed.

Advances in Mucosal Immunology, Edited by
J. Mestecky *et al.,* Plenum Press, New York, 1995

IL-6 Assay

The subclone B9, derived from cell line B13.29, was used for IL-6 determinations. B9 cells were harvested from tissue culture flasks, seeded into microtiter plates (Nunc, Roskilde, Denmark) and serum samples were added. [^{3}H]thymidine was added after 68 h of culturing, and the cells were harvested 4 h later. The samples were tested in twofold dilutions and compared with an IL-6 standard. One unit of IL-6 is the concentration required for half-maximal proliferation of B9 cells.

TNF Assay

The MTT tetrazolium (Sigma Chemical Co., St. Louis, MO) cytotoxicity assay was used to measure levels in serum of TNF with clone 13 of the WEHI 164 cell line as target cells. In brief, target cells were seeded in complete medium in flat-bottom microtiter plates at a concentration of 2 x 10^4 cells per well. Different dilutions of serum were added, in triplicate, to the wells. After 20 h of incubation at 37° C, 10 µl of MTT tetrazolium at a concentration of 5 mg/ml in phosphate-buffered saline was added, and further incubation for 4 h at 37° C was performed. After aspiration of 100 µl of serum from the wells, 100 µl of 95 % ethanol was added to the wells, and incubation was performed for 10 min. The absorbance at 570 nm was registered on a microtiter plate reader. The concentration of TNF in the samples were obtained from a standard curve produced with known concentrations of recombinant TNF.

Immunohistochemistry

Sections of the intestine containing Peyer's patches, mesenteric lymph nodes (MLN) and spleens were snap-frozen and stored at -70° C. Staining of the sections were done with monoclonal antibodies directed against CD4, CD8, TNF, Ig, IL-1 and MAC1 and a rabbit anti-rat immunoglobulin antibody conjugated to peroxidase.

Blood from germ-free mice or from animals carrying a normal intestinal microflora did not have any detectable levels of either IL-6 or TNF. Eight hours after colonization of the germ-free mice the mean serum concentration of IL-6 increased to 12 pg/ml (range 5 pg/ml to 70 pg/ml). Sixteen hours later the concentration of IL-6 was again below detection limit (Fig. 1). TNF was not present in sera from germ-free or conventionally reared mice. Eight hours after colonization in 2 of 6 animals serum TNF was detected. Sixteen hours later all animals had increased levels of serum TNF, mean concentration 13 pg/ml (range 7 pg/ml to 19.5 pg/ml) (Fig. 2). Immunohistochemistry did not reveal any difference in number of CD4$^+$, CD8$^+$, TNF$^+$, Ig$^+$, 1L-1$^+$ and MAC1$^+$ cells during the 24 hour observation period in the spleen, the MLN or the intestine.

DISCUSSION

In the present study we have shown that bacterial colonization of the intestine of germ-free mice results in the production of IL-6 and TNF, cytokines which are closely associated with inflammatory reactions. IL-6 or TNF were not present in the blood from germ-free or conventional mice.

The intestinal mucosa in conventional animals is constantly exposed to live bacteria and bacterial products in different forms but it does not normally show any obvious signs of inflammation. Thus it seems as if the immune system is adapted to the intestinal microflora and actively down-regulates the production of inflammatory mediators. In the present study this was especially evident for IL-6 which during the short 24 hour observation period peaked and disappeared. TNF was still present after 24 hours but must apparently also be down-regulated since animals carrying a normal intestinal flora did not have detectable levels of TNF.

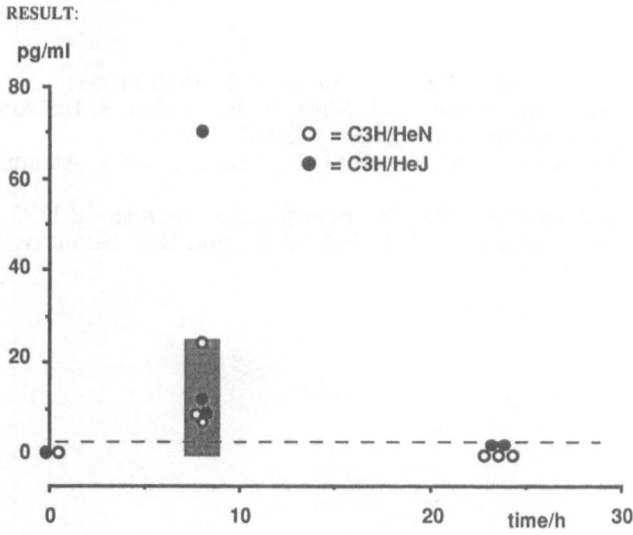

Figure 1. Concentration of serum IL-6 in mice at different times after colonization with *E. coli* OCK13.

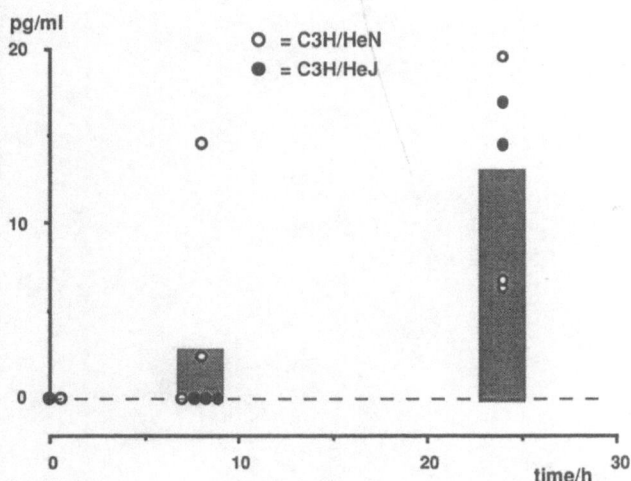

Figure 2. Concentration of serum TNF in mice at different times after colonization with *E. coli* 06K13.

The very early appearance of IL-6 preceding the appearance of TNF is different from what is normally expected when animals are exposed to LPS intravenously.[2] Administration of LPS leads to production of TNF which induces IL-1 which in turn upregulates IL-6. Several of the biological effects ascribed to TNF and IL-1 has been shown to be mediated by IL-6.[3] Recent studies have indicated that IL-6 can be produced by epithelial cells after stimulation with bacteria *in vitro*.[4] The precedence of serum IL-6 compared to TNF in the current study might be due to bacterial stimulation of the epithelial cells in the mucosa inducing synthesis and release of IL-6 by these cells. The production of TNF would then be a result of translocation of the bacteria over the intestinal mucosa and stimulation of hosts immune system.[5]

REFERENCES

1. G. Evans and S. Zuckerman, *Eur. J. Immunol.* 21:1973 (1991).
2. J. K. McIntosh, D. M. Jablons, J. J. Mule, R. P. Nordan, S. Rudikoff, M. T. Lotze, and S. A. Rosenberg, *J. Immunol.* 143:162 (1989).
3. R. Neta, R. Perlstein, S. N. Vogel, G. D. Ledney, and J. Abrams, *J. Exp. Med.* 175:689 (1992).
4. S. Hedges, M. Svensson, and C. Svanborg, *Infect. Immun.* 60:1295 (1992).
5. C. L. Wells, S. L. Maddaus, S. L. Erlandsen, and R.L. Simmons, *Infect. Immun.* 56:278 (1988).

STIMULATION OF INTESTINAL IMMUNE CELLS BY *E. coli* IN GNOTOBIOTIC PIGLETS

Leos Mandel,[1] Ilja Trebichavsky,[1] Igor Splichal,[1] and Jürgen Schulze[2]

[1]Institute of Microbiology, Czechoslovak Academy of Sciences, Department of Immunology and Gnotobiology, Prague, Czech Republic; and [2]Ardeypharm Labotratory Herdecke, Germany

INTRODUCTION

Enterobacteriaceae, among them various *E.coli* strains, exhibit two contradictory effects: stimulation of local and systemic immunity on the one hand[1], and threatening the life by penetration and even bacteriaemia and death if the immune system is impaired on the other hand.[2] Therefore, an effort has been made to find a non-pathogenic *E. coli* strain which has the ability to suppress unwanted intestinal bacteria. Highly infection-prone colostrum-deprived gnotobiotic piglets were used for association experiments. Results obtained with the promising *E. coli* strain 06:K5:H1 (Nissle 1917) are reported.

METHODS

Ten day old germfree piglets were mono- or diassociated *per os* with 10^8 viable bacteria/2 ml PBS. The following bacterial strains were used: *E. coli* 06:K5:H1 Nissle 1917; *E. coli* hemolytic, resistant to colicins, Czechoslovak State Collection of Microorganisms #Ec 542/88; *Salmonella typhimurium* LT_2R_4. The number of bacteria (CFU) in feces were counted using cultivation methods and differentiated by agglutination with specific antisera. On days 1,3,7, and 12, 2 associated piglets were sacrificed and the numbers of CFU of respective bacteria in various organs (mesenteric lymph nodes, spleen, lungs, liver) determined after homogenization of samples and cultivation. On day 7, IgA+, IgG+, IgM+, and MHC Class II+ (SLA-D+) lymphocytes in the intestinal mucosa were observed using immunofluorescent methods.

RESULTS

Gnotobiotic piglets monoassociated with *E. coli* Nissle 1917 showed no apparent impairment in health. The number of bacteria found in mesenteric lymph nodes was highest on day 3 after monoassociation, and, unlike in animals monoassociated with *E. coli* Ec 542/88 or with some other *E. coli* strains previously tested[3,4] decreased to zero after day 7. This time period corresponded with the occurrence of large amounts of IgA+ and IgG+ cells in lamina propria of the intestinal villi and in the intestinal lymphatic follicles (Fig.1). No IgA+ and IgG+ cells could be detected in comparable tissues of control germfree piglets (Fig.2). IgM+ cells were frequent in both groups of animals. In all mucosal and submucosal tissues, SLA-D+ cells (MHC class II+) were observed in piglets monoassociated with *E. coli* Nissle 1917 strain. This *E. coli* strain was also found to suppress the hemolytic strain Ec 542/88 and *Salmonella hyphimurium* LT2R4 in diassociation experiments. This observed competitive effect of *E. coli* Nissle 1917 is attributed to a microcin with a bacteriostatic activity.[5] The results indicate a possible use of the *E. coli* 06:K5:H1 strain Nissle 1917 for stimulation of local and systemic immunity and for modulation of intestinal microflora in cases where the intestinal microecology is disturbed.

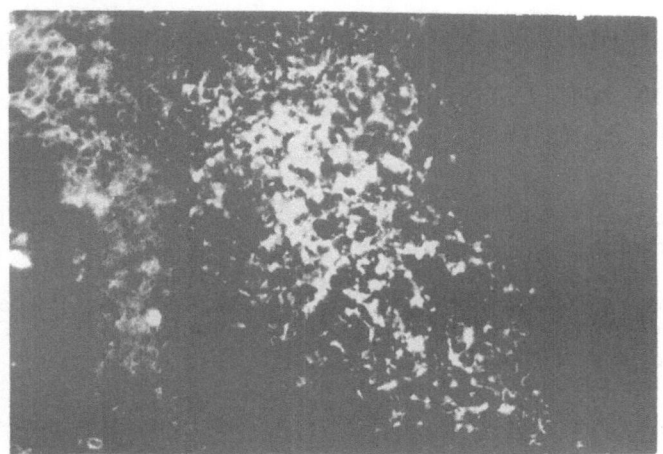

Figure 1. Ileal villus, piglet 18 days old, 7 days after monoassociation with *E. coli* 06:K5:H1 Nissle 1917. Specific immunofluorescence showing large numbers of IgA$^+$ cells. (250x)

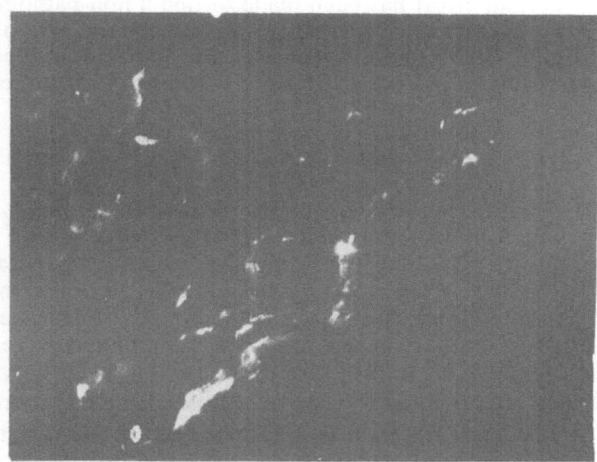

Figure 2. Ileal villus, germfree piglet 18 days old. No IgA$^+$ cells visible by specific immunofluorescence. (250x)

REFERENCES

1. H. Tlaskalova, L. Mandel, P. Rossmann, I. Trebichavsky, J. Kopecek, and J. Rejnek, *in:* "Bacteria and the Host", M. Ryc, ed., p. 355, Avicenum & Med. Press, Prague (1986).
2. I. Brook, *in:* "Treatment of Radiation Injuries", D. Browne, ed., p. 227, Plenum Press, New York (1983).
3. V. Dlabac, M. Talafantova, and L. Mandel, *in:* "Bacteria and the Host", M. Ryc, ed., p. 61, Avicenum & Med. Press, Prague (1986).
4. M. Talafantova, L. Mandel, and I. Trebichavsky, *Microecol.Therapy* 18:311 (1989).
5. F. Baquero and F. Moreno, *FEMS Microbiol. Letters* 23:117 (1984).

IMMUNOMODULATION OF THE GNOTOBIOTIC MOUSE THROUGH COLONIZATION WITH LACTIC ACID BACTERIA

Harriet Link, Florence Rochat, Kim Y. Saudan,
and Eduardo Schiffrin

Nestlé Research Centre
Vers-Chez-les-Blanc
1000 Lausanne 26
Switzerland

INTRODUCTION

The contribution of the intestinal microflora and microbial antigens to the development of the mucosal immune system has been recognized. The germ-free rodent is the main model to study the influence of the microflora in theontogeny of this system. The aim of this work was to detect if different strains of lactic acid bacteria or their soluble products could produce different patterns of immunomodulation in the host. Some bacteriological and immunological parameters were chosen: bacterial colonization, bacterial translocation, white blood cell counts, serum immunoglobulins, and the intraepithelial lymphocyte (IEL) population of the gut.

MATERIALS AND METHODS

Germ-free OFI mice (24) were randomly assigned to four experimental groups. Group I was maintained germ-free throughout the experiment. Group I was gavaged with, *Lactobacillus acidophilus* Lal (7 x 108 CFU) on days one and four. Group III was gavaged with *Lactobacillus casei* GG (3x108 CFU) on the same days. Group IV was treated orally with a sterile hydrolysate of *Lactobacillus bulgaricus* administered in the drinking water. Three animals of each group were sacrificed on day four and the rest on day eight.

Bacterial colonization was studied in the feces and jejunum. Bacterial translocation from the gastrointestinal tract to extraintestinal tissues was determined in the mesenteric lymph nodes (MLN) and liver.

Serum immunoglobulins (IgG, IgM and IgA) and secretory IgA in 5 ml intestinal lavages were quantified by radial immunodiffusion. The proportion of different types of leucocytes was determined in peripheral blood.

Morphological studies were performed in formalin-fixed and frozen tissues. Particular emphasis was placed on the number of intraepithelial lymphocytes (IEL). At least 250 IEL were counted per tissue sample and their phenotype determined using immunoperoxidase staining of frozen sections. Monoclonal antibodies recognizing CD3, Thy 1-2, CD8 and CD4 were employed.

RESULTS

The level of fecal colonization was higher in the group treated with L. casei than in the group treated with *L. acidophilus* (6.2 x 108 vs 2.7 x 107 CFU/g). On the other hand no differences were observed in the jejunal colonization by the different strains (1.4 x 105 and 4.3 x105 CFU/g for *L. acidophilus* and *L. casei*, respectively). The incidence of bacterial translocation to MLN and liver was similar for both strains at day four. Despite the low number of animals, it seems that at day 8 the incidence diminished for L. acidophilus but remained high for *L. casei* GG (Table 1).

Table 1. Incidence of bacterial translocation to mesenteric lymph nodes and liver 4 and 8 days post-colonization.

Day	Tissue	*L. acid*	*L. casei*
4	MLN	2/3*	2/3
4	Liver	2/3	1/3
8	MLN	0/3	3/3
8	Liver	1/3	1/3

* Number of positive animals per goup of three.

Serum immunoglobulin levels were IgG 437 ± 145 µ/ml and IgM 90 ± 14 µ/ml for the control group, and changed only in the group colonized with *L. acidophilus* (633 ± 62 and 877 ± 114 for IgG respectively on days 4 and 8; 124 ± 18 and 160 ± 8 for IgM on days 4 and 8 respectively). Serum IgA (<50 µg/ml) and intestinal IgA (874 ± 179 µg/ml) did not change with any of the treatments.

An increased proportion of polymorphonuclear leucocytes was observed in the leucocyte compartment particularly in the animals treated with *L. casei* GG (Table 2).

Table 2. Percentage of polymorphonuclear leucocytes in peripheral blood total leucocytes.

Day	GF	*L. acid*	*L. casei*	Hydrolysate
4	12 (+/-6)*	16 (+/-12)	30 (+/-15)	12 (+/-5)
8	12 (+/-6)	19 (+/-8)	47 (+/-15)	23 (+/-5)

* Average per group +/- SD.

TheIEL lymphocyte population showed a quick expansion after bacterial colonization with both strains and some expansion was evident after the treatment with the hydrolysate (Fig. 1). The phenotype of the IELs in the germ-free animals was Thy 1.2⁻, CD3⁺, CD8⁺ and CD4⁻. During the length of the study, even though an increased quantity of IELs was detected, the phenotype remained the same.

CONCLUSIONS

This study showed that in germ-free mice there were 4 IEL/100 epithelial cell in agreement with previous studies.[1] After monocolonization with *L. acidophilus* or *L. casei* an early augmentation of IELs was detected (9 IEL/100 epithelial cells). The phenotype of IELs in germ-free mice was Thy 1.2⁻, CD3⁺, CD8⁺, CD4⁻ (probably / + T cells as shown by Goodman[2]). No change in the phenotype was observed after the expansion of this

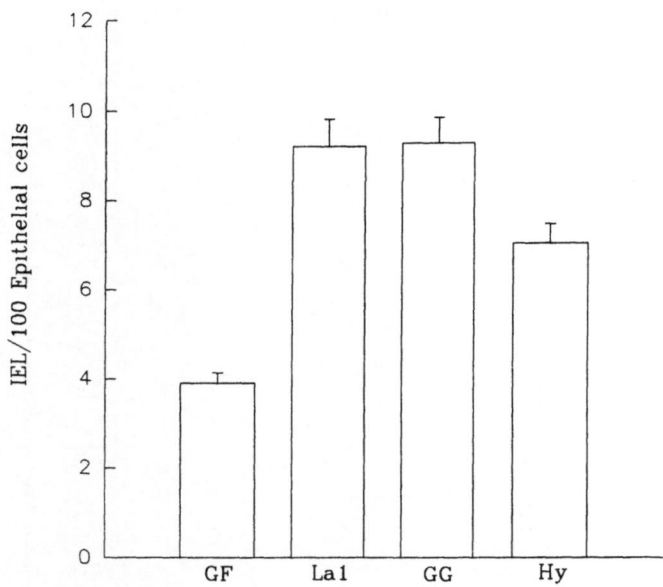

Figure 1. Intraepithelial lymphocytes eight days post-colonization.

population. Bacterial antigens have been reported to produce an augmentation of IEL with the appearance of Thy 1.2^+ phenotype.[3,4] Up to this moment, no studies have considered the variation of IEL population in the monocolonized animals or what is happening in the first days post-colonization. Here it is shown that the first event in the kinetics of that expansion depends on an increased number of $CD3^+$, Thy 1.2^-, $CD8^+$, the same phenotype present in germ-free animals. It is necessary to examine monocolonized animals for a longer period of time to determine whether monocolonization can lead to the full development of the IEL compartment, or whether different individual strains or a synergy of several strains pertaining to the normal intestinal microflora is required. Finally, a higher level of translocation occurred in the group colonized with *L. casei*, concomitant with an increase in circulating polymorphonuclear leucocytes.

REFERENCES

1. J. F. Glaister, *Int. Arch. Allergy* 45:719 (1973).
2. T. Goodman and L. Lefrancois, *Nature* 333:855 (1988)
3. L. Lefrancois and T. Goodman, *Science* 243:1716 (1989)
4. A. Bandeira, T. Mota-Santos, S. Itohara, S. Degermann, C. Heusser, S. Tonegawa, and A. Coutinho, *J. Exp. Med.*, 172:239 (1990).

EXPRESSION OF INTERLEUKIN-2 RECEPTOR ON ENTEROCYTES IN CONVENTIONAL AND GERM-FREE RATS AFTER STIMULATION WITH GLIADIN

Jiří Sinkora and Renata Stepánková

Institute of Microbiology, Czech Academy of Sciences
Department of Immunology and Gnotobiology
Prague, Czech Republic

INTRODUCTION

The high affinity interleukin-2 receptor (IL-2R) is a membrane-bound heterodimer resulting from the noncovalent association of at least 2 subunits. Both α (p55, Tac protein) and β (p75) chains bind interleukin-2 (IL-2) with low and intermediate affinity, respectively.

IL-2 was originally described as an autocrine and paracrine T cell growth factor and the IL-2R complex was detected on the surface of activated T cells. Recently it has been shown that both subunits of IL-2R can be expressed on several cell types of lymphoid and non-lymphoid origin (macrophages, monocytes, NK and B cells as well as endothelial and thymic stromal cells).[1] In addition to its ability to increase the affinity of IL-2R for the ligand, p55 glycoproteins can be released from the plasma membrane and thus generate a soluble form of IL-2-binding protein. While the α chain is important for efficient IL-2 binding and does not seem to play a role in signal transduction or IL-2 internalization, the β subunit of the receptor is clearly involved in the IL-2:IL-2R signaling pathway although its cytoplasmic domain does not contain a kinase motif. A unique serine-rich cytoplasmic region is responsible for signal transduction after receptor-ligand complex formation. Cell lineage-dependent tyrosine kinases of the non-receptor membrane-associated src family become associated with the β chain following activation, which results in phosphorylation of several cytoplasmic protein targets.[2,3]

The expression of both chains on the surface of different cell lineages is controlled independently, varies during ontogeny of organisms, and depends on physiological or pathological activation of the immune system (infection, autoimmune disorders).[1,4,5]

We used germ-free (GF) and conventional (CV) rats for studying age-dependent IL-2R-α chain occurrence in enterocytes. Intragastrically treated animals were used to investigate the ability of a food antigen (gliadin) to induce IL-2R-α chain expression.

MATERIALS AND METHODS

Animals

Inbred F (87) rats of the Wistar AVN strain, were used as conventional and GF animals. The 8th generation of GF rats reared in plastic isolators and fed a granulated diet sterilized by autoclaving was used in our experiments.[6]

Advances in Mucosal Immunology, Edited by
J. Mestecky *et al.*, Plenum Press, New York, 1995

Intragastric Immunization

Immunization of GF animals with gliadin was performed by intragastric administration of 0.5% gliadin using a rubber tube 3 times a week from birth until the age of 2 months as described elsewhere.[7]

Immunohistochemical Methods

Rats were sacrificed and the proximal and medial parts of the jejunum were frozen in liquid nitrogen. 6-7 μm thick tissue sections were stained with primary anti-IL-2R-α antibody (MRC OX 39, Serotec) and secondary fluorescein-conjugated sheep anti-mouse polyclonal immunoglobulin (Amersham) without blocking endogenous enzyme activities. Photomicrographs were made using a Leitz Orthoplan microscope equipped with a fluorescence illuminator, interference beam-splitting filters for fluorescein absorption and emission, and an Orthomat-E camera (see Stepankova *et al.* in this issue).

Table 1. IL-2R-α expression on enterocytes in GF and CV animals.

Animals	CV		GF		GFx
Months	2	12	2	12	2
IL-2R-alph	++	-	+	++	+++

-	negative
+	weakly positive in villi
++	positive in villi
+++	bright positivity in villi and crypts
GFx	animals treated with gliadin as described in methods

RESULTS

Table 1 and Fig. 1 show the level and localization of IL-2R-α detected by the immunofluorescence technique on enterocytes of CV and GF animals. We found different levels of IL-2R-α expression in the Golgi region of enterocytes in both 2 and 12 month-old GF animals as well as in young CV rats. No α chain was found in enterocytes in adult (1 year old) CV rats. After long-lasting administration of gliadin to GF rat pups, much more IL-2R-α could be seen in the Golgi region of enterocytes and the α chain seemed to reach the microvillar zone of the enterocytes. In our experimental model of coeliac disease, for unknown reasons, GF animals appear to be more convenient for the induction of coeliac-like symptoms by gliadin than CV animals. Our preliminary unpublished results suggest that, in comparison to CV rats, there is a higher level of IL-2 gene expression in GF animals. It remains to be shown whether the difference in IL-2 and IL-2R expression between GF and CV rats is associated with their divergent responses to gliadin.

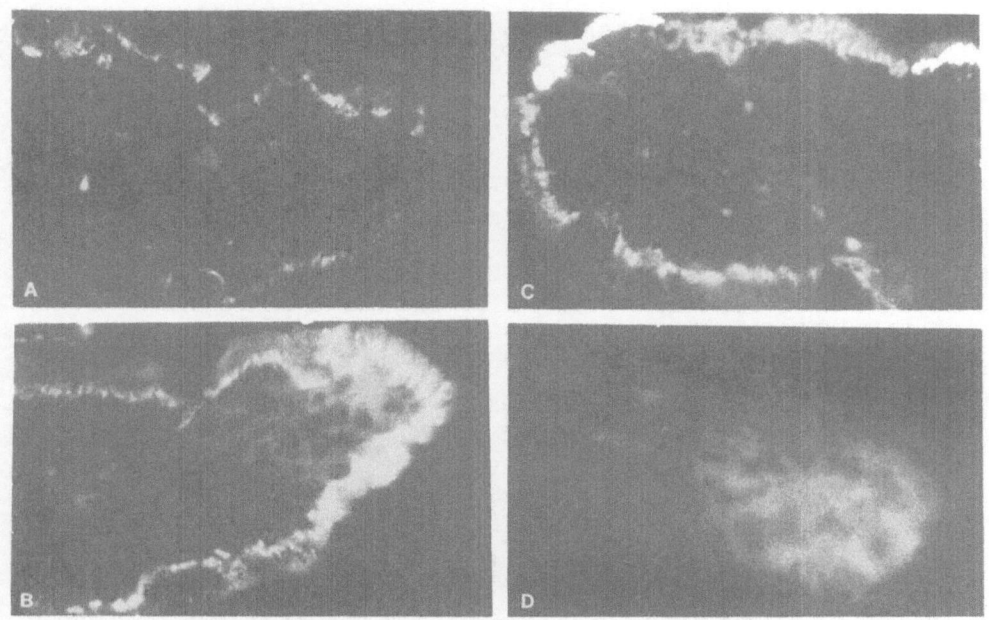

Figure 1. IL-2R-α chain detection in enterocytes of GF and CV rats
A, B: 2-month old GF animals untreated and treated by gliadin, respectively
C, D: 1-year old GF and CV rats, respectively

REFERENCES

1. G. Kromer and G. Wick, *Immunol. Today* 10:246 (1989).
2. T. Torigoe, H. Uri Saragovi, and J. C. Reed, *Proc. Natl. Acad. Sci.* USA 89:2674 (1992).
3. M. Hatakeyama, T. Kono, N. Kobayashi, M. Tsudo, S. Minamoto, T. Kono, T. Doi, T. Miyata, M. Miyasaka, and T. Taniguchi, *Science* 244:551 (1989).
4. M. Lipoldova, A. Zajicova, and V. Holan, *Immunology* 71:497 (1990).
5. M. A. Callgiuri, A. Zmuidzinas, T. J. Manley, H. Levine, K. A. Smith, and J. Ritz, *J. Exp. Med.* 171:1509 (1990).
6. R. Stepankova, *Folia Microbiol.* 24:11 (1979).
7. R. Stepankova, P. Tlaskalova, I. Fric, and I. Trebichavsky, *Folia Biol.* 35:19 (1989).

FACTORS INFLUENCING THE FATE OF *ESCHERICHIA COLI* AND *SALMONELLA TYPHIMURIUM* IN GERM-FREE PIGLETS AND RATS

V. Dlabac, M. Talafantová, L. Mandel, and R. Stepánková

Departments of Immunology and Gnotobiology
Institute of Microbiology
Czech Academy of Sciences Prague, Czech Republic

INTRODUCTION

Germ-free (GF) animals are an excellent model for studies of the interactions between intestinal bacteria and the host. Such animals can be easily colonized with various strains of enteric bacteria and this facilitates studies of the fate of microorganisms in various compartments of the host on one side and the immune response on the host on the other. GF piglets represent a model exceptionally suited for such studies because only extremely low concentrations of antibodies are present in their sera as a result of a special type of placentation in this animal species.[1,11] In addition, higher weight of organs and mesenteric lymph nodes (MLN) facilitates quantitative estimation of translocated bacteria.

To elucidate some of the factors influencing the fate of the bacteria in the intestinal tract and those involved in the control of bacterial translocation from the intestinal tract to MLN and other organs, a collection of smooth and rough strains of *E. coli* and *S. typhimurium* with known structures of their LPS was used for peroral infection of GF piglets and rats.[2]

RESULTS

Changes of the Microbial Population in the Intestine of GF Piglets and Rats Infected with Smooth and Rough Strains of *S. typhimurium*

After peroral infection all strains used, irrespective of the chemotype, readily colonized the intestinal tract of GF animals reaching maximal counts within 1-2 days. As seen in Fig. 1 in all groups of piglets infected with smooth and rough strains of *S. typhimurium* a proportion of animals died as a result of septicemia, the mortality ratio and length of survival being dependent on the type of mutation. However, microbiological examination of isolates from organs of animals dying from infection with strain LT2M1 (Rc chemotype) and TV 148 (Rb chemotype) revealed only smooth revertants with characteristics of the original wild type strain. This corresponded to the findings in the intestinal content. Taken together these results indicate that the pathogenicity of these two rough mutants is probably very low, the cause of death of infected animals being probably due solely to rapid reversion to a virulent smooth strain. The much lower mortality ratio (about 40%) in these two groups of animals may be due to stimulated occurrence of reversion. It is interesting that the surviving animals can withstand the presence of a pure culture of smooth virulent *S. typhimurium* in their intestinal content for appreciably long time. It should be stressed, however, that all surviving piglets suffered from chronic diarrhoea and chronic inflammation of the intestinal mucosa.

Most piglets infected with original wild type smooth strain *S. typhimurium* LT2 died within 1-2 days after infection as a result of septicemia (see Fig. 1). One animal survived the full length of the experiment (7 weeks). In this case the isolates at the end of the experiment were found to belong to the Ra chemotype by immunochemical analysis of the isolated LPS. This finding might be revelant to progressive changes to rough mutants described by Sack and Miller[3] in GF mice infected with a smooth strain of *Vibrio cholera* and with the progressive frequency of rough mutants of *E. coli* 086 in the stool and intestinal content of GF rats originally colonized with a smooth strain (unpublished). As shown in Fig. 1, rough strain R4 of the Ra chemotype was relatively stable *in vivo* and the occurrence of smooth revertants in the intestine was observed only after prolonged colonization. It should be stressed that in organs and blood of the animals dying usually within 1 week as a result of infection with this mutant, only colonies corresponding to the Ra mutant used for infection were found.

Immune Response Against Intestinal Bacteria and the Effect of Antibodies and Peroral Immunization on the Outcome of Infection and Passage of Intestinal Bacteria to Regional Mesenteric Lymph Nodes and Organs

The colonization of GF piglets leads to a profound local and systemic immune response.[13] Using immunofluorescent staining, IgG- and IgA-producing cells were regularly found in the mucosa of GF piglets colonized with various strains of *S. typhimurium* and *E. coli*. The systemic antibody response against smooth and rough LPS estimated in sera of GF piglets colonized with smooth and rough strains of *S. typhimurium* is shown in Fig. 2. It is apparent that irrespective of the strain used for colonization and the speed of reversion, the first antibodies observed during the first week were directed against rough LPS, whearas the antibody response against smooth LPS occured much later. Similar results were obtained in GF piglets colonized with *E. coli* 086 (data not shown). It should be stressed, however, that treatment of rough LPS with alkali completely abolished the reactivity with these early antibodies directed against rough LPS in contrast to the reactivity with hyperimmune anti-rough sera, where alkali treatment of rough LPS led to elevated reactivity.

As shown earlier[8] and confirmed in this study, colonization of GF piglets with strains of *S. typhimurium* and *E. coli* is accompanied by the appearance of bacteria in MLN and/or spleen and liver. In nonpathogenic strains bacterial counts in MLN and organs reached a maximum within the first week after colonization, followed by a continuous decline during subsequent weeks. This process is usually not accompanied by pathological symptoms. However, irradiation with a dose of 8 gy of GF piglets 3 days after colonization with the same strain led to septicemia and death in spite of the fact that this dose does not appreciably alter the morphological and functional integrity of the intestinal mucosa. Irradiation of piglets with the same dose 30 days after colonization, when living bacteria could no longer be detected in the MLN and organs, did not lead to the reappearance of bacteria in MLN or organs, or to septicemia. These results are in agreement with a view that the translocation of intestinal bacteria to MLN is controlled by mechanisms of local immunity.

The coincidence of the onset of antibody response with the development of protective immunity as a result of colonization of GF piglets with rough strains of *S. typhimurium* prompted us to study the effect of antibodies and peroral vaccination on the course of intestinal infection with various pathogenic strains of *E. coli* and *S. typhimurium*. However, hyperimmune sera from colonized piglets applied perorally were repeatedly found to be ineffective in preventing the penetration of intestinal bacteria.[8] Even peroral application of bacteria killed by different procedures (heat, X-irradiation) and applied in different doses and schedules did not prevent translocation of the corresponding strain in spite of prominent local and systemic antibody production.

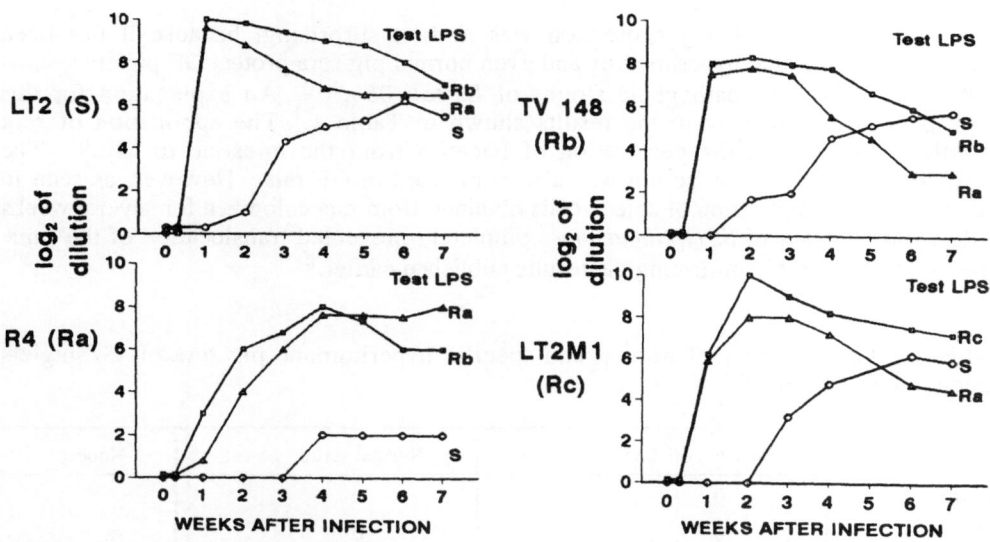

Figure 1. Changes of the microbial population in the stool of GF piglets infected with *S. typhimurium.*

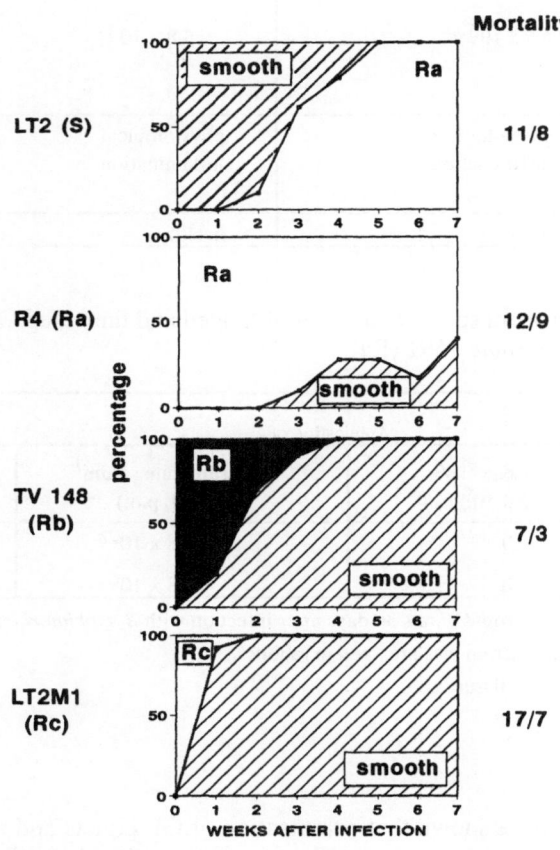

Figure 2. Antibody response to LPS in GF piglets infected with *S. typhimurium.*

The absence of any protection was at first surprising, because it has been repeatedly shown, that hyperimmune and even normal pig sera protect GF piglets against septicemia caused by pathogenic strains of *E. coli* 055.[9,10] An explanation for this discrepancy is evident from the results shown in Table 1. The application of sera completely prevented the penetration of bacteria from the intestine to MLN. The ineffectiveness of this antiserum was also confirmed in GF rats. However, as seen in Table 2, the i.v. application of spleen cells obtained from rats colonized for several weeks with a rough strain of *S. typhimurium* completely prevented translocation of the same strain into MLN, thus confirming the results published earlier.[8]

Table 1. The influence of normal and specific hyperimmune pig sera on GF piglets infected with *E. coli* 055.

		Anti-*E. coli* 055	Normal serum	None
Blood		0	0	
Spleen		0	0	s
				e
Liver		0	0	p
				t
M	jejunal	1.7×10^4	8.1×10^3	i
				c
L	ileal	5.4×10^4	2.8×10^3	a
				e
N	colic	6.5×10^5	5.8×10^4	m
				i
				a
Intestinal histology		physiological inflammation	physiological inflammation	destruction of intestinal mucosa
Clinical state		healthy	healthy	death

Table 2. Bacterial counts in spleen and MLN of treated and untreated GF rats 2 days after infection with *S. typhimurium* 1591 (Ra).

	Application of		
Bacterial counts in	Spleen cells[a] $(1.2 \times 10^9$ iv.)	Hyperimmune serum[b] (1 ml p.o.)	Control
Spleen	0	3.8×10^{2c}	5.9×10^3
MLN	0	3.8×10^6	5.4×10^6

a   Spleen cells were prepared from GF rats 30 days after infection with *S. typhimurium* 1591

b   Anti-*S. typhimurium* 1591 serum was prepared in rabbits

c   Bacterial counts per gram of tissue

CONCLUSION

Presented data have shown that colonization of GF piglets and rats results in both local and systemic antibody responses. The specificity of early antibodies in serum is directed to an alkali-labile epitope of LPS present in both smooth and rough LPS.

However, the protective role of antibodies could be demonstrated only in the case of enteropathogenic *E. coli* 055. In the case of *Salmonella* infection, the most important defense mechanism seems to be the function of the mucosal barrier that presents the passage of intestinal bacteria into regional lymph nodes and organs. The fact that colonization of GF animals with rough strains has a protective effect against infection with smooth virulent strains indicates that the protective antigen may be different from the polysaccharide part of smooth LPS.

REFERENCES

1. J. Sterzl, J. Rejnek, and J. Trávnicek, *Folia Microbiol.* 11:7 (1966).
2. V. Dlabac, *Folia Microbiol.* 13:439 (1968).
3. R. B. Sack and C. E. Miller, *J. Bacteriology* 99:688 (1969).
4. V. Dlabac, J. Klepalová, and L. Mandel, *Folia Microbiol.* 16:533 (1971).
5. V. Dlabac, M. Talafantová, and L. Mandel, *in:* "Bacteria and the Host", M. Ryc and J. Franek ed., p. 351, Avicenum, Praha (1986).
6. T. Staley, E. W. Jones, and L. D. Corley, *Am. J. Pathol.* 56:371 (1969).
7. A. Takeuchi, *Curr. Topics Pathol.* 54:1 (1971).
8. H. Tlaskalová-Hogenová, J. Sterzl, R. Stepánková, V. Dlabac, V. Vetvicka, P. Rossmann, L. Mandel, and J. Rejnek, *Ann. N.Y. Acad. Sci.* 409:96 (1983).
9. J. Rejnek, J. Trávnicek, J. Kostka, J. Sterzl, and A. Lanc, *Folia Microbiol.* 13:36 (1966).
10. H. Tlaskalová-Hogenová, J. Rejnek, J. Trávnicek, and A. Lanc, *Folia Microbiol.* 15:372 (1969).
11. V. Dlabac, I. Miller, J. Kruml, F. Kováru, and M. Leon, *in:* "Developmental Aspects of Antibody Formation and Structure", J. Sterzl and I. Riha, eds., p. 105, Academia, Prague, (1970).
12. R. G. Berg and A. W. Garlington, *Infect. Immun.* 23:403 (1979).
13. H. Tlaskalová-Hogenová, J. Sterzl, M. Pospisil, and J. Hofman, *in:* "Developmental Immunobiology", J. B. Solomon and J. D. Horton, eds., p. 355, Elsevier, North-Holland, Amsterdam, (1977).

BIFIDOBACTERIA AND *ESCHERICHIA COLI* TRANSLOCATION IN GNOTOBIOTIC MICE

Vladimír Kmet,[1] Marta Kmetová,[2] Michel Contrepois[3], and Yves Ribot

[1]Institute of Animal Physiology, Slovak Academy of Sciences, Hlinkova 1/B, 040 01 Kosice; [2]Department of Medical Microbiology, Medical Faculty, Srobárova 56,040 01 Kosice, CSFR; and [3]Laboratory of Microbiology I.N.R.A., 63122 St. Genés Champanelle, France

INTRODUCTION

Bacterial translocation is the passage of viable bacteria from the gut through the mucosa to the internal organs and it depends mainly on the host, on the pathogenicity of *E. coli* strains, and on the level of colonization achieved by the bacteria in the gut.[1]

The purpose of the work reported here was to study the interactions between bifidobacteria with iron intake activity and the translocation activity of aerobactin-producing *E. coli* in animal experiments.

There is iron-intake interaction between the normal and pathogenic gut microflora. Bifidobacteria, as indigenous microflora, are capable of accumulating large quantities of iron, if it is present in the ferrous state, and can serve to maintain a relatively iron-free environment in the gut, depriving pathogens of ferrous ions.[2]

Pathogenic bacteria, especially invasive *E. coli*, synthesize ferric iron chelators (aerobactin) known generally as siderophores. Aerobactin influences the extent of bacterial translocation from the intestinal tract or of bacterial multiplication in tissues following translocation, or both.[3]

MATERIAL AND METHODS

Bacteria and Media

Bifidobacterium sp. No. 904 isolated from calf faeces (eightfold dilution), was obtained from the Institute of Animal Physiology and Nutrition, Polish Academy of Sciences (Dr. A. Ziolecki †). Strain No. 904 fermented only galactose, melibiose, and raffinose, and the ratio of acetic acid/lactic acid formation from glucose was 5.4. Bifidobacterium was grown in a carbon dioxide atmosphere at 37°C in TPY medium[4]: trypticase, 10g; phytone, 5g; glucose, 5g; yeast extract, 2.5g; Tween 80, 1 ml; cysteine hydrochloride, 0.5g; K_2HPO_4, 2g; $MgCl_2.6H_2O$, 0.5g; $ZnSO_4.7H_2O$, 0.25g; $CaCl_2$, 0.15g; $FeCl_3$ a trace; agar, 15g; distilled water to 1,000 ml. Final pH was about 6.5 after autoclaving at 121°C.

E. coli strain R 15, aerobactin positive, was isolated from poultry.

Advances in Mucosal Immunology, Edited by
J. Mestecky *et al.*, Plenum Press, New York, 1995

Iron Intake Activity

Bifidobacterium was grown in one liter of TPY medium in the presence of iron. Iron intake was measured in the microbial pellets by atomic absorbance spectrophotometry (Perkin-Elmer 5000).

Production and Breeding of Mice

A breeding colony of axenic mice was established from outbred axenic Swiss type breeders, strain OF1, supplied at the age of eight weeks. The mice were raised in sterile isolators and given ad libitum RO3, 4MR concentrate (UCAR, Epinay-sur-Orge), sterilized by irradiation at 4 megarads. The animals were subjected to a nycthemeral cycle of 12/12 h.

Inoculation of Mice

Experimental and control mice were inoculated orally with 0.25 ml of fresh bacterial suspensions of Bifidobacterium sp. No. 904 and of *E. coli* R15 after one day, or with suspension of *E. coli* R15 only by a 1 ml syringe fitted with a probe (10 to 12 mm long and 0.45 in diameter), both sterilized beforehand.

Measurement of Translocation

The mice were removed from the isolators after 24 h or 48 h and killed by cervical dislocation. The thorax and abdomen were opened and samples were taken in the following order: about 0.3 ml of blood by intracardiac puncture, the spleen, kidneys, liver, and mesenteric lymph nodes. Tenfold dilutions in sterile saline were made and the viable bacteria counted by seeding on desoxycholate agar (Diagnostic-Pasteur). The results were expressed as the number of colony forming units (CFU) per gram of organ or caecal content and per ml of blood.

The results are given as arithmetic means $\pm$ S.E.M. Differences between groups were calculated using the t test.

RESULTS AND DISCUSSION

It is generally known that intestinal lactic acid bacteria (lactobacilli, enterococci, and also bifidobacteria) suppress the numbers of pathogenic bacteria by producing antibacterial substances. Primary metabolites, such as organic acids and hydrogen peroxide, are known to be effective *in vitro*.[9] Another possibility for the inhibition of *E. coli* by bifidobacteria is through its iron intake activity. Bifidobacterium sp. No 904 had such activity[5], which was expressed as 5 - 7 nanomols of iron per bacterial pellet.

The quantity of *E. coli* aerobactin influences the extent of bacterial translocation also.[3] In our previous *in vitro* experiments[5] on the cocultivation of pectate gel-immobilized bifidobacteria and aerobactin-producing *E. coli*, a decrease of aerobactin production was achieved. Therefore, we expected that the translocation of *E. coli* would be inhibited by bifidobacteria.

The results of experiments on the interaction between bifidobacteria and the translocation of pathogenic aerobactin-producing *E. coli* in gnotobiotic mice are in Table 1 (24 hours after inoculation of *E. coli*) and Table 2 (48 hours after inoculation of *E. coli*).

There was significant inhibition of numbers of *E. coli* in the caecum in both groups of experimental gnotobiotic mice by bifidobacteria. However, there was great variability in translocation activity of *E. coli* in experimental and control animals, for example (Table 1) the number of *E. coli* in the mesenteric lymph nodes was higher in experimental mice.

We assume that *E. coli* translocation in our experiments was not dependent on the *E. coli* counts in the gut after their inhibition by bifidobacteria. It will be necessary to prolong

Table 1. The effect of Bifidobacterium sp No 904 on *E coli* numbers (log $\pm$ SEM) in gnotobiotic mice after 24 hours

Organ	Experiment	Control
Caecum	7 35 ± 0 13	8 62 ± 0 09[a]
Mesenteric LN	2 85 ± 0 24[b]	1 86 ± 0 20
Blood	2 60	2 0
Liver	2 55 ± 0 07	1 17 ± 0 05
Spleen	2 58 ± 0 31	2 76 ± 0 18
Kidney	2 08 ± 0 24	2 59 ± 0 48

[a]P<0 001 [b]P<0 05

Table 2. The effect of Bifidobacterium sp No 904 on *E coli* numbers (log $\pm$ SEM) in gnotobiotic mice after 48 hours

Organ	Experiment	Control
Caecum	7 54 ± 0 08	8 92 ± 0 05[a]
Mesenteric LN	1 76 ± 0 63	3 60 ± 0 59
Blood	1 84 ± 0 08	2 26 ± 0 34
Liver	1 86 ± 0 38	0 84 ± 0 15
Spleen	2 67 ± 0 09	2 20 ± 0 36
Kidney	2 00 ± 0 05	2 48 ± 0 65

[a]P<0 001

the time after the inoculation (minimum five or ten days) of bifidobacteria to permit the bacterial translocation Lactobacilli can translocate as shown by Berg[7] and can survive for many days in the spleen, liver, and lungs [8] Positive translocation of *Lactobacillus acidophilus* or *L casei* was achieved after four or eight days [6]

Little work has been done on the relation between the incidence of translocation and the degree of virulence of different strains of the same bacterial species The invasiveness of the *E coli* strain R 15 has already been studied by Buzoni-Gatel[10] after subcutaneous or intravenous inoculation The *E coli* R 15 strain used in our study has been shown to belong to the 078 Col V clone [11] Ribot and Contrepois (unpublished results) showed direct correlation between the translocation of the *E coli* and their ability to survive outside the gut The internal organs were colonized soon after inoculation of the bacteria, between two and four hours On a practical level, a high septicaemic potential of the *E coli* that colonize the animal gut, should be considered as a predisposing factor in the development of disease

REFERENCES

1   R D Berg, *Infect Immun* 33 854 (1981)
2   A Bezkorovainy, in "Biochemistry and Physiology of Bifidobacteria", A Bezkorovainy, and R Miller, eds , p 147, Catchpole CRC Press, Boca Raton, Florida (1989)
3   M Der Vartanian, B Jaffeux, M Contrepois, M Chavarot, J P Girardeau, Y Bertin, and C Martin, *Infect Immun* (1992) in press

4. V. Scardovi, *in:* "Bergeys Manual of Systematic Bacteriology", Vol. 2, P.H.A. Sneath, ed., p. 1418, Williams & Wilkins, Baltimore (1986).
5. V. Kmet, J. Cizmarova, and M. Kmetova, *Biologia* (Bratislava) 47:767 (1992).
6. H. Link, F. Rochat, Ky. Saudan, and E. Schiffrin, *in:* "Abstract Book of 7th Inter. Congress of Mucosal Immunology", Czechoslovak Immunol. Soc., ed., p. 143, Prague, Czechoslovakia (1992).
7. R. Berg, *in:* "Human Intestinal Microflora in Health and Disease", D.J. Hentges, ed., p. 333, London Academic Press, London, England (1983).
8. N. Bloksma, H. Ettekoven, F. M. Hothuis, L. van Noorle-Jansen, M. J. De Revuer, J. G. Krefleenberg, and J. M. Willers, *Med. Microbiol. Immunol.* 170:45 (1981).
9. R. Fuller, *J. Appl. Bacteriol.* 66:365 (1989).
10. D. Buzoni-Gatel, *Ann. Inst. Pasteur/Microbiol.* 135:B323 (1984).
11. A. Dassouli-Mrani-Belkebir, M. Contrepois, J. P. Girardeau, and M. Der Vartanian, *Vet. Microbiol.* 17:345 (1988).

EFFECTS OF NOCARDIA-DELIPIDATED CELL MITOGEN ON INTESTINAL MUCOSA AND SPLEEN LYMPHOCYTES OF GERM-FREE RATS

Hana Kozáková,[1] Renata Stépánková,[1] Helena Tlaskalová[1]
Rita Barot-Ciorbaru[2], and Jirina Kolinská[3]

[1]Institute of Microbiology, Department of Immunology and Gnotobiology, Czechoslovak Academy of Science, Videnská 1083, CS-142 20 Prague 4, Czech Republic; [2]Institut de Biochemie, Université Paris-Sud, 91405 Orsay, France; and [3]Institute of Physiology, Czechoslovak Academy of Science, Videnská 1083, CS-142 20 Prague 4

INTRODUCTION

Many compounds of bacterial origin can modulate basic physiological parameters of the mammalian organism including the immune system. One of them is Nocardia-delipidated cell mitogen (NDCM) which was isolated by delipidation from *Nocardia opaca*.[1] NDCM stimulates proliferation of small resting human B lymphocytes and their differentiation into Ig-secreting cells.[2] The mucosa of small intestine, especially the enterocytes, cells with digestive and absorptive function, produce a number of glycohydrolases. The disaccharidase (sucrase, lactase, and glucoamylase) activities of brush border membrane vesicles (BBMV) of enterocytes after a short-term NDCM-treatment have not been studied. Measurement of lymphocyte proliferation is an established method of quantifying the immune response to foreign antigens. Antigenic stimulation of human peripheral blood lymphocytes by NDCM was measured by Barot-Ciorbaru.[2] Neither ^{3}H-TdR-nor ^{3}H-UdR-uptake by T cells has been measured after NDCM stimulation.

MATERIAL AND METHODS

Animals

Two-month old male rats of the Wistar AVN strain (inbred F, generation 89) were used. Some of the rats were reared as germ-free (GF) for 8 generations in plastic isolators and fed with same sterilized granulated pellets; other animals were conventional (CV).[3]

Treatment of GF and CV Rats with NDCM

NDCM (gift of Dr. Barot) was suspended at a concentration of 100 µg/ml in a phosphate buffered saline (PBS), homogenized and sterilized. Two-month old males were deprived of food for 24 h and then given a single intragastric dose of NDCM (200 µg in 2 ml PBS) via a silicon rubber cannula. The animals were sacrificed 4 days later.

Determination of the Level of Immunoglobulins in Sera

Immunoglobulins IgA and IgG in sera of NDCM-treated rats were measured by ELISA using specific anti-isotype sera (Inst. Sera and Vaccines, Prague) as described for human Igs.[4]

Lymphocyte Preparation

Lymphocytes were obtained by washing from the spleen using syringe with bent needle, containing RPMI 1640 medium supplemented with antibiotics. The concentration used for the experiments was 5×10^5 lymphocytes/100 µl/well.

Lymphocyte Proliferation

A computer program was used for planning and evaluating our microplate experiment[5] and for the lymphocyte proliferation assay.[6] All measurements were done with four parallel samples in all experiments both with and without Con A (1.5 µg/ml)(Pharmacia). Incubation proceeded in humidified atmosphere (37°C, 5% CO_2) for 48 h, at which time each well was pulsed with 10 µl (37 kBq) ^{3}H-thymidine. Cultures were then incubated for additional 18 h. ^{3}H-UdR incorporation was determined in parallel with ^{3}H-TdR incorporation. After 17 hour incubation 10 µl (37 kBq) ^{3}H-UdR were added, the cultures were incubated for additional 31 h and measured using beta-counter Rackbeta 1214 (LKB).

MHC Class II Antigen Expression

Cryo-slides from the jejunum were first incubated with the murine monoclonal antibody MRCOX6 to MHC Class II (Serotec). The anti-mouse FITC antibody (Amersham) was used as a secondary reagent.

Preparation of Brush-Border Membrane Vesicles (BBMV)

The intestinal jejunum of rats was rinsed in cold saline, the mucosal layer was gently scraped off, immediately weighed and frozen in liquid nitrogen and placed into deep-freeze until BBMV preparation. The procedure for isolation of BBMV was adapted from the divalent cation precipitation method.[7]

Protein and Enzyme Determination in BBMV

Protein concentrations were determined by the method of Lowry[8] using bovine serum albumin as a standard. Sucrase, lactase and glucoamylase activities were measured.[9] The liberated glucose was estimated[10] with the Tris-glucose oxidase-peroxidase reagent (Koch -Light).

RESULTS AND DISSCUSION

We studied local and systemic changes on day 4 after NDCM intragastric application in both GF and CV rats. Sucrase is located at the brush border membrane surface. Its activity is often used as an indicator of differentiated enterocytes. Table 1 shows that the sucrase and lactase activity in BBMV from both GF and CV increased after a short-term treatment with NDCM, while the activity of glucoamylase was activated only in GF rats. We assumed that the activation of these BBMV enzymes after NDCM treatment of GF animals is linked with an adaptive mechanism in the absence of normal bacterial microflora in the intestine.

Table 1. Effect of NDCM on the sucrase, lactase and glucoamylase activities of BBMV from GF and CV rats (U/mg protein).

	Sucrase	Lactase	Glucoamylase
Control. GF	3.012 ± 0.743	1.059 ± 0.041	0.302 ± 0.055
GF + NDCM	5.650 ± 0.450	1.495 ± 0.275	1.355 ± 0.403
Control CV	0.772 ± 0.178	0.148 ± 0.038	0.677 ± 0.150
CV + NDCM	1.274 ± 0.051	0.251 ± 0.002	0.504 ± 0.100

Values are expressed as mean ± S.E.M. in U/mg protein for n=4 rats per group. P<0.05 vs. control.

We examined the changes on MHC Class II antigen expression after pretreatment of both GF and CV rats with NDCM. In comparison with control GF rats, MHC Class II antigen was detected on enterocytes from GF rats on day 4 after NDCM application (Figure 1B).

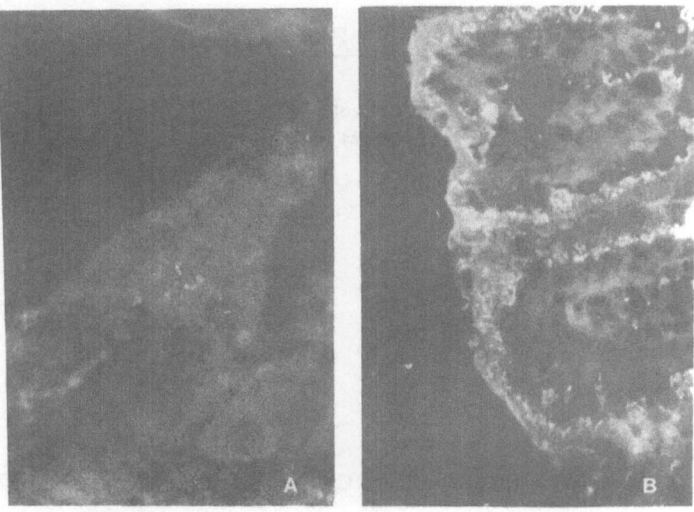

Figure 1. Immunofluorescence staining of small intestinal epithelium of rat. A) 2-month old GF rat. MHC class II molecules in the villus epithelium are negative (x500). B) 2-month old GF rat. Expression of MHC class II molecules in villus epithelium was detected on day 4 after NDCM intragastric application (x500).

The proliferative response was estimated as both ^{3}H-thymidine and ^{3}H-uridine incorporation in spleen lymphocytes from both GF and CV rats after intragastrically applied NDCM (Tables 2, 3). T cells encountering the first antigen stimulus (GF) seem to be

Table 2. Effect of NDCM and GF state on the ^{3}H-TdR incorporation of Con A stimulated spleen lymphocytes of rats

	-Con A (c p m)	+Con A (p m)	S I
Control GF	1, 689 ± 382	113, 671 ± 26 300	67 3
GF + NDCM	1, 715 ± 295	187, 769 ± 15 300	110 0
Control CV	4, 774 ± 1220	229, 119 ± 19 300	48 0
CV + NDCM	5, 420 ± 1150	270, 233 ± 38 500	49 9

Values are expressed as mean ± S E M in c p m from 8 samples   P<0 05 vs control

Table 3. Effect of NDCM and GF state on the ^{3}H-TdR incorporation of Con A stimulated spleen lymphocytes of rats

	-Con A (c p m)	+Con A (c p m)	S I
Control GF	85 ± 22	1, 172 ± 205	13 8
GF + NDCM	180 ± 48	1, 521 ± 95	8 5
Control CV	147 ± 33	900 ± 382	6 1
CV + NDCM	211 ± 28	1, 112 ± 180	5 3

Values are expressed as mean ± S E M in c p m from 8 samples*   P<0 05 vs control

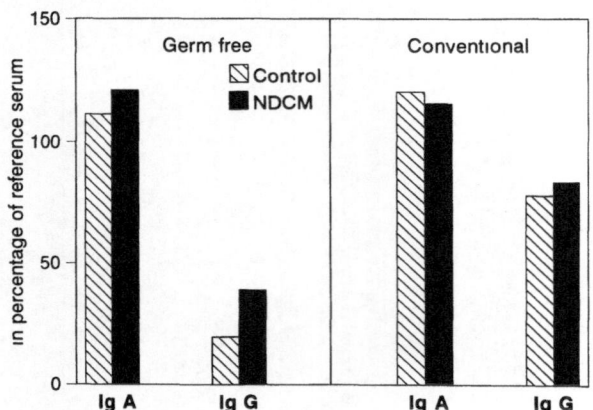

Figure 2. Level of IgA and IgG in sera from GF and CV rats after intragastric application of NDCM

induced to high DNA synthesis while RNA synthesis was decreased   IgG levels in sera of GF rats were markedly decreased in comparison with CV rats   IgG levels in NDCM-treated GF rats were stimulated Fig 2

Previous NDCM studies document its profound immunostimulatory effects [2] Here we present evidence that NDCM activates the rat intestinal mucosa after a short-term application and that GF rats are a convenient and very sensitive animal model It is expected that NDCM will soon be introduced into preclinical trials because of its possible role in immunodiagnostic and immunotherapeutic procedures. The knowledge of the relationship between the structure and biological effects of this biologically active compound is therefore highly desirable.

REFERENCES

1. R. Ciorbaru, A. Adam, J. F. Petit, E. Lederer, C. Bona, and L. Chedid, *Infect. Immun.* 11:257 (1975).
2. R. Barot-Ciobaru, J. Brochier, T. Miyawaki, L. Preud'homme, J. F. Petit, C. Bona, N. Taniguchi, and J. P. Revillard, *J. Immunol.* 135:3277 (1985).
3. R. Stepánková, *Folia Microbiol.* 24:11 (1979).
4. H. Tlaskalová-Hogenová, J. Bártová, L. Mrklas, P. Mancal, Z. Broukal, R. Barot-Ciorbaru, M. Novák, and M. Hanikyrová, *Folia Microbiol.* 30:258 (1985).
5. P. Siman, *J. Immunol. Methods* 146:1 (1992).
6. H. Kováru, M. Pospisil, *Lymphology* 13:30 (1980).
7. J. Schmitz, H. Preisner, D. Maestracci, B. K. Ghosh, J. Cerda, and R. K. Crane, *Biochim. Biophys. Acta* 323:98 (1973).
8. O. H. Lowry, N. J. Rosenbrough, A. L. Farr, and R. J. Randall, *J. Biol. Chem.* 193:265 (1951).
9. J. Kolinská and J. Kraml, *Biochim. Biophys. Acta* 284:235 (1972).
10. A. Dahlquist, *Anal. Chem* . 7:18 (1964).

STIMULATION OF GALT BY *Nocardia* DELIPIDATED CELL MITOGEN (NDCM) IN IRRADIATED GERMFREE PIGLETS

Ilja Trebichavsky,[1] Leos Mandel,[1] Helena Tlaskalová,[1] Igor Splichal,[1] and Rita Barot[2]

[1]Institute of Microbiology, Departments of Immunology and
  Gnotobiology, 142 20 Prague, Czech Republic
[2]Université Paris Sud, Department of Biochemistry, Orsay,
  France

INTRODUCTION

Ionizing radiation seriously damages gastrointestinal compartments of the mucosal immune system. Even small doses lead to mitotic disturbances, death of stem cells in crypts of Lieberkuhn, injury of the epithelium and endothelium and extensive edema and hemorrhages. Low doses of ionizing radiation affect mostly lymphocytes which are extremely radiosensitive cells. Their morphology is influenced by doses as low as 0.25 Gy after 2 h. Lymphocytes which are in the interphase of cell cycle are killed after irradiaton by interphase death, a kind of programmed cell death.

Different populations of immune cells reveal different levels of radiosensitivity. The most radioresistant cells of the immune system are macrophages and plasma cells. Increasing radiosensitivity is expressed by NK and ADCC cells, cytotoxic T lymphocytes, activated lymphocytes, lymphoblasts, various progenitors and stem cells, helper T cells, pre-B cells, and virgin B cells with membrane IgM (the most sensitive immune cells).

Do (the dose resulting in 37% survival of irradiated cells *in vitro*) of human lymphocytes increases in PHA activated T cells from 0.5 Gy to 2.9 Gy and in B cells from 1 Gy to 1.5 Gy. Generally, activation of lymphocytes with antigens, mitogens, target and allogenic cells increases their radioresistance. The same effect is found after any immunization, application of lipopolysaccharide, IL-1, IL-3, interferon gamma, TNF and other immunomodulators.[1]

Substances isolated from some pathogenic microorganisms are frequently shown to be potent immunomodulators. The cells of *Nocardia opaca* are a source of potent immunostimulating substances which possess various properties such as:

1. adjuvanticity
2. antitumor activity
3. induction of interferon and interleukins
4. activation of NK cells
5. polyclonal stimulation of B cells
6. upregulation of MHC class II antigens

Advances in Mucosal Immunology, Edited by
J. Mestecky *et al.*, Plenum Press, New York, 1995

The aim of this study was to stimulate the gut-associated lymphatic system with NDCM *in vivo* and to decrease the radiation effect after sublethal irradiation with a [60] Co source. Germfree animals have become an important tool in radiobiology because of the possibility to exclude the influence of associated microflora on the course of the radiation - induced changes.[3] The germfree miniature pig is an ideal immunological model because its immune system in the postnatal period does not contain maternal immune molecules. This model is also similar to humans with respect to physiological and radiobiological parameters.

MATERIALS AND METHODS

Animals

Germfree piglets of miniature Minnesota breed of both sexes were obtained by hysterectomy on the 110th day of gestation. Piglets were fed by sterile milk diet.

Irradiation

Animals were irradiated ([60] Co Chizotron) in a sterile transport isolator by single whole-body dose of 2.5 Gy. They were immobilized before irradiation by 30 mg of Ketalar/kg body weight, given i.m. The time of irradiation was 1 Gy/5 min 40 sec.

NDCM

NDCM or Nocardia-delipidated cell mitogen[2] was obtained by delipidation of *N. opaca* cells after acetone extraction of fresh microorganisms (Pasteur Institute, Paris) for 24 h in a Soxhlet apparatus with acetone, ether, chloroform, methanol and methanol/ chloroform. Delipidated *Nocardia* cells were suspended in ethyl alcohol for 24 h and dried. NDCM contain cell wall peptidoglycan and Cy1 fraction derived from cytoplasmic compartment.

Immunomodulation

NCDM was applicated intragastrically in a plastic tube (1mg/kg of body weight). NDCM was dissolved in saline (0.5 mg/ml). NDCM was applied either 4 days before or 2 h after the irradiation. As controls, animals of the same age were used. Some controls were neither irradiated nor stimulated, some were only stimulated or only irradiated.

Immunology

Immunoglobulin content was measured in the sera and intestinal lavages using ELISA method with monoclonal antibodies directed against immunoglobulin isotypes (porcine peroxidase labeled swine anti-mouse Ig. Immunofluorescence was accomplished with monoclonal antibodies reacting with pig IgM, IgG, IgA, CD2 and SLA-D porcine class II MHC) followed by FITC-labeled anti-mouse Ig. Cryostat sections were cut from snap frozen terminal ileum, spleen, and mesenteric lymph nodes.

Electron Microscopy

Samples fixed in paraformaldehyde/glutardehyde and osmium tetroxide were dehydrated and embedded in Vestopal W resin. Ultrathin sections were observed in a Tesla BS500 (Brno, Czech Republic) transmission electron microscope.

RESULTS

Gut-associated lymphatic tissue of NDCM treated and irradiated animals was less damaged than the same tissue of irradiated but non-stimulated animals. Similar findings were obtained in samples of spleen and mesenteric lymph nodes.

NDCM applied 2 h after irradiation protected lymphatic tissues significantly more than NDCM applied before irradiation. The humoral immune system expressed the highest protection, T cells moderate protection. NDCM protection of irradiated animals is summarized as follows:

Ileum

1. Lesser content of mucus (increased secretion is the hallmark of gut injury)
2. Smaller number of goblet cells and fibroblasts
3. Less collagen in lamina propria mucosae
4. No edema and no extravasation of IgG
5. IgM$^+$ B cells and CD2$^+$ T cells survive in intestinal villi
6. The presence of cells containing IgM, IgG and IgA in follicles

Lymph Nodes

Great follicles containing IgM$^+$ and IgG$^+$ cells were conserved.

Spleen

Great follicles with B cells and white pulp IgM$^+$ cells were conserved. On the other hand, irradiated animals not treated with NDCM had lymphatic tissues depleted and replaced by fibrotic tissue; only the spleens contained rare B cells. The protective effect of NDCM was confirmed also by ELISA. Irradiated animals were almost completely devoid of serum immunoglobulins whereas NDCM-treated irradiated animals displayed Ig levels 10 times higher (Table 1). Intestinal IgA was also elevated (Table 2).

Table 1. Serum immunoglobulins levels ($\mu g/ml$) in germfree pigs, day 8 after whole-body gamma irradiation (2.5 Gy).

Group	IgM	IgG	IgA
Control	18.40	3.79	2.60
2.5 Gy	0.92	0.70	0.34
NDCM	9.09	12.50	3.21
NDCM before 2.5 GY	8.51	5.11	2.80
NDCM after 2.5 Gy	8.31	8.32	3.02

Table 2. Immunoglobulin levels in the ileum (mg/ml), day 8 after whole-body gamma irradiation (2.5 Gy).

Group	IgM	IgG	IgA
Control	5.6	–	1.0
2.5 Gy	2.4	–	1.1
NDCM	10.3	–	57.0
NDCM before 2.5 GY	2.7	–	12.5
NDCM after 2.5 Gy	2.9	–	24.5

CONCLUSION

NDCM was shown to protect the immune system against the most deleterious effects of gamma rays and this effect was elicited even if NDCM was applied 2 h after irradiation.

Note In Proof

Brook et al.[4] recently described the treatment of gamma-irradiated mice with the immunodulator trehalose dimycolate (TDM) to be effective against endogeneous infections. Glycolipid TDM is present in cell walls of *Mycobacteria, Nocardia* and *Corynebacteria.* It is nontoxic and stimulates macrophages and antibacterial activity.

ACKNOWLEDGMENT

We gratefully acknowledge Mrs. Marie Zahradnickova, Jaroslave Steparová, Gita Hanikyrová, and Iva Mnuková, for their excellent technical assistance.

REFERENCES

1. L. Mandel and I. Trebichavsky, *in*: "Clinical Radiobiology", L. Navratil, ed., p. 48 (in Czech), SPN Prague (1990).
2. R. Barot-Ciorbaru, J. Brochier, T. Miyawaki, J. L. Preud'homme, J. F. Petit, C. Bona, N. Teniguchi, and J. P. Revillard, *J. Immunol.* 135:3277 (1985).
3. L. Mandel, F. Moravek, and I. Trebichavsky, *Fol. Microbiol.* 24:107 (1979).
4. I. Brook, G. D. Ledney, G. S. Madonna, R. M. DeBell, and R. I. Walker, *Milit. Med.* 157:130 (1992).

MUCOSAL ADAPTATION AND DESTRUCTION IN RESPONSE TO LAMINA PROPRIA T CELL ACTIVATION IN EXPLANTS OF HUMAN FETAL INTESTINE

Paolo Lionetti, Simon H. Murch, Jacqueline Taylor, and Thomas T. MacDonald

Department of Paediatric Gastroenterology
The Medical College of St Bartholomews Hospital
West Smithfield
London EC1A 7BE

INTRODUCTION

T cell-mediated hypersensitivity is probably important in the immunopathogenesis of some of the important gastrointestinal diseases in man.[1,2] By immunohistology, activated T cells can be seen in the mucosa of active Crohn's disease[3], coeliac disease[4] and intractable diarrhoea of infancy.[5] In biopsies from patients with active Crohn's disease there are increased levels of mRNA transcripts for interleukin-2[6], interferon-gamma mRNA is detectable by Northern blots and the frequency of IL-2 and interferon-gamma secreting T cells is increased.[7] Mononuclear cells isolated from the mucosa of patients with active Crohn's disease spontaneously secrete higher levels of interferon-gamma than cells from control patients.[8] We have shown previously that activation of lamina propria T cells in explant cultures of human fetal small bowel by pokeweed mitogen results in villous atrophy and crypt cell hyperplasia.[9,10] In these previous experiments, the fetal gut explants were cultured in medium which contained hydrocortisone[11], an important growth and maturation factor for fetal human intestine.[12] However hydrocortisone is also a potent immunosuppressant and hence the present series of experiments were carried out to determine the effects of steroids and another immunosuppressive agent (FK506) on the immunopathology caused by the activation of lamina propria T cells.

MATERIALS AND METHODS

Organ Culture of Fetal Human Small Intestine

Human fetal small intestine was obtained within 2 h of surgical termination from the Medical Research Council Foetal Tissue Bank, The Brompton Hospital, London. This study also received ethical approval from the City and Hackney Health Authority. The small intestine from fetuses of 13-16 weeks gestation was dissected into 2 mm square explants and these were then cultured (15-20 per dish) at 37°C in a 95% oxygen, 5% CO_2 atmosphere in 7 ml of serum-free CMRL-1066 medium (Flow Laboratories Inc, McLean, VA), modified according to Autrup et al.[11], but with the omission of hydrocortisone. At various times thereafter, the explants were examined by phase contrast microscopy for gross evidence of damage and then snap frozen in liquid nitrogen and stored at -70°C.

Advances in Mucosal Immunology, Edited by
J. Mestecky et al., Plenum Press, New York, 1995

Activation of Mucosal T Cells and Immunosuppressive Agents

Mucosal T cells were activated by adding pokeweed mitogen (PWM, Sigma, 7 μg/ml unless otherwise specified) FK 506 (10-30 nM, Fujisama Pharmaceutical Co, Osaka, Japan) was used to inhibit mucosal T cell activation Dexamethasone (Sigma Chemical Co, Poole, Dorset) was used at a concentration of 0 5-1 μM

Immunohistochemistry

Frozen sections (6 μm) of the explants were stained immunohistochemically using the immunoperoxidase technique as described previously [13] The monoclonal antibody used was Ki67 (which recognizes a nuclear antigen present in all dividing human cells, Dako Ltd, High Wycombe, Bucks)

Assessment of Mucosal Damage by Inverted Phase Contrast Microscopy

Individual explants could easily be visualised by nverted phase contrast microscopy and changes in gross morphology assessed As shown previously[10], control explants showed long villi and there was little surface debris The epithelial layer was clearly visible and the phase-dense villus core easily identifiable Two other types of explants were seen in cultures stimulated with PWM The first type showed some debris on the top of the villi and the villi were short Immunohistology (see results) showed that these explants characteristically showed partial villus atrophy and crypt hyperplasia Thus, it was considered that these explants showed adaptive changes following T cell activation Other explants however showed a different appearance In those, no villi were apparent By histology it was seen that only the muscularis externa remained, covered by a thin layer of lamina propria, overlain patchily by epithelium Thus, it was considered that T cell activation had led to destruction of the mucosa in these explants

RESULTS

Analysis of the gross morphology of a large number of explants from fetuses of different ages showed that the age of the fetal intestine and the type of immunosuppressive added to the culture had dramatic effects on the mucosal damage produced by PWM In young fetal guts (13-15 weeks old) about half of the explants were destroyed by PWM, whereas the other half showed adaptive changes (Table 1) In explants from older fetuses however virtually all of the explants were destroyed

Table 1. The effect of fetal age and FK506 and dexamethasone on the consequences of PWM-induced T cell activation in fetal human intestine

Fetal age		# of explants	Morphological Appearance		
			NORMAL	ADAPTED	DESTROYED
13 14 weeks	CONTROL	85	85(100%)		
	PWM	83	0(0%)	43(52%)	40(48%)
	PWM+FK 506	82	69(84%)	11(13%)	2(3%)
	PWM+DM	98	12(12%)	85(87%)	1(1%)
15 16 weeks	CONTROL	61	61(100%)		
	PWM	59	1(2%)	5(8%)	53(90%)
	PWM+FK 506	50	20(40%)	11(22%)	19(38%)
	PWM+DM	26	2(8%)	22(84%)	2(8%)

Immunohistology revealed that destroyed explants had no villi and only the remnants of crypts (Fig. 1A), whereas explants showing adaptive changes had short villi with long crypts containing large numbers of dividing epithelial cells (Fig. 1B).

When FK506 was added to the cultures, it almost completely blocked the pathological effects of PWM in the younger specimens (Table 1) but it was less effective in the older tissues. Dexamethasone on the other hand produced the same effects in explants of all ages in that the majority of explants showed adaptive changes.

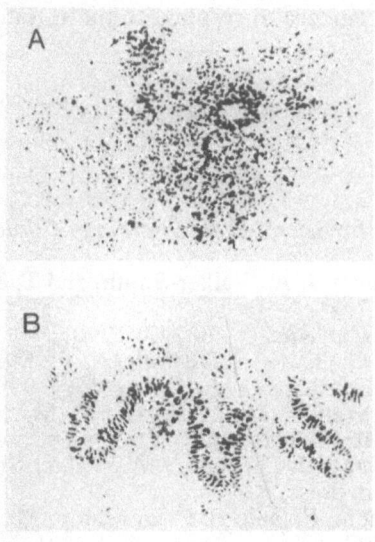

Figure 1. After the addition of PWM to explants of fetal small intestine, two types of pathology are seen. Some of the explants are completely destroyed (A) containing only a few crypts with dividing cells, whereas others show adaptive changes with crypt cell hyperplasia (B). Immunoperoxidase with Ki67, original magnification x 100.

DISCUSSION

The results obtained in this study differ from our previously published studies in that it is now clear the immunosuppressive effects of steroids were largely responsible for producing the crypt hyperplasia and villous atrophy seen in explants of fetal gut after activation of lamina propria T cells with PWM. In older fetal gut explants of the age used in our previous studies, instead of T cell activation leading to mucosal adaptation, it leads to mucosal destruction. Clues to the reason for these differences come from studies of very young fetal intestine, in which about half the explants showed adaptive changes whereas the rest were destroyed. We have previously shown that around 13-14 weeks gestation aggregates of T cells appear in fetal small intestine[14,15], so that the distribution of T cells is patchy. In the next few weeks the number of lamina propria T cells increases. Thus, we would conclude that the outcome for an individual explant is a stochastic event which will depend on the number of T cells in the explant at the onset of culture. There is no means of determining this prior to culture, except that older intestine will contain more T cells than younger specimens. Studies are ongoing to determine if individual explants which are destroyed contain more T cells than those in whom an adaptive response is seen. If this is the case then it implies that a strong local cell-mediated immune response can destroy the tissues but that the mucosa can adapt to a lesser inflammatory response with structural changes.

As expected FK506 was effective at limiting gut damage due to it's direct effects on inhibiting T cell activation.[16] Surprisingly however, when dexamethasone was added to the PWM-stimulated cultures, virtually all showed adaptive changes, confirming our previous observations.[9,10] Dexamethasone can down-regulate cell mediated immune reactions by decreasing T cell lymphokine production and cytokine production by activated macrophages.[17-20] These results indicate that there is a quantitative relationship between the extent of the local cell-mediated response and the capacity of the epithelial and stromal elements in the mucosa to respond to the inflammation with alterations in mucosal morphology.

REFERENCES

1. S. P. James, W. Strober, T. C. Quinn, and S. H. Danovitch, *Digestive Dis. Sci.* 32:1297 (1987).
2. T. T. MacDonald and J. Spencer, *Gastroenterology Clinics of North America* 21:367 (1992).
3. M.-Y. Choy, P. I. Richman, J. A. Walker-Smith, and T. T MacDonald, *Gut* 31:1365 (1990).
4. T. T. MacDonald, *Clin. Exp. Allergy* 20:247 (1990).
5. B. Cuenod, N. Brousse, O. Goulet, S. De Potter, J.-F. Mougenot, C. Ricour, D. Guy-Grand, and N. Cerf-Bensussan, *Gastroenterology* 99:1037 (1990).
6. G. E. Mullin, A. J. Lazenby, M. L. Harris, T. M. Bayless, and S. P. James, *Gastroenterology* 102:1620 (1992).
7. E. Breese, C. P. Braegger, C. J. Corrigan, J. A. Walker-Smith, and T. T. MacDonald, *Immunology* 1992 (in press).
8. S. Fais, M. R. Capobianchi, F. Pallone, P. Di Marco, M. Boirivant, F. Dianzani, and A. Torsoli, *Gut* 32:403 (1991).
9. T. T. MacDonald and J. M. Spencer, *J. Exp. Med.* 167:1341 (1988).
10. R. daC. Ferreira, L. A. Forsyth, P. I. Richman, C. Wells, J. Spencer, and T. T MacDonald, *Gastroenterology* 98:1255 (1990).
11. H. Autrup, L. A. Barrett, F. E. Jackson, M. L. Jesudason, G. Stoner, P. Phelps, B. F.Trump, and C. C. Harris, *Gastroenterology* 74:1248 (1978).
12. P. Arsenault and D. Menard, *J. Paed. Gastro. Nutr.* 4:893 (1985).
13. P. G. Isaacson and D. H. Wright, *in:* "Immunocytochemistry Practical Applications in Pathology and Biology", J. M. Polak and S. van Noorden, eds., p. 249,J ohn Wright & Sons, Bristol UK (1983).
14. J. Spencer and T. T. MacDonald, *in:* "Ontogeny of the Intestinal Immune System", T. T. MacDonald, ed., p. 23-50, CRC Press, Boca Raton, FL, 1990.
15. J. M. Spencer, T. T. MacDonald, T. T. Finn, and P. G. Isaacson, Clin. Exp. Immunol. 64:536 (1986).
16. M. J. Tocci, D. A. Matkovich, K. A. Collier, P. Kwok, F. Dumont, S. Lin, S. Degudicibus, J. J. Siekierka, J. Chin, and N. I. Hutchinson, *J. Immunol.* 143:718 (1989).
17. S. K. Arya, F. Wong-Staal, and R. C. Gallo, *J. Immunol.* 133:273 (1984).
18. B. Beutler, N. Krochin, I. W. Milsark, C. Luedke, and A. Cerami, *Science* 232:977 (1986)
19. T. R. Cupps and A. S. Fauci, *Immunol. Reviews* 65:133 (1982).
20. J. C. Reed, A. H. Abidi, J. D Alpers, R. G. Hoover, R. J. Robb, and P.C. Nowell, *J. Immunol.* 137:150 (1986).

DEVELOPMENT OF IgE ANTIBODIES AND T CELL REACTIVITY AGAINST INTESTINAL BACTERIAL ANTIGENS IN RATS AND THE INFLUENCE OF FEEDING

Anna Dahlman,[1,2] Lars Å. Hanson,[1] Esbjorn Telemo,[1] and Ulf I. Dahlgren[1,2]

[1]Departments of Clinical Immunology; and [2]Endodontics/Oral Diagnosis, University of Goteborg, Sweden

INTRODUCTION

One difficult task for the immune system of the gastrointestinal (GI) tract is to discriminate between dietary and bacterial antigens. An illustration of this ability is that rats monocolonized with *Escherichia coli* and fed ovalbumin (OA)-containing diet develop antibodies against the bacterial but not against the dietary antigens.[1] Instead rats which are fed OA become tolerant against the protein.[2] Colonisation of rats with a bacterial strain producing OA alters the immune response profoundly and the animals respond with biliary IgA antibodies against OA.[3] We have also shown that conventional rats colonised from birth with the OA producing *E. coli*, get an enhanced delayed type hypersensitivity (DTH) reaction to OA after immunization.[4] They also develop IgE and IgG antibodies against OA but not against the fimbrial and the lipopolysaccharide (LPS) antigens of the bacteria. Thus when OA is presented for the immune system of young rats in a bacterial context, it induces both a T cell response as well as an antibody response while feeding rats a diet containing OA induces tolerance.

METHODS

Animals

Eight week old Sprague-Dawley rats were obtained from ALAB, Stockholm, Sweden. Rats of the same strain, born by dams which had been colonised with the genetically manipulated *E. coli* were also raised in our own animal house.

Expression of Ovalbumin in *E. coli*

In the present study the pOMP21 plasmid[5,6] was introduced into the *E. coli* strain 06K13H1.[3] The plasmid also carries a gene which confers ampicillin resistance to the bacteria.

Bacterial Colonization

Eight week old female rats were given streptomycin (Evans Med. Ltd, Langhurst, Horsham, England) in the drinking water (5 g/l) for one day. They were starved for 24 hours and then given 10^{10} genetically manipulated *E. coli* in 1 ml bicarbonate buffer (0.2 M), through a stomach tube. The rats were given water containing ampicillin (0.5 g/l, Doktacillin, Astra, Sodertalje, Sweden) *ad libitum* during the whole period of colonisation.

Faecal cultures were done on ampicillin agar at regular intervals. The rats were mated immediately after colonization.

Experimental Procedures

The experiment included five groups of rats. Group numbers one, two and three were delivered by conventional dams colonised with the OA producing bacteria. The pups born by these dams were followed weekly to collect fecal samples that were cultured on ampicillin agar. The fourth and fifth groups were born by conventional rat dams carrying a normal intestinal bacterial flora. Group number one, two and five were weaned at 3 weeks of age onto a standard rat diet without OA and were used as controls in the DTH experiment. The pups in group numbers three and four were weaned at 21 days of age onto an OA contain ing diet[1] fed ad libitum for 5 weeks. The intake of OA from this diet was estimated to be 0.8 g per rat per day. When the rats in group numbers 2, 3, 4 and 5 were 8 weeks old they were given a subcutaneous (s.c.) injection with 50 ml OA (20 mg/ml), (Sigma, St. Louis, Mo, U.S.A) emulsified in Freund's complete adjuvant. Two weeks later, all rats in groups 1-5 were given a challenge intradermally in one ear with 20 ml OA (2.5 mg/ml) in phosphate-buffered saline. The increase in ear thickness was measured 24 h later with an Oditest (Kroplin, Hessen, Germany).

Antibody Determinations

Serum samples were analysed with an enzyme linked immunosorbent assay (ELISA).

RESULTS

DTH Response

The results of colonisation with the genetically manipulated *E. coli* on the DTH reaction against OA is shown in Figure 1. The rats in group 2 which were colonised with the genetically manipulated OA producing bacteria had a significantly greater increase in ear thickness (p = (),0018) after immunisation than the controls, group number 5, harbouring a conventional bacterial flora. Feeding the colonised rats OA diet resulted in a down regulation of the immune response against OA (p = 0.0001).

Figure 1.
Delayed type hypersensitivity reaction against ovalbumin (OA) measured as increase in ear-thickness in rats colonized with an *Escherichia coli* strain producing OA (O), and in control rats (O). The animals were immunised intracutaneously with OA in Freund's complete adjuvant two weeks (+) before challenge with OA in the ear (group 1-5). Some rats (+) were fed OA-containing diet. Each symbol represents one rat and the bars denote the medians.

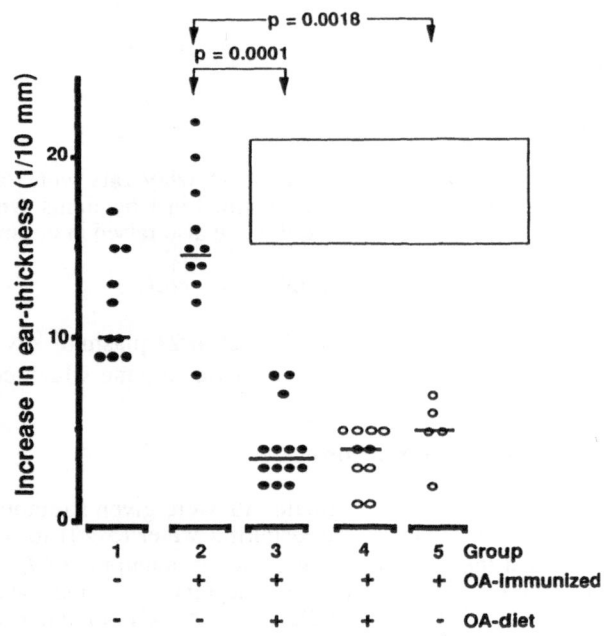

498

The IgE Antibody Response

Immunization of the colonised rats with OA resulted in a significantly higher level (p = 0.038) of IgE antibodies than in the control rats (group 2 versus 5). In group 3 where the colonised rats were eating OA containing diet, most of the animals did not develop any detectable IgE antibodies to OA. The rats in group 1, which were colonised with the OA-producing *E. coli* bacteria and fed conventional diet, did not develop IgE antibodies against OA.

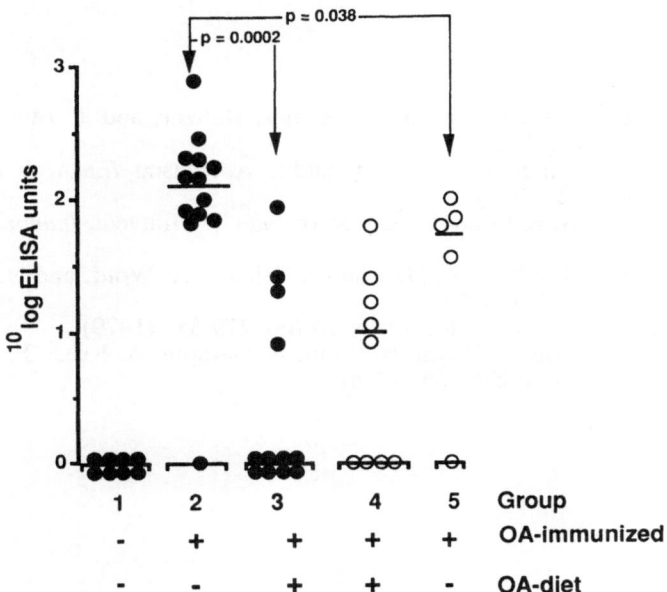

Figure 2. IgE antibodies in serum against ovalbumin (OA) recorded by ELISA in 10 weeks old rats immunised intracutaneously with OA in Freund's complete adjuvant two weeks earlier (+). Filled circles (O) are rats colonised with an *Escherichia coli* strain producing OA and empty circles (O) are control rats with a conventional intestinal bacterial flora. Some rats (+) were fed OA-conlaining diet after weaning. Each symbol represents one rat and the bars denote the medians.

The young animals colonised from birth did not develop any IgE anti-LPS or anti-fimbrial antibodies. However, the dams which were colonised as adults developed both IgE anti-LPS and anti-fimbrial antibodies.

DISCUSSION

The results demonstrate that colonisation with an OA producing bacteria by itself results in increased T cell reactivity to OA and enhanced priming for IgE anti-OA antibodies whereas feeding colonised rats antigen abrogated the intestinally induced immune response to OA. The OA produced by the bacteria is retained in the periplasmic space and not secreted, so there will be a very limited amount of free OA available in the intestine of rats fed standard diet. The rats eating OA diet will have a much more OA present in the intestine, probably resulting in passage of antigen over the entire intestinal epithelium. On the other hand, the bacterial antigens may be taken up by and processed in the Peyer's patches. This difference in antigen uptake and presentation for the immune system might explain the difference in the immune response to dietary and bacterial OA.

Another interesting finding in the present study was that the young animals did not produce any antibodies against the bacterial antigens (LPS or Fimbriae), while the adult animals did. We postulate that the lack of IgE response in the neonatally colonised animals is due to the induction of tolerance during early life because the rats colonised as adults developed IgE antibodies against LPS.

ACKNOWLEDGEMENTS

The study was supported by grants from the Swedish Medical Research Council and Nestec Laboratories.

REFERENCES

1. A. Wold, U. Dahlgren, L. Hanson, I. Mattsby-Baltzer, and T. Midtvetdt, *Infect. Immun.* 57:2666 (1989).
2. A. E. Wold, U. I. Dahlgren, S. Ahlstedt, and L. Å. Hanson, *Int. Arch. Allergy Appl. Immunol.* 84:332 (1987).
3. U. I. H. Dahlgren, A. E. Wold, L. Å. Hanson, and T. Midtvedt, *Immunology* 73:394 (1991).
4. A. Dahlman, S. Ahlstedt, L. Å. Hanson, E. Telemo, A. Wold, and U. I. Dahlgren, *Immunology* 76:225 (1992).
5. O. Mercereau-Puijalon and P. Kourilsky, *Nature* 279:647 (1979).
6. O. Mercereau-Puijalon, A. Royal, B. Cami, A. Garapin, A. Krust, F. Gannon, and P. Kourilsky, *Nature* 275:505 (1978).

EXPRESSION OF MHC CLASS II ANTIGENS AND ENZYMATIC ACTIVITY IN ENTEROCYTES OF GERM-FREE AND CONVENTIONAL RATS: DEPENDENCE ON NUTRITIONAL FACTORS DURING SUCKLING

Renata Stepánková,[1] P Kocna [2] Jaroslav Sterzl,[1] and Boshuslav Dvorák[1]

[1]Institute of Microbiology, Czech Academy of Sciences, Department of Immunology and Gnotobiology, Prague
[2]Laboratory of Gastroenterology, Faculty of General Medicine, Charles University, Prague, Czech Republic

INTRODUCTION

MHC class II molecules are involved in a variety of immune functions, including presentation of antigens to CD 4^+ T cells and regulation of T cell activity [1] Columnar absorptive cells of the small intestinal villous epithelium have been shown to express class II antigens of the major histocompatibility complex (MHC) in the mouse and the rat [2] In the normal adult rat their expression is restricted to fully differentiated columnar epithelial cells on the upper two-thirds of the villus and they are not normally expressed on immature crypt cells [2] The expression of class II antigens on epithelial cells can be altered in immune processes, this suggests that they may participate in immune reactions in the gut mucosa [3] The epithelial cells may function as antigen-presenting cells The exogenous growth factor for lymphocytes, interleukin-2 (IL-2), is produced by T lymphocytes, and its receptor has been found in both lymphoid and non-lymphoid cells [4] In the present study we determined the expression of MHC class II antigens in enterocytes of germfree (GF) and conventional (CV) rats, and also in rats after a prolonged intragastric immunization by gliadin

MATERIALS AND METHODS

Animals

Inbred F (89) rats of Wistar AVN strain were reared as an 8th generation of germfreeas (GF) animals in plastic isolators and they were fed a sterilized granulated diet [5] Artificially reared infant rats were obtained by hysterectomy and fed through a rubber tube introduced into the esophagus until the age of 16 days The composition of the artificial milk diet used was different in the quality of essential fatty acids [6] [8]

Immunohistochemical Methods

Rats were killed under anesthesia by bleeding Proximal and medium parts of their jejunum were frozen in liquid nitrogen, cut into 6-7 μm slices and used for immunofluorescence staining, after dewaxing followed by incubation with murine monoclonal antibodies MRC OX 39 (IL-2R), and MRC OX 8 (CD 8[+]). Endogenous enzyme activity was blocked with hydrogen peroxide (6%) 5 min, periodic acid (2.5%) 5 min, potassium borohydrate (0.02%) 2 min before incubation with murine monoclonal antibodies MRC OX 6 (MHC class II). All monoclonal antibodies were from Serotec. The secondary reagent was fluorescein-labelled sheep anti-mouse Ig (Amersham).

Microscopy and Photography

An immunofluorescence microscope (Leitz Orthoplan, Wetzlan, Germany) equipped with a vertical illuminator with interference filters for selective observation of fluorescein (green) was used.

Determination of Sucrase Activities

Collection of Tissue. The small intestine was removed, cleaned of adhering tissue, and flushed with normal saline to remove the luminal contents. The resulting preparation was then cut into two halves of equal length. The proximal half was designated as jejunum and the distal half as ileum. The mucosal layer was gently scraped off, immediately weighed and frozen in liquid nitrogen. Sucrase activity was determined as glucose liberated from sucrose, measured according to Dahlqvist by the GOD/POD method with o-dianisidine as chromogen, and evaluated photometrically at 540 nm.[9] Enzyme activities were calculated as mU/mg protein determined according to Lowry with albumin as standard.

RESULTS AND DISCUSSION

The distribution of MHC class II antigens is not restricted to bone marrow derived cells, but other tissues including the gastrointestinal epithelium can also express these antigens.[12] The sole recognized function of MHC class II molecules is participation in the presentation of antigens.

In 2-month old conventional rats MHC class II was found to be expressed on apical parts of the villi (Table 1), whereas no expression of MHC class II was observed in enterocytes from 2-month or 12-month old GF rats (Table 1, Fig. 1A). A conspicuous expression of MHC class II by enterocytes was found in GF rats subjected to prolonged intragastric immunization with 0.5% gliadin (Fig. 1 B). Immunization of the GF rats was carried out 3 times a week from birth until the age of 2 months.[10] These longitudinally immunized GF animals expressed MHC class II in the apical parts of both the villi and crypts (Table 1, Fig. 1 B), a situation found in the coeliac disease.[12]

Our results are in keeping with the finding that neither GF nor young rats show MHC class II on their intestinal epithelial cells while CV adult healthy rats express these antigens.[11] Our experiments did not reveal any expression of MHC class II in entrocytes of both 2- and 12-months old GF rats, but expression of IL-2R on enterocytes was found in 2-month-old GF animals after intragastric immunization with gliadin. IL-2R was found in the Golgi region and microvillous zone of the enterocytes (Fig. 1D). These events depend on the quantity of intraepithelial lymphocytes, in accordance with the notion that intraepithelial lymphocytes are likely to be involved in the modulation of intestinal MHC class II expression.[12] Old GF rats were found to exhibit lymphocytes with CD 8[+] and CD 4[+] characteristics only in the lamina propria, not in the epithelium. Old CV rats exhibited no CD 8[+] lymphocytes in the lamina propria, while these lymphocytes were found in the epithelium; this is similar to GF rats after prolonged intragastric immunization with gliadin.[10]

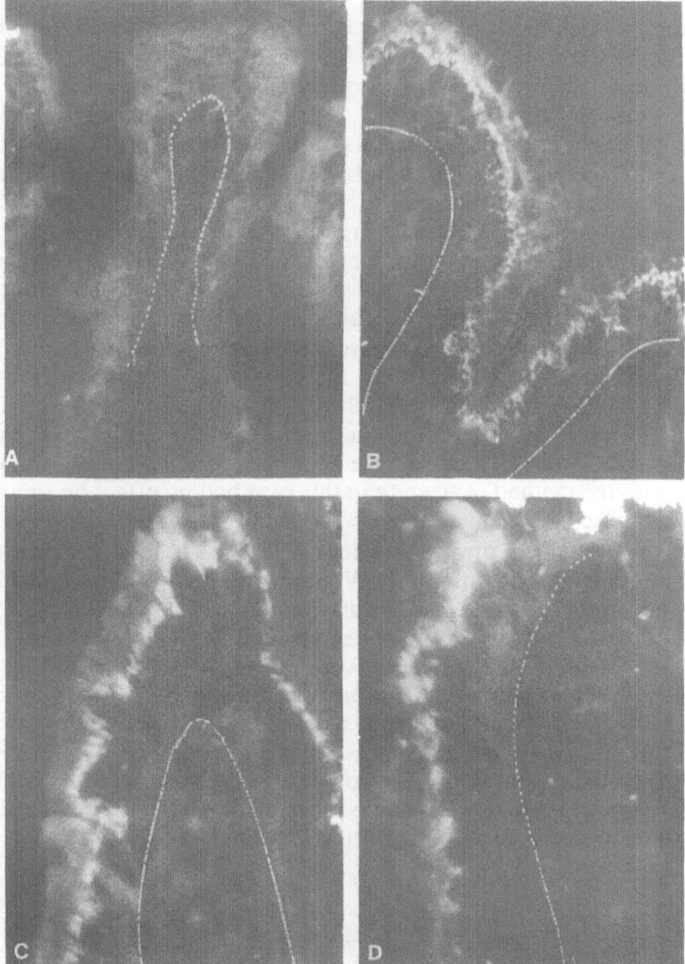

Figure 1. Immunofluorescence staining (FITC) of small intestinal epithelium of rat.

A) 2-month old GF rat. MHC class II molecules in the crypt and villus epithelium are negative (x 500).

B) 2-month old GF rat. The rat was longitudinally intragastrically immunized with gliadin. The expression of MHC class II molecules takes place in crypt and villus epithelium (x 500).

C) 2-month old conventional rat. MHC class II expression is confined to the Golgi region (x 800).

D) 2-month old GF rat. The rat was longitudinally intragastrically immunized with gliadin. The expression of IL-2R in enterocytes was confined to the Golgi region and the microvilli (x 800).

Basement membrane indicated by dotted line.

Table 1. MHC II expression on jejunal enterocytes of germfree and conventional rats of different age.

| | CV | | GF | | GF |
Months	2	12	2	12	2*
MHC class II	++	0	0	0	+++

*   intragastrically immunized with 0.5% gliadin
GF   = germ-free   + +   = positivity in villi
CV   = conventional   + + + = bright positivity in villi and crypts
                                                            0   = no positivity

Nutritional studies included the use of milk diets from birth until the age of 16 days. We used a fatty acid-deficit diet (EFA-D) and another with excess essential fatty acids (EFA-R). Rats reared artificially on the EFA-D milk diet had a low level of prolactin in the sera and low numbers of cells producing hemolytic antibodies after immunization with sheep red blood cells, as compared with rats reared artificially on the EFA-R milk diet or suckled.[6,7] These nutritional models were used to determine the biochemical and immunological characteristics of the intestinal mucosa. This included the detection of sucrase activities in the tissue in animals of different ages. Sucrase is one of several disaccharidases localized at the surface of the brush border membrane. Its activity is often used as an indicator of the functional state of the intestinal epithelium. In suckling rats the activity is dependent on glucocorticoids and the administration of hydrocortisone during the second postnatal week causes a precocious appearance of sucrase activity.[14] In our experiments with suckled young rats, an increase in sucrase activity was observed no sooner than during weaning, i.e. on days 18-20. Young rats reared on the EFA-D diet exhibited high sucrase levels already at the age of 14 days (Fig. 2). Young rats fed artificially on maternal milk evinced sucrase activities comparable with those of suckled animals (Fig. 2), i.e., the method of feeding in this case had no effect on sucrase levels. As stated above, the disaccharidase activities in the small intestine of developing mammals depend strongly on the effects of glucocorticoids.[13-15] These effects are thought to be associated with biochemical maturation. We expected to find also a difference in the expression of MHC class II between artificially reared and suckled animals, especially in cases where an increased sucrase level had been found, i.e. in animals kept on the EFA-D diet. However, the MHC class II antigens were not expressed in 7-, 14- and 21-day old rats, irrespective of whether they were kept on any of the artificial diets or suckled. Thus, MHC class II molecules appear to be absent from the epithelium throughout fetal development of the rat; the expression begins only at the age of 3-4 weeks[16] independent of the biochemical maturation of the enterocytes. The expression of MHC class II in enterocytes is influenced by IFN-γ and some other cytokines, but biochemical maturation has no direct influence on this expression during suckling.

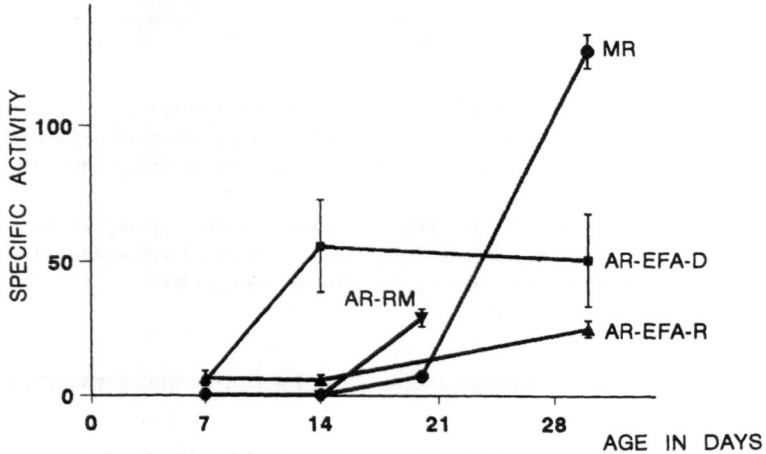

The jejunal sucrase activity (U/g protein). Each data point represents the mean ± S.E. of 3 to 7 animals. 1U=1μmol products formed per minute.

 AR - artifficially reared

  diet EFA-R = high proportion EFA (49.5% EFA)
  diet EFA-D = low proportin EFA (4,7% EFA)
   RM = rat milk
 MR - mother reared

Figure 2. Developmental changes in enzymatic activity in mother (MR) and artificially reared rats (AR).

REFERENCES

1. E. Thorsby, *Transplant Proc.* 19:29 (1987).
2. G. Mayrhofer, C. W. Pugh, and A. N. Barclay, *Eur. J. Immunol.* 13:112 (1983).
3. P. W. Bland and L. G. Warren, *Immunology* 58:1 (1986).
4. T. Diamantstein and H. Osawa, *Immunol. Rev.* 92:5 (1986).
5. R. Stepankova, *Folia Microbiol.* 24:11 (1979).
6. R. Stepankova, J. Sterzl, and I. Trebichavsky, *Physiol. Bohemoslov.* 34:49 (1985).
7. R. Stepankova, B. Dvorak, J. Sterzl, and I. Trebichavsky, *Physiol. Bohemoslov* 39:185 (1990).
8. B. Dvorak and R. Stepankova, *Prostagl. Leukotrien Essent. Fatty Acids* 46:183 (1992).
9. P. Kocna, P. Fric, J. Slaby, and E. Kasafirek, *Sbornik Lek.* 88 (1986).
10. R. Stepankova, H. Tlaskalova, P. Fric, and I. Trebichavsky, *Folia Biol.* 35:19 (1989).
11. N. Cerf-Bensussan, A. Quaroni, J. T. Kurnick, and A. T. Bhan, *J. Immunol.* 132:2244 (1984).
12. H. Scott, L. M. Sollid, O. Fausa, P. Brandtzaeg, and E. Thorsby, *Scand. J. Immunol.* 26:567 (1987).
13. R. G. Doell and N. Kretchmer, *Science* 143:42 (1964).
14. O. Koldovsky, J. Jumawan, and M. Palmieri, *J. Endocrinol.* 66:31 (1975).
15. J. Neu, W. N. Crim, and N. C. Hoge, *Pediatr. Res.* 29:109 (1986).
16. P. Bland, *Immunol. Today* 9:174 (1988).

QUANTIFICATION OF SALIVARY, URINARY AND FECAL SECRETORY IgA, AS WELL AS IN SALIVA TITERS AND AVIDITIES OF IgA ANTIBODIES IN CHILDREN LIVING AT DIFFERENT LEVELS OF ANTIGENIC EXPOSURE AND UNDERNUTRITION

A. T. Nagao,[1,3] MIDS. Pilagallo,[2] A. B. Pereira,[2] MMS. Carneiro-Sampaio,[1] and L. A. Hanson[3]

[1] Dept. of Immunology, Universidade de Sao Paulo, Brazil
[2] Dept. of Medicine, Escola Paulista de Medicina, Brazil
[3] Dept. of Clinical Immunology, University of Gothenburg, Sweden

INTRODUCTION

The prevalence of malnutrition, often associated with parasitic, bacterial and viral infection, is high in Brazil.[1-3] Since the mucosa is an important access of pathogens, one supposes that different levels of antigenic exposure can influence secretory IgA (sIgA) production and secretion.

STUDY GROUPS

Group I (N=80) was composed of 37 one to three and 43 four to six year-old well-nourished children attending a kindergarten located in a residential middle to upper class area in Sao Paulo, Brazil, with municipal water and sewage system.

Group II (N=62) was composed of 53 well-nourished (subgroup IIA) and 9 malnourished (subgroup IIB) children. Subgroup IIA was composed of 28 one to three and 25 four to six year-old children, and subgroup IIB, had 9 one to three year-old children.With respect to subgroup IIB, 7 children present mild undernutrition and 2, moderate undernutrition. All the children attended a day nursery located in a slum in Sao Paulo with municipal water, but without a sewage system.

The reference standard for weight evaluation was that from National Center of Health Statistics.[4] Children with weight for age equal to or higher than the tenth percentile were considered well-nourished; those lower than the tenth percentile were considered undernourished.

Non-stimulated saliva, urine and feces were collected from the children, after consent from the parents. An enzyme-linked immunosorbent assay (ELISA), as described elsewhere[5], was employed to quantify sIgA in the samples. We also collected saliva samples from children of the same slum and from undernourished children assisted at a Health Center in Sao Paulo to measure titers[6] and avidities[7] of IgA antibodies against beta lactoglobulin, against a pool of the most common *E. coli* serotype O antigens and against poliovirus type III. 59 well-nourished (group III) and 18 undernourished children (group IV) were studied. With respect to the undernourished children only 3 presented moderate form (second degree) of malnutrition, and 2 of them had chronic malnutrition; the other 15 presented a mild form of malnutrition.

For statistical analysis, nonparametric Mann-Whitney test was employed for comparison of two groups and Kruskal-Wallis and Dunn's tests, for comparison of three or more groups. Statistical differences were considered at the level of $p \leq 0.05$.

RESULTS

Presence of Parasites in Feces

Among the one to three year-old children, 10.7% in Group I and 71.9% in Group II presented enteroparasites. Among the four to six year-old children, 14.7% in Group I and 100% in Group II presented enteroparasites. The frequency of polyparasitism (two or more different parasites in the same sample) was in Group I, 0%, and in Group II, 50 % among the younger children and 85% among the older ones.

Comparison Between Fecal, Salivary and Urinary sIgA Levels of Well-Nourished Children in Group I and Subgroup IIA

For fecal sIgA, the levels for the children from subgroup IIA were significantly lower than those from Group I (Figs. 1 and 2, p<0.05); fecal sIgA was undetectable in 14 of 41 children in subgroup IIA (33.3%) and in no child in Group I.

For salivary sIgA, the levels for the four to six year-old children from subgroup IIA were significantly higher than those from Group I (Fig. 2, p<0.05).

For urinary sIgA, no statistical difference was found between subgroup IIA and Group I (Figs. 1 and 2, p>0.05).

Comparison Between Fecal, Salivary and Urinary sIgA Levels of Well-Nourished and Undernourished Children

No statistically significant difference was found comparing fecal, salivary and urinary sIgA levels of children from subgroup IIB (undernourished, slum) with those from subgroup IIA (well-nourished, slum) or from Group I (Fig. 1, p>0.05).

Comparison Between Titers and Avidities of Salivary IgA Antibodies to a Bacterial, Viral and Food Antigen of Well-Nourished and Undernourished Children

We did not find any statistical difference between undernourished and well-nourished children from Groups IV and III, respectively, for the antibodies against a "pool" of the most common *E. coli* serotype O antigens, poliovirus type III and beta-actoglobulin, in respect to titers and avidities in the saliva samples (Figs. 3, 4 and 5, p>0.05).

SUMMARY

We can speculate that children in Group II (slum) suffer of "environmental enteropathy."[8,9] This entity is associated with heavy antigenic exposure particularly among people living in tropical countries. The major characteristics are an alteration in D-xylose absorption and/or histological abnormalities of jejunal mucosa.[8] Also, an overgrowth of bacteria in the gastrointestinal tract can occur.[10] In these situations, secretion and/or production of sIgA could be altered, as well as an increase of IgA digestion by bacterial proteases.

well-nourished vs undernourished

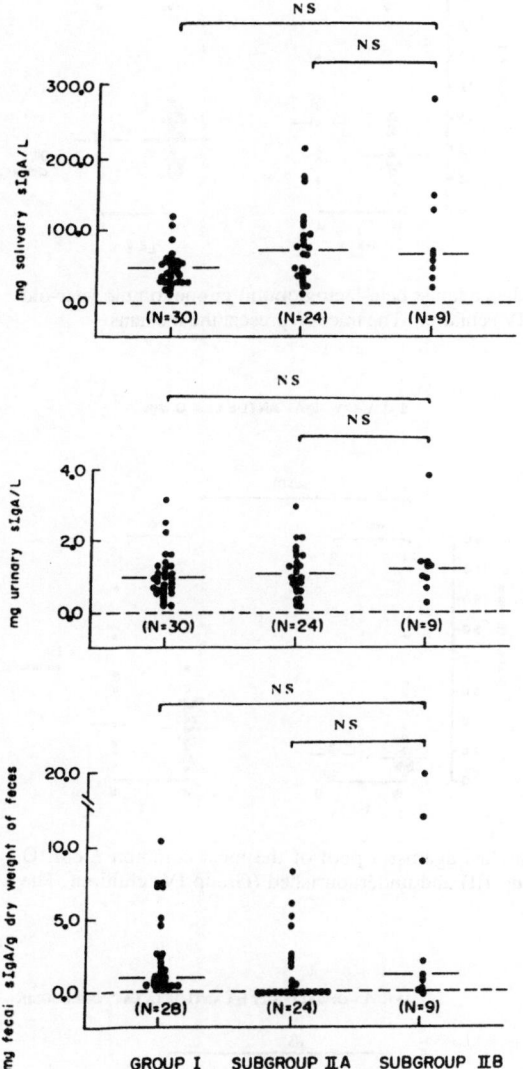

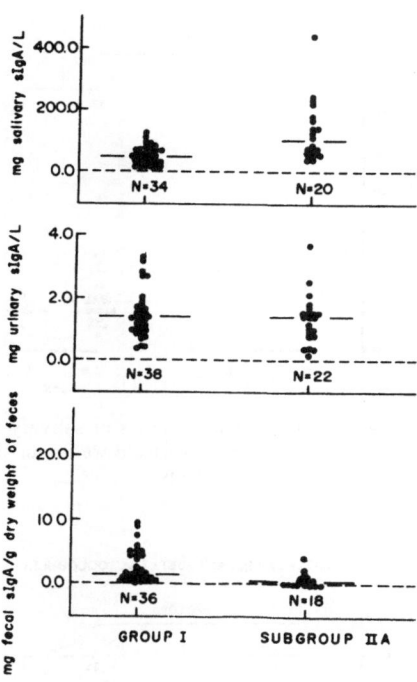

Figure 1. Levels of salivary, urinary and fecal sIgA in one to three year-old undernourished children from subgroup IIB and in well-nourished children from Group I and subgroup IIA. The traces represent the medians.

Figure 2. Levels of salivary, urinary and fecal sIgA in four to six year-old well-nourished children from Group I and subgroup IIA. The traces represent the medians.

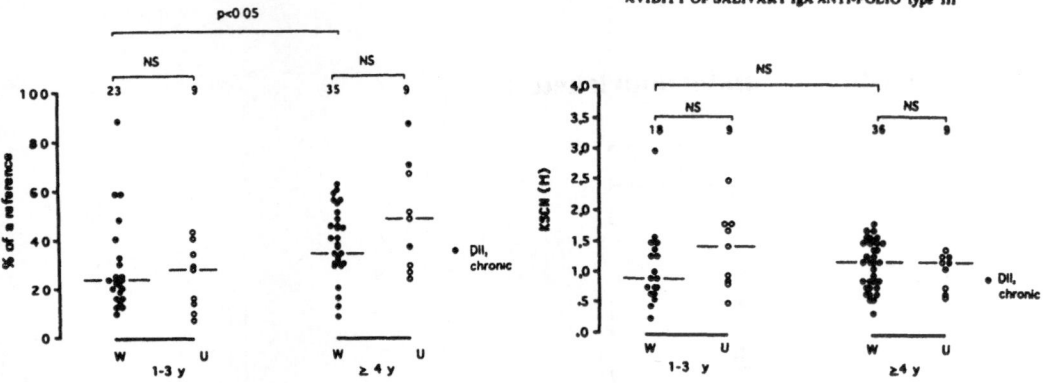

Figure 3 Titers and avidities of salivary IgA antibodies against beta-lactoglobulin in one to nine year-old well-nourished (Group III) and undernourished (Group IV) children The traces represent the medians

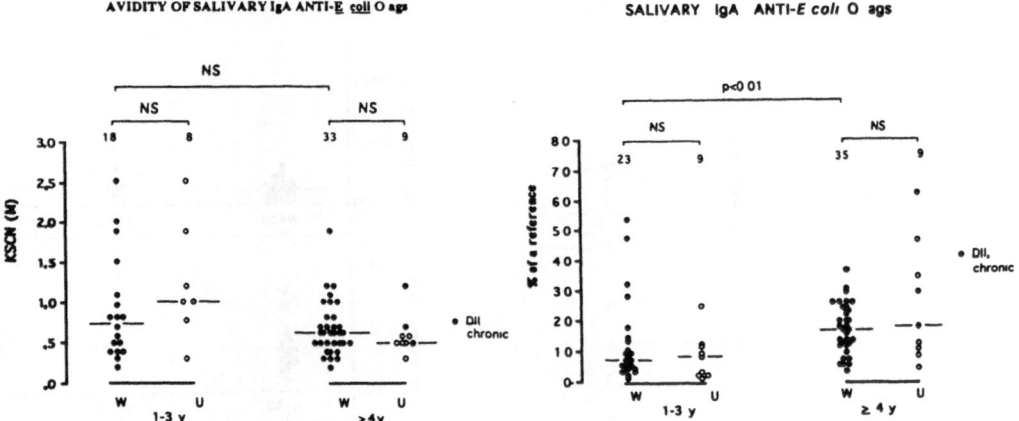

Figure 4. Titers and avidities of salivary IgA antibodies against a pool of the most common *E coli* O antigens in one to nine year-old well-nourished (Group III) and undernourished (Group IV) children The traces represent the medians

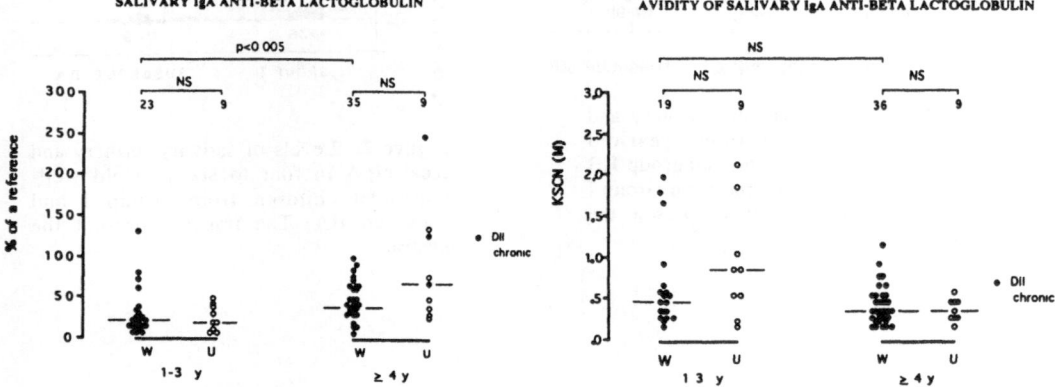

Figure 5. Titers and avidities of salivary IgA antibodies against poliovirus type III in one to nine year-old well-nourished (Group III) and undernourished (Group IV) children The traces represent the medians

510

With respect to the salivary sIgA, the fact that subgroup IIA (slum), the children, particularly those between four to six years old, presented higher levels than those in Group I agrees with data obtained by Mellander et al.[6], comparing salivary sIgA levels between Pakistani and Swedish infants. According to them, the higher the antigenic exposure, the earlier the maturation process of the mucosal system.

The mucosal sIgA immune system was probably not impaired in our undernourished children. This point is important since rather mild undernutrition commonly occurs in the slums in Sao Paulo, Brazil.

ACKNOWLEDGMENT

This study was supported by the Fundacao de Amparo a Pesquisa do Estado de Sao Paulo (Brazil), the Swedish Medical Research Council (No. 215) and the Ellen, Lennart, Walter Hesselman Foundation (Sweden).

REFERENCES

1. A. L. Torres, Estudo microbiologico da diarreia aguda da crianca em uma comunidade de favelados da cidade de Sao Paulo. Sao Paulo, p. 79 (Master thesis, Escola Paulista de Medicina, Brazil) (1984).
2. C. A. Monteiro, M. H. D'Aquino-Benicio, H. P. P. Zuniga, and S. C. Szarfarc, *Rev. Saude Publ.,* 20:446 (1986a).
3. C. A. Monteiro, P. P. Chieffi, M. H. D'Aquino-Benicio, R. M. S. Dias, D. M. A. G. V. Torres, and A. C. S. Mangini, *Rev. Saude Publ.,* 22:8 (1988).
4. World Health Organization.- Medicion del cambio del estado nutricional. Genebra, OMS, p. 105 (WHO/FAP/79.1) (1983).
5. A. T. Nagao, O. L. Ramos, and A. B. Pereira, *Braz. J. Med. Biol. Res.,* 23:211 (1990).
6. L. Mellander, B. Carlsson, F. Jalil, T. Soderstrom and.L. A. Hanson, *J. Pediatr.,* 107:430 (1985).
7. D. M. Roberton, B. Carlsson, K. Coffman, M. Hahn-Zoric, F. Jalil, C. Jones, and L. Å. Hanson, *Scand. J. Immunol.,* 28:783 (1988).
8. U. Fagundes-Neto, T. Viaro, J. Wehba, F. R. S. Patricio, and N. L. Machado, J. Ped., 54:313 (1983).
9. M. C. V. Martins - Enteropatia ambiental: alteracoes funcionais e morfologicas da mucosa jejunal decorrentes do ambiente desfavoravel. Sao Paulo, p. 93 (Master thesis, Escola Paulista de Medicina, Brazil) (1988).
10. M. Gracey, M. R. A. C. P. Suharjono, M. D. Sunoto, and D. E. Stone, *Am. J. Clin. Nutr.,* 26:1170 (1973).

ALTERATIONS OF GALT DUE TO MALNUTRITION AND DECREASE IN THE SECRETORY IMMUNE RESPONSE TO CHOLERA TOXIN

Juan Fló, Ruben Benedetti, Catalina
Feledi,Silvia G.Ariki*, Hebe Goldman,
María E.Roux* and Ernesto Massouh

Laboratory of Cellular Immunology*, FFyB
Laboratory of Immunochemistry, FCEyN,
Ciudad Universitaria, UBA
1428 Cap. Fed., Argentina

INTRODUCTION

Generalized malnutrition affects all aspects of host immunity, the impact being greatest on T and B cell function and cell mediated immunity.[1] Moreover, in an animal model of immunodeficiency due to severe protein deprivation at weaning, B and T cell differentiation in Peyer's patches (PP) and in mesenteric lymph node (MLN) showed severe impairment, which persisted after protein refeeding.[2] In this report we present data that define the impact of malnutrition during the suckling period on the gut associated lymphoid tissues (GALT) .

MATERIALS AND METHODS

Rats of the Wistar strain were suckled in groups of 14-16 pups per mother from birth until weaning (21 days of age) to achieve undernutrition (weight less than 60% of the control).[3] Well-nourished controls were suckled in litters of 6-8 pups. At weaning both groups were fed stock diet.

Three doses of cholera toxin (CT) (10, 7.5, and 7.5 µg) were given intragastrically at intervals of one week, the first when rats were 28 days old. Six days after the third dose (50 days of age), rats were bled and intestinal fluid from lavage of the small intestine was collected.

Surface antigenic determinants were characterized in a suspension of living cells from PP and MLN by immunofluorescence. The following affinity-purified antibodies were used: goat anti-rat IgM (µ chain specific) and goat anti-rat IgA (α chain specific), and the fluorescein conjugated IgG fraction of rabbit anti-goat IgG and goat anti-mouse IgG (Organon Tecknika Co., Cappel Division, Durham, NC). T cell population was characterized using the OX19 MAb (CD5), T helper subset (CD4) with the W3/25 MAb and the non helper subset (CD8) with the OX8 MAb (Seralab, Westbury, NY, USA)

Indirect ELISA was used to measure antitoxin IgA and sandwich ELISA for whole IgA in intestinal fluid. IgA concentrations (µg/ml) were interpolated from a standard rat IgA curve . Intestinal ligated loop test was performed as described by Lange and Holmgren.[4]

Advances in Mucosal Immunology, Edited by
J. Mestecky *et al.*, Plenum Press, New York, 1995

The weight per length ratio (gm/cm) was compared among normal, malnourished and nonimmunized control rats.

RESULTS

Rats 50 days old, fed for 30 days with stock diet displayed altered percentages of cells bearing IgM (sμ⁺) and IgA (sα⁺) in GALT (Fig. 1). The sα⁺ cells from PP and MLN were significantly diminished in the malnourished group while the sμ⁺ cells were increased in both organs. CD5⁺ T cells in PP were diminished in the malnourished group as well as the CD4⁺ and CD8⁺ subset (Fig. 1). The same alterations were found in MLN (Fig. 1). Moreover the levels of IgA in the intestinal fluid were decreased (683 ± 101 μg vs 480 ± 111 μg, $p < 0.0002$).

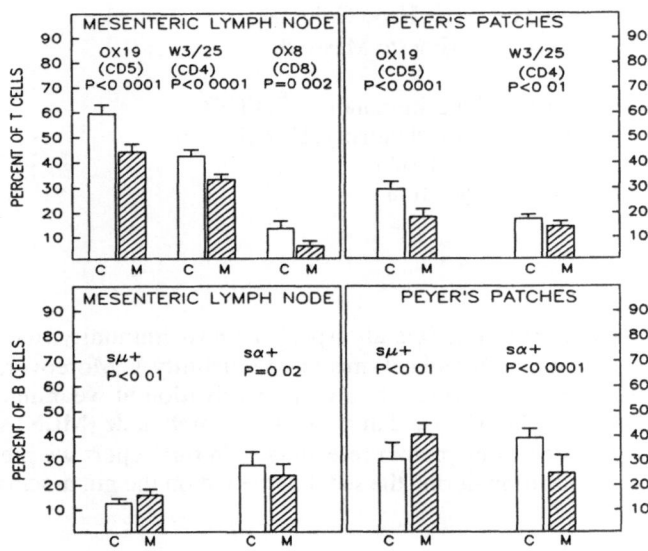

Figure 1. Percent of T and B cells from PP and MLN of control (C) and malnourished during lactation (M) 50 day old rats. Bars represent the mean data obtained on 9 - 16 animals. Vertical lines indicate the SD. The Student's t test was used.

In the malnourished rats reciving three doses of CT the specific IgA production in the intestinal fluid was abrogated, while the control group presented titers from 1/800 to 1/3200 (Table 1). Protection against CT challenge in the ligated intestinal loop was also studied (Table 1). Whereas control animals demonstrated partial ability to neutralize the CT, no antitoxin activity was observed in the malnourished group.

Table 1. Specific IgA anti CT titers in intestinal fluid and results of intestinal ligated loop test (ILLT).

	CONTROL RATS	NORMAL IMMUNE RATS	MALNOURISHED IMMUNE RATS
ILLT (grs/cm)	0.34 ± 0.035*	0.22 ± 0.021[a]	0.314 ± 0.019[b]
IgA titers		2080 ± 1070	< 100

*All values are means ± SD, n = 5. One μg of CT was used as challenge in the ILL. [a] $p < 0.001$ when compared to the control. [b] $p = 0.285$ when compared to the control. The statistical test used was the Tukey's T test.

DISCUSSION

The impact of malnutrition during lactation affects body weight, organ weight and cell number in different organs as well as lymphoid tissues such as spleen and thymus.[3] In an animal model of immunodeficiency due to severe protein deprivation at weaning, a severe impairment of B and T cell differentiation in PP and MLN has been described, which persists even after protein refeeding.[4] Recently, it has been shown that there is also an impairment in the terminal differentiation of IgA-Bcells in the lamina propria.[5]

The results reported in this communication show that malnutrition induced during suckling until weaning provokes severe alterations in GALT. Subsequent feeding on stock diet for 30 days is unable to revert the damage, and the percentage of CD5+ T cells in PP and in MLN remain decreased as well as the CD4+ T cell subset. The above findings correlate with what has been observed in the thymus.[6] Concomitantly, a diminished percentage of B cells bearing IgA with accumulation of B cells bearing IgM was observed in the PP. These results seem to indicate that the switch from IgM to IgA is altered, perhaps due to the diminished numbers of the CD4+ T cell subset. CD4+ T cells through soluble factors would play the main role in the isotype switching.[7,8]

The fall in the content of IgA in the intestinal fluid may be a consequence of the diminished number of B cells bearing IgA in PP and MLN since it is generally thought that IgA cells in intestinal lamina propria are derived from precursor cells located in GALT. Furthermore, in the rat the other pathway through which polymeric IgA can reach the gut (via hepatocytes and biliary duct) may be also affected.[9]

The observed alterations due to this type of malnutrition affecting GALT B and T cell populations inhibit the triggering of a specific immune response to orally administered CT. This results in low titers of both IgA in the intestinal fluid and IgG in serum (data not shown), and also in the incapability to neutralize CT in the ligated intestinal loop test. According to recent studies CT is a thymus dependent antigen which requires functional CD4+ T cells in GALT for the development of a host defense mechanism against it.[10,11] We may assume that the diminished percentages of CD4+ T cells is the main cause of the low response in the malnourished rats. Furthermore, in a model of severe protein deprivation at weaning pre-B cells were found[2] in PP; thus, we propose that perhaps in the present model pre-B cells would persist at 28 days of age,and would not be in condition to be triggered by antigenic stimulus.

ACKNOWLEDGMENTS

We thank the skilfull assistance of Mr. G. Assad Ferek. This work has been supported by grants from CONICET (PID 300700/88) and UBA (EX 148).

REFERENCES

1. R. K. Chandra, *Nutr. Rev.* 39:225 (1981).
2. M. C. Lopez and M. E. Roux, *Devel. Comp. Immunol.* 13:253 (1989).
3. M. Winick and A. Noble, *J. Nutr.* 89:300 (1966).
4. S. Lange and J. Holmgren, *Acta Pathol. Microbiol. Scand. Sect.* C 86: (1978).
5. S. Gonzalez Ariki, M. C. Lopez, and M. E. Roux, *Reg. Immunol.* 4:41 (1992).
6. H. N. Slobodianik, A. N. Pallaro, I. Fernandez, M. E. Roux, and M. E. Rios, *Com. Biol.* 9:313 (1991).
7. R. L. Coffman, A. L. Deborah, and B. Shrader, *J. Exp. Med.* 170:1039 (1989).
8. E. Sonoda, Y. Hitoshi, N. Yamaguchi, T. Ishii, A. Tominaga, S. Araki, and K. Takatsu, *Cell. Immunol.* 140:158 (1992).
9. L. Villalon, B. Tuchweber, and I. M. Yousef, *J. Nutr.* 117:678 (1987).
10. N. Lycke, L. Eriksen, and J. Holmgren, *Scand. J. Immunol.* 25:413 (1987).
11. E. Hörnqvist, T. J. Goldschmidt, R. Holmdahl, and N. Lycke, *Infect. Immun.* 59:3630 (1991).

DIETARY ANTIGEN HANDLING BY MOTHER AND OFFSPRING IN A TWO GENERATION STUDY

Esbjorn Telemo, Ulf Dahlgren, Lars Å. Hanson, and Agnes Wold

Department of Clinical Immunology
University of Göteborg, Sweden

INTRODUCTION

The neonatal period seems to be of major importance in the establishment of tolerance to harmless environmental antigens . In most species studied it is possible to induce a long lasting specific tolerance to orally administered antigens during this period,[1] however antigens introduced to rodents before one week of age have been shown to cause priming rather than tolerance.[2,3] It has also been shown that factors transferred from the mother can specifically suppress or enhance the response of her offspring to such antigens.[4-6] In the present experiment we studied the response of the offspring to an antigen introduced very early in life to the mother.

METHODS

The first generation of Sprague-Dawley rats were fed β-lactoglobulin (β-LG) either from 7 days or from 21 days of age, OvA was first introduced at 21 days of age as a control antigen. Feeding of β-LG and OvA continued either for 2 w or until mating (9 w) or until the 2nd generation was weaned at 3 w of age (Fig. 1).

The second generation was immunised with β-LG and OvA in ICFA at 8 w of age, and the serum was monitored for IgG and IgE (data not shown) anti- β-LG and anti- OvA responses (Fig. 1).

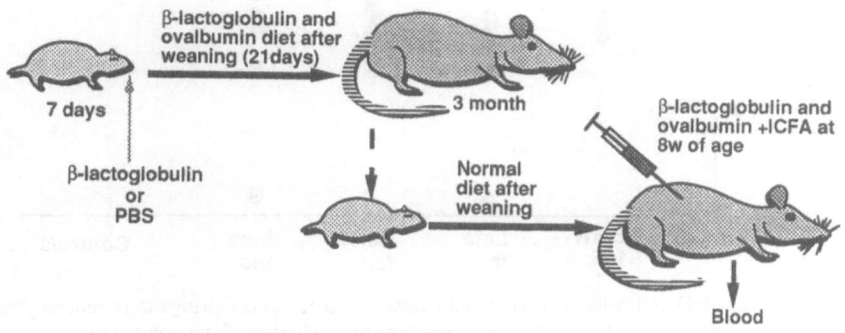

Figure 1. Experimental design.

Advances in Mucosal Immunology, Edited by
J. Mestecky *et al.*, Plenum Press, New York, 1995

RESULTS

There was a significant suppression (p<0.05, Mann-Whitney U-test) of the response to both antigens (OVA data not shown) in all groups compared with controls from non-fed mothers (Fig. 2). In addition, within the groups there was a significantly more pronounced suppression in rats from mothers fed from 21 d than from 7 d of age (Fig. 3).

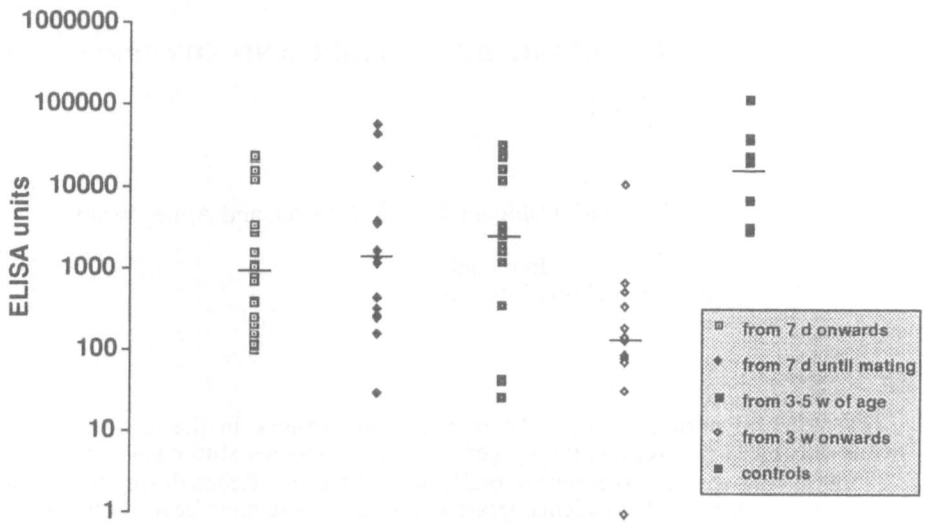

Figure 2. Serum IgG anti β-lactoglobulin titre in offsprings from mothers fed the antigen as indicated.

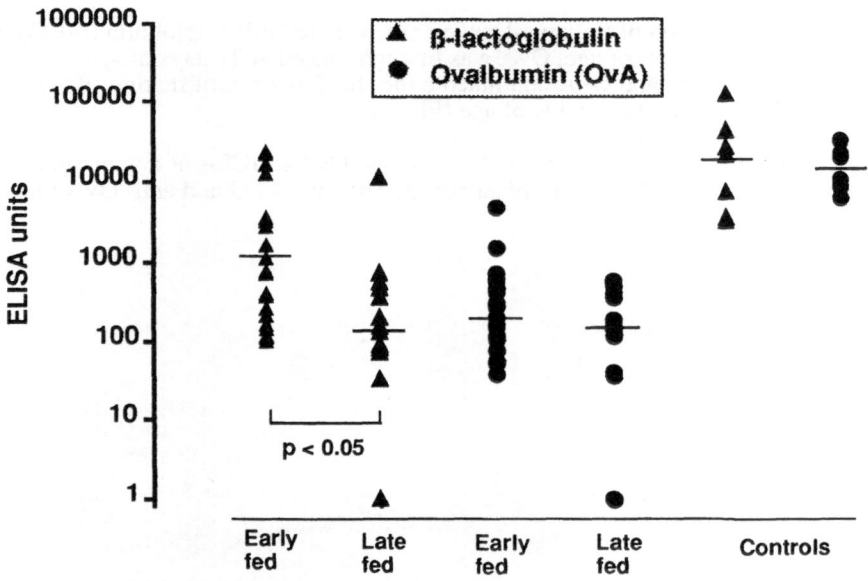

Figure 3. Serum IgG anti β-lactoglobulin and ovalbumin titres in offsprings from mothers fed β-lactoglobulin from 7 d (early fed) or 21 day of age (late fed). Controls from non-fed mothers.

DISCUSSION

As shown in this study very early antigen experience by the mother can alter the immunological handling of this antigen in her offspring resulting in a disturbed tolerance development. When the same antigen was introduced to the mother at normal weaning age she produced a tolerant offspring. In addition it was demonstrated that feeding an antigen from weaning and only for two weeks was sufficient to transfer tolerance to the next generation. This implies an education of the immune system of the offspring in avoiding hypersensitivity reactions to "harmless" environmental antigens.

ACKNOWLEDGEMENTS

The skillful technical assistance of Ms Helena Kahu is very much appreciated. This study was supported by the Swedish Medical Research Council.

REFERENCES

1. A. E. Wold, U. I. H. Dahlgren, S. Ahlstedt, and L. A. Hanson, *Monogr. Allergy* 24:251 (1988).
2. S. Strobel and A. Ferguson, *Pediatr. Res.* 18:588 (1984).
3. D. G. Hanson, *J. Immunol.* 127:1518 (1981).
4. E. Telemo, M. Bailey, B. G. Miller, C. R. Stokes, and F. J. Bourne, *Scand. J. Immunol.* 34:689 (1991).
5. E. Telemo, 1. Jacobsson, B. Westrom, and H. Folkesson, *Immunology* 62:35 (1987).
6. C. Pathriana, N. J. Goulding, M. J. Gibney, M. Jennifer, P. J. Gallagher, and T. Taylor, *Int. Archs. Allergy Appl. Immunol.* 66:114 (1981).

THE EFFECT OF CALORIC SUPPLEMENTATION ON LEVELS OF MILK IgA ANTIBODIES AND THEIR AVIDITIES IN UNDERNOURISHED GUATEMALAN MOTHERS

M.V. Herías,[1,2] J. R. Cruz,[2] T. González-Cossío,[2] F. Nave,[2] B. Carlsson,[1] and L.Å. Hanson[1]

[1]Department of Clinical Immunology, University of Gotegorg, Goteborg, Sweden; [2]Institute for Nutrition of Central America and Panama, INCAP, Guatemala City, Guatemala

INTRODUCTION

Undernutrition affects the immune response and, therefore, might worsen the outcome of infections.[1,2] This is especially important in developing countries where the prevalence of infectious diseases is high.

Human milk is necessary for the young infant because it provides a well balanced nutrition and protects against infections.[3-7] Some of the major protective factors are secretory IgA (S-IgA) and lactoferrin. It has not been proven that these factors are negatively affected in the malnourished mother's milk.

In this study, we investigated the effect of two different levels of caloric supplementation on the expression of antibodies against three microbial antigens and the concentration and daily output of total S-IgA and lactoferrin in the milk of moderately undernourished mothers.

MATERIALS AND METHODS

Milk samples were obtained from moderately undernourished mothers living in the poorest sections of a region in the western Guatemalan highlands. Altogether, 70 women were in a random blind design divided in two groups. Group A received a high caloric supplement (500 kcal/day) and Group B received a low caloric one (140/kcal/day). The supplement was given for 20 weeks, beginning at 5 weeks post-partum. Milk samples were obtained at week 5, 10, 20 and 25, that is 0, 5, 15, and 20 weeks after the supplementation was given.

We evaluated the level and avidity of specific antibodies against tetanus toxoid, *E. coli* 06 and a pool of 10 common *E. coli* O antigens, as well as the concentration and daily output of lacteoferrin and total S-IgA. The determinations were done using the ELISA in various modifications, except for lactoferrin which was quantified by single radial immunodiffusion.

Advances in Mucosal Immunology, Edited by
J. Mestecky *et al.,* Plenum Press, New York, 1995

RESULTS

In the low-calorie supplemented group the S-IgA concentration decreased significantly form week 5 to 25, but remained within the normal range. Furthermore, the S-IgA concentration as well as the 24 h output differed significantly between the two groups, being lower in group B (Table 1). On the other hand, the concentration of lactoferrin and the avidity and antibody levels against the antigens tested, showed no significant effect of the supplementation (Table 1). Fluctuations during the study period could be seen for the antibody levels in both groups, especially a significant increase in the low-caloric group towards *E.coli* 06. However, no significant differences could be observed between the groups.

Table 1. The effect of supplementation on the milk factors evaluated within each group and between groups. Group A represents the high-calorie supplemented group and group B represents the low-calorie supplemented groups.

| | | | Effect of supplementation | |
| | | | Within the group | Between A and B |
Milk factors	Group	n	p value	p value
S-IgA				
g/L	A	30	NS	p<0.05
	B	25	p<0.05	
g/24 h	A	30	NS	p<0.05
Lactoferrin				
g/L	A	31	p<0.05	NS
	B	25	p<0.05	
g/24 h	A	31	NS	NS
	B	25	NS	
***E. coli* 06**				
Level	A	39	NS	NS
	B	27	p<0.05	
Avidity	A	30	NS	NS
			p<0.05	
Tetanus				
Level	A	29	p<0.05	NS
	B	25	NS	
Avidity	A	26	NS	NS
	B	19	NS	

DISCUSSION

Decreased S-IgA concentration and daily S-IgA output were the only findings associated with the low- versus the high-caloric supplementation treatment, although both groups of women showed S-IgA levels within the normal range.[3,8] There were significant fluctuations for antibodies to all three antigens studies, which may be due to several factors beyond the nutritional situation, including varying antigen exposure, variations in milk volume, the state of hydration of the mother, effects of intercurrent infections or varying bacterial colonization of the mother's gut. The antibody levels showed no significant differences between the two supplemented groups. A similar pattern of fluctuation was also observed in earlier studies.[9] The avidity indices against the pool of

10 *E. coli* antigens was lower in both groups compared to the levels reported by Roberton *et al.*[10] in Pakistani and Swedish mothers. The exact reason for the lower avidity are not known. On the other hand, the mean avidity for anti-*E. coli* 06 were comparable to the ones obtained from well nourished Costa Rican and Swedish mothers in a recent study.[11] The basis for the observed fluctuations remain unclear to us, although it may relate e.g., to geographical differences in exposure to the *E. coli* serotypes.

CONCLUSIONS

We found that S-IgA was lower in the low- compared to the high-supplemented group, but we were unable to show an impairment in the antibody response to specific antigens. Supplementation did not enhance the levels of antibodies to the specific antigens, but prevented the decrease in the content of total S-IgA in milk. We consider it of importance that the specific antibody levels were not hampered and could even increase in spite of the mother's poor nutritional condition.

ACKNOWLDGEMENTS

We thank Monika Ericsson-Schmidt, Lisbeth Larsson and Behrit Bahr for their assistance in past of the laboratory determinations. The study was supported by the Swedish Agency for Research Cooperation with Developing Countries (SAREC). the Ellen, Walter and Lennart Hesselman's Foundation and by the Regional Office for Central America and Panama (ROCAP) of the US Agency for International Development.

REFERENCES

1. R. K. Chandra, *Acta Paediatr. Scand.* 69:137 (1979).
2. A. M. Tomkins, *Proc. Nutr. Soc.* 45:289 (1986).
3. B. Lonnerdal, *Am . J. Clin. Nutr.* 42:1299 (1985).
4. A. Prentice and A. A. Paul, *in* :"Human Lactation 4. Breastfeeding, Nutrition, Infection and Infant Growth in Developed and Emerging Countries", S. A. Atkinson, L. Å. Hanson, and R. K. Chandra, eds., pp. 87-101, ARTS Biomed Publ Distr Ltd, St. John's, Newfoundland (1990).
5. G. M. Ruiz-Palacios, J. J. Clava, L. K. Pickering, Y. Lopez-Vidal, P. Volkow, H. Pezzarossi, and M. Stewart West, *J. Pediatr.* 116:707 (1990).
6. C. G. Victoria, J. P. Vaughan, C. Lombardi, S. M. C. Fuchs, L. P. Gigante, P. G. Smith, L. C. Nobre, A. M. B. Teixeira, L. B. Moreira, and F. C. Barros, *Lancet* ii:319 (1987).
7. R. I. Glass, A. M. Svennerholm, B. J. Stoll, M. R. Khan, K. M. Belayet Hossain, M. I. Huq, and J. Holmgren, *N. Engl. J. Med.* 308:1389 (1983).
8. A. S. Goldman, C. Garza, B. L. Nichols, and R. M. Goldblum, *J. Pediatr.* 100:563 (1982).
9. J. R. Cruz, C. Arevalo, and L. Å. Hanson, *in*: "Human Lactation 2. Maternal and Enviromental Factore", M. Hamosh and A. S. Glodman, eds., pp.569-79, Plenum Press, New York, (1986).
10. D. M. Roberton, B. Carlsson, K. Coffman, M. Hahn-Zoric, F. Jalil, C. Jones, and L. Å. Hanson, *Scand. J. Immunol.* 28:783 (1988).
11. L. Å. Hanson, F. Jalil, R. Ashraf, S. Bernini, B. Carlsson, J. Cruz, T. Gonzalez, M. Hahn-Zoric, L. Mellander, Y. Minoli, G. Moro, F. Nave, S. Zaman, L. Matta, J. Karlberg, and B. Lindblad, *in*: "Immunology of Milk and the Neonate", J. Mestecky, C. Blair, and P.L. Ogra eds., pp.1-15, Plenum Press, New York, (1991).

THE EFFECT OF AN ELEMENTAL DIET ON GUT-ASSOCIATED LYMPHOID TISSUE (GALT) IN RATS

Soichiro Miura, Hiroshi Serizawa, Hirokazu Tashiro,
Hiroyuki Imaeda, Hiroshi Shiozaki, Nobuyuki Ohkubo,
Shin Tanaka and Masaharu Tsuchiya

Department of Internal Medicine, School of Medicine
Keio University, Tokyo, Japan

INTRODUCTION

Intestinal gut-associated lymphoid tissue (GALT) IS constantly exposed to various antigenic stimuli in the gut lumen. Various dietary substances are known to have a great influence on the intestinal lymphatic system which plays an important role in intestinal mucosal immunity. For example, in our previous study,[1] we reported that a remarkable increase in lymphocyte transport was induced during lipid adsorption in rat mesenteric lymphatics.

An elemental diet (ED) mainly consists of non-antigenic amino acids and oligosaccharides. ED has been known to be effective for maintenance of the nutritional state in patients with malabsorption syndrome, especially for Crohn's disease in inducing a remission by "resting bowel".[2,3] There is the possibility that ED may have some beneficial effect on immunological disturbances in Crohn's disease. However, there is little experimental information as to how ED directly affects the immunological function of the intestine.

In this study, we therefore investigated the effect of ED on GALT, especially lymphocyte transport, in intestinal lymph and IgA production by intestinal mucosa in rats.

METHODS

Male Wistar rats weighing 180 g were used for the experiments. Rats were fed with ED (Elemental®, Morishita Pharmaceutical Co. Ltd. Tokyo) for 4 weeks. Conventional diet (CRF-1®, Charles River Japan Co. Ltd). was given to the control group. Composition of ED was: amino acids 16.4 %, oligosaccharide 70.4 %, linoleic acid 0.3%, minerals, vitamins and trace elements. Free access to water was permitted.

After 4 weeks, intestinal lymphatics near cisterna chyli were cannulated according to Bollman et al.[4] under anesthesia with intraperitoneal injection of sodium pentobarbital. Intestinal lymph was collected hourly, lymph flow and lymphocyte output in intestinal lymph were measured. The number of lymphocytes was counted by Bürker-Türke plate.

Blastogenesis of lymphocytes in intestinal lymph was compared in both groups. Lymphocytes were suspended in RPMI 1640 medium with 10 % fetal calf serum at the concentration of 1.0×10^6 cells/ml. Two hundred μl of cell suspension was added to individual wells of microtiter plate with PHA (Sigma Chemical Co. Ltd., USA) at a concentration of 10 μg/ml. The cell cultures were incubated in 95 % O_2 and 5 % CO_2 at 37°C for 72 h. [^{3}H]-methyl thymidine (New England Nuclear Co., USA) was added and cells were harvested by multiple cell harvester 5 h later. TCA-precipitated radioactivity was measured and stimulation index (S.I.) was expressed as follows;

S.I. = dpm with stimulation of PHA/dpm without stimulation of PHA

Four weeks later, morphological alteration of intestinal mucosa was examined. Rats were killed, small intestine was resected and cut off along the mesentery. The number of visible Peyer's patches was counted and total area of Peyer's patches was measured. The tissues were fixed in PLP (periodate, lysine-paraformaldehyde) solution, embedded and frozen in dry-ice and acetone before sections were cut by Cryostat. Sections were stained for IgA-containing cells by direct immunoperoxidase technique using horseradish peroxidase-labeled sheep anti-IgA (Fc) (Binding Site Ltd., England). After counterstaining of nuclei with methyl-green, the slides were observed under a light microscope and peroxidase-positive cells were counted per 1000 mononuclear cells in the lamina propria.

RESULTS AND DISCUSSION

The amount of calorie intake per day was not different between ED and control groups. Body weight gain showed no difference between two groups during 4 weeks as shown in Table 1. No significant morphological alteration appeared to occur in the intestinal mucosa of ED-fed rats. When size of villus height to crypt depth was compared in both groups there was no significant change in this ratio. The number and total area of Peyer's patches also showed no significant alterations between ED and control groups.

Table 1. Comparison of body weight gain, villus and crypt ratio and area of Peyer's patch between ED-fed and control rats.

	Control	Element Diet
Body weight gain (for 4 weeks) (g)	169.4 ± 10.5[1]	157.9 ± 8.7
Area of Peyer's patches of ileum (mm^2)	150.6 ± 57.3	132.5 ± 54.1
Villus/crypt ratio	1.64 ± 0.22	1.65 ± 0.55

[1]Mean ± S. D., N.S.: not significant

As shown in Fig. 1, the intestinal lymph flow rate was not changed, however, lymphocyte transport through intestinal lymph was significantly decreased in the ED-fed group. Stimulation index of lymphocytes from intestinal lymphatics was 28.4 ± 6.0 in control rats and 24.7 ± 7.4 in ED-fed rats. There was no significant difference in spontaneous blastogenesis and stimulation index between the two groups.

From these observations, the significant decrease in lymphocyte transport through intestinal lymph appeared not to be secondary to the morphological changes in the intestinal mucosa itself.

As shown in Table 2, the number of IgA-containing cells in the lamina propria of the ileum was significantly decreased in ED-fed rats by morphometrical analysis.

We have previously reported that rats on total parenteral nutrition for 2 weeks, there is a remarkable reduction in lymphocyte transport and a decrease in IgA-containing cell number in the lamina propria of intestine with significant atrophic changes of Peyer's patches.[5] When compared with total parenteral nutrition, decreases in lymphocyte transport and in IgA-containing cell numbers in the mucosa were relatively moderate in ED-fed rats, and atrophy of Peyer's patches not occur in ED-fed rats. In the case of ED, mitogen response of lymphocytes in intestinal lymph was also well maintained. It is conceivable that ED-treatment will be effective for intestinal inflammation in case of Crohn's disease to ameliorate the excessive immune response in the gut by its moderate immunological inhibition.

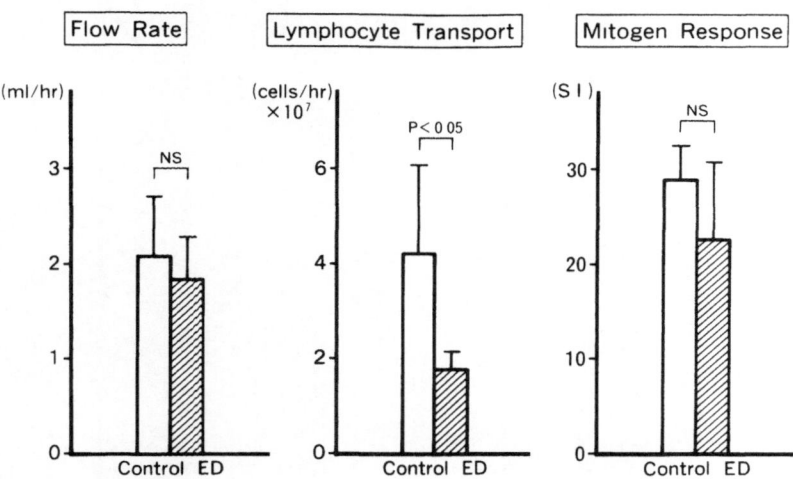

Figure 1. Effect of elemental diet (ED) feeding on flow rate, lymphocyte transport number and lymphocyte blastogenesis in response to PHA in rat intestinal lymphatics.

Table 2. IgA-containing cell number in the lamina propria of ileum in ED-fed rats.

	Control	Element Diet
IgA-containing cells (per 1000 cells)	210.0 ± 20.0	154.1 ± 30.2[1]

[1]p < 0.05 as compared with controls. Mean ± S.D. of 7 animals.

Absence or decrease in dietary stimulation to intestinal mucosa could induce significant changes in GALT, especially by inhibition of lymphocyte transport from intestinal mucosa and the appearance of IgA-containing cells in the intestinal mucosa. As we have previously demonstrated, fat absorption is thought to be an important factor influencing the lymphocyte migration through GALT.[6] Inhibitory effect of ED feeding on intestinal lymphocyte traffic and IgA production may be related to the low concentration of lipids in ED-diets. Further investigations are necessary to elucidate the exact mechanisms of immunological effect of ED on GALT.

REFERENCES

1. S. Miura, E. Sekizuka, H. Nagata, C. Oshio, H. Minamitani, M. Suematsu, M. Suzuki, Y. Hamada, K. Kobayashi, H. Asakura, and M. Tsuchiya, *Am. J. Physiol.* 253:G596 (1987).
2. C. O'Morain, A. W. Segal, and A. J. Levi, *Br. Med. J.* 288:1859 (1984).
3. S. J. D. O'Keefe, J. Ogden, and J. Dicker, *J. parent. nutri.* 13:455 (1989).
4. J. L. Bollman, J. C. Cain, and J. H. Grindlay, *J. Lab. Clin. Med.* 33:1349 (1948).
5. S. Tanaka, S. Miura, H. Tashiro, H. Serizawa, Y. Hamada, M. Yoshioka, and M. Tsuchiya, *Cell Tissue Res.* 266:29 (1991).
6. K. Kobayashi, S. Miura, Y. Hamada, H. Asakura, and M. Tsuchiya, *In*: "Progress in Lymphology X". J. R. Casley-Smith and N. B. Piller, Eds., p. 94, University of Adelaide Press, Adelaide, Australia (1985).

EFFECT OF BACTERIAL ANTIGENS ON LOCAL IMMUNITY

Ch. Ruedl, G. Wick and H. Wolf

Institute for General and Experimental Pathology,
University of Innsbruck, Medical School, Fritz-Pregl
Strasse 3, A-6020 Innsbruck, Austria

INTRODUCTION

The ability of intestinal lymphoid cells to migrate to, and localize in, other mucosal tissues, e.g. the respiratory and genitourinary tract, and other secretory glands, underlines the relevance of the gut-associated lymphoid tissue (GALT) in the local immune response. Following antigen presentation, activated cells from the GALT pass from the lymphatics to the blood circulation via the ductus thoracicus, disseminate into distant mucosal tissues where the cells further proliferate and differentiate into memory and effector cells. In fact, it is known that enteric ingestion of microbial antigens induces S-IgA responses in remote external secretions.[1, 2]

Based on this concept, our interest is to induce an immune response at mucosal surfaces where the infection actually occurs, in the respiratory tract. For this purpose we applied a bacterial lysate (LUIVAC®), consisting of components of common respiratory microorganisms, as an immunomodulator.

We have demonstrated previously that oral application of LUIVAC® leads to a primed state of polymorphonuclear leukocytes (PMN) for increased oxidative burst activity in rabbits.[3] Moreover, we showed an increase of serum IgA and IgA-producing cells in Peyer´s patches (PP) of Swiss and nude mice after oral administration of this bacterial lysate.[4] The aim of this study was to follow the kinetics of lamina propria lymphocytes (LPL) isolated from the small intestine of orally immunized mice. Using fluorescence or radioactively labelled donor cells, we were able to monitor the migration and tissue distribution *in vivo* after i.v. injection of these cells into syngeneic recipients.

MATERIAL AND METHODS

Antigens and Immunization Protocol

LUIVAC®, kindly provided by M. Ellenrieder, Luitpold-Pharma GmbH, Munich, FRD, contains lysates of at least 1010/g of each of the following heat-inactivated lyophilized bacteria: *Staphylococcus aureus, Streptococcus pyogenes, Streptococcus mitis, Streptococcus pneumoniae, Haemophilus influenzae, Branhamella catarrhalis and Klebsiella pneumoniae.*

Female BALB/c mice (5-7 weeks old) were immunized orally for 5 consecutive days with 15 mg of the bacterial lysate LUIVAC®/kg body weight. The first group of animals was sacrificed at day 6. After a 10-day interval the immunization was repeated for another 5 days and then the second group was killed on day 21.

Isolation of Intestinal Lamina Propria Lymphocytes

Lymphocytes were isolated from the small intestinal lamina propria by a modification of the method previously described by Davies and Parrot.[5] Splenocytes from the same animals served as control. The lumen of the intestines was washed with Hanks' salt solution w/o Ca^{++} and Mg^{++} and the gut was cut into small pieces. After 2 h incubation with EDTA- (5mM) to remove the epithelium, the pieces were enzymatically digested with 10 U/ml of purified collagenase from *Cl. histolyticum* (CLSPA, Worthington Biochemical Corp., Freehold, NJ) to obtain the final cell suspension.

Cell Phenotype Analysis

To characterize the cell subpopulations of isolated LPL we examined the expression of different surface markers, e.g.Th1.2, CD4, CD8, B220, sIg and adhesion molecules of the ß2-integrin family (LFA-1, yMAC-1, p150-95, CD18) by FACS-analysis. The monoclonal antibodies used as culture supernatant were generously provided by N. Romani, Department of Dermatology, University of Innsbruck, Medical School, Austria.

In vivo Lymphocyte Migration

To assess lymphocyte migration *in vivo* two different methods were used: one was based on radioactive cell labelling (^{51}Cr), the other used fluorescence (supravital nuclear fluorochrome Hoechst H33342). Ten million labelled cells were injected into the tail vein of each syngeneic mouse. The recipients were sacrificed after 2 hours, and organs, such as spleen, liver, lungs, PP, mesenteric lymph nodes (MLN) and kidneys were harvested and screened for the presence of labelled cells. In case of ^{51}Cr-labelled cells, the radioactivity was counted and calculated as cpm/g organ or tissue; the fluorochrome stained cells were counted on 4-mm thick frozen sections in a fluorescence microscope.

Characterization of Fluorescent Stained Cells in Lung Sections

To characterize the H33342-stained cells on lung frozen sections a double immunofluorescence staining was performed using a biotinylated Thy1.2 to identify T-cells and a FITC-conjugated anti-mouse Ig to identify B-cells.

RESULTS AND CONCLUSIONS

The main cell populations found in the lamina propria of small intestine are lymphocytes (65%); $67 \pm 9\%$ of these were T-cells and $15 \pm 3\%$ B-cells, respectively. Within the T cell population equal numbers of $CD4^+$ and $CD8^+$ stained cells were identified; $85 \pm 5\%$ expressed the α-chain CD11a of LFA-I and about $80 \pm 7\%$ expressed the ß-chain CD18. The non-lymphocyte population consists of $35 \pm 3\%$ MAC-1+ and $20 \pm 2\%$ of these expressed the α-chain CD11c of the integrin p150-95. With the exception of an increased presence of B-cells (30%), no difference of surface marker expression on lymphocytes prepared from immunized mice compared to the untreated control could be observed.

The distribution of radiolabelled LPL and splenocytes in different organs is shown in Fig. 1. Two hours after i.v. injection control splenocytes were preferentially localized in the spleen (40% ; Fig. 1b), whereas LPL were found in the lung (55% ; Fig 1a). Labelled cells were detected also in liver (20%), but only a small proportion reached PP, MLN and kidneys. Furthermore, the radioactivity recovered in the lung of animals that received LPL isolated from immunized donors was increased as compared to the untreated controls. In fact, significant rises of 25% and 35% ($p<0.015$ and $p<0.005$, respectively) were observed in both immunized groups.

A similar migratory pattern could be demonstrated with fluorochrome-labelled cells, as illustrated in Fig. 2. In lungs, a significantly increased number of LPL obtained from immunized mice in comparison to control groups was observed: 23 and 34 cells/viewing field of both immunized groups compared to the 12 cells/viewing field of the untreated group (Fig. 2a). No difference in homing pattern of untreated and stimulated cells could be detected in the case of splenocytes (Fig. 2b).

The fluorochrome-stained LPL localized in the lung after i.v. injection could be characterized as 60% Thy 1.2+, 20% sIg+, and 20% remaining unstained with these antibodies.

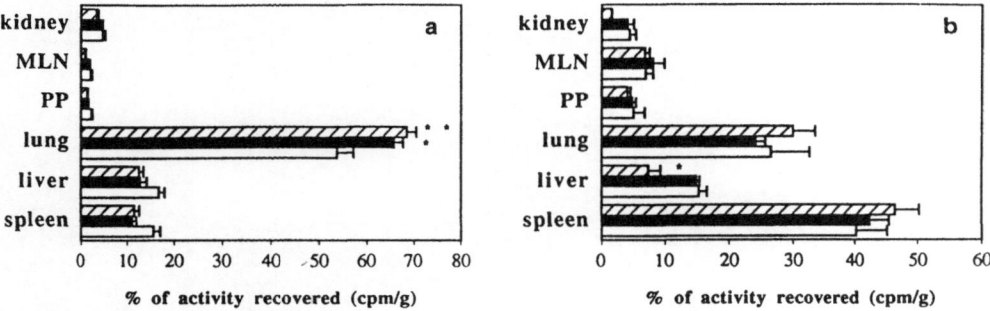

Figure 1. Distribution of lamina propria (a) and spleen cells (b) isolated from control and immunized mice in syngeneic recipients using radiolabelling technique.
Black bars : control; hatched bars: 15 mg LUIVAC®/kg (one vaccination period); white bars: 15 mg LUIVAC®/kg (two vaccination periods).
Bars represent the percentage of recovered radioactivity (cpm/g) and indicate mean ± SEM of 7 mice per group of three separate experiments. * and ** indicate values significant different from controls, $p < 0.015$ and < 0.005, respectively.

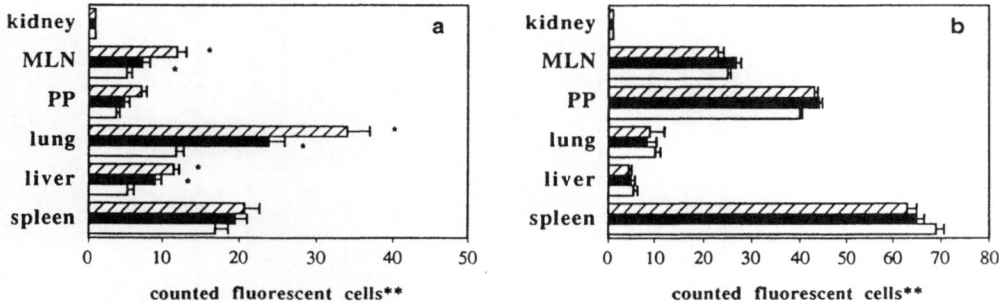

Figure 2. Distribution of lamina propria (a) and spleen cells (b) isolated from control and immunized mice in syngeneic recipients using fluorochrome H33342 labelling technique. Black bars : control; hatched bars: 15 mg LUIVAC®/kg (one vaccination period); white bars: 15 mg LUIVAC®/kg (two vaccination periods).
** 10 microscopic fields were counted and the mean calculated. Bars indicate mean ± SEM of 8 mice per group of three separate experiments.
* indicates values significantly different from controls, $p < 0.001$.

The aim of this study was to investigate whether single or repeated oral application of bacterial lysate from pneumotropic bacteria influences the migratory behaviour of intestinal LPL. Our results indicate a different localization pattern *in vivo* of LPL isolated from immunized mice compared to the untreated controls, shown as an increased presence of labelled cells in the lung.

In conclusion, this study shows that LPL migration *in vivo* can be modulated by oral administration of LUIVAC®. These observations may have important practical and clinical implications for the development of new vaccination strategies for protection against respiratory infections.

REFERENCES

1.   J. R. McGhee, J. Mestecky, M. T. Dertzbaugh, J. H. Eldridge, M. Hirasawa, and H. Kiyono, *Vaccine* 10:75 (1992).

2.   J. Holmgren, C. Czerkinsky, N. Lycke, and A. Svennerholm, *Immunobiology* 184:157 (1992).
3.   A. Helmberg, G. Böck, H. Wolf, and G. Wick, *Infect. Immun.* 57:3576 (1989).
4.   M. Frühwirth, C. Ruedl, G. Wick, and H. Wolf, (submitted for publication).
5.   M. D. J. Davies, and D. M. V. Parrot, *Gut* 22:481 (1981).

INTESTINAL AND SYSTEMIC IMMUNE RESPONSES IN RATS TO DIETARY LECTINS

Enrique Gómez,[1] Víctor Ortiz,[1] Javier Ventura,[2]
Rafael Campos,[2] and Héctor Bourges[1]

[1]Departmento de Fisiología de la Nutrición, Instituto
Nacional de la Nutrición; and [2]Escuela Superior de
Medicina, Instituto Politécnico Nacional México, DF
Mexico

INTRODUCTION

Common beans *(Phaseolus vulgaris)* constitute a staple food for most of the World's population. Nevertheless, it has been known for a long time that common beans contain a wide variety of substances which may be considered as "anti-nutritional", unless they are properly inactivated or destroyed.[1-3] Among substances that are inactivated by the cooking process are the lectins, a especial type of proteins that display sugar-binding properties.[4-6]

The bean lectin, phytohemagglutinin (PHA), administered in a raw form to animals, causes multiple physiological and metabolic disturbances, such as disruption of the intestinal microvilli, malabsorption of some nutrients by a mechanism of non-especific interference, decreased activity of membrane-associated enzymes, negative nitrogen balance, an increase in the size and weight of different organs, growth retardation or failure to thrive, and even death.[7-12] It has been shown that PHA is resistant to the digestive secretions of the gastrointestinal tract since it has been recovered in an intact form from feces of animals fed PHA.[10,12]

From the immunological point of view, the gastrointestinal tract plays an important role in the body's mechanisms of defence, since it limits the entry of many potentially antigenic substances without interfering with the normal processes of digestion and absorption.[13]

Due to the high resistance of many lectins to thermal inactivation, the diet represents the main source of exposure to these proteins.[14] There is a possibility that PHA is not completely destroyed during cooking and may produce different disturbances to the gut-associated immune system.

The main objective of this study was to describe the *in vivo* effects of PHA on the cellular and humoral immune responses both at the systemic and at the mucosal levels, as well as to describe its binding pattern in the intestine, and to determine the effect of heat treatment (autoclaving at $121°C$, 15 lb/in^2, for 30 min) in these biological activities.

MATERIALS AND METHODS

Common bean seeds *(Phaseolus vulgaris)* of the Brown variety were obtained from a local market in the México city area. They were grounded to a fine powder and its soluble components were extracted by overnigth stirring in PBS at 4°C. The suspension was filtered through cheese cloth and centrifuged at 16,000xg for 2 h. Part of this extract was autoclaved (121°C, 15 lb/in², 30 min) to obtain the "cooked" extract; the other part of the extract was filtered through 0.22μm Millipore membranes to obtain the "raw" extract. The total protein in both extracts was quantified by the Lowry method and adjusted to 5 mg/ml.

Female Wistar rats (2-3 months old) received either the raw or cooked extracts (1 ml) daily by gastric intubation with a "balloon tipped" canulae for 3, 7 and 14 days. Control rats received PBS only.

After completion of each period, blood was obtained by cardiac puncture under ether anesthesia and animals were killed by cervical dislocation. The abdominal region was exposed and the intestine, mesenteric lymph nodes, and spleen were obtained. Lymphocytes were isolated from the intestine (I), Peyer's patches (PP), mesenteric lymph nodes (MN) and the spleen (S) by standard methods.[15]

Serum IgG and IgM anti-PHA antibodies were assayed by an indirect ELISA method in polyestyrene microtitration plates, coated with commercially purchased purified PHA. Rabbit anti-rat IgG and IgM-HRP conjugates were used to detect rat IgG and IgM, respectively.

Lymphoid cells from the I, PP, MN and S, were cultured without the addition of PHA, and the uptake of tritiated thymidine (^{3}H-TdR) in 3-day-cultures was determined. Cells cultured without PHA reflected their degree of *in vivo* activation.

PHA-binding pattern along the intestine was assessed by an immunohistochemical peroxidase technique. Different sections of the intestine were tested: proximal (duodenum), medial (jejunum) and distal (ileum).

RESULTS

Serum IgG and IgM anti-PHA antibodies appeared in a time-dependent fashion. IgM anti-PHA antibodies appeared at day 3 and reached its maximal at day 7, while IgG anti-PHA antibodies, appeared at day 7 and remained constant until day 14 (Figs. 1,2).

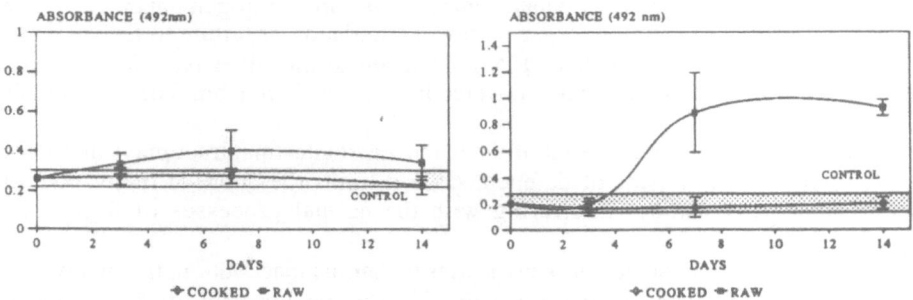

Figure 1. IgM anti-PHA antibodies   **Figure 2.** IgG anti-PHA antibodies

Total mononuclear cells (TMNC) obtained from the intestine (I), Peyer's patches (PP), mesenteric lymph nodes (MN) and spleen (S) of rats fed raw extracts, were activated *in vivo* by the PHA (Fig. 3). No activation was found in rats fed PBS nor in rats fed cooked extract (Fig. 4).

TMNC were activated on day 3 in both the PP and MN, and also at day 7 in the S; on day 14 TMNC were activated in all organs.

PHA was bound only in the proximal region (duodenum) of the small intestine and in the upper third of the intestinal villi, in rats fed raw extracts (Table 1). No PHA was detected in the intestine of rats fed PBS or cooked extracts.

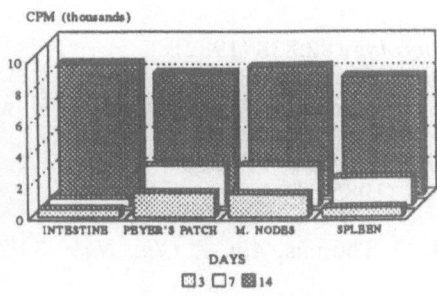

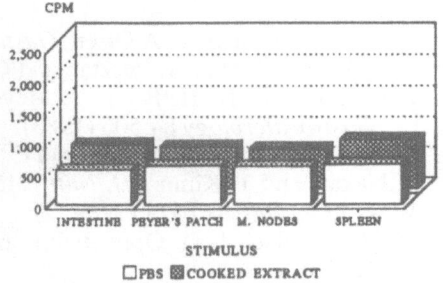

Figure 3. Cellular proliferation of TMNC rats fed raw extracts

Figure 4. Cellular proliferation of TMNC rats fed PBS or cooked extract

Table 1. Binding pattern of PHA to the intestine of rats fed raw saline extracts.[1]

TIME OF ADMINISTRATION (Days)	Duodenum	Jejunum	Ileum
3	+++	-	-
7	++	-	-
14	+	-	-
Control	-	-	-

[1] Intensity of staining was defined as negative (-), weak (+), positive (++) and strongly positive (+++)

DISCUSSION

PHA administered to rats in the form of raw saline extracts induced a specific immune response, both at the systemic and mucosal levels.

Serum IgG and IgM anti-PHA antibodies appeared only in rats fed raw extracts in a time-dependent fashion. This finding indicates that PHA can cross the intestinal barrier, as previously reported.[16]

The activation pattern observed in TMNC from rats fed raw extracts is the same for the well-known "Common Mucosal Immune System".[13,16,17]

PHA in the raw extracts was found only in the proximal region of the small intestine (duodenum). PHA-receptors are present only at this site and are restricted to mature epithelial cells (at the tip of the villi).

Heat treatment (autoclaving) was effective in destroying all the biological activities displayed by PHA.

REFERENCES

1. I. E. Liener, *Am. J. Clin. Nutr.* 11:281 (1962).
2. W. G. Jaffé, *in:* "Toxic Constituents of Plant Foodstuffs", I.E. Liener, ed., p. 73, Academic Press (1980).
3. W. G. Jafféy and M. E. Flores, *Arch. Latinoamer. Nutr.* 25:79 (1975).
4. N. Sharon and H. Lis, *Science* 177:949 (1972).
5. H. Lis and N. Sharon, *Annu. Rev. Biochem.* 55:35 (1986).
6. N. Sharon and H. Lis, *FASEB J.* 4:3198 (1990).
7. A. Pusztai, E. M. Clarke, and T. P. King, *Proc. Nutr. Soc.* 38:115 (1979).

8. V. Lorenzsonn and W. A Olsen, *Gastroenterology* 82:838 (1982).
9. J. T. A. de Oliveira, A. Pusztai, and G. Grant, *Nutr. Res.* 8:943 (1988).
10. J. G. Banwell, D. H. Boldt, J. Meyers, F. L. Weber, B. Miller, and R. Howard, *Gastroenterology* 84:506 (1983).
11. D. A. Donatucci, I. E. Liener, and C. J. Gross, *J. Nutr.* 117:2154 (1987).
12. S. Nakata and T. Kimura, *J. Nutr.* 115:1621 (1985).
13. W. F. Doe, *Gut* 30:1679 (1989).
14. M. S. Nachbar, J. D. Oppenheim, and J. O. Thomas, *Am. J. Clin. Nutr.* 33:2338 (1980).
15. P. J. van der Heijden and W. Stock W, *J. Immunol. Meth.* 103:161 (1987).
16. H. J. de Aizpurua and G. J. Rusell-Jones, *J. Exp. Med.* 167:440 (1988).
17. C. O. Elson, M. F. Kagnoff, C. Fiocchi, A. D. Befus, and S. Targan, *Gastroenterology* 91:746 (1986).

THE EFFECT OF AN ANTI-GLUCOCORTICOID (ZK 98299) ON THYMUS EVOLUTION AND ON HYDROCORTISONE-INDUCED THYMOLYSIS, INTESTINAL BRUSH-BORDER ENZYMES AND THEIR DESIALYLATION IN SUCKLING RATS

Jirí Kraml,[1] Jirina Kolínská,[2] Libuse Kadlecová,[1] Marie Zákostelecká,[2] Dana Hirsová,[1] and Vratislav Schreiber[1]

[1]The First Faculty of Medicine, Charles University
[2]Institute of Physiology, Czech Academy of Sciences
 Prague, Czech Republic

INTRODUCTION

The postnatal developmental changes in the activities of intestinal brush-border enzymes and the desialylation of intestinal brush-border proteins in infant rats are enhanced or precociously induced by glucocorticoids. During ontogeny the physiological changes in the small intestine are linked to the period of weaning, starting at the end of the 3rd week of life.[1-3] Thymolysis is induced by glucocorticoid hormones.[4] Sialylation of the intestinal surface may be related to mucosal immunology, because the bound sialic acid serves as a receptor for some viruses and bacteria or may exhibit a masking effect for membrane receptors. Thymus function may also indirectly influence the immunological mechanisms of intestinal mucosa. In both cases, i.e. for the mucosal enzymes or proteins and for thymocytes we face a similar problem related to the glucocorticoid hormones. If these hormones enhance the physiologically occurring phenomena during postnatal development, are the endogeneously secreted glucocorticoids also responsible for the physiological regulation of these postnatal changes?

The answer to this problem may be partially found in the effect of anti-glucocorticoids, which would not only suppress the effect of exogeneously given hormones, but they would act also as antagonist agents supressing the postnatal development of some functions in which the involvement of endogeneously secreted glucocorticoids is assumed.

ZK 98299 (Onapristone, Schering, Berlin) belongs to 11-beta-aryl-substituted, 4,9-estradiene compounds with anti-glucocorticoid and anti-progestational effects.[4] It is analogous to the better known drug RU 38486 (Mifepristone, Roussel-Uclaf, France); however, it possesses a weaker anti-glucocorticoid activity. We have tested the anti-glucocorticoid activity of Onapristone on the following parameters of infant male rats:

1. Antagonistic effect on the precocious induction of intestinal brush-border enzymes by hydrocortisone (HC), and influence of the drug (ZK) on the physiological development of these enzymes during the early postnatal period.

2. Antagonistic effect on the enhancement of desialylation of brush-border proteins and enzymes by HC, which we have discovered before[3], and the influence of the drug on the physiological desialylation during the early postnatal period.

3. The antagonistic effect on thymolysis induced by HC and the influence of the drug on thymus growth during postnatal development.

Advances in Mucosal Immunology, Edited by
J. Mestecky *et al.*, Plenum Press, New York, 1995

HC was injected to infant male rats s.c. in a dose 1 mg on day 9 and 0.5 mg on day 10. Onapristone was given on day 9, 10 and 11 either alone, or simultaneously with HC (the latter on day 9 and 10). The animals were killed either on day 12 for the evaluation of the short-term effect, or on day 16-17 for the evaluation of the delayed effect.

RESULTS

Intestinal Brush-Border Enzymes

Table 1 shows induction and marked increase of sucrase and glucoamylase 3 days after the first HC injection, and suppression of this effect by the anti-glucocorticoid. However, the activity of glucoamylase was suppressed also by Onapristone alone. Intestinal lactase showed a similar pattern.

Dipeptidyl peptidase IV (DP IV) was suppressed by Onapristone and the drug exhibits an antagonistic effect on the induction of gamma-glutamyl transferase (GMT) by hydrocortisone. In the majority of enzymes tested, the antiglucocorticoid not only suppresses the increase of activities after hydrocortisone injection, but given alone it also decreases the activities of several brush-border enzymes during the short-term effect.

Table 1. Enzyme activities in the homogenates of the small-intestinal mucosa of 12-day-old male rats (nkat/mg prot.).

Enzyme	Controls	HC	ZK	HC+ZK
Sucrase	0.069 ± 0.051	** 1.257 ± 0.451	0.039 ± 0.038	** 0.814 ± 0.297
Amylase	0.295 ± 0.082	** 1.850 ± 0.373	** 0.181 ± 0.037	** + 1.397 ± 0.267
Lactase	2.146 ± 0.730	** 3.493 ± 0.900	** 1.130 ± 0.099	+ 2.316 ± 0.327
DP IV	0.463 ± 0.086	0.523 ± 0.145	0.359 ± 0.080	* 0.595 ± 0.079
GMT	0.731 ± 0.129	** 3.17 ± 0.688	0.720 ± 0.164	* ++ 1.300 ± 0.388

t-test vs. controls	*	$p<0.05$	HC = hydrocortisone
	**	$p<0.01$	ZK = anti-glucocorticoid
HC+ZK vs. HC	+	$p<0.05$	ZK 98299
	++	$p<0.01$	

The effect of Onapristone on sucrase and glucoamylase activities, however, was not apparent 7 days after the beginning of the experiment, when the drug was omitted after the first 3 days of its administration. On the other hand, the delayed effect of injected HC and its suppression by anti-glucocorticoid was convincing, even 3 days after the treatment was stopped (Table 2).

Desialylation of the Brush-Border Proteins and Enzymes During Ontogeny and After Hydrocortisone (Table 3).

The effect of HC on the content of membrane-bound sialic acid was delayed and became apparent only on day 15, although we know that its effects start earlier in the differentiating enterocytes of the Lieberkuehn crypts.[5]

Table 2. Changes of sucrase and glucoamylase activities in intestinal brush-borders of 16-17-day-old male rats (nkat/mg protein).

Experiment	Sucrase	Glucoamylase
Controls (16)	0.128	0.84
Controls (17)	0.224	0.62
HC (16)	10.410	4.86
HC (17)	9.500	2.21
ZK 98299 (16)	0.202	1.01
HC + ZK (16)	1.084	1.08
HC + ZK (17)	1.04	1.06

Table 3. Sialic acid content in the brush-border fraction of rat small intestine (umol/g membrane protein).

	12-day-old (3 days after HC)	15-day-old (6 days after HC)
Controls	78+11	93+10
HC	72+10	** 51+3.5
ZK	93+18 (n.s.)	
HC + ZK	45+31 (n.s.)	
t-test vs. controls	** p<0.01 n.s. (nonsignificant)	

While following the pI changes of several brush-border enzymes during the postnatal development of the rat by analytical isoelectric focusing, we observed a shift to more basic values due to the loss of bound sialic acid.[2] Most pronounced changes were observed for DP IV and GMT. Therefore DP IV may serve as a marker of the desialylation processes in the brush-border during ontogeny and after HC treatment.

In the delayed effect, DP IV revealed a shift to more basic pI values after HC; this effect was suppressed by the anti-glucocorticoid. However, no effect was observed after Onapristone alone (Table 4).

Table 4. pI-changes of DP IV in 16-day-old male rats.

Experiment	pI
Controls	4.2 - 5.0
HC	4.5 - <u>5.6</u>
ZK	4.2 - 5.0
HC + ZK	<u>4.2</u> - 5.6

Changes in Thymus Mass

Thymolysis induced by HC and its inhibition by anti-glucocorticoids have been observed and used as a test for anti-glucocorticoid action. We could not observe any inhibition of thymolysis induced by HC in a short-term and at the dose used. However, in 12-day-old male rats we found a marked, 2-fold increase of thymus mass after the previous 3-day treatment by Onapristone; this increase in thymus mass exceeded the relative increase of the body mass (Table 5). At the same time the DNA/protein ratio increased to a 2-fold value (Table 6).

Table 5. Changes in the body weight and in the weight of thymus (12-day-old male rats).

Experiment	n		Body weight (g)		Thymus mass (mg)		(mg/kg b.w.)
Controls							
group A	8		26.5±3.3		73.5±27.8		2750.8±930.9
group B	8		23.4±1.6		59.4±18.9		2511.7±709.6
HC	8		26.0±1.3	**	27.4±6.5	**	1057.5±249.2
ZK	6	**	36.6±1.1	**	135.4±17.9	*	3709.7±594.4
HC + ZK	7		23.6±1.1	**	28.4±8.3	**	1197.8±329.7

vs. controls * $p<0.05$ ** $p<0.01$

Table 6. Effect of ZK 98299 (3x1 mg a day s.c.) on thymus of 12-day-old male rats, DNA/protein.

Experiment	n	Body weight (g)		Thymus DNA (ug/mg prot.)
Controls	7	31.6±1.2		26.5± 3.9
ZK	8	33.6±2.8	**	50.9±12.5

** $p<0.01$

In the delayed effect, 7 days after the beginning of the experiment, when the anti-glucocorticoid was omitted after the first 3 days of its administration, increase of the thymus mass was almost the same in controls as in the Onapristone-treated animals and was more or less proportional to the increase in body weight during postnatal development. On the other hand, the inhibitory (40%) effect on the HC-induced thymolysis became apparent, although he difference was not statistically significant (n.s.), as shown in Table 7.

Table 7. Effect of ZK 98299 injected on day 9,10,11 (3x1 mg a day s.c.) on thymus mass of 16-day-old male rats.

Experiment	n		Body weight (g)		Thymus mass (mg)		(mg/kg b.w.)
Controls	4		36.5±0.4		115.6±28.2		3171.7±801.1
HC	4	**	25.5±2.4	**	40.4±26.3	*	1537.8±875.1
ZK	4		37.1±1.0		110.8±38.0		3291.5± 67.7
HC + ZK	4		30.2±2.8	+	70.1±12.4	+	2310.2±249.6

HC vs. controls   * $p<0.05$ HC+ZK vs. HC + n.s.
                   ** $p<0.01$

SUMMARY

1.   The action in Onapristone infant male rats displays short-term and delayed effects.

2.   Suppression of intestinal brush-border enzymes and increase of thymus mass were observed only immediately after 3-day treatment with Onapristone. After an additional 3 days its effect disappeared. There was no immediate or delayed effect of Onapristone on the desialylation of brush-border enzymes.

3.   In the short-term and delayed effects, Onapristone suppressed the HC-provoked induction of several intestinal brush-border enzymes, especially α-glycosidases. In the delayed effect the drug also suppressed thymolysis induced by the exogeneously given glucocorticoid, and suppressed the HC-induced desialylation of a brush-border enzyme DP IV, which serves as a marker of the desialylation process.

4.   These experiments seem to support a conclusion that the postnatal development of intestinal brush-border enzymes and the development of thymus in infant rats are controlled by endogeneously secreted glucocorticoids.

5.   The control of sialylation of intestinal brush-border proteins by endogeneously secreted glucocorticoids during the postnatal development of the rat remains debatable.

ACKNOWLEDGMENT

We thank Dr. Ekkehard Schillinger from Schering AG for providing us with the drug Onapristone.

REFERENCES

1.   N. Kretchmer, J. S. Latimer, F. Raul, K. Berry, C. Legum, and H. L. Sharp, Ciba Foundation Symposium 70 (new series), 117 (1979).

2.   J. Kraml, J. Kolínská, L. Kadlecová, M. Zákostelecká, and Z. Lojda, *FEBS Lett.* 151:193 (1983).

3.   J. Kraml, J. Kolínská, L. Kadlecová, M. Zákostelecká, and Z. Lojda, *FEBS Lett.* 172:25 (1984).

4.   D. Henderson, *in:* "Pharmacology and Clinical Uses of Inhibitors of Hormone Secretion and Action", B.J.A. Furr and A.E. Wakeling, eds., p.184, Bailliere Tindall, London (1987).

5.   J. Kolínská, J. Kraml, M. Zákostelecká, and L. Kadlecová, INSERM Symp. 26, 381 (1986).

A COMPARATIVE STUDY OF THE IMMUNE RESPONSE TO POLY-[α(2→8)-N-ACETYL NEURAMINIC ACID]

J. Diaz Romero[1] and I.M. Outschoorn[2]

[1]Inmunologia, Centro Nacional de Microbiologia; and
[2]C.N. Biologia Celulary Retrovirus, Instituto de Salud
Carlos III, Majadahonda, 28220 Madrid, Spain

INTRODUCTION

The capsular polysaccharides from *Escherichia coli* K1, a normal inhabitant of the human gastrointestinal tract, and *Neisseria meningitidis* serogroup B, resident in the human nasopharynx, are [α(2→8)-N-acetylneuraminic acid] polymers (p/α(2→8)) structurally and serologically indistinguishable.[1] Both organisms are causative agents of meningitis in humans and the capsule acts as an essential virulence factor and protective antigen for both pathogens. For the development of meningitis the capsule appears to be critical.[2]

A hypothetical immunoepidemiologic model for meningococcal disease advocates a link between capsule-associated cross-reactivity and susceptibility to disseminated disease.[3] The mucosal induction of anticapsular serum IgA by an enteric (*E. coli* K1) priming organism would block the initiation of immune lysis of a cross-reactive organism (group B *N. meningitidis*) mediated by antibodies of identical specifity elicited by the second organism. IgA blocking includes two different mechanisms in relation to the isotype involved.[4] Whereas blocking of IgG-initiated immune lysis is a function of the ratio of IgA to IgG, blocking of IgM is a noncompetitive, linear function of the ratio of IgA to antibody binding sites and independent of the concentration of IgM.

The poor immunogenicity of purified capsular polysaccharide makes the production of capsule-specific antibodies extremely difficult and prevents the formulation of a comprehensive polysaccharide vaccine against these pathogens. For these reasons, protocols using hyperimmunization with whole bacteria, mainly *N. meningitidis* group B, have been usually employed to obtain polyclonal as well as monoclonal antibodies with p/α(2→8) specificity[5], and non-covalent complexes between p/α(2→8) and outer membrane proteins have been proposed as potential vaccines.[6] In both cases variable results have been obtained, perhaps due to the absence of a standard immunoassay for the quantitation of total levels of antibodies elicited. Like other purified carbohydrates, p/α(2→8) fails to adhere efficiently and reproducibly to a solid phase.

MATERIALS AND METHODS

In order to study the induction of a specific response to p/α(2→8) in different animal species, an ELISA was employed, recently developed in our laboratory (submitted). This assay uses a streptavidin solid phase and biotinylated purified polysaccharide, with biotinylated bovine serum albumin as a non-specific binding control (Fig. 1). Anti-p/α(2→8) sera from rabbit, hamster, and mouse were titred by this assay. Hamster, an animal model not previously used in such a system, offers the advantage of quick separation and assay of individual isotypes. The isotype predominance in response to different

Advances in Mucosal Immunology, Edited by
J. Mestecky *et al.*, Plenum Press, New York, 1995

immunogens could be assayed by the separation of IgG subclasses on protein A Sepharose columns[7] followed by protamine separation of the excluded IgM and IgA.[8] Only two subclasses of IgG have been described[9], but hyperimmune sera from Syriam hamsters on gradient elution frequently show a third peak eluting after the characteristic IgG2 and IgG1 peaks, while this peak is rarely seen in Chinese hamsters where elution patterns are in the reverse order. Additionally, a horse group B meningococcal antiserum, H46[10], was titred by haemagglutination. The preparation of immunogens used to elicit specific antibodies to $p/\alpha(2\rightarrow 8)$, formalinized whole bacteria, purified polysaccharide and the non-covalent complex of polysaccharide and outer membrane proteins (kindly provided by Dr. Robert Lifely, Wellcome Res. Labs., Beckenham, England), has been previously described.[11,12]

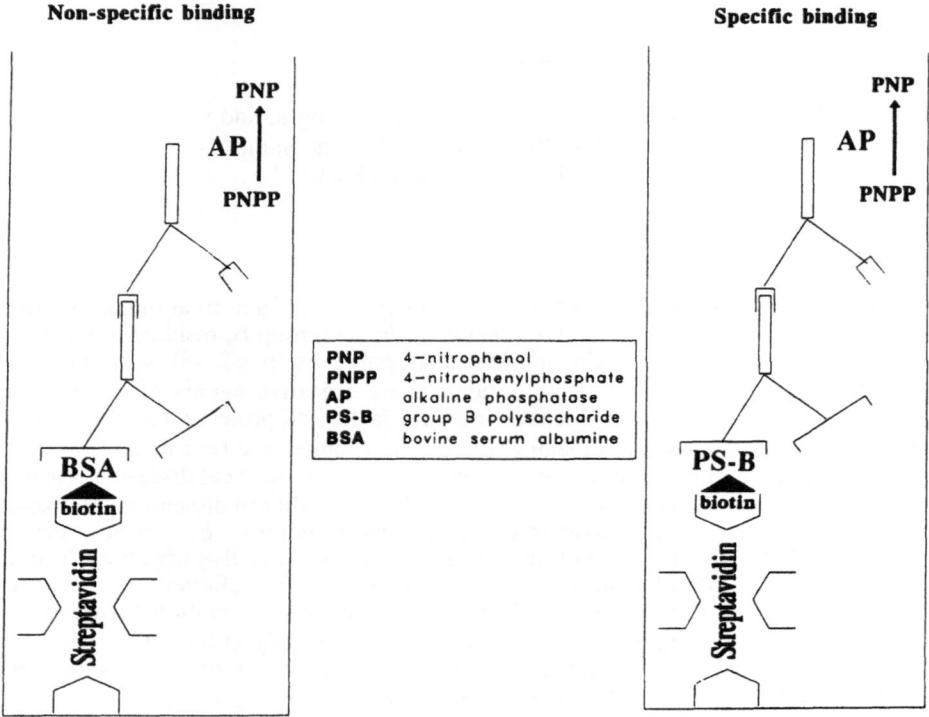

Figure 1. Diagram of anti-$p/\alpha(2\rightarrow 8)$ ELISA.

RESULTS

First, we evaluated the titre and specificity of a commercial rabbit immune serum produced by hyperimmunization with whole bacteria (*N. meningitidis* group B), and currently used in diagnosis for serogroup identification, comparing it to similar sera specific for groups A and C. The results of an ELISA assay of these sera against homologous polysaccharide are shown in Fig. 2. In contrast to groups A and C, group B *N. meningitidis* fails to elicit a specific response. No significant differences were obtained using sera from 2 commercial sources (Difco and Pasteur) nor when ELISA data were compared by haemaglutination as an alternative assay (data not shown), although the latter was less sensitive. In contrast, a horse group B meningococcal antiserum obtained by hyperimmunization showed a high haemagglutination titre, (as previously described[13]).

In an attempt to elucidate these data, a temporal follow up of the specific response to $p/\alpha(2\rightarrow 8)$ using a hyperimmunization protocol was carried out in Syriam hamsters. Figure 3 shows that hyperimmunization produced a poor response, which was irregular and not maintained; 4 immunizations were required to obtain a significant response, and titres fell after the fifth immunization.

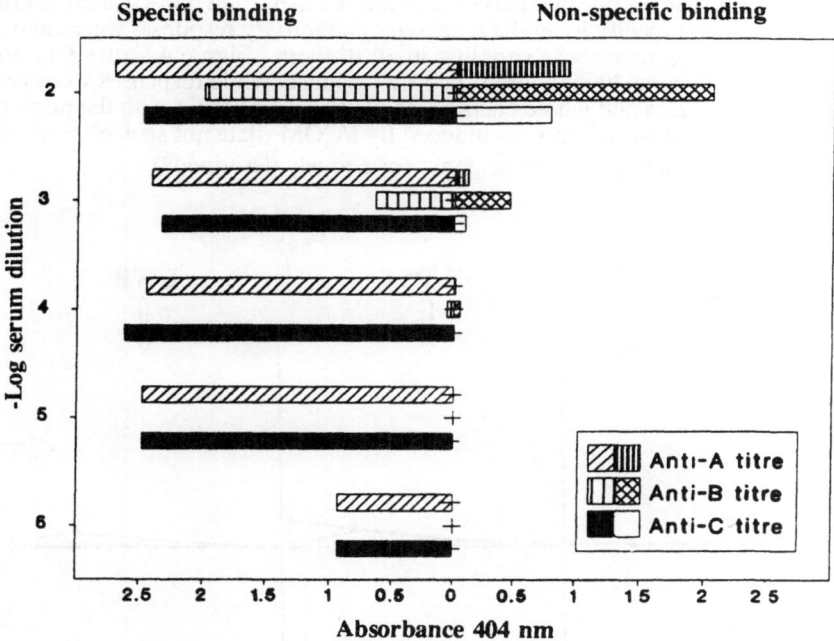

Figure 2. Evaluation of titre and specificity of serogrouping rabbit antisera

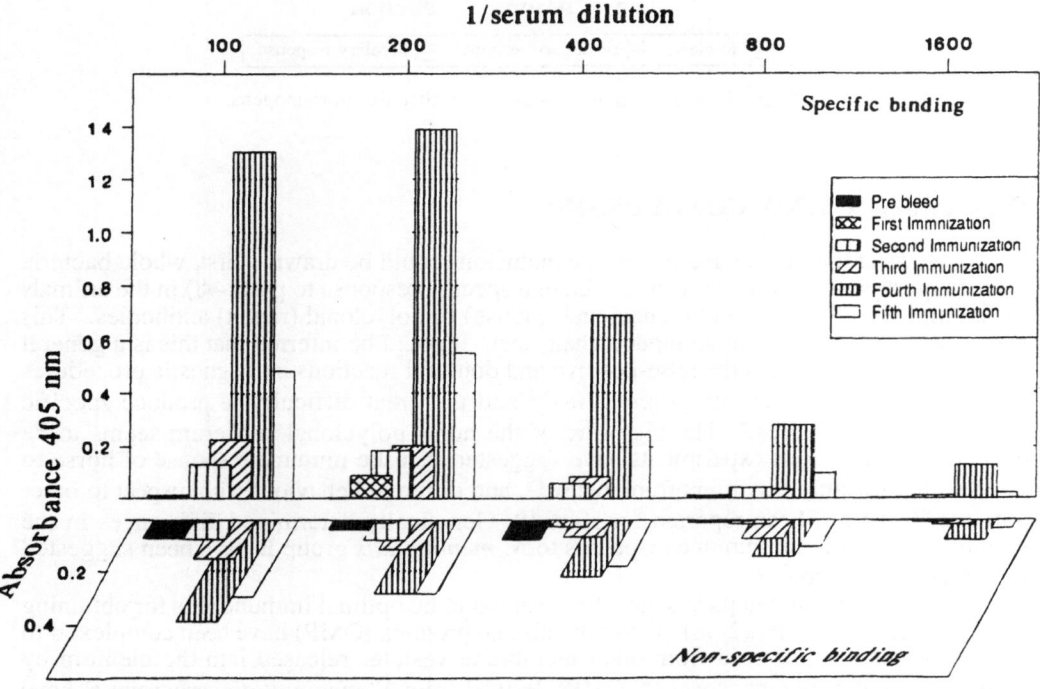

Figure 3. Temporal follow up of the specific response in hamster by hyperimmunization with *N meningitidis* serogroup B

Finally, mouse sera were produced using three diferent immunogens: immunization with whole bacteria (I/WB), purified polysaccharide (I/PS-B), and non-covalent complexes (I/COM). The titre and specificity of the secondary and tertiary responses were analysed by ELISA. The primary response was very low in all of them. Figure 4 shows that only the complex induced a moderately high titred specific response, while responses were very low to purified polysaccharide and whole bacteria, the latter overlapping with the non-specific binding curves. An analysis of isotypes induced by I/COM (data not shown) confirmed an almost exclusively IgM anti-p/α(2→8) response (previously described[6]).

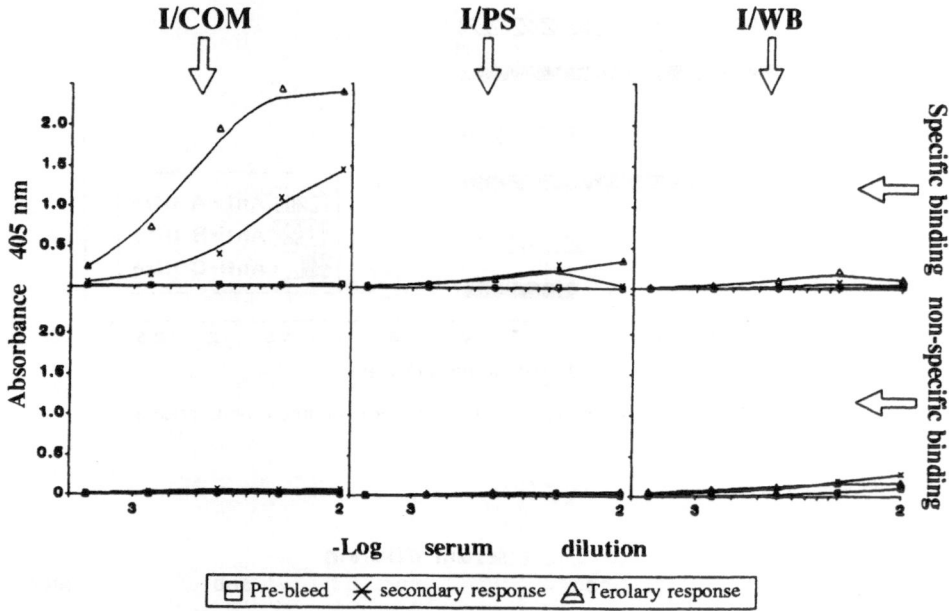

Figure 4. Mouse immunization with different immunogens.

DISCUSSION AND CONCLUSIONS

Several theoretical and practical conclusions could be drawn. First, whole bacteria were inappropriate as immunogens to obtain a specific response to p/α(2→8) in the animals most employed for obtaining monoclonal (mouse) or polyclonal (rabbit) antibodies. This was also true for other animal models (hamster). It could be inferred that this is a general defect and would explain the false-positive and doubtful reactions in diagnostic procedures that use polyclonal reagents to p/α(2→8)[14] and the great difficulty to produce specific monoclonal antibodies.[5] The high titre of the horse polyclonal antiserum seems to be exceptional. Previous experiments have suggested that the immune response of horse to polysaccharide antigens differs from mouse[15], and different behaviour with respect to other anti-p/α(2→8) has been reported for H-46.[16] Genetically determined differences in the magnitude of humoral immune responses to *N. meningitidis* group B have been suggested in humans and mice.[5,17]

Non-covalent complexes have been shown to be optimal immunogens for obtaining specific responses to p/α(2→8). Outer membrane proteins (OMP) have been complexed to B polysaccharide prepared from outer membrane vesicles released into the medium by actively growing meningococci.[18] OMPs derived from *N. meningitidis* serogroup B have shown mitogenic activity for lymphocytes and this ability has been exploited in conjugate vaccines to induce strong IgG responses against the capsular polysaccharide from *Haemophilus influenzae* type b[19], but imputing the enhanced anti-p/α(2→8) response to the complex as being exclusively due to the carrier is questionable. Whole meningoccoci also have been shown to be mitogenic for human peripheral blood lymphocytes[20] but in these experiments failed to induce an anti-p/α(2→8) response, and the response induced by I/COM is restricted to IgM. The enhanced immunogenicity of p/α(2→8) in mice occurred with an increase in the degree of binding of the polymer to OMPs in non-covalent complexes.[21] In our studies, we used complexes prepared by the method of Moreno *et al.*[12] (to whom we are also indebted for advice and materials) which showed a high ratio of p/α(2→8):OMP. Non-covalent complexes with a low ratio of these components or covalent complexes that use one molecule of tetanus toxoid as carrier induced poor anti-p/α(2→8) responses.[21,22]

What could explain the immunogen-associated differences in the immune response to p/α(2→8)? Both the activation mechanism of B cells to T cell-independent antigens as well as the epitope nature asociated with this antigen could provide the answer. Cross-linking of the antigen receptors, IgM and IgD, on mature B-lymphocytes by multivalent ligands plays a critical role in the inductive phase of the humoral response.[23,24] In relation to this, it has been postulated (the "immunon" model[25,26]) that a minimum specific number of antigen receptors must be connected together as a spatially compact cluster (or "immunon") before an immunogenic signal is delivered to the receptor cell. This also predicts a requirement for a minimum number of spatially linked haptens for immunogenicity. Molecules below threshold values both of hapten (epitope) valence and of molecular mass were inferred to be unable to form an aggregate of antigen receptors sufficient for immunon formation. It has also been shown that capsular polysaccharides are T-independent antigens which are composed of a few repetitive epitopes, enabling them to crosslink the specific receptors in B cells and trigger the humoral immune response without T cell help, although a modulatory role has been postulated for the latter[27] and a cross-linking mechanism (mediated by IgD)[28] could, at least partially, explain this process.

In the case of the p/α(2→8) molecule, the postulated presence of a conformational epitope[16] probably results in a low number of epitopes/molecule. So, the epitope presented in the native purified polysaccharide or in an equimolecular complex to carrier[22] is unable to trigger the specific immune response because the antigen structure is below the epitope valence threshold. By analogy, when the epitope is presented on a suface, as a bacterium or on outer membrane vesicles, the epitope spacing is critical to induce a specific response. Only a close epitope grouping, directly dependent on the antigenic density in this case, can work as a stimulatory signal. Exclusively high ratio p/α(2→8):OMP seems to fulfill this requirement and low ratio p/α(2→8):OMP as well as whole bacteria fail to do so. In contrast, whole bacteria are good immunogens for the induction of a specific response against capsular polysaccharides from *N. meningitidis* group A and group C - the latter a polymer of N-acetylneuraminic acid α(2→9) - which have conventional linear and repetitive epitopes, and for which the epitope grouping exists on the antigen molecule itself. Additional factors such as antigen stability and concentration, specific receptor number, affinity and mobility, must also play an important role in the antigen-B cell interaction, and any of them could be the basis of the differential equine anti-p/α(2→8) response.

Additional functional implications of the nature of the p/α(2→8) epitope can be postulated. The density of antigen determinants on the target cell can influence the efficiency of antibody mediated effector functions: a low density would hinder the obligatory multivalent binding of IgM - the isotype induced by p/α(2→8) - suggested for immune lysis[29], and would facilite the IgA-mediated IgM blocking, directly dependent on the ratio of IgA to the number of specific binding sites.[4] This model is supported by the bactericidal inefficiency of anti-p/α(2→8) antibodies[6,21] and would explain the failure of the bactericidal activity of these antibodies in conjunction with human serum as complement source.[30] This failure is partially reversed by the use of serum from an individual with agammaglobulinemia[30], very low IgA levels (<3 mg/ml) and hence a diminished blocking capacity.

REFERENCES

1. H. V. Raff, D. Devereux, W. Shuford, D. Abbott-Brown, and G. Maloney, *J. Infect. Dis.* 157:118 (1988).
2. K. S. Kim, H. Itabashi, P. Gemski, J. Sadoff, R. L. Warren, and A. S. Cross, *J. Clin. Invest.* 90:897 (1992).
3. J. M. Griffiss, *Rev. Infect. Dis.* 4:159 (1982).
4. J. M. Griffiss, *Ann. N.Y. Acad. Sci.* 409:697 (1983).
5. F. Nato, J. C. Mazie, J. M. Fournier, B. Slizewicz, N. Sagot, M. Guibourdenche, D. Postic, and J. Y. Riou, *J. Clin. Microbiol.* 29:1447 (1991).
6. M. R. Lifely, S. C. Roberts, W. M. Shepherd, J. Esdaile, Z. Wang, A. Cleverly, A. A. Aulaqi, and C. Moreno, *Vaccine.* 9:60 (1991).
7. J. E. Coe, P. R. Coe, and M. J. Ross, *Mol. Immunol.* 18:1007 (1981).
8. L. Hudson and F. C. Hay, *in*: "Practical Immunology", p. 319, Blackwell Scientific Publication, Oxford (1989).
9. J. E. Coe, *J. Immunol.* 100:507 (1968).
10. P. Z. Allen, M. Glode, R. Schneerson, and J. Robbins, *J. Clin. Microbiol.* 15:324 (1982).
11. C. Moreno, J. Hewitt, K. Hastings, and D. Brown, *J. Gen. Microbiol.* 129:2451 (1983).
12. C. Moreno, M. R. Lifely, and J. Esdaile, *Infect. Immun.* 47:527 (1985).
13. F. Ørskov, I. Ørskov, A. Sutton, R. Schneerson, W. Lin, W. Egan, G. E. Hoff, and J. Robbins, *J. Exp. Med.* 149:669 (1979).
14. R. C. Tilton, F. Dias, and R. W. Ryan, *J. Clin. Microbiol.* 20:231 (1984).
15. B. N. Manjula and C. P. J. Glaudemans, *Immunochemistry* 15:269 (1978).
16. M. R. Lifely and J. Esdaile, *Immunology* 74:490 (1991).
17. J. P. Pandey, W. D. Zollinger, H. H. Fudenberg, and C. B. Loadholt, *J. Clin. Invest.* 68:1378 (1981).
18. C. E. Frasch and M. S. Peppler, *Infect. Immun.* 37:271 (1982).
19. M. A. Liu, A. Friedman, A. I. Oliff, J. Tai, D. Martinez, R. R. Deck, J. T. C. Shieh, T. D. Jenkins, J. J. Donnelly, and L. A. Hawe, *Proc. Natl. Acad. Sci. USA.* 89:4633 (1992).
20. J. Melancon, R. A. Murgita, and I. W. Devoe, *Infect. Immun.* 42:471 (1983).
21. M. R. Lifely and Z. Wang, *Infect. Immun.* 56:3221 (1988).
22. H. J. Jennings and C. Lugowski, *J. Immunol.* 127:1011 (1981).
23. J. G. Monroe and V. L. Seyfert, *Immunol. Res.* 7:136 (1988).
24. J. Hombach, T. Tsubata, L. Leclercq, H. Stappert, and M. Reth, *Nature* 343:760 (1990).
25. R. Z. Dintzis, B. Vogelstein, and H. M. Dintzis. *Proc. Natl. Acad. Sci. USA* 79:884 (1982).
26. R. Z. Dintzis, M. Okajima, M. H. Middleton, G. Greene, and H. M. Dintzis. *J. Immunol.* 143:1239 (1989).
27. P. J. Baker, *Infect. Immun.* 58:3465 (1990).
28. R. F. Coico, G. W. Siskind, and G. J. Thorbecke, *Immunol. Rev.* 105:45 (1988).
29. T. Borsos, R. M. Chapuis, and J. J. Langone, *Mol. Immunol.* 18:863 (1981).
30. W. D. Zollinger and R. E. Mandrell, *Infect. Immun.* 40:257 (1983).

IN VIVO ACTIVATION OF EXTRATHYMIC T CELLS IN MICE BY STAPHYLOCOCCAL ENTEROTOXIN B

Kazuo Ohtsuka,[1,2] Hisami Watanabe,[1] Hitoshi Asakura,[2] and Toru Abo[1]

[1]Department of Immunology, [2]Department of 3rd Internal Medicine, School of Medicine, Niigata University, Niigata, Japan

INTRODUCTION

Extrathymic T cell differentiation was recently known to occur in the sinusoids of the liver,[1,2] and in the epithelial region of the intestine.[3] Such extrathymic T cells in the liver and intraepithelial lymphocytes (IEL) in the intestine both have common and distinct properties with each other. Both extrathymic T cells become predominant in corresponding organs of aged humans and animals with the involved thymus, and comprise a considerable proportion of self-reactive forbidden clones, possibly due to the lack of negative selection systems for such clones. Staphylococcal enterotoxin B (SEB) that activates T cells is known as one of the super-antigens.[4,5] We investigated herein how extrathymic T cells in the liver and intestine were activated when SEB was administered in mice.

METHODS

Mice

C3H/HeN mice were originally purchased from Jackson Laboratory, Inc., Tokyo, Japan, and maintained in the Animal Facilities of Niigata University under specific pathogen-free conditions. These mice were used at the indicated ages from 8 to 10 wks.

Cell Preparations

Hepatic mononuclear cells (MNC) were isolated as follows.[6] Anesthetized mice were sacrificed, and the liver was removed, pressed through 200-gauge stainless steel mesh, and then suspended in PBS (0. 1 M, pH 7.2). After being washed once with PBS, MNC were isolated from hepatocytes by the Ficoll-Isopaque density (1.090) gradient centrifugation. The collected MNC from the interface was then suspended in MEM medium supplemented with 2 % fetal calf serum. Spleen cells were also collected by Ficoll-Isopaque method, whereas thymocytes were obtained by forcing the thymus through 200-gauge steel mesh. IEL in the intestine were collected as follows. Small intestine was removed and was flushed with PBS to eliminate lumen contents. Mesenterium and Peyer's patches were then removed. The intestine was longitudinally opened and cut into fragments 1 cm long. These were incubated for 15 min in 20 ml Ca^{++}- and Mg^{++}-free HBSS containing 5 mM EDTA in 37° C shaking water bath. Supernatant containing cells was collected.

Advances in Mucosal Immunology, Edited by
J. Mestecky *et al.*, Plenum Press, New York, 1995

Injection of SEB

SEB was intraperitoneally or perorally inoculated (100 μg/mouse). Mice were sacrificed on days 1, 3 and 7, and MNC were prepared from the liver, spleen, thymus and epithelia of the small intestine.

Immunofluorescence Test

The surface phenotype of cells was analyzed by using monoclonal antibodies (mAb) in conjunction with a two-color immunofluororescence test. The fluorescence-positive cells were analyzed by a FACScan (Becton-Dickinson & Co., Palo Alto, CA).

RESULTS

After intraperitoneal inoculation of SEB, the numbers of MNC in the liver and spleen were increased on day 3, while those of thymus were decreased (Fig. 1). The quantity of IEL remained constant. On the other hand, after oral inoculation, the numbers of MNC in the liver were increased up to three times on day 3, while those of MNC in the spleen, thymus, and IEL remained constant.

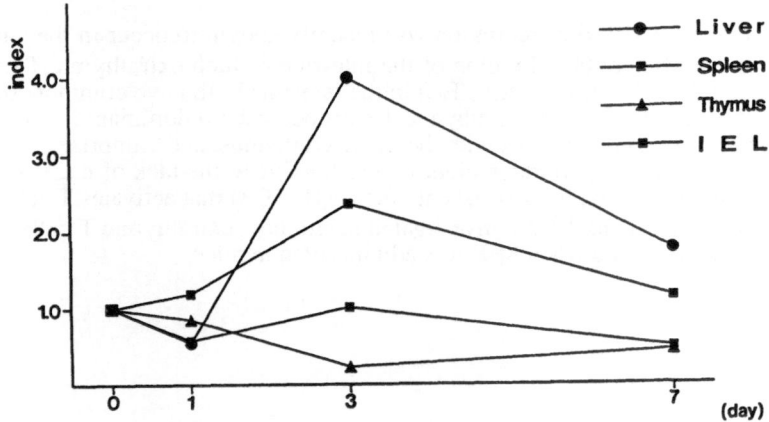

Figure 1. Numbers of MNC when intraperitoneally inoculated.

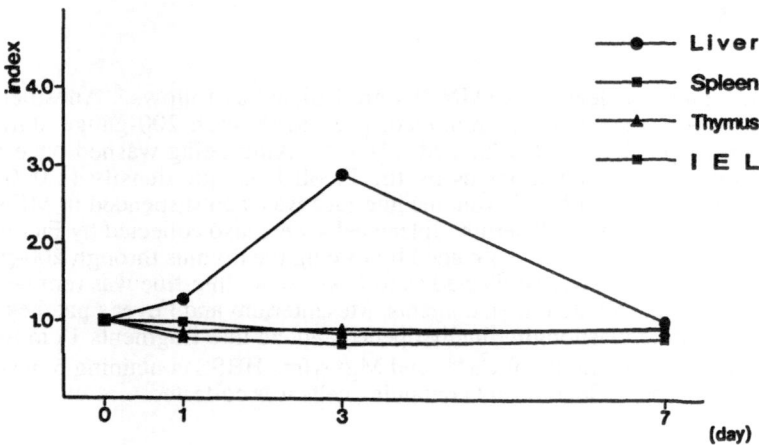

Figure 2. Numbers of MNC when orally inoculated.

The two-color staining for CD3 and IL-2 receptor β chain (IL-2Rβ) in normal mice clearly identified four distinct lymphocyte populations, (1) CD3-IL-2Rβ B cells, (2) CD3-IL2Rβ+ : NK cells, (3) CD3+ IL-2Rβ-:bright TCR cells, and (4) CD3+ IL-2Rβ +:intermediate TCR cells (Fig. 3). In the thymus, three populations were identified, (1) CD3-IL-2R-:immatured T cells, (2) CD3+ IL-2Rβ -:mature T cells, and (3) CD3+ IL-2R {3 +:intermediate TCR cells. IEL composed of populations, (1) CD3-IL-2Rβ+: NK cells, (2) CD3+ IL-2Rβ -:bright TCR cells, (3) CD3+ IL-2R +:extrathymic T cells.

After oral inoculation on day 3 (Fig. 3), the most striking difference was an increase in the proportion of intermediate TCR cells in the liver. On day 7, bright TCR cells also increased. The distribution patterns of MNC in the spleen, thymus and IEL showed no significant change. On the other hand, in the intraperitoneally inoculated mice, intermediate TCR cells in the liver increased day by day. In the spleen and thymus, intermediate TCR cells had increased by day 3, then decreased.

As already reported, in normal mice only IEL consisted of double-positive CD4+ CD8+ cells, but hepatic or splenic MNC did not (Fig.4). After the oral inoculation, CD8+ cells increased in the liver, while the distribution patterns of single-positive CD4+ or CD8+ cells in MNC of the spleen, and of double-positive cells in the thymus showed no significant change. After intraperitoneal SEB inoculation, CD8+ cells increased in the liver and spleen, while the distribution patterns of double-positive cells in the thymus decreased on day 3, and returned to starting levels on day 7.

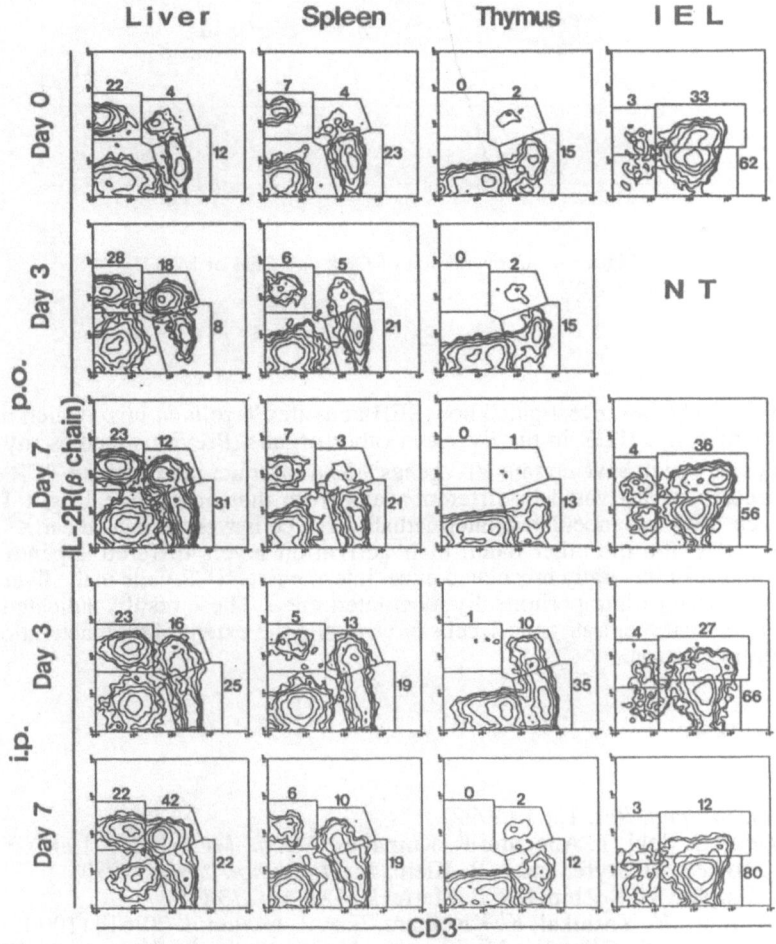

Figure 3. Coexpression of CD3 and IL-2Rβ of MNC.

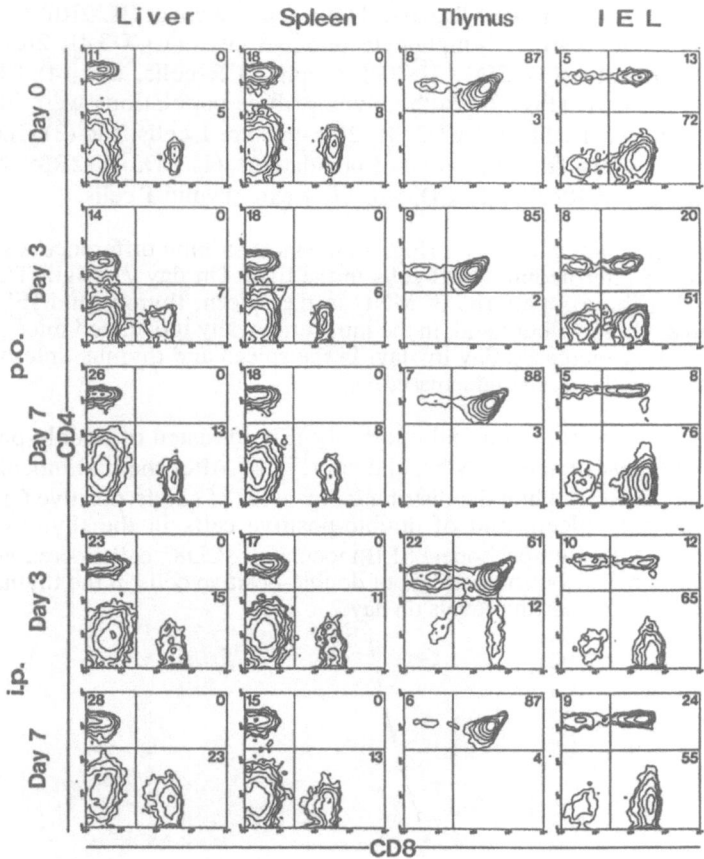

Figure 4. Coexpression of CD8 and CD4 of MNC.

DISCUSSION

In this study, we investigated how SEB activates T cells *in vivo* which have either intermediate or bright TCR in the liver and other organs. Previous studies say that SEB activates T cells which have certain Vβ3 genes.[6] Extrathymic, intermediate TCR cells were demonstratred to be activated in different ways from that of regular, bright TCR cells. Namely, when SEB was inoculated, intermediate TCR cells were activated earlier than were bright TCR cells. On the other hand, their activation levels differed depending on the inoculated routes. In the orally inoculated mice, intermediate TCR cells in the liver increased earlier than did that of intraperitoneally inoculated mice. These results indicated that SEB preferentially activates extrathymic T cells *in vivo*, and the extent of their activation owe the routes of administration.

REFERENCES

1.   T. Abo, *Biomed. Res.* 13: 1 (1992) .
2.   T. Ohteki, S. Seki, T. Abo, and K. Kumagai, *J. Exp. Med.* 172:7 (1990).
3.   R. L. Mosley, D. Styre, and J. R. Klein, *Int. Immunol.* 2:361 (1990).
4   J. W. Kappler, N. Roehm, and P. Marrack, *Cell* 49:273 (1987).
5.   H. Takimoto, Y. Yoshikai, K. Kishihara, *Eur. J. Immunol.* 20:617 (1990).
6.   H. Watanabe, K. Ohtsuka, M. Kimura, Y. Ikarashi, K. Ohmori, A. Kusumi, S. Ohteki, Seki, and T. Abo, *J. Immunol. Methods* 146:145 (1991)

INTERACTION OF A SULFHYDRYL ANALOGUE OF VASOACTIVE INTESTINAL PEPTIDE (VIP) WITH MURINE LYMPHOCYTES

C.A. Ottaway

Department of Medicine
University of Toronto, Toronto, Ontario, Canada, M5S 1A8

INTRODUCTION

Vasoactive intestinal peptide (VIP) is a 28 amino acid neuropeptide with potent immunoregulatory properties.[1] The cDNA of the VIP receptor (VIPR) of NALM-6 cells predicts a heptahelical polypeptide chain homologous to other G-protein coupled neurotransmitter receptors.[2] The predicted structure of VIPR includes two cysteine residues which are conserved in this receptor family. These are represented at positions 117 and 196 of the VIPR[2] and positions 106 and 184 of the hamster beta-adrenergic receptor (BAR).[3] Selective replacement of either of these cysteine residues[4] in BAR alters the ligand binding properties of the receptor, and a disulfide bond between these residues on adjacent extracellular loops has been implicated in the maintenance of the conformational and functional integrity of the BAR ligand binding site.[3,4] The purpose of this work was to examine the ability of lymphocytes to recognize and respond to an analogue of VIP in which cysteine was substituted for serine in the second position of the neuropeptide.

METHODS

VIP and the VIPR antagonist[5] [4Cl-D-Phe^6, Leu^{17}]VIP were purchased from Peninsula Laboratories (Belmont, CA, USA). The analogue [Cys^2] VIP was synthesized by solid-phase methods and purified by HPLC. Mesenteric lymph node lymphocyte suspensions (MLN) were prepared from BALB/c mice as previously described.[6,7] VIP binding studies were carried out using 7.5×10^{-11}M ^{125}I-VIP (Amersham Canada Ltd.) in RPMI-1640 containing 5% BSA as previously described.[6] Specific binding of ^{125}I-VIP was determined by the difference between the cell-associated radioactivity found in the absence and the presence of 10^{-7}M unlabeled VIP after 60 min at 37°C. Cyclic AMP determinations were performed in RPMI-1640 containing 5% BSA and 2mM 3-isobutyl-1-methylxanthine (Sigma). MLN were incubated at 37°C and reactions initiated by the addition of peptides in the same medium. Reactions were terminated at 10 min by transferring aliquots to ice-cold 100% ethanol. Cell lysates were centrifuged (6500xg, 5 min) and supernatants dried in a vacuum oven and then acetylated and assayed for cyclic AMP by scintillation proximity radioimmunoassay (Amersham). The effect of peptides on thymidine incorporation by Concanavalin A (ConA) stimulated cultures was assessed as previously described[6,7] using 2.5×10^5 cells/ml in 0.2 ml volumes with and without ConA, 2.5 ug/ml.

Advances in Mucosal Immunology, Edited by
J. Mestecky *et al.*, Plenum Press, New York, 1995

RESULTS

As expected, VIP competed effectively for the binding of ^{125}I-VIP to MLN lymphocytes (Fig. 1A). The sulfhydryl analogue, [Cys^2]VIP, was less effective than VIP as a competitor, but more effective than the VIPR antagonist ([4Cl-D-Phe^6, Leu^{17}]VIP) introduced by Pandol and coworkers[5] (Figure 1A). Both of these VIP derivatives were able to compete at high concentrations, however, and the VIPR antagonist displaced virtually all of the lymphocyte-bound ^{125}I-VIP when present at micromolar concentrations (Fig. 1A).

VIP stimulated intracellular cyclic AMP in MLN suspensions (Fig. 1B). Exposure of MLN to [Cys^2]VIP was able to stimulate intracellular cyclic AMP of the lymphocytes to a similar extent, but only at much higher peptide concentrations (Fig. 1B). In contrast, the VIPR antagonist had no effect on intracellular cyclic AMP, even at concentrations that could occupy VIP binding sites of the lymphocytes (Figs. 1A, 1B). The cyclic AMP dose-response to both VIP and [Cys^2]VIP was shifted to higher peptide concentrations, however, in the presence of micromolar concentrations of the VIP antagonist (Figs. 2, 3). The proliferative response of MLN cultures stimulated with ConA was inhibited by VIP in a dose-dependent manner (Fig.e 1C). In the presence of [Cys^2]VIP, the response of this peptide competed for the inhibitory effect of both VIP and [Cys^2]VIP on the proliferation of the MLN cultures (Figs. 2, 3).

DISCUSSION

MLN lymphocytes recognize and respond to a simple sulfhydryl-containing analogue of VIP, but the molar potency of [Cys^2]VIP as a competitor for the binding of radiolabeled VIP to MLN, as a stimulator of MLN intracellular cyclic AMP, and as an inhibitor of the ConA-stimulated proliferation of MLN is two orders of magnitude less than that of the intact neuropeptide. Although [Cys^2]VIP appears to exert its effect on MLN by means of the neuropeptide receptors, its ability to occupy and activate VIPR is markedly disrupted. Why should this substitution result in such a marked difference with respect to the interaction of the VIP analogue with its receptor?

Substitution of a cysteine side chain for that of a serine is usually considered to be the most conservative possible replacement of amino acid residues in a protein.[8] The -CH$_2$SH side chain is only slightly larger than -CH$_2$OH and the bond lengths and angles are only slightly different in the two residues.[8] Deletion of the N-terminal residue[9] of VIP, i.e. histidine[1], or steric substitution of serine[2] with its D-isomer[10], results in about an order of magnitude decrease in the binding of the peptide to its receptors on other tissues. For MLN lymphocytes, however, removal of all of the first 9 amino acids from the N-terminal only impairs the ability of the peptide to bind to the cells, and to inhibit their response to ConA, by one order of magnitude.[6] If the cysteines on the second and third extracellular domains of the VIPR represented by those at positions 117 and 196 of the cloned NALM-6 VIPR cDNA[2] form a disulfide bridge that helps to maintain the integrity of the ligand binding domain of the receptor, then [Cys^2]VIP may interfere with this bond and produce a conformational disruption of the ligand-binding site.

This derivative of VIP adds to the armamentarium of available probes of VIP-VIPR interactions *in vitro*. It may be possible also to target such a modification of the neuropeptide *in vivo*. Agoston and coworkers[11] have shown that a minimal 5.2 kb flanking sequence of the VIP gene is all that is required to target the expression of the VIP-gene to the intestine in transgenic mice. Similar insertion of a modified VIP gene to encode for [Cys^2]VIP might alter local neurotransmission and neuropeptide-mediated immunoregulation in the intestinal mucosa *in vivo*.

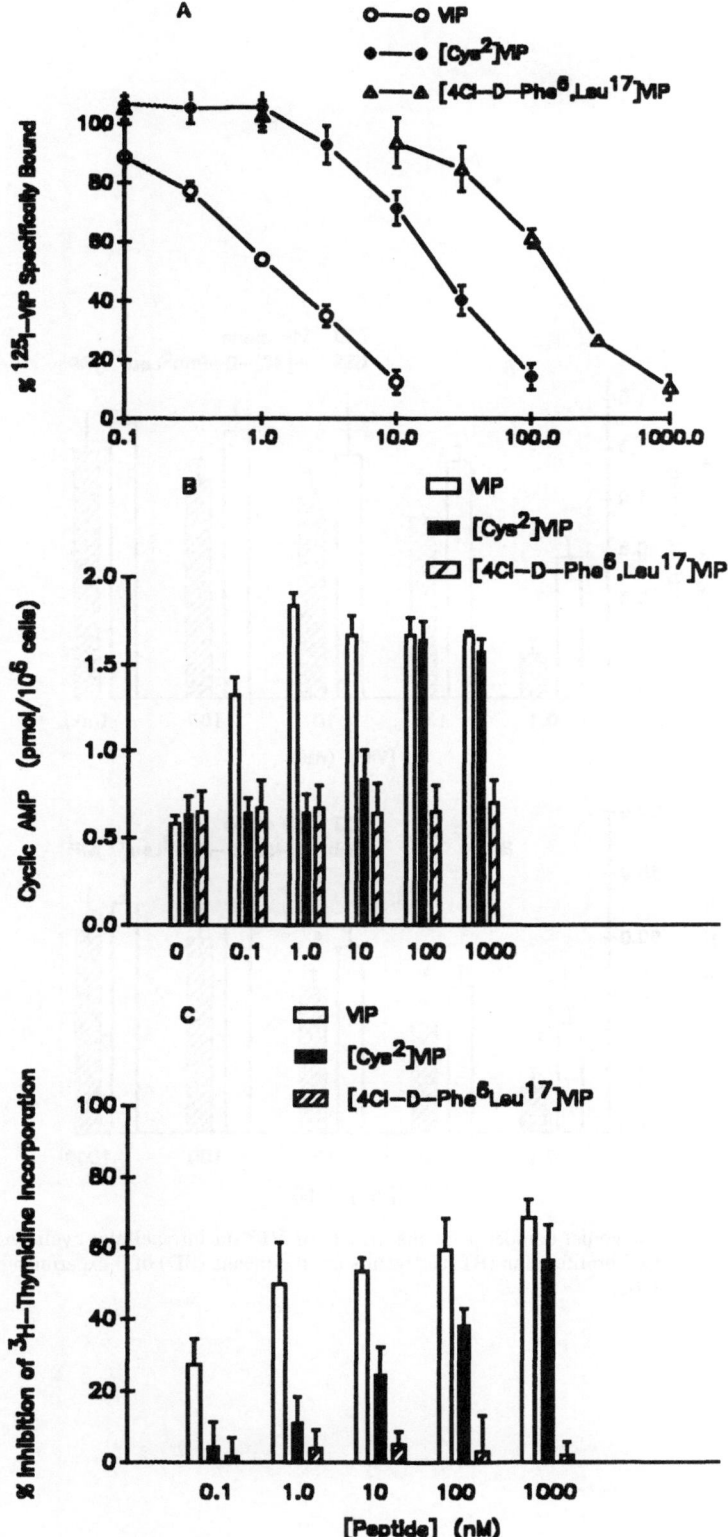

Figure 1. Effects of VIP, [Cys^2]VIP, and [4Cl-D-Phe^6, Leu^{17}]VIP on: A) Specific binding of VIP; B) Intracellular cyclic AMP; and C) Inhibition of ConA proliferation (means (SD) of 3 experiments).

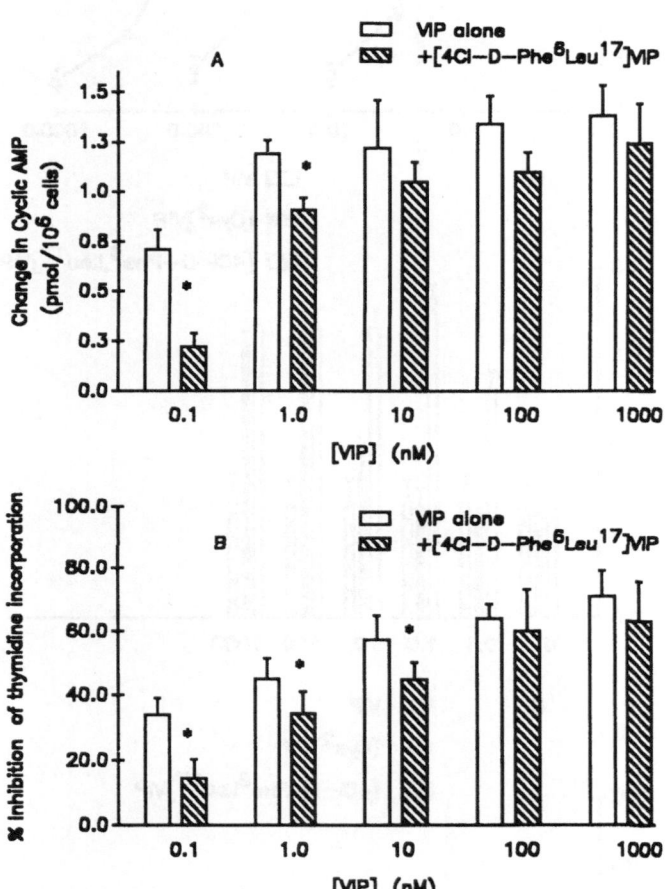

Figure 2. The VIPR antagonist competes for the effects of VIP on intracellular cyclic AMP (A) and inhibition of ConA induced proliferation (B). The results are the means (SD) of 3 experiments, * indicates significant difference p<0.05.

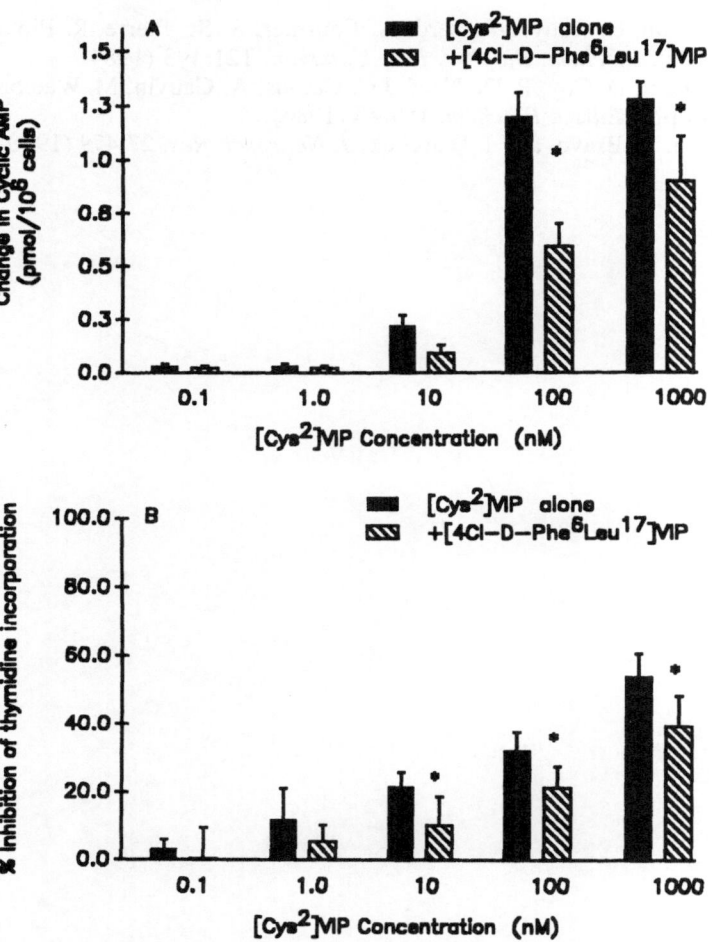

Figure 3. The VIPR antagonist competes for the effects of [Cys^2] VIP on intracellular cyclic AMP (A) and inhibition of ConA induced proliferation (B). The results are the means (SD) of 3 experiments, * indicates significant difference p<0.05.

REFERENCES

1. C. A. Ottaway, *in:* "Psychoneuroimmunology", second edition, R. Ader, N. Cohen, and D. Felten ed., Academic Press, Inc., New York, p. 225, (1991).
2. S. P. Sreedharan, A. Robichon, K. E. Peterson, and E. J. Goetzl, *Proc. Natl. Acad. Sci. USA* 88:4986 (1991).
3. C. D. Strader, I. S. Sigal, and R. A. F. Dixon, *FASEB* J. 3:1825 (1989).
4. R. A. F Dixon, I. Sigal, I. E. Rands, R. B. Register, M. R.Candelore, C. D Blake, and C. D. Strader, *EMBO J.* 6:3269 (1987).
5. S. J. Pandol, K. Dharmsathaphorn, M. S. Schoeffield, W. Vale, and J. Rivier, *Am. J. Physiol.* 250:G553 (1986).
6. C. A. Ottaway and G. Greenberg, *J. Immunol.* 132:417 (1984).
7. C. A. Ottaway, *Immunology* 62:291 (1987).
8. J. J. He and F. Quiocho, *Science* 251:1479 (1991).

9. A. Couvineau, C. Rouyer-Fessard, A. Fournier, S. St. Pierre, R. Pipkorn, and M. Laburthe, *Biochem. Biophys. Res. Commun.* 121:493 (1984).
10. P. Robberecht, D. Coy, P. De Neef, J.C. Camus, A. Cauvin, M. Waelbroeck, and J. Christophe, *Eur. J. Biochem.* 159:45 (1986).
11. D. Agoston, D. Bravo, and J. Waschek, *J. Neurosci. Res.* 27:479 (1990).

ANALYSIS OF SP/VIP FIBER ASSOCIATION WITH T4 AND T8 LYMPHOCYTES IN NORMAL HUMAN COLON

P. Weisz-Carrington, N. Nagamoto, M. Farraj, R. Buschmann, E.B. Rypins, and A. Stanisz

Departments of Pathology and Surgery,West Side VAMC and University of Illinois at Chicago, USA and McMaster University, Ontario, Canada

INTRODUCTION

The possibility that neuropeptides control many facets of the secretory immune system is now an established fact.[1] In addition, a variety of receptors for neuropeptides have been described on lymphocytes.[2] A close contact between specialized nerve fibers and lymphocytes has been demonstrated in the rat spleen.[3] This neighboring contact suggests that for neuropeptides to act on immunocytes, this close range between fibers and cells is required. Others have also observed a proximity between Substance P and IgA producing lymphocytes.[4] In other work, we have demonstrated that IgA cells have a highly significant relationship by proximity, to SP fibers rather than VIP fibers in the colonic mucosa.[5] Since such proximity does not appear to occur randomly, we posed the question that SP-fibers might also be associated with lymphoid cells, such as T cells known to influence differentiation and maturation of B cells in the lamina propria of mucosal surfaces.

MATERIALS AND METHODS

Sections of morphologically normal human colon were obtained fresh from surgical specimens, fixed in 10% buffered formalin and embedded in paraffin. The paraffin embedded blocks were then cut into 5μm sections and mounted on poly-L-lysine-coated slides. Double-enzyme immunostaining techniques were modified from method described by Nakane & Pierce.[6] Paraffin embedded sections were deparaffinized and then rinsed with 0.05M Tris-HCl pH 7.6 for 5 min twice, followed by an incubation with 0.1% trypsin/Tris for 10 min.[7] The sections were then washed with deionized water (3 times for 5 min) and immersed in 0.5% periodic acid solution for 10 min to block endogenous peroxidase activity. After washing with PBS 5 min three times, sections were immersed in 0.3% Triton-X 100/PBS for 1 hr to obtain good antibody penetration. The sections were then incubated with 5% bovine serum albumin/PBS for 10 min to reduce background and then washed with PBS, 3 times for 5 min. Then the slides were incubated with mouse anti-VIP monoclonal antibody at a 1:400 dilution obtained from DAKO Laboratories, or rabbit anti-SP antibody 1:500 dilution (Milab/Malmo, Sweden) and incubated simultaneously with peroxidase-conjugated sheep anti-human CD4 or CD8 antibodies at a 1:200 dilution (DAKO).

The slides were then incubated in a humid chamber at room temperature overnight followed by washing with PBS, 3 times for 5 min. A DAB substrate kit (Vector Laboratories, Burlingame, CA) was then used for 10 min to detect anti-immunoglobulin antibody activity followed by washing with deionized water and then PBS, 3 times for 5 min. Sections were then incubated with biotinylated antibody for 30 min, washed with PBS, 3 times for 5 min and then incubated with ABC-AP Reagent (Vectastain ABD-AP kits, Vector Laboratories Inc., Burlingame, CA) for 1 h followed by washing with PBS as above. Alkaline Phosphatase Substrate Kit III (Vector Labs, Inc., Burlingame, CA) was then applied for 30 min to detect antineuropeptide antibody activity. After washing with tap water, the sections were counterstained with eosin, de-hydrated with stepwise increases of ethanol, cleared with xylene and mounted in non-aqueous solution. Sections were then photographed at 400x magnification and cells scored per field. Single fibers were counted per field, and fibers associated with either CD4-positive cells or CD8-positive cells. Ten fields per case were scored. ANOVA and "t" tests were performed between groups. Step slides were also stained by immunoperoxidase for macrophage-specific antigen (DAKO). Total number of macrophages per section were subtracted from totals because these cells express CD4.

RESULTS

Sections of normal mucosa, showed substance P fibers in both mucosal and submucosal tissues. SP fibers reached high into the epithelium and T4 cells were seen in close proximity to SP-positive fibers. In contrast to SP fibers, VIP fibers were seen mostly in submucosal nerve fibers and nerve ganglions. Few T4 or T8 positive cells were observed in the vicinity of these fibers, some of T4 cells had abundant cytoplasm and were not enumerated. In addition, Mac+ in CD4-stained and CD8-stained step sections were subtracted from totals. More T4 cells were associated with SP fibers than did T8 cells. A 3.7 fold difference between T4 and T8 cells associated with SP fibers was observed. VIP fiber/T-cell association was minimal.

The data summarized in the table, show a significant difference (p[<0.003) between the percentages of T4 cells and T8 cells associated with SP fibers. There was, on the other hand, no difference between the T4 and T8 cells in their association with VIP fibers.

Table 1. Association of SP and VIP fibers with T4 or T8 in normal colon.

	SP-fiber associated n=4	VIP-fiber associated n=4
% T4 Cells	66.7 ± 3.9	0.67 ± 0.4
% T8 Cells	17.6 ± 2.7	1.3 ± 0.8
p ≤	0.003	n.s.

CONCLUSIONS

We have found that a close proximity of T4 lymphocytes with SP fibers exists, and is more significant than the association of T8 cells with these nerve endings. Our findings suggest a role for these fibers in the migration of T4 cells to these locations and/or the maturation of T cell-dependent B cells in the lamina propria of the gut. A heterogeneity of T helper cells has been described in humans and in the mouse (Th1 and Th2).[8] Th2 cells have been shown to synthesize Il-5, a critical lymphokine for the production of IgA by IgA-committed B cells.[9] Though we did not test for these subpopulation sin the present study, this obviously represents potential future studies in lymphocyte neuropeptide-fiber correlation.

SP belongs to a peptide family termed tachykinins derived as the gene product of preprotochynin gene.[10] Prior work has shown that SP-innervation in mesenteric lymph nodes (MN), occurs in a scant manner and mainly in T cell rather than B cell areas[11] but in Peyer's Patches (PP), nerve fibers are in close apposition to both T and B cells.[4] In addition, receptors for SP exist on both T and B cells.[12] SP in the presence of Con A is capable of stimulating IgA production 300% in PP compared to 40% in MN.[13] This suggests that SP is involved in the switching process of lymphocytes from gut-associated lymphoid tissue. Our finding of this close proximity of SP fibers and T4 cells, and the finding by us and others of a close proximity between SP fibers and IgA-bearing lymphoid cells suggests at least two hypothetical mechanisms for the role of SP on the regulation of IgA production. In one SP could be involved in directly stimulating, along with other mediators, the production of IgA, and in another, SP could directly stimulate T4 cells that would effect neighboring IgA lymphoblasts, to either further differentiate or produce increased amounts of IgA.

REFERENCES

1. K. Croiton, P. B. Ernst, A. M. Stanisz, and J. Bienenstock, *in:* "Immunology and Immunopathology of the Liver and Gastrointestinal Tract", Shanahan and Torgan, eds., p. 183, Igaku Shoin Tokyo (1989).
2. K. L. Bost, *Prog. Allerg.* 43:68 (1988).
3. S. Y. Felten and J. Olschowka, *J. Neurosci. Res.* 18:37 (1987).
4. R. Stead, J. Bienenstock, and A. M. Stanisz, *Immunol. Rev.* 100:333 (1987).
5. N. Nagamoto, J. Cintron, A. M. Stanisz, R. Buschmann, E. B. Rypins, and P. Weisz-Carrington, *in:* "Recent Adv. Mucosal Immunol.", J.R. McGhee, J. Mestecky, J. Sterzl, and H. Tlaskalova, eds., Abs., Prague, Czechoslovakia (1992).
6. P. K. Nakane and G. B. Pierce, *J. Cell. Biol.* 33:307 (1967).
7. R. C. Curren and J. Gregory, *J. Clin. Pathol.* 31:974 (1978).
8. T. R. Mosmann and R. L. Coffman, *Immunol. Today* 8:233 (1987).
9. R. L. Coffman, B. Shrader, J. Carty, T. R. Mosmann, and M. W. Bond, *J. Immunol.* 139:3685 (1987).
10. H. Nawa, H. Kotani, and S. Nakanishi, *Nature* 12:729 (1984).
11. P. Popper, C. R. Mantyh, S. R. Vigna, J. E. Maggio, and P. W. Mantyh, *Peptides* 9:257 (1988).
12. A. M. Stanisz, R. Scicchitano, P. Dazin, J. Bienenstock, and D. G. Payan, *J. Immunol.* 139:749 (1987).
13. A. M. Stanisz, D. Befus, and J. Bienenstock, *J. Immunol.* 136:152 (1986).

CHOLECYSTOKININ-OCTAPEPTIDE (CCK-OP) AND SUBSTANCE P (SP) INFLUENCE IMMUNE RESPONSE TO CHOLERA TOXIN IN LIVE ANIMALS

Sana Jarrah,[1] Maya Eran,[1] Serem Freier,[1] and
Etan Yefenof[2]

[1]Shaare Zedek Medical Center
[2]Department of Immunology
Hebrew University Medical School, Jerusalem, Israel

INTRODUCTION

The immune response in the intestine is initiated by antigen uptake at the level of Peyer's patches (PP)[1] and along the small intestine.[2,3] A series of initiating and enhancing signals are required to effect T cell activation by polypeptide antigens. A degree of antigen processing by the antigen-presenting cell (APC) to reveal epitopes reactive with the T cell receptor is usually required.[4] Interaction of the processed antigen with the T cell receptor is regulated by products of the major histocompatibility complex residing in the APC surface membrane.[5] Release by the APC of the cytokine IL-1 is required to induce the synthesis of receptors for IL-2 on the T cell.[6] These processes regulate the subsequent production of specific antibodies by B cells. It may be desirable, in certain circumstances, to enhance the immune response. An obvious example is the administration of oral vaccines. On the other hand, there may be occasions when it is desired to abrogate the immune response, for instance, in order to prevent hypersensitivity responses. One method of enhancing the immune response in the intestine is by the concomitant administration of the antigen with an appropriate adjuvant. In this report, we wish to investigate the potential adjuvant effect of some peptide hormones in regulating the immune response to foreign antigens in the intestine. We chose the peptide hormones, substance P (SP) and cholecystokinin-octapeptide (CCK-OP) as well as their respective antagonists, spantide and L 316,718. In addition, we investigated the effect of capsaicin, which is a toxin to SP containing-nerve fibers.

METHODOLOGY

The experiments were performed on C57/black mice. Miniosmotic pumps (Alzet model 2001, Alza Corp. Palo Alto, Cal 94303) were implanted subcutaneously under anesthesia. They contained the test substance and were designed to deliver it at the rate of 1 μl/h for one week. The pump was changed weekly for a total of 3 weeks. On the day of the first implantation, the mice were immunized orally with whole cholera toxin (CT) in bicarbonate buffer 10μg/0.5 ml. A booster immunization was given at 14 days. After three weeks the mice were purged and the stools collected for measurement of IgA CT-specific antibodies.

In the control group CT immunization was administered as above and empty pumps were inserted subcutaneously.

In a third group, capsaicin was injected subcutaneously to a group of mice between 2-6 days old in a dose 50mg/kg. At the age of 6 weeks they received CT immunization in the schedule described in the previous paragraph. Three weeks later, they were purged and the stool IgA CT-specific antibody estimated by ELISA assay.

Groups:

1. Control-CT immunization only
2. CT immunization + SP by miniosmotic pump
3. CT immunization + SP antagonist, spantide, by miniosmotic pump
4. CT immunization + CCK-OP by miniosmotic pump
5. CT immunization + CCK antagonist L 314,718 by miniosmotic pump
6. CT immunization + capsaicin at birth5

RESULTS

The administration of the SP antagonist, spantide, (Peninsula Laboratories, Belmont, CA 94002), caused a depression of intestinal antibodies to CT. This was significantly different from the SP-treated group ($p<0.001$) (Fig. 1).

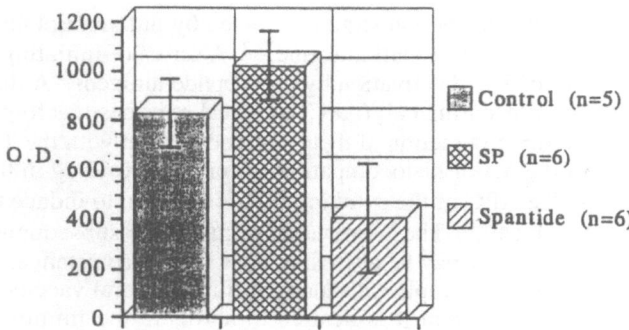

Figure 1. The SP antagonist, spantide, caused a significant ($p<0.001$) depression of intestinal CT antibodies compared to the SP-treated group (results are expressed in OD units).

The administration of the CCK antagonist L 364,718 (Merck, Sharp and Dohme, Rahway, NJ) caused a significant depression of intestinal antibodies to CT when compared with the CCK-OP treated group ($p<0.001$) (Fig. 2).

The injection of capsaicin (Sigma Laboratories, St. Louis, MO) to neonatal mice caused a significant depression of the intestinal immune response to CT (Fig. 3).

DISCUSSION

The reasons for choosing SP as a potential candidate for modulating the intestinal immune response were as follows: SP is a brain gut peptide that is widely distributed in the gastrointestinal tract. *In vitro*, SP has been shown to cause proliferation of PP cells with or without co-stimulation with PHA or ConA and to produce increased secretion of IgA and IgM.[7] The same holds true for CCK-OP (Jarrah S. *et al.*, to be published). In the case of

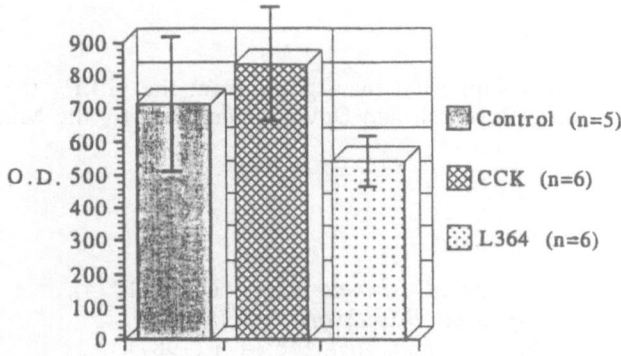

Figure 2. The CCK antagonist, L364,718, caused a significant (p<0.001) depression of intestinal CT antibodies compared to the CCK-OP group (results are expressed in OD units).

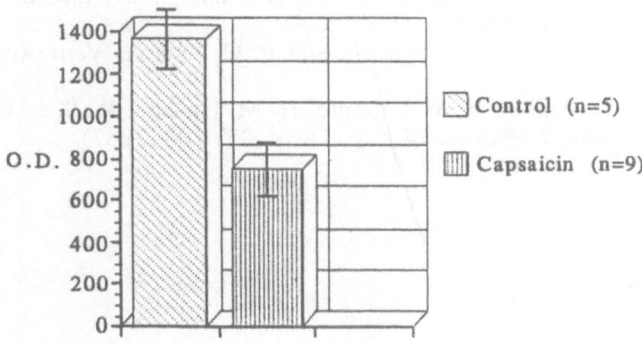

Figure 3. The injection of capsaicin to neonatal mice caused a significant depression of the intestinal immune response to CT (p<0.001).

SP, the increased production of IgA and IgM is associated with α-chain and μ-chain production. Prolonged stimulation with SP also augments Ig-production in PP cells when these are subsequently grown in tissue culture.[8] Receptors for SP have been shown on PP cells.[9] These cells also have receptors for CCK-OP (Jarrah S. *et al.*, to be published.)

Capsaicin is a toxin to nerves containing SP, causing their degeneration. When administered at the age and dose used by us it is specific to these nerves.[10]

Spantide is an NK 1 receptor antagonist and L 364,718 is a non-peptide, peripheral CCK antagonist.[11]

Our data show that the two peptide hormones studied by us play a role in the regulation of the intestinal immune response to foreign antigens. The prolonged administration of these peptides at the time of immunization with CT only slightly raised the production of specific antibodies. On the other hand, the administration of the respective antagonists or of the SP nerve toxin significantly depressed the production of antibody. These findings suggest that optimal amounts of SP and CCK were present and participated in promoting the immune response. The absence of these peptides impaired this response.

ACKNOWLEDGMENT

This research was supported by a grant from the G.I.F., the German-Israeli Foundation for Scientific Research and Development, and by the Miriam Coven-Fish Foundation.

REFERENCES

1. R. L. Owen and A. L. Jones, *Gastroenterology* 66:189 (1974).
2. P. W. Bland, *Adv. Exp. Med. Biol.* 216A:219 (1987).
3. Ll. Mayer and R. Schlien, *J. Exp. Med.* 166:1471 (1987).
4. K. Ziegler and E. Unanue, *J. Immunol.* 127:1869 (1986).
5. K. C. Lee, A. Wilkinson, and M. Wong, *J. Cell. Immunol.* 48:79 (1979).
6. J. Kaye, S. Gillis, S. B. Mizel, E. M. Shevach, T. R. Males, C. A. Dinarello, L. B. Lachman, and C. A. Janeway, *J. Immunol.* 133:1339 (1984).
7. A. M. Stanisz, D. Befus, and J. Bienenstock, *J. Immunol.* 136:152 (1986).
8. R. Scicchitano, J. Bienenstockl, and A. M. Stanisz, *Immunology* 63:733 (1988).
9. A. M. Stanisz, R. Scicchitano, P. Dazin, J. Bienenstock, and D. G. Payan, *J. Immunol.* 139:794 (1987).
10. J. L. Nagy, S. P. Hunt, L. L. Iversen, and P. C. Emson, *Neuroscience* 6:1923 (1981).
11. V. J. Lotti, R. G. Pendelton, R. J. Gould, H. M. Hanson, R. S. L. Chang, and B. V. Clineschmidt, *J. Pharmacol. Exp. Thera.* 231:103 (1987).

NEURITE OUTGROWTH INDUCED BY RAT LYMPHOID TISSUES *IN VITRO*

Y. Kannan,[1,2] R. H. Stead,[1,2] C. H. Goldsmith,[3] and J. Bienenstock[2,4,5]

[1]Intestinal Disease Research Unit, [2]Pathology, [3]Clinical Epidemiology and Biostatistics, [4]Medicine, and [5]Molecular Virology and Immunology Program, McMaster University, Hamilton, Ontario, L8N 3Z5, Canada

INTRODUCTION

Growing evidence for sympathetic innervation of primary and secondary lymphoid organs suggests an important functional link between the nervous and immune systems. Histochemical studies have shown that noradrenergic sympathetic fibers abundantly innervate the vasculature and parenchymal fields of lymphocytes, and are directed particularly into zones of T cells.[1]

To observe target-induced extension of neurites, several co-culture systems have been reported,[2,3] and preferential neurite outgrowth has been found towards tissues that are directly innervated *in situ*, compared with normally sparsely innervated tissues. Several neurite outgrowth-inducing factors have been suggested to be present in nerve target organs. Nerve growth factor (NGF) is the best known neurotrophic factor affecting sympathetic and sensory neurons.

The present study was carried out to quantitate the extent and relative efficacy of the neurostimulators from several lymphoid organs. This was accomplished by coculturing rat superior cervical ganglia (SCG) with tissue explants in basement membrane Matrigel. We also examined the role of NGF in the stimulation of neurite outgrowth by lymphoid tissues, using affinity purified rabbit anti-NGF.

METHODS

Explants

SCG were isolated from neonatal Wistar rats. Thymus, spleen and mesenteric lymph nodes (MLN) were isolated from adult male Wistar rats. MLN were also isolated from rats after Nb-infection (Nb-MLN 1 week (W), 2W, 3W, or 4W). MLN 2W to 4W were macroscopically noted to be increased in size. Spinal cord (SC, as negative control) and heart (as positive control) were isolated from neonatal rats.

Cultures

Three-dimensional cultures were made using basement membrane Matrigel in 24well culture plates. Matrigel (0.3 ml) diluted with medium containing 0.1% bovine serum albumin, was added to each well. SCG and various target tissue explants were set 1 mm apart in Matrigel and cultured at 37° C for 48 hr. In one series of cultures (Series 1), one explant was surrounded by four SCG for assessment of the stimulation on neurite outgrowth by the various tissues. In another series of cultures (Series 2), one ganglion was surrounded by three different tissue explants to confirm the relative efficacy of neurite outgrowth induction.[4] To determine the role of NGF in the target tissues, 10 µg/ml of polyclonal rabbit anti-NGF[5] or rabbit serum Ig fraction were added to the co-cultures. This amount of anti-NGF was found to inhibit totally the dense nerve growth from rat SCG observed in cultures with 1 - 10 ng/ml NGF.

Quantitative Estimation

The cultures were examined on an inverted microscope with phase contrast optics. Neurite outgrowth was scored after 12, 24, and 48 hr in culture by counting the number of nerve fibers crossing an eyepiece graticule positioned at 0.5 mm between the ganglion and the target tissues, according to Ebendal.[6] The mean value and the standard error of the mean were determined. In series 1, one-way analysis of variance (ANOVA) was performed to determine significant differences of neurite numbers growing towards different explants at 12, 24, and 48 hr in culture. When these were found to be significant, independent two-sample Student's t-tests were used to assess the differences between the mean neurite number induced by each explant, compared to the negative control (SC). In series 2, ANOVA of the neurite number growing towards the different explants was performed. The level of significance was $p < 0.05$ and no adjustment was made for multiple comparisons.

RESULTS

Neurite Outgrowth Induced by Lymphoid Tissue Explants in Co-Culture

In Series 1 (Table 1, A), although variable neurite outgrowth was observed towards targets, ANOVA of the number of neurites did not reveal a statistically significant difference at 12 hr ($F_{(4,122)} = 1.01$, $p = 0.403$). At 24 hr, neurite outgrowth was greater than 12 hr, and there was a significant difference in the density of outgrowth towards targets ($F_{(4,124)} = 4.38$, $p < 0.010$). Neurite outgrowth towards heart 07 < 0.001), thymus, $p < 0.001$) or spleen $p < 0.050$), but not MLN, were significantly greater than towards SC. At 48 hr, there was greater variation of neurite outgrowth ($F_{(4,126)} = 10.21$, $p < 0.001$) compared to 24 hr culture. The number of neurites towards heart, thymus and spleen, but not MLN, were still significantly greater than SC. In Series 2, the relative effects seen in Series 1 were confirmed when SCG were surrounded by SC, heart and thymus, or thymus, spleen and MLN (data not shown).

Effect of Nb-Infected MLN Explants on Neurite Outgrowth

In Series 1 (Table 1, B), there was a post-infection time-dependent increase of neurite outgrowth towards Nb-infected MLN explants, especially after 48 hr of culture. The maximum outgrowth was observed towards Nb-MLN 3W explants. When a single SCG was cultured with MLN from uninfected, 2W and 4W post-infection animals in Series 2, the extent of neurite outgrowth towards Nb-MLN 2W was greater than uninfected MLN or Nb-MLN 4W (data not shown).

Table 1. Neurite outgrowth induced by various target explants (Series 1).[1]

	Target explants	12h	24h	48h
A[2]	SC 4 + 2-1(33)	22 + 4.0(34)	22 + 3.5(34)	
	Heart	13 + 4.2(30)	53 + 7-4(30)**	65 + 6.6(30)***
	Thymus	16 + 3.1(29)	48 + 5-1(29)+*	64 + 9.3(28)***
	Spleen	9 + 2.7(20)	45 + 11.0(20)*	49 + 6-5(20)"*
B	MLN	13 + 7.2(15)	27 + 8.0(16)	30 + 5.0(19)
	B MLN	13 + 7.2(15)	27 + 8.0(16)	30 + 5.0(19)
	Nb-MLN 1w	4 + 23(15)	29 + 7.6(15)	43 + 7.2(15)#
	Nb-MLN 2w	14 + 4.0(21)	33 + 4-4(21)	54 + 7-6(21)^
	Nb-MLN 3w	8 + 1.4(28)	35 + 4-4(26)	60 + 8.9(26)*"
	Nb-MLN 4w	8 ± 2.4(12)	20 ± 3.9(12)	38 ± 7.6(12)*

[1]Values represent the mean numbers of neurites + SEM, calculated from the number of SCG, indicated in parentheses. *$p < 0.05$, **$p < 0.01$, and ***$p < 0.001$, compared with SC by independent two sample Student's t-tests.
[2]A: normal tissues; B: Nb-infected tissues

Blockade of Lymphoid Tissue-Induced Neurite Outgrowth by Anti-NGF

When 10 µg/ml of anti-NGF was added to co-cultures of SCG and heart, thymus or spleen, there was a significant inhibition of neurite outgrowth towards these tissues (Table 2). At 24 hr, anti-NGF inhibited neurite outgrowth by 85 %, 80 % and 77% compared to rabbit serum treated heart, thymus and spleen. At 48 hr, anti-NGF still significantly inhibited neurite outgrowth induced by these three tissues (86%, 70% and 75% inhibition, respectively). However, there was an increased in neurite outgrovth towards anti-NGF treated thymus and spleen in 48 hr cultures compared to 24 hr (39% and 28% increase, respectively), in contrast to anti-NGF treated heart (0 % increase). These results suggest the presence of other molecules than NGF that affect nerve growth in thymus and spleen.

Table 2. Effect of anti-NGF on neurite outgrowth in co-culture.[1]

	24h			48h		
Targets	Control	Anti-NGF	Inhibition	Control	Anti-NGF	Inhibition
Heart	59 + 9.4(18)	9 + 2.7(22)	85%	63 + 10.3(14)	9 + 2.8(16)	86%
Thymus	55 + 9.9(20)	11 + 2.5(25)	80%	61 + 8.1(19)	18 + 3.9(21)	70%
Spleen	57 ± 9.1(14)	13 ± 2.9(22)	77%	73 ± 7.5(21)	18 ± 4.4(21)	75%

[1]10 µg/ml polyclonal anti-NGF or rabbit serum were added to co-cultures of SCG and heart, thymus or spleen. The mean numbers of neurites + SEM were calculated from the number of SCG, indicated in parentheses.

DISCUSSION

Our data suggest the existence of neurotrophic/tropic factors in lymphoid tissues whose levels are modified during an inflammatory response. It appears that the predominant neurite growth factor in thymus and spleen is NGF, as it is in the heart,[7] although indirect stimulation of NGF production in SCG by other factor(s) can not be excluded. NGF synthesis in lymphoid tissues has not yet been shown, but relatively high concentrations of

NGF have been detected in rodent spleen.[8] We are now trying to determine the key cells that produce NGF and modulate its synthesis. We are also investigating whether the major neurotrophic factor involved in Nb-MLN is NGF or other cytokines released by activated lymphocytes, since several lymphokines are known to modulate the function and maintenance of the nervous system, including interleukins 1, 2, 3 and 6, and granulocyte/macrophage colony stimulating factor.[9-13]

REFERENCES

1. D. L. Felten, S. Y. Felten, D. L. Bellinger, S. L. Carlson, K. D. Ackerman, K. S. Madden, J. A. Olschowki, and S. Livnat, *Immunol. Rev.* 100:225 (1987).
2. T. Lahtinen, and O. Eranko, *Dev. Biol.* 62:189 (1984).
3. M. Tessier-Lavigne, and M. Placzek, *Trends in Neurosci.* 14:303 (1991).
4. T. Ebendal, and C-O. Jacobson, *Exp. Cell. Res.* 105:379 (1977).
5. J. Diamond, M. Holmes, and M. Coughlin, *J. Neurosci.* 12:1454 (1992).
6. T. Ebendal, *Dev. Biol.* 72:276 (1979).
7. B. B. Randon, *Dev. Brain. Res.* 59:49 (1991).
8. R. Kato-Semeda, R. Semeda, S. Kashiwamata, and K. Kato, *J. Neurochem.* 52:1559 (1989).
9. T. Satoh, S. Nakamura, T. Taga, T. Matsuda, T. Hirano, T. Kishimoto, and Y. Kaziro, *Mol. Cell. Biol.* 8:3546 (1988).
10. P. K. Haugen, and P. C. Letourneau, *J. Neurosci. Res.* 25:443 (1990).
11. M. Kamegai, Y. Konishi, and T. Tabira, *Brain Res.* 532:323 (1990).
12. M. Kamegai, K. Niijima, T. Kunishita, M. Nishizawa, M. Ogawa, M. Araki, A. Ueki, Y. Konishi, and T. Tabira, *Neuron* 2:429 (1990).
13. R. A. Gadient, K. C. Cron, and U. Otten, *Neurosci. Lett.* 117:335 (1990).

EXERCISE, STRESS AND MUCOSAL IMMUNITY IN ELITE SWIMMERS

M. Gleeson,[2] W. A. McDonald,[1] A. W. Cripps,[2] D. B. Pyne,[1]
R. L. Clancy,[2] P. A. Fricker,[1] and J. H. Wlodarczyk

[1]Australian Institute of Sport, Canberra, ACT, Australia
[2]Hunter Immunology Unit, Hunter Area Pathology Service
and Faculty of Medicine, University of Newcastle, Newcastle, NSW,
Australia

INTRODUCTION

Elite athletes have been reported to be susceptible to upper respiratory tract infections (URTI), particularly during the period immediately prior to major competitions.[1] Studies of the effects of exercise on immune parameters have shown that alterations in systemic immunity and cytokine levels are related to the intensity of the exercise and fitness of the athlete.[2] In elite athletes decreases in salivary IgA levels have been observed following intense endurance exercise[3,4] but it is not clear whether the changes are associated with an increased incidence of URTI.[5] Psychological stress has also been shown to decrease salivary IgA levels,[6] but the relevance of this observation to the immune fitness following fatiguing exercise is unclear. This prospective study assessed the impact of long term exercise (physical stress) and psychological stress on systemic and mucosal immunity and the relationship to URTI in a cohort of elite swimmers.

SUBJECTS

Athletes and Controls

The 1990 Australian Institute of Sport (AIS) swimming squad consisted of 27 elite swimmers (16 males, 11 females) aged 16-24 years. The athletes were training 25 h per week in preparation for the World Championship Trials. Twelve Australian Institute of Sport staff (7 males, 5 females) aged 19-41 years, who were involved in regular but moderate exercise programmes acted as environmental controls.

STUDY DESIGN

Long Term Exercise Effects

The athletes and controls were studied monthly from March to October 1990. Blood for lymphocyte subsets and IgG subclasses was collected at the beginning of the study period and the lymphocyte subsets were repeated at the end of the study. A psychological stress test was completed and serum and saliva were collected for Ig and albumin prior to the testing session each month and saliva again immediately after the training session. An assessment of the exercise intensity for the training session and the distances swam for the session and during the previous week and previous month were recorded. A throat swab was collected to determine the quantitative routine bacterial carriage rates.

Infection Episodes

Every infection episode was assessed clinically by the medical team. A psychological stress score was completed, serum and saliva were collected for immunoglobulins and albumin and a throat swab taken for culturing during each URTI.

METHODS

White Cell Count and Lymphocyte Subsets

The total white cell count (WCC) and percentage (%) of lymphocytes were measured on a Model Stks Coulter Counter (Coulter Electronics, USA). Commercial monoclonal markers (Coulter Electronics, USA) were used to determine numbers and percentages of T-Cells (CD3, CD4, CD8), B-Cells (CD19), NK cells (CD56) and an activation marker (HLA-DR) by flow cytometry using an EPICS II Flow Cytometer (Coulter Electronics, USA).

Serum and Salivary Immunoglobulins and Albumin

IgA, IgG, IgM and albumin were measured in serum by rate nephelometry using a Beckman ARRAY analyser and Beckman antisera and calibration material (Beckman, USA). The same proteins were measured in saliva by electroimmunodiffusion[7] using commercially prepared IgA-specific antisera (Tago, USA) and IgG, IgM and albumin specific antisera (Dako - Immunoglobulins, Denmark), calibrated with Standard Human Serum (Behringwerke, Germany).

Throat Swabs

Calcium alginate throat swabs were collected into transport media and cultured on blood agar and chocolate bacitracin agar. The quantitative carriage rate[8] was determined for β-haemolytic streptococci, ß-haemolytic streptococci, non-haemolytic streptococci, *Streptococcus pneumoniae*, Corynebacterium sp., Neisseria sp., Haemophilus sp., Staphylococcus sp., *Pseudomonas aeruginosa*, coliform organisms, and yeast.

Psychological Stress Score and Exercise Intensity

The Spielberger State-Trait Anxiety Inventory STAI Form Y self evaluation questionnaire[9] was used to determine the stress score. The reference range stated for this age group was 26-46.

Each study session was graded according to physical intensity based on an aerobic/anaerobic classification[10] and on the distance swum in the study session, during the previous 7 days and previous 4 weeks. The phase of training was classified as Endurance, Quality, Recovery, Taper and Competition.[10]

Statistics

For comparisons between athletes and controls and between infection and routine samples repeated measurements on an athlete were assumed to be independent.[11] Changes during the training session were assessed by taking the difference between last and first measurement for each athlete. Non-parametric methods were used for all comparisons.

EFFECT OF LONG TERM EXERCISE TRAINING ON SYSTEMIC IMMUNE RESPONSES

Serum

Mean **serum IgA, IgG and IgM** were lower in athletes compared with controls (Table 1).

Table 1. Mean serum immunoglobulin and albumin levels for athletes and controls.

Serum (g/L)	Athletes (n=220)	Controls (n=83)	Significance Level
IgA	1.60	1.83	p=0.0120
IgG	10.36	10.90	p=0.0264
IgM	1.17	1.48	p=0.0001
Albumin	42.83	43.08	p=0.3418

Serum IgG and IgM levels did not change significantly during the study period in athletes or controls. Serum albumin and IgA levels were slightly higher at the end of the study period in athletes. The mean increase for serum albumin in athletes was 1.2 g/L [2.8%] and for serum IgA was 0.07 g/L [4.4%].

Lymphocyte Subsets

There was a significant decrease in the total numbers (0.068×10^9 cells [51 %]) and percentages (2.1% of cells [35%]) of NK cells in athletes but not controls at the end of the study period. There were no changes in total WCC, B cell and T cell subsets or activation markers in athletes or controls.

EFFECT OF LONG TERM EXERCISE TRAINING ON LOCAL IMMUNE RESPONSES

Saliva

There was a significant increase in pre-exercise (mean = +17.3 mg/L [47 %] p=0.003) and post-exercise (mean = +16.0 mg/L [36%], p=0.019) salivary albumin levels over the study period in athletes. There was a substantial decrease in the pre-exercise salivary IgA levels (mean = -10.8 mg/L, 17%, p=0.054) over the study period in athletes. The post-exercise decrease in salivary IgA (mean = -10.1 mgL, 16%) was not statistically significant (p=0.141). There was a decrease in the ratio of salivary IgA/albumin in pre-exercise (-1.54) and post-exercise (-1.03) saliva in athletes over the study period. There were no trends in either pre-exercise or post-exercise salivary IgG levels in athletes or controls. There were no changes in the pre-exercise salivary IgM levels or the detection rates over the study period in athletes. The detection rate for salivary IgM increased in the post-exercise saliva collected in October (pre-competition) in athletes. There were no trends observed in controls for either pre or post-exercise salivary IgM.

EXERCISE LOAD - STRESS SCORE - INFECTION RATE

There was an inverse relationship between stress score and exercise load as measured by physical intensity and session volume (kms swam at session). The maximum monthly stress score correlated with the number of infections. A stress score threshold of 45 appeared to be associated with an increase in infection rate in athletes and controls. The stress score was significantly higher (p = 0.0001) in infection-prone periods (mean score = 43) compared to non-infection periods (mean score = 34). There was less salivary IgM detected in both pre and post-exercise samples collected from athletes during infection periods compared with non-infection periods. There was no significant differences in bacterial carriage rates between athletes and controls throughout the study period for any organism and no significant changes in infection periods in either athletes or controls.

CONCLUSIONS

The trend of lower serum Ig levels in athletes compared to controls prior to commencing the study may indicate chronic systemic immune suppression in elite athletes after years of intense training. The significant decreases in both pre and post exercise salivary IgA levels in athletes over the 8 months training period indicate a chronic local immune suppression with intensive exercise training. The increase in salivary albumin over the study period and the changes in IgA/albumin ratios indicate that the decrease in salivary IgA was not due to alterations in hydration or mucosal transudation. The positive correlation between stress score and infection rate suggests neuroendocrine influences may play a role in the exercise/stress induced alterations of local immune function. The inverse correlation between detection of salivary IgM and infection rate may indicate that during periods of IgA suppression after intensive exercise training, compensation with salivary IgM may have a protective effect against the development of respiratory infections. The significant decrease in NK cells over the training period may leave athletes susceptible to infection.

ACKNOWLEDEGMENTS

The authors wish to thank members and coaches of the 1990 AIS Swimming Squad and the AIS staff who participated in this study, in particular Sr Sue Beasley. The assays were performed by the staff of the Hunter Immunology Unit at Royal Newcastle Hospital and the Newcastle Mater Misericordiae Hospital. The project was funded by the Australian Sports Commission through the Australian Institute of Sport, Department of Sports Medicine.

REFERENCES

1. J. Weidemann, *et al., Today's Life Sci.* 4 (7): 24 (1992).
2. L. Fitzgerald, *et al., Immunol. Today* 9: 337 (1988).
3. T. B. Tomasi, *et al., Clin. Immunol.* 2: 173 (1982).
4. L. T. Mackinnon, *et al., Sports Training Med. Rehab.* 1: 1 (1989).
5. L. T. Mackinnon, *et al., In*: Behaviour and Immunity. Ed. A J Husband p. 169. (1992).
6. J. B. Jemmott, *et al., Lancet.* i:1400 (1983).
7. M. Gleeson, *et al., Aust. N. Z. J. Med.* 12: 255 (1982).
8. H. L. Butt, *et al., Pathology* 20: 253 (1988).
9. C. D. Spielberger, *et al., In*: Manual for State-Trait Anxiety Inventory, Consulting Psychologists Press Inc, California. (1983).
10. D. B. Pyne *et al., Excel* 5:9 (1988).
11. M. Gleeson, *et al., Scand J. Immunol.* 33: 533 (1991).

STRUCTURE OF IgA: FACTS AND GAPS IN OUR DATA ON DISULFIDE BONDS

Jeike Biewenga

Department of Cell Biology, Medical Faculty
Vrije Universiteit, Amsterdam, The Netherlands

INTRODUCTION

The mucosa is provided with a unique immune system that can generate specific IgA antibodies against pathogens and unresponsiveness against non-pathogenic substances. The function of the specific IgA antibodies is to protect the mucosal surfaces against further pathogen infiltration. IgA is quantitatively the major Ig isotype in man and laboratory animals. Nevertheless, the structure and function of IgA are not known in as much detail as those of IgG.

AMINO ACID SEQUENCE

The complete amino acid sequences of the α chains of human IgA1 and IgA2 as well as of mouse and rabbit IgA have been determined. When comparing these sequences it is obvious that the constant domains of the α chains are well preserved and that homology increases from CH1 to CH2 and CH3 domains. The hinge regions show little homology (Table 1), which is in accordance with the major role of the constant domains, not the hinge region, in the function of IgA.

Table 1. Amino acid sequence homology between human, mouse and rabbit IgA

Species compared	Homology			
	CH1 domain (176-223)	CH2 domain	CH3 domain	Hinge region
Human vs mouse	51%	55%	69%	18%
Human vs rabbit	35%	50%	70%	9%
Rabbit vs mouse	30%	69%	59%	5%
Human/mouse/rabbit	21%	40%	52%	0%

Adapted from Knight et al.[1]

Advances in Mucosal Immunology, Edited by
J. Mestecky et al., Plenum Press, New York, 1995

DISULFIDE BONDS

Most the half-cystines in α chains are conserved in man, mouse, and rabbit, especially those forming the intra-domain bonds that stabilize the domain structures (Table 2). The α chains of human, mouse, and rabbit IgA are linked by two interchain disulfide bonds. One is at position 242, the other probably at 301. The function of further half-cystines in the CH1 and CH2 domains is not clear.

Table 2. Half-cystines in the CH1 domain, hinge region and CH2 domain, except for the residues involved in intra-domain bonds

	Chain region and amino acid position											
	CH1		HR						CH2				
Protein	196	220	223	224	229	230	232		241	242	299	301	311
Human IgA1/IgA2[a]	+	+	−	−	−	−	−		+	+	+	+	+
Murine IgA1[b]	+	+	−	−	−	−	−		−	+	+	+	+
Rabbit IgA-g[c]	+	−	+	+	+	+	+		−	+	−	+	−

[a] from Putman et al.[2], Torano and Putman[3]; [b] from Robinson and Appella[4]; [c] from Knight et al.[1]

In the CH1 domain human and mouse α chains contain two half-cystines at positions 196 and 220 which form a disulfide bond. Cys196, but not Cys220, is present in rabbit α chains. It could be that in rabbit α chains the Cys196 residue is linked to one of the five hinge region half-cystines.

In the CH2 domain human α chains contain three half-cystines of which Cys 241 is missing in mouse and rabbit. In addition, the Cys residues at positions 299 and 311 are missing in α chains of the rabbit IgA-g subclass. Cys311 may be present in the rabbit IgA-f subclass. The presence of this residue seems to be associated with binding of SC (see below). The function of Cys241 and Cys299 is completely unknown.

The hinge regions of human and mouse α chains are devoid of half-cystines. The rabbit α chain hinge region contains 5 Cys residues of which 4 are presumed to form inter-heavy chain disulfide bonds (Fig. 1). The fifth may be linked to Cys196 as mentioned.[1]

Residue:	223	230	238
Human IgA1:	Pro-Ser-Thr-Pro-Pro-Thr-Pro-Ser-Pro-Ser-Thr-Pro-Pro-Thr-Pro-Ser		
Human IgA2:	Pro-Pro-Pro		
Murine IgA:	Pro-Thr-Pro-Pro-Pro-Ile-Thr		
Rabbit IgA-g	Cys-Cys-Pro-Ala-Asn-Ser-Cys-Cys-Thr-Cys-Pro-Ser-Ser-Ser-Arg-Asn-Lys		

Adapted from Knight et al.[1]; numbering according to the a chain Bur (Putman et al.[2]).

Figure 1. Hinge region amino acid sequences of human, murine, and rabbit α chains

STABILITY OF DISULFIDE BONDS

The presence of free -SH groups, and the stability of disulfide bonds in proteins can be measured by blockage of -SH groups with 5,5'-dithio-(2,2'dinitro-)benzoate (DTNB) in unreduced or reduced proteins, respectively. In a recent study[5] we demonstrated that human IgA1 myeloma proteins and human S-IgA1 contain a similar number of free -SH groups as in rabbit S-IgA[6], with one free -SH per dimer or 0.5 free -SH per monomer. Under denaturing conditions this number increases slightly in human IgA1, but in rabbit IgA (IgA-g and IgA-f) an average of 7.5 free -SH groups was found after denaturation. Moreover, in rabbit IgA these free -SH groups were located in the $F(ab')_2$ fragments, whereas $F(ab')_2$ fragments of human IgA1 contain little free -SH, even after denaturation. Since the main difference between human and rabbit F(ab')2 fragments are the disulfide bonds in the hinge region (Table 2), these data indicate that the hinge region disulfide bonds in rabbit IgA are highly susceptible to reduction in the unfolded molecule.

We further studied the stability of the disulfide bonds in human IgA1, human IgA2 of both allotypes, and rat IgA by reduction with (a) glutathione (0.48 mg/ml), which exposes labile disulfide bonds, (b) glutathione and 0.01M DTT which exposes weak bonds, and (c) glutathione and DTT in the presence of 2% SDS to expose weak bonds in denatured IgA. The results (Table 3) showed that 4-6 labile half-cystines are present in human IgA1, including an IgA1 cryoglobulin, in human IgA2 of the A2m(2) allotype, and in rat IgA. The addition of DTT exposed 2-7 more half-cystines in these IgA proteins. It is likely that all reduced half-cystines were located in the α chains, since no half-cystines were reduced under these conditions in Fab fragments of human IgA1. The latter also shows that the Cys196-Cys220 bond is rather strong. In contrast, the inter-α chain bonds in human IgA1 are relatively weak as demonstrated by SDS-PAGE analysis of IgA1 proteins reduced with glutathione and DTT.[5] Reduction with glutathione alone exposed more half-cystines in IgA2 A2m(1) than in other IgA molecules, but the addition of DTT did not increase exposure of half-cystines in IgA2 A2m(1). This may be associated with the lack of H-L bonds in human IgA2 A2m(1). It would be interesting to analyse rabbit IgA molecules under non-denaturing conditions for labile and weak disulfide bonds, as these differ in hinge region disulfide bonds, and rabbit IgA-g molecules also lack H-L disulfide bonds.[1]

Table 3. Number of half-cystine residues in IgA proteins reduced with glutathione, glutathione and DTT, or glutathione + DTT + SDS.

Protein	Reduction with:		
	Glut.	Glut. + DTT	Glut. + DTT + SDS
Human IgA1[a]	3,8	9.6	18.6
Human IgA1, cryoglobulin[b]	5.3	8,4	13,2
Human IgA2 A2m(1)[c]	11.6	9.7	n.d.
Human IgA2 A2m(2)[c]	5.6	13.8	n.d.
Rat IgA[d]	5.7	7.7	17.7

[a] mean values, from Biewenga and van Run[5]; [b] protein kindly provided by Dr. R. Oosterom, Rotterdam; [c] 3 proteins obtained from Dr. J. Mestecky, Birmingham, Alabama; [d] monomeric protein, kindly provided by Dr. J.P. Vaerman, Brussels.

BINDING OF SECRETORY COMPONENT

The binding of SC to IgA dimers is conserved in man, mouse, and rabbit. SC is covalently bound in human IgA, mouse IgA, rabbit IgA-f, and non-covalently in rabbit IgA-g. The presence of J chain in the dimers is important for efficient binding, but it may not be a prerequisite for SC binding.[7] J chain may stabilize the structure of dimeric IgA and thereby enhance its SC binding capacity. J chain is disulfide linked to the penultimate Cys471 residues of one α chain of either IgA molecule of the dimer.[8] Besides binding the J chain, the CH3 domains seems to add little to the binding of SC, since a deletion of 36 amino acids from the CH3 domain of a mutant murine IgA dimer that contained J chain, did not abolish SC binding.[9]

Cys311 in the CH2 domain is most likely the half-cystine residue involved in covalent linkage of SC.[10] In rabbit IgA-g molecules Cys311 and also Cys299 are missing. Whether they are present in rabbit IgA-f molecules remains to be determined. Therefore, further studies on the mechanisms of SC-binding and the involved amino acids are needed. Underdown and coworkers[9] have approached this question by studying SC binding to murine IgA mutants with amino acid substitutions at different positions in the CH2 domain.

As shown for human dimeric IgA, the SC binding site is sensitive to physical and chemical manipulation e.g., to autoradiation and exposure to oxidizing agents.[7] This is not surprising because the binding of SC requires a disulfide interchange. The disulfide bond formed between SC and the α chain, however, is rather strong. Moreover the binding of SC protects the two monomers in S-IgA against reduction of the weaker inter-α chain bonds.[5] It would be interesting to study the effect of the amino acid substitutions in the CH2 domain on the stability of the aforementioned murine mutants.

FUNCTIONAL IMPLICATIONS

The capacity of Ig to bind antigen and to exert effector functions depends on an intact tertiary structure that has sufficient segmental flexibility. For instance, the angle between Fab fragments, which themselves are rigid structures, can vary significantly due to segmental flexibility. This enables the Fab fragments of a single molecule to bind to spacially close or more distant epitopes. For IgG, segmental flexibility is associated with the length and amino acid composition of the hinge region.[12] As for human IgA, the flexibility of IgA1 and IgA2 molecules seems to be similar, despite the shorter hinge region of the α2 heavy chain, suggesting that the lack of hinge region disulfide bonds in human IgA molecules is more important for segmental flexibility than the length of the hinge region. In this respect it would be interesting to compare human IgA1 and rabbit IgA-g, since both molecules have an extended hinge region, but rabbit IgA-g has 4 disulfide bonds in the hinge region, which are missing in human IgA1.

Complement activation is another feature that requires an intact tertiary structure. In studies on human IgA, Hiemstra et al.[13] showed that IgA1 and S-IgA1 when directly coated onto ELISA plate, activate the complement system only through the alternative pathway. This activation is F(ab')$_2$ mediated. Mild reduction of the IgA1 preparations abolished complement activation. Alternative, but not classical pathway, complement activation was also demonstrated for chimeric rabbit/mouse IgA antibodies bound to ELISA plates that were coated with specific antigen.[14] These chimeric IgA antibodies had rabbit α chains and mouse light chains.[15] Classical pathway activation was demonstrated by Jarvis and Griffiss[16] using human IgA1 antibodies bound to bacterial outer membrane proteins. Human IgA2 was not tested in these studies. Further studies are warranted to delineate the structural requirements for classical and alternative pathway complement activation by IgA proteins of different species and molecular forms.

IgA molecules vary strongly in susceptibility to proteolytic degradation. Mild reduction increases proteolytic degradation by increasing the accessibility of susceptible sites. S-IgA is less susceptible to proteolysis than monomeric IgA, which is attributed to blockage of proteolytic sites by SC. The fact that rabbit S-IgA-g is more susceptible to papain cleavage[17] than rabbit sIgA-f may be due to the fact that in S-IgA-g SC is bound non-covalently whereas in S-IgA-f it is bound by a disulfide bond. Moreover, IgA-g molecules lack an intra-domain bond in the CH2 domain, which is present in human IgA and possibly also in rabbit IgA-f molecules.[15] Degradation by specific bacterial IgA proteases has been investigated for human IgA by Kilian and coworkers.[18] These enzymes cleave IgA1 or IgA1 and IgA2 in the hinge region in a site-specific manner.

CONCLUSION

Further studies are warranted to determine the position of half-cystine residues in rabbit IgA-f and to delineate the function of the half-cystine residues in the hinge region (rabbit) and the CH2 domain of the IgA molecules of the different species.

REFERENCES

1. K. L. Knight, C. L. Martens, C. M. Stoklosa, and R. D. Schneiderman, *Nucleic Acid Res.* 12:1657 (1984).
2. F. W. Putnam, Y. S. V. Liu, Y-S. V. Low and T. L. K. *J. Biol. Chem.* 254:2865 (1979).
3. A. Torano and F. W. Putnam, *Proc. Natl. Acad. Sci. USA*, 75:966(1978).
4. E.A. Robinson and E. Appella, *Proc. Natl. Acad. Sci. USA* 77:4909 (1980).
5. J. Biewenga and P. E. M. van Run, *Mol. Immunol.* 29:327 (1992).
6. B. W. Elliott, B. Friedenson, and K. L. Knight, *J. Immunol.* 125:1611 (1980).
7. J. M. Schiff, M. M. Fisher, and B. J. Underdown, *Mol. Immunol.* 23: 45 (1986).
8. A. Garcia-Pardo, M. E. Lamm, A. G. Plaut, and B. Frangione, *J. Biol. Chem.* 256:11734 (1981).
9. I. C. Switzer, G. M. Loney, D. S. C. Yang, and B. J. Underdown (this volume) (1994).
10. A. Garcia-Pardo, M. E. Lamm, A. G. Plaut, and B. Frangione, *Mol. Immunol.* 16:477 (1979).
11. C. W. Hanly, C. H. Chang, and M. Schiffer, *Fed. Proc.* 44:1299 (abstract 5192) (1985).
12. R. Nezlin, *Adv. Immunol.* 48:1 (1990).
13. P. S. Hiemstra, J. Biewenga, A. Gorter, M. E. Stuurman, A. Faber, L. A. Van Es, and M. Daha, *Mol. Immunol.* 25:527 (1988).
14. R. D. Schneiderman, T. F. Lint , and K. L. Knight, *J. Immunol.* 145:233 (1990).
15. R. D. Schneiderman, W. C. Hanley, and K. L. Knight, *Proc. Natl. Acad. Sci. USA* 86:7561 (1989).
16. G. A. Jarvis and J. M. Griffiss, *J. Immunol.* 143:1703 (1989).
17. T. R. Malek, W. C. Hanly, and K. L. Knight, *Eur. J. Immunol.* 4:692 (1974).
18. M. Kilian and J. Reinholdt, *in:* "Medical Microbiology", C.S.F. Easmon and J. Jeljaszewicz, eds., Vol 5, p. 173, Academic Press, London (1986).

INTRA- AND INTER-CHAIN DISULFIDE BRIDGES OF J CHAIN IN HUMAN S-IgA

A. Bastian, H. Kratzin, E. Fallgren-Gebauer,
K. Eckart, and N. Hilschmann

Max-Planck-Institute for Experimental Medicine
Hermann-Rein-Straße 3
W3400 Göttingen, Germany

INTRODUCTION

S-IgA is synthesized in two steps. First the IgA dimer is linked to the J chain, and afterwards the $(IgA)_2$ J complex is linked to the Secretory Component (SC). Both linkages are accomplished by disulfide exchange and occur at different sites. The linkage of the J chain to IgA is the pre-requisit for the linkage of the SC to the IgA dimer.[1]

Here we describe the linkage of the J chain to IgA.

RESULTS AND DISCUSSION

S-IgA was purified from human colostrum according to Pincus et al.[2] In order to remove the Fab-fragments, S-IgA was split by *Streptococcus sanguis* protease prepared according to Plaut et al.[3], and the digest was separated by gel filtration on Sephacryl S-300. To remove indigested material, intact S-IgA was adsorbed by antibodies against κ or λ type L chains.

The purified $(Fc)_2 \cdot J \cdot SC$ fragment was cleaved with CNBr and a J chain containing 28 kDa fragment was isolated by FPLC on Superose 6, equilibrated in 25 mM Tris HCl, 150 mM NaCl and 5 M guanidine HCl.

Amino-terminal sequencing of this 28 kDa product showed that it contained the J chain and a C-terminal octapeptide of α chains. The yield of the α chain octapeptide was about twice as high as that of the J chain. From this we concluded that one J chain is linked to two α chains.

The 28 kDa product was digested with trypsin in a 0.1 M ammonium bicarbonate buffer containing 0.1 M guanidine HCl and 20% methanol. The resulting peptides were separated by reversed phase HPLC and the cystine-containing peptides were characterized by gas phase sequencing and plasma desorption mass spectrometry. This way all intra-J chain, the two J chain α chain peptides and one inter-α chain peptide could be isolated and the disulfide bridges determined (Tab. 1).

Advances in Mucosal Immunology, Edited by
J. Mestecky *et al.*, Plenum Press, New York, 1995

Table 1. Properties of peptides from digested S-IgA

calculated mass	measured mass	sequence	Cys-Cys
1201.3	1203.5	AEVDGTCY | CAR	S-IgA | J:Cys2
2077.3	2078.8	FVYHLSDLCK | AEVDGTCY	J:Cys3 | S-IgA
1294.4	1297.5	ETCYTYDR | CK	J:Cys6 | J:Cys1
2042.1	2044.8	ATQSNICDEDSAT | KCDPTE	J:Cys5 | J:Cys4
2990.4	2994.3	CYTAVVPLVYGGETK VETALTPDACYPD	J:Cys7 | J:Cys8
1709.8	1712.4	AEVDGTCY | AEVDGTCY	S-IgA | S-IgA

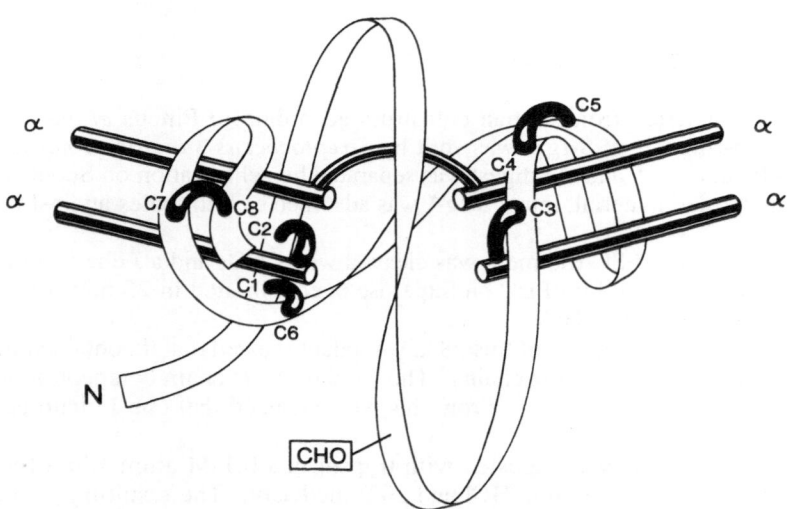

Figure 1. Intra- and inter-chain disulfide bridges of the J chain

Figure 1 summarizes these results in which the two IgA monomers are linked tail to tail in two ways are summarized: 1. by the original Cys 17-Cys 17 disulfide bridge, which connects the two IgA monomers in dimeric IgA[4], and 2. by Cys 2 and Cys 3 of the J chain, which link the J chain to the ultimate Cys of the other two α chains. (Numbering of J chain Cys according to E. E. Max and S. J. Korsmeyer[5]). Since Cys 1 and Cys 6, Cys 4 and Cys 5 and Cys 7 and Cys 8 of the J chain are also linked by disulfide bridges, a pseudo-symmetric figure is obvious, which bridges two α chains by one big and one small loop. This way the $(IgA)_2 \cdot J$ molecule gets a V shaped bend which sometimes can be seen in the electron microscope.[6] This bend might enhance the binding of SC to the $(IgA)_2 \cdot J$ complex. The linkage of SC to IgA has recently been determined by us.[7]

Our results disagree with all previous models proposed for the intra- and inter-J chain disulfide bridges in S-IgA.[8,9,10]

CONCLUSIONS

S-IgA was isolated from human colostrum and cleaved with IgA1 specific protease, CNBr, and trypsin. Trypsin cleavage was performed in a methanol and guanidine hydrochloride-containing buffer for unfolding the J chain under non-reducing conditions. Tryptic peptides were isolated by RP-HPLC and characterized by amino acid analysis, plasma desorption mass spectrometry, and automated gas phase sequencing. Based on these data intra-J chain disulfide bridges were assigned to cysteines C1-C6, cysteines C4-C5, and cysteines C7-C8. Cysteines C2 and C3 of the J chain are linked to one ultimate cysteine of each IgA monomer in dimeric S-IgA. The remaining chains of the two IgA monomers are linked by their ultimate cysteines. These data result in a model which for the first time includes both correct intra J chain and S-IgA J chain interchain disulfide bridges.

ACKNOWLEDGMENT

We thank Mrs. M. Praetor and Mrs. D. Hesse for technical assistance and Mrs. U. Brockhaus and Mr. L. Kolb for the illustrations.

REFERENCES

1. P. Brandtzaeg and H. Prydz, *Nature* 311:71 (1984).
2. C. S. Pincus, M. E. Lamm, and V. Nussenzweig, *J. Exp. Med.* 133:987 (1971).
3. A. G. Plaut, R. J. Genco, and T. B. Tomasi, Jr., *J. Immunol.* 113:289 (1974).
4. Ch. Y. Yang, H. Kratzin, H. Totz, and N. Hilschmann, *Hoppe-Seyler's Z. Physiol. Chem.* 360:1919 (1979).
5. E. E. Max and S. J. Korsmeyer, *J. Exp. Med.* 161:832 (1985).
6. R. B. Doumashkin, G. Virella, and R. M. E. Parkhouse, *J. Mol. Biol.* 56:207 (1971)
7. E. Fallgren-Gebauer, W. Gebauer, A. Bastian, H. Kratzin, H. Eiffert, B. Zimmermann, M. Karas, and N. Hilschmann, These Proceedings.
8. G. M. Cann, A. Zaritsky, and M. E. Koshland, *Proc. Natl. Acad. Sci. USA* 79:6656 (1982).
9. J. Zikan, J. Novotny, T. L. Trapane, M. E. Koshland, D. W. Urry, J. C. Bennett, and J. Mestecky, *Proc. Natl. Acad. Sci. USA* 82:5905 (1985).
10. R. S. H. Pumphrey, *Immunol. Today* 7:206 (1986).

CHARACTERIZATION OF IgA1, IgA2 AND SECRETORY IgA CARBOHYDRATE CHAINS USING PLANT LECTINS

Agnes E. Wold,[1] Cecilia Motas,[2] Catherina Svanborg,[3] Lars Ake Hanson,[1] and Jiri Mestecky[4]

[1]Department of Clinical Immunology, University of Goteborg, Sweden; [2]Department of Biochemistry, University of Bucharest, Romania; [3]Department of Clinical Immunologyk University of Lund, Sweden; and [4]Department of Microbiology, University of Alabama, Birmingham, AL, USA

INTRODUCTION

The glycosylation pattern of immunoglobulins varies according to class and subclass. IgA1 and IgD, but not the other Ig isotypes, are substituted with O-linked oligosaccharide chains, which are linked to serine or threonine in the polypeptide chain. N-linked oligosaccharide chains are, in contrast, found in all isotypes, but their number varies, from one per heavy chain in IgG to six per heavy chain in IgE.[1] N-linked oligosaccharide chains can be either of the "complex" type (Fig. 1a), or of the "oligomannose" type (Fig. 1b). The N-linked oligosaccharide chains are attached to asparagine residues in the polypeptide backbone of the protein.

The carbohydrate structural variability is increased enormously by a phenomenon termed microheterogeneity. This means that a single type of oligosaccharide chain can show considerable heterogeneity between different individual molecules of the same protein. Thus, in a complex type of oligosaccharide chain, the carbohydrate chain may be complete, ending with sialic acid on both branches, or terminate at either of the positions indicated by arrows in the figure (Fig. 1a). This premature termination of the oligosaccharide chain probably depends on a limited capacity of different glycosylation enzymes. These enzymes put on the peripheral sugar residues, one after another, during the passage of the protein through the endoplasmic reticulum and Golgi apparatus.[2] There is one specific enzyme for each binding.

Microheterogeneity is present in Ig carbohydrates, as well as in other glycoproteins. Thus, a large number of different oligosaccharide chains are found in a single IgA myeloma protein.[3]

There are indicatins of an overall differences in the glycosylation pattern between IgA1 and IgA2, though both contain N-linked oligosaccharide chains. Thus, chemical analyses have shown a difference in the percentage of different monosaccharides between IgA1 and IgA2 myeloma proteins.[4] Functionally, IgA2 proteins had a higher tendency to function as receptors for type 1-fimbriated *Escherichia coli*, although there were large individual differences between the individual myeloma proteins.[3]

Advances in Mucosal Immunology, Edited by
J. Mestecky *et al.,* Plenum Press, New York, 1995

Plant lectins may be used as tools to determine the oligosaccharides exposed on glycoproteins.[5,6] In this study, we have investigated the capacity of IgA1 and IgA2 myeloma proteins, free secretory component, and secretory IgA, to precipitate plant lectins with specificity for different aspects of N-linked oligosaccharide chains.

a) Complete type:

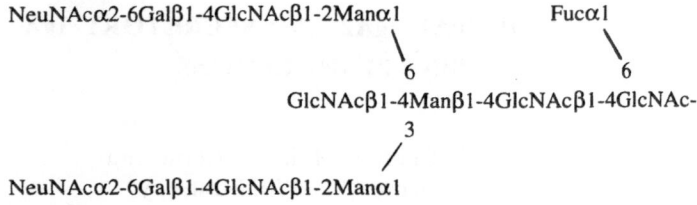

b) Oligomannose type:

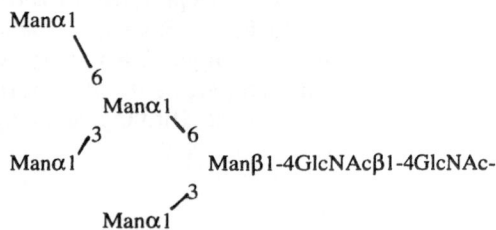

Figure 1. N-linked oligosaccharide chains

MATERIALS AND METHODS

Lectins were purchased from Sigma Chemical Co., St. Louis, MO and used at a concentration of 1 g/l. Secretory IgA (S-IgA) and secretory component (SC) were purified from human colostrum as earlier described.[7] S-IgA consisted of about equal proportions S-IgA1 and S-IgA2. Myeloma proteins of the IgA1 and IgA2 isotype were purified from serum.[7] Twenty-three IgA1 and seven IgA2 myeloma proteins were tested. All IgA2 proteins were polymeric; half of the IgA1 ones were monomeric and half were polymeric. All preparations were used at 1 g/l.

The IgA preparations were titrated in twofold steps and mixed with equal volumes of lectin and PBS (or inhibiting sugar at 2.5% in PBS) in U-shaped 96 well hemagglutination plates. The plates were shaken, incubated overnight, and evaluated in a plate microscope for the highest dilution giving a visible precipitate. The reciprocal of this dilution was recorded as the precipitation titre. The absence of any precipitation in the presence of the inhibiting sugar proved that the precipitation occurred as a result of a lectin-carbohydrate reaction, and not because of any antibody reaction against the lectin.

RESULTS

Mannose-Specific Lectins

Mannose-specific lectins bind to the tri-mannan sequence at the branching point of complex type oligosaccharide chains (Fig. 2). They are all blocked by mannose. Despite their common binding site, they all have different requirements for the structural details surrounding the binding site. For example, Con A, which is the best defined, binds to oligo-mannose oligosaccharide chains, and to complex type chains, preferably if they terminate with G1cNAc or mannose.

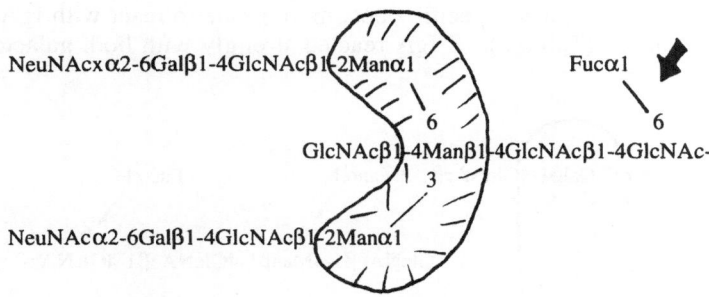

NeuNAcxα2-6Galβ1-4GlcNAcβ1-2Manα1

Fucα1

GlcNAcβ1-4Manβ1-4GlcNAcβ1-4GlcNAc-

NeuNAcα2-6Galβ1-4GlcNAcβ1-2Manα1

Figure 2. Binding site for mannose-specific lectins.

When tested for precipitation with IgA1 and IgA2 myeloma proteins, ConA precipitated IgA2 more efficiently than IgA1 myeloma proteins (Table 1). S-IgA showed a very high precipitation titre, free SC a moderate titre.

The lectins from *Pisum sativum* and *Lens culinaris* also bind to the core tri-mannan sequence. However, they will only bind if fucose is linked to the innermost G1cNAc of the core (Fig. 2, arrow). Probably, the presence of this fucose alters the conformation of the region where the lectin binds. Binding is not inhibited by monomeric fucose, but by mannose, showing that the binding site is, indeed, the tri-mannan sequence.

The IgA2 myeloma proteins were more readily precipitated by the *Pisum sativum* and *Lens culinaris* lectins (Table 1). S-IgA reacted strongly with both lectins.

The detailed structural requirements of the mannose-specific lectin of *Lathyrus odoratus* are, to our knowledge, not known. None of the 23 IgA1 myeloma proteins were precipitated by this lectin, but 5/7 of the IgA2 myeloma proteins, and S-IgA were (Table 1).

In summary, all mannose-specific lectins showed a greater reactivity with IgA2 than with IgA1.

Table 1. Mannose-specific lectins: ConA, *Pisum sativum*, *Lens culinaris*, and *Lathyrus odoratus*. Precipation titres with IgA1 myeloma proteins (monomeric and polymeric) IgA2 myeloma proteins (polymeric), SC and S-IgA.

	ConA		Pisum		Lens		Lathyrus	
	median	range	median	range	median	range	median	range
IgA1:								
mono	1	<1-8	8	<1-16	<1	<1-<1	<1	<1-<1
poly	1	<1-64	<1	<1-16	<1	<1-4	<1	<1-<1
IgA2:								
poly	16	4-64	16	16-16	16	<1-32	4	<1-16
SC	8		16		8		<1	
S-IgA	64		16		64		4	

Galactose-Specific Lectins

The galactose-specific lectins *Ricinus communis* agglutinin-I and *Abrus precatorius* lectin bind to terminal galactose, or galactose substituted with sialic acid (NeuNAc) as shown in Fig. 3. They are inhibited by monomeric galactose. These lectins were, in contrast to the mannose-specific ones, more prone to react with IgA1 than with IgA2 myeloma proteins (Table 2). S-IgA reacted strongly with both galactose-specific lectins.

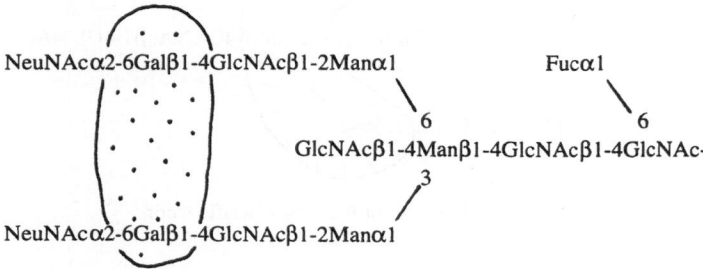

Figure 3. Binding site of galactose-specific lectins.

Table 2. Galactose-specific lectins:*Ricinus communis* agglutinin I (RCA-I) and *Abrus precatorius*. Precipitation titres with IgA1 myeloma proteins (monomeric and polymeric) IgA2 myeloma proteins (polymeric), SC and S-IgA.

	RCA-I		Abrus	
	median	range	median	range
IgA1:				
mono	16	4-32	4	<1-16
poly	16	4-32	16	<1-16
IgA2:				
poly	8	2-8	8	4-16
SC	8		8	
S-IgA	32		16	

DISCUSSION

These results show that mannose-specific lectins preferentially precipitate IgA2 myeloma proteins, while galactose-specific lectins prefer IgA1. This is in accordance with our earlier findings of a greater reactivity of IgA2 than IgA1 with the bacterial mannose-specific lectin of type 1 fimbriae.[3] Most mannose-specific lectins prefer an incomplete oligosaccharide chain ending with GlcNAc or mannose on one or both branches. The findings, thus, suggest that the N-linked oligosaccharide chains of IgA2 in general are shorter than those of IgA1.

The greater reactivity of IgA2 than IgA1 with the mannose-specific lectins from *Pisum sativum* and *Lens culinaris*, which requires core fucose to be present, further suggest that the N-linked oligosaccharide chains of IgA2 are more fucosylated than those of IgA1. This is in accordance with chemical analyses.[4]

The glycosylation pattern of immunoglobulin molecules may influence a number of functions. Thus, phagocytes have lectin-like receptors for mannose as well as galactose. Catabolism by the liver depends on a recognition of galactose residues on glycoproteins through the asialoglycoprotein receptor. It is also possible that the difference in glycosylation pattern of the different IgA subclasses may be involved in the greater tendency of IgA1 to be deposited in the glomeruli in IgA nephropathy.

S-IgA reacted very strongly with all lectins tested, showing that the N-linked oligosaccharide chains are indeed very favorably exposed in this glycoprotein. S-IgA is thus well equipped to serve as a blocking agent for attaching microorganisms by providing carbohydrate-binding sites.

REFERENCES

1. A. Torano, Y. Tzuzukida, Y.-S. V. Liu, and F. W. Putnam, *Proc. Natl. Acad. Sci. USA* 74:2301 (1977).
2. H. Iwase, *Int. J. Biochem.* 20:479 (1988).
3. A. E. Wold, J. Mestecky, M. Tomana, A. Kobata, H. Obayashi, T. Endo, and C. Svanborg Eden, *Infect. Immun.* 58:3073 (1990).
4. M. Tomana, W. Niedermeyer, J. Mestecky, and F. Skvaril, *Immunochemistry* 13:325 (1976).
5. H. Lis and N. Sharon, *Annu. Rev. Biochem.* 55:35 (1986).
6. R. K. Merkle and R. D. Cummings, *Meth. Enzymol.* 138:233 (1987).
7. J. Mestecky and M. Kilian, *Meth. Enzymol.* 116:37 (1985).

HUMAN IgA1 AND IgA2 HAVE DISTINCT SPECTROTYPES BUT DISPLAY SUBCLASS SIMILARITIES BETWEEN INDIVIDUALS

Peggy A. Crowley-Nowick, John E. Campbell,
Alan L. Mullins, and Susan Jackson

Department of Microbiology, The University of
Alabama at Birmingham, Birmingham, AL

INTRODUCTION

Isoelectric focusing (IEF) is a powerful tool for the examination of protein heterogeneity. A variety of IEF techniques have been employed to analyze the spectrotypes of total IgA from humans, rats and rabbits.[1,2,3] In all of these studies however, purified IgA was used, with the potential of having an artificially altered pI range.

The present study was designed to compare the spectrotypes of IgA1 and IgA2 in normal human sera without prior manipulation of the proteins. A combination of Immobiline® isoelectric focusing and western blotting and densitometry analysis allowed for the examination of discrete differences between the two subclasses and demonstrated distinct contrasts in spectrotypes between polyclonal IgA1 and IgA2.

MATERIALS AND METHODS

Polyacrylamide Immobiline® (Pharmacia) isoelectric focusing gels were mixed and poured between glass plates in a manner similar to the product specification sheet to produce a gel 0.25 cm thick. After polymerization, the glass plates were separated. Individual serum samples from 20 healthy individuals were loaded directly onto the basic end of the gel and electrophoresed for 5 hr. Protein transfer to Immobilon-P filter membranes (Millipore, Medford, MA) was accomplished on a semi-dry electroblotter. The membrane was removed from the gel immediately after transfer, blocked for one hr, and incubated consecutively with biotinylated antibody overnight, followed by 1.0 mg/ml of extra-avidin alkaline phosphatase (Sigma, St. Louis, MO) and subsequently developed with nitro blue tetrazolium/5-bromo-4-chloro-3-indolyl phosphate substrate (BioRad, Richmond, CA).

All blots were scanned using a Silver Scan Capture, then analyzed using Image Analyst Software (Research Services Branch, NIMH) to assign x, y coordinates to each spectrotypic band. pI estimations were based on broad range standards (FMC, Rockland, ME) run on each gel. Differences between the density or area under the curve for the spectrotypic IgA1 and IgA2 bands were determined by non-parametric analysis using the Mann-Whitney test.

Advances in Mucosal Immunology, Edited by
J. Mestecky *et al.*, Plenum Press, New York, 1995

RESULTS

The total spectrotypic range of IgA1 and IgA2 was determined using three pH unit gels, spanning the range of pH 4.5 to pH 7.5 and pH 5.5 to pH 8.5. Although IgA2 was detected in the extremes of the basic and acidic range (data not shown), the major banding occurred within the narrow range of pH 4.7 to pH 6.3. It is possible to visually detect the diversity between IgA1 and IgA2 (Fig. 1); however analysis by denisitometric scan (Fig. 2) is a more sensitive way to recognize these differences.

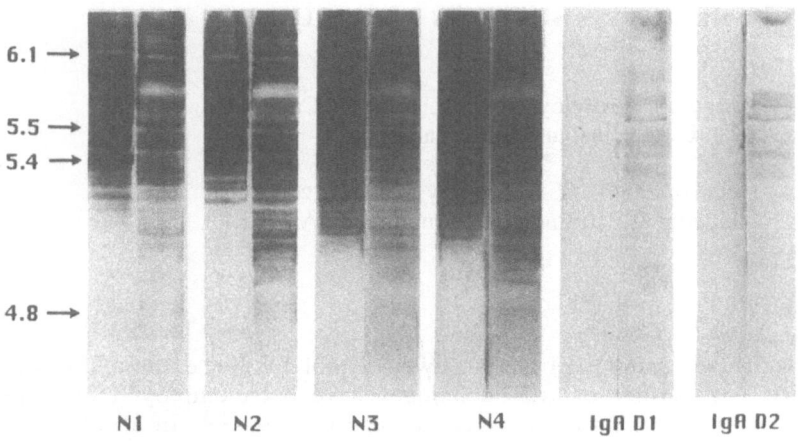

Figure 1. Four representative spectrotypes between the range of pH 4.5 to 6.5 of IgA1(1) and IgA2 (2) from sera of healthy controls (N1-N4) and two IgA deficient subjects are depicted.

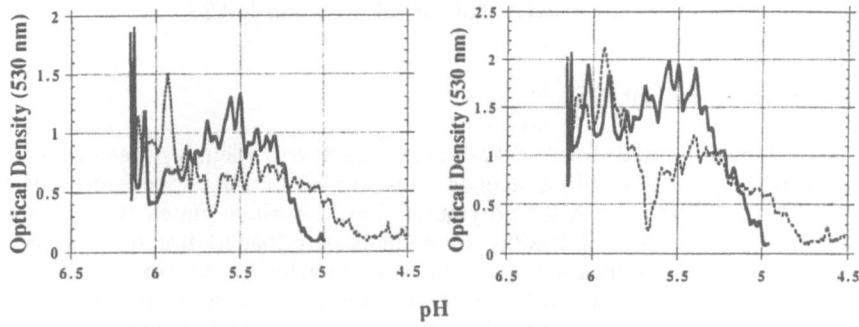

Figure 2. A graphic description of the densitometry scans obtained from the spectrotypes of two subjects shown in Fig. 1 is depicted. The solid and dashed lines represent IgA1 and IgA2, respectively.

Examination of narrow pH changes illustrates the significant pI differences between IgA1 and IgA2. These differences were not due to variance between gels; individual samples run on different gels showed no differences in intensity or spectrotypic pattern. Although both molecules focused over the range of pH 4.5 to 6.5, the major bands of IgA1 were located between pH 5.4 - 5.7. A large proportion of IgA2 also focused here, but it was significantly less than that of IgA1 ($p=0.0001$-$p=0.02$). IgA2 displayed a much wider range of banding, spanning from pH 4.7 to >6.3. The major IgA2 bands were from pH 5.8 to 6.3 and higher. However, IgA2 also showed significantly more focusing in the acidic range from 4.7 to 5.1 than did IgA1 ($p=0.0007$-$p=0.03$).

DISCUSSION

The data presented here represent an extensive evaluation of the spectrotypes of serum IgA1 and IgA2 from healthy individuals. In agreement with previous studies on IgA of rabbit and rat, human IgA was found to focus over a wide pH range. However, IgA1 was not found to focus in the acidic range below pH 4.5 as described previously[5]. Our results indicated that IgA1 focused over a range of pH 4.9 to 7.5 whereas IgA2 focused from pH 4.6 to >8.0. This is a wider range and more basic than that described previously (pH 4.6-6.7)[4] and demonstrates the limited spectrotypic range of IgA1 compared to IgA2. The size of the population in this study, the sensitivity of our methods, and the use of whole serum verses purified IgA may explain the differences in our results.

The data presented in this report further describe molecular differences between the IgA subclasses. Although the two molecules have similar amino acid structures, there are significant differences in the charge heterogeniety that can be differentiated by isoelectric focusing. These differences may reflect the vairability in carbohydrate structures of the molecules, or antigen specificity of the two subclasses. Analysis of charge differences between IgA1 and IgA2 as determined by a comprehensive analysis of serum IgA from healthy subjects provides a reference point for future studies of IgA subclasses in disease.

ACKNOWLEDGMENTS

Supported by USPHS Grants AI 10854 and DK 40117

REFERENCES

1. D. E. Jackson, C. A. Skandera, J. Owen, E. T. Lally, and P. C. Montgomery, *J. Immunol.* 36:315 (1980).
2. R. J. Mairs and J. A. Beeley, *Arch. Oral Biol.* 32:873 (1987).
3. P. C. Montgomery, A. Ayyildiz, I. M. Lemaitre-Coelho, J.-P. Vaerman, and J. H. Rockey, *Ann. N. Y. Acad. Sci.* 428-439 (1983).
4. K. B. Elkon, *J. Immunol. Methods* 66:313(1984).

PEPTIC FRAGMENTS OF RAT MONOMERIC IgA

J.P. Vaerman,[1] A. Langendries,[1] C. Van der Maelen,[1]
J.P. Kints,[2] F. Cormont,[2] F. Nisol,[2] and H. Bazin[2]

[1]Unit of Experimental Medicine; and [2]Experimental
Immunology Catholic University of Louvain and
Institute of Cell Pathology, B-1200 Brussels, Belgium

INTRODUCTION

The structure of rat IgA is still poorly known, despite the existence of rat monoclonal
IgA myeloma and hybridoma proteins. This paper describes the peptic digestion fragments
of two monomeric myeloma and two monomeric hybridoma (anti-dinitrophenyl [DNP]) rat
IgA proteins. In the case of one hybridoma (LO-DNP-67), but not in the three other cases,
the peptic digest comprised, besides the expected F(ab')α-like fragments, also an Fcα-like
fragment.

MATERIALS AND METHODS

Anti-DNP IgA monoclonal antibodies (Mabs) [LO-DNP-67 ("67") and LO-DNP-64
("64")] were purified from ascites by dinitrophenol elution from DNP-lysine-Sepharose and
separated into polymers and monomers of IgA by gel filtration on Ultrogel AcA34.[1]
Monomers of IgA myeloma proteins (IR-22 and IR-1060) were similarly separated from
polymers and further purified by preparative electrophoresis on Pevikon blocks.[2] Pepsin
digestion was performed for 24h at 37°C with 2% pepsin at pH 4.5; digestion was stopped
by raising the pH to 8.0. The peptic digests were analyzed and fractionated by agarose gel-
electrophoresis (AE), immunoelectrophoresis (IEP) using anti-rat α-chain and anti-rat-
F(ab')$_2$γ antisera, and gel-filtration on Ultrogel AcA44 (GF). SDS-PAGE of the various
fractions, with an without reducing agent (2-mercaptoethanol), was run in 4-30% gradient
gels or in 12% gels. Anti-DNP antibody activity of fractions was tested by affinity
chromatography on a DNP-lysine-Sepharose column and elution with 0.1 M dinitrophenol,
pH 8.0.

RESULTS

When analyzed by AE, only the "67" hybridoma IgA monomer yielded in addition to
a band with the same mobility as the original IgA, one more cationic band and two bands
with higher anionic mobility. Neither of the monomers from "64", IR-699 or IR-1060
yielded these more anionic bands after peptic digestion and AE, although they both yielded
the original and more cationic bands. The four bands observed after AE of the peptic digest
of "67" will hereafter be called AE-1 to AE-4, from anode to cathode respectively (Fig. 1).

Advances in Mucosal Immunology, Edited by
J. Mestecky *et al.,* Plenum Press, New York, 1995

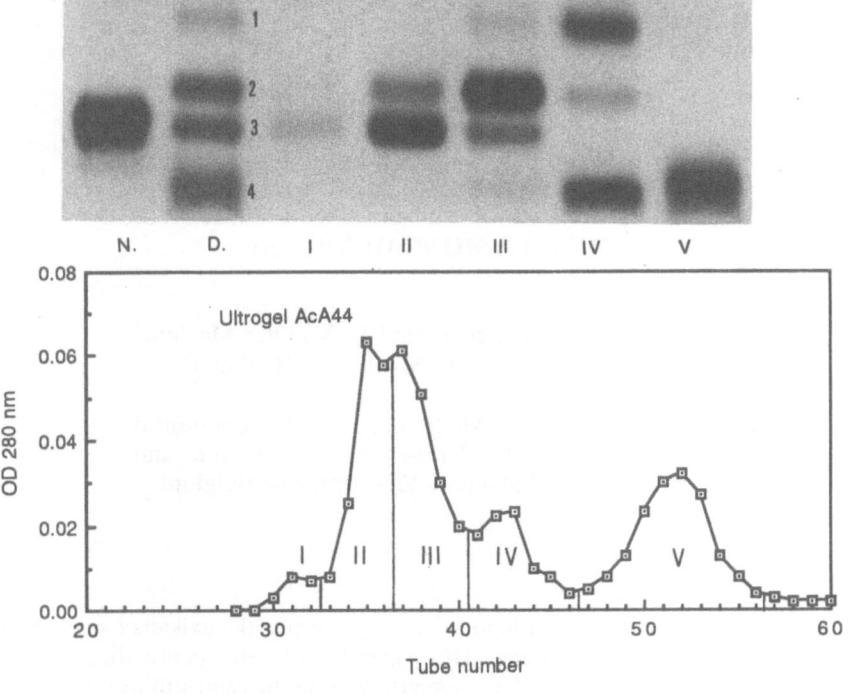

Figure 1. Agarose-gel electrophoresis (upper panel) of native IgA "67" (N), its peptic digest (D), and the five gel filtration fractions (I to V) of the digest gel filtered in non-dissociating solvent (lower panel) Numbers 1 to 4 indicate the 4 agarose-gel electrophoretic bands of the digest, from the anode to the cathode.

When the digest was analyzed by IEP (not shown), AE-1 reacted only with anti-α antiserum, AE-4 only with the anti-F(ab')$_2$γ antiserum, whereas AE-2 and AE-3 reacted with both antisera. This was confirmed after preparative purification of the 4 AE fractions (not shown). Therefore, AE-1 was thought to represent Fcα-like fragments, and AE-4 F(ab')α-like fragments. Because AE-3 had the same mobility as the native IgA and reacted with both antisera, it was suggested to be intact monomeric IgA, and this was confirmed later by SDS-PAGE.

When the digest was analyzed by GF on Ultrogel AcA44, 5 peaks were obtained, hereafter called GF-I to GF-V (Fig. 1). GF-I to FG-V were also analyzed by AE (Fig. 1). The major component of GF-I and GF-II migrated as AE-3, and was thus thought to represent intact IgA. GF-II also contained a significant proportion of AE-2, which predominated in GF-III, which also contained some AE-1 and AE-4. GF-IV, like GF-II, was contaminated with some AE-2, but its main components were AE-1 and AE-4. Finally, GF-V only contained AE-4, although with a more diffuse pattern than AE-4 from GF-IV.

SDS-PAGE was run with AE-purified AE-1 (from GF-IV), purified AE-2 and AE-3 (both from pooled GF-II plus GF-III), and AE-4 (either purified from GF-IV or directly as GF-V), both unreduced and reduced (Fig. 2). In the unreduced state, AE-1 yielded mostly a broad ± 68 kDa band, which upon reduction became mostly a ± 34 kDa band, with a minor ± 31 kDa band. Unreduced AE-2 yielded a major ± 100 kDa band with smaller ± 23 and ± 130 kD bands; after reduction, 3 bands were observed, ± 64 kDa, ± 29 kDa, and ± 34 kDa, respectively. Unreduced AE-3 (like intact IgA) yielded ± 130 and ± 50 kDa bands, thought to be dimeric α- and dimeric light-chains, respectively. AE-4 from GF-IV and AE-4 in GF-V were identical upon reduction (2 strong bands of about equal intensity of ± 29 and ± 32 kDa). When unreduced, AE-4 from GF-IV displayed mainly 4 bands of ± 50, 66, 32, and 60 kDa, with decreasing intensity, whereas GF-V consisted of 3 strong bands of ± 32, ± 60, and ± 23 kDa.

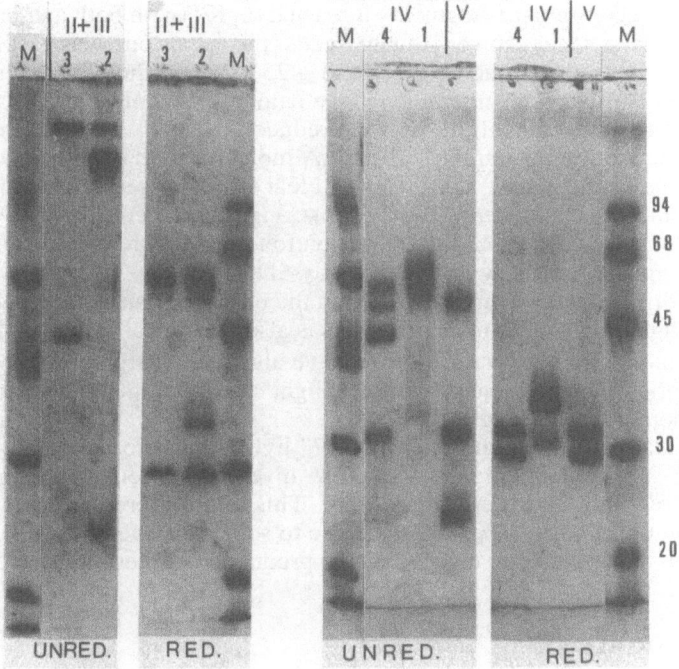

Figure 2. SDS-PAGE, reduced or unreduced, of AE-3 (3) and AE-2 (2) purified from GF-II plus GF-III (II + III), on the left part, and of AE-1 (1) and AE-4 (4) purified from GF-IV, and of FG-V (V), on the right part. Molecular weight markers (M) are indicated by appropriate kDa values along the right side.

GF-IV, containing AE-1, AE-2, and AE-4, was passed on DNP-lysine-Sepharose and separated into a fall-through fraction (FT) and an hapten eluate (EL). The FT contained only AE-1, whereas the EL contained both AE-2 and AE-4, indicating that only AE-2 and AE-4 had anti-DNP activity, in contrast to AE-1. When the whole "67" digest was passed on DNP-lysine-Sepharose, AE-3 was shown to also have antibody activity (as expected) in addition to AE-2 and AE-4. When the peptic digest of "64" was similarly analyzed (not shown), both AE-bands obtained were retained and hapten-eluted.

DISCUSSION

This article describes peptic fragments from monomeric IgA anti-DNP "67". The most anionic fragment (AE-1) must represent Fcα because it reacts only with the anti-α-chain and not with the anti-F(ab')$_2\gamma$ antiserum; it does not bind to DNP-Sepharose, and it has a size of ± 68 or ± 34 kDa in unreduced or reduced state, respectively. Two types of F(ab')α-fragments, with anti-DNP activity, were apparently produced, which had similar mobility on AE (AE-4 from GF-IV and AE-4 in GF-V), but differed by size upon GF in a non-denaturing solvent. These fragments also had different SDS-PAGE patterns in the unreduced state, but appeared identical in SDS-PAGE after reduction, with ± 29 and ± 32 kDa bands representing, respectively, light chains and Fdα. One could suggest that AE-4 from GF-IV was the F(ab')α dimer whereas AE-4 in GF-V could be its monomer. However, these unreduced fragments also had complex but different SDS-PAGE patterns. Non-reduced AE-4 from GF-IV contained ± 66 and ± 50 kDa bands not found in AE-4 from GF-V. Further studies with Western blots might help in identification of these various bands AE-2 was also a unique fragment it reacted on IEP with both antisera, but on SDS-PAGE (in unreduced state) showed 3 main bands, the major one was $\pm$ 100 kDa and was not identified The 2 weaker bands, $\pm$ 130 and $\pm$ 23 kDa, are thought represent dimeric α-chains and monomeric light chains, although the latter needs confirmation Upon reduction, AE-2 showed 3 bands with mobilities of the reduced α-chain, reduced Fcα, and reduced light chains The possibility that AE-2 is an IgA molecule with one intact α-chain and one (half) Fcα will be investigated Also, it is not clear why apparently only monomeric light chains were found in unreduced AE-2 The strong unreduced $\pm$ 130 kDa band will be purified and rerun in SDS-PAGE after reduction to unravel its reduced chain composition When AE-3 or intact IgA "67" were analyzed by SDS-PAGE, $\pm$ 130 kDa dimeric α-chains with $\pm$ 50 kDa dimeric light chains were seen in the unreduced state, and $\pm$ 68 kDa α-chains with $\pm$ 29 kDa light chains after reduction This is also the typical pattern of monomeric IgA in BALB/c myeloma monomeric IgA [3] We have also repeatedly observed, as others did[4], that rat IgA-hybridoma proteins release dimeric light chains without reduction after treatment with SDS or another dissociating solvent

It remains to be seen why only the ' 67" hybridoma monomeric IgA, and not the 64 nor the IR-22 and IR-1060 monomeric IgA myeloma proteins, were able to yield Fcα and the AE-2 fragment upon peptic digestion This might represent an unknown type of heterogeneity in rat IgA, but it could also be due to some *in vivo* and/or *in vitro* cleavage of IgA occurring in the ascitic fluid induced for the production of these IgA molecules

ACKNOWLEDGMENT

Supported by the Belgian State (Science Policy Programming, Office of the Prime Minister, PA1 7bis and AC 88-93-122)

REFERENCES

1   M Rits, F Cormont, H Bazin, R Meykens, and J P Vaerman, *J Immunol Methods* 89 81 (1986)
2   J P Vaerman, D L Delacroix, and C Moussebois, *J Immunol Methods* 54 95 (1982)
3   C A Abel and H M Grey, *Biochemistry* 7 2682 (1968)
4   M Rits, *in* "IgA Immune Complexes are Efficient Activators of the Alternate Pathway of Complement Studies in a Homologous Rat System", pp 1-195, Ph D Thesis, C I P Koninklijke Bibliotheek Albert I, Berchem, Belgium (1989)

ANTIGENIC AND GENETIC HETEROGENEITY AMONG *HAEMOPHILUS INFLUENZAE* AND *NEISSERIA* IgA1 PROTEASES

Hans Lomholt, Knud Poulsen, and Mogens Kilian

Institute of Medical Microbiology, The Bartholin
Building, University of Aarhus, DK-8000 Aarhus C,
Denmark

INTRODUCTION

Bacterial immunoglobulin A1 (IgA1) proteases are extracelluar enzymes which specifically cleave the hinge region of human IgA1 including secretory IgA1 (S-IgA1) resulting in intact monomeric Fab and Fc fragments (or dimeric Fc coupled to the secretory component and J chain in the case of S-IgA1) (for reviews see 1-3). The Fab fragments have retained antigen binding capacity while Fc mediated effector functions are lost. S-IgA1 is the dominant Ig isotype protecting the mucosal membranes in the upper respiratory tract suggesting that specific cleavage of this antibody is an important factor in the process of bacterial colonization and infection at this location. IgA1 protease is secreted by a number of human pathogens which colonize at mucosal sites where they are capable of penetrating the mucosal barrier. Notably, the three leading causes of bacterial meningitis *Haemophilus influenzae, Neisseria meningitidis,* and *Streptococcus pneumoniae* all produce IgA1 protease, whereas closely related nonpathogenic species lack this enzyme.

IgA1 cleaving activity has evolved convergently in different bacteria through at least three independent evolutionary lines. Thus, the *N. meningitidis, N. gonorrhoeae,* and *H. influenzae* IgA1 proteases are all serine proteases[4], whereas those from *Streptococcus sanguis* and *Prevotella (Bacteroides) melaninogenica* are metalloproteases and cysteine proteases, respectively.[5,6] Cloning and sequencing of *iga* genes encoding IgA1 proteases from *N. gonorrhoeae, H. influenzae,* and *S. sanguis* have confirmed these evolutionary relationships.[5,7,8] Thus, the deduced amino acid sequences of the gonococcal and *H. influenzae* IgA1 proteases are approximately 50 % identical. In contrast, there is no homology between the primary structure of these two serine proteases and the *S. sanguis* IgA1 metalloprotease.

The amino acid sequences together with protein analyses have revealed that for secretion the IgA1 serine proteases have a unique mechanism involving autoproteolytic maturation of the pre-protease.[7,8] The pre-protease of approximately 170 kDa consists of at least three functionally distinct domains: (1) the N-terminal signal peptide, (2) the protease domain of approximately 110 kDa, and (3) the C-terminal β-domain which is supposed to create a pore in the outer membrane for secretion of the protease domain and which is autoproteolytically cleaved off during the secretion and remains integrated in the outer membrane. The autoproteolytic site(s) separating the protease domain and the β-domain is characterized by amino acid sequences rich in prolines and very similar to the target site in human IgA1.

The substrate specificity of the IgA1 proteases implies that no relevant animal model exists, so direct evidence for the role of these proteases as virulence factors in the infection process is still lacking. We have proposed a hypothetical model for invasive infections like meningitis due to IgA1 protease-producing bacteria.[9] According to that model these bacteria may evade the immune system and penetrate the mucosal barrier by becoming coated with Fab fragments generated by IgA1 protease cleavage of preexisting IgA1 antibodies to bacterial constituents other than the IgA1 protease. These antibodies could be induced as a result of prior mucosal colonization by cross-reactive bacteria. We are currently engaged in evaluating this model which suggests that the IgA1 proteases are essential virulence factors in mucosal infections and that neutralizing mucosal antibodies to the proteases secreted by the bacterium confer protection from invasive disease.

The hypothetical model implies that, theoretically, the IgA1 proteases may be used as immunogens for vaccination against, e.g., bacterial meningitis. In this context we have examined the variation of the IgA1 serine proteases among *H. influenzae* and the pathogenic *Neisseria*.[10,11] Since the putative protection by an IgA1 protease-containing vaccine relies on neutralizing antibodies on the mucosal surfaces we have used an assay system based on enzyme neutralization. In short, rabbit antisera raised against selected IgA1 proteases were tested for inhibition of IgA1 cleaving activity of standardized preparations of homologous as well as heterologous IgA1 proteases. Interestingly, the antisera did not inhibit all the heterologous IgA1 proteases from the same bacterial species. Using this assay system we were able to identify serologically distinct versions termed "inhibition types" among the IgA1 proteases. This result should be considered when studying the relations between host and pathogen in infection as well as recurrent colonizations with these bacteria. In addition, the results from genetic studies on distinct *H. influenzae iga* genes have added to our understanding of the mechanisms operating in the molecular evolution of this bacterial species.

H. INFLUENZAE IgA1 PROTEASES

We have previously examined IgA1 proteases secreted by 155 *H. influenzae* strains including isolates from patients with meningitis, epiglottitis, conjunctivitis and various respiratory infections, as well as isolates from healthy individuals.[10] By the use of nine selected antisera in enzyme neutralization assays, a surprising antigenic heterogeneity was revealed as fifteen inhibition types with only limited cross-reactivity could be defined. A close correlation was found between IgA1 protease inhibition type and capsular serotype.[10] Four inhibition types were defined among *H. influenzae* type b strains, the serotype responsible for the vast majority of invasive disease caused by this species.[12,13] Interestingly, an antiserum raised against one type neutralized the enzyme activity of IgA1 proteases from three of the four types accounting for more than 98% of the *H. influenzae* type b population.

It is not known what role this antigenic variation among *H. influenzae* IgA1 proteases plays for the *in vivo* colonization and penetration of human mucosal surfaces by these bacteria. One possibility is that it serves an immune escape purpose to retain IgA1 protease activity. This could function in at least two ways. One could be a continuous exchange of clones secreting IgA1 proteases of different inhibition types. This would serve as an immune escape for the species as a whole and additionally, the single clones would benefit from a less frequent induction of neutralizing antibodies to their particular IgA1 protease inhibition type. Another possibility could be that the immunologic pressure might, by selection, force the bacteria to change the antigenic properties of the IgA1 protease secreted as it has previously been suggested for, e.g., the outer membrane protein

P2.[14] To evaluate this possibility we studied sequential isolates of *H. influenzae* from three healthy children. Pharyngeal swabs were taken every two months during a study period of one to two years and the clonal types of the *H. influenzae* isolates were determined by a combination of biotyping, analysis of outer membrane protein patterns by SDS-gel electrophoresis, and genomic DNA fingerprinting. The clonal type of strains was compared to the IgA1 protease type determined by a combination of inhibition typing with eleven antisera and restriction fragment length polymorphism (RFLP) analysis of the *iga* gene region. In addition, we raised an antiserum to the initial isolate from each person to analyze neutralization of IgA1 proteases secreted by strains subsequently colonizing the person. In this study we observed a frequent clonal exchange with each new clone secreting a new inhibition type of IgA1 protease and during the study period we did not see the same type to reappear. A change of IgA1 protease type in sequential isolates assigned to the same clonal type was not observed. Thus, exchange of clones expressing antigenically different IgA1 proteases is the principal mechanism operating *in vivo* to evade the host immune response against IgA1 proteases. The results support, though do not prove, that IgA1 protease activity is important for the survival of these bacteria on mucosal surfaces.

We have used RFLP analysis of the *iga* gene region to estimate the degree of genetic variation underlying the observed antigenic variation of the expressed enzyme. Among *H. influenzae* type b organisms four *iga* RFL types were found corresponding to the four antigenic types of the IgA1 protease.[12] Among 30 isolates of other capsular serotypes and non-capsular strains we identified more than twenty different *iga* gene region types and found that IgA1 proteases of the same inhibition type could be encoded by *iga* regions that produced very different RFL patterns. In general, variation was limited among strains of the same capsular serotype whereas among non-capsular isolates no two *iga* RFL pattern were found to be identical.

In order to study the molecular and genetic basis for the serological variation among the IgA1 proteases we have cloned and sequenced four *H. influenzae iga* genes encoding IgA1 proteases of different inhibition types.[13] A comparison of the deduced amino acid sequences of the four IgA1 protease domains revealed that they are all very similar and that the sequence homologies are characterized by conserved regions interrupted by short, highly variable sequences. Notably, the signal peptide as well as the sequence around the active serine and the proposed autoproteolytic sites are conserved among the four pre-proteases. The areas constituting the neutralizing epitopes on the IgA1 proteases have not yet been identified. However, our results suggest that the observed variation in inhibition by neutralizing antibodies is caused by epitopes of the discontinuous conformational type.

In relation to evolution of the *H. influenzae* population the pattern of homology among the four *iga* genes reveals two interesting features. First, the N-terminal part of the secretory β-domain, which is assumed to function only as a hydrophillic spacer region, is highly variable with several reminiscents of ancient insertions, deletions, and duplications. This suggests that such genomic rearrangements are relatively frequent *in vivo* since they accumulate when tolerated without selective disadvantage. Secondly, the pattern of homology among the sequences in the variable regions differ in different variable regions. A similar mosaic-like organization of a part of the *iga* gene has been reported for *N. gonorrhoeae*.[15] Such an organization indicates a process of horizontal genetic exchange between the *iga* genes of independent strains. This is confirmed by the observation that the nucleotide sequence of the *iga* gene in two *H. influenzae* strains was identical in a 2.4 kb region in contrast to the remaining 2.5 kb which were very different in the two strains.[13] The mechanisms creating the horizontal transfer of *iga* gene fragments is presumably transformation and subsequent homologous recombination. This assumption is supported by the finding that copies of the *H. influenzae* DNA uptake signal, which is involved in the naturally competence of *H. influenzae*, surround the *iga* gene.

NEISSERIA IgA1 PROTEASES

The extensive antigenic variation found among *H. influenzae* IgA1 proteases prompted us to examine the corresponding relations among IgA1 proteases in the population of *N. meningitidis*. Using rabbit antisera raised against three selected IgA1 proteases, five different antigenic types could be defined among IgA1 proteases of meningococci in a collection of 133 isolates representing major epidemic and carrier strains from 19 countries during 40 years.[11] In contrast to the *H. influenzae* IgA1 protease inhibition types a high degree of cross-reaction was found between these five types and interestingly, an antiserum to a single protease strongly neutralized the enzyme activity of all IgA1 proteases in the *N. meningitidis* collection. Preliminary results indicate that a similar limited variation in neutralizing epitopes on the IgA1 protease is found among *N. gonorrhoeae* strains. Among meningococci no association was found between capsular serotype and IgA1 protease inhibition type except for group A strains, which in general appear to be very homogenous.[16]

RFLP analysis of the *iga* gene region revealed more than twenty different patterns among the 133 *N. meningitidis* isolates and the variation in DNA fragment sizes suggests that the meningococcal *iga* gene is situated in a highly variable area of the chromosome. However, within most clonal types determined by multilocus enzyme electrophoresis the isolates showed an identical *iga* RFLP pattern. In addition, all group A strains showed very similar patterns.[11] The homogeneity of group A strains may suggest that this part of the *N. meningitidis* population is deficient in horizontal genetic exchange.

RELATIONSHIPS BETWEEN IgA1 PROTEASES FROM DIFFERENT SPECIES

The serine IgA1 proteases of *H. influenzae* and pathogenic neisseria are derived from a common ancestor as the amino acid sequences are homologous.[7,8] Based on this close relationship we wanted to analyze if cross-reactions between species are found in epitopes recognized by neutralizing antibodies. A collection of IgA1 proteases and corresponding antisera including all five inhibition types of meningococci, most of the *H. influenzae* inhibition types, and four types of gonococci were studied in enzyme neutralization assays. We also included representatives of the streptococcal IgA1 metalloproteases to determine whether different classes of IgA1 proteases cross-react. From this study we conclude that extensive cross-reactions are found between IgA1 proteases of the two *Neisseria* species. In addition, few examples of cross-reactions of various extents are seen between *H. influenzae* and neisseria IgA1 proteases. No cross-reactions were found between IgA1 serine and metalloproteases. The observed cross-reactions suggest that mucosal colonization with one bacterial species may, in some cases, affect colonization by another species.

The relatively low degree of variation in inhibition type of the IgA1 protease within and among the two *Neisseria* species compared to *H. influenzae* is paralleled in variable degree of homology among the corresponding *iga* gene sequences. Thus, the four partial gonococcal *iga* sequences described by Halter *et al.*[15] are very homologous with few silent site substitutions. In addition, we have cloned and sequenced a meningococcal *iga* gene and found it to be very similar to the gonococcal genes. These observations are in contrast to the four known *H. influenzae iga* gene sequences which are characterized by a higher degree of variation including a large number of silent site substitutions. The reason for this different degree of variation among IgA1 proteases of the three species is not known. It is not a mere reflection of a more homogeneous population structure among neisseria since population genetic studies have shown a degree of diversity in meningococci at least as great as for the *H. influenzae* population.[16] Also, meningococci have developed specific mechanisms for variation of other virulence factors such as pili and certain outer membrane proteins.[17] Interestingly, oral streptococci which colonize the human oral mucosal surfaces for the whole life almost lack antigenic diversity in neutralizing epitopes on the IgA1 protease.[18] Further studies on the variation of bacterial IgA1 proteases and the underlying genetic mechanisms may teach us more about the complex interplay between these bacteria and the human immune system in the colonization and infection process.

ACKNOWLEDGMENTS

This study was supported by the Danish Medical Research Council Grant 12-9505.

REFERENCES

1. A. G. Plaut, *Annu. Rev. Microbiol.* 37:603 (1983).
2. M. H. Mulks, *in:* "Bacterial Enzymes and Virulence", I. A. Holder, ed., p. 81, CRC press, Boca Raton, Florida (1985).
3. M. Kilian and J. Reinholdt, *in:* "Medical Microbiology", C.S.F. Easmon and J. Jeljaszewicz, ed., p. 173, Academic Press, London (1986).
4. W. W. Bachovchin, A. G. Plaut, G. R. Flentke, M. Lynch, and C. A. Kettner, *J. Biol. Chem.* 265:3738 (1990).
5. J. V. Gilbert, A. G. Plaut, and A. Wright, *Infect. Immun.* 59:7 (1991).
6. S. B. Mortensen and M. Kilian, *Infect. Immun.* 45:550 (1984).
7. J. Pohlner, R. Halter, K. Beyreuther, and T. F. Meyer, *Nature* 325:458 (1987).
8. K. Poulsen, J. Brandt, J. P. Hjorth, H. C. Thøgersen, and M. Kilian, *Infect. Immun.* 57:3097 (1989).
9. M.Kilian and J. Reinholdt, *Adv. Exp. Med. Biol.* 216B:1261 (1987).
10. M. Kilian and B. Thomsen, *Infect. Immun.* 42:126 (1983).
11. H. Lomholt, K. Poulsen, D. A. Caugant, and M. Kilian, *Proc. Natl. Acad. Sci. USA* 89:2120 (1992).
12. K. Poulsen, J. P. Hjorth, and M. Kilian, *Infect. Immun.* 56:987 (1988).
13. K. Poulsen, J. Reinholdt, and M. Kilian, *J. Bacteriol.* 174(9):2913 (1992).
14. K. Groeneveld, L. van Alphen, C. Voorter, P. P. Eijk, H. M. Jansen, and H. C. Zanen, *Infect. Immun.* 57(10):3038 (1989).
15. R. Halter, J. Pohlner, and T. F. Meyer, *EMBO J.* 8:2737 (1989).
16. D. A. Caugant, L. F. Mocca, C. E. Frasch, L. O. Frøholm, W. D. Zollinger, and R. K. Selander, *J. Bacteriol.* 169(6):2781 (1987).
17. T. F. Meyer, C. P. Gibbs, and R. Haas, *Annu. Rev. Microbiol.* 44:451 (1990).
18. J. Reinholdt, M. Tomana, S. B. Mortensen, and M. Kilian, *Infect. Immun.* 58:1186 (1990).

TITRATION OF INHIBITING ANTIBODIES TO BACTERIAL IgA1 PROTEASES IN HUMAN SERA AND SECRETIONS

Jesper Reinholdt,[1] and Mogens Kilian[2]

[1]Royal Dental College, [2]Institute of Medical Microbiology, Faculty of Health Sciences, University of Aarhus, DK-8000 Aarhus C, Denmark

INTRODUCTION

Bacterial IgA1 proteases cleave human IgA1 and S-IgA1 molecules in the hinge region hereby leaving them as Fcα and antigen-binding, but functionally incompetent, Fabα fragments. IgA proteases are produced by several overt pathogens and also by members of the indigenous oral flora involved in the formation of dental plaque.[1]

As putative virulence factors with a considerable molecular polymorphism, IgA1 proteases of pathogenic bacteria have been found to be a target for the immune response of the host.[2,3] However, methodological problems associated with the titration of IgA1 protease-inhibiting antibodies in humans may be the reason why the occurrence and significance of such antibodies have not been studied in any detail. To titrate protease-inhibiting antibodies, Gilbert et al.[2] incubated serial dilutions of serum or purified colostral S-IgA with a fixed amount of IgA1 protease and then identified persisting protease activity, if present, by its capacity to cleave a small amount of subsequently added, radiolabeled myeloma IgA1 acting as indicator. The authors speculated this this method is at risk of overestimating titers as the inherent IgA1 of the inhibitor might reduce the cleavage of indicator IgA1 by a competitive mechanism independently of the activity of enzyme-neutralizing antibodies. Here we present an assay for the titration of inhibiting antibodies which takes this possibility into account. With this assay, we have quantitated inhibiting activities in human sera, saliva, and colostral S-IgA against IgA1 proteases from pathogenic strains of *Haemophilus influenzae, Neisseria meningitidis,* and *Streptococcus pneumoniae.* Inhibiting activities against IgA1 proteases of oral streptococci were also determined. Except for selected results obtained with the protease of *Streptococcus sanguis,* however, the data for these proteases will be presented elsewhere.

MATERIALS AND METHODS

Serum and Secretions

All eight donors of serum and saliva, and two donors of colostrum were healthy adults with no recorded history of infections due to IgA1 protease-producing pathogenic bacteria. Saliva was collected with a minimum of bacterial contamination at the orifice of the submandibular duct, cleared by cold centrifugation after light sonication on ice, and used immediately. Colostral S-IgA was purified as described.[4]

Advances in Mucosal Immunology, Edited by
J. Mestecky *et al.,* Plenum Press, New York, 1995

Calibration of IgA1 Proteases

Partly purified IgA1 proteases from *H. influenzae* serotype b, strain HK 368, *N. meningitidis* group B, nontypeable, strain HF 161, and *S. pneumoniae* serotype 14, strain 800/91, and *S. sanguis* ATCC 10556 were prepared as described.[5] The proteases were calibrated on the basis of activity. Serial two-fold dilutions (25 µl) of proteases in PBS, pH 7.4, containing 0.1% BSA (PBSA) were incubated with 25 µl of myeloma IgA1 substrate, 0.5 mg/ml, for 6 h at 35°C in microtiter wells. IgA1 incubated with PBSA, served as intact substrate control. Subsequently, cleavage of IgA1 in each well measured by analysis of contents, diluted 100-fold, in the ELISA depicted in Fig. 1. With this assay, increase in cleavage of IgA1 is reflected as decrease in OD. Based on dose-response curves obtained by fitting a four parameter sigmoid model (*Fig P*, Biosoft, Cambridge, UK), the proteases were adjusted to an activity causing 50% cleavage of IgA1 as reflected by the OD obtained for protease-incubated relative to control substrate.

Inhibition Assay

Two identical two-fold dilution series, each in triplicate, of test serum or secretion were made, volumes of 25 µl being dispensed in microtiter plates. In one of the series, each well then received 25 µl of IgA1 protease. In the other series, providing uncleaved substrate controls matched with respect to IgA1 contributed by diluted inhibitor, wells received 25 µl of PBSA. Following incubation for 1 h at room temperature, all wells received 25 µl of myeloma IgA1 at 20 µl/ml and the temperature was raised to 35°C. After incubation for 6 h at 35°C, IgA1 in protease-containing and corresponding control wells was assayed (at 100-fold dilution) for cleavage by the ELISA depicted in Fig. 1. The titer of the inhibitor was defined as the dilution at which 25% cleavage of IgA1 was observed. After fitting regression functions (1st order monoexponential decay with residual) to means of ID for protease-containing and control wells, respectably, the titer was calculated as the solution of an equation setting the ratio of the functions equal to 0.75 (*Fig P*, Biosoft).

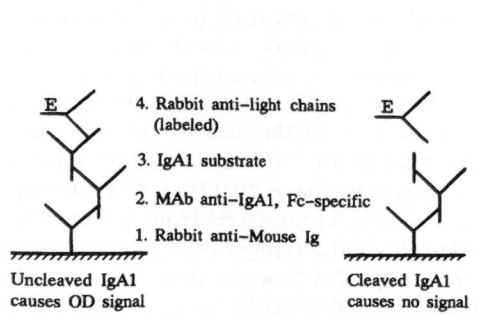

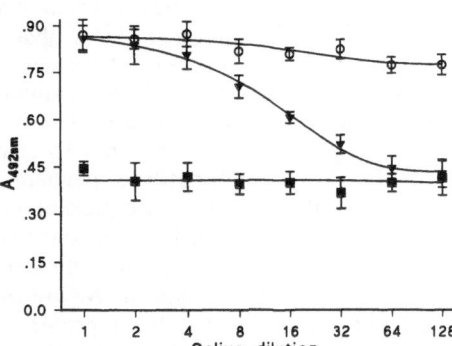

Figure 1. Immunochemical principle of ELISA for Ug/a1 oritease-induced cleavage of IgA1.

Figure 2. ELISA data obtained by titration of inhibiting activity against two IgA1 proteases in saliva of subject C. OD regression curves for wells receiving *S. sanguis* protease, *N. meningitidis* protease, and PBSA (controls) are shown.

Saliva Fractionation Experiment

A 2 ml sample of saliva was fractionated on a size-exclusion column (1.6 x 50 cm) of Superose 6 (Pharmacia) equilibrated in PBS. Eluent fractions were assayed for content of S-IgA by sandwich ELISA[6] and for inhibiting activity against four IgA1 proteases as described above.

RESULTS

Due to the inherent IgA1 of the inhibitor (serum, saliva, or colostral S-IgA), the concentration of substrate was higher in wells representing initial as compared to terminal dilution steps. However, ELISA data obtained with the *S. sanguis* protease (Fig. 2), demonstrated that IgA1 proteases calibrated as described were potentially capable of cleaving the same proportion (50%$) of IgA1 in "initial" as in "terminal" wells. Thus, the differential concentrations of substrate did not invalidate the assay.

Table 1. Inhibiting activity against 3 bacterial IgA1 proteases measured in serum and secretions of healthy subjects.

| | | Source of IgA1 protease | | |
	Donor	*H. influenzae* HK368	*N. meningitidis* HF161	*S. pneumoniae* Pn800
	A	512.7 / 4.3[a]	2582.3 / 12.2	362.0 / 2.2
	B	656.2 / 6.1	1456.4 / 11.9	432.3 / 8.0
	C	142.4 / 1.5	1281.7 / 14.5	79.2 / 1.8
	D	122.8 / 1.4	976.8 / 6.1	962.7 / 49.3
Serum/saliva	E	1152.3 / 1.2	832.7 / 32.2	476.5 / 7.6
	F	279.1 / 0	2217.8 / 14.1	189.3 / 4.2
	G	319.0 / 0	512.3 / 2.2	736.4 / 3.8
	H	2204.6 / 8.3	1297.9 / 16.5	814.3 / 8.2
	Median	319.0 / 1.5	1289.8 / 13.5	454.4 / 5.9
		r^b = 0.47	r = 0.12	r = 0.78[c]
Colostral S-IgA 0.2 mg/ml	J	8.7	98.3	28.2
	K	12.6	15.5	4.5

[a] Titer of serum / titer of saliva (means of triplicate determinations)
[b] Spearman's r for titer in serum vs titer in saliva
[c] p<0.05

To examine the reproducibility of the assay, 8 sera, 2 saliva samples, and 2 samples of colostral S-IgA were titrated against 4 different IgA1 proteases twice on different days. The mean deviation of the highest relative to the lowest of duplicate titers was 36.6% (range 10.2-98.4%), 100% corresponding to one doubling dilution.

Inhibiting activity against the IgA1 proteases from *H. influenzae, N. meningitidis,* and *S. pneumoniae,* was detected in serum, colostral S-IgA, and saliva samples, however at highly variable titers (Table 1). The overall level of titers in saliva and in colostral S-IgA (0.2 mg/ml) was roughly two orders of magnitude lower than the level in serum (Table 1). Linear regression analysis revealed variable correlation of homologous titers in serum and saliva, Spearman's ranging from 0.12 to 0.78 (Table 1).

By size-exclusion chromatography of saliva, IgA1 protease-inhibiting substances co-eluted with S-IgA (data not shown).

DISCUSSION

The finding of inhibiting activity in serum and colostral S-IgA against IgA1 proteases of pathogenic bacteria is in accordance with observations by others.[2,3] These investigators identified the activity as being due to enzyme-specific antibodies, which in serum represented the IgG as well as the IgA isotypes. The present study has demonstrated that IgA1 protease-inhibiting activity may be expressed also in saliva. The observation that the activity copurified with colostral salivary S-IgA suggests that inhibition by saliva is mediated by enzyme-specific antibodies resulting from stimulation of the mucosal immune system. The overall level of titers in saliva was lower than in serum, but should be evaluated in view of a roughly 1 to 70 ratio of the Ig contents of saliva and serum.

A considerable amount of direct and indirect evidence indicate that IgA1 proteases of pathogenic bacteria are important virulence factors.[1] The pronounced antigenic diversity of these enzymes[1,7], along with the response of the human host with inhibiting antibodies as demonstrated here, support this notion. That antibodies to IgA1 proteases play a role in the defence against the bacteria is suggested by a recent study involving clonal analysis of *H. influenzae* isolates sequentially sampled from the pharyngeal mucosa of infants.[8] In each infant we observed a frequent exchange of clones, each new clone producing an inhibition type of IgA1 protease different from that of previous isolates. Thus, it seems that IgA1 protease is among the target antigens for an immunological selection pressure on protease-producing mucosal pathogens in humans. In our view, therefore, it is likely that protection against infection due to IgA1 protease-producing pathogens can be obtained by stimulation of protease-inhibiting antibody responses at relevant mucosal sites. The inhibition assay described may be of value in an evaluation of this hypothesis.

REFERENCES

1. M. Kilian and J. Reinholdt, *in*: "Medical Microbiology", C.F.S. Easmon and J. Jeljaszewicz, ed., vol.5, p. 173, Academic Press Inc., London (1986).
2. J. V. Gilbert, A. G. Plaut, B. Longmaid, and M. E. Lamm, *Molec. Immunol.*, 20:1039 (1983).
3. K. Kobayashi, Y. Fujiyama, K. Hagiwara, and H. Kondoh, *Microbiol. Immunol.* 31:1097 (1987).
4. J. Mestecky and M. Kilian, *Methods Enzymol.*, 116:37 (1985).
5. J. Reinholdt, M. Tomana, S. B. Mortensen, and M. Kilian, *Infect. Immun.*, 58:1186 (1990).
6, T. Ahl and J. Reinholdt, *Infect. Immun.* 59:3619 (1991).
7. H. Lomholt, K. Poulsen, D. Caugant, and M. Kilian, *Proc. Natl. Acad. Sci. USA*, 89:2120 (1992).
8. H. L. Lomholt, K. Poulsen, and M. Kilian, *Recent Adv. Mucosal Immunol.*, this volume.

CLEAVAGE OF IgG AND IgA *IN VITRO* AND *IN VIVO* BY THE URINARY TRACT PATHOGEN *PROTEUS MIRABILIS*

M.A. Kerr, L.M. Loomes, and B.W. Senior

Departments of Pathology and Medical Microbiology
Dundee University Medical School
Ninewells Hospital
Dundee DD1 9SY, Scotland

Pathogenic microorganisms have evolved a number of strategies for avoiding the immune system of the host. For example, *Staphylococcus aureus* and some *Streptococcus* spp. express surface proteins which are able to bind immunoglobulins by the Fc region thus eliminating the Fc-mediated effector functions of the immunoglobulins. Some other pathogens secrete proteinases which are able to degrade immunoglobulins (reviewed by Senior *et al.*[1]). These enzymes include a number of highly specific IgA1 proteinases produced by pathogens of mucosal surfaces such as *Neisseria meningitidis*, *Neisseria gonorrhoeae*, *Haemophilus influenzae* and several *Streptococcus* spp. There is indirect evidence[2-4] that these proteinases might be involved directly in the pathogenicity of the microorganisms. A wider range of microorganisms including strains of *Proteus* spp. and *Pseudomonas aeruginosa* and *Serratia marcescens* secrete proteinases of broader specificity which are able, *in vitro*, to cleave both IgA and IgG and other non-immunoglobulin proteins.

We have shown that *Proteus mirabilis*, a common pathogen of the urinary tract produces a proteinase which is able to cleave IgA1, AgA2 and IgG but not IgM.[5] Active enzyme and fragments of IgA produced by the enzyme have been shown to be present in the urine of infected patients.[6] Subsequent studies[7] showed that the enzyme could be purified rapidly by affinity chromatography on Phenyl-Sepharose. The purified enzyme was complex in that it appeared as a number of closely spaced bands on SDS-PAGE. Each of the bands had proteinase activity and was inhibited by EDTA and other inhibitors of metalloproteinases. The molecular mass of the protein suggested that it was monomeric and that the different forms detected on electrophoresis were the result of post-translational modification.[7]

An analysis was made of the proteinase of 18 diverse strains of *P. mirabilis*. All of the strains had typical biochemical characteristics of *P. mirabilis* and all showed proteolytic activity on gelatin-CLED agar. For each strain, the active proteinase in the supernate of nutrient broth cultures incubated for 24h at 37°C was detected as a doublet of Mr approximately 50 kDa when subjected to electrophoresis in polyacrylamide-gelatin gels and subsequent development. Incubation of ^{125}I-IgG with culture supernates followed by SDS-PAGE and autoradiography resulted in each case in cleavage of the heavy chain to fragments visible as a broad band of Mr ~30 kDa. The immunoglobulin light chain was not cleaved. Each of the strains also cleaved the heavy chain of ^{125}I-IgA. The rate of cleavage of IgA by any strain was always greater than that of IgG. All proteolytic activity was inhibited by 10

Advances in Mucosal Immunology, Edited by
J. Mestecky *et al.*, Plenum Press, New York, 1995

mM EDTA. There was a clear correlation between the ability to cleave IgG and IgA and the azocaseinase activity of culture supernates.

P. mirabilis strain 64676 was chosen for further study and its proteinase purified by affinity chromatography on Phenyl-Sepharose.[7] When the purified P. mirabilis proteinase (PMP) at different dilutions was incubated with radiolabelled immmunoglobulins, IgA cleavage was more easily detected at lower concentrations of enzyme than IgG cleavage. Thus for one preparation, whereas IgA cleavage was clearly detected at dilutions down to 1:128, IgG cleavage was detected only at dilutions down to 1:8. Consient with our previous observations,[5,8] IgA heavy chain was cleaved to yield fragments of 47 kDa. Greater amounts of enzyme produced a fragment of 34 kDa. However the IgG heavy chain was cleaved to give fragments of 31 kDa and a low molecular weight fragment of ~14 kDa. In neither case was the immunoglobulin light chain cleaved.

The specificity of cleavage of IgG was studied in more detail using purified PMP. The fragments produced from cleavage of unlabelled IgG were identified after separation on SDS-PAGE and their composition determined by immunoblotting with anti-heavy (γ) - and anti-light chain (λ and κ) - specific antisera. The results confirm and extend those obtained previously[5] using radiolabelled IgG incubated with culture supernates or purified enzyme from the same strain. IgG digested with limiting amounts of PMP gave fragments of 130 kDa and 110 kDa corresponding to F(abc)'$_2$ and F(ab)'$_2$. When digested with greater amounts of enzyme, fragments of 50 kDa and 43 kDa were produced. The 50 kDa fragment contained only heavy chain determinants; the 43 kDa contained only light chain determinants. This suggests that they correspond to Fc and Fab fragments respectively and that the anti-heavy chain antiserum recognises predominantly determinants in the CH2 and CH3 domains.

The amounts of the 130 kDa and 110 kDa fragments produced did not seem to vary with enzyme concentration, suggesting that these fragments were not further degraded into the smaller fragments.

In contrast to the fragments detected when radiolabelled IgG was digested, there was never any low molecular weight (14 kDa) fragment formed which could be detected by immunoblotting or by staining of blots with gold stain when unlabelled IgG was digested. This suggests that iodination might protect the small fragment from further degradation.

Comparison with the fragments generated by cleavage of IgG with pepsin or papain confirmed that PMP generated some F(ab)'$_2$ fragments, but in the presence of more enzyme, the major fragments were Fab and Fc. The loss of effector functional activity of IgG resulting from cleavage by PMP, pepsin or papain was studied by measuring the ability of the fragments produced to induce a burst of chemiluminescence when incubated with purified human neutrophils. IgG, when aggregated by binding to plastic luminometer plates, was able to produce a respiratory burst from the neutrophils as described previously.[9] This ability of IgG to elicit a respiratory burst rapidly diminished upon incubation of the IgG with pepsin. Analysis of the incubation mixtures by SDS-PAGE showed that the loss of activity was coincident with cleavage of IgG to produce F(ab)'$_2$ fragments. Incubation of IgG with papain or PMP resulted in a marked decrease in activity down to a level approximately one-half of that stimulated by intact IgG. Analysis of the incubations by SDS-PAGE showed that the IgG had in each case been cleaved to Fab and Fc fragments.

It is clear therefore that in this experimental system, the Fc fragment produced by PMP is functionally equivalent to that produced by papain. The new lower level of chemiluminescent activity corresponds to the activity of the Fc fragment bound to the microtitre plate and aggregated by the binding because purified papain derived Fc, but not Fab, fragments when bound to microtitre plates, elicited a similar response.

When urine specimens from a number of patients with a P. mirabilis urinary tract infection were analyzed by SDS-PAGE, many showed the presence of serum proteins including IgG and IgA detectable by immunoblotting. A number of these urine specimens were seen to contain fragments of IgA typical of those produced by cleavage of IgA by PMP. The amount of IgA did not reflect the amount of proteinuria or haematuria. In a number of urine specimens where the SDS-PAGE showed clearly the presence of plasma proteins, the IgG was partially cleaved into Fab and Fc fragments. These results show clearly that the secreted proteinase is active *in vivo*.

The presence of IgA and IgG fragments in urine suggests that the proteinase might act as a virulence factor through by-passing the immune effector functions of immunoglobulins. Cleavage of IgG will yield Fab fragments, which, if they are of sufficient affinity, will still bind to the microorganism. These fragments will not be able to elicit effector functions mediated by the Fc part of the molecule. They might, however, block the binding of intact IgG molecules and thereby further limit the effectiveness of the immune response. Although the Fc fragments can be aggregated artifically *in vitro*, allowing them to mediate effector functions, there is no evidence that such aggregation can occur *in vivo* and therefore the cleaved IgG can be assumed to be inactive.

Although the mechanisms of effector functions mediated by IgA are less fully understood, our evidence (manuscript in preparation) suggests that the F(abc)'$_2$ produced on cleavage of IgA by the proteinase is unable to bind to Fcα receptors on human myeloid cells. The detection of fragments of IgG and IgA as well as active enzyme[6] in *P. mirabilis* infected urine containing plasma proteins, suggests that *P. mirabilis* proteinase is not efficiently inhibited by any of the many proteinase inhibitors present in plasma.

REFERENCES

1.   B. W. Senior, L. M. Loomes, and M. A. Kerr, *Revs. Med. Microbiol.* 2:200 (1991).
2.   M. H. Mulks and A. G. Plaut, *N. Engl. J. Med.* 299:973 (1978).
3.   M. Blake, K. K. Holmes, and J. Swanson, *J. Infect. Dis.* 139:89 (1979).
4.   R. A. Insel, P. Z. Allen, and I. D. Berkowitz, *Semin. Infect. Dis.* 4:225 (1982).
5.   L. M. Loomes, B. W. Senior, and M. A. Kerr, *Infect. Immun.* 58:1979 (1990).
6.   B. W. Senior, L. M. Loomes, and M. A. Kerr, *J. Med. Microbiol.* 35:203 (1991).
7.   L. M. Loomes, B. W. Senior, and M. A. Kerr, *Infect. Immun.* 60:2267 (1992).
8.   B. W. Senior, M. Albrechtsen, and M. A. Kerr, *J. Med. Microbiol.* 24:175 (1987).
9.   W. W. Stewart and M. A. Kerr, *Immunology.* 71:328 (1990).

CLEAVAGE OF HUMAN IMMUNOGLOBULINS BY PROTEINASE FROM *STAPHYLOCOCCUS AUREUS*

Ludmila Prokesová,[1] Bela Potuzníková,[1] Jan Potempa,[3] Jirí Zikán,[2] Jiri Radl,[4] Zofia Porwit-Bóbr,[3] and Ctirad John[1]

[1]First Medical Faculty, Charles University, [2]Institute of Microbiology, Czech Academy of Sciences, Prague, Czech Republic; [3]Jagiellonian University, Cracow, Poland; and [4]Vascular Research, Leiden, The Netherlands

INTRODUCTION

Many pathogenic bacteria release proteinases that cleave Ig.[1] The proteolytic damage of antibodies by bacterial exoproducts is usually taken to be a factor of virulence which compromises the immunological protection of the host against infection. Most existing studies of bacterial proteinases have concentrated on the cleavage of human IgA because of its important role in defense on mucosal surfaces that are the entrance point for most infections. Bacterial proteinases usually cleave IgA1 selectively in the hinge region. A more extensive cleavage or the cleavage of both IgA subclasses occurs only rarely.[1,2] The cleavage of more immunoglobulin classes has so far been described only in *Pseudomonas aeruginosa*[3], *Serratia marcescens*[4], and *Proteus mirabilis*.[5]

S. aureus is a conspicuous human pathogen. The only report on Ig cleavage by staphylococcal proteinase has been that by Rousseaux *et al.*[6] which describes the cleavage of rat IgG2b. *S. aureus* secretes 3 main proteinases: serine proteinase (SP), metalloproteinase (MP) and thiol proteinase (TP).[7] TP is produced very rarely while SP and MP are produced by some two-thirds of *S. aureus* strains and therefore have to be taken into account in the pathogenesis of staphylococcal infections. We provided evidence that the stimulation of human lymphocytes in culture is affected by these enzymes[8,9] and decided to test also their effect on human Ig. MP does not cleave human Ig whereas SP exhibits substantial cleavage.

MATERIALS AND METHODS

SP produced by *S. aureus* strain V8 was isolated according to Drapeau.[10] The resulting enzyme was homogeneous during SDS-PAGE. IgG and IgM were isolated from human serum, SIgA from human colostrum. Human myeloma IgA1 and IgA2 were provided by Prof. J. Mestecky.

Ig at a concentration of 1 mg/ml (in 0.02 M Na phosphate buffer with 0.15 M NaCl, pH 7.8) was incubated with SP for 4 h or 24 h at 37°C. The cleavage was carried out at different enzyme/substrate ratios – 1:5 (I), 1:20 (II) and 1:100 (III). Ig incubated in the absence of the enzyme served as control (C).

Advances in Mucosal Immunology, Edited by
J. Mestecky *et al.*, Plenum Press, New York, 1995

SDS-PAGE was carried out in a 10% or gradient (5 - 20%) polyacrylamide gel under reducing conditions according to Laemmli.[11]

ELISA was performed in polystyrene microtitration plates in modification according to Tlaskalova *et al.*[12] After termination of incubation with SP the samples were supplemented with the enzyme inhibitor diisopropylfluorophosphate at a final concentration of 1 mM. Ig was determined either as an antigen on plates coated with anti-Ig antibodies or as antibodies on plates coated with antigen. The antigen was a sonicate from *S. aureus* strain Wood 46 (which does not produce protein A) because normal Ig contains a considerable proportion of antistaphylococcal antibodies.

RESULTS AND DISCUSSION

On incubating human Ig with SP a marked cleavage of all three Ig classes is observed. The cleavage intensity depends on the enzyme/substrate ratio and on the length of incubation.

SDS-PAGE under reducing conditions demonstrated the cleavage of individual peptide chains of the Ig molecule (Fig. 1,2). Incubation with the highest enzyme concentration (I) causes a complete disappearance of γ, μ, and α chain zones and of SC zone of the colostral IgA. IgM and SIgA exhibit also attenuation of the L chain bands. In the case of IgG this is apparently masked by fragments of the γ chains. At a lower enzyme concentration the cleavage is less pronounced but it is detectable even at an enzyme/substrate ratio of 1:100 (III). Cleavage of the myeloma IgA1 and IgA2 demonstrated the sensitivity of both IgA subclasses to SP. Based on the analysis of several monoclonal IgA proteins the λ chains appear to be more readily degraded than the α chains. The cleavage of Ig by SP yields a considerable number of proteolytic fragments with a molecular weight range of 42 to <12 kDa. The low molecular-weight fragments are more conspicuous after a longer incubation with the enzyme. The formation of a whole spectrum of fragments implies that staphylococcal SP does not cleave Ig only in the hinge region as described for the majority of bacterial proteinases, but acts on multiple sites of the Ig molecule.

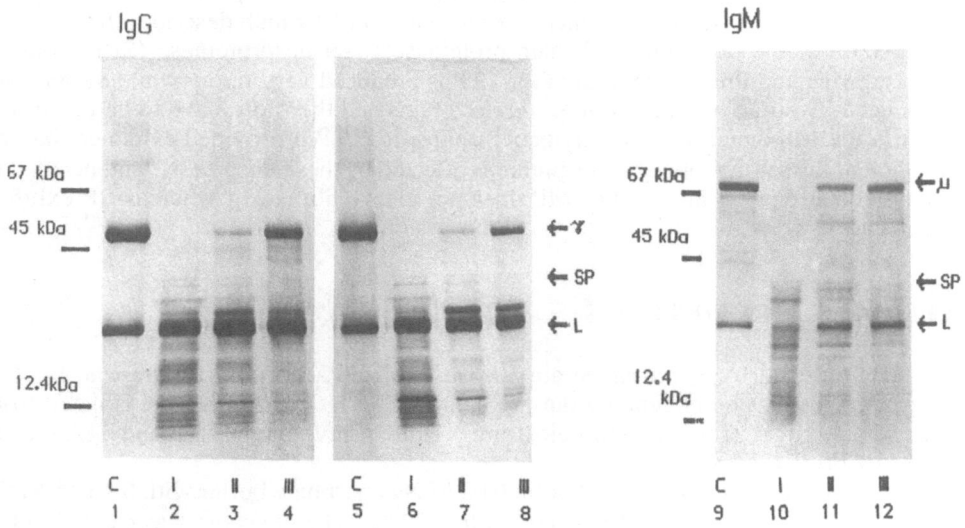

Figure 1. SDS-PAGE of serum IgG and IgM after cleavage with SP. Gradient gel in reducing conditions. C-controls; I, II, III - enzyme/substrate ratio 1:5, 1:20, and 1:100, respectively. Lanes 1-4 and 9-12, 4 h incubation; lanes 5-8, 24 h incubation.

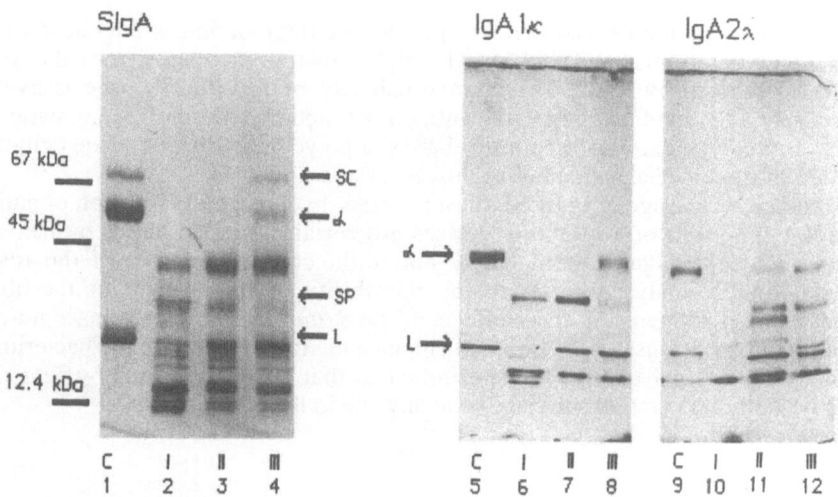

Figure 2. SDS-PAGE of colostrum SIgA and of myeloma IgA1 and IgA2 after 4 h cleavage with SP. Reducing conditions. Lanes 1-4, gradient gel; lanes 5-12, 10% gel. C-controls; I, II, III - enzyme/substrate ratio 1:5, 1:20, and 1:100, respectively.

Table 1. ELISA – Detection of Ig after 4 h and 24 h incubation with SP of *S. aureus*. Results are expressed as per cent (Ig concentration in the control incubated in the absence of the enzyme = 100%).

	Enzyme : substrate	IgG 4 h	24h	IgM 4 h	24 h	SIgA 4 h	24 h
C	0 : 1	100	100	100	100	100	100
I	1 : 5	70	38	78	76	86	78
II	1 : 20	72	52	79	70	93	91
III	1 : 100	90	62	93	75	104	91

Table 2. ELISA – Binding of Ig incubated 4 h and 24 h with serine proteinase to antigens of *S. aureus*. Results are expressed as per cent (amount of the uncleaved Ig control that binds with *S. aureus* sonicate = 100%)

	Enzyme : substrate	IgG 4 h	24h	IgM 4 h	24 h	SIgA 4 h	24 h	SIgA1 4 h	SIgA2 4 h
C	0 : 1	100	100	100	100	100	100	100	100
I	1 : 5	28	4	92	104	39	36	47	68
II	1 : 20	64	10	96	90	60	47	–	–
III	1 : 100	98	36	120	98	64	63	–	–

As shown by the ELISA method, proteolytic degradation is the cause of loss of some antigenic determinants (Table 1). The high sensitivity to cleavage by the SP in this respect is evidenced by IgG. The large sensitivity of IgG to cleavage is even more prominent when the binding of Ig to the antigen is tested by an immunoenzymatic method (Table 2). Proteolytic damage to both subclasses of polyclonal SIgA has been demonstrated with the aid of monoclonal antibodies against IgA1 and IgA2.

Extensive cleavage of Ig by SP must necessarily damage the function of antibodies. The effector functions of all three Ig classes are certain to be damaged by heavy chain cleavage. The cleavage of both heavy and light chains, along with the results of immunoenzymatic analyses points to the possibility of impairment of the ability of antibodies to bind antigen. All these effects of the *S. aureus* serine proteinase may impair the immunological defense of the host and enhance the invasiveness of the bacterium. The high sensitivity of IgG to SP cleavage indicates that the enzyme may affect immune reactions not only on mucosal surfaces but at any site in the organism.

CONCLUSION

Serine proteinase is seen to be one of the many virulence factors of *S. aureus* which can, apart from a direct damage to tissues, lower the defense of the host against infection.

REFERENCES

1. M. Kilian, B. Thomsen, T. E. Petersen, and H. S. Bleeg, *Ann. NY Acad. Sci.* 409:612 (1983).
2. Y. Fujiyama, K. Kobayashi, S. Senda, Y. Benno, T. Bamba, and S. Hosoda, *J. Immunol.* 134:573 (1985).
3. G. Döring, H.-J. Obernesser, and K. Botzenhart, *Zbl. Bacteriol. Microbiol. Hyg.* 249:89 (1981).
4. A. Molla, T. Kagimito, and H. Maeda, *Infect. Immun.* 56:916 (1988).
5. L. M. Loomes, B. W. Senior, and M. A. Kerr, *Infect. Immun.* 58:1979 (1990).
6. J. Rousseaux, R. Rousseaux-Prévost, H. Bazin, and G. Biserte, *Biochim. Biophys, Acta* 748:205 (1983).
7. S. D. Arvidson, *in:* "Staphylococci and Staphylococcal Infections", C.S.F. Easmon and C. Adlam, eds., Vol. 2, p. 780, Academic Press, London (1983).
8. L. Prokesová, Z. Porwit-Bóbr, K. Baran, J, Potempa, and C. John, *Immunol. Lett.* 19:127 (1988).
9. L. Prokesová, Z. Porwit-Bóbr, K. Baran, J. Potempa, M. Pospísil, and C. John, *Immunol. Lett.* 27:225 (1991).
10. G. R. Drapeau, *Can. J. Biochem.* 56:534 (1978).
11. U. K. Laemmli, *Nature* 227:680 (1970).
12. H. Tlaskalová-Hogenová, J. Bártová, L. Mrklas, P. Mancal, Z. Broukal, R. Barot-Ciorbaru, M. Novák, and M. Hanikyrová, *Folia Microbiol.* 30:258 (1985).

CLONING, CHROMOSOMAL LOCALIZATION, AND LINKAGE ANALYSIS OF THE GENE ENCODING HUMAN TRANSMEMBRANE SECRETORY COMPONENT (THE POLY-Ig RECEPTOR)

Peter Krajci, Dag Kvale, and Per Brandtzaeg

Laboratory for Immunohistochemistry and Immuno-
pathology (LIIPAT), Institute of Patholology,
University of Oslo, The National Hospital,
Rikshospitalet, Oslo, Norway

INTRODUCTION

The human transmembrane secretory component (SC) acts as the epithelial poly-Ig receptor (PIGR), mediating the translocation of J-chain-containing polymeric IgA (poly-IgA) and pentameric IgM into exocrine secretions.[1] SC thus exerts a key role in antibody-mediated protection of mucosal surfaces.[2] Primary SC deficiency has not been convincingly documented[3]; this might be explained by the crucial functional role of this receptor protein. However, it is also possible that its gene might be under positive selection pressure exerted by tight linkage to some other essential gene(s) whose absence or dysfunction is incompatible with survival of the species. Such linkage was examined in the present study. Moreover, characterization of the SC gene, in addition to its chromosomal localization, was performed to supplement present knowledge about its structural organization as part of the Ig supergene family.[4]

METHODS AND RESULTS

Genomic Organization and Polymorphism

Twenty-two clones, hybridizing with our human transmembrane SC cDNA probe[5,6] were isolated from a human genomic leucocyte library in EMBL3. Southern blot analysis revealed[7] three overlapping clones, which were characterized by automated sequencing. The composite clone was shown to span a total length of 19 kb of the gene encoding human SC (Fig. 1).

Altogether eleven exons were isolated and their exon-intron junctions were characterized. Exons 1 and 2 were shown to be separated by an intron with a length of approximately 6 kb. The translational start and stop codons were found in exon 2 and 11, respectively (Fig. 2).

The exon-intron gene structure *versus* the domain structure of the SC protein and possible polymorphism[6] were studied (Fig. 3). The signal peptide was encoded by both exons 2 and 3. No introns were observed between the codons for the paired cysteine residues comprising disulfide bridges: exons 3, 5 and 6 each encoded a single Ig-related

Advances in Mucosal Immunology, Edited by
J. Mestecky *et al.*, Plenum Press, New York, 1995

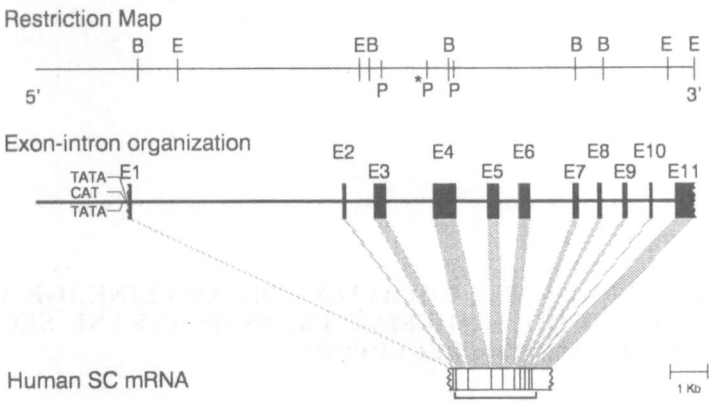

Figure 1. Schematic representation of the gene encoding human transmembrane SC. *Top:* Partial restriction map: (B) *Bam* HI; (E) *Eco* RI; (P) *Pvu* II (only the three sites involved in *Pvu* II RFLP are indicated; the polymorphic site is labelled by asterisk). TATA and CAT boxes are indicated. *Middle:* Exon-intron organization: E1-E11 (–■–). *Bottom:* Schematic representation of SC mRNA with coding region indicated (└──┘) (Adapted from Krajci *et al.*[7]).

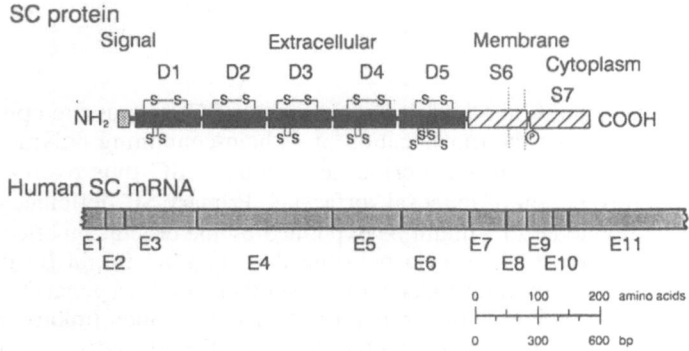

Figure 2. *Top:* Putative extracellular homologous human SC domains (D1-D5), segment including the membrane-spanning portion (S6), and cytoplasmic segment (S7) are indicated. (-S-S-), disulfide bridges; (Ⓟ), the phosphorylated Ser residue in rabbit. *Bottom:* Schematic representation of the organization of exons (E1-E11) in SC mRNA (Adapted from Krajci *et al.*[7]).

domain (D1, D4 and D5, respectively), whereas exon 4 encoded two domains (D2 and D3). The segment including the membrane-spanning portion (S6) was encoded by E7. Exon 8 coded for the second half of this segment, which includes the membrane-spanning part of SC. The 103-amino-acid long cytoplasmic tail was encoded by the four distal exons (E8-E11). A polymorphic *Pvu* II site was detected in the intron between E3 and E4.

Chromosomal Localization

Southern blot analysis of human-rodent cell-hybrid DNAs assigned the human transmembrane SC gene to chromosome 1 (0% discordancy).[6] Regional mapping was performed on somatic cell hybrids containing different translocations of this chromosome(Fig. 4). Human-specific hybridization was obtained with DNA of the hybrid cell lines F15 RAG 6-1-8 (p31-qter) and 12CB-17B (p32-qter), thus assigning the human SC gene to 1q31-q42.

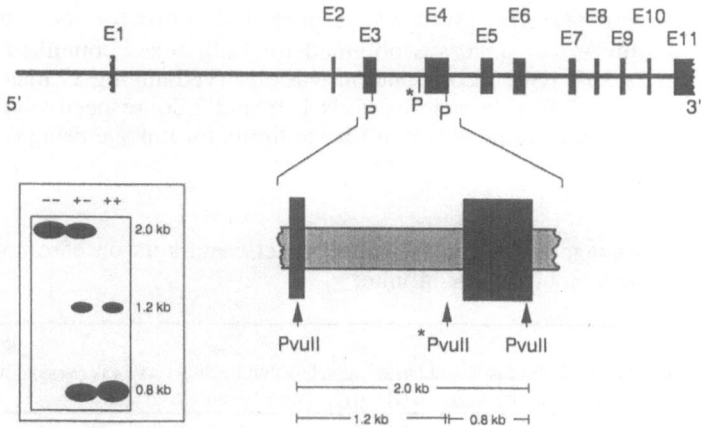

Figure 3. Schematic representation of the gene encoding human transmembrane SC. *Top:* Exon-intron organization, with three restriction sites for *Pvu* II (P) involved in RFLP, is indicated. The polymorphic site is labelled by asterisk. *Bottom left:* Southern blot of *Pvu* II-digested genomic DNA hybridized with the 0.67-kb *Pvu* II cDNA probe; heterozygotic (+-) or homozygotic for the presence (++) or absence (--) of the polymorphic cleavage site. The estimated DNA fragment sizes are indicated on the right. *Bottom right:* Schematic enlargement of the 2.0-kb *Pvu* II fragment of the gene showing the suggested location of the polymorphic site. (Adapted from Krajci *et al.*[7])

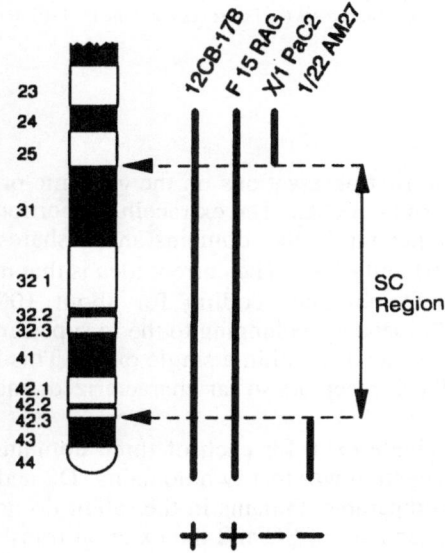

Figure 4. Schematic depiction of the regional mapping of the SC gene on human chromosome 1. *Vertical lines:* retained parts of chromosome 1 in the somatic cell hybrids used for regional sublocalization of the human SC gene. The presence (+) or absence (-) of the human SC hybridizing fragments in each hybrid is noted at the *bottom*. (Adapted from Krajci *et al.*[6])

Linkage Analysis

A 0.67-kb *Pvu* II fragment from the human SC cDNA[5,6] and a 5.0-kb *Msp* I fragment from the polymorphic DNA sequence pYNZ23 (locus D1S58)[8] were used for the analysis of linkage relations between SC (locus PIGR), F13B and D1S58 among twenty-six Norwegian family groups.[9] A peak lod score (or lods) of +5.06 at recombination fraction $\theta_{max} = 0.06$ was obtained for both sexes combined (Table 1)[9] between PIGR and D1S58. One recombination was observed among 17 male (including three phase-known) and 17 female meioses (lods 1.98 and 2.86, respectively), indicating linkage between these two loci (the 95% confidence limits for linkage being $\theta_1 = 0.01$ and $\theta_1 = 0.18$).

Table 1. Pairwise linkage relation of PIGR to genetic markers on chromosome 1. R: recombinant. NR: nonrecombinant. ∞: infinite.

Relation	Sex*	Phase known R	Phase known NR	Two-generation Fam	Two-generation Children	Lods at recombination fraction (θ_1) 0.00	0.05	0.10	0.20	0.30	Peak (θ_{max}) lods	95% confidence interval given linkage
PIGR-F13B	M	4				1.20	1.12	1.02	0.82	0.58	1.20 (0.00)	
	F		4	4	34	- ∞	-3.23	-1.10	0.37	0.61	0.62 (0.28)	
	I		1	1	4	- ∞	0.05	0.20	0.21	0.12	0.23 (0.15)	
	M+F	4	5		38	- ∞	-2.06	0.12	1.40	1.31	1.46 (0.25)	0.12-0.43
PIGR-D1S58	M	3	5		14	- ∞	1.98	1.93	1.51	0.94	1.99 (0.06)	0.01-0.29
	F		2		17	- ∞	2.86	2.78	2.27	1.52	2.86 (0.06)	0.00-0.25
	I		1	1	2	0.22	0.30	0.15	0.05	0.01	0.30 (0.00)	
	M+F	3	8		33	- ∞	5.06	4.86	3.83	4.47	5.06 (0.06)	0.01-0.18
F13B-D1S58	M		1		3	0.60	0.54	0.47	0.32	0.17	0.60 (0.00)	
	F		4		15	- ∞	1.70	1.67	1.28	0.73	1.72 (0.07)	0.00-0.31
	M+F		5		18	- ∞	2.24	2.14	1.60	0.90	2.24 (0.06)	0.00-0.25

*Male and female backcrosses and double intercrosses (I). For sexes combined (M+F), $\theta_m = \theta_f$

(Adapted from Krajci et al.[9]).

DISCUSSION

Here we summarize the first observations on the genomic organization of the gene encoding the human transmembrane SC. The extracellular portion of SC is organized in five covalently stabilized repeating Ig-like domains and it shares homology with other gene products of the Ig superfamily.[10,11] The current idea is that molecules of this family have evolved from a primordial gene coding for about 100 amino acids.[12] A characteristic feature of most members belonging to the Ig superfamily family is that each domain sequence tends to be encoded within a single exon. This has been demonstrated for all domains of Ig and T-cell receptors so far characterized, and also for MHC class I and II molecules.[4]

Our study revealed a single exon for each of three domains (D1, D4 and D5) of human SC, but a notable exception was that two domains (D2 and D3) were encoded by the same exon. The two comparable domains in the rabbit do not seem to be of major importance for the binding capacity of SC and the external translocation of poly-Ig; the corresponding exon message tends to be eliminated by alternate splicing[13], thus resulting in a small variant of SC mRNA which, however, encodes a functional translational product.

In this report we have chosen to focus on the organization of the SC gene in relation to putative structural and functional regions of the receptor protein. Exons 2 to 11 display lengths of 67-655 nucleotides and encode the entire open reading frame of the SC cDNA clone.[5,6] The membrane-spanning part and cytoplasmic tail of SC (encoded by exons 8 to 11) are highly conserved between the rabbit[10], human[5,6] and rat[14] species. In this region the human receptor protein shares an amino-acid homology of ~85% with the rat and ~70% with the rabbit counterpart. By contrast, the comparable amino-acid similarities for the entire proteins are about 65% and 55%, respectively. Extensive studies on mutant rabbit SC as reviewed by Bomsel and Mostov[15] have documented that various regions of the cytoplasmic tail are responsible for the intracellular sorting of this receptor protein Our study showed that exon 8 of the human SC gene encodes the C-terminal half of segment 6, including the proposed 23-amino-acid long membrane-spanning region Furthermore, a 14-residue segment, just downstream to the membrane-spanning part (rabbit pos 655-668), directs the receptor to the basolateral cell surface The corresponding human segment is encoded by exons 8 and 9, whereas the serine residue (human pos. 655), whose negative charge after phosphorylation appears to be a signal for transcytosis[15], is confined to exon 9 (Fig. 5) This exon also encodes a region corresponding to the part of rabbit SC (pos 670-707) involved in protecting it from lysosomal degradation [15] The 59-bp-long exon 10 is the shortest exon of the human SC gene, it encompasses the region corresponding to the rabbit pos 708-725 No definite function has been assigned to this segment The upstream extension of exon 11 encodes the 28 C-terminal residues of human SC It corresponds to the 30 amino acid portion of the cytoplasmic tail of rabbit SC that has been shown to be necessary for rapid endocytosis of the poly-Ig receptor at the basolateral cell surface [15]

Our findings therefore suggest a striking correspondence between exon boundaries and the putative functional regions of the encoded human SC protein This would be in keeping with the remarkable homology of the cytoplasmic SC tail observed between different species as described above and with the theory that exons in general correspond to structural domains [4] Further studies on the genomic organization of other genes in relation to the function of the protein segments they encode, will improve our understanding of the development of genes and gene families

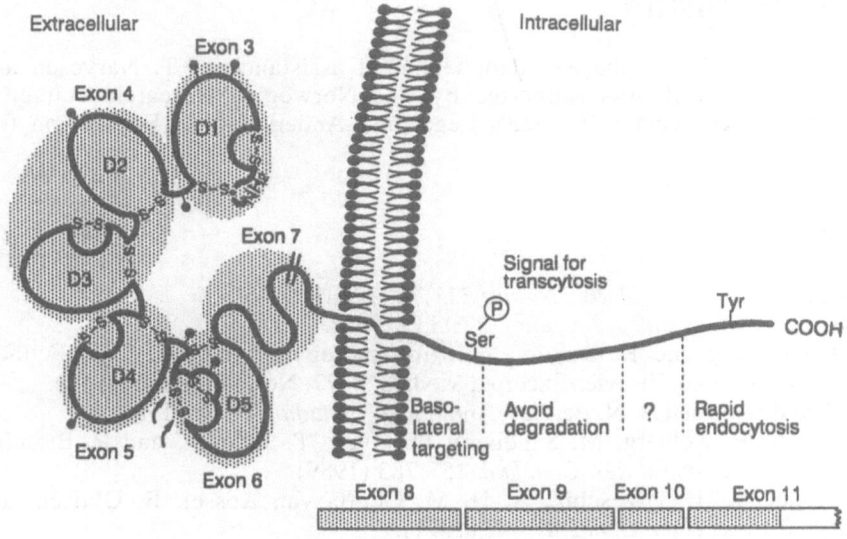

Figure 5. Putative extracellular homologous human SC domains (D1-D5) including the membrane spanning portion and cytoplasmic tail The shaded areas represent extensions of exons (Exons 3 6) encoding for the respective domains Exon 7 encodes the N-terminal half of segment 6 (☐), exons 8-10 coding for the different functional regions of the cytoplasmic tail of rabbit SC, (–●), carbohydrate-binding sites, (-S-S-), disulfide bridges, (–●–), the amino-acid stretch (Ser-Val-Ser-Ile-Thr-Cys-Tyr-Tyr-Pro) possibly involved in the noncovalent binding to poly-Ig, (/), the cysteins probably, involved in the formation of disulphide bridges created after binding to poly-IgA, (//), conserved cleavage site proposed for the generation of free and poly-Ig-bound SC, the sorting signals involved in the targeting of SC are also indicated (Ⓟ) the phosphorylated Ser residue in rabbit SC Tyr, this residue in 6 SC apparently contributes to rapid endocytosis, (?), no definite function has yet been assigned to this segment (Adapted from Krajci et al [7])

We have previously demonstrated[6] that the human SC gene exhibits a two-allele RFLP for *Pvu* II. Our present study[7] demonstrated that this RFLP is indeed due to a polymorphic restriction site located in intron 3, in agreement with the allele A1 pattern (Fig. 3). This fact enables designation of PCR products as new markers for allelic typing of individuals in population and linkage studies.

The SC gene had earlier been provisionally localized to the q31-q41 region.[16] Our independent results, based on the use of a separate probe and a different method, confirmed the originally proposed assignment.

The *a priori* probability of linkage between PIGR and F13B was high in view of their partially overlapping physical assignment. D1S58 was selected[8] as an additional marker that could be tested on the Southern blots made for PIGR. Multipoint linkage analyses[17,18] have determined the following order of these loci: cen-F13B-REN-D1S58-DAF-CR1/CR2/CR3-qter. Linkage analysis of the D1S58-PIGR relation demonstrated a combined lod score for both sexes of +5.06 at $\theta_{max} = 0.06$. The 95% confidence limits for linkage being $\theta_l = 0.01$ and $\theta_l = 0.18$, suggesting close linkage between PIGR and D1S58 (Table 1). The progeny of a triply heterozygotic female segregant indicated a cen-F13B-D1S58-PIGR-qter gene sequence on human chromosome 1. This would place PIGR close to the regulator of complement action (RCA) cluster in 1q32. This cluster includes the loci for decay-accelerating factor (DAF), C3b/C4b receptor (CR1), and C3d/Epstein Barr virus receptor (CR2). The proteins encoded by these loci serve a very important role in the control of the complement cascade.[19] Other genes of immunological interest that have been localized to this chromosome region are Factor H (HF), C4-binding protein (C4BP) and the CD 45 complex.

ACKNOWLEDGMENTS

We are grateful for the excellent technical assistance of T. Narvesen and B. Simonsen. This work was supported by The Norwegian Research Council, The Norwegian Cancer Society, Torsted's Legat and Anders Jahre's Foundation for the Promotion of Science.

REFERENCES

1. P. Brandtzaeg and H. Prydz, *Nature* 311:71 (1984).
2. P. Brandtzaeg, *Scand. J. Immunol.* 22:111 (1985).
3. P. Brandtzaeg and K. Baklien, *in:* Immunology of the gut. Ciba Foundation Symposium 46: Elsevier/Excerpta Medica, p. 77, North-Holland (1977).
4. A. F. Williams and A. N. Barclay, *Annu. Rev. Immunol.* 6:381 (1988).
5. P. Krajci, R. Solberg, M. Sandberg, O. Øyen, T. Jahnsen, and P. Brandtzaeg, *Biochem. Biophys. Res. Commun.* 158:783 (1989).
6. P. Krajci, K. H. Grzeschik, A. H. M. Geurts van Kessel, B. Olaisen, and P. Brandtzaeg, *Hum. Genet.* 87:642 (1991).
7. P. Krajci, K. Taskén, D. Kvale, and P. Brandtzaeg, *Eur. J. Immunol.* 22:2309 (1992).
8. Y. Nakamura, M. Culver, P. O'Connell, M. Leppert, G. M. Lathrop, J.-M. Lalouel, and R. White, *Nucl. Acids Res.* 15:9620 (1987).
9. P. Krajci, B. Olaisen, T. Gedde-Dahl, B. Høyheim, S. Rogde, B. Olaisen, and P. Brandtzaeg, *Hum. Genet.* 90:215 (1992).
10. K. E. Mostov, M. Friedlander, and G. Blobel, *Nature* 308:37 (1984).
11. H. Eiffert, E. Quentin, J. Decker, S. Hillemeir, M. Hufschmidt, D. Klingmüller, M. H. Weber, and N. Hilschmann, *Hoppe-Seyler's Z. Physiol. Chem.* 365:1489 (1984).
12. R. L. Hill, R. Delaney, R .E. Fellows, and H. E. Lebowitz, *Proc. Natl. Acad. Sci. USA* 56:1762-1769 (1966).

13. D. L. Deitcher and K. E. Mostov, *Mol. Cell Biol.* 6:2712 (1986).
14. G. Banting, B. Brake, P. Braghetta, P. J. Luzio, and K. K. Stanley, *FEBS Lett.* 254:177 (1989).
15. M. Bomsel and K. Mostov, *Current Opinion Cell Biol.* 3:647 (1991).
16. M. K. Davidson, M. M. Le Beau, R. L. Eddy, T. B. Shows, L .A. DiPietro, M. Kingzette, and W. C. Hanly, *Cytogenet. Cell. Genet.* 48:107 (1988).
17. P. O'Connell, G. M. Lathrop, Y. Nakamura, M. L. Leppert, R. H. Ardinger, J. L. Murray, J.-M. Lalouel, and R. White, *Genetics* 4:12 (1989).
18. K. L. Buetow, D. Nishimura, P. Green, Y. Nakamura, O. Jiang, and J. C. Murray, *Genomics* 8:13 (1990).
19. T. E. Mollnes and P. J. Lachmann, *Scand. J. Immunol.* 27:12 (1988).

THE COVALENT LINKAGE OF THE SECRETORY COMPONENT TO IgA

E. Fallgren-Gebauer, W. Gebauer, A. Bastian, H. Kratzin,
H. Eiffert, B. Zimmermann, M. Karas, and N. Hilschmann

Max-Planck-Institute for Experimental Medicine
Hermann-Rein-Straße 3
W3400 Göttingen, Germany

INTRODUCTION

Secretory immunoglobulin A (S-IgA) is the most important Ig in human secretions.[1] It consists of four proteins, two IgA monomers, SC and the J chain.[2]

IgA dimers, bound to J chain, are produced in lymphoid cells. SC is synthesized as a membrane protein in epithelial cells where it acts as a specific receptor for polymeric Ig and mediates its transport through the cell.[3]

The primary structures of IgA1 and SC, including the arrangement of the disulfide bonds, have been determined in our laboratory.[4-6] In this paper we describe the covalent linkage between IgA1 and SC.

MATERIAL AND METHODS

S-IgA was prepared from human colostrum according to Pincus et al.[7], and was cleaved in the hinge region with an IgA1-specific protease prepared from the supernatant of a *Streptococcus sanguis* culture as described by Plaut et al.[8] The resulting $(Fc)_2 \cdot J \cdot SC$ fragment was purified by gel-chromatography on Sephacryl S-300 superfine. CNBr cleavage (Sigma, St.Louis, USA) was performed in 70% formic acid and the sample was fractionated on Superose 12 in 30% formic acid. The major component was cleaved by pepsin (Boehringer Mannheim, FRG) and the resulting mixture chromatographed on Superose 12 (Pharmacia, Sweden).

A 46 kDa protein, obtained by peptic digestion and determined to consist of the fifth homology region of SC and the CH2 region of IgA monomer, was subjected to thermolysin cleavage (Boehringer Mannheim, FRG).

The generated peptides were separated by two dimensional RP-HPLC, first on Aquapore C8 followed by chromatography on Kromasil 100, C8, 5 μm with the TFA/acetonitrile system.

The amino acid sequences of the thermolysin peptides were determined by automated gas-phase sequencing (Model 470A, Applied Biosystems, Weiterstadt, FRG). Molecular masses were confirmed by plasma desorption mass spectrometry (Applied Biosystems, Weiterstadt, FRG).

SDS-PAGE was performed according to Laemmli.[9] Separated proteins were transferred to PVDF membranes (Millipore, Eschborn, FRG) by semi-dry blotting as described by Hirano.[10] Electroblotted proteins were subjected to amino terminal sequencing according to Kratzin *et al.*[11]

RESULTS

S-IgA was treated with an IgA1 specific protease to specifically cleave IgA1 in the hinge region between Pro 227 and Thr 228. The resulting 220 kDa peptide lacking the Fab fragments was purified by gel-chromatography. By the following CNBr cleavage the IgA1 fragment was cleaved in the third constant region of IgA1. SC lacks methionine and therefore remains complete. The resulting 130 kDa polypeptide showed that the SC is bound to only one monomer of S-IgA.

For further fragmentation this peptide was cleaved by pepsin. The resulting product was purified by gel-filtration and was characterized by SDS-PAGE, automated gas-phase sequencing and laser desorption mass spectrometry indicating a mass of about 46 kDa. We determined this peptic fragment to consist of the second constant domain of IgA1 and the fifth homology region of the secretory component.

In the final enzymatic cleavage of the 46 kDa fragment thermolysin was used. The generated peptides were separated by RP-HPLC.

A 3189 Da peptide was identified. It consists of peptides derived from IgA1 and the secretory component. It was characterized by automated gas-phase sequencing and plasma desorption mass spectrometry under reducing and non-reducing conditions (Fig. 1).

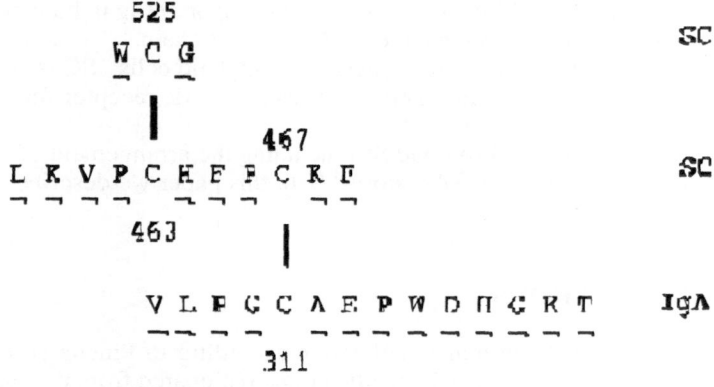

Figure 1. SC-IgA link in S-IgA.

Thereby Cys 311 of IgA1 and Cys 467 of SC were found to be responsible for the covalent linkage between IgA and SC in sIgA (Fig. 1).

Further analysis by plasma desorption mass spectrometry showed that cysteines are linked covalently via SS bridges to cysteine 311 of the other α chain and to cysteine 501 of the secretory component.

DISCUSSION

Cys 311 in serum IgA, not associated with SC is known to form an interchain disulfide bond between the two chains.[5] In S-IgA we were able to show that Cys 311 of one α chain is linked to SC. Cys 311 from the other α chain of the same monomer is blocked by a cysteine. The penultimate cysteine residues are responsible for the covalent linkage of the two IgA monomers in S-IgA and it has been demonstrated that the J chain is attached to both subunits of S-IgA at this site.[12]

For the secretory component a 5 domain structure has been described.[7] Compared to the 4 preceding regions there is an extra-disulfide bridge in the fifth homology region. It has been suggested earlier that these cysteine residue(s) are responsible for the covalent linkage to IgA.[13] This assumption is proved by these results.

Our investigations show that Cys 467 of SC alone is involved in a covalent bond between the two molecules. Cys 501 from SC that becomes free by disulfide exchange is blocked by cysteine (Fig. 2).

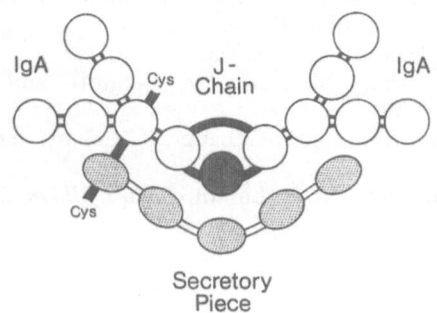

Figure 2. Structure of S-IgA.

The location of the SC-IgA linkage in the fifth domain of SC migth be a result of an interaction betweeen each of the five SC domains with the five domains of the CH2-CH3-J-CH3-CH2 structure.

CONCLUSIONS

S-IgA was isolated from human colostrum and cleaved in the hinge region. The resulting $(Fc)_2 \cdot J \cdot SC$ fragment was purified by gel-chromatography and subsequently cleaved with CNBr and pepsin. A fragment was obtained which consisted of the fifth homology region of the secretory component (SC) and the CH2 region of a single IgA monomer covalently linked by one disulfide bond.

Peptides generated by subsequent thermolysin cleavage were separated by RP-HPLC and characterized by automated gas-phase sequencing and plasma desorption mass spectrometry.

SC was found to be covalently linked to IgA by a single disulfide bond between Cys 467 of the SC and Cys 311 of one α chain. Cys 311 of the other α chain and Cys 501 of SC are blocked by cysteines.

ACKNOWLEDGMENT

We thank Mrs. M. Praetor and Mrs. D. Hesse for technical assistance and Mrs. U. Brockhaus and Mr. L. Kolb for illustrations.

REFERENCES

1. J. F. Heremans, *in*: "The Antigens", Vol. 2, M. Sela, ed., p. 365, Academic Press, New York (1974).
2. J. Mestecky and J. R. McGhee, *Adv. Immunol.* 40:153 (1987).
3. R. Solari and J. P. Kraehenbuhl, *Immunol. Today* 6:17 (1984).

4. H. Kratzin, P. Altevogt, A. Kortt, E. Ruban, and N. Hilschmann, *Hoppe Seyler's Z. Physiol. Chem.* 359:1717 (1978).

5. C. Yang, H. Kratzin, H. Götz, and N. Hilschmann, *Hoppe Seyler's Z. Physiol. Chem.* 360:1919 (1979).

6. H. Eiffert, E. Quentin, J. Decker, S. Hillemeir, M. Hufschmidt, D. Klingmüller, M. Weber, and N. Hilschmann, *Hoppe Seyler's Z. Physiol. Chem.* 365:1489 (1984).

7. C. S. Pincus, M. E. Lamm, and V. Nussenzweig, *J. Exp. Med.* 133:987 (1971).

8. A. G. Plaut, R. J. Genco, and T. B. Tomasi, *J. Immunol.* 113:289 (1974).

9. U. K. Laemmli, *Nature* 227:680 (1970).

10. H. Hirano, *J. Prot. Chem.* 8:115 (1989).

11. H. D. Kratzin, J. Wiltfang, M. Karas, V. Neuhoff, and N. Hilschmann, *Anal. Biochem.* 183:1 (1989).

12. A. Bastian, H. Kratzin, E. Fallgren-Gebauer, K. Eckart, and N. Hilschmann, These Proceedings.

13. C. Cunningham-Rundles and M. E. Lamm, *J. Biol. Chem.* 250:1987 (1975).

MOUSE SECRETORY COMPONENT

Pascal G. Pierre,[1] Xavier B. Havaux,[2] Agnes Langendries,[1] Pierre J. Courtoy,[3] Kunihiko Goto,[1] Paul Maldague,[2] and Jean-Pierre Vaerman[1]

[1]Catholic University of Louvain, Unit of Experimental Medicine
[2]Unit of Experimental Pathology and Cytology
[3]Unit of Cell Biology, Institute of Cell Pathology, B-1200 Brussels, Belgium

INTRODUCTION

The receptor for polymeric immunoglobulins (pIg-R) has already been extensively studied in many species, but not much in mice[1], the most common laboratory animals. This study was conducted to isolate and characterize mouse secretory component (SC) and/or pIg-R and to produce an antiserum allowing immunolocalization of mouse SC/pIg-R as well as the study of its expression modulated by various cytokines.

MATERIALS AND METHODS

Sample Collection

Fresh mouse (NMRI) milk was of commercial source (Proefdieren Centrum, KUL, Leuven, Belgium). Bile was obtained by cannulation of the common bile duct and blood by bleeding the retroorbital plexus. Liver fragments obtained after removing blood by PBS-perfusion, were homogenized in PBS; bacterial growth and proteolytic activities were inhibited by addition to all samples of a cocktail of 1mM benzamidine, 1mM iodoacetamide, 2mM EDTA, 0.1% sodium azide, 0.05% soybean trypsin inhibitor, and 0.0175% PMSF. Milk was delipidated by high speed centrifugation and casein was removed by acidification to pH 3.5. All samples were frozen until use.

Purification of Mouse Free SC (FSC)

Mouse FSC was purified from mouse lactoserum. Briefly, 80 ml of lactoserum was concentrated 5 times and gel-filtered on a column of Ultrogel AcA22. Eluates containing FSC and SC bound to IgA (sIgA) were identified by immunodiffusion using rabbit anti-rat SC and anti-mouse IgA antisera produced in our laboratory. The FSC-containing fraction was dialyzed against PBS with 5 mM EDTA and iodoacetamide, and 0.1% NaN$_3$. Then, this fraction was passed at 4°C on a column of human polymeric IgA myeloma proteins coupled to Sepharose 4B (Pharmacia, Uppsala). After thorough washing, bound proteins were eluted at 4°C with PBS containing 1M NH$_4$SCN, 5mM EDTA and iodoacetamide, and thereafter dialyzed against PBS. Rat FSC was similarly purified from rat bile.[2]

Production of Rabbit Anti-Mouse-SC Antiserum

A rabbit (L881) was injected subcutaneously every two weeks with 0.3 mg of mouse FSC in Freund's complete adjuvant and bled. The antiserum was solid-phase absorbed on normal mouse serum, mouse IgA (MOPC-315), and on those mouse milk proteins that were not retained on the human pIgA-Sepharose column used above; all proteins used for absorption of the antiserum were coupled to Sepharose 4B. Then, the antiserum was concentrated by precipitation with neutralized ammonium sulfate at 45% of saturation.

Sodium Dodecylsulfate Polyacrylamide Gel Electrophoresis (SDS-PAGE) and Western Blot

Two μg of purified mouse FSC and 5 μg of purified rat FSC were applied on a 12% SDS-PAGE gel. The gel was stained with 0.25% Coomassie blue R 250 (Merck, Darmstad).

Various amounts of bile, serum, lactoserum and liver homogenate were applied on an 8% SDS-PAGE gel. After migration, the proteins were electrotransferred to a PVDF membrane (Millipore). SC/pIg-R was revealed with the rabbit anti-mouse-SC antiserum, followed by biotinylated goat anti-rabbit antibodies and Extravidin-peroxidase (Sigma). The latter was revealed with 4-chloro-1-naphthol (Sigma).

Immunolocalization of Mouse SC/pIg-R

After perfusion of the anesthetized mouse with 5% paraformaldehyde, liver, small intestine and a salivary gland were dissected, excised and further immersed in 2.3 M sucrose. Blocks were frozen in liquid nitrogen for transfer to a Reichert FC4 cryo-ultramicroto me (0.5 μm sections). Rabbit anti-mouse SC antibodies (1/100) were applied and revealed with sheep FITC-IgG antibodies against rabbit IgG (Organon Technika, Durham, NC). Mouse IgA was directly revealed with rhodamine-labelled sheep anti-mouse-IgA (The Binding Site, Birmingham, U.K.). Controls included omission of the anti-SC antibodies and their replacement by normal rabbit serum.

RESULTS

Molecular Characterization of Mouse SC/pIg-R

Mouse SC was unambigously identified in mouse lactoserum fractions by means of its cross-reactivity with anti-rat SC. The elution position of mouse milk FSC on gel-filtration demonstrates that it has the same molecular size as other mammalian FSC (elution between IgG and albumin, result not shown). Mouse FSC was, as expected, able to bind to a column of insolubilized human pIgA myeloma proteins. By SDS-PAGE, the thiocyanate eluate of this column, corresponding to highly purified mouse FSC, yielded two close bands with apparent Mr of 86 and 88 kD, a mass slightly larger than that of rat FSC (Fig. 1,A).

By Western blot analysis (Fig. 1, B), the absorbed rabbit anti-mouse SC antiserum only reacted with mouse FSC and sIgA. FSC was a doublet in milk and bile (Mr 86 and 88 kD), with an additional 95 kD band in bile, as well as a probable degradation product. In the liver extract, four bands were identified (Mr of ± 90, 95, 97 and 122 kD). The latter probably corresponds to the mature pIg-R. No SC-band was detected in serum (result not shown).

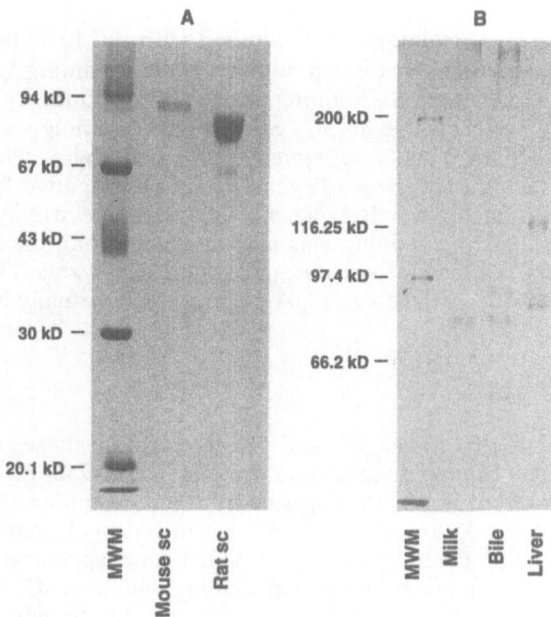

Figure 1. SDS PAGE of pure mouse and rat FSC (A) and Western blot (B) of mouse milk, bile and liver extract revealed with anti mouse SC antibodies MWM=molecular weight markers

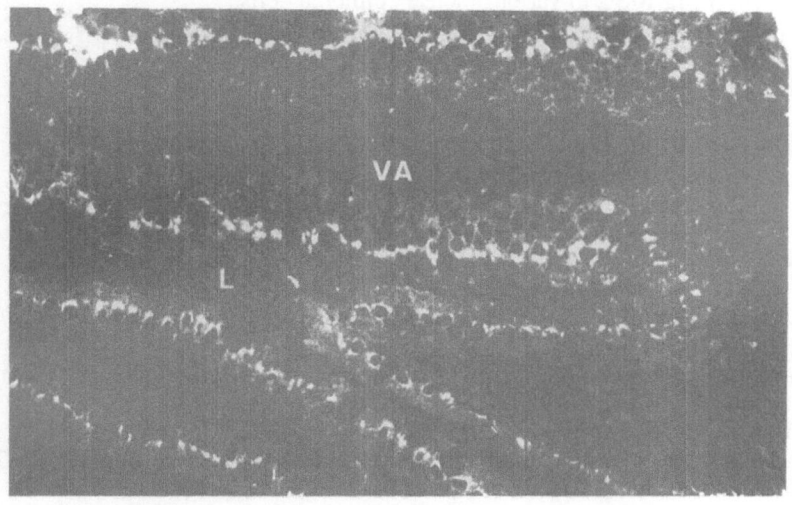

Figure 2. Immunofluorescence of mouse SC/pIgR in mouse small intestine   Note strong supranuclear signal L=intestinal lumen, VA=villus axis

Immunolocalization of Mouse SC/pIg-R

The antiserum strongly and specifically reacted with intestinal (Fig 2) and common bile duct (not shown) epithelium and, weaker, with hepatocytes and salivary gland acini (not shown). When green SC-staining was combined with red IgA-staining on an intestinal section, the classical picture[3] was found, with bright red staining IgA-plasma cells in the lamina propria, whereas green SC staining was confined to the epithelium (not shown). Mucous goblet cells were negative for SC. Strongest SC-staining was typically observed in enterocytes just above their nuclei, obviously corresponding to what was called "Golgi-staining". The epithelium covering a Peyer's patch was negative for SC, but the Peyer's patch dome area contained a few IgA-plasmacytes (not shown). When only SC-staining was used, weak but specific staining was also detected along the basolateral enterocyte membrane, with some reinforcement in the apical cytoplasmic zone. The lamina propria was SC-negative. Controls for both SC- and IgA-staining were virtually black (not shown).

CONCLUSIONS

Our data show that mouse SC and pIg-R resemble those of other mammals. A slightly larger apparent Mr could correspond to small interspecies glycosylation differences. The reactivity with anti-rat-SC demonstrates many common epitopes between SC of rats and mice. As described in other rodents, pIg-R is produced by secretory epithelial cells and hepatocytes. In mouse enterocytes, the "Golgi" zone appeared as a bright granular supranuclear cluster after immunofluorescent staining with anti-SC. Mouse pIg-R may also mature by incorporation of carbohydrates, as suggested by bands of different Mr in the perfused liver extract. Such an intracellular maturation process by sugar addition was already shown for rat and rabbit pIg-R.[4,5]

These reagents will be of great value in establishing the true sIgA nature of mouse IgA antibodies detected in mouse intestinal and bronchial secretions after various immunization protocols.[6] They could also help in establishing which cytokines modulate the expression of pIg-R/SC in various mouse epithelial cells.

ACKNOWLEDGMENTS

Supported by the Belgian State (Science Policy Programming, Office of the Prime Minister, PAI 7bis and AC 88-93-122).

REFERENCES

1. I. Lemaître-Coelho, C. André, and J. P. Vaerman, *Protides Biol. Fluids* 25:891 (1977).
2. G. Acosta Altamirano, C. Barranco-Acosta, E. Van Roost, and J.P. Vaerman, *Mol. Immunol.* 17:1525 (1980).
3. P. Brandtzaeg, *J. Immunol.* 112:1553 (1974).
4. J. P. Buts, J. P. Vaerman, and G. Lescoat, *Gastroenterology* 102:949 (1992).
5. K. E. Mostov and G. Blobel, *J. Biol. Chem.* 257:11816 (1982).
6. M. Van Damme, M. P. Story, T. Biot, J. P. Vaerman, and G. Cornelis, *Infect. Immun.* 103:520 (1992).

ANTIBODY PRODUCTION TO SECRETORY COMPONENT (SC) USING RECOMBINANT SC FRAGMENT

Miwako Kamei,[1] Takashi Iwase,[1] Peter Krajci,[2] Per Brandtzaeg,[2] and Itaru Moro[1]

[1]Department of Pathology, Nihon University School of Dentistry, Tokyo 101, Japan; and [2]LIIPAT, Institute of Pathology, University of Oslo, Rikshospitalet, N-0027 Oslo, Norway

INTRODUCTION

Secretory component is a glycoprotein produced by glandular epithelia and localized on the basolateral membrane of epithelial cells. SC functions as a receptor for J chain-containing polymeric IgA and IgM, and facilitates the transport of polymeric immunoglobulins through the epithelium to the lumen. It has been reported that some antisera to human SC cross-react with galactosyltransferase (GT) because of copurification of SC and GT from human milk as previously described.[1,2] In this study, we have attempted to produce an antibody to SC without crossreactivity to GT using a part of human SC cDNA.

MATERIALS AND METHODS

Construction of SC Expression Plasmid

A part of human SC cDNA (*Pst* I digested fragment[3], position 789-1455) was ligated into the *Pst* I digested vector pEX1[4] which expresses a fusion protein with β-galactosidase (β-gal). The plasmid construct was used to transform competent *E. coli POP2136* cells as described by Hanahan[5] in which all incubations were carried out at 30°C (heat shock at 37°C). The colonies obtained were grown up in L-broth (1% tryptone, 0.5% yeast extract, 1% NaCl) containing 50 µg/ml ampicillin and harvested, then their plasmid DNA was analyzed with a DNA sequencer (373A, Applied Biosystems, CA) using fluorescence labelling.

Preparation and Identification of SC-β-Gal Protein

Bacteria containing recombinant plasmid were grown with gentle shaking overnight at 30°C in 5 ml of L-broth containing 50 µg/ml ampicillin. Following induction of SC-β-gal protein at 42°C for 120 min with vigorous agitation, the bacteria were pelleted by centrifugation. The pellet was solubilized by boiling for 5 min in 500 µl of lysis buffer (5% SDS, 50 mM Tris-HCl, pH 8.0, 15 mM 2-mercaptoethanol). In order to identify SC-β-gal protein, 10 µl of solubilized pellet was subjected to 7.5% SDS-polyacrylamide gel electrophoresis (SDS-PAGE).

Immunization Protocol

Rabbits were immunized twice subcutaneously using SC-β-gal protein emulsified in Freund's complete adjuvant with a one-week interval. After three weeks, the immune response was boosted again using SC-β-gal protein emulsified in Freund's incomplete adjuvant. Rabbits were bled at one week after the last immunization.

Identification of Antibody to SC-β-Gal Protein

Human SC and secretory IgA (SIgA) purified from human milk were subjected to 7.5% SDS-PAGE; thereafter the proteins were transferred to membrane (Immobilon-pSQ; Millipore) for Western blotting. In brief, the membrane was incubated for 2 h with blocking solution (4% BSA/PBS) and for 1 h with the antibody to SC-β-gal protein as a 1:50 dilution in blocking solution. After washing 3 times for 15 min with PBS containing 0.05% Tween 20, horseradish peroxidase-conjugated anti-rabbit IgG (diluted 1:500 in blocking solution) was applied on the membrane for 1 h. The membrane was washed again and stained by an immunoperoxidase method with 50 mg 3,3'-diaminobenzidine tetrahydrochloride in 100 ml 0.5 M Tris-HCl, pH 7.5 containing 30 µl of 30% H_2O_2.

Immunohistochemical Staining

Human salivary gland and small intestinal tissues were embeded in O.C.T. compound (Tissue Tek II) and frozen in dry ice-acetone. Cryostat sections were fixed in cold acetone for 10 min before staining. Cultured glioblastoma cell line used as SC-negative control was fixed in cold acetone for 10 min and washed in PBS. Specimens were treated with 10% normal horse serum in PBS for 10 min and incubated with an antibody to SC-β-gal protein at various dilutions in 1% BSA/PBS for 1 h. After washing in PBS for 1 h, specimens were incubated with FITC-conjugated anti-rabbit IgG (diluted 1:40 in 1% BSA/PBS) for 1 h and mounted in PBS-glycerol.

RESULTS AND DISCUSSION

Construction and Identification of SC-β-Gal Protein

The expression of cloned eukaryotic genes in E. coli has greatly facilitated basic studies of a large number of proteins. When the expression of a native, unfused protein is unsatisfactory, the expression of protein or parts of proteins as hybrids fused with E. coli polypeptides such as β-gal, has often been more successful. Furthermore, the fusion of a foreign polypeptide with β-gal appears to enhance significantly the stability of the polypeptide in E. coli; a variety of biochemical procedures can be employed to isolate the hybrid product for use in functional studies or as an immunogen.

SC consists of 764 amino acids residues and contains five extracellular domains, a membrane-anchoring region and an intracellular tail. In this study, we used the SC-encoding fragment (extracellular domains 3-5) that has low homology with GT. To confirm the correct direction of insert, the SC-β-gal plasmids obtained were analyzed with a DNA sequencer using fluorescence labelling. DNA sequencing revealed complete accordance with a part of the SC cDNA sequence. SDS-PAGE revealed the extracted SC-β-gal protein as a 150 kDa band, confirming fusion of SC with β-gal.

Analysis of Antibody to SC

The IgG fraction of the rabbit antiserum was applied on β-gal protein affinity column. Its specificity for SC was defined by binding to free SC and SIgA in Western blotting. Free SC on the membrane reacted with the antibody and also SIgA was weakly reactive. Commercially available antibody to SC was used as positive control and reacted as expected with free SC and SIgA.

Immunohistochemical Performance of Antibody to SC

In salivary gland and small intestine, immunofluorescence with antibody to SC was displayed in the cytoplasm of acinar cells and the luminal contents. Commercially available anti-SC showed the same localization of SC. Cultured glioblastoma cells were negative with all anti-SC reagents.

CONCLUSIONS

This study demonstrated the possibility of raising an antibody to recombinant SC, thus avoiding the problem with contaminating antibody to GT inherent in the traditional methods used to produce antisera against purified SC.

ACKNOWLEDGMENTS

We are grateful for the contribution of Dr. T. Koga (National Institute of Health, Japan) in the generation of antisera. This work was supported by a Grant-in-Aid for Scientific Research from Ministry of Education (No. 034052) the Norwegian Research Council for Science and the Humanities, and the Norwegian Cancer Society.

REFERENCES

1. E. J. McGuire, J. J. Cebra, and S. Roth, *Monogr. Allergy* 24:81 (1988).
2. E. J. McGuire, R. Kerlin, .J. J. Cebra, and S. Roth, *J. Immunol.* 143:2933 (1989).
3. P. Krajci, R. Solberg, M. Sandberg, O. Øyen, T. Jahnsen, and P. Brandtzaeg, *Biochem. Biophys. Res. Commun.* 158:783 (1989).
4. K. K. Stanley and J. P. Lucio, *EMBO. J.* 3:1429 (1981).
5. D. Hanahan, *J. Mol. Biol.* 166:557 (1983).

A LACK OF RELATIONSHIP BETWEEN SECRETORY COMPONENT AND GALACTOSYLTRANSFERASE IN HUMAN MILK

Kunihiko Kobayashi,[1] Naoki Mafune,[1] Hisashi Narimatsu,[2] Hirohisa Nakao,[3] and Naoyuki Taniguchi[3]

[1]Department of Laboratory Medicine, Hokkaido University School of Medicine, Sapporo 060; [2]Institute of Life Science, Soka University, Tokyo 192; and [3]Department of Biochemistry, Osaka University Medical School, Osaka 530, Japan

INTRODUCTION

Secretory component (SC), a glycoprotein produced by mucosal epithelial cells and expressed on their cell surface, is known to be the receptor of polymeric immunoglobulins (Ig) of IgA and IgM. SC on the basolateral surface of epithelial cells binds polymeric Igs, and undergoes endocytosis and transcellular transport to apical surfaces for the release of secretory Igs which retain the major portion of the SC receptor.[1] During this process, a considerable amount of SC, unbound to polymeric Igs, is also released from the apical surfaces and present in various secretions as a soluble form of SC which has identical structure to the SC attached to the secretory Ig molecules.[2]

β-(1-4)-galactosyltransferase (β-GT) is an enzyme which transfers galactose from UDP-galactose to non-reducing N-acetylglucosamine residues of oligosaccharides on various glycoproteins. It is expressed in the plasma membrane of various cells in membrane bound form and is also present in a soluble form in various secretions as well as in serum.[3] The β-GT is considered to have affinity for glycoproteins through binding to their oligosaccharides. In this context, the β-GT is similar to SC in its molecular characteristics; both molecules are present in two forms, membrane bound and soluble forms, and have affinity for glycoproteins including immunoglobulins. In fact, it was reported that β-GT purified specifically by UDP-galactose affinity chromatography was associated with SC antigen and was claimed to be related to, or identical to SC.[4,5]

The present study was initiated to clarify whether the above claim is true or not.

MATERIALS AND METHODS

SC, Secretory IgA (sIgA) and Corresponding Antibodies

SC and sIgA were purified from human milk by salting out, ion-exchange chromatography and gel filtration, and their antibodies were prepared in rabbits and goats as described before[2]. Mouse monoclonal antibodies (MAbs) to SC were provided by Dr. I. Moro, Nihon University, Tokyo and by Dr. J.H. Perez, University of Cambridge, Cambridge.

Anti-ß-Galactosyltransferase (ß-GT) MAb

A mouse MAb8628 to human β-GT was established by immunization with human GT which was purified from serum using α-lactalbumin, and anti-human immunoglobulin affinity chromatography. MAbs were screened with purified antigen. The antigenic epitope recognized by the MAb8628 was defined using recombinant proteins directed by deletion mutants of human β-GT cDNA. It was located at the stem region of the soluble form of β-GT.

ß-GT Assay

A reversed-phase HPLC was employed using pyridylaminated agalactosyl non-bisected biantennary complex as galactose acceptor as described in detail by Fujii et al.[6]

Recombinant Human ß-GT

The cDNA encoding human β-GT was cloned from a human placental cDNA library using bovine cDNA as a probe.[7] A plasmid containing full-length human β-GT-cDNA was digested with AccIII and EcoRI to excise a fragment encoding the soluble form of β-GT. The fragment was inserted into AvaI and EcoRI sites of the pUC19 vector. The resulting plasmid (pHGT832) encoded a fusion protein in frame comprising 18 amino acid residues of β-galactosidase derived from the lac Z gene of the vector and 320 residues of the soluble form of β-GT. An E. coli strain, XL1-Blue, transformed with pHGT832 was grown in broth containing 50μg/ml of ampicillin. After induction of the recombinant protein by addition of isopropyl β D-thiogalacto-pyranoside (IPTG) at a final concentration of 2 mM, the cells were collected by centrifugation and washed once with 10 mM Tris-HCl buffer, pH 7.4, containing 10 mM $MnCl_2$. They were suspended in 0.065 M Tris-HCl buffer, pH6.8, with 2% SDS and 5% 2-mercaptoethanol and lyzed for SDS-PAGE.

Affinity Chromatography

UDP-galactose (UDP-hexanolamine-Sepharose, Sigma) affinity chromatography (1.5 X 5 cm) was done essentially the same as that by McGuire et al.[5] An anti-SC antibody immunoaffinity column (1.5 X 6 cm) was prepared by coupling goat anti-SC to Sepharose 4B by CNBr. The column operation was done at 4 °C with 10 mM cacodylate buffer, pH 7.6.

Ion-Exchange and Gel Chromatography

DEAE cellulose column (DE 52, Whatman) was carried out in Na-phosphate buffer by stepwise elution.[2] Gel filtration was done by FPLC using Superose 12 HR 10/30 (Pharmacia).

Immunoblot .ing

Western blot procedures, including SDS-PAGE and electrotransfer to nitrocellulose, were carried out as described by Laemmli[8] and Towbin et al.[9], respectively. Immunostaining of a single sample with various different antibodies was performed using the Cross Blot System (SEBLA, France) with biotin-conjugated second antibodies and peroxidase-labeled avidin. Dot blot procedure was employed for the detection of antigens in chromatographic separation. The procedure for immunostaining the dot-blot was the same as that for Western blotting.

RESULTS AND DISCUSSION

UDP-Galactose Affinity Chromatography

Human milk, precipitated with ammonium sulfate at 50% saturation, was applied to a UDP-galactose column. The column was washed extensively with initial buffer and proteins adsorbed were eluted with 2 M NaCl in the initial buffer. When the column effluent was monitored for ß-GT, SC, and α-chains by dot blot procedure with appropriate antibodies, these three antigens were all eluted together in the same fractions, indicating that ß-GT was copurified with SC as well as α-chain antigens (possibly sIgA), as mentioned by McGuire et al.[5](not shown).

Anti-SC Affinity Chromatography

The preceding results indicated the possible association of SC or sIgA with β-GT. We then tried whether or not the SC and α-chains are separable from β-GT by anti-SC antibody affinity chromatography. Fractions of the UDP-galactose column, rich in β-GT, SC and α-chains, were loaded on an anti-SC antibody column. After break-through fractions were obtained, the column was eluted with NaCl at 0.5, 1.0 and 2.0 M in the initial buffer, and finally with 8 M urea (Fig. 1). Each fraction was monitored for β-GT, SC, and α-chains by dot-blotting. As can be seen in Fig. 1, β-GT and α-chains but not SC were found in the break-through fraction, and exclusively β-GT in the 0.5 M NaCl fraction. Most of the SC and α-chains were eluted in 8 M urea fractions, although some β-GT was also found in this fraction. Nevertheless, it is apparent that β-GT devoid of SC was obtained in the break-through and 0.5 M NaCl fractions by anti-SC antibody chromatography. This strongly indicates that SC and β-GT are independent molecules. The co-elution of some β-GT antigen with SC by 8 M urea seemed to be due to non-specific interaction between the β-GT and the anti-SC antibody column, but not to antigenic cross-reactivity of the anti-SC antibody. To prove this, we then tested the cross-reactivity of the anti-SC antibody with β-GT, and also that of anti-β-GT with SC, by Western-blot procedure.

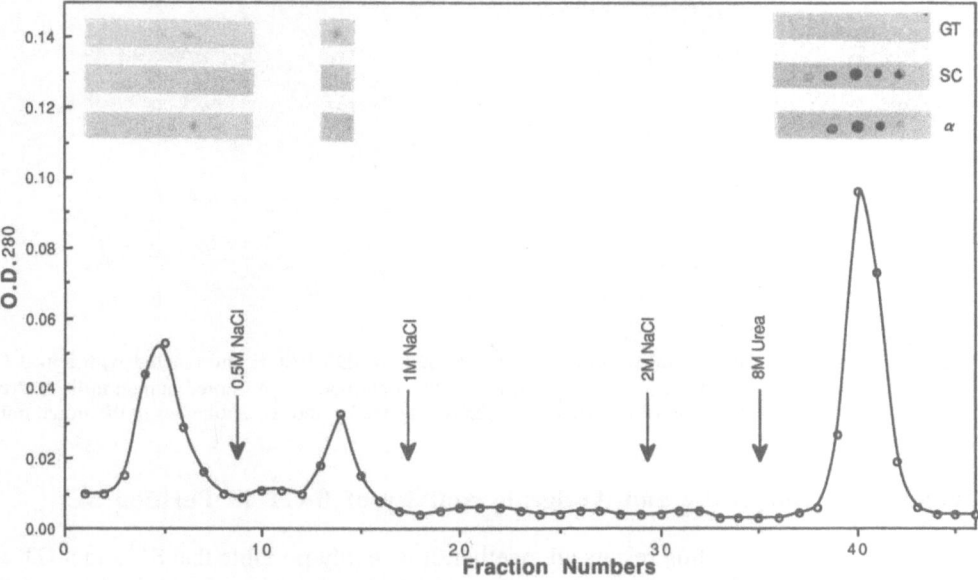

Figure 1. Separation of β-GT, SC, and α-chains by anti-SC antibody affinity chromatography. Photos in the figure represent dot-blot analysis obtained by anti-β-GT monoclonal antibody (GT), anti-SC rabbit antiserum(SC), and anti-α-chain rabbit antiserum(α).

639

Tests for Immunological Cross-Reactivity Between SC and ß-GT

Ammonium sulfate precipitates of human milk containing both β-GT and SC were analyzed by Western blotting using anti-β GT antibody and several different preparations of anti-SC antibodies As can be seen in Fig 2A, the anti-β-GT detected a single band at around 45 kD which is the proper position of native β-GT in milk, while each anti-SC antibody detected two bands, one major band at 75 kD and the other a very faint band at around 45 kD (Fig 2A arrows) The band at 75 kD corresponded to that of native SC, while the other at around 45 kD was reminiscent of β-GT in molecular weight It has been reported by us that SC is labile to proteolytic degradation and converts into fragment of around 40-45 kD [10] In this connection, it was conceivable that the 45 kD fragment detected by every anti-SC antibody, was not from β-GT but SC, possibly due to degradation of SC during storage of milk To prove this, freshly collected milk was then analyzed in the same manner, and was found to have no band other than 75-kD SC (Fig 2B), indicating that the 45 kD could be derived from fragmentation of SC molecules but not due to cross-reaction of anti-SC antibodies with β-GT To ensure this further, we then analyzed the reactivity of anti-SC antibodies with recombinant β-GT The cell lysate of *E coli* with the β-GT gene, expressing recombinant β-GT, and one without the β-GT gene were electrophoresed and analyzed as shown in Western blot of Fig 3 Although a number of bands appeared in each *E coli*-lysate with anti-SC antibodies, banding patterns between *E coli* with and without β GT gene were almost identical, thus indicating that the bands appeared through non specific interactions between some cellular elements of *E coli* and the reagents, including the first and second animal antibodies and/or peroxidase-labeled avidin

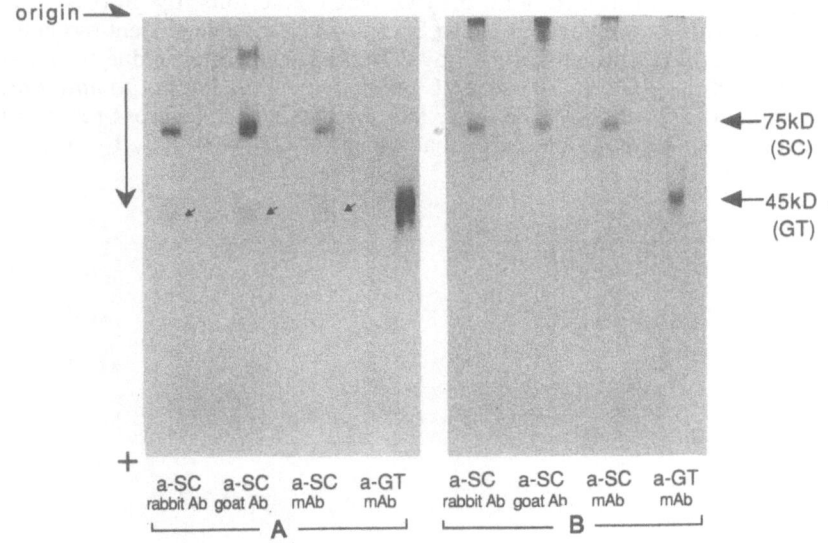

Figure 2. Western blotting of human milk electrophoresed on SDS PAGE and reacted with anti β GT monoclonal antibody and anti SC polyclonal and monoclonal antibodies A stored human milk B fresh human milk Arrows in the photo (A) indicate faint bands detected by anti SC antibodies in the stored milk

Detection of Enzymatic and Antigenic Activity of ß-GT in Purified SC

Taking the preceding results all together, it is highly possible that SC and β-GT are each independent molecules and that no immunological cross-reaction is present between the two In order to verify this further, conventionally purified SC from milk was assessed for β-GT enzymatic activity As shown in reversed-HPLC charts in Fig 4, the pyridylaminated agalactosyl oligosaccharide (formula I, in Fig 4) incubated with as much as 250 µg of SC in

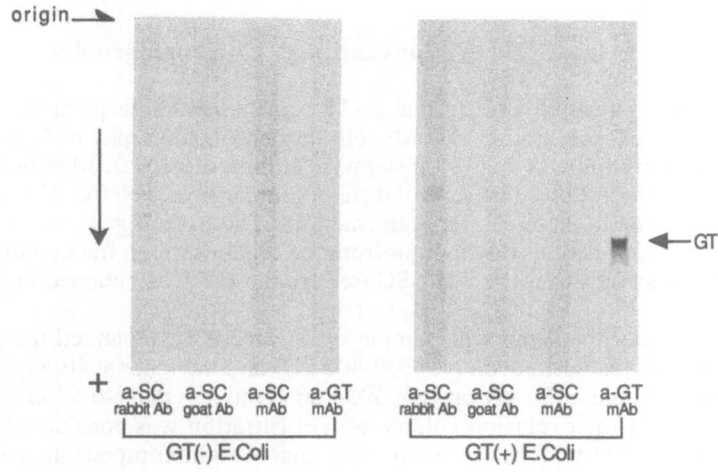

Figure 3. Western blotting of *E coli* lysate expressing recombinant β GT (GT(+) *E coli*) and one without the β-GT gene (GT(-) *E coli*) reacted with several different anti-SC antibodies and an anti-β-GT monoclonal antibody Anti-β-GT antibody clearly stained a band of recombinant β-GT in *E coli* with β-GT gene No such band corresponding to the recombinant β-GT was detected by any anti-SC antibodies, although they detected several undefined bands almost equally in *E coli* both with or without the β-GT gene

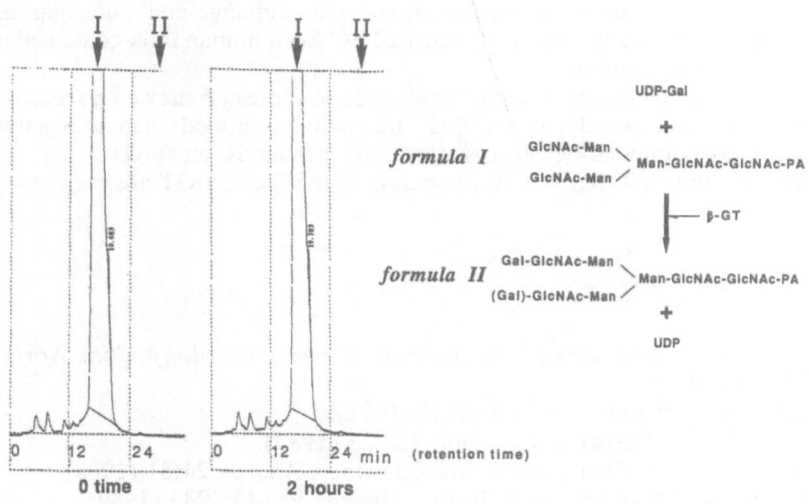

Figure 4. Detection of β-GT enzymatic activity in conventionally purified SC Pyridylaminated agalactosyl non-bisected biantennary complex (*formula I*) was mixed with 250 µg of purified SC in the presence of UDP-galactose A portion of the mixture before incubation at 37°C (0 time) and that after 2 h incubation (2 h) were analyzed by reversed-phase HPLC Both samples at 0 time and 2 h revealed equally huge peaks at the retention time for *formula I* (I), and none corresponding to the galactosylated complex (*formula II*) was detected at its retention time (II) even after 2h incubation

the presence of UDP-galactose showed no conversion of formula I into galactosyl oligosaccharide (formula II), confirming a lack of GT activity in the SC preparation. Detection of β-GT antigen in the purified SC by Western blot analysis with anti-β-GT antibody also gave a negative result (not shown).

Separation of ß-GT from SC by Conventional Chromatography

The preceding results indicate that β-GT can be separable from SC somewhere during conventional SC preparation. DEAE cellulose chromatography of 50% ammonium sulfate precipitates of milk was done with stepwise elution using 0.01 M phosphate buffer, pH 7.6, as initial buffer followed by several different molarities of NaCl. The β-GT antigen was exclusively eluted in the break-through fraction. The SC antigen, on the other hand, was eluted in the wide fractions distributing from the break-through fraction to 0.05 M and 0.1 M NaCl fractions (not shown). Thus SC free from β-GT was obtained in 0.05 and 0.1 M NaCl fractions.

Gel filtration of the same milk sample on Superose 12 separated the β-GT into 2 fractions, one near the exclusion volume (300-400 kD) and the other at 40-50 kD, while the SC antigen came out quite diffusely ranging 300-400 kD to less than 40-50 kD (not shown). The β-GT eluted near the exclusion volume on gel filtration was considered to be β-GT aggregated or associated with certain high molecular weight components, such as sIgA. The binding affinity of β-GT to sIgA has been reported before.[5] The β-GT at 40-45 kD should easily be defined as the soluble form of β-GT. Thus fractions around at 70-90 kD, where the native SC comes out, consists of SC devoid of β-GT.

CONCLUSIONS

SC and β-GT were copurified by UDP-galactose affinity chromatography as reported by McGuire et al..[4,5] However these two components were separable either by anti-SC antibody affinity chromatography or conventional ion-exchange and molecular exclusion chromatography. In fact, conventionally purified SC from human milk contained no β-GT antigenic and enzymatic activity.

Anti-SC polyclonal or monoclonal antibodies so far tested showed no reactivity with recombinant β-GT. Conversely, an anti-β-GT monoclonal antibody had no reactivity with SC. Thus there is no immunological cross-reaction between SC and β-GT.

Taking these result together, it is apparent that SC and β-GT are each independent molecules.

REFERENCES

1. J. Mestecky, C. Lue, and M. W. Russell, *Gastroenterology Clin. North Amer.* 20:441 (1991).
2. K. Kobayashi, *Immunochemistry* 8:785 (1971).
3. J. Roth and E. G. Berber, *J. Cell Biol.* 92:223 (1982).
4. E. J. McGuire, J. J. Cebra, and S. Roth, *Monogr. Allergy* 24:81 (1988).
5. E.J . McGuire, R. Kerlin, and S. Roth, *J. Immunol.* 143:2933 (1989).
6. S. Fujii, T. Nishiura, A. Nishikawa, R. Miura, and N. Taniguchi, *J. Biol. Chem.* 265:6009 (1990).
7. H. Narimatsu, S. Sinha, K. Brew, H. Okayama, and P.K. Qasba, *Proc. Natl. Acad.Sci.USA.* 83:4720 (1986).
8. U. K. Laemmli, *Nature* 227:680 (1970).
9. H. Towbin, T. Staehelin, and J. Gordon, *Proc. Natl. Acad. Sci. USA.* 76:4350 (1979).
10. K. Kobayashi, J.-P. Vaerman, and J. F. Heremans, *Immunochemistry* 10:73 (1973).

SUBSTANCE P ACCELERATES SECRETORY COMPONENT-MEDIATED TRANSCYTOSIS OF IgA IN THE RAT INTESTINE

Dennis McGee,[1] Maya Eran,[2] Jerry R. McGhee,[1]
and Serem Freier[2]

[1]Department of Microbiology, University of
Alabama at Birmingham, USA;
[2]Laboratory of Mucosal Immunology, Shaare Zedek
Medical Center, Jerusalem, Israel

INTRODUCTION

We have previously shown that cholecystokinin (CCK) causes secretion of specific IgA antibodies in the rat intestine.[1] The objective of the experiments described below was to establish if another gut-brain peptide, substance P (SP), also accelerates secretion of IgA. The reason for choosing this peptide was that SP is functionally similar to CCK, though chemically quite unrelated. Once having established that SP, indeed, causes secretion of IgA specific antibodies, we proceeded to obtain data on the molecular size of the IgA, on its probable mode of transport and on the absolute amounts secreted.

METHODS

Estimation of Specific IgA Antibodies

Rats of the Hooded-Lister strain were immunized with ovalbumin (OVA) by an intraperitoneal injection of 250 µg in 0.1 ml 0.9% saline mixed with an equal amount of Freund's complete adjuvant. Fourteen days later, a booster injection of 10 µg OVA in 0.1 ml 0.9% saline mixed with an equal amount of Freund's complete adjuvant was administered intraperitoneally. On the 21st day the rats were anesthetized and a loop of intestine 10 cm long was isolated 10 cm distal to the pylorus, and perfused with normal saline at a rate of 0.5 ml/2.5 min. After flushing the loop for 30 min, aliquots were collected every 2.5 min. After 10 min, SP was injected intravenously and the collection continued. Specific IgA antibodies to OVA were estimated as previously described by us.[1] As basal antibody levels varied between rats, all results were expressed as percentages. The mean optical density (OD) reading of the first four, 2.5 min aliquots (the base line period) was designated 100%. Subsequent readings in OD were compared with this baseline and expressed as a percentage of it. The control group was immunized but did not receive the peptide.

Estimation of IgA

Rats of the Hooded-Lister strain, weighing 180-200 g were anesthetized, a loop of intestine was isolated, perfused, and aliquots collected as described above. Total and secretory IgA (S-IgA) were estimated by ELISA: Total IgA was estimated with a monoclonal mouse anti-rat IgA (Zymed Laboratories, San Francisco, CA). Standard curves were generated using a rat Ig reference serum (Miles Laboratories, Elkhart, IN). S-IgA was estimated with a biotin conjugated F(ab')$_2$ fraction of rabbit anti-rat SC antibody (kindly provided by Dr. Brian Underdown, McMaster University, Hamilton, Ontario).

Secretion of Albumin

In one group of animals, the secretion of albumin was measured. These rats received 0.5 μCi of ^{125}I-human serum albumin intravenously, 45 min before the iv injection of 25 μg of SP. The perfusate was precipitated with trichloacetic acid and the precipitable radioactivity was measured.

Student's t test was used for statistical analysis.

RESULTS

The injection of SP produced increased secretion of IgA specific OVA antibodies. This was significant already at 2.5 min and remained so for a total of 5 min (Fig. 1).

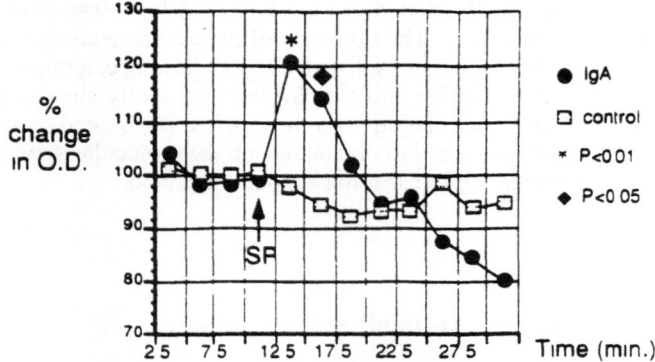

Figure 1. The rise in IgA specific OVA antibody secretion, following the injection of substance P.

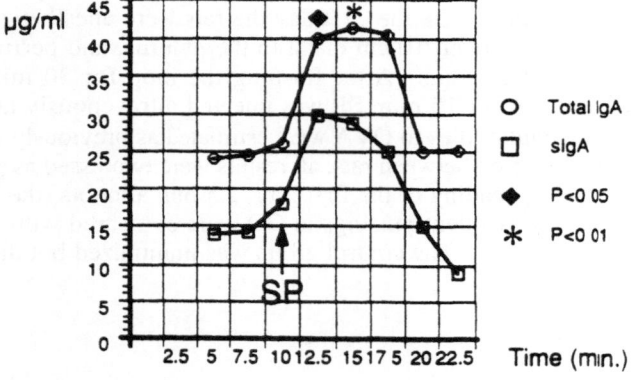

Figure 2. Rise in total IgA and S-IgA following the injection of SP.

The amount of S-IgA constituted about two-thirds of the total IgA in the intestinal fluid at rest. Following the injection of SP, there was a significant rise of sIgA and of total IgA (Fig. 2). The rise of total IgA was almost the same as that of S-IgA, indicating that the SP-induced rise of IgA was almost entirely sIgA (Table 1 and Fig. 2).

Substance P caused increased the secretion of albumin which remained significant for 10 min (Fig. 3).

Table 1. Secretion of total IgA and S-IgA into the lumen of rat intestine following the iv injection of substance P (results are expressed as µg/2.5 min/10 cm intestine).

| | Time (min) | | | |
	0	2.5'	Increment	P
Total IgA	12.4 ± 0.5	20 ± 6.3	7.6	<0.05
S-IgA	7.9 ± 1.2	15 ± 7.4	7.1	<0.05

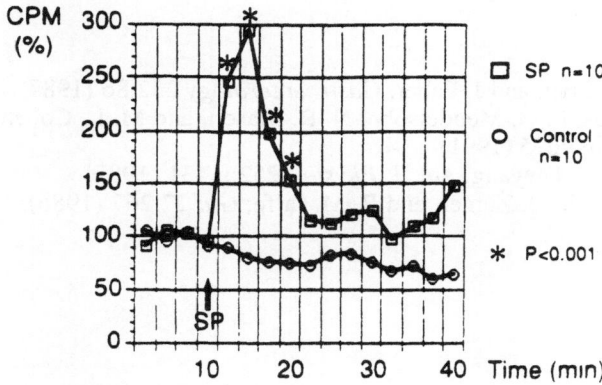

Figure 3. The injection of SP causes a 3-fold rise in albumin secretion into the intestine.

DISCUSSION

We have previously shown that CCK-octapeptide (CCK-OP) can bring about the secretion of IgA isotype antibodies in the intestine.[1] As there are CCK receptors on SP-containing nerves[2] and as in certain situations SP has been shown to mediate the effects of CCK[3], it was reasonable to suspect that SP, like CCK, may bring about the secretion of IgA. We show here that our surmise was correct. Specific IgA antibodies to OVA are secreted into the rat intestine in response to stimulation with SP. Our quantitative studies presented above show that in the resting rat intestine, about two-thirds of the total IgA is S-IgA. Following the injection of SP, the total IgA rises by 7.6 µg/2.5 min/10 cm of intestine. This rise was almost entirely due to a rise of S-IgA. As S-IgA transported is driven by microtubules, we suggest that SP accelerates the intracellular transport of S-IgA. It seems likely that the S-IgA that was transported was already present within the cell and that SP caused its depletion within the cell. We suggest this as the most likely explanation of the effect of SP, in view of the rapid, but short-lived, response to the injection of the peptide. It is known, however, that SP also causes secretion of electrolytes and water in the gut.[4] Furthermore, we have shown above, that SP also causes secretion of albumin into the rat intestine (Wilschanski *et al.*, to be published). These facts raise the question whether microtubular translocation is the only mode of transport induced by SP. It is generally accepted that 80% of intestinal secretion of water and electrolytes takes place along the intercellular pathway, across the tight junction. We cannot, therefore, exclude the possibility that some or most of the albumin and electrolyte transport induced by SP takes place along this route. Substance P may, therefore, induce both intracellular and intercellular transport.

CONCLUSION

We would like to stress what seems to us the most important of the observations here described, namely that the intravenous injection of SP almost doubles the intestinal output of S-IgA for a limited period of time.

ACKNOWLEDGMENT

This research was supported by a grant from the G.I.F., the German-Israeli Foundation for Scientific Research and Development, and the Miriam Coven-Fish Fund.

REFERENCES

1. S. Freier, M. Eran, and J. Faber, *Gastroenterology* 232:86 (1987).
2. U. L. Lucaites, L. G. Mendelsohn, N. R. Mason, and M. L. Cohen, *J. Pharma. Exp. Thera.* 2565:695 (1991).
3. J. Wiley and L. Owyang, *Am. J. Physiol.* 252:g 431 (1987).
4. D. McFadden, M. J. Zinner, and B. M. Jaffe, *Gut* 27:267 (1986).

NEW FUNCTIONS FOR MUCOSAL IgA

Michael E. Lamm, Mary B. Mazaneca, John G. Nedrud
and Charlotte S. Kaetzel

Institute of Pathology and Department of Medicine
Case Western Reserve University
Cleveland, Ohio, USA

INTRODUCTION

The concept of a mucosal barrier function for secretory IgA is well established.[1-3] In this role lumenal IgA serves to prevent exogenous antigens, including microorganisms, from attaching to and penetrating the epithelial lining of mucous membranes and helps to promote their degradation and excretion. Our recent work[4,5] *in vitro* has provided evidence suggesting that mucosal IgA can function in host defense at two additional levels: (a) intracellularly within the mucosal epithelium and (b) within the lamina propria. In the former locus, as IgA passes through the epithelial cell on its way to the secretions, if it should encounter the specific antigens of an intracellular pathogen, it has the potential to neutralize the pathogen. In the latter locus, IgA antibody bound to specific antigen within the lamina propria can excrete the antigen as an immune complex by following the same transepithelial route as free dimeric IgA. Here we shall present some of the kinds of data that support these concepts.

MATERIALS AND METHODS

The principal experimental system used in these studies employs a polarized monolayer of epithelial cells that express the polymeric Ig receptor, of which the cleaved extracellular domain is secretory component. The cells are grown on a nitrocellulose filter that separates upper and lower fluid compartments that can be sampled independently. Tight junctions between adjacent cells prevent the diffusion of macromolecules across the monolayer in either direction. Since the cells express the polymeric Ig receptor on their basolateral surface, they can take up dimeric IgA by receptor mediated endocytosis and transport it across the cell to the apical surface, where secretory IgA is released into the upper fluid compartment after proteolytic cleavage of the polymeric Ig receptor at the junction of the transmembrane and extracellular domains. To study the ability of IgA to neutralize virus intracellularly, the polarized epithelial cells are infected via the apical surface and specific antibody is introduced at the basolateral surface, providing an opportunity for IgA antibody to combine with viral antigen inside the cell. To study the ability of specific IgA antibody to transport antigen across the monolayer, soluble immune complexes are added to the lower fluid compartment and their appearance in the upper fluid compartment is then assessed.

Advances in Mucosal Immunology, Edited by
J. Mestecky *et al.*, Plenum Press, New York 1995

RESULTS

A typical experiment for showing the ability of specific IgA antibody to neutralize viruses within epithelial cells expressing polymeric Ig receptor on the basolateral surface is shown in Table 1. MDCK cells transfected with cDNA for rabbit polymeric Ig receptor provided by K. Mostov[6] were grown as polarized monolayers on a membrane separating apical and basal compartments. The cells were infected via the apical surface with Sendai virus. Afterwards free virus was removed from the apical compartment, and IgA or IgG monoclonal antibodies against the viral HN glycoprotein were added to the lower compartment.

Table 1. IgA-mediated intracellular virus neutralization.

	Virus Titer[a]	
Antibody[b]	Intracellular	Supernatant
IgA	2.73q.24	3.07q.35
IgG	4.16q.14	4.47q.05
None	4.43q.25	4.97q.28

[a]mean q S.D. log10 pfu/ml. Neutralization by IgA vs. IgG or no antibody control is statistically significant (p<.001) for both cell lysates and apical supernatants.
[b]anti-viral monoclonal antibodies (IgA #37, IgG #20)

A day later viral titers were measured in both cell lysates and apical supernatants. The data show that IgA antibody decreases virus titers in both locations compared to specific IgG, irrelevant IgA (data not shown) or no antibody. Additional controls employing antibody conditioned medium confirmed that virus neutralization was occurring intracellularly.[5] Our original studies on intraepithelial cell neutralization of viruses by IgA antibody were done, as illustrated above, with Sendai virus, a rodent parainfluenza virus. More recently we have obtained similar data with human influenza virus (see Mazanec, *et al.,* this volume). Radioiodinated IgA monoclonal antibody against DNP, either dimeric or monomeric, was used to form soluble immune complexes with DNP-BSA antigen. Complexes were then added to the compartment below a polarized monolayer of MDCK epithelial cells expressing polymeric Ig receptor. The quantity of radioactive immune complexes transported to the upper compartment was measured. The cells are able to transport immune complexes when the antibody is dimeric IgA but not when it is monomeric IgA. When, however, mixed immune complexes were prepared by allowing both monomeric and dimeric IgA antibody molecules to bind to the same multivalent antigen, the entire complex, including the monomeric IgA, was transported (not shown). IgG antibodies could also be transported in complexes containing dimeric IgA (not shown). Other experiments (not shown) have demonstrated that dimeric IgA immune complexes are not transported by wild- type MDCK cells not expressing polymeric Ig receptor. We also have preliminary evidence that IgA, but not IgG antibodies can similarly promote the passage of "antigens" as large as intact, live bacteria, specifically Salmonella typhimurium, across epithelial cells bearing the polymeric Ig receptor (data not shown).

DISCUSSION

Our *in vitro* experimental results strongly suggest that mucosal IgA functions more broadly in host defense than simply as a lumenal barrier to exclude foreign matter from the body proper. The fact that specific IgA antibody introduced at the basolateral cell surface can neutralize virus within cells that were infected through the apical surface indicates that the intracellular pathway for transepithelial migration of dimeric IgA at some point intersects the pathway for synthesis and assembly of virions. We do not yet know where this intersection occurs; indeed, it could vary according to the viral component recognized by a given IgA antibody. From the viewpoint of host defense intracellular neutral- ization of viruses by IgA antibody has several implications. First, in the general sense it implies that antibody, as well as cell mediated immunity, can promote recovery from viral infection, in contrast to the traditional view that antibody can prevent viral infection but cannot cure it and that recovery from infection by intracellular microbial pathogens depends on cell mediated immunity. Second, in a primary infection in which the barrier function of IgA was not available because there was no pre-existing antibody, as synthesis of specific IgA begins in response to the infection, it can promote recovery. Third, since IgA has a relatively low inflammatory potential compared to other Ig isotypes, it can potentially neutralize an intracellular pathogen without destroying the infected cell and damaging the integrity of the mucosal lining, in contrast to cell mediated immunity in which T cells or NK cells act by killing infected cells. Fourth, our preliminary data with IgA antibody and salmonellae suggest that an intraepithelial role for IgA in host defense can apply to a variety of microbes, not just viruses. Transport of immune complexes from the basolateral to apical surface of mucosal epithelial cells also has important implications for host defense. Since the mucosal epithelium is literally in contact with the external environment from which antigens can penetrate to some extent and since the lamina propria contains abundant IgA-producing plasma cells, there is a significant potential for immune complexes that contain IgA antibodies to form in the lamina propria. Transport of such immune complexes directly across the mucosal epithelium by polymeric Ig receptor-mediated transcytosis into the external secretions could serve to spare the systemic circulation from an excessive load of foreign matter and immune complexes. Again, the low inflammatory potential of IgA would be helpful in that formation and excretion of immune complexes composed solely of IgA could occur without inciting local tissue damage. Since monomeric IgA can be transported in an immune complex that also contains dimeric IgA, the full output of local IgA plasma cells is potentially available to play a role in defense, in contrast to the barrier function where only oligomeric IgA has an opportunity to act. Since IgA antibody against a bacterial surface antigen appears capable of promoting the transport of bacteria across an epithelium, IgA antibodies in the lamina propria would seem to be of potential importance in defense against antigens as large as intact bacteria that have penetrated a mucous membrane.

CONCLUSIONS

We envision that mucosal IgA functions in host defense in a three-tiered mode. First is the traditional lumenal barrier function to exclude antigens from the body proper. Second is intraepithelial cell neutralization of intracellular pathogenic microbes, which can occur if IgA, during its transepithelial passage to the external secretions, meets an antigen of an invading microorganism. Third is to complex antigens within the mucosal lamina propria and excrete them into the lumen via the same transport route taken by free dimeric IgA.

ACKNOWLEDGMENTS

This work was supported NIH grants AI-26449, HL-37117, HL- 02002, AI-11949, CA-51998, DK-43999 and an American Lung Association research grant.

REFERENCES

1. B. J. Underdown and J. M. Schiff, Ann. Rev. Immunol., 4:389 (1986).
2. J. Mestecky and J.R. McGhee, *Adv. Immunol.*, 40:153 (1987).
3. P. Michetti, M.J. Mahan, J.M. Slauch, J.J. Mekalanos and M.R. Neutra, Infect. Immun., 60:1786 (1992).
4. C. S. Kaetzel, J. K. Robinson, K. R. Chintalacharuvu, J. -P. Vaerman and M. E. Lamm, *Proc. Natl. Acad. Sci. USA,* 88:8796 (1991).
5. M. B. Mazanec, C. S. Kaetzel, M. E. Lamm, D. Fletcher and J. G. Nedrud, *Proc. Natl. Acad. Sci.* USA 89: (1992).
6. K. E. Mostov and D. L. Deitcher, *Cell* 46:613 (1986).

INTRACELLULAR NEUTRALIZATION OF SENDAI AND INFLUENZA VIRUSES BY IgA MONOCLONAL ANTIBODIES

Mary B. Mazanec,[1,2] Charlotte S. Kaetzel,[2] Michael E. Lamm,[2]
David Fletcher,[2] Janet Peterra,[1] and John G. Nedrud[2]

[1]Department of Medicine; and [2]Department of Pathology Case
Western Reserve University School of Medicine Cleveland,
Ohio USA

INTRODUCTION

Secretory IgA (S-IgA), a polymeric immunoglobulin, plays a central role in protecting the mucosal surfaces of the body against foreign pathogens.[1,2] Resistance to viral infections correlates with the presence of specific IgA antibody in mucosal secretions.[3] Polymeric IgA is transported into mucosal secretions by the receptor for polymeric immunoglobulins, which is expressed on the surface of many mucosal epithelial cells.[4,5] Traditionally, S-IgA is thought to neutralize virus extracellularly by binding to virions and preventing their attachment and entry into epithelial cells.[6] Since IgA is transported through epithelial cells and since viruses are obligate intracellular parasites requiring the host cell's metabolic machinery to replicate, we hypothesize that IgA may be able to interfere with virus assembly by binding to newly synthesized viral proteins within the infected epithelial cell. Thus, to examine the possible role of IgA in intracellular neutralization of viruses, we employed polarized monolayers of Madin-Darby canine kidney (MDCK) epithelial cells stably transfected with the cDNA encoding the receptor for polymeric immunoglobulin.[7] Cells were infected on the apical surface with either Sendai or influenza virus and subsequently treated on the basal surface with viral specific monoclonal antibody (mAB). Using this *in vitro* epithelial model, we showed that anti-viral IgA mAB co-localizes with viral protein within the cell.[8] In concert, viral titers in both apical supernatants and cell lysates were reduced from monolayers treated with viral-specific IgA but not from those receiving viral-specific IgG or an irrelevant IgA.[8] Based on these observations, we conclude that IgA, as it is transported through the epithelial cell, is capable of neutralizing microbial pathogens intracellularly.

MATERIALS AND METHODS

Polarized Epithelial Cell Monolayers

MDCK cells transfected with the cDNA coding for rabbit polymeric immunoglobulin receptor (obtained from Keith Mostov, University of California, San Francisco)[7] were cultured on nitrocellulose filters in microwell chambers. Confluency was determined by the ability of a monolayer to sustain a transmembrane electric potential difference. The ability of a monolayer to selectively transcytose polymeric IgA was documented as described previously.[8]

Intracellular Neutralization of Virus by IgA

At confluence, MDCK cell monolayers were infected with either Sendai or influenza virus (1 plaque forming unit (pfu)/cell) placed in the apical chamber for 60 min at 37°C. The cell layers were washed, and 6-8 h later ascites containing viral specific IgA or IgG or an irrelevant IgA mAB was added to the lower compartment. The monolayers were incubated with antibody for 4 h, after which the antibody was removed and the monolayer washed. Twenty-four h after infection, viral titers were determined by plaque assay in the supernatants and cell lysates.[9]

Intracellular Co-localization of IgA and Viral Protein

MDCK cells were examined for intracellular co-localization of mAB and viral protein. MDCK cell monolayers were cultured and infected as above. Eight h after infection, ascites containing mAB was added to the lower compartment. Twenty-four h after the addition of antibody, cells were detached with trypsin, cytocentrifuged onto glass slides and fixed with acetone. Two-color immunofluorescence was used to detect viral glycoprotein and IgA simultaneously.[8]

RESULTS

Two lines of evidence support the hypothesis of intracellular neutralization of virus by IgA. First, by two-color immunofluorescence, anti-Sendai HN IgA co-localized within epithelial cells with Sendai HN, the viral surface glycoprotein responsible for cell attachment.[8] This finding suggests that anti-HN IgA antibody contacted newly synthesized viral HN protein. In addition, every cell that stained for intracellular anti-HN IgA also demonstrated HN protein in the same location.[8] In contrast, infected cells treated with an irrelevant IgA did not show intracellular accumulation of IgA antibody.[8] Thus, only transport of anti-viral IgA is impeded. Therefore, the accumulation of anti-viral IgA within an infected cell is not due to the viral infection nonspecifically interrupting IgA transport.

Table 1. Treatment of Sendai virus-infected MDCK cells with anti-HN IgA reduces intracellular and extracellular virus.

Antibody	Mean virus titer ($\log_{10}$)	
	Apical supernatant	Cell lysate
Anti-HN IgA	3.2 ± .6*	4.3 ± .6*
Anti-HN IgG	6.4 ± .1	5.6 ± .1
Irrelevant IgA	6.2 ± .1	5.3 ± .1

Anti-HN IgA (mAb 380) or IgG (mAb 20) or irrelevant IgA (MOPC-315) was added for 4 h to the compartment below the infected monolayers. After 24 h, virus titers were determined in both apical supernatants and cell lysates.[8]

*Virus titers (pfu/ml) in apical supernatants and cell lysates from cells exposed to IgA anti-HN are significantly reduced ($p \leq 0.0001$) compared to those from cells exposed to anti-HN IgG or irrelevant IgA. Data are means ± S.D. (n = 4).

Second, treatment of either Sendai or influenza virus infected MDCK cell monolayers with the respective anti-viral IgA mAB reduced mean viral titers up to 1000 times in both the supernatants and the cell lysates as compared to monolayers treated with either anti-viral IgG or an irrelevant IgA monoclonal antibody (Tables 1 and 2). In both experiments, the viral-specific IgA mAB was directed against the surface glycoprotein (Sendai HN or influenza HA) responsible for virus adherence to epithelial cells. The results of these experiments indicate that viral-specific IgA antibody, as it is transported through the epithelial cell, can interact with viral glycoproteins, effectively disrupting viral replication and egress.

Table 2. Anti-HA IgA antibody neutralizes influenza virus during transcytosis.

Antibody	Mean virus titer ($\log_{10}$)	
	Apical supernatant	Cell lysate
Anti-HA IgA	4.3 ± .1*	3.2 ± .5*
Irrelevant IgA	7.1 ± .3	6.1 ± .2
No Antibody	7.3 ± .3	6.1 ± .1

Anti-HA IgA mAb (obtained from Walter Gerhard, The Wister Institute) or an irrelevant IgA (MOPC-315) was added for 4 h to the compartment below the infected monolayers. After 24 h virus titers were determined in both apical supernatants and cell lysates.

* Virus titers (pfu/ml) in apical supernatants and cell lysates from cells exposed to anti-HA IgA are significantly reduced ($p \leq 0.0001$) compared to controls, suggesting intracellular neutralization of virus by IgA.

DISCUSSION

S-IgA plays a central role in defending the mucosal surfaces of the body against invading pathogens.[1,2] The current studies provide evidence for a previously undescribed mechanism by which IgA interacts with foreign antigens at the epithelial surface. Since IgA is selectively transported through the epithelial cell, it is uniquely poised to interact with intracellular pathogens. Thus, intracellular neutralization, along with the more traditional extracellular actions of IgA including the recently described excretory function,[10] maximizes the efficiency of the mucosal immune system in limiting the spread of virus.

In order for IgA to complex with viral proteins within the epithelial cell, the exocytic and transcytotic pathways must intersect. Other studies have demonstrated intersection of these pathways, resulting in mixing of endocytic and exocytic vesicles.[11,12] Although there are several steps during viral replication where IgA might be able to neutralize virus intracellularly, including uncoating of the capsid, transcription and translation of the genome, post-translational modification of protein, and assembly and budding of virions, we suspect that this is most likely to occur during virion assembly. It is at this phase of virus replication, as viral proteins mature, that the exocytic and transcytotic pathways are most likely to cross.

Intracellular interactions between IgA antibody and antigen would have the greatest utility in defending against intracellular pathogens, such as viruses, which invade the body at a mucosal surface. In addition to the traditional role assigned to cell mediated immunity for protection against intracellular pathogens, the ability of IgA to complex with viral proteins within cells and hence abort viral assembly and release represents another mechanism for recovery from virus infections. Consistent with this hypothesis, other investigators have demonstrated that mice deficient in cytotoxic T cells are able to recover from influenza infection.[13,14,15] An explanation for their observations, would be that virus-specific IgA antibody prevents viral replication within the epithelial cell, promoting virus clearance and recovery.

CONCLUSION

During its transport into mucosal secretions, IgA is able to interact with intracellular antigens. This additional role of IgA in mediating viral protection limits virus spread while possibly preserving the integrity of the mucosa. Thus, as opposed to cell-mediated immunity, IgA, which is intrinsically less phlogistic than IgG, may be able to eradicate viral infection without destroying the host cell.

REFERENCES

1. M. E. Lamm, *Adv. Immunol.* 22:223 (1976).
2. J. Mestecky and J. R. McGhee, *Adv. Immunol.* 40:153 (1987).
3. F. Y. Liew, S. M. Russell, G. Appleyard, C. M. Brand, and J. Beale, *Eur. J. Immunol.* 14:350 (1984).
4. R. Solari and J. P. Kraehenbuhl, *Immunol. Today* 6:17 (1985).
5. M. A. Fiedler, C. S. Kaetzel, and P. B. Davis, *Am. J. Physiol.* 261:L255 (1991).
6. M. C. Outlaw and N. J. Dimmock, *J. Gen. Virol.* 71:69 (1990).
7. K. E. Mostov and D. L. Deitcher, *Cell* 46:613 (1986).
8. M. B. Mazanec, C. S. Kaetzel, M. E. Lamm, D. Fletcher, and J. G. Nedrud, *Proc. Natl. Acad. Sci. USA* 89:6901 (1992).
9. K. Sugita, M. Maru, and K. Sato, *Jpn. J. Microbiol.* 18:262 (1974).
10. C. S. Kaetzel, J. K. Robinson, K. R. Chintalacharuvu, J. P. Vaerman, and M. E. Lamm, *Proc. Natl. Acad. Sci. USA* 88:8796 (1991).
11. R. J. Youle and M. Colombatti, *J. Biol. Chem.* 262:4676 (1987).
12. W. Stoorvogel, H. J. Geuze, J. M. Griffith, and G. J. Strous, *J. Cell Biol.* 106:1821 (1988).
13. R. M. Kris, R. A. Yetter, R. Cogliano, R. Ramphal, and P. A. Small, Jr., *Immunology* 63:349 (1988).
14. P. Scherle, G. Palladino, and W. Gerhard, *J. Immunol.* 148:212 (1992).
15. M. Eichelberger, W. Allan, M. Zijlstra, R. Jaenisch, and C. Doherty, *J. Exp. Med.* 174:875 (1991).

THE BINDING OF MONOMERIC IgA TO MYELOID FcαR: EVIDENCE FOR RECEPTOR RE-CYCLING AND DETERMINATION OF ITS AFFINITY

Wilson W. Stewart and Michael A. Kerr

Department of Pathology
University of Dundee
Ninewells Hospital Medical School
Dundee, DD1 9SY, Scotland

INTRODUCTION

Receptors for the Fc region of immunoglobulins (FcRs) have been identified on a variety of cell types and have been shown to promote multiple cellular functions.[1,2] They are differentially expressed on different types of leucocyte and show slightly different subclass specificity.[3] Only FcγRI binds to monomeric IgG with sufficient affinity that binding is likely to occur at physiological concentrations. Efficient binding of IgG to FcγRII and FcγRIII only occurs when the IgG is aggregated. Binding of aggregated IgG causes clustering of Fc receptors which triggers a biological response by the cell.

In humans, the amount of IgA-produced in total is more than the production of all the other immunoglobulin classes combined. It therefore seems reasonable to expect that receptors for the Fc region of IgA will have functional activities which are of equal importance as FcγRs. Fc receptors for IgA (FcαR) are expressed on human neutrophils, monocytes, macrophages, and eosinophils as well as certain populations of lymphocytes.[4] It has been demonstrated that interaction of IgA with myeloid cells expressing FcαR stimulates a range of functions such as phagocytosis[5], degranulation[6], superoxide release and chemiluminescence production[7,8] as well as ADCC.[9] FcαRs havebeen purified from monocytes[10] and neutrophils[11] by affinity chromatography using IgA-Sepharose and have been characterised as heavily glycosylated proteins of Mr 50-70kDa. The FcαR purified from neutrophils binds monomeric IgA1 and IgA2 with similar high affinity.[12] This affinity suggests that *in vivo*, the receptors would be saturated with IgA.

RESULTS AND DISCUSSION

We have previously shown that purified serum or secretory IgA of either subclass, when bound to microtitre plates, was able to elicit a respiratory burst from PMN.[8] The chemiluminescent burst was similar in size and duration to that elicited by IgG. My43, a monoclonal IgM antibody which recognises an FcαR on monocytes and neutrophils, has also been shown to elicit a chemiluminescent burst from purified PMN when cross-linked on their surface.[13]

Advances in Mucosal Immunology, Edited by
J. Mestecky *et al.*, Plenum Press, New York, 1995

Cross-Linking of PMN FcαR with Monoclonal Antibody (My43)

When purified PMN were added to microtitre plates which had been coated with My43, bound via F(ab')$_2$ anti-mouse IgM, a chemiluminescence burst was elicited, the size of which was dependent on the amount of My43 present. No repiratory burst was observed when PMN were added to plates coated with monoclonal antibody MC2, an IgM antibody recognising the PMN surface carbohydrate CD15. The response elicited by My43 was inhibitable by pre-incubating the PMN with human IgA at a concentration greater than 1mg/ml. The Ki for inhibition was low, consistent with our earlier determination of receptor affinity.[12] Pre-incubation of the PMN with IgG did not inhibit the My43 chemiluminescence burst.

Cross-Linking of PMN Monomeric and Dimeric Cytophilic IgA

When freshly purified PMN were incubated on F(ab')$_2$ anti-human IgA coated chemiluminescence plates, no respiratory burst was elicited. However when the same PMN were pre-incubated with IgA at 4°C and then after washing at 4°C were added to the same plates, respiratory bursts were observed which were dependent on the amount of IgA added. Incubation with IgG resulted in no burst on anti-human IgA plates. The size of the bursts elicited in IgA coated cells suggest saturation of the receptor at concentrations greater than 1mg/ml. The results show that unaggregated serum IgA is able to bind to the receptor but crosslinking is necessary to elicit a respiratory burst. When the same IgA coated cells were washed after incubating for 1 hour at room temperature, no bursts were obtained when the cells were added to F(ab')$_2$ anti-human IgA plates, indicating loss of IgA from the PMN surface at this temperature. This explains the inability of F(ab')$_2$ anti-human IgA to stimulate freshly purified PMN since the cells are routinely prepared at room temperature.

IgA1 purified from normal serum was subjected to gel filtration chromatography to separate monomeric IgA1 from dimeric and any other higher aggregated forms of IgA1. Immunoblotting of the resulting fractions, run on SDS-PAGE, with anti-human IgA, suggested that only dimeric and monomeric forms were present. When PMNs were added to chemiluminescence plates coated with these fractions, the absence of polymeric IgA1 was again indicated by the fact that significant bursts were only elicited by dimer and monomer IgA1 fractions. Purified washed PMNs were incubated at 4°C with the fractions from gel filtration. After washing at 4°C the cells were added to F(ab')$_2$ anti-human IgA-coated chemiluminesence plates. Bursts were elicited by cross-linking both surface bound monomer and dimer IgA1. Even though the concentration of dimer was much lower than that of the monomer, the size of the respiratory burst elicited by both was similar.

The chemiluminescence data clearly show that monomeric and dimeric IgA bind to the FcαR without eliciting a response but crosslinking of the receptor triggers a strong respiratory burst.

Cytophilic IgA on Myeloid Cells in Whole Blood

To further study the binding of unaggregated IgA to neutrophils, we developed a method for the measurement of IgA bound to the surface of leucocytes in whole blood or on washed cells. When whole heparinized blood was cooled to 4°C for 1 hour then washed free of plasma at 4°C, surface bound IgA could be measured on PMN and monocytes by flow cytometry following staining at 4°C using FITC-F(ab')$_2$ anti-human IgA. Analysis of blood from different individuals showed that the amount of surface IgA on PMN correlated well with the plasma IgA concentration as measured by radial immunodiffusion. A much poorer correlation was observed between the amount of surface IgA on monocytes and the plasma IgA concentration.

Endocytosis via FcαR

Markedly less surface IgA could be detected on cells of whole blood which had been washed free of plasma at room temperature then stained using using FITC-F(ab')$_2$ anti-human IgA at 4°C. Similarly, PMN purified by conventional techniques at room temperature lacked surface IgA (see above). However, when the cells were permeabilised by treatment with 0.2% Triton/PBS for 10 minutes, internal IgA could be detected by flow cytometry following staining with the same FITC-F(ab')$_2$ anti-human IgA. The presence of internalised IgA was confirmed by confocal microscopy and by immunoblotting of detergent extracts of normal PMN using alkaline phosphatase conjugated anti-α chain antisera. These results suggest that the depletion of surface IgA at room temperature was due to internalision via FcαR mediated endocytosis rather than the release of IgA from the receptor.

This was confirmed by following the fate of purified serum IgA which had been bound to the surface of PMN or monocytes by incubating with washed whole blood cells at 4°C. Loss of surface IgA was faster at 37°C but was negligible at 4°C as measured by flow cytometry. The internalisation of this re-bound IgA could be detected by using confocal microscopy to follow the internalization of surface bound FITC-IgA.

Estimation of FcαR Affinity on Cells in Whole Blood

The binding of IgA by FcαR of PMN and monocytes could be followed by incubating whole blood which had been washed at room temperature to remove cytophilic IgA, with purified normal plasma IgA at 4°C. From the resulting association curves values for half saturation of FcαR on PMN and monocytes could be estimated to be 0.7 mg/ml for PMN (N=6) and 0.44 mg/ml for monocytes (N=6). Since these concentrations are similar to normal plasma concentrations, the data are consistent with the detection of IgA on cells in whole blood which had been chilled to 4°C. Similar results were obtained using directly conjugated FITC-IgA.

We have previously predicted that IgA would be associated with the surface of PMN *in vivo* since the affinity of the purified PMN FcαR has been estimated to be 2 x 10^{7}M and the serum concentration of IgA is normally much greater than this. Thus, it would be expected that this receptor would be permanently occupied with IgA in the circulation. In these investigations we have demonstrated that no cytophilic IgA was present on freshly purified PMN as incubation of these cells on chemiluminescence plates coated with anti- IgA did not elicit a respiratory burst. However, our results also show that this was the result of endocytosis of the IgA from the surface of the cells during preparation at room temperature (see above).

CONCLUSIONS

Taken together, the results suggest that monomeric or dimeric IgA in plasma will bind to PMN and probably to monocyte FcαR. The binding does not elicit a response from the cells but crosslinking of the IgA using anti-IgA antibody fragments will elicit a strong response. It is therefore suggested that crosslinking of IgA by polyvalent antigen would have a similar effect. The significance, *in vivo* of the internalisation of IgA bound to receptors which we have studied in vitro has yet to be determined.

REFERENCES

1.   I. Mellman, *Curr. Opin. Immunol.* 1:16 (1988).
2.   M. W. Fanger, L. Shen, R. F. Graziano, and P. M. Guyre, *Immunology Today* 10:92 (1989).

3. J. G. J. Van de Winkel and C. L. Anderson, *J. Leuk. Biol.* 49:511 (1991).
4. M. A. Kerr, *Biochem. J.* 271:285 (1990).
5. A. Gorter, P. S. Hiemstra, P. C. J. Leijh, M. E. Van Der Sluys, M. T. Van Den Barselaar, L. A. Van Es, and M. R. Daha, *Immunology* 61:303 (1987).
6. G. R. Yeaman and M. A. Kerr, *Clin. Exp. Immunol.* 68:200 (1987).
7. W. W. Stewart and M. A. Kerr, *Immunology* 71:328 (1990).
8. L. Shen and J. Collins, *Immunology* 68:491 (1989).
9. M. W. Fanger, S. N. Goldstine, and L. Shen, *Ann. N. Y. Acad. Sci.* 409 (2):552 (1983).
10. R. C. Monteiro, H. Kubagawa, and M. D. Cooper, *J. Exp. Med.* 171:597 (1990).
11. M. Albrechtsen, G. R. Yeaman, and M. A. Kerr, *Immunology* 64:201 (1988).
12. R. L. Mazengera and M. A. Kerr, *Biochem. J.* 272:159 (1990).
13. L. Shen, *J. Leuk. Biol.* 51:373 (1992).

SIGNAL TRANSDUCTION VIA Fc RECEPTORS; INVOLVEMENT OF TYROSINE KINASE AND REDOX REGULATION BY ADF

Satoshi Iwata,[1] Mitsuhiro Matsuda,[1] Katsuji Sugie,[2]
Yasuhiro Maeda,[1] Takumi Kawabe,[1] Hajime Nakamura,[1]
Hiroshi Masutani,[1] Toshiyuki Hori[1], and Junji Yodoi[1]

[1]Department of Biological Responses, Institute for Virus
Research, [2]Department of Late Effect Studies, Radiation
Biology Center, Kyoto University, Sakyo-ku, Kyoto, Japan

INTRODUCTION

Low affinity Fc receptors for IgE (FcεRII) were initially described on the surface of peripheral blood B lymphocytes[1,2] and turned out to be identical to CD23, one of the B cell activation antigens. Cloning of FcεRII cDNA revealed a unique primary structure compared to other Fc receptors such as FcγRI~III or FcεRI, consisting of a cytoplasmic N-terminal domain of 23 amino acids, a transmembrane domain of 21 amino acids, extracellular C-terminal domain of 321 amino acids and no signal sequence.[1] FcεRII belongs to the c-type animal lectin superfamily, whereas FcεRI and FcγRI~III belong to the immunoglobulin superfamily.

FcεRII is expressed in a variety of hemopoietic cells, which includes macrophages, eosinophils, and platelets. T lymphocytes are also known to express FcεRII in the case of abnormal transformation by HTLV-I or in some allergic disease patients. Recently, we have demonstrated that FcεRII is induced in peripheral blood T lymphocytes stimulated by PHA and this induction is enhanced by IL-2 or IL-4.[3] Expression of FcεRII is markedly up-regulated in Epstein-Barr virus (EBV) mediated B lymphocyte transformation. It has been reported that introduction of EBNA-2 and LMP-(latent membrane protein)-1 DNA causes B cell immortalization and FcεRII induction. The function of EBNA-2 and LMP-1 in B cell transformation or FcεRII induction is still unclear.

In these virus-mediated transformed lymphoid cell lines, some activation antigens on the cell surface are up-regulated. For example, FcεRII, as mentioned above, and the α chain of the IL-2 receptor complex (IL-2R)/p55(Tac) are enhanced. Actually, abnormal levels of IL-2R/p55 are expressed in HTLV-I transformed T cell lines and it has been suggested that dysregulation of IL-2R/p55 may be involved in HTLV-I mediated T cell transformation. Adult T cell leukemia-derived factor (ADF) was initially described as an IL-2R/p55 inducing activity in the culture supernatant of HTLV-I transformed T cell line, ATL-2. A cDNA cloning and deduced amino acid sequence revealed marked homology with E.coli thioredoxin.[4,5] Thioredoxin of *E. coli* is a heat-stable small protein with a molecular weight of approximately 12kDa and is a powerful reducing agent with a dithiol-mediated active center (-Cys-Gly-Pro-Cys-), which is also conserved in ADF. Recombinant ADF exhibits disulfide reducing activity equivalent to *E. coli* thioredoxin. ADF/human thioredoxin is highly expressed by virus-transformed host cells not only in the case of HTLV-I but also EBV. Wakasugi *et al.*[6] reported autocrine growth factor produced by EBV-transformed B

Advances in Mucosal Immunology, Edited by
J. Mestecky *et al.*, Plenum Press, New York, 1995

cell line (3B6-IL-1), which turned to be identical to ADF/thioredoxin. ADF/thioredoxin promotes growth and FcεRII expression of EBV-transformed B cell lines.[6] Thus, involvement by ADF/thioredoxin has been suggested in virus-mediated lymphocyte transformation.

Here we report the FcεRII mediated signal transduction pathway for IL-2R/p55 induction and the association of FcεRII with protein tyrosine kinase/Fyn in YTSER cells. We also show the involvement of dithiol-mediated redox-regulation by ADF in IL-2R/p55 gene activation and the possible role of ADF in FcεRII/CD23 gene regulation of EBV-transformed B lymphocytes.

SIGNAL TRANSDUCTION THROUGH FcεRII/CD23

There is evidence showing that FcεRII might mediate signal transduction. Anti-FcεRII antibody (H107) is known to enhance the expression of FcεRII in a EBV-transformed B cell line[8] and HTLV-I transformed T cell line (ED).[2,3]

**Possible Mechanism of IL-2R/p55(Tac) and FcεR2/CD23
Gene Activation in Virus-Induced Lymphocyte Transformation**

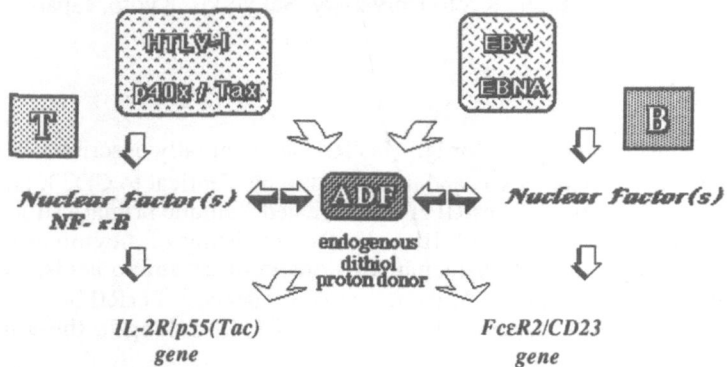

Figure 1. Activation antigens induced in virus-mediated transformation of T and B lymphocytes. Both HTLV-I transformed T cell lines and EBV-transformed B cell lines express high level of ADF.

The molecular cloning of FcεRII cDNA revealed that this molecule possessed a short cytoplasmic domain and had no consensus sequence for a kinase domain such as receptor-type tyrosine kinase. To elucidate this signal transduction mechanism through FcεRII, we introduced the cDNA of FcεRII (pSV2neoSER) into YT cells and obtained a stable transformant of FcεRII named YTSER under a selection medium containing G418. YT cells were originally derived from a 15 year old patient suffering from thymic lymphoma which had LGL (large granular lymphocyte) like morphology and activity. Fresh leukemic cells were also obtained from a patient with B-CLL (B-cell chronic lymphoblastic leukemia), which were known to express high levels of endogenous FcεRII. In these cells, IL-2R/p55 expression was induced when cells were triggered by anti-FcεRII antibody cross-linked by goat-anti mouse IgG polyclonal antibody. This IL-2R/p55 induction was also detected with Northern blotting as well as by surface staining. Using control YT cells (YT neo) or control triggering antibody, induction of IL-2R/p55 was not observed . Therefore this IL-2R/p55 induction was thought to be specific for FcεRII.

Molecules associated with this FcεRII mediated signal transduction pathway were analysed by immunoprecipitaion with H107 antibody followed by immuno-blot using anti-phosphotyrosine polyclonal antibody. The immunoprecipitant of YTSER contained a phosphoprotein of molecular weight about 60kDa. This phosphoprotein associated with FcεRII proved to have the same digestion pattern with V8 protease as

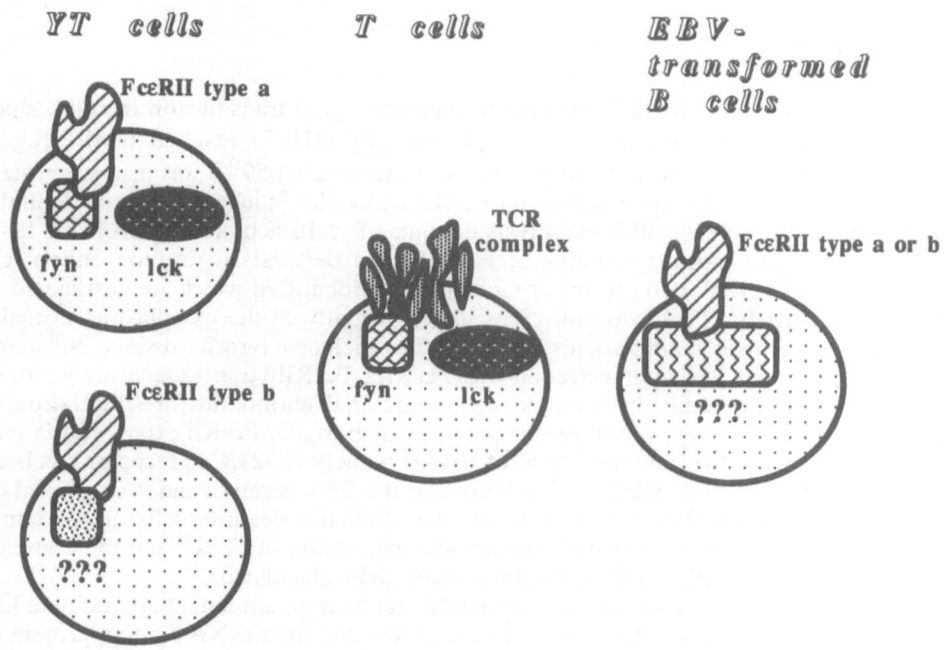

Figure 2. Schematic model of association of protein tyrosine kinases with FcεRII in various cell types.

p59^{fyn}. The p59^{fyn} is a membrane-anchoring protein tyrosine kinase and is reported to be associated with the T cell receptor pathway. It was thus suggested that at least transfected FcεRII might associate with p59^{fyn} and this interaction might play an important role in FcεRII mediated signal transduction in YTSER.

DITHIOL-MEDIATED REDOX REGULATION BY ADF/HUMAN THIORE-DOXIN

Co-expression of FcεRII and IL-2R/p55 has been reported in both B cell and T cell lines. EBV-transformed B cell lines and HTLV-I transformed T cell lines often express both FcεRII and IL-2R/p55. Indeed, ADF/human thioredoxin was purified and cloned from both EBV-transformed B cell lines and HTLV-I transformed T cell lines independently[4-6] and was thought to be involved in both FcεRII and IL-2R/p55 induction.

While the effect of ADF/thioredoxin on gene regulation of FcεRII has not been analyzed in detail, the analysis of the IL-2R/p55 gene regulation mechanism by ADF has been studied. The 5' flanking region of or IL-2R/p55 genome is known to contain various consensus sequences required for interacting with transcriptional activators, which are κB-like elements, Sp-1-like sequence, UE-1 and so on. First of all, we analyzed the interaction of the κB-like element of IL-2R/p55 and ADF/thioredoxin. Nuclear extracts of YT cells were semi-purified with a heparin-agarose column and subjected to mobility shift assays using a DNA oligomer with the κB-sequence as a probe. *In vitro* treatment with diamide, which is an oxidizing agent of thiol-group, completely abolished DNA binding activity of NF-κB. This loss of DNA binding activity of NF-κB was markedly restored by addition of ADF/thioredoxin with thioredoxin reductase and NADPH. Co-transfection analysis of ADF cDNA and IL-2R/p55-CAT (consists of the CAT gene flanked upstream by non-translated regions of IL-2R/p55 gene) demonstrated that ADF might enhance transcriptional activation of the IL-2R/p55 gene by regulating the activity of NF-κB.[8,9]

DISCUSSION

In our study, FcεRII/CD23 seemed to mediate signal transduction in cells, since stimulation of FcεRIII$^+$ cells by anti-FcεRII antibody (H107) resulted in IL-2R/p55 induction. Association of FcεRII and protein tyrosine kinase/p59^{fyn} was also suggested. Other groups have recently reported that anti-FcεRII antibodies induce a rise in intracellular calcium and polyphosphoinositide hydrolysis in human B-cells activated by SAC and IL-4. Their results suggest indirect coupling of FcεRIIb and pertussis toxin-independent Gp protein. In human FcεRII, two species of molecules are identified which are designated as FcεRIIa and FcεRIIb. These two molecules differ partially in the cytoplasmic domain. FcεRIIb lacks C-terminal 6 amino acids of FcεRIIa that include a tyrosine residue. Since our YTSER cells were obtained by transfection with FcεRIIa, FcεRIIb transfectants are yet to be examined. As these studies both employed monoclonal antibodies for stimulation of FcεRII, further elucidation of physiological stimuli which trigger FcεRII expression *in vivo* is required. Very recently, another ligand of FcεRII namely CD21/C3d receptor has been reported by Bonnefoy *et al.* CD21 is also known as the EBV receptor and is expressed on the cell surface of mature B lymphocytes, B cell lines, follicular dendritic cells and certain T cell lines. EBV infection is known to enhance the expression of FcεRII and the complex relationship between FcεRII, CD21 and EBV remains to be elucidated.

There has been increasing evidence that this redox regulation might participate in a variety of signal transduction pathways. AP-1 (jun/fos) and the mRNA binding protein of IRE (iron-responsive element) have been found to be regulated by thiol-mediated redox control mechanisms. The activity of certain protein tyrosine kinases is also regulated by thiol-specific oxido-reduction.

The evidence suggests the possibility that signal transduction via FcεRII might be regulated by an oxido-reduction system including ADF/thioredoxin, through the modulation of DNA-protein interactions or protein tyrosine kinase activity.

REFERENCES

1. K. Ikuta, M. Takami, C. W. Kim, T. Honjo, T. Miyoshi, Y. Tagaya, T. Kawabe, and J. Yodoi, *Proc. Natl. Acad. Sci. USA* 84:819 (1987).
2. T. Kawabe, M. Takami, M. Hosoda, Y. Maeda, S. Sato, M. Mayumi, H. Mikawa, K. Arai, and J. Yodoi, *J. Immunol.* 141:1376 (1988).
3. T. Kawabe, N. Maekawa, Y. Maeda, M. Hosoda, and J. Yodoi, *J. Immunol.* 147:548 (1991).
4. Y. Tagaya, Y. Maeda, A. Mitsui, N. Kondo, H. Matsui, J. Hamura, N. Brown, K. Arai, T. Yokota, H. Wakasugi, and J. Yodoi, *EMBO J.* 8:757 (1989).
5. Y. Tagaya, H. Masutani, H. Nakamura, S. Iwata, A. Mitsui, N. Wakasugi, S. Fujii, A. Uchida, H. Wakasugi, T. Tursz, and J. Yodoi, *Mol. Immunol.* 27:1279 (1990).
6. N. Wakasugi, Y. Tagaya, H. Wakasugi, A. Mitsui, M. Maeda, J. Yodoi, and T. Tursz, *Proc. Natl. Acad. Sci. USA* 87:8282 (1990).
7. A. Yamauchi, H. Masuani, Y. Tagaya, N. Wakasugi, A. Mitsui, H. Nakamura, T. Inamoto, K. Ozawa, and J. Yodoi, *Mol. Immunol.* 29:263 (1992).
8. J. Yodoi and T. Tursz, *Adv. Cancer Res.* 57:381 (1992).
9. J. Yodoi and T. Uchiyama, *Immunol.Today* 13:379 (1992).

A NEW IgA RECEPTOR EXPRESSED ON A HIGHLY ACTIVATED MURINE B CELL LYMPHOMA

Ana M.C. Faria, T. Dharma Rao, and Julia M. Phillips-Quagliata

Pathology Department and Kaplan Cancer Center
New York University School of Medicine
New York, NY 10012

INTRODUCTION

The predominance of IgA plasma cells within the mucosal immune system has been known for many years but the regulation of IgA responses at mucosal sites is still poorly understood. The tendency of mucosal B cells to switch to IgA is thought to be influenced both by local bacterial antigens and mitogens and by locally produced lymphokines that promote IgA responses. The B cell populations on which these stimulants act are, however, already distinct from those elsewhere in the body, having been selected for their ability to migrate to and within the mucosal immune system.

One feature of the mucosal milieu that might help to regulate local IgA responses is IgA itself. IgA and IgA immune complexes are known to bind to lymphocytes bearing receptors for IgA and IgA receptor activity is increased when receptor-bearing cells are exposed to IgA both *in vivo* and *in vitro*.[1-6] Both T and B lymphocyte populations in the gut-associated lymphoid tissue (GALT) are enriched in cells bearing IgA receptors (IgAR) and these may well play a role in regulating the responses of local immunocytes. Indeed, it is already known that receptors specific for the Fc portion of IgA (FcαR) aid in focusing GALT helper T cells onto B cells that bear surface IgA.[7] Very little is, however, known about the possible role of IgAR on B cells in regulating their behavior. Interaction of IgA or IgA immune complexes with receptors for IgA on GALT B cells could conceivably regulate B cell proliferation and/or differentiation or influence the movement of B cells within and between the mucosae.

It would be difficult to study the function of IgAR on normal B cells because only small percentages of them express IgAR, at least as detected by binding assays: IgA-induced modifications in the responses of IgAR-bearing B cells would be hard to discern against the background of responses by B cells lacking the receptors. For this reason, we have characterized an IgAR-bearing B lymphoma, T560, which arose in GALT of a (B10 X B10.H-2^{a}H4^b) F1 hybrid mouse.[8] T560 appears to represent a malignant, highly activated GALT B cell which has retained several B cell functions. It can potentially serve as a model for investigation of the function of IgA receptors on this type of B cell.

CHARACTERISTICS OF T560

T560 cells are IgG2aκ+, Ia+, B220+, J11d.2+, CD3-, CD4-, CD5-, Mac 1-, Mac 2-, non-specific esterase-. They bind bromelain-treated mouse red blood cells (Br-MRBC) in a phosphorylcholine (PC) chloride-inhibitable manner but do not bind sheep or ox red blood cells (ORBC) or 2, 4, 6, trinitrophenyl (TNP) coupled-ORBC. They constitutively secrete IL-1, IL-4 and IL-6 but not IL-2, IL-5 or TGF-β and they present antigen to T cells. The efficiency with which they do this is enhanced approximately 30 fold by coupling PC to the antigen, indicating that the PC-binding surface moieties function in the uptake of antigen for processing.[8]

Although T560 cells bear surface IgG2a, upon activation with either LPS or PC-keyhole limpet hemocyanin (PC-KLH) and irradiated, KLH-reactive, cloned T cells, they secrete predominantly IgA rather than IgG2a (Fig. 1). In LPS-stimulated cultures, the amount of IgA produced is not altered by addition of any of a variety of lymphokines at concentrations known to influence IgA production in other systems[9-13] and secretion of IgG2a is not enhanced by addition of either IFN-α or IFN-γ, reported to enhance IgG2a production by activated, normal B cells[14,15] (Table 1). T560 cells do not secrete IgE when they are stimulated with LPS alone or together with IL-4 at a range of concentrations (5-160 u/ml) known to promote production of IgE by LPS-activated normal B cells.[16]

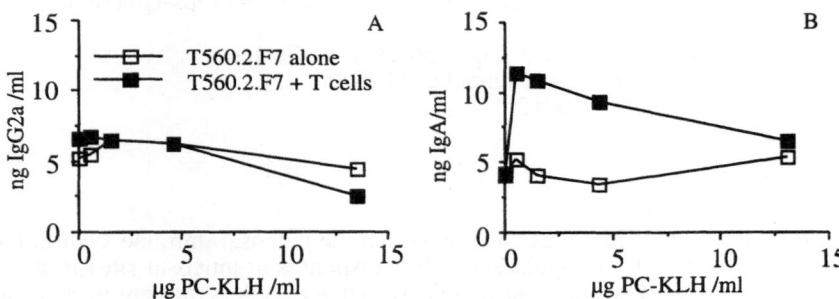

Figure 1. IgG2a (A) and IgA (B) measured by ELISA in the supernatants of T560.2.F7 cells cultured at 2.5 X 10^5/ml alone or together with 2 X 10^5/ml γ-irradiated KLH-reactive T cells of clone K28.E6 for 96 h.

Table 1. Lack of effect of lymphokines on IgA and IgG2a production by T560.2.F7 cells.[a]

Lymphokine	Concentration	IgA	IgG2a	References
IL-2	12.5 - 400 u/ml	unchanged	unchanged	9
IL-4	5 - 160 u/ml	"	"	10
IL-5	5 - 160 u/ml	"	"	10
IL-2 (200 u/ml) + IL-4	2.5 - 80 u/ml	"	"	9, 10
IL-2 (200 u/ml) + IL-5	2.5 - 80 u/ml	"	"	9, 10
IL-4 (80 u/ml) + IL-5	2.5 - 80 u/ml	"	"	9, 10
IL-6	0.16 - 10 ng/ml	"	"	11
TGF-β	0.16 - 10 ng/ml	"	"	12
Nerve growth factor	3.1 - 100 ng/ml	"	"	13
IFN-γ	6.25 - 200 u/ml	"	"	14
IL-2 (200 u/ml) + IFN-γ	6.25 - 200 u/ml	"	"	9, 14
IL-4 (80 u/ml) + IFN-γ	6.25 - 200 u/ml	"	"	10, 14
IL-5 (80 u/ml) +IFN-γ	6.25 - 200 u/ml	"	"	10, 14
IFN-α	3.1 - 100 ng/ml	"	"	15

[a]IgA and IgG2a present in supernatants of T560.2.F7 cells cultured at 2.5 X 10^5/ml with 50 μg/ml LPS from *E. coli* 0111:B4 in the presence or absence of the lymphokines for 24, 48 and 96 h were assayed in isotype-specific ELISA.

Two lines and several clones of T560 cells exist. As determined in rosette assays using indicator erythrocytes coated with monoclonal IgG2b antibody or MOPC-315 IgA myeloma protein, T560.1 cell populations express moderately high amounts of the low affinity IgG2b receptor, FcγRII, and small amounts of a receptor for IgA.[17] By contrast,

T560.2 cell populations express moderately high amounts of a receptor for IgA and low amounts of the FcγRII. To point up the similarities and differences between the T560 IgAR and other IgAR, the properties of the other receptors are summarized in Table 2.

The T560 IgAR is like the murine FcαR described by others in being trypsin-sensitive, neuraminidase-resistant[18] and in being up-regulated by exposing the cells overnight to high concentrations of polymeric IgA.[1-4] It is, however, unlike the normal spleen cell FcαR[3] in binding at pH 4.0 but not at pH 8.0 and unlike other IgAR in being inhibited not only by IgA but also by low concentrations of IgM. This cross-reactivity suggests that the T560 IgAR might bear more similarity to the poly-Ig receptor, also known as secretory component (SC), expressed on secretory epithelial cells[19], but it does not react specifically with a polyclonal antibody to murine SC. Furthermore, the T560 IgAR, unlike most T cell FcαR or the poly-Ig receptor, is also inhibited by high concentrations of murine IgG2a and IgG2b. We note that IgAR up-regulated on lymphoid cells from IgA myeloma-bearing mice are partially inhibitable by IgG2a and IgG2b.[3] The T560 IgAR is not a galactosyl transferase and does not seem to bind to IgA through its carbohydrate (CHO) residues. Unlike the human monocyte IgAR, the T560 IgAR is down-regulated by activation of protein kinase C (PKC) and is sensitive to phosphatidylinositol-specific phospholipase C (PI-PLC), indicating that it might be glycosyl phosphatidylinositol-linked to the cell membrane. The effect of PI-PLC is, however, blocked by the PKC inhibitors, staurosporine[20] and calphostin C (data not shown), suggesting that PI-PLC may actually take its effect by releasing diacylglycerol, which activates PKC, rather than by cleaving the IgA receptor from the membrane.

A rat IgM monoclonal antibody (mAb) that blocks IgA rosette-formation has been prepared. This immunoprecipitates a 36 kDa molecule from both NP-40 and digitonin lysates of T560.2 cells as do both MOPC-315 and TEPC-15 IgA-coated beads (data not shown). The MW of the T560 IgAR is comparable to the MWs reported for the protein cores of the heavily glycosylated human IgAR (32 and 36 kDa)[6,38] and to that reported for the mouse T cell IgAR (38kDa).[23]

DISCUSSION

The origin of T560 in GALT, coupled with its ability to present antigen and to differentiate into an IgA-secreting cell, support the idea that T560 is a malignant representative of a type of highly activated, IgA-committed GALT B cell. In its ability to bind Br-MRBC and PC it resembles the CD5- members of the B1 cell population prominent in the mouse peritoneal cavity[42], but, unlike them, it lacks Mac 1. The possible relationship of T560 to the peritoneal B1 cell population is of interest since the latter has been shown to contain precursors capable of giving rise to at least some of the IgA plasma cells in GALT[43]. Normal B cell precursors of IgA-producing plasma cells in Peyer's patches (PP) usually express either IgM or IgA and it was somewhat unexpected to find that this IgG2a-bearing cell line could give rise to IgA-secreting cells. The surface IgM-bearing CH12[44,45] and I.29[46] lymphomas both switch to IgA secretion but the frequency with which they do so can be influenced by lymphokines. Presumably the T560 progenitor cell came under whatever GALT-derived extrinsic influences were necessary for the hypomethylation of the Cα gene required for switching[47] before the oncogenetic event, but its transformed progeny retained their ability to make the membrane-bound form of IgG2a. T560 secretes lymphokines such as IL-4 and IL-6 and additional lymphokines seem to be unnecessary for its further differentiation. CH12 switches to several isotypes besides IgA but not, apparently, to IgG2a.[45] I.29 switches not only to IgA, but also to IgE and, rarely, to IgG2a.[47] Direct evidence of switching from IgM to IgG2a to IgA by the I.29 cell line has not, however, been documented. We have not yet explored the mechanism of switching in T560 and do not know whether the V-regions and idiotypes of the IgG2a and IgA made by T560 cells are the same. Regardless, however, of the mechanism of switching, it is an intriguing idea that commitment to production of IgA and the activity of a surface receptor for IgA might in some way be coupled in the mucosal B cells represented by T560. One possibility is that binding of IgA immune complexes by the IgAR might lead to up- or down-regulation of IgA production by the B cell. Another possibility is that IgAR up-regulated on normal, activated mucosal B cells might act as homing receptors or as receptors for chemotactic signals, capable of promoting migration of IgA plasma cell precursors into

Table 2. Characteristics of IgAR on various cell types.

Species	Cell Type	Specificity of IgAR	Properties and function of IgAR	Biochemistry of IgAR
Mouse	Blood T[2], thymic and PP T[22], splenic T1, 3, 21 and splenic B[3,4,21] cells.	Binds dimeric better than monomeric IgA[3]; not inhibited by IgG2a or IgG2b[3,4,21], IgG1 or IgM1,4,22 or IgG3[4]. Up-regulated receptors are inhibited 12% by IgG2a and 20% by IgG2b[3].	Up-regulated by IgA *in vivo*[1-3] or *in vitro*[3,4]. Expressed on large T, induced by activation of small T cells[22].	Pronase-sensitive[1,21] but recovers after overnight culture. Protein synthesis required for recovery[3].
	PP-derived TH clones[7], T-T hybridomas[18,23].	Binds monomeric, dimeric and polymeric IgA with increasing affinity[23]. Not inhibited by IgM, IgG1, IgG2a, IgG2b or IgG3[7]. 20% inhibited by IgG2a and IgG2b18. Up-regulated IgAR increase in number and/or affinity for IgA[5]; i.e., there is more than one kind of receptor.	Up-regulated by IgA, down-regulated by IFN β1[8]. Helps focus T help on to IgA-bearing B cells[7].	Trypsin-sensitive[18]; neuraminidase treatment increases IgA-binding[18]; sensitive to acid pH[3]. MW 38 kDa[23].
Rabbit	Normal T and B cells from spleen, PP, appendix, mesenteric and popliteal lymph nodes[24,25].	Blocked by rabbit monomeric and dimeric IgA, by human IgA2, by the Cα3-deficient, IgA1 Wal protein and by human SC; not inhibited by rabbit IgG or IgM. PP IgAR bind better to rabbit IgAg than to IgAf, splenic IgAR, the reverse, implying that there are two types of IgAR[25].	Up-regulated by anti-rabbit Ig antibody. Detected immediately after isolation only in spleen cells. Appears after culture on cells from other tissues[24].	Pronase-sensitive, but recovers after 18 hrs culture[24]. Nothing known about structure.
Human	Normal PMNs[26-35].	Binds secretory IgA (sIgA) and polymeric but not monomeric IgA1 and IgA2[6,27]. Not inhibited by IgG, IgG4, IgM, or IgD[6]. IgA1-binding 43% inhibited by IgG1, 16% by IgG2[7].	Binding IgA : inhibits serum-induced chemotaxis and chemokinesis[33-35] and IgG-mediated phagocytosis[31]; induces respiratory burst and phagocytosis[29,30] and degranulation in PMNs[32] and eosinophils[26].	Trypsin-resistant, pronase-sensitive[28]; neuraminidase treatment increases IgA binding; insensitive to PI-PLC; heavily glycosylated; MW 60kDa [6,29]; gene encodes a 30kDa peptide[36].
	Monocytes and myelo-monocytic cell lines[6,37].	Binds sIgA and polymeric, but not monomeric, IgA1 and IgA2; not inhibited by IgG or IgM[6].	Inhibits casein-induced chemotaxis[35]. Up-regulated by PMA or polymeric IgA[6].	Pronase sensitive ; MW 60kDa; removal of multiple N-linked CHO chains reveals 2 peptide cores of 32 and 36 kDa[6].
	Normal T and B and CLL cells[38-41].	Binds IgA2 better than IgA1[39-41] but does bind Wal protein; binds monomeric and dimeric IgA with similar avidity; not inhibited by IgG or IgM.	Number of rosette-forming cells increases after overnight culture[38,41].	Nothing known.

lamina propria where IgA immune complexes are abundant. The observed down-regulation of the T560 IgAR in response to activation of PKC would be consistent with such a function. It would allow the cell to stop moving if PKC were activated following interaction between the cell surface Ig molecules and specific antigen.

The T560 IgAR bears some similarity to the IgAR described by others, but these are clearly heterogeneous. Data summarized in Table 2 indicate that in mice, humans and rabbits, more than one type of IgAR exists. The cross-reactivity of the T560 IgAR with IgG2a and IgG2b suggests that it might represent the receptor up-regulated in a sub-set of cells from mice bearing IgA plasmacytomas[3]. Its cross-reactivity with IgM suggests it could be related to the poly-Ig receptor but its MW, 36 kDa, is much lower. Perhaps it has homologous domains. Murine M cells bear a receptor capable of transporting IgA, IgM and IgG immune complexes across the GALT epithelium[48]; the T560 IgAR might, alternatively, be related to this receptor. Ongoing studies with our monoclonal antibodies should clarify its function and relationships.

REFERENCES

1. R. G. Hoover and R. G. Lynch, *J. Immunol.* 125:2580 (1980).
2. R. G. Hoover, B. K. Dickgraeffe, and R. G. Lynch, *J. Immunol.* 127:1560 (1981).
3. J. Yodoi, M. Adachi, and T. Masuda, *J. Immunol.* 128:888 (1982).
4. K. Tsujimura, Y. -H. Park, M. Miyama-Inaba, K. Meguro, T. Ohno, M. Ueda, and T. Masuda, *J. Immunol.* 144:4571 (1990).
5. M. Sandor, T. J. Waldschmidt, K. R. Williams, and R. G. Lynch, *J. Immunol.* 144:4562 (1990).
6. R. C. Monteiro, H. Kubagawa, and M. D. Cooper, *J. Exp. Med.* 171:597 (1990).
7. H. Kiyono, J. O. Phillips, D. E. Colwell, S. M. Michalek, W. J. Koopman, and J. R. McGhee, *J. Immunol.* 133:1087 (1984).
8. T. D. Rao, A. A. Maghazachi, A. M. C. Faria, R. S. Basch, and J. M. Phillips-Quagliata, *Internat. Immunol.* 4:107 (1992).
9. P. D. Murray, S. Swain, and M. F. Kagnoff, *J. Immunol.* 135:4015 (1985).
10. S. L. Tonkonogy, D. T. McKenzie, and S. Swain, *J. Immunol.* 142:4351 (1989).
11. D. Y. Kunimoto, R. P. Nordan, and W. Strober, *J. Immunol.* 143:2230 (1989).
12. R. L. Coffman, D. A. Lebman, and B. Schrader, *J. Exp. Med.* 170:1039 (1989).
13. H. Kimata, A. Yoshida, C. Ishioka, T. Kusunoki, S. Hosoi, and H. Mikawa, *Eur. J. Immunol.* 21:137 (1991).
14. S. H. Stein and R. Phipps, *J. Immunol.* 147:2500 (1991).
15. F. D. Finkelman, A. Svetic, I. Gresser, S. Clifford, J. Holmes, P. P. Trotter, I. M. Katma, and W. C. Gause, *J. Exp. Med.* 174:1179 (1991).
16. C. M. Snapper, F. D. Finkelman, D. Stefany, D. H. Conrad, and W. E. Paul, *J. Immunol.* 141:489 (1988).
17. T. D. Rao, A. A. Maghazachi, A. V. González, and J. M. Phillips-Quagliata, *J. Immunol.* 149:143 (1992).
18. J. Yodoi, M. Adachi, K. Teshigawara, T. Masuda, and W. H. Fridman, *Immunology* 48:551 (1983).
19. K. E. Mostov, and G. Blobel, *J. Biol. Chem.* 257:11816 (1982).
20. J. M. Phillips-Quagliata, T. D. Rao, A. A. Maghazachi, A. V. González, and A. M. C. Faria, *Immunol. Res.* 10:432 (1991).
21. W. Strober, N. E. Hague, L. G. Lum, and P. A. Henkart, *J. Immunol.* 121:2440 (1978).
22. D. Char, W. K. Aicher, J. R. McGhee, J. H. Eldridge, K. W. Beagley, and H. Kiyono, *Regional Immunol.* 3:228 (1990/91).
23. W. K. Aicher, M. L. McGhee, J. R. McGhee, J. H. Eldridge, K. W. Beagley, T. F. Meyer, and H. Kiyono, *J. Immunol.* 145:1745 (1990).
24. H. A. Stafford and M. W. Fanger, *J. Immunol.* 125:2461 (1980).
25. H. A. Stafford, K. L. Knight, and M. W. Fanger, *J. Immunol.* 128:2201 (1982).
26. R. I. Abu-Ghazaleh, T. Fujisawa, J. Mestecky, R. A. Kyle, and J. Gleich, *J. Immunol.* 142:2393 (1989).
27. D. A. Lawrence, W. O. Weigle, and H. L. Spiegelberg, *J. Clin. Invest.* 55:368 (1975).

28. M. Fanger, J. Pugh, and G. M. Bernier, *Cell. Immunol.* 60:324 (1981).
29. M. Albrechtsen, G. R. Yeaman, and M. A. Kerr, *Immunology* 64:201 (1988).
30. A. Gorter, P. S. Hiemstra, P. C. J. Leijh, M. E. van der Sluys, M. T. van den Barselaar, L. A. van Es, and M. R. Daha, *Immunology* 61:303 (1987).
31. J. M. A. Wilton, *Clin. Exp. Immunol.* 34:423 (1978).
32. P. M. Henson, H. B. Johnson, and H. L. Spiegelberg, *J. Immunol.* 109:1182 (1972).
33. D. E. Van Epps and R. C. Williams, *J. Exp. Med.* 144:1227 (1976).
34. S. Ito, H. Mikawa, K. Shinomiya, and T. Yoshida, *Clin. Exp. Immunol.* 37:436 (1979).
35. A. S. Kemp, A. W. Cripps, and S. Brown, *Clin. Exp. Immunol.* 40:388 (1980).
36. C. R. Maliszewski, C. J. March, M. A. Schoenborn, S. Gimpel, and L. Shen, *J. Exp. Med.* 172:1665 (1990).
37. A. Chevailler, R. C. Monteiro, H. Kubagawa, and M. D. Cooper, *J. Immunol.* 142:2244 (1989).
38. M. W. Fanger and P. M. Lydyard, *Mol. Immunol.* 18:189 (1981).
39. L. G. Lum, A. V. Muchmore, D. Keren, J. Decker, I. Koske, W. Strober and R. M. Blaese, *J. Immunol.* 122:65 (1979).
40. L. G. Lum, A. V. Muchmore, N. O'Connor, W. Strober, and R. M. Blaese, *J. Immunol.* 123:714 (1979).
41. S. Gupta, C. D. Platsoucas, and R. A. Good, *Proc. Natl. Acad. Sci. USA* 76:4025 (1979).
42. L. A. Herzenberg, A. M. Stall, P. A. Lalor, C. Sidman, W. A. Moore, D. R. Parks, and L. A. Herzenberg, *Immunol. Rev.* 93:81 (1986).
43. F. G. M. Kroese, E. C. Butcher, A. M. Stall, P. A. Lalor, S. Adams, and L. A. Herzenberg, *Internat. Immunol.* 1:76 (1989).
44. L. W. Arnold, T. A. Grdina, A. C. Whitmore, and G. Haughton, *J. Immunol.* 140:4355 (1988).
45. D. Y. Kunimoto, G. R. Harriman, and W. Strober, *J. Immunol.* 141:713 (1988).
46. J. Stavnezer, S. Sirlin, and J. Abbott, *J. Exp. Med.* 161:577 (1985).
47. J. Stavnezer-Nordgren and S. Sirlin, *EMBO J.* 5:95 (1986).
48. R. Weltzin, P. Lucia-Jandris, P. Michetti, B. N. Fields, J. P. Kraehenbuhl and M. R. Neutra, *J. Cell. Biol.* 108:1673 (1989).

ANTI-INFLAMMATORY CAPACITIES OF HUMAN MILK: LACTOFERRIN AND SECRETORY IgA INHIBIT ENDOTOXIN-INDUCED CYTOKINE RELEASE

Lars Å. Hanson,[1] Inger Mattsby-Baltzer,[2] Inga Engberg,[1]
Anca Roseanu,[3] Johan Elverfors,[1] Cecilia Motas[3]

[1]Department of Clinical Immunology and
[2]Department of Clinical Bacteriology,
University of Göteborg, Göteborg, Sweden and
[3]Institute of Biochemistry,
Bucharest, Rumania

INTRODUCTION

To study various immunological functions, passive transfer experiments are often crucial, both for humoral and cellular activities. They can readily be performed in animals, but less commonly in man. There is, however, one situation which provides good possibilities to study passive transfer of mucosal immunity in man: the role of maternal milk for the protection of the breast fed infant. Very strong effects have been found, so that the risk of dying of diarrhoea for a non-breast fed baby in a poor country is 25 times that of an exclusively breast fed.[1]

We have found, surprisingly, that even partial breast feeding can prevent 70-80 % of diarrhoeas during the first several weeks of life in infants in under-privileged populations of Pakistan.[2] Furthermore, breast feeding decreases the risk of attracting neonatal septicemia 18 times compared with non-breast fed in the same populations.[3] This disease, with a mortality of about 40%, is second only to diarrhoea as a cause of death in early infancy in that part of the world.[4] Undoubtedly secretory IgA antibodies are important for this protection as they have been proven to be for the prevention of cholera, *Campylobacter* and ETEC infections in breast-fed infants.[5,6,7] However, it is likely that many other components and functions of human milk may play a role for the well-being of the infant.

A few years ago, we presented the hypothesis that maternal milk provides protection via non-inflammatory mechanisms, in contrast to tissue immunity where, for example, complement and cytokine enhancement of inflammatory cells play an important role.[8] There is little IgG and IgM and complement factors in milk, which instead is rich in the non-inflammatogenic secretory IgA antibodies, as well as other factors, which block complement and responsiveness of neutrophils.

The newborn is colonized with microorganisms, including endotoxin-carrying gram-negatives, shortly after birth. Infants in a developing country started to be colonized already the first day of life with gram-negatives, whereas Swedish infants were colonized with such bacteria only some days later.[9] We were wondering how the neonate can manage this early exposure to endotoxin-producing, -carrying and -releasing bacteria. Would that not cause production of cytokines from the intestinal mucosa, affecting the well-being of the off-spring? We even proposed that the so called "physiological weight loss of the newborn" is not really physiological, but due to cytokine release in the gut induced by the intestinal flora, e.g., the gram-negative bacteria.[10] In agreement with our hypothesis is the fact that this weight loss is mainly seen in non-breast fed infants. It is significantly less, or non-existent in exclusively breast fed infants.[11,12] Against this background, we wanted to search for components in human milk with the capacity of preventing the release of cytokines from cells exposed to endotoxin.

METHODS AND MATERIAL

Lactoferrin and secretory IgA were isolated from human milk as previously described.[13,14] Human lactoferrin was also purchased from Sigma, St Louis, USA. Lipopolysaccharide from *Escherichia coli* 06K13H1 was prepared by the hot phenol-water extraction method.[15] The cell lines HT-29 of human colon epithelium, U-937 of human macrophage-monocyte, a murine macrophage-monocyte cell line, as well as fresh murine spleen cells were used.

The cells were exposed to endotoxin in appropriate amounts with and without one or the other of the milk proteins in different combinations in time and concentration.

The amount of interleukin-6 (IL-6) in the cell cultures was determined with a bioassay using the B9 cell line, dependent on IL-6 for its growth.[16]

RESULT

The experiments showed that lactoferrin was capable of inhibiting the release of IL-6 from the cell lines HT-29 and U-937. The effect was best obtained by the addition of lactoferrin after the endotoxin to the cell culture (Fig. 1). Inhibition of 40-100% was obtained depending on the conditions of the experiment. Different lactoferrin preparations seemed to vary in their inhibitory capacity.

Purified secretory IgA fractions as well were able to reduce the IL-6 release from the endotoxin exposed human cells up to 100%. This was noted also using spleen cells from C3H/HeN mice (Fig. 2).

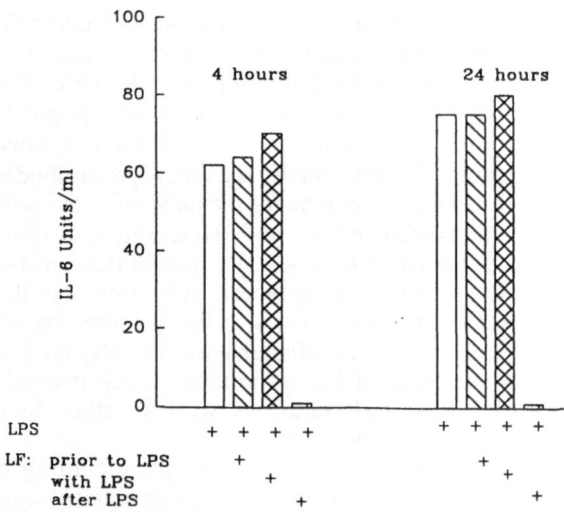

Figure 1. HT-29 colon cell line stimulated with E. coli O6 LPS (10 µg/ml). Human lactoferrin (LF) added (100 µg/ml); 30 min prior, at the same time, or 30 min after the addition of LPS. The IL-6 concentrations of the supernatants were determined by the B9-cell bioassay; results after 4 and 24 hr of incubation are given. Lactoferrin alone did not induce IL-6 production.

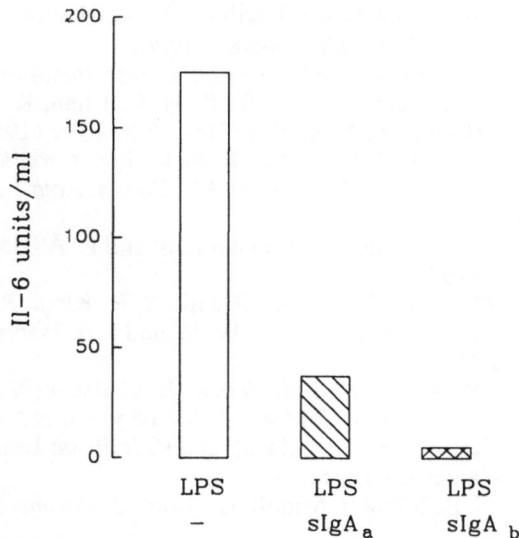

Figure 2. Inhibition of the LPS induced IL-6 response in spleen cells from C3H/HeN mice incubated *in vitro* for two hrs. LPS (10 μg/ml) was added to the cells alone or in combination with secretory IgA (sIgA) (100 μg/ml). Two different sIgA fractions were analysed ($sIgA_a$, $sIgA_b$). No IL-6 was observed with medium, or the sIgA fractions alone.

DISCUSSION

Human milk transfers a remarkable protective capacity to the infant. A mother can be infected and have symptoms of *Shigella* and her breast-fed infant may be infected, but have no symptoms.[17] Our finding of protection against some 70-80% of diarrhoeas in young Pakistani infants even by partial breast feeding is also notable.[2] Breast feeding is not known to protect well against rotavirus-induced diarrhoea, which is a common cause, and still this high efficiency is reached. It is likely that the maternal milk contains protective components in addition to the secretory IgA, but this has never been formally proven. Lactoferrin has been mentioned as one interesting candidate, because of its iron-binding capacity giving a bacteriostatic effect, synergistic with that of secretory IgA *in vitro*.[18] Lately, more attention has been given to its bactericidal capacity.

In this study, we found that lactoferrin could block the release of IL-6 from human and murine cells being exposed to endotoxin. A similar effect was seen with secretory IgA isolated from human milk.

The intestinal flora containing many gram-negatives might, especially in the neonate with immature host defense, expose the intestinal mucosa and initiate cytokine production. That might be followed by symtoms which could be considered to be signs of infection, such as poor appetite, fever and malaise. The affected appetite might be one explanation for the "physiological weight loss", which is mainly seen in formula-fed infants[11,12] which miss the anti-inflammatory capacities of human milk, including inhibition of cytokine release. It will be of interest to try to define the possible significance of such an activity in *in vivo* experiments.

REFERENCES

1.   R. A. Feachem, and M. A. Koblinsky, *Bull. WHO* 62:271 (1984).
2.   F. Jalil, A. Mahmud, R. N. Ashraf, S. Zaman, J. Karlberg, L. Å. Hanson, and B. S. Lindblad, *Acta Paediatr.* (in manuscript).

3.   R. N. Ashraf, F. Jalil, S. Zaman, J. Karlberg, S. R. Khan, B. S. Lindblad, and L.Å. Hanson, *Arch. Dis. Child.* 66:488 (1990).

4.   S. R. Khan, F. Jalil, S. Zaman, and J. Karlberg, *Acta Paediatr. Suppl.* (in press).

5.   R. I. Glass, A. M. Svennerholm, B. J. Stoll, M. R. Khan, K. M. B. Hassain, M. I. Huq, and J. Holmgren, *N. Engl. J. Med.* 308: 1389 (1983).

6.   G. M. Ruiz-Palacios, J. J. Calva, and L. K. Pickering, *J. Pediatr.* 116:707 (1990).

7.   J. R. Cruz, L. Gil, F. Cano, P. Caceres, and G. Pareja, *Acta Paediatr Scand.* 77:658 (1986).

8.   A. S. Goldman, L. W. Thorpe, R. M. Goldblum, and L. Å. Hanson, *Acta Paediatr Scand.* 75:689 (1986).

9.   I. Adlerberth, B. Carlsson, P. de Man, F. Jalil, S. R. Khan, P. Larsson, L. Mellander, C. Svanborg-Edén, A. E. Wold, and L.Å. Hanson, *Acta Paediatr Scand.* 80:602 (1991).

10. L. Å. Hanson, R. N. Ashraf, M. Hahn-Zoric, B. Carlsson, V. HÇrias, U. Wiedermann, U. Dahlgren, C. Motas, I. Mattsby-Baltzer, T. Gonzales-Cossrio, J.R. Cruz, J. Karlberg, B.S. Lindblad, and F. Jalil. *in*: Immunophysiology of the Gut, A. Walker, ed. (In press).

11. D. K. Rassin, N. C. R. Räihä, I. Minoli, G. Moro, *J. Parent. Ent. Nutr.* 14:4 (1990).

12. T.A. Picone, J.D. Benson, G. Moro, I. Minoli, F. Fulconis, D. K. Rassin, and N. C. R. Räihä, *J. Pediatr. Gastroenterol. Nutr.* 9:351 (1989).

13. M. Boesman, and R. A. Finkelstein, *FEBS Letter* 144:1 (1982).

14. T. W. Hutchens, S. S. Magnusson, T. T. Yip, *Pediatr. Research* 26:623 (1989).

15. O. Westphal, and K. Jann, *Methods Carbohydr. Chem.* 5:83 (1965).

16. M. Helle, L. Boeije, and L. A. Aarden, *Eur. J. Immunol.* 18:1535 (1988).

17. L. J. Mata, "The Children of Santa Maria Cauqué, *A Prospective Field Study of Health and Growth*". The MIT Press, Cambridge (1978).

18. J. J. Bullen, H. J. Ryers, and L. Leigh, *Brit. Med. J.* 1:69 (1972).

INHIBITION OF ENTEROPATHOGENIC *ESCHERICHIA COLI* (EPEC) ADHERENCE TO HeLa CELLS BY HUMAN COLOSTRUM. DETECTION OF SPECIFIC sIgA RELATED TO EPEC OUTER-MEMBRANE PROTEINS

Lilia M. Câmara,[1] Solange B. Carbonare,[1] Isabel C. A. Scaletsky,[2] M. Lourdes Monteiro da Silva,[2] and Magda M. S. Carneiro-Sampaio[1]

[1]Department of Immunology, University of São Paulo; and
[2]Department of Microbiology, Escola Paulista de Medicina, São Paulo, Brazil

INTRODUCTION

Enteropathogenic *Escherichia coli* (EPEC) is the main cause of acute diarrhea in infants up to one year old in many developing countries.[1] EPEC includes several specific serotypes, 0111:H-, 0111:H2 and 0119:H6 being the most frequent agents of diarrhea found in Brazil.[2] *In vivo* EPEC strains adhere intimately to cuplike projections of the apical enterocyte membrane causing localized destruction of brush border microvilli, described as an attaching and effacing lesion.[3] *In vitro* EPEC strains attach to HeLa[4] or HEp-2 cells[5] in a specific pattern named localized adherence (LA).[4]

Breastfeeding has been found to be an important protective factor against intestinal and respiratory infections in infants.[6] Silva and Giampaglia demonstrated that adhesion of EPEC of different serotypes to HeLa cells is strongly inhibited by human colostrum and milk.[7] The purpose of the present work was to study the immunological mechanisms that interfere in adherence inhibition of EPEC to HeLa cells by colostrum.

MATERIALS AND METHODS

Colostrum samples were obtained within the two first days after delivery after informed consent from 25 healthy mothers who had given birth to normal term babies. Samples were pooled, delipidated and kept frozen at -20°C until use.

Adherence assays were performed using HeLa cells grown into Leighton tubes as described by Scaletsky *et al.*[4] An inoculum of 260ll of an exponential-phase culture of *E.coli* 0111:H- LA+ 00041-1-85 was used in each assay. Unattached bacteria were removed by washing six times with phosphate-buffered saline (PBS) after 30 min of incubation at 37°C. Sebsequently, a new culture medium was added for an additional incubation period of 3h. The cells were then washed, fixed with methanol, stained with May-Grünwald and Giemsa as previously described[4] and mounted on glass slides for observation under light microscope.

To study the effect of colostrum upon EPEC adherence to HeLa cells the assays were performed in the presence of colostrum or its fractions diluted 1:2 in cell culture medium (Dulbecco's modified Eagle medium supplemented with 2% fetal bovine serum) in the first 30 min incubation period of assay. The effect of colostrum on bacterial adhesion was determined by calculating the percentage of cells with six or more attached bacteria in relation to the control, carried in same conditions without colostrum. All the experiments were performed at least in quintuplicate and 300 cells or more were observed in each preparation.

RESULTS AND DISCUSSION

Delipidated pooled colostrum strongly inhibited bacterial adherence to HeLa cells, (mean adherence=13%), compared to control experiments, (mean adherence=100%). In order to study the effect of heat-treatment, we heated pooled colostrum at 56ºC for 30 min. The percentage of inhibition observed was similar to that of whole untreated colostrum, as showed in Fig. 1. These results suggest that heat-labile factors, such as the complement system, probably do not interfere in adherence inhibition in this experimental system.

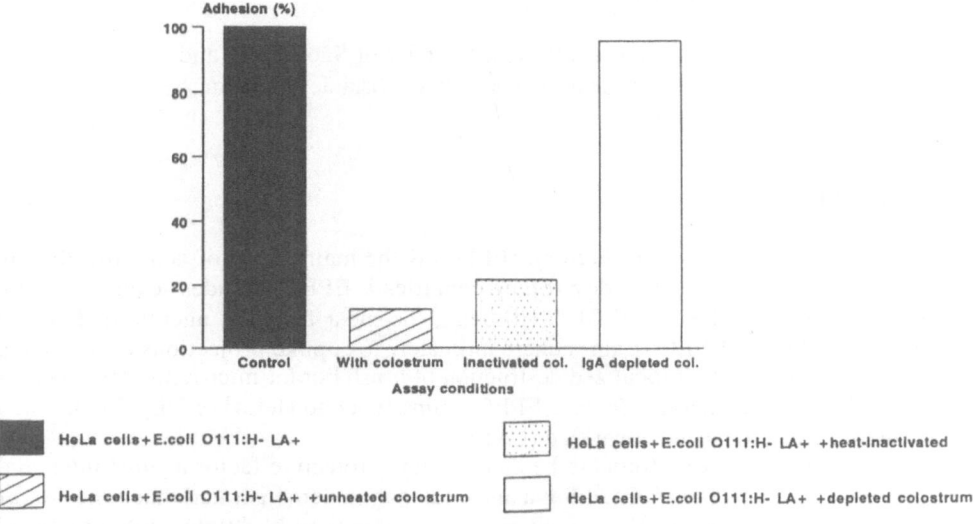

Figura 1. Inhibitory effect of colostrum on EPEC adherence to HeLa cell.

The second approach to study the role of colostrum components in the inhibition phenomenon was to separate fractions by filtering colostrum out through a dialysis membrane which retains molecules of molecular weight >14,000Da. The fractions obtained were called high (>14,000Da) and low (<14,000Da) molecular weight fractions, respectively HMWF and LMWF. The technique used was based on that described by Andersson *et al.*[8]

Fig. 2 shows that adherence tests performed in the presence of HMWF had a significant inhibition of bacterial adhesion (80.9%), equivalent to whole untreated colostrum (87.5%), while the LMWF showed no inhibitory effect. These results suggest that colostrum oligosaccharides do not interfere with inhibition of bacterial adherence in the present experimental system.

The role of specific antibodies against EPEC adhesins was studied by the absorption of whole colostrum as well as its fractions with the LA+ EPEC strains as described by Pál *et al.*[9] The control experiment was to absorb pooled colostrum and fractions with a non-adherent EPEC strain (LA-) of same serotype, *E. coli* 606-54 CDC.

The results of the assays performed in the presence of previously absorbed colostrum and fractions with LA+ and LA- strains are presented in Figure 2. Whole colostrum showed no more significant inhibition and HMWF had a significant loss of inhibitiory activity after absorption with LA+ strain. LMWF showed similar results in adherence assays before and after the absorption procedure. Adhesion assays with materials previously treated with LA- strain showed high inhibition levels in the presence of colostrum and HMWF and virtually no alteration in the presence of absorbed LMWF. These data suggest that specific antibodies against EPEC adhesins have a significant role on the inhibition of bacterial adherence to Hela cells.

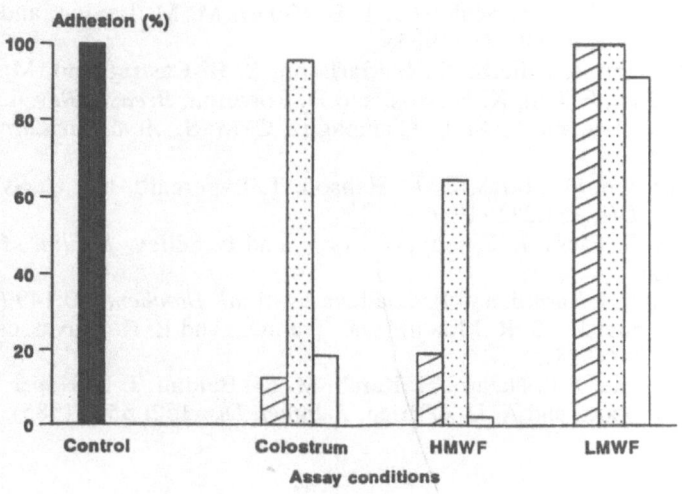

Figura 2. Effect of previous absorption of colostrum and its fractions with LA+ and LA- EPEC strains

In order to verify if IgA was the most important class of antibodies involved in this phenomenon, colostrum was depleted of IgA through passage over a 4B Sepharose affinity column linked to sheep anti-human alpha chain, as described by March et al.[10] IgA concentration of absorbed colostrum was 0.2g/l, about 10 fold less than the untreated material, 1.7g/l. IgA-depleted colostrum did not show inhibitory effect on bacterial adherence to HeLa cells (Fig. 1), suggesting that IgA antibodies are important in human milk inhibitory activity on EPEC adherence to this human epithelial cell lineage.

Finally, an EPEC outer-membrane complex (OMC) extraction was performed as described by Scaletsky et al.[11] to identify which surface antigens would be recognized in a western blotting assay by the antibodies detected in the previous experiments. OMC was separated by SDS-PAGE electrophoresis using a 10% gel and transfered onto nitrocelulose sheets to react with the different samples of treated colostrum and fractions. Antibody-antigen complexes were detected using peroxidase-labelled goat anti-human alpha chain conjugate (Sigma).

Using the above technique, a 94 KDa outer-membrane protein was recognized by both whole untreated colostrum and HMWF, both of which show strong inhibitory effect on bacterial adherence. LA+ absorbed colostrum and HMWF were not able to recognize the same bacterial surface protein as well as LMWF. A 94KDa OMP has been previously related to EPEC adherence by Levine et al.[12]

According to the results obtained in this work we may conclude that in the present experimental system the main factor of adherence inhibition appears to be represented by specific IgA antibodies that recognize an EPEC surface protein of 94 KDa.

REFERENCES

1. M. M. Levine, *J. Infect. Dis.* 155, 377 (1987).
2. M. R. F. Toledo, M.C.B. Alvariza, J. Murahovischi, S.R.T.S. Ramos, and L.R. Trabulsi, *Infect. Immun.* 39, 586 (1983).
3. H. W. Moon, S. C. Whipp, R. A. Argenzio, M. M. Levine, and R. A. Gianella, *Infect. Immun.* 41:1340 (1983).
4. I. C. A. Scaletsky, M. L. M. Silva, and L. R. Trabulsi, *Infect. Immun.* 45, 534 (1984).
5. J. P. Nataro, I. C. A. Scaletsky, J. B. Kaper, M. M. Levine, and L. R. Trabulsi, *Infect. Immun.* 48:378 (1985).
6. L. A. Hanson, I. Adlerberth, B. Carlsson, S. B. Castrignano, M. Hahn-Zoric, U. Dahlgren, F. Jalil, K. Nilsson, and D. Roberton, *Breastf. Rev.* 13:19 (1988).
7. M. L. M. Silva, and C. M. S. Giampaglia, C. M. S.. *Acta Paediatr. Scand.* 81: 266 (1992).
8. B. Andersson, O. Porras, L. A. Hanson, T. Lagergard, and C. Svanborg-Eden, *J. Infect. Dis.* 153:232 (1986).
9. T. Pál, A. S. Pácsa, L. Emody, S. Voros, and E. Sélley, *J. Clin. Microbiol.* 21:415 (1985).
10. S. C. March, I. Tarikii, and P. Cuatrecasas, *Anal. Biochem.* 60:149 (1977).
11. I. C. A. Scaletsky, S. R. Milani, L. R. Trabulsi, and L. R. Travassos, *Infect. Immun.* 56:2979 (1988).
12. M. M. Levine, J. P. Nataro, H. Karch, M. M. Baldini, J. B. Kaper, R. E. Black, M. L. Clements, and A. D. O'Brien, *J. Infect. Dis.* 152: 550 (1985).

SALIVARY SPECIFIC ANTIBODIES IN RELATION TO ADHESION OF *STREPTOCOCCUS PYOGENES* TO PHARYNGEAL CELLS OF PATIENTS WITH RHEUMATIC FEVER AND RHEUMATIC HEART DISEASE

K. S. Nanda Kumar,[1] N. K. Ganguly,[2] Y. Chandrashekher,[3]
I. S. Anand[3] and P. L. Wahi[3]

[1]National Facility For Animal Tissue
and Cell Culture, JOPASANA,85/1, Puad Road,Kothrud,
Pune -29; [2]Department of Experimental Medicine
[3]Cardiology, Postgraduate Institute of Medical Education
and Reseach, Chandigarh-12, India

INTRODUCTION

The association of *Streptococcus pyogenes* with human pharyngeal mucosa and the development of pharyngitis and its sequelae rheumatic fever and rheumatic heart disease (RF/RHD) are well established.[1-3] However only < 3% of pharyngitis cases ultimately develops RF.[4,5] The reason for this remains unknown. Streptococcal surface molecules like lipoteichoic acid (LTA)[6] and epithelial components like fibronectin (Fn), collagen and other proteins seem to mediate the adherence of the organism to oral mucosa.[7,8] Patients with RF/RHD exhibit a heightened antibody response to streptococcal antigens which are found to cross-react with host tissue molecules.[9] The present study was aimed to determine the mucosal antibody reactive with LTA, Fn and collagen in the saliva of patients with RF/RHD and streptococcal pharyngitis. The antibody levels were then correlated with the quantum of streptococci adhered to the isolated pharyngeal cells of these patients.

METHODS

Patient groups consisted of 21 cases of acute recurrent rheumatic fever (ARRF), 33 cases of chronic rheumatic heart disease (CRHD) and 12 children with streptococcal pharyngitis. RF/RHD patients were followed-up at one month, six months and 12 months after day zero (before starting any medication on the day patients presented to the hospital). Only six of the pharyngitis children were studied after one month. Normal controls consisted of 19 subjects for ARRF group and 16 subjects for CRHD group and seven children for pharyngitis group.The characteristics of the patients have been described previously.[3] LTA was extracted and purified from M type 6 streptococci.[10] Fn was from Boehringer Mannheim (Germany) and collagen type III and IV from Sigma (USA).ELISA was employed to determine the salivary specific antibodies with the help of HRP cojugated rabbit anti-human IgA (Dako).

RESULTS

Salivary Antibody

In ARRF patients anti-collagen III antibody was elevated at six and 12 months of the study as compared to the respective control (p<.001, Table 1) whereas in other two patient groups there was no difference. In all the patient groups there was no significant level of anti-collagen IV antibody during the period of the study.In ARRF group anti-LTA antibody was higher than that of controls (Table 1) except at six months. Whereas in CRHD patients anti-LTA antibody was elevated during the course of the study as compared to the controls (p<.001). No significant level of antibody reactive with Fn was observed in any of the patient groups.

Table 1. Salivary antibodies in patient and control groups.[1]

	ARRF (21) Follow up					CRHD (33) Follow up					SD (12) Follow up		
	NC19	0	1	6	12	NC16	0	1	6	12	NC7	0	1(6)
Anti-III	0.8 ± 0.3	0.7 ± 0.4	1.0 ± 0.5	1.7^2 ± 0.3	1.3^4 ± 0.3	0.9 ± 0.3	1.1 ± 0.5	1.1 ± 0.4	1.1 ± 0.6	1.0 ± 0.4	0.9 ± 0.6	1.0 ± 0.3	0.9 ± 0.3
Anti-IV	1.0 ± 0.2	0.9 ± 0.4	0.8 ± 0.2	0.9 ± 0.4	0.9 ± 0.3	1.0 ± 0.2	1.3 ± 0.6	1.2 ± 0.6	1.0 ± 0.5	1.2 ± 0.3	1.1 ± 0.5	1.2 ± 0.3	1.1 ± 0.3
Anti-Fn	1.2 ± 0.3	1.0 ± 0.4	1.0 ± 0.4	1.2 ± 0.5	1.1 ± 0.5	1.2 ± 0.3	1.1 ± 0.6	1.1 ± 0.5	1.4 ± 0.5	1.1 ± 0.4	1.0 ± 0.6	0.9 ± 0.4	1.1 ± 0.4
Anti-LTA	0.2 ± 0.03	0.4^4 ± 0.04	0.3^4 ± 0.03	0.3 ± 0.07	0.3^4 ± 0.05	0.2 ± 0.04	0.4^4 ± 0.03	0.3^4 ± 0.03	0.4^4 ± 0.06	0.4^4 ± 0.06	0.2 ± 0.02	0.3^4 ± 0.02	0.3^3 ± 0.03

[1]Values represent mean ELISA value (A_{492}) ± S.D. For details, see the foot note of Table 2.
^{2}p < .05,
^{3}p < .01,
^{4}p < .001 as compared to the respective control group (NC). Anti-III, Anti-Collagen Type III ; Anti-IV, Anti-Collagen Type IV.

Streptococcal Adhesion and Salivary Antibody

The number of M type 5 streptococci that adhered to the isolated pharyngeal and buccal cells of the patients differ from the controls during the one year period of the study (Table 2).[3] Therefore, it was of interest to look if the streptococcal adherence altered the production of naturally occuring salivary antibodies in patients. But correlation analyses did not show any significant relation between the adhesion of streptococcal to pharyngeal and buccal cells and the the level of antibodies determined in the saliva of the patients studied over a period of one year.

Table 2. Mean number of M type 6 streptoocci adhered to PEC and BEC of patients and controls.[1]

	ARRF (21) Follow up					CRHD (33) Follow up					SD (12) Follow up		
	NC19	0	1	6	12	NC16	0	1	6	12	NC7	0	1(6)
PEC	17.8 ± 1.2	31.2 ± 3.3	19.8 ± 1.3	22.0 ± 1.0	21.7 ± 1.1	16.9 ± 1.8	24.4 ± 1.1	23.2 ± 1.2	23.1 ± 0.6	22.1 ± 1.2	37.4 ± 2.6	39.0 ± 1.7	20.9 ± 1.0
BEC	16.9 ± 1.7	19.6 ± 1.6	14.5 ± 0.9	15.1 ± 0.9	15.25 ± 0.7	18.7 ± 2.1	23.7 ± 1.1	20.7 ± 1.2	21.9 ± 1.0	18.0 ± 1.3	28.0 ± 2.8	28.4 ± 2.2	18.5 ± 2.6

[1]Values are Means ± S.E. ; ARRF, acute recurrent rheumatic fever ; CRHD, chronic rheumatic heart disease ; SP, streptococcal pharyngitis ; PEC, pharyngeal epithelial cells ; BEC, buccal epithelial cells ; NC, normal control ; values in parentheses are number of subjects ; follow-up of patients at zero day (O), one month (1), six months (6) and twelve months (12); for details see reference 3.

DISCUSSION

Secretory antibody (S-IgA) generated in the mucosal in the host is important in preventing the bacteria from attaching to the mucosal cells.[11] Such antibodies may be directed against the bacterial adhesins or to the complementary molecules expressed on the epithelial cells thereby inducing steric hinderance in the bacterial adhesion process. Hence, in the present study we determined the S-IgA specific antibodies for LTA, Fn and collagen of in the saliva of patients with RF/RHD and strep pharyngitis during a period of one year. Because of the lack of salivary anti-Fn and anti-collagen antibody the adhesion of streptococci to the isolated pharyngeal and buccal cells to Fn and collagen was not altered. Pretreatment of oropharyngeal cells of healthy subjects with anti-Fn antibody (1:10) was capable of blocking the adhesion (data not shown). Although salivary anti-LTA antibody was elevated in the patient groups, there was no interfering effect of this antibody on the adherence of streptococci to the isolated oropharyngeal cells. This may be due to the fact that salivary anti-LTA may not react with the epithelial cells of the patients and therefore fail to interfere with the adhesion.Pharyngeal cells have been found to be more susceptible to streptococcal adhesion than buccal cells.[12] The findings of the present study suggest that the preferential adherence of S.pyogenes to the pharyngeal cells of RF/RHD patients[3] was not altered by the naturally occuring anti-LTA or anti-Fn or anti-collagen antibodies. The decrease in the adherence of S.pyogenes to the oropharyngeal cells of the patients may be due to other host related factors.[3]

REFERENCES

1. G. H. Stollerman, "Rheumatic Fever and Streptococcal Infections", Grune and Stratton, New York(1975).
2. D. S. Selinger, N. Julie, W. P Reed, and W. P. Williams, Jr, *Science* 201:455 (1977).
3. K. S.Nanda Kumar, N. K. Ganguly, I. S. Anand, and P. L.Wahi, *APMIS* 100:353 (1992).
4. A. L. Bisno, I. Pearce, H. P. Wall, D. Moody, and G. H. Stollerman, *N. Eng. J. Med.*, 283:561 (1970).
5. L. W. Wannamaker, *N. Eng. J. Med.*, 282:23 and 78 (1970).
6. E. H. Beachey and H. S. Courtney, *Respiration* 55 (Supl.I):33 (1989).
7. H. Poly, M. Couble, D. J. Hartman, M. Faure, and H. Magloire, *J. Periodont. Res.* 23:252 (1988).
8. M. Kostrzynska, C. Schalam, and T. Wadstrom, *FEMS Microbiol. Let.* 59:229 (1989).
9. J. B. Zabriskie, *Circulation* 71:1079 (1985).
10. I. Ofek, E. H. Beachey, W. Jefferson, and G. C. Campbell, *J. Exp. Med.* 141:990 (1975).
11. S. N. Abraham, and E. H.Beachey, *In*: "Advances in Host Defense Mechanisms", J. I. Gallin, and A. S.Fauci, (ed.), p.63, Raveen Press, New York(1985).
12. K. S. Nanda Kumar, N. K. Ganguly, I. S. Anand, and P. L. Wahi, *Microbiol. Immunol.* 35:1029 (1991).

ENTAMOEBA HISTOLYTICA ADHERENCE: INHIBITION BY IgA MONOCLONAL ANTIBODIES

O. Leyva, G. Rico, F. Ramos, P. Moran, E. I. Melendro, and C. Ximenez

Experimental Medicine Department, Faculty of Medicine
UNAM. Mexico City, Mexico

INTRODUCTION

Adhesion of *Entamoeba histolytica* to target cells in the host intestine, is the first of three consecutive steps (adhesion, cytolytic effect and phagocytosis) involved in the invasion of colonic tissues.[1] The present study investigated the role of the local secretory immune response in the-interference with this early host-parasite relationship. IgA monoclonal anti-*E. histolytica* antibodies were produced. One of the clones obtained (FlPlD5) had been tested in its capacity to block the adhesion process *in vitro* with two different epithelial cells (MDCK and HT-29 cell lines), and *in situ* using colonic mucosa from BALB/c mice or gerbils (*Meriones unquiculatum*), which differ in susceptibility to *Entamoeba* experimental infection.

METHODS AND RESULTS

Inhibition of Trophozoite Adherence to Epithelial Cells *In Vitro*

Adherence of *E. histolytica* HM1: IMSS (10^4) to MDCK or HT-29 cells (2×10^5) was studied by rosette formation assays as previously described;[2] trophozoites were incubated with specific monoclonal IgA during 30 min at 4° C, then trophozoites were washed with D-MEM without fetal bovine serum, (GIBCO BRL, Grand Island, NY, USA). Trophozoites and MDCK or HT-29 cells were mixed (1 ml total volume) and after incubation (2 hrs 4° C) the supernatant (0.8 ml) was removed and the pellet was gently stirred. Percentage of amoeba with 3 or more cells attached was calculated. Results show that FlPlD5 IgA monoclonal antibody inhibited rosette formation very efficiently (85 %) compared with control assays, where rosette formation was almost complete (Fig. 1). There are differences between inhibition adherence with MDCK compared to HT-29 human adenocarcinoma cell line (Table 1).

Inhibition of Adherence to Colonic Mucosa Assay

Two rodent species with different susceptibility to amoebic infection, were used to evaluate adherence of trophozoites to colonic mucosa. In both models adherence of axenically cultured [^{3}H]-TdR labeled trophozoites,[3] to 3 mm sections of fixed glutaraldehyde (2.5 %) colonic mucosa was 15 % after 30 min at 37° C. Subsequently we evaluated whether anti-*E. histolytica* IgA monoclonal antibody could inhibit amoebic adherence in the *in situ* model. As may be observed in Table 2, FlPlD5 antibody, interferes with the adherence process between 50 % and 70 % compared with control assays in the abscence of specific monoclonal antibody (non-opsonized trophozoites). However, hyperimmune mouse anti-amoebic serum shows no greater inhibitory activity.

Advances in Mucosal Immunology, Edited by
J. Mestecky *et al.,* Plenum Press, New York, 1995

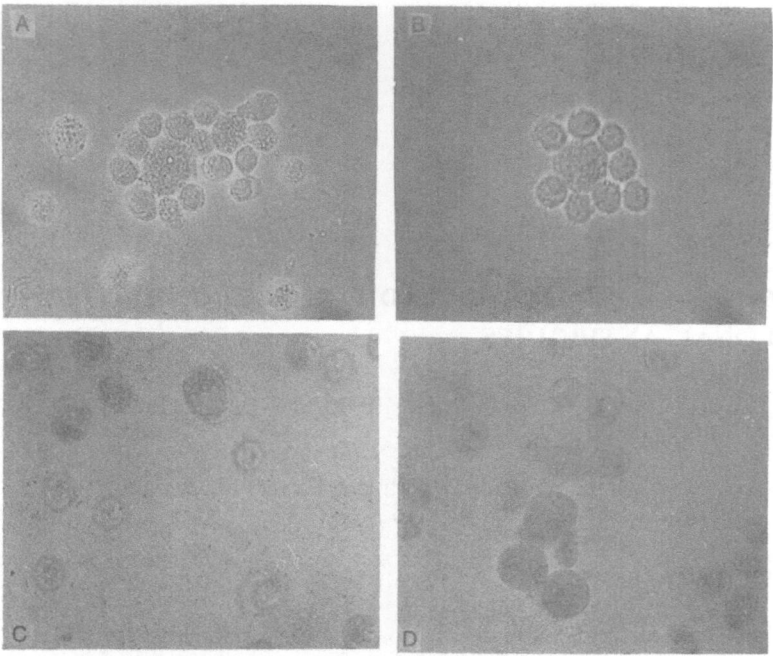

Figure 1. A) Rosette formation with *E. histolytica* HM1:IMSS trophozoites and HT-29 cells; B) in the presence of Sp2/0-Ag14 (C) and F1P1D5 (D) anti-*E. histolytica* monoclonal IgA.

Table 1. Inhibition of *E. histolytica* adherence to MDCK or HT-29 cell lines.

Cell line	Antibody	Adherence (%)	Inhibition adherence (%)
MDCK	-	80	-
	F1P1D5	13	85
	Hiperimmune Serum	0	100
HT-29	-	90	-
	F1P1D5	33	64
	serum	36	60

Affinity purity anti-monoclonal F1P1D5 was used at a protein concentration of 100 µg/ml. Ascitic fluid from Balb/c mice inoculated with Sp2/0-Ag14 cells, previously precipitated with saturated $(NH_4)_2 SO_4$ solution (50% v/v) was used at 100 µg/ml protein concentration. Anti-amoeba hyperimmune serum was diluted 1:10. 100 µl volumes were used in each case.

Table 2. Inhibition of *E. histolytica* adherence to fixed or mice gerbils cononic mucosa.

Inhibiton of Adherence (%)

Animal	Antibody			
	Ascitic[a] fluid	F1P1[b] D4	F1P1[b] D5	Hyperimmune serum[c]
BALB/c	-	40	49	74
Gerbils	-	41	67	48

a) Ascitic fluid treated as was indicated in Table 1, adjusted to a protein concentration of 90 μg/ml.
b) Affinity purify anti-*E. histolytica* monoclonal IgA (F1P1D5) (90 μl/ml).
c) Anti-amoeba hyperimmune serum was used as indicated in Table 1.

DISCUSSION

In the present study, experimental conditions were kept, similar to those described in other *in vitro* target cell systems.[3,4] The HT-29 cell line was used as an *in vitro* adherence model since this line offers adventages over other epithelial cell lines. HT-29 cells are physiologically closer to normal target cells for amoebae.[5] Our results indicate that this may be a useful model for adhesion studies of *E. histolytica* to human cells. There is a correlation between the *in vitro* model and the adherence model of trophozoites to glutaraldehyde fixed colonic mucosa, either from BALB/c mice or gerbils. Though these two species are different in suceptibility to amoebic infection *in vivo*,[6,7] in the *in situ* model of adherence differences were not evident.

Our results suggest that although the biological significance of the secretory immune response in amoebiasis remains obscure, a reasonable possibility is that local immunity plays a critical role in mantaining the balance between intestinal and luminal amoebiasis. This may contribute to the differences in world prevalence and morbidity of the disease.

ACKNOWLEDGMENT

This work was supported by the Agency for International Development US AID throught the grant No.936-5542-02-5239, 130.

REFERENCES

1. A. Martinez-Palomo, A. Gonzalez-Robles, and B. Chavez, Cell Biology of the Cytopathic Effect of *Entamoeba histolytica. In:* Ambiasis. Infection and Disease by *Entamoeba histolytica.* Kretschmer, Ed. Boca Raton, FL, C.R.C. Press Inc., pp. 43-58, 1990.
2. J. I. Ravdin and R. Guerrant, *J. Clin. Invest.* 68:1305 (1981).
3. J. I. Ravdin, J.E. Jhon, L. I. Johnson, D. J. Onnes and K. L. Guerrant, *Infect. Immun.* 48:292 (1985).
4. E. Orozco, A. Martinez-Palomo, A. Gonzalez Robles, G. Guarneros,, and J. Mora-Gobiedo, *Arch. Invest. Med.* (Mexico) 13:159 (1982).
5. M. Rousset, *Biochem.* 68:1035 (1986).
6. F. Anaya-Velazauez and B. J. Underwon, *Arch. Invest. Med.* (Mexico) 21:53 (1990).
7. K. Chadee and E. Mecrovith, *Am. J. Trop. Med. Hyg.* 34:283 (1985).

BIOLOGICAL PROPERTIES OF YOLK IMMUNOGLOBULINS

Anna K. Janson, C. I. Edvard Smith, and Lennart Hammarström

Dept of Clinical Immunology
Karolinska Institute at Huddinge Hospital
S-141 86 Huddinge, Sweden

INTRODUCTION

Selected immundeficient patients suffer from gastrointestinal infections probably due to defective local production of antibodies. The patients are usually treated with immunoglobulins given intramuscularly, intravenously or subcutaneously. There is normally a leakage of antibodies to the gut, but in some patients the fraction of immunoglobulins that reach the intestines after systemic application of antibodies is not enough. In these patients it would be desirable to introduce the immunoglobulins directly into the gastrointestinal tract.

In recent years, immunoglobulin produced in different species has been used successfully in human as oral prophylaxis[1-6] and/or treatment[7-15] against diarrhea caused by a number of microorganisms. The rationale for using immunoglobulins for this purpose is based on the observation that breast milk containing secretory immunoglobulins protects infants against diarrhea[16-18]. In some cases when other chemotherapeutic drugs have proved ineffective, which is the case for Cryptosporidiosis in AIDS and immunologically compromised patients[9-11], passive oral immunization has proven successful. Passive immunization is also to be preferred, for example, in protection against traveler's diarrhea since no side effects, such as development of resistance in the bacteria which is the case in large scale use of prophylactic antimicrobial agents, are induced. It is important to find an antibody source with sufficient levels of antibodies and to assess whether the immunoglobulins are tolerated orally, and furthermore, to investigate the protective mechanisms of the antibodies in the gastrointestinal tract.

Hens can be readily immunized with enteropathogenic microorganisms and specific antibodies can then be purified from serum or yolk. IgG in chicken is effectively transferred *in vivo* to the yolk and may reach levels comparable to serum.[19] In an average egg there is about 100-400 mg antibodies packaged in a convenient and sterile form. Yolk antibodies (IgY) from immunized hens have also been shown to protect infant mice and cats against diarrhea due to rotavirus challenge[20-22] and have been used effectively as oral treatment of piglets suffering from diarrhea due to *E.coli.*[23-24]

The aim of this study was to purify antibodies from eggs, laid by hens immunized with *Shigella flexneri* and *Campylobacter jejuni,* to measure the specific antibody levels in the yolk and to investigate the biological properties of yolk antibodies.

MATERIALS AND METHODS

Immunization Schedule

In our study, the hens were vaccinated orally with an ARO-mutant of *S. flexneri* (obtained from A. Lindberg, Dept. of Bacteriology, Huddinge Hospital) but since we could not detect the bacteria in the feces of the hens (probably due to effective gastrointestinal killing), we vaccinated them 5 times intramuscularly either with heatshocked *S. flexneri* (SFL124-Y obtained from A Kärnell, Dept. of Bacteriology, Huddinge Hospital) or *C. jejuni* (CCUG;12074, 12070, 19506, 12085, 12066, CCUG, obtained from Dept. of Clin. Bacteriology, University of Göteborg, NCTC-1135, JY-3483 obtained from K.Karlsson, Dept. of Bacteriology, Huddinge Hospital) killed by formalin (all strains were cultured and prepared by K.Karlsson). Complete Freund's adjuvant was used in the first vaccinations and incomplete Freund's adjuvant was used in the first booster dose in the hens vaccinated with SFL. The other booster vaccinations were performed without any adjuvant. The eggs were collected before and after each vaccination.

ELISA

To determine the amount of IgY, rabbit-anti chicken antiserum (IMS,Uppsala, Sweden) was coated in wells of microplates. A sample of purified antibodies (IMS) from yolk of nonimmunized hens (2.34 mg/ml) was used as standard. For determination of specific IgY, the wells were coated with either whole *S. flexneri* 124 Y, *S. flexneri* lipopolysaccharide (LPS- obtained from Dr. A. Kärnell, Dept. of Bacteriology, Huddinge Hospital, Huddinge) or whole *C. jejuni* 12074 and 12078. Alkaline phosphatase labeled rabbit anti-chicken immunoglobulin (IMS), rabbit anti-chicken (IMS or Sigma, St. Louis, MO) or a mouse monoclonal anti-chicken IgG antibody (Janssen, Biochemica, Beerse, Belgium) followed by alkaline phosphatase-labeled rabbit anti-mouse (Dako, Denmark) was then added and the plates were developed with p-nitrophenyl phosphate-disodium (Sigma).

FACScan

S. flexneri was mixed either with IgY from prevaccinated or vaccinated hens and incubated for 20 min. Fluorescein isothiocyanate (FITC)-conjugated rabbit anti -chicken antibodies (IMS diluted 1:20) was added and the samples were incubated again for 20 min. Between each step the samples were washed twice with PBS to remove unreacted components. The intensity of FITC-labeled antibodies bound to the bacteria was measured in a FACScan flow cytometer.

Hemagglutination

25 µl of human bloodgroup 0 Rh- (1.5% red blood cells), coated with *S. flexneri* LPS (0.5mg/ml) or left unreacted, was added into microtiter wells. The IgY samples and positive controls (rabbit anti -Shigella antiserum and human anti -Shigella patient antiserum from Dr. A Kärnell) were titrated in twofold dilutions in 25 µl PBS containing 1% fetal calf serum (FCS, Kemila, Sollentuna, Sweden) and incubated at room temperature for 24 hours.

Inhibition of Invasion

1-2 x 10^5 HeLa cells were coated in wells (Costar, tissue culture plates) and incubated overnight at 37° C in 7 % CO_2. The bacteria, *S. flexneri* (Y SFL1) and *E. coli* K-12 with the invasive plasmid of *S. flexneri* 5 (SP10) (grown in brain heart infusion up to early log phase and then diluted to OD(600)= 0.1 in RPMI 1640) and the IgY samples were added to the wells and incubated for 90 min. at 37° C in 7 % CO_2. The wells were washed 3 times with PBS containing 20 µg/ml Gentamicin to kill extracellular bacteria. 1ml RPMI 1640 (+ 30 µg/ml Gentamicin) was added to each well and incubated for 30 minutes at 37° C in 7 % CO_2. The cells were then treated with trypsin, lysed by sodium-deoxycholate, diluted, plated and CFU were counted.

Phagocytosis Test

0.5 ml of human granulocytes from normal healthy donors (5×10^6/ml purified with Macrodex, 60 mg/ml with Nacl Pharmacia, diluted in Hanks, a balanced salt solution, SBL) was mixed with 0.1 ml of *S. flexneri* (1×10^8/ml in Hanks) and 0.4 ml of IgY (2 mg/ml in Hanks) from either prevaccinated hens or hens vaccinated with *S. flexneri* 124 or *S. flexneri* 124-20. The total number of live bacteria (intra+extracellular) were counted before and after 37° C incubation by lysing the granulocytes in distilled water and then plated to count CFU.

RESULTS

Estimation of Specific Antibodies

The antibody concentration in the IgY samples from eggs laid by hens vaccinated 1-4 times and prevaccinated hens, was measured by ELISA and adjusted to approximately 2 mg/ml before being tested against different antigens. The specific antibody levels increased according to the number of vaccinations (Figure 1).

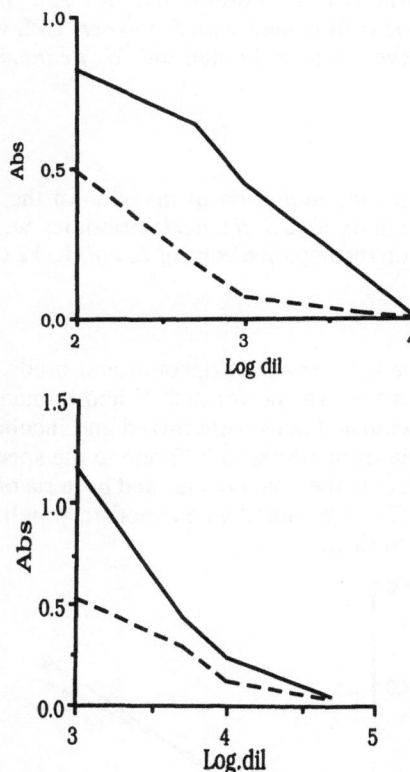

Figure 1. The top figure shows the results of the specific IgY antibodies against SFL, the bottom figure shows the results of the specific antibodies against *Campylobacter*, in the ELISA assay. There is an increase of specific antibodies in the samples from hens boosted (___) compared to samples from hens vaccinated once (_ _ _).

FACS is another method to measure the intensity of antibodies bound to different antigens on whole cells. Figure 2 shows that yolk antibodies from *S. flexneri* vaccinated hens binds more intensely to the bacteria than yolk antibodies from prevaccinated hens.

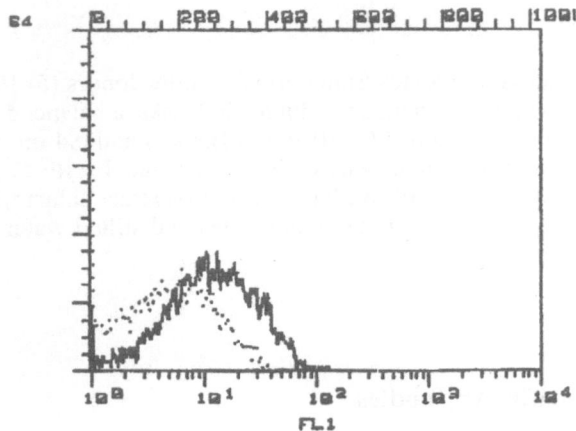

Figure 2. The figure shows that IgY from hens vaccinated with SFL (___) binds more intensely to the bacteria than IgY from prevaccinated hens (_ _ _)

Hemagglutination

The *S. flexneri* specific IgY antibodies, but not IgY from prevaccinated hens, agglutinated human red blood cells coated with *S. flexneri* LPS with a titer of 1/64 which is comparable to the positive controls human-anti *S. flexneri* 1/128 and rabbit anti *S. flexneri* 1/128.

Inhibition of Invasion

There was an increase rather than inhibition of invasion of the HeLa cells of *S. flexneri*, when the IgY samples containing anti-*S. flexneri* antibodies were addded in the invasion assay. No effect was noted on the negative control *E. coli* K-12 (data not shown).

Phagocytosis

To investigate whether IgY serves as opsonin and binds to human phagocytosing cells, a granulocyte-function test was performed. When human granulocytes, *S. flexneri* and IgY from *S. flexneri*-vaccinated hens were mixed and incubated, there was a decrease in the amount of live bacteria from 100 % to 2 % due to the specific IgY antibodies, since no effect was achieved when only the granulocytes and bacteria or the granulocytes and IgY were mixed (Figure 3), no effect was noted when another, rough strain, of *S. flexneri* (SFL 124-20) was used (data not shown).

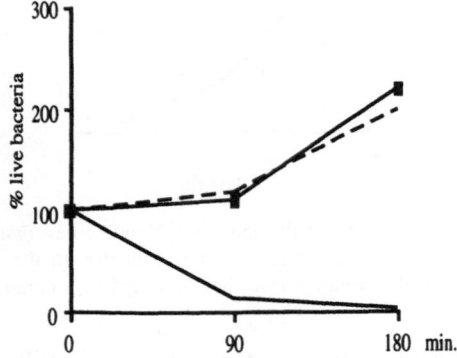

Figure 3. The figure shows a decrease of live Shigella from 100% to 2% when IgY from SFL vaccinated hens is added together with human granulocytes (___) compared to Shigella mixed with only the IgY (-n-) or human granulocytes (_ _ _)

688

DISCUSSION

In this study, an increased specific antibody level in yolk from eggs laid by hens vaccinated with *S. flexneri* or *C. jejuni* was found, using ELISA as well as agglutination and FACS techniques. It still remains to be determined whether these antibodies would be effective as treatment/prophylaxis in humans. Chicken yolk immunoglobulin do not fix either chicken[25] or mammalian complement,[26] but the antibodies bind to bacteria, agglutinate *S. flexneri* in an indirect hemagglutination assay, opsonize the bacteria and bind to human granulocytes as evidenced by *in vitro* phagocytosis of *S. flexneri*. The latter is in contrast to earlier suggestions since the yolk antibodies do not bind to protein A.[27] Unexpectedly, the chicken yolk antibodies increased the adhesion of *S. flexneri* to HeLa cells. However, this does not rule out an *in vivo* effect since the cells used are different from the cells affected *in vivo* and, since the specific yolk antibodies agglutinate the bacteria and therefore promote efficient contact between bacteria and cultured cells in this infection model. This has previously also been suggested in a similar experimental system.[28] Our studies indicate that chicken antibodies could be of potential interest for oral use in humans. The cost also makes it an interesting alternative for human use.

REFERENCES

1. G. P. Davidsson *et al.*, *Lancet* 2:709 (1989).
2. T. Ebina *et al.*, *Med. Microbiol. Immunol.* 174:177 (1985).
3. C. O. Tacket *et al.*, *N. Engl. J. Med.* 318:1240 (1988).
4. C. O. Tacket, S. B. Binion, E. Bostwick, G. Losonsky, M. J. Roy, and R. Edelman, *Am. J. Trop. Med. Hyg* (in press) (1992).
5. G. L. Barnes *et al.*, *Lancet* 1:1371 (1982).
6. M. M. Eibl, H. M. Wolf, H.F. Fürnkranz and A. Rosenkranz, *N. Engl. J. Med.* 319:1 (1988).
7. A. K. Mitra, D. Mahalanabis, H. Ashraf, S. Tzipori, L. Unicomb, R. Eeeckels, Abstract 27th Joint Conference on Cholera and Related Diarrheal Diseases, Virginia (1991).
8. R. Lodinová-Zádníkova, B. Korych, and Z. Baatáková, Nahrung 31 (5-6):465 (1987).
9. S. Tzipori, D. Roberton, and C. Chapman, *Br. Med. J.* 293:1276 (1986).
10. S. Tzipori, D. Roberton, D.A. Cooper, and L. White, *Lancet* 2:344 (1987).
11. B. L. P. Ungar, D. J. Ward R. Fayer, and C. A. Quinn, *Gastroenterology* 98:486 (1990).
12. R. Arndt *et al.*, Abstract 5. Frühjahrstagung der Gesellschaft für Immunologie, Freiburg (1989).
13. A. Guarino, S. Guandalini, F. Albano, A. Mascia, G. DeRitis, and A. Rubino, *Pediatr. Infect. Dis. J.* 10:612 (1991).
14. S. M. Borowitz and F. T. Saulsbury, *Clin. Lab. Obs.* 119:593 (1991).
15. I. Melamed, A. M. Griffiths, and C. M. Roifman, *J. Pediatr.* 119:486 (1991).
16. A. S. Cunningham, *J. Pediatr.* 90:726 (1977).
17. S. A. Larsen, Jr. and D. R. Homer, *J. Pediatr.* 92:417 (1978).
18. R. I. Glass *et al.*, *N. Engl. J. Med.* 308:1389 (1983).
19. R. Kühlmann, V. Wiedemann, P. Schmidt, R. Wanke, E. Linckh, and U. Lösch, *J. Vet. Med.* 35:610 (1988).
20. C. R. Bartz, R. H. Conklin, C. B. Tunstall, and J. H. Steele, *J. Infect. Dis.* 142:439 (1980).
21. R. H. Yolken, F. Leister, S.-B. Wee, R. Miskuff, and S. Vonderfecht, *Pediatrics* 81(2):291 (1988).
22. C. Hiraga, Y. Kodama, T. Sugiyama, and Y. Ichikawa, *J. Jpn. Assoc. Infect. Dis.* 64:118 (1990)
23. V. Wiedemann, E. Linckh, R. Kühlmann, P. Schmidt, and U. Lösch, *J. Vet. Med.* B 38:283 (1991).
24. H. Yokayama, R. C. Peralta, R. Diaz, S. Sendo, Y. Ikemoria, and Y. Kodama, *Infect. Immun.* 60:998 (1992).

25. E. Orlans, *Immunology* 12:27 (1967).
26. H. N. Benson, H. P. Brumfield, and B. S. Pomeroy, *J. Immunol.* 87:616 (1961).
27. G. Kronvall, U. S Seal, J. Finstad, and R. C. Williams Jr, *J. Immunol.* 104:140 (1970).
28. T. L. Hale and P. F. Bonventre, *Infect. Immun.* 24:879 (1979).

ROLE OF ALPHA-HAEMOLYSIN IN RESISTANCE OF *ESCHERICHIA COLI* STRAINS TO BACTERICIDAL ACTION OF HUMAN SERUM AND POLYMORPHONUCLEAR LEUKOCYTES

Leonard Siegfried,[1] Jozef Filka,[2] and Hana Puzová[1]

[1]Institute of Medical Microbiology, Faculty of Medicine,
P.J. Safárik University, Alzbetina 2, 040 00 Kosice;
[2]Department of Pediatrics, University Hospital, Trieda
SNP 1, 040 66 Kosice, Slovak Republic

INTRODUCTION

It is known that *Escherichia coli* strains isolated from intestinal and extraintestinal infections possess various virulence traits that make them injurious to their host. These properties include capabilities to adhere and alter physiology of host cells, to invade host cells and tissues and to resist host defense mechanisms.

It has been shown that several properties of the bacterial cell wall complex of *E. coli* function as determinants of resistance to the bactericidal effect of serum, e.g., O-antigen polysaccharide side chains, capsular polysaccharides, and outer membrane proteins.[1] Capsular antigens also play an important role in virulence of *E. coli* by inhibiting phagocytosis.[2]

Another factor promoting virulence of *E. coli* seems to be alpha-haemolysin (AH). The activity of AH is multifactorial. It is based on the release of iron from erythrocytes and on direct toxicity to host tissues.[3] Moreover, it has been shown that AH is involved in dysruption of phagocyte function.[4]

The aim of this work was to evaluate a possible role of AH for resistance of *E. coli*, both to serum bactericidal activity and intracellular killing in polymorphonuclear leukocytes (PMNL).

MATERIAL AND METHODS

Bacterial Strains

A total of 213 clinical isolates of *E. coli* was studied, 109 alpha-haemolytic and 104 non-haemolytic strains. They were isolated from faeces of infants with diarrhoea (106 isolates), or from extraintestinal sites (107 isolates). Of the latter isolates, 82 were from urine of infants with urinary tract infections, 13 from pharyngeal exudate of infants with pharyngitis and 12 from mothers with vaginal discharge.

Serum and PMNL

For serum bactericidal test, blood from one healthy donor was obtained. Serum was collected, aliquoted, and stored at -70°C until needed.

PMNL for tests of intracellular killing of *E. coli* were used from fresh, nonheparinized blood of healthy donors.

Advances in Mucosal Immunology, Edited by
J. Mestecky *et al.,* Plenum Press, New York, 1995

Quantification of AH Production

The test was a modification of the method of Tabouret et al.[5]

Serum Bactericidal Assay

The assay was a slight modification of a method described by Miler et al.[6] and Benge.[7] Susceptibility of bacteria was evaluated after the treatment of serum for 180 min serum. Strains were designated serum-sensitive if the viable count dropped to 1% of the initial value and resistant if 90% of organisms survived after 180 min. Strains that gave results between these values were considered of intermediate sensitivity.

Intracellular Killing in PMNL

The intracellular killing of bacteria by human PMNL was determined microscopically with a slightly modified acridine orange staining method described by Pantazis and Kniker.[8] The viability of organisms in a total of 100 PMNL was estimated. Red organisms were regarded as non-viable, green and orange ones as viable. The percentage killed was the percentage of intracellular organisms that were non-viable, i.e.,

$$\text{percent killed} = \frac{\text{total number of non-viable bacteria in 100 PMNL}}{\text{total number of intracellular bacteria}} \times 100$$

RESULTS

By comparing the response of alpha-haemolytic and non-haemolytic E. coli strains, a significantly greater proportion of resistant isolates was found among the former compared to the latter (47% and 28% after 100% serum treatment, $x^2=8.11$, $p<0.01$) (Table 1). By comparing the intracellular killing of E. coli strains in PMNL, a significantly lower intracellular killing was found in alpha-haemolytic isolates compared to non-haemolytic ones (78.6% and 84.4%, t=3.26, p<0.01).

Table 1. Resistance of alpha-haemolytic and non-haemolytic E. coli strains to human serum bactericidal activity.

Haemolytic status	Number (%) of strains that were resistant after 100% serum treatment
Hly+ (n=109)	51 (47%)
	p < 0.01
Hly− (n=104)	29 (28%)

Hly+ - isolates showing production of alpha-haemolysin; Hly− - non-haemolytic isolates

Table 2. Intracellular (i.c.) killing of alpha-haemolytic and non-haemolytic E. coli in human PMNL.

Haemolytic status	Percentage i. c. killing in PMNL		
	Mean	(SD)	Range
Hly+ (n=43)	78.6	(8.9)	54 - 92
	p < 0.01		
Hly− (n=37)	84.4	(6.9)	64 - 94

DISCUSSION

Several reports have shown that there are various virulence determinants at the cell wall surface that may play a role in the resistance of E coli strains to the complement mediated serum killing and to intracellular killing in PMNL [9] [10]

Our results indicate that AH could be another factor that increases serum resistance of E coli This suggestion arises from significantly greater proportion of alpha-haemolytic isolates showing serum resistance compared to non-haemolytic ones Until now only Hughes et al [11] suggested possible association of haemolysin production by E coli with serum resistance, and recently Kubens and Opferkuch[12] showed that the loss of secretion in mutant E coli strains compared to their haemolytic parental wild-type strain was accompanied by a decrease in serum resistance

Our results also indicate the importance of AH for the extent of E coli intracellular killing in PMNL We found significantly higher resistance of alpha-haemolytic isolates to intracellular killing in PMNL than was non-haemolytic isolates These results are in agreement with those of Gadeberg et al [13]

Taking into account the results obtained in this study, we consider that AH contributes both to the resistance of E coli to serum killing and intracellular killing in PMNL and thus may be a tool which allows E coli strains to resist immune mechanisms operating at the mucosa and enables them to invade host

REFERENCES

1   J R Johnson, *Clin Microbiol Rev* 4 80 (1991)
2   I S Roberts, F K Saunders, and G J Boulnois, *Biochem Soc Trans* 17 462 (1989)
3   S J Cavalieri, G A Bohach, and I S Snyder, *Microbiol Rev* 48 326 (1984)
4   O V Gadeberg and B Mansa, *Zbl Bakt* 273 492 (1990)
5   M Tabouret, J De Rycke, A Audurier, and B Poutrel, *J Med Microbiol* 34 13 (1991)
6   I Miler, J Vondracek, L Hromádková, *Folia Microbiol* 24 143 (1979)
7   G R Benge, *J Med Microbiol* 27 11 (1988)
8   C G Pantazis and W T Kniker, *RES* 26 155 (1979)
9   K N Timmis, G J Boulnois, D Bitter-Suermann, and F C Cabello, *Curr Top Microbiol Immunol* 118 197 (1985)
10 H Leying, S Suerbaum, H -P Kroll, D Stahl, and W Opferkuch, *Infect Immun* 58 222 (1990)
11 C Hughes, R Phillips, and A P Roberts, *Infect Immun* 35 270 (1982)
12 B S Kubens and W Opferkuch, *Zbl Bakt Hyg* A270 52 (1988)
13 O V Gadeberg, J Hacker, and I Orskov, *Zbl Bakt* 271 205 (1989)

OPSONO-PHAGOCYTOSIS OF NON-ENCAPSULATED *HAEMOPHILUS INFLUENZAE*

Liesbeth Vogel,[1] Loek van Alphen,[1] Forien Geluk,[1]
Henk Jansen,[2] and Jacob Dankert[1]

Departments of [1]Medical Microbiology, University of
Amsterdam and [2]Pulmonology, Academic Medical
Center, Amsterdam, The Netherlands

INTRODUCTION

Non-encapsulated *Haemophilus influenzae* causes serious respiratory tract infections in adults with chronic obstructive pulmonary disease (COPD).[1] These bacteria persist despite the presence of abundant numbers of polymorphonuclear leukocytes (PMNs), complement, and antibodies specific for the infecting organism in the sputum.[1,2,3,4] Opsono-phagocytosis, an important immune defense mechanism toeliminate bacterial intruders[3], is also operating in the bronchial tree. The basic mechanisms for persistence of H.influenzae in the bronchial tree are unknown, buthost and bacterial factors are likely to contribute[1,5,6], one of them might be adisturbed opsono-phagocytosis.

These observations were the impetus to investigate the opsono-phagocytosis of non-encapsulated *H. influenzae*. Flow cytometry[7,8] was adapted to study opsono-phagocytosis quantitatively. This method combines the specific advantages of the determination of opsono-phagocytosis by colony counting and binding of radioactive bacteria. Bacteria are labeled with fluorescamine for the detection in the flow cytometer. We compared the opsono-phagocytosis by PMNs of non-encapsulated *H. influenzae*, isolated from the sputum of a COPD patient, and of Staphylococcus aureus to evaluate whether inadequate opsono-phagocytosis is one of the reasons for the persistence of *H. influenzae* in the bronchial tree.

MATERIALS AND METHODS

Bacteria

H. influenzae strain d3, isolated from the sputum of COPD patient T, and *S. aureus* (Wood 46) were used for the experiments. *Candida albicans* was included as internal standard for the quantification of phagocytosis. Bacteria were grown late exponentially while shaking at 120 rpm, in Brain Heart Infusion broth (Difco laboratories, Detroit, Mich., USA) supplemented with NAD and haemin (10 mg/l each). After centrifugation the bacteria were washed and resuspended in 20 mM HEPES, containing 0.1% glucose, 5mM Ca^{2+}, pH 7.4 and incubated with fluorescamine isothiocyanate (FITC) in a concentration of 0.015 mg/ml for 15 min at 37°C under shaking.

Isolation of Polymorphonuclear Leukocytes

The method described by Roos and De Boer was used.[9] After isolation the cells were washed and resuspended in 20 mM HEPES - 5mM Ca^{2+} - 0.1% gelatine - 0.1% glucose - 0.1% Human Serum Albumin (Central Blood Transfusion Laboratory, Amsterdam, The Netherlands). The viability of the PMNs was always greater than 95% as determined with the lactate dehydrogenase activity.

Opsono-Phagocytosis of Bacteria

FITC labeled *H. influenzae* strain d3 bacteria were incubated with heat-inactivated Human Pooled Serum (HPS) and complement (antibody-free serum of an agammaglobulinemic patient), both in saturating concentrations for opsonization, and with PMNs for 30 min at 37°C in a shaking waterbath. Controls were always run in parallel in which the sera or PMNs were omitted from the incubation mixture to determine killing or agglutination of the bacteria in the absence of PMNs and phagocytosis of unopsonized bacteria. The phagocytosis was interrupted by five-fold dilution of aliquots of the incubation mixture into ice-cold HEPES buffer. A standard amount of FITC-labeled *C. albicans* was added to each sample as an internal standard for quantitative analysis. Paraformaldhyde (PFA) was added to a final concentration of 1% to fix the reaction mixture. Samples were analyzed with a FACScan flow cytometer (Becton Dickinson, Heidelberg, Germany). Data were acquired by using an instrument status with a logarithmic data mode for forward scatter (FSC), side scatter (SSC) and for FITC fluorescence, as measured in the FL1 channel (525-545 nm). A live gate was set around the bacteria, yeast cells and PMNs. Routinely 4000 events were counted. Phagocytosis was quantified as the decrease in the percentage of bacteria not associated with PMNs.

ELISA

With the whole cell ELISA as previously described[11], antibody titers of sera were determined using two-fold dilutions of antisera. The titer was the inverse of the dilution, which results in an optical density of 0.2 at 405 nm under the experimental conditions.[2]

RESULTS AND DISCUSSION

The ELISA titers to *H. influenzae* strain d3 as antigen before and after FITC labeling were compared in HPS. The titers were $1.6 \ 10^3$ and $2.5 \ 10^3$ respectively, indicating that the titers did not decrease by the labeling. The FITC labeling was sufficient to discriminate positive bacteria and PMNs from negative bacteria and PMNs. Opsono-phagocytosis of non-encapsulated *H. influenzae* strain d3 was measured after opsonization with HPS in the presence or absence of antibody-free serum of an agammaglobulinemic patient as source of complement. During opsono-phagocytosis of *H. influenzae* strain d3 a gradual decrease in the number of free bacteria was observed with time as determined with *C. albicans* as internal standard. The fluorescence of the PMNs increased accordingly (Fig. 1). Quantitative analyses of the opsono-phagocytosis are summarized in Fig. 2. HPS promoted opsono-phagocytosis poorly since it was $30 \pm 36\%$ in the presence and $23 \pm 36\%$ in the absence of complement. Opsono-phagocytosis was not promoted by antibodies either in the presence or absence of complement. The percentage of phagocytized bacteria was not higher than the percentage of phagocytized bacteria in the absence of opsonins. To demonstrate that the poor opsono-phagocytosis was a property of *H. influenzae*, similar experiments were performed with *S. aureus*, opsonized with HPS, as a source of antibodies and antibody-free serum from an agammaglobulinemic patient as a source of complement. The decrease of the number of free *S. aureus* by phagocytosis was $90 \pm 5\%$ (n=4). These results indicate that the assay condition were optimal for quantitative analysis of opsono-phagocytosis.

Flow cytometric quantification of phagocytosis combines the advantages of measuring phagocytosis by means of colony counting or binding of radioactive labeled bacteria to PMNs, because association with PMNs and the reduction in the percentage 'free' bacteria are measured. We have demonstrated that non-encapsulated *H. influenzae* is poorly phagocytized compared to *S. aureus*. The opsono-phagocytosis of *S. aureus* is in agreement with data from the literature[7,10,] indicating that the low percentage phagocytosis of *H. influenzae* is not due to inappropriate assay conditions. Hansen *et. al.*[12] and Garofalo *et. al.*[13] described almost 100% and 90% phagocytosis of non-encapsulated *H. influenzae* respectively. However they measured phagocytosis by means of colony counting or binding of radioactive labeled bacteria and they did not control for bacterial aggregation or bactericidal activity of the serum.

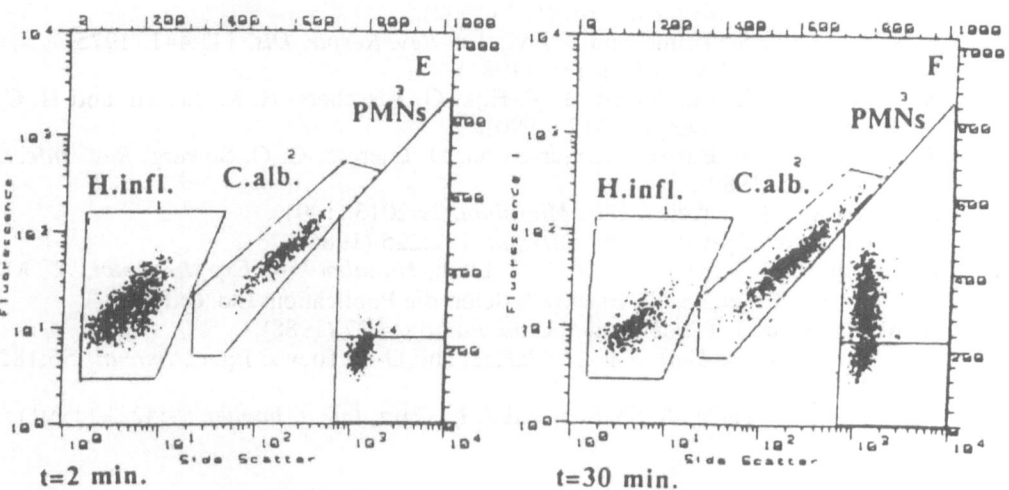

Figure 1. Flow cytometric anaalysis of opsono-phagocytosis of non-encapsulated *H. influenzae* strain d3. A decrease in the number of free bacteria and an increase of the fluorescence of the PMNs was observed with time.

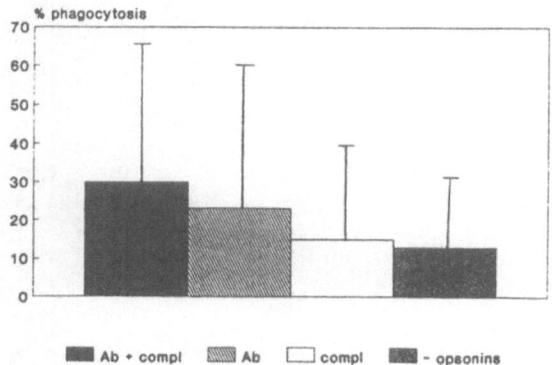

Figure 2. Phagocytosis of non-encapsulated *H. influenzae* strain d3 opsonized with antibodies and complement. Antibody (Ab) source: Human Pooled Serum. Complement (Compl) source: antibody free serum of an agammaglobulinemic patient.
n=7

We conclude that with flow cytometry opsono-phagocytosis of non-encapsulated *H. influenzae* can be analysed quantitatively and that non-encapsulated *H. influenzae* is poorly opsono-phagocytized. The poor opsono-phagocytosis may contribute to persistence of non-encapsulated *H. influenzae* in the presence of antibodies and complement in COPD patients.

ACKNOWLEDGMENTS

This work was supported by the Nederlands Asthma Foundation. We thank Drs. S. Bhakdi and E. Martin for their help in the development of the assay.

REFERENCES

1. T. F. Murphy and M. A. Apicella, *Rev. Infect. Dis.* 9:1 (1987).
2. K. Groeneveld, P. P. Eijk, L. van Alphen, H. M. Jansen, and H. C. Zanen, *Am. Rev. Respir. Dis.* 141:1316 (1990).
3. P. Brandtzaeg, *J. Infect. Dis.* 165:S167 (1992).
4. E. Kagan, C. L. Soskolme, and S. Zwi, *Am. Rev. Respir. Dis.* 111:441 (1975).
5. D. C. Turk, *J. Med. Microbiol.* 18:1 (1984).
6. K. Groeneveld, L. van Alphen, P. P. Eijk, G. Visschers, H. M. Jansen, and H. C. Zanen, *J. Infect. Dis.* 161:512 (1990).
7. R. Bjerknes, C. F. Bassoe, H. Sjursen, O. D. Laerum, C. O. Solberg, *Rev. Infect. Dis.* 11:16 (1989).
8. E. Martin and S. Bhakdi, *J. Clin. Microbiol.* 29:2013 (1991).
9. D. Roos and M. de Boer, *Meth. Enzymol.* 132:225 (1986).
10. R. van Furth, T. L. van Zwet, P. C. J. Leyh, *Handbook of Exp. Immunol,*. D. M. Weir, ed., 3rd ed., p. 32, Blackwell Scientific Publication. Ltd. Oxford.
11. H. Abdillahi and J. T. Poolman, *Microb. Pathog.* 4:27 (1988).
12. E. J. Hansen, D. A. Hart, J. L. McGeHee, and G. B.Toews, *Infect. Immun.* 56:182 (1988).
13. R. Garofalo, H. Faden, S. Sharma, and P. L. Ogra, *Infect. Immun.* 59:4221 (1991).